CRYSTALS THAT FLOW

THE LIQUID CRYSTALS BOOK SERIES

Edited by

G.W. GRAY, J.W. GOODBY & A. FUKUDA

The Liquid Crystals book series publishes authoritative accounts of all aspects of the field, ranging from the basic fundamentals to the forefront of research; from the physics of liquid crystals to their chemical and biological properties; and, from their self-assembling structures to their applications in devices. The series will provide readers new to liquid crystals with a firm grounding in the subject, while experienced scientists and liquid crystallographers will find that the series is an indispensable resource.

CRYSTALS THAT FLOW

Classic papers from the history of liquid crystals

Compiled with translation and commentary by

Timothy J. Sluckin, David A. Dunmur and Horst Stegemeyer

CRC Press
Taylor & Francis Group
Boca Raton London New York

CRC Press is an imprint of the
Taylor & Francis Group, an **informa** business
A TAYLOR & FRANCIS BOOK

First published 2004 by Taylor & Francis

Published 2022 by CRC Press
Taylor & Francis Group
6000 Broken Sound Parkway NW, Suite 300
Boca Raton, FL 33487-2742

First issued in paperback 2022

© 2004 Timothy J. Sluckin, David A. Dunmur and Horst Stegemeyer
CRC Press is an imprint of Taylor & Francis Group, an Informa business

No claim to original U.S. Government works

ISBN-13: 978-0-415-25789-3 (hbk)
ISBN-13: 978-1-03-240262-8 (pbk)

DOI: 10.1201/9780203022658

Typeset in Times by Integra Software Services Pvt. Ltd,
Pondicherry, India

Publisher's Note
The publisher has gone to great lengths to ensure the quality of this reprint but points out that some
imperfections in the original copies may be apparent.

Every effort has been made to ensure that the advice and information in
this book is true and accurate at the time of going to press. However,
neither the publisher nor the authors can accept any legal responsibility
or liability for any errors or omissions that may be made. In the case
of drug administration, any medical procedure or the use of technical
equipment mentioned within this book, you are strongly advised to
consult the manufacturer's guidelines.

British Library Cataloguing in Publication Data
A catalogue record for this book is available
from the British Library

Library of Congress Cataloging in Publication Data
A catalog record for this book has been requested

**Visit the Taylor & Francis Web site at
http://www.taylorandfrancis.com**

**and the CRC Press Web site at
http://www.crcpress.com**

CONTENTS

CONTENTS

Section C
The modern physical picture   **333**

x

CONTENTS

PREFACE

Science, unlike many other human activities, progresses through the published word, so in some sense the history of science is simply the published scientific literature. Nevertheless, amongst the vast accumulation of papers, many do not really contribute to scientific progress and some are just plain wrong. There are, however, occasional gems of inspiration, serendipitous discoveries of important new phenomena and more rarely landmarks of understanding. These might reasonably be called "classic papers", and a collection of them focussed on a particular theme provides some sort of history of an aspect of science. This volume of classic papers attempts to tell the story of the development of liquid crystal science through the words of the scientists themselves, as reported in their original publications. But the recorded history of a scientific idea often hides important pointers towards a deeper understanding, not only of the underlying science, but also of its interaction with the greater world outside. Thus the reprinted papers are augmented with a series of essays, which place the papers in context, and give some flavour of the scientific milieu of the period during which they were published.

The science of liquid crystals began more than a century ago with a baffling observation of two melting points in a single pure substance by a German botanist (Friedrich Reinitzer). The botanist passed his results to a physicist (Otto Lehmann), who realised that he was studying something interesting, but misidentified the physical phenomenon as crystalline in origin. Despite his misidentification, Lehmann gave us a name that stuck: **liquid crystals**, and our subject was born. Later a combative French crystallographer (Georges Friedel) realised that the materials were not crystals at all, but special liquids, albeit exhibiting some properties akin to those of crystals, and put the subject on a firmer footing. The early story of liquid crystals involves botanists, chemists, physicists, mineralogists and mathematicians mainly from Europe, who pursued their interests through a turbulent period of history involving wars (the Great War and the Second World War) and revolutions (the birth of communism). The events of history took their toll on our scientists (Vsevolod Konstantinovich Frederiks published and perished in one of Stalin's gulags), but did nothing to stem their fervour or determination to promote their ideas. After the Second World War, in common with many other physical sciences, liquid crystals benefited from a burgeoning in both fundamental science and applications. The former resulted in a Nobel Prize for one of the scientists featured in our collection (Pierre-Gilles de Gennes), while the latter led to the birth of a multi-billion dollar display industry, and fame and fortune for other players in the story of liquid crystals. This scientific odyssey is mapped by the classic papers we have selected, but we intend that our essays of introduction and the biographies of the principal authors will reinforce the human perspective of the story. Despite the historical content of this book, we cannot claim to have produced a definitive history of liquid crystal science. We have had access to some original documents, including of course the selected papers, but there is a wealth of information still to be assessed. We are optimistically confident that further study will not change the conclusions we have reached, but there are still many interesting stories to be told in the history of liquid crystals.

This book is divided into five separate sections, each headed by an essay of introduction. The papers of the first three sections and their commentaries are arranged chronologically, and these record important experiments and theoretical developments concerning the thermotropic liquid crystals discovered by Reinitzer. These papers take us from the early scientific arguments about the existence of liquid crystals to an increasingly sophisticated interpretation of their properties, and an enormous increase in the number of materials and phases available for study. This formative period for liquid crystals ends, in the context of this volume, around 1970. Such an early cut-off should not be seen to diminish the importance of subsequent studies, but these are more appropriate for scholarly scientific review than for this history. This volume is not the first to attempt a history of liquid crystals, and we refer the interested reader to the publications listed at the end of this Preface.

Our fourth section takes a different time-slice through the subject of liquid crystals, and its focus is the development of the liquid crystal display, which is now familiar to all. This part of our story ends rather arbitrarily around 1980, and so neglects many more recent important contributions, which have transformed liquid crystal displays from novelties to high-value, high-quality mass-produced electronic goods. In particular, the imaginative technical developments in displays that have come from Japanese scientists have largely taken place since 1980. The fifth and final section deals with lyotropic and polymeric liquid crystals, the origins of which are sufficiently different from thermotropic liquid crystals to justify a separate treatment. Indeed, lyotropic liquid crystals, so-named because their liquid crystallinity involves principally a reduction in water content of a solution rather than a reduction in temperature, have their origins in the soaps of the ancients. Furthermore, polymer liquid crystals in biology were recorded in the scientific literature, even before Reinitzer's studies, but have taken much longer to understand. This part of our story takes us back to the early part of the twentieth century, and Nobel-prize winner Richard Zsigmondy, and his studies of gel phases. Since that time, the understanding of lyotropics progressed largely in parallel with that of thermotropic liquid crystals, with occasional points of contact. Theoretical developments also emerged from different sources, with important contributions from other Nobel prize winners, Paul Flory and Lars Onsager. We finish this part of the story in the early 1980s, with the emergence of polymer liquid crystals and their applications. With hindsight, we can now see that the stories of lyotropic, polymer and thermotropic liquid crystals are all part of the same scientific picture, but history saw otherwise.

From the hundreds of papers that might be regarded as 'classic', we have been forced to make a selection. Some of the papers included (for example, those of Lehmann, Reinitzer and Georges Friedel) would appear in any compilation. In the case of others, the choice is not so obvious. Those who know this literature well will have their own favourites, and some of these may not be the same as ours. The science of liquid crystals crosses disciplinary boundaries, and to reflect this, we have tried to choose representative papers from the major strands contributing to the current state of liquid crystal knowledge.

Much of the early work on liquid crystals is published in German or French (and occasionally in Russian), and in order to make this accessible to a readership around the world, we have translated a number of the selected papers into English. We apologise to French and German readers in particular for this necessity, for taken as literature, the original is surely better than a translation. Apart from the papers by Reinitzer and Saupe, new translations have been made by the present authors for this volume. We hope that the translated papers give a faithful representation of not only the meaning but also the spirit of the originals. Inevitably readers will find minor errors: for these we apologise.

Two authors require particular mention. To remark that the writing style of Otto Lehmann was ponderous would be to indulge in weighty understatement. We found the translation of Lehmann almost impossible, for his sentences were longer than some people's books! Nevertheless the task is worth the effort, for the content is remarkable, both from a historical and from a scientific viewpoint. Georges Friedel, although eleven years younger than Lehmann, was his scientific colleague and rival. His 1922 paper was a classic which leaves a mark on the science of liquid crystals to this day.

The genesis of this volume lies in an inaugural lecture given by one of the present authors (TJS) in March 2000, celebrating his election to the Chair of Applied Mathematical Physics at the University of Southampton. The title of this lecture "Fluids with attitude – the story of liquid crystals from oddity to technology" was an attempt to review of the origins of the science of liquid crystals for a general university audience, and to show how a multi-billion dollar industry developed from Reinitzer's esoteric observations on the biochemistry of carrots. The original hope was that the lecture could be made available to a wider audience through publication. The single lecture has matured into the present volume of more than seven hundred pages. To do proper justice to the subject of liquid crystals, which embraces many aspects of science including chemistry, physics, mathematics, biology, materials science, and electronic engineering, has required the involvement of two other authors (DAD and HS).

A volume such as this has required input from many sources. Many colleagues have provided advice and guidance concerning both the choice of papers, and their significance. Peter Knoll was kind enough to provide copies of the unpublished monographs he co-authored with Hans Kelker on the early history of liquid crystals. Peter Raynes and Cyril Hilsum suffered the indignity of formal interviews in sharing their memories of the early days of the development of liquid crystal displays. Jacques Friedel recalled his grandfather Georges, and also his own studies in the liquid crystal field, while Yves Bouligand pointed us towards the long-ignored theoretical contribution of François Grandjean. The late Frank Leslie carefully explained the background to his work on the continuum theory of liquid crystals, but unfortunately did not live long enough to peruse the result. Tomas Carlsson also transmitted some of his insights in this area, as well as information on Björndåhl and Oseen. Maxim Tomilin and Victor Reshetnyak were extremely helpful in putting us in touch with the Russian literature, and in transmitting memories from the Frederiks and Tsvetkov groups. Hiro Kawamoto and Kazu Toriyama helped us to reach some appreciation of the great contributions of Japanese science to the development of displays, and we are grateful to them. The authors also greatly appreciated the advice of Mark Warner concerning the contributions of Paul Flory to the theory of liquid crystals. The biographies of scientists whose work is included have been compiled from many different sources. We also thank all authors who have been kind enough to send us brief descriptions of their lives; we hope that our (minor) rewriting of their life stories meets with their satisfaction.

Preliminary versions of essays or translations have been read by Heinrich Arnold, Joe Castellano, Noel Clark, Jerry Ericksen, Heino Finkelmann, Pierre-Gilles de Gennes, George Heilmeier, Maurice Kléman, Sven Lagerwall, Vittorio Luzzati, Bob Meyer, Peter Raynes, Martin Schadt, John Seddon, Joachim Stauff, Richard Williams and Louis Zanoni. Karl Hiltrop was generous with helpful advice concerning the history of lyotropic liquid crystal research. We are most grateful for their comments, but of course absolve them of any responsibility for errors of fact or interpretation, which remains exclusively that of the authors. We are also indebted to the families of the late Horst Sackmann and Hans Kelker for useful comments on the circumstances of their investigations on liquid crystals.

This volume appears in the Liquid Crystal Series, for which the Series Editors are George Gray and John Goodby. We are grateful to both Series Editors for their support and encouragement in guiding this book to publication. We must thank particularly George Gray who has encouraged the authors during the gestation of this volume. He has been generous with his recollections, and vigorously contributed ideas for the book. He has read many drafts of the essays and translations, finding and correcting innumerable errors and infelicities. For all this, we thank him, but emphasise that the final responsibility for remaining errors rests with the authors. Our 'minder' at Taylor and Francis over the period of the preparation of this collection has been Janie Wardle, and we thank her also for her patient help in guiding us towards a final product. We are also extremely grateful to Anita Ananda and her production team at Integra-India for their effort in typesetting an extremely non-standard text.

TJS finally wishes to thank his wife Celia and his children Ben and Rachel for their patience in allowing family time to be diverted for so long into the task of compiling this collection. Well might they protest that this was a never-ending task whose fruits they might never live to see. It is to be hoped that this prediction at least will turn out to be false! DAD, as one of the 'co-opted' authors, wishes to thank the initiating author Tim Sluckin for the opportunity to participate in this particular literary venture. These thanks are for the chance to have spent many leisure hours translating, writing and searching the stacks of his two universities of Southampton and Sheffield. Time thus spent might otherwise have been spent with his family, to whom thanks are also due for their understanding and forbearance. HS thanks his family and friends for mental support and tolerating his absence during his dive into the interesting history of liquid crystals.

Southampton, Münstertal, January 2003

General historical references

A.S. Sonin, *Doroga dlinuyu v vyek: iz istorii nauki o zhidkikh kristallov* (A road reaching back a century: on the history of liquid crystal science) (Nauka Publishing House, Moscow, 1988).

P.M. Knoll and H. Kelker, *Otto Lehmann: Erforscher der flüssigen Kristalle: Eine Biographie mit Briefen an Otto Lehmann* (Otto Lehmann, discoverer of liquid crystals: a biography of Otto Lehmann including letters) (privately published) (contains a lot of biographical detail on Lehmann, with reproductions of many letters, in German).

H. Kelker and P.M. Knoll, *Die Erforschung der flüssigen Kristalle zu Beginn des zwanzigsten Jahrhunderts: Ein Kapitel deutsch-französischer Wissenschaftsgeschichte* (Liquid crystal research at the beginning of the twentieth century: a chapter in Franco-German scientific history (privately published) (more biographical detail, concentrating on Lehmann's interaction with the French school, again in German).

A.S. Sonin and V.Ya. Frenkel, *Vsevolod Konstantinovich Freédericksz (1885–1944)* (Nauka, Moscow 1995) (in Russian, contains a lot of scientific detail, including equations).

H. Kelker, *History of liquid crystals*, Molecular Crystals Liquid Crystals **21**, 1–48 (1973).

H. Kelker, *Survey of the Early History of Liquid Crystals*, Molecular Crystals Liquid Crystals **165**, 1–42 (1988).

H. Kelker and P.M. Knoll, *Some pictures of the history of liquid crystals*, Liquid Crystals **5**, 19–42 (1989).

H. Sackmann, *Smectic liquid crystals, a historical review*, Liquid Crystals **5**, 43–55 (1989).

I. Korenic, *A millennium of liquid crystals*, Optics and Photonics News (Feb 2000) 16–22.

C. Hilsum, *The anatomy of a discovery – biphenyl liquid crystals*, in *Technology of chemicals and materials for electronics* ch. 3, in (ed.) E.R. Howells (ed.) (Ellis Horwood, Chichester 1991).

V. Vill, *Early history of liquid crystalline compounds*, Condensed Matter News **1**, No. 5, 25–28 (1992).

H. Kawamoto, *The history of liquid-crystal displays,* Proc. IEEE **90**, 460–500 (2002).

ACKNOWLEDGMENTS

The editors and publishers are pleased to acknowledge the following organizations for permission to reproduce selected papers. Every effort has been made to contact and acknowledge copyright holders. In some cases the time lapse is great and we have been unable to locate the copyright holder, unable to determine whether papers or material are protected by copyright, or unable to obtain replies from a presumed copyright holder. If any errors or omissions have been made we would be pleased to correct them at a later printing.

Taylor & Francis, as publishers, and the editors of this volume are most grateful to all who have given permission for their published work to be reproduced here.

Section B: The inter-war period: anisotropic fluids or mesomorphic phases?

B1. Les états mésomorphes de la matière, G. Friedel, *Annales de Physique*, **18**, 273–474 (1922) Copyright Masson Editeur.

B2. La diffraction des Rayons X par les corps smectiques, M. de Broglie and E. Friedel, *Comptes rendus de l'Académie des Sciences*, **176**, 738–40 (1923). Reproduced by permission of Institut de France.

B3. On the use of a magnetic field in the measurement of the forces tending to orient an anisotropic liquid in a thin homogeneous layer, V. Fréedericksz and V. Zolina, *Transactions of the American Electrochemical Society*, **55**, 85–96 (1929). Reproduced by permission of The Electrochemical Society Inc.

B4. Extracts from General Discussion on liquid crystals, P.P. Ewald *et al.*, *Zeitschrift für Kristallographie*, **79**, 269–347 (1931). Reproduced by permission of Oldenbourg Wissenschaftsverlag.

B5. The theory of liquid crystals, C.W. Oseen, *Transactions of the Faraday Society*, **29**, 883–900 (1933). Reproduced by permission of The Royal Society of Chemistry.

B6. New arguments for the swarm theory of liquid crystals, L.S. Ornstein and W. Kast, *Transactions of the Faraday Society*, **29**, 930–44 (1933). Reproduced by permission of The Royal Society of Chemistry.

B7. The effect of a magnetic field on the nematic state, H. Zocher, *Transactions of the Faraday Society*, **29**, 945–57 (1933). Reproduced by permission of The Royal Society of Chemistry.

B8. Extracts from General Discussion, J.D. Bernal *et al.*, *Transactions of the Faraday Society*, **29**, 1060–85 (1933). Reproduced by permission of The Royal Society of Chemistry.

B9. Influence of a magnetic field on the viscosity of *para*-azoxyanisole, M. Mięsowicz, *Nature*, **136**, 261 (1936). Reprinted by permission from *Nature*, **136**, 261 (1936), Macmillan Publishers, Ltd.

B10. The three coefficients of viscosity of anisotropic fluids, M. Mięsowicz, *Nature*, **158**, 27 (1946). Reprinted by permission from *Nature*, **158**, 27 (1946), Macmillan Publishers, Ltd.

B11. Sur l'orientation des cristaux liquides par les surface frottées, P. Chatelain, *Comptes rendus de l'Académie des Sciences*, **213**, 875–76 (1941). Reproduced by permission of Institut de France.

B12. Sur l'orientation des cristaux liquides par les surface frottées, P. Chatelain, *Bulletin de la Société française de Minéralogie*, **66**, 105–30 (1944). Reproduced by permission of European Journal of Mineralogy.

Section C: The modern physical picture

C1. Sur l'application de la théorie du magnétisme aux liquides anisotropes, F. Grandjean, *Comptes rendus de l'Académie des Sciences*, **164**, 280–83 (1917). Reproduced by permission of Institut de France.

C2. Über die Molekülanordnung in der anisotropy-flüssigen Phase, V.N. Tsvetkov, *Acta Physicochimica (URSS)*, **16**, 132–47 (1942).

C3. Eine einfache molekulare Theorie des nematischen kristallinflüssigen Zustandes, W. Maier and A. Saupe, *Zeitschrift Naturforschung*, **13a**, 564–66 (1958). Reproduced by permission of Verlag der Zeitschrift fur Naturforschung.
C3. Translation. A simple molecular theory of the nematic liquid-crystalline state, W. Maier and A. Saupe, first published in *Dynamics and defects in liquid crystals* (eds P.E. Cladis and P. Palffy-Muhoray, 1998). Reproduced by permission of Taylor & Francis.

C4. On the theory of liquid crystals, F.C. Frank, *Discussions of the Faraday Society*, **25**, 19–28 (1958). Reproduced by permission of The Royal Society of Chemistry.

C5. Isomorphiebeziehungen zwischen kristallin-flüssigen Phasen. 4. Mitteilung: Mischbarkeit in binären Systemen mit mehreren smektischen Phasen, H. Arnold and H. Sackmann, *Zeitschrift für Elektrochemie*, **11**, 1171–77 (1959). Reproduced by permission of Wiley-VCH.

C6. Some constitutive equations for liquid crystals, F.M. Leslie, *Archives for Rational Mechanics and Analysis*, **28**, 265–83 (1968). Copyright Springer-Verlag.

C7. Short-range order effects in the isotropic phase of nematics and cholesterics, P.G. de Gennes, *Molecular Crystals and Liquid Crystals*, **12**, 193–214 (1971). Reproduced by permission of Taylor & Francis.

Section D: The development of liquid display crystal technology

D1. Improvements in or relating to light valves, B. Levin and N. Levin (Marconi's Wireless Telegraph Company), British Patent 441, 274 (1936). Copyright Her Majesty's Stationary Office.

D2. Domains in liquid crystals, R. Williams, *Journal of Chemical Physics*, **39**, 384–88 (1963). Reprinted with permission from Journal of Chemical Physics, American Institute of Physics.

D3. Dynamic scattering: a new electro-optic effect in certain classes of nematic liquid crystals, G.H. Heilmeier, L.A. Zanoni and L.A. Barton, *Proceedings of the Institution of Electronic and Electrical Engineers*, **56**, 1162–71 (1968). © 1968 Institution of Electronic and Electrical Engineers. Reprinted, with permission.

D4. A liquid-crystalline (nematic) phase with a particularly low solidification point, H. Kelker and B. Scheurle, *Angewandte Chemie International Edition*, **8**, 884–85 (1969). Reproduced by permission of Wiley-VCH.

D5. Distortion of twisted orientation patterns in liquid crystals by magnetic fields, F.M. Leslie, *Molecular Crystals and Liquid Crystals*, **12**, 57–72 (1970). Reproduced by permission of Taylor & Francis.

D6. Voltage-dependent optical activity of a twisted nematic liquid crystal, M. Schadt and W. Helfrich, *Applied Physics Letters*, **18**, 127–28 (1971). Reprinted with permission from Applied Physics Letters, American Institute of Physics.

D7. New family of nematic liquid crystals for displays, G.W. Gray, K.J. Harrison and J.A. Nash, *Electronic Letters*, **9**, 130–31 (1973). © 1973 IEE. Reprinted, with permission.

D8. Ferroelectric liquid crystals, R.B. Meyer, L. Liébert, L. Strzelecki and P. Keller, *Journal de Physique*, **26**, L69–71 (1975). Reproduced by permission of EDP Sciences.

D9. Submicrosecond bistable electro-optic switching in liquid crystals, N.A. Clark and S.T. Lagerwall, *Applied Physics Letters*, **36**, 899–901 (1980). Reprinted with permission from Applied Physics Letters, American Institute of Physics.

Section E: Lyotropic and polymeric liquid crystals

E2. Liquid crystalline substances from virus-infected plants, F.C. Bawden, N.W. Pirie, J.D. Bernal and I. Fankuchen, *Nature*, **138**, 1051–52 (1936). Reprinted by permission from *Nature*, **138**, 1051–52 (1936), Macmillan Publishers, Ltd.

E3. Die Mizellarten wässeriger Seifenlösungen, J. Stauff, *Kolloid–Zeitschrift*, **89**, 224–33 (1939). Reproduced with permission from Steinkopff Verlag, Springer Verlag.

E4. The effects of shape on the interaction of colloidal particles, L. Onsager, *Annals of the New York Academy of Sciences*, **51**, 627–59 (1949). © 1949. New York Academy of Sciences, USA. Reprinted, with permission.

E5. Phase equilibria in solutions of rod-like particles, P.J. Flory, *Proceedings of the Royal Society of London*, **A234**, 73–89 (1956). Reproduced by permission of The Royal Society of London.

E6. La structure des colloides d'association. I. Les phases liquide-crystallines des systèmes amphiphile-eau, V. Luzzati, H. Mustacchi, A. Skoulios and F. Husson, *Acta Crystallographica*, **13**, 660–67 (1960). Reproduced by permission of the International Union of Crystallography.

E7. Mesophasic structures in polymers. A preliminary account on the mesophases of some poly-alkanoates of *p,p'*-di-hydroxy-α,α'-di-methyl-benzalazine, A. Roviello and A. Sirigu, *Journal of Polymer Science B: Polymer Letters Edition*, **13**, 455–63 (1975). Reproduced by permission of Wiley-VCH.

E8. Model considerations and examples of enantiotropic liquid crystalline polymers, H. Finkelmann, H. Ringsdorf and J.H. Wendorff, *Makromolekulare Chemie*, **179**, 273–76 (1978). Reproduced by permission of Wiley-VCH.

Photographs

We gratefully acknowledge the following sources for photographs which appear in this volume:

The Deutsche Bunsen Gesellschaft (Kelker collection): (F. Reinitzer, O. Lehmann, L. Gattermann, R. Schenck, D. Vorländer, E. Bose, C.W. Oseen, L. Ornstein, V.N. Tsvetkov, W. Maier, G. Friedel, R.A. Zsigmondy);
Dr Lutz Rohrschneider (H. Kelker);
Société française de Minéralogie et Cristallographie (C. Mauguin);
© Acarologia (F. Grandjean);
© Académie française (M. de Broglie, P. Châtelain);
Professor J. Friedel (E. Friedel);
Professor G. Meier (W. Kast);
Mrs E. Maier (W. Maier);

Mrs E. Leslie (F.M. Leslie);
Sackmann family (H. Sackmann);
© Godfrey Argent Studio (F.C. Bawden, N.W. Pirie);
Russian Academy of Sciences (J.D. Bernal)
University of Bristol Physics Department (F.C. Frank);
Atomic Weapons Establishment Aldermaston (N. Levin);
Brooklyn Polytechnic Institute (I. Fankuchen);
© Nobel Foundation (L. Onsager, P. Flory)
Liquid Crystals Today (© Taylor & Francis) (M. Mięsowicz, V.K. Frederiks,
    A. Skoulios)
Molecular Crystals and Liquid Crystals (© Taylor & Francis) (H. Zocher)
Dynamics and Defects in Liquid Crystals (© Taylor & Francis) (A. Saupe)
Zeitschrift für Elektrochemie (© John Wiley) (W. Bachmann)

We thank other authors of papers whose photographs appear in this volume and
who have provided us with photographs of themselves.

Section A

THE EARLY PERIOD: LIQUID CRYSTALS OR ANISOTROPIC LIQUIDS?

THE EARLY PERIOD: LIQUID CRYSTALS OR ANISOTROPIC LIQUIDS?

Introduction

The foundation of liquid crystal science is traditionally set in the year 1888, with the work of Friedrich Reinitzer. Reinitzer is commonly termed a botanist, although in modern terms he would perhaps be thought of more as a biochemist. He was at the time 30 years old and assistant to Professor Weiss at the Institute of Plant Physiology at the German University of Prague. We remind readers that Prague was then the capital of the province of Bohemia in the Austro-Hungarian empire, but that the university in Prague was highly prestigious within the German-speaking world. Reinitzer's experiments involved extracting cholesterol from carrots in order to determine its chemical formula, which at that time was unknown. He thought that cholesterol was chemically related to carotene (the red pigment) and thus to chorophyll. At the same time, cholesterol had been observed to occur in the cells of many animals, and it was of some interest to determine whether this was exactly the same cholesterol, or whether there were a number of closely related compounds. He presented his results to the Vienna Chemistry Society at its monthly meeting on 3 March 1888. This paper is article A1 in our collection.

In the paper Reinitzer examined the physico-chemical properties of various derivatives of the carrot cholesterol. Most of his results are not specifically of interest in a liquid crystal context. Naturally there was much discussion concerning speculations on the exact chemical formula of cholesterol; is it $C_{26}H_{44}O$ as suggested by Gerhardt?[1] He noted that a number of previous workers, such as Raymann,[2] Löbisch[3] and Planar,[4] had observed some dramatic colour effects on cooling cholesteryl acetate or related compounds just above the solidification temperature. He himself found the same phenomenon both in cholesteryl acetate and in cholesteryl benzoate, which has the chemical formula $C_{27}H_{45} \cdot C_7H_5O_2$.

But the coloured light show near the solidification of cholesteryl benzoate was not its most peculiar feature. Reinitzer found, to his amazement, that this compound does not melt like other compounds. Cholesteryl benzoate appeared to have *two* melting points. At 145.5 °C the solid melted into a cloudy liquid. The cloudy liquid lasted up to 178.5 °C, at which point the cloudy liquid suddenly became clear. Furthermore, the phenomenon appeared to be reversible. Near both transition points the system exhibited some dramatic colours.

Both the colours and the double melting were worthy of note. What was going on? There were several different solid modifications, and the colours suggested to Reinitzer that some form of physical isomerism was occurring. Reinitzer sought help from Dr Otto Lehmann, a well-known crystallographer, then the assistant of Professor Wüllner at the Polytechnical School of Aachen. The expectation was that Lehmann's polarising microscope might clarify the situation.

There followed an exchange of letters, and presumably of samples as well, throughout March and April of 1888. Lehmann examined the intermediate cloudy fluid, and reported that he had seen crystallites. When the exchange of letters ended on April 24, although definitive answers to the nature of the cloudy phase had not been elicited, Reinitzer felt that he had enough to publish. The important point here is that these first observations of liquid crystals (although not yet recognised as such) were a serendipitous by-product of an apparently unrelated and unprofound piece of research. Neither for the first nor for the last time, Nature had a sprung a surprise on an unprepared investigator.

Reinitzer's first letter to Lehmann was sent on 14 March 1888. It was 16 pages long, and handwritten in Gothic characters. An extract is shown in Fig. A1. In it Reinitzer relates to Lehmann most of the content of article A1. The colour phenomenon in particular is of interest to the modern observer. When he cooled cholesteryl benzoate below its second melting point at 178.5 °C (later called by Lehmann and others the *clearing point*), he observed that

> ...violet and blue colours appear, which rapidly vanish with the sample exhibiting a milk-like turbidity, but still fluid. On further cooling the violet and blue colours reappear, but very soon the sample solidifies forming a white crystalline mass.

Reinitzer observed the appearance of colours twice! However, in the case of cholesteryl acetate with its *monotropic* cholesteric phase, with a melting point 114.3 °C and a clearing point 94.8 °C, Reinitzer observed the appearance of colours only once on cooling. At that time the mere existence of the double-melting and the colours was sufficient to excite interest. In fact, nowadays we are also able to understand why in one material two sets of colours were seen, and in the other material only one. Furthermore, it turns out that this is not an accident, and indeed a tribute to the exactness of Reinitzer's experimental method that he observed and recorded rather subtle phenomena whose significance could not

4

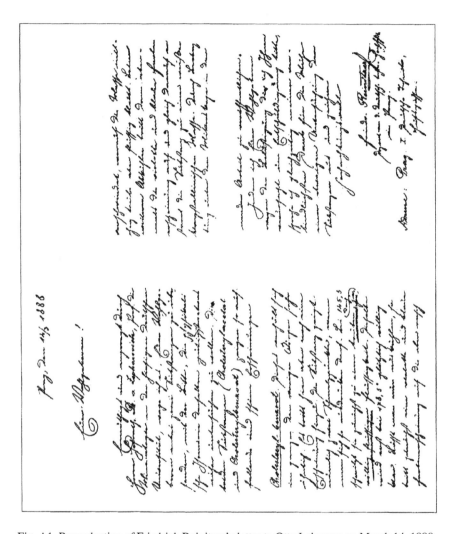

Fig. A1 Reproduction of Friedrich Reinitzer's letter to Otto Lehmann on March 14, 1888, in which he sought help in identifying the properties of cholesteryl benzoate. Note the Gothic script which is very difficult to disentangle.

have been understood at the time. The explanation itself is more sophisticated and involves concepts which are of extremely recent origin.*

Reinitzer wrote a little extra on his work on cholesteryl benzoate,[5] and then more or less disappears from this narrative. His one further contribution to the story comes some 20 years later, in a rather unedifying episode which we shall come to presently.

The mantle is taken up by Otto Lehmann, who both realised that he had come across a new phenomenon, and was in a position to launch a research programme to investigate it. In 1888, Lehmann was 33. Up till 1888 Lehmann had enjoyed a rather peripatetic life. His postdoctoral years had been spent building up expertise in crystallography. The principal weapon in his scientific arsenal was experimental microscopy, for which he was well prepared because his schoolmaster father had been an amateur microscopist before him. Lehmann was known to be a coming man, and even at a relatively young age in 1889 he was elected as professor of physics at the Technical University of Karlsruhe, as the successor to Heinrich Hertz (1857–94), who had lately demonstrated experimentally Maxwell's theory of electromagnetism.

It was Lehmann's jealously guarded and increasingly prestigious microscope, not yet available off the shelf, which led Reinitzer to approach him for help. Lehmann was not only able to make observations in polarised light, but also, and this was a key advantage, his microscope possessed a hot stage enabling *in situ* high temperature observations. With Reinitzer's peculiar double-melting liquid, a problem in search of a technique had met a scientist in search of a problem.

Lehmann immediately launched a vigorous programme of intense investigation into the new phenomenon. Already by the end of August 1889 he had his first article ready for submission to the *Zeitschrift für Physikalische Chemie* (Journal of Physical Chemistry). This is article A2 in our collection.

The reader will observe that, even in the imperfect translation presented in this volume, Lehmann's language is particularly flowery. The article is entitled 'On flowing crystals'. What emerges in article A2 is that the cloudiness of the intermediate fluid occurs when what we would now call nucleating droplets merge, and that sometimes the individual droplets exhibited a black cross when viewed between crossed nicols. The cloudiness itself was the macroscopic manifestation of 'large star-like radial aggregates of needles'.

---

* In order to appreciate the origin of the colours we have to move forward to the 1970s. They result from selective reflection of circularly polarised light from helically structured chiral liquid crystals. In a careful reinvestigation of selective reflection in cholesteryl benzoate one of the present authors (H. Stegemeyer and K. Bergmann, Springer Ser. Chem. Physics 11 (1980), p. 161ff) was able to show that whereas the low temperature colours are caused by the *cholesteric* phase (a fluid which is nevertheless crystallographically a one-dimensional 'solid'), the high temperature colours are due to selective reflection from a *blue phase* (a fluid which is crystallographically a three-dimensional solid)! Cholesteryl acetate does not exhibit a blue phase, and as a consequence Reinitzer observed colours only once in this compound. It is amazing that the first study of liquid crystals already revealed a structure as complex as a blue phase, whose cubic structure was only disentangled some hundred years later.

Lehmann was certain that the cloudy liquid has all the attributes of a crystal and those of a liquid. He believed truly to have discovered 'crystals that flow'. Much of the rest of the article is concerned with advocating the point of view that the properties of liquidity and crystallinity could indeed coexist, and is not without rhetorical flashes. Given the effort he went to in order to justify his picture, he must clearly have expected to meet with a good deal of opposition.

In a series of papers over the period 1890–1900,[6,7] Lehmann made exhaustive studies of the phenomenon. Because the essence of the phenomenon seemed to occur in droplets, he made a virtue out of necessity and often deliberately prepared fluid mixtures from which the intermediate phase would then precipitate in droplet form. We show in Fig. A2 a series of coloured images of droplets, taken from Lehmann's review article published in 1900.[7] These dramatic pictures, originally taken only in black and white, have been coloured by hand so as to resemble what Lehmann saw under the microscope.

Lehmann found materials some of which exhibited, as in cholesteryl benzoate, two melting points, and some of which even exhibited *three* melting points. He found a phase which he called *Fliessende Kristalle* (flowing crystals) or *Schleimig flüssige Kristalle* (slimy liquid crystals), and another which he named *Kristalline Flüssigkeit* (crystalline fluid) or *Tropfbar flüssige Kristalle* (liquid crystals which form drops). If both phases existed in the same material the latter was always the higher temperature phase. The latter was cloudy, but the former was clear, although very viscous. All this culminated in an enormous and generously illustrated tome, simply entitled 'Liquid Crystals',[8] published in Leipzig in 1904.

Lehmann's first article quickly elicited a response from his scientific colleagues. On 14 March 1890, Ludwig Gattermann of the University of Heidelberg wrote to Lehmann:[9]

> It was with great interest that I read your article on flowing crystals in Zeitschrift für physikalische Chemie. For some time I have had several substances here which also exhibit the same properties. To begin with I thought I was considering mixtures of several materials, but the properties remained unchanged after several crystallisation cycles. Following your article I am now clear as to what is going on.

Later that year Gattermann and his student Ritschke published the first report of the complete synthesis of one of the new substances. This is article A3 in our collection 'On azoxyphenol ethers'. It seems that as with so many organic chemists, the immediate spur to Gattermann's work lay in industrial concerns, for the initial step of the work involved the reduction of *p*-nitrophenetole provided by the Bayer & Co. dyeworks in Elberfeld. The ultimate goal of Bayer was evidently to produce better dyes for textiles and other industrial goods.

In 1890 Gattermann was a 30-year-old Assistant Professor at the University of Heidelberg. Later on he became Full Professor at the University of Freiburg and a

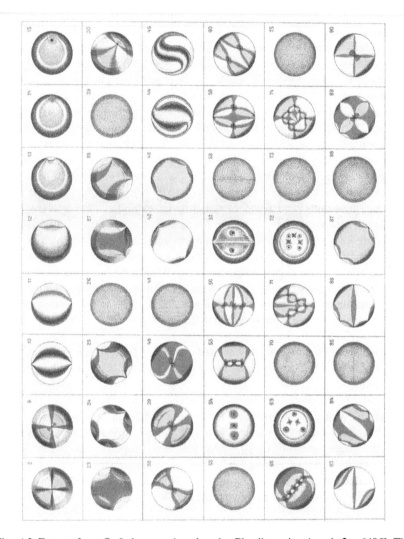

Fig. A2 Extract from O. Lehmann, Annalen der Physik, series 4, vol. **2**, p649ff. The Schlieren texture – the *stains* that so struck the early workers in the field of liquid crystals – can be seen in particular in the images that show what Gattermann and Lehmann called *copulating* drops, in which two drops are amalgamating (e.g. images 55, 56, 59, 74). See Plate I.

famous organic chemist, well-known to generations of chemistry students by his textbook '*Die Praxis des Organischen Chemikers*' (The practice of organic chemistry).

The aim of their investigations was a real problem of organic chemistry, namely to test if there existed two or more different isomers (i.e. molecules with the same chemical formula but different structure) of phenol ethers, as was claimed in the literature in the case of azoxytoluenes. In the process of synthesising azoxyphenetole they were indeed able to isolate two different reaction products with quite different crystal structures. However, by skilful chemical analysis they found out that the two substances were not isomers, but rather different compounds which differ in their alkyl groups: *p*-azoxyanisole with a methyl group, and *p*-azoxyphenetole with an ethyl group. Using ingenious chemical procedures, they found evidence that an exchange of alkyl groups had taken place during the synthesis in methanol as solvent (ethyl- *vs.* methyl-group). Consequently, they solved a typical problem of organic chemistry – the replacement of one alkyl group by another.

During the procedure, however, they made an unusual and extremely interesting observation. All the three compounds under investigation (azoxyanisole, azoxy-phenetole, anisole azoxyphenetole) exhibited a sharp melting point in the usual way, but were not transformed into the usually observed transparent fluid phase. Rather they gave rise to cloudy liquids, which resembled the cloudy liquid produced by Reinitzer's cholesteryl benzoate.

From our point of view Gattermann and Ritschke's article presents a number of interesting features. For brevity we have omitted some of the technical details not directly relevant to the liquid crystal story. The article reports in particular the synthesis of para-azoxyanisole, which formed a cloudy liquid phase at 116 °C, and what soon became known as a *clearing point* at 134 °C. This compound was to become the material of choice for liquid crystalline studies. The method of synthesis was well defined and relatively easy, and the temperature range over which the anomalous properties were manifest was rather more accessible than in the case of cholesteryl benzoate. In this article too we find the first known reference to the term *Flüssige Kristalle* ('liquid crystals'), although not yet in any precise fashion. The paper emphasises that, although the three compounds under discussion possess different chemical structures and quite different crystalline properties, they exhibit totally the same physical properties above the 'third melting point', i.e. in the liquid crystalline state. Another interesting snippet from Gattermann's report concerns his observations of liquid crystal droplets. These resembled oil droplets, apart from some peculiar 'stains' in the middle of the droplets. The stains, or *Schliere*, have entered the canon as the familiar Schlieren texture.

At the end of his paper Gattermann described a proposal for the structure of the isotropic liquid. He believed that the isotropic, non-birefringent liquid phase of those compounds exhibiting liquid crystalline phases should possess a regular (i.e. cubic) crystal structure. We now know that this speculation went well beyond the facts available and is false. Paper A3 was Gattermann's pioneering publication in the liquid crystal field. Despite the importance of this paper for the development

9

of the subject, only once did he return to liquid crystals. This was in an extensive paper in 1906,[10] in which he briefly mentioned that the condensation products of aromatic aldehydes with benzidine resulted in Schiff bases, which exhibited the properties of Lehmann's 'flowing crystals'. Gattermann's contribution is remarkable not only because it was an assay by a well-known scientist into this field, but also because it provided a solid foundation for much of the work which came later.

Lehmann's slimy liquid crystals were obviously solid-like, if only because of their reluctance to flow. The drop-like variety also showed one physical property which had hitherto been uniquely associated with solidity, that of *birefringence*, which explained the peculiar dark crosses seen through the polarising microscope in droplets in what Lehmann came to call the *erste Hauptlage*, or 'first principal position'. In other circumstances the droplets could be found in the *zweite Hauptlage*, or second principal position. In this case the droplets acted like a lens, with an apparently greater refractive index in the centre of the droplet.

Lehmann continued to insist on his interpretation of his microscope observations as representing materials combining all the properties of fluidity and crystallinity, while freely admitting his ignorance of the precise molecular explanation. By 1900, he was prepared to subsume all the new phenomena under the more general classification of *Flüssige Kristalle*.

It is worth pausing a moment to consider the enormity of the step that Lehmann felt he had to take when he proclaimed the existence of liquid crystals. The distinction between solid, liquid and gas is evident even to the non-scientist. The earliest theories of matter in ancient Greek times emphasise the physical form of matter (under the guise of earth, water and air) over its chemical basis.

Chemistry, by contrast, only slowly developed in the seventeenth and eighteenth centuries out of the medieval art of alchemy. Although atomic ideas had been around since Democritus and Leucippus in the fifth century BC, it was not until John Dalton (1766–1844) published his seminal work, *A New System of Chemical Philosophy*, in 1808, that it was established that chemical compounds came from constituent materials in definite proportions. This gave strong, but not yet decisive, evidence of the existence of atoms.

Similarly by the end of the nineteenth century a basic understanding of continuum mechanics had already developed. Solids possess *displacement* (we would say shear) elasticity, whereas liquids do not. Many solids exhibit crystallinity, which at this time usually meant that under the microscope crystal facets were observed. Crystals were often optically birefringent (i.e. refracted light differently depending on their orientation). Birefringent materials were said to be *anisotropic*, i.e. to possess different properties in one direction as compared to another.

The current picture of crystallinity in terms of periodic atomic lattices is due to the French crystallographer René-Just Haüy (1743–1822). Although this picture had gained wide acceptance by the end of the nineteenth century, it did not yet rest on firm foundations. The debate between Ludwig Boltzmann (1844–1906) and Ernst Mach (1838–1916) – between materialism and idealism – on the atomic nature of matter was still in full swing. Definitive evaluation of the size of the Avogadro number (or equivalently Boltzmann's constant) had to wait for the experiment of Jean Perrin[11] (1870–1942) in 1908, and the definitive experiments in which the crystal lattice was directly observed in X-ray scattering[12] were carried out by Max von Laue (1879–1960) in 1912 and the father–son team of W.H. Bragg (1862–1942) and W.L. Bragg (1890–1971) in 1913.

In any event, notwithstanding the lack of proof of its essential nature, crystal-linity certainly seemed incompatible with fluidity. Perhaps not surprisingly given the circumstances and intellectual climate of the day, Lehmann's work elicited not a little scepticism from his scientific peers concerning the very existence of the liquid crystal phenomenon.

Throughout the 1890s, even up till 1905, the physical chemists Georg Quincke (1834–1924)[13] and Gustav Tammann (1861–1938)[14] insisted that the observations were more parsimoniously explained by supposing that the substances which Lehmann labelled as liquid crystals were mixtures of some sort. The multiple melting phenomena must then be explicable in terms of separate transitions for each quasi-separate component of this mixture. Thus, averred Quincke and the theoretical physicist Wulff (later to become famous for explaining why macro-scopic solids take their characteristic crystalline shape), liquid crystals must really be colloidal. According to Quincke and Wulff, the existence of a solid component would explain the birefringence. The strong light scattering (i.e. the turbid appearance) in colloids such as white paint results from strong scattering by indi-vidual colloidal particles, which are just the right size for maximal light scattering.

An alternative but related picture, supported by Tammann and Walther Nernst (1864–1941), proposed that the so-called liquid crystal phase was rather a colloidal emulsion, akin to milk or vinaigrette, in which droplets of one liquid are suspended in another.

Colloid science was at that stage relatively young, though no longer in its infancy. The founding father is generally reckoned to be the Scottish chemist Thomas Graham (1805–69). It is to Graham that we owe many of the important terms in colloid science, including the appellation itself, *colloid* from the Greek word κολλα, meaning glue.

Tammann's colloidal picture was not supported by detailed theory, but seemed more plausible than the alternative of a liquid which was simultaneously crystal-line. Nernst later became extremely famous, winning the Nobel prize in 1920 for discovering the third law of thermodynamics (that the entropy of a material goes to zero at the absolute zero of temperature). But on this occasion he was not right.

The controversy between Tammann and Lehmann was particularly intense, and led to a bitter exchange in the pages of the *Annalen der Physik* in the early years of the twentieth century.[15] Tammann was an Estonian physical chemist who was appointed as Professor of Inorganic Chemistry in Göttingen in 1903. In 1907 he succeeded Nerst as Professor of Physical Chemistry. He is best known now for his contributions to the thermodynamics of metallurgy and glass-formation.

Twice, in 1901 and again in 1902, Tammann wrote articles in *Annalen der Physik* entitled *Über die sogenannten flüssigen Krystalle* (On the so-called liquid crystals). Lehmann was quick to make a response. By the end of 1901 his article *Flüssige Krystalle: Entgegnung auf die Bemerkungen des Herrn Tammann* (Liquid crystals: a rebuttal of Herr Tammann's comment) had already appeared. The second article received similar short shrift.

Tammann's scepticism was at the very least understandable, given the knowledge of thermodynamics at the time. Moreover, the colloidal hypothesis was the obvious parsimonious explanation of the turbidity data. The most difficult question concerned the apparent anisotropic optical properties of the droplets. Tammann was inclined to consider this problem a detail which would be resolved in the course of time, but Lehmann regarded it (correctly, as it turned out) as a central question. The elimination of the colloidal hypothesis, which involved careful experimentation, was an essential step in the understanding of the new phenomenon.

In the end careful experiments by a number of workers, including Lehmann, and, in particular, the physical chemist Rudolf Schenck of Marburg, demonstrated that liquid crystallinity persisted in the limit of chemical purity, the sceptics' contrary prediction notwithstanding. Already in 1904, Both Coehn,[16] and separately Bredig and Schukowsky,[17,18] had attempted to verify the emulsion hypothesis by means of electrophoresis. This effect, sometimes also known as cataphoresis, involves using an electric field to remove colloidal particles from the 'emulsion'. No effect was observed, and the authors concluded that liquid crystals were not emulsions.

The more complete robust demonstration by Schenck of the falsity of the colloidal picture is included in our collection as article A4. This *coup de grace* was administered at a meeting of the *Deutsche Bunsengesellschaft* (the German Physical Chemistry Society), held in Karlsruhe (Lehmann's home territory), in the afternoon of 3 June, 1905. The proceedings of this meeting are reported in the 1905 edition of the *Zeitschrift für Elektrochemie*. Schenck's article (a presumably rather faithful written version of his oral presentation) is designed as a brief overview of the liquid crystal field to prepare the audience for Herr Geheimrat Lehmann's kind demonstrations of liquid crystal behaviour, due to take place later in the afternoon.

Schenck's[19] article emphasised that attempts to obtain the phase separation which must occur if the turbidity were a colloidal phenomenon have failed. Observations in the literature, he claimed, were in fact due to impurities in the system. He himself had personally shown this using the experimental samples of, for instance, Rotarski,[20] who had claimed evidence for phase separation. He had found no evidence of phase separation at the onset of the anisotropic phase, and indeed found discontinuities in density and viscosity. He repeated the experiment for water–phenol mixtures and found an entirely different signature at the transition. He used an elaborate argument due to Eötvos – omitting the source of the argument! – to show that in the clear liquid there could be no isomerism because the specific heat obeys a law of corresponding states.

We also include Tammann's restatement of his position immediately after Schenck's talk. This statement was only barely disguised as a question. The final statement of the official report of the meeting (also included) is replete with irony, as the session chair, the well-known physical chemist van't Hoff, made brave attempts at compromise in order to avoid the real possibility of physical violence. This was socially necessary, but with the benefit of hindsight, scientifically not so. The controversy gradually died down. The most plausible explanations had clearly failed. Nevertheless, there remained neither microscopic explanation nor macroscopic description of the nature of the phenomenon, even if some of the basic elements were now becoming clearer.

The big debate as to the nature of liquid crystallinity may overshadow another fundamental question which still remained in the early years of the twentieth century, and to which the discovery of liquid crystals led to much confusion. This concerned the relationship between thermodynamic phase and chemical constitution. In 1821 the German chemist Eilhard Mitscherlich discovered that many crystal compounds could appear in different solid forms, while retaining identical liquids and vapours. This phenomenon is known as polymorphism. The natural conclusion was that the molecules of the polymorphic modifications are chemically identical, and that polymorphism is only caused by different space lattice structures of the same molecules.

The discovery of *liquid* crystals muddied this picture. For now the lack of a unique liquid modification gave rise to doubts about the original interpretation of polymorphism. Lehmann, who believed firmly in the *physical* existence of his liquid crystals, called for a revision of the dogma. In his 1904 book, he argued:

> The behaviour of flowing and liquid crystals demonstrates that the molecules of polymorphic modifications are unambiguously different from each other. It follows that the optical properties of a substance are not determined by the type of molecular aggregation . . . but must be caused by the special molecular structure.

Shortly afterward he claimed:

> . . . my previous investigations on polymorphism, especially on liquid crystals, have shown that enantiotropic transformations are not caused by the aggregation of the same molecules into another space lattice. Rather changes in the molecules themselves must occur.

The fundamental problem of different molecular structures in different polymorphic phases remained open into the early fifties. Kast,[21] in 1939 claimed:

> . . . the molecules will lose their extended structure [*in liquid crystalline phases*] at the clearing point . . .

and even as late as 1955[22] he maintained the point of view that:

> . . . free rotation of alkoxy side groups does not occur until one arrives at the clearing point.

13

Eventually, however, the first infrared spectroscopic investigations on liquid crystals, carried out by Maier and Englert[23] in 1957, told a different story:

> The infrared spectrum of a cr.-l. compound is identical with that of the isotropic liquid phase in the band position, in intensity, and in half-width. No significant changes in the structure of a single molecule appear during the phase transition. As a result, the earlier idea — that there is a "thawing" of the intramolecular rotational degrees of freedom at the isotropic-crystalline-liquid phase transition — no longer holds.

The story has a mixed moral. Certainly the vociferousness and vituperation with which Tammann pursued his liquid crystalline quarry turned out to be at the very best misplaced. Indeed Lehmann's liquid crystals did have some liquid and some crystal qualities. For these qualities the term 'liquid crystal' was not wholly inappropriate, at least at the time. We shall see in article B1 that Georges Friedel criticised this terminology with arguments that were formulated in a much better manner than Tammann's. But too literal an attachment to the concept of a *liquid crystal* led to errors which took more than half a century to correct.

Let us now return to 1905. The search for the microscopic origin of liquid crystallinity was now really on, as the number of research groups studying the phenomenon began to increase, though still not transcending the boundaries of the German-speaking world. A prerequisite for any detailed picture involves knowing what kind of materials are able to form liquid crystals. New materials are provided by synthetic chemists, and in this case the synthetic chemists are usually organic chemists, in that liquid crystals without a large number of carbon atoms seemed rare. Gattermann had lost interest in liquid crystals, but a new champion emerged in Daniel Vorländer of the University of Halle.

To begin with liquid crystals (or *crystalline liquids*, as Vorländer insisted on calling them right through his long career) cannot have seemed very promising. The first paper which appeared from the Halle group on this subject (which we do not reproduce) was signed only by his graduate students F. Meyer and K. Dahlem.[24] The paper was entitled 'Azo-and Azoxybenzoic acid esters', and was published in the *Annalen der Chemie* in 1903. In a rather detailed commentary on the oxidation of an intermediate chemical, they mention, almost in passing, that they synthesise a material which

> ...exhibits a double melting point,...(which)...reminds one of the strange materials investigated by O. Lehmann and R. Schenck.

Indeed just to check, they asked Schenck to repeat the observation. However, these remarks occupy less than half a page in a fifteen page article, whose main thrust concerns more technical aspects of organic synthesis.

What is noteworthy in this article to the historian of science is that it is not co-authored by Vorländer himself. It was an unwritten but nevertheless extremely rigid convention of the day that the Director of an institute publish the results of his co-workers under his own name. Only in passing would the Director have mentioned that these results had been obtained (in this case) 'together with F.M and K.D'. It stretches plausibility to suppose that Vorländer omitted his own name out of a spirit of generosity, in order to promote his subordinates' careers. Much more likely is that Vorländer missed the significance of these observations and, furthermore, aware of the controversial nature of the crystalline liquid hypothesis, was simply keeping his distance from a possibly dangerous scientific controversy.

Whatever his motivation in 1903, the Halle experiments continued, and Vorländer finally realised that he was tapping a rich vein. Three years later he returned to the subject, this time in an article authored by himself alone, in the premier German journal *Berichte der Deutschen chemischen Gesellschaft*. So influential was this journal in its day, that references to articles in it were often abbreviated just by the letter *B*! We include this paper, entitled simply 'On crystalline-fluid substances', as article A5 in our collection.

By now it was becoming clear that this was a *class* of substances, and not just a peculiarity. Starting with Reinitzer's cholesteryl benzoate, and Gattermann's *p*-azoxyanisole, Schenck recorded the existence of 24 crystalline liquid compounds; by the time Vorländer wrote his article in the following year, the number had increased to 35, and that did not include oleates.

Article A5 is significant for a number of reasons. It is Vorländer's first major contribution to a subject in which for the next 30 years he was the dominant synthetic chemist. At least partly, Vorländer is simply staking a claim: crystalline fluid research is happening here in Halle. But beyond this, Vorländer is beginning to apply the tools of a trained organic chemist to the problem in hand.

What are the essential molecular properties for a compound to form a crystalline liquid? How can one change a singly-melting fluid into a double-melting fluid? Paper A5 is the first systematic study of how to manipulate organic molecules in order to favour or disfavour liquid crystal properties. He emphasises:

> ... we have been guided by the idea that the formation of the crystalline-liquid state could be caused or promoted by the same atomic groups which also influence other physical properties, such as light refraction, colour, rotatory power, and so on.

He makes tables of compounds from a homologous series, in order to see which of them possesses a crystalline liquid phase, and which do not; which has the highest first melting point, and which the highest second melting point. Gattermann had found that para-azoxyanisole forms a crystalline liquid. Vorländer finds further that that the para-position of the substituents is important; the analogous ortho- and meta- compounds (with the azoxy group placed differently, and at an angle, with respect to the rest of the molecule) do not give rise to crystalline liquids. However, in this article Vorländer has not yet reached any definitive conclusions from his observations.

15

It was his next article, published in the following year in *Berichte*, which we include as article A6, which was to have the really long-term influence on the subject. In it he discerned the first important clue to the real nature of liquid crystals. This very short article is entitled 'On the influence of molecular shape on the crystalline liquid state' and it reported that most liquid crystalline materials were constructed from molecules with a strongly *rod-like* structure. Over the years Vorländer and his students synthesised hundreds of liquid crystalline compounds. An interesting discovery was that amongst the slimy liquid crystals were many soaps and soap-like compounds. And in 1908, he detected for the first time that a given substance may exhibit more than one liquid crystal phase.[25] Already by 1908 he had enough material to feel able to report his accumulated results, not just in a series of papers, but in a book, which he entitled *Kristallinisch-flüssige Substanzen* (Crystalline-liquid Substances).[26]

One unintended consequence of the book was an unedifying battle, fought out on the pages of the *Annalen der Physik*,[27] concerning priority over who had *really* discovered liquid crystals. We quote the first paragraph of Lehmann's opening gambit in his battle to defend *his* discovery:

The history of liquid crystals
O. Lehmann

In a book published recently, the origin of the liquid crystal concept has been described. The book contains many new observations and is a valuable addition to the *chemical* liquid crystal literature. However it presents a picture of the origin of the liquid crystal concept which is seriously misleading. One might suspect from this book that Herr Fried. Reinitzer, who is professor in the Botanical Institute at the University of Graz, had actually discovered the phenomenon in 1888, and I had merely renamed it. I present here, for the first time, the full story.

Reinitzer replied some months later in *Annalen der Physik*. He set the record straight as follows:

From these arguments it is indisputably clear that the unambiguous concept of flowing crystals was recognised by Lehmann first of all by studying my derivatives. Furthermore, it is obvious that the perception is due to Lehmann, but that I also contributed considerably in this matter.... One should admit that the credit of discovering the phenomenon ought to be attributed to me.

Reinitzer did not continue his investigation of cholesteryl derivatives. This plea for scientific recognition was his last contribution to the liquid crystal story. His career continued as University Rector and Director of the Botanical Institute in Graz.

It was not only against Reinitzer that the barbs of Lehmann's pen were applied. Here he is again, in 1914, in a direct reproach to Vorländer:[28]

> Mr. D. Vorländer believes he should reproach me for some errors, but in reality these errors do not exist. However, his remarks are of value. Misunderstandings between the points of view of Vorländer and myself have often arisen by confusion. Hopefully these misunderstandings can be put behind us. Vorländer has obviously only concerned himself with the investigation of which chemical constitution of a substance is necessary for the appearance of a liquid-crystalline modification. In this endeavour he has achieved great success. However, he has only rarely been involved with physical and crystal optical investigations. As a result it is not surprising that my ideas seem extremely unfamiliar to him.

We shall return in the next section to Vorländer and also to further debates of this type!

A full study of Lehmann's scientific work and impact goes well beyond the scope of this book. One further aspect of his scientific interests of particular interest to the modern reader concerns his correspondence with the zoologist and natural philosopher Ernst Haeckel (1834–1919). Haeckel was an ultra-Darwinist who attempted to combine the natural laws of organic and inorganic matter in the context of the same set of physical laws. This contrasted with traditional ideas of a 'vital force' breathed (by the almighty or otherwise) into living beings.

In many ways he was before his time, in that almost all of the fundamental science which would have enabled him to achieve this task was lacking in the mid-nineteenth century. His programme was premature by at least a hundred years and maybe by a good deal more. Despite Haeckel's commitment to the scientific method, the result was a set of speculations about natural relationships many of which were not in any way empirically based. Some were correct – for example he was the first to suggest that the seat of inheritance was in the nucleus of the cell. Haeckel has developed a reputation as a bit of a crank because of the tenacity with which he held to implausible ideas.

Lehmann was attracted to Haeckel's ideas. Even while at school he had studied Haeckel's books and looked for a link between the areas of minerals, plants and animals. Haeckel was most fascinated by Lehmann's liquid crystal studies and believed liquid crystals to be a missing link between inorganic and living systems. Their correspondence lasted between 1906 and Haeckel's death in 1919, although they never met personally. Lehmann was sufficiently influenced to write a book on the topic.[29] Haeckel's last book *Kristallseelen – Studien über das anorganische Leben* (Crystal souls – Studies on inorganic life), published in 1917, included a chapter entitled *Rheokristalle* (Rheocrystals), devoted to Lehmann's observations of liquid crystals.

Here he discusses the entire life of these rheocrystals, which he thought of as an intermediate stage between simple materials and life itself:

> By critical comparison between spherical rheocrystals (myelin spherules) and primitive cytodes (*chroococcus*) the traditional as well as artificial border between inorganic and organic nature is finally removed.

As we have seen, Lehmann's ideas about the nature of his liquid crystals did not meet with uncritical acceptance by his peers. His links with Haeckel were a further hindrance to his credibility. Indeed, following one of his lectures on liquid crystals a colleague ironically asked him, 'What's this about your liquid crystals? Can they now eat?'

What was lacking now was any input from theoretical physics. In the years 1907–9 the first serious attempt at a mathematical theory of liquid crystals was made by Emil Bose from the Physical Chemistry department of the University of Danzig (now Gdansk in Poland, but then a free German-speaking city). He published three papers[30] in the *Physikalische Zeitschrift*. The first of these was entitled 'For and against an emulsion structure for crystalline fluids'. The basic conclusion (perhaps for the first time, but certainly not for the last time, in the history of the subject!) is that rather than talk of liquid crystals or crystalline fluids, it would be better to refer to *anisotropic* fluids, for there is no real crystal structure. At any rate, following Schenck, Bose finds against the emulsion picture despite having been initially attracted by it.

We include Bose's second paper in our collection as article A7. This is the most influential of the three articles; the third is an attempt to compare theory with experiment. At this time there were really only two sets of statistical mechanical models on which to base a theory. One of these strands was van der Waals's 1873 model of the fluid equation of state.[31] This leads naturally to the idea of a liquid–gas phase transition. The other strand included Pierre Weiss's (very new) model of ferromagnetism,[32] now known as the Curie-Weiss theory, as well as the ideas of Paul Langevin.[33]

Bose tried, not entirely successfully, to draw eclectically from both of these. He was led to introduce the idea of molecular *swarms (Molekülschwärmen)*. We include this paper, despite the eventual failure of Bose's ideas, because this idea remained influential in liquid crystal science for a long time. For many years the literature contained earnest but awkward discussions about the difference between a *chemical* molecule (the real one!) and the *physical* molecule. The latter was supposed to be the swarm of molecules pointing more or less in the same direction. In any event the swarm theory, as it became known, passed, albeit temporarily, into the canon. As late as 1957, it was quoted approvingly in the major review article of the day in *Chemical Reviews*.[34]

What is wrong with Bose's paper is that although his physical motivation was correct (he is searching for a theory of anisotropic fluids), his mathematical starting point was not, in that he started out from the van der Waals theory of fluids. He was compelled to introduce molecular anisotropy in a forced, unnatural and essentially phenomenological manner. The result was that he missed the point, despite realising that the important physics lay in the Weiss and Langevin theories of magnetism. Shortly afterwards Bose died tragically young at only 37. Had he lived, it seems likely that he would have been able to reformulate his ideas and obtain a good molecular field theory of liquid crystals before the First World War.

In the spring of 1909 Lehmann, by now 54 and an established figure, visited Geneva and Paris, and at each venue gave a long seminar accompanied by experimental demonstrations.[35] The visits seem to have been a success, for they inspired the formation of a French school of liquid crystal science which has remained influential to this day.

His host in Paris was the eminent crystallographer and member of the Academy of Sciences Fréderic Wallerant, who held a chair at the École Normale Supérieure. Amongst those influenced by Lehmann's visit were Charles Mauguin, Georges Friedel and François Grandjean. Mauguin was at the time Wallerant's assistant in Paris, Friedel was director of the School of Mines at St Étienne, with Grandjean his assistant. Lehmann's lecture turned Mauguin toward a study of liquid crystals. He must have established a warm relationship with Lehmann, for two years later he paid him a return visit in Germany. Both groups started working in liquid crystals.

We include two of Mauguin's early papers, published in 1911 as articles A8 and A9 in our collection. It is noteworthy that in fact a digest of this work was also published in the *Physikalische Zeitschrift*, presumably at the behest of Lehmann, who felt that they should be available in the German literature as well. Mauguin's studies were carried out using Gattermann's by-now-standard liquid crystals azoxyanisole and azoxyphenetole. The more substantial of these, article A9, simply entitled 'On Lehmann's liquid crystals', was published in the *Bulletin de la société française de mineralogie*.

Mauguin concentrated on the behaviour of a liquid crystal confined between plates in thin layers, of thickness between 10 and 150 microns, i.e. roughly of the dimensions on which present-day studies are carried out. This is to a certain extent in contrast to Lehmann's work, in that much of Lehmann's studies concerned droplets. His studies involved both parallel light and converging light, and of course he used a polarising microscope. The use of converging light – conoscopy – was necessary in order to detect birefringence in the direction of view, for otherwise the sample appeared isotropic.

A number of important ideas appear in Mauguin's article. The extinction reappears when the sample is heated and subsequently cooled. In modern language, it is the director whose orientation is retained by the surface. The *memory effect*, as it has come to be known, was thought by Lehmann to depend on films of oriented molecules adhering to the glass slides. Mauguin further noticed that the optical phenomena persisted in a moving liquid, a fact apparently inconsistent with Lehmann's conception of a liquid crystal as a crystal which flows. He was able to calculate the degree of optical birefringence from his experiments.

Most profoundly for the subsequent history of the subject, he examined 'birefringent liquid films with a helicoidal structure'. These were films with non-coincident surface films which no longer extinguished light between crossed polarisers. Mauguin found that under certain circumstances the polarisation of incident light is twisted in such a sample, and in general that an incident linearly polarised beam exits the sample elliptically polarised. Furthermore he was able to show theoretically that this is the consequence of what we would now call a twisted nematic cell. He demonstrated that if the ratio of the twist pitch to the wavelength of light is long, the polarisation does indeed follow the twisting birefringence. Following these observations, to this day, light undergoing weak polarisation rotation under this circumstance is said to be in the *Mauguin régime*.

Article A9, which appeared in the premier French scientific journal, the *Comptes rendus de l'Académie des Sciences*, is but a brief note, but once again of great importance. The problem he had had in his previous work had always been to obtain sufficiently large so-called crystalline (and in reality oriented liquid crystalline) domains. In this paper he reported observations that large domains could be obtained using magnetic fields. In other words, fields orient liquid crystals. He then went on to discuss the competing effects of glass plates orienting in the plane of the sample and a magnetic field orienting in the plane perpendicular to it, and vice versa. Extinction between crossed nicols could be induced or destroyed by the application of the field.

His final observation is particularly significant with hindsight. The initial sample is set up with what we would call random planar boundary conditions, so that (as we now know) the static fluctuations are sufficient to cause the sample to appear cloudy. A magnetic field is applied perpendicular to the sample. The molecules reorient giving the '... equivalent of a film perpendicular to the axis, except for a thin layer next to the glass plates ....'

The experimental programme of Friedel and Grandjean was contemporaneous with that of Mauguin. There was a healthy rivalry between the two research groups. In part this was born from the competition between Parisian patricians and provincial practical men. Reading between the lines it is clear that experimental progress ran in parallel, and that there was an understandable unwillingness by either group to grant priority to the other (this spirit is also observable in early German work!). Perhaps this purely French rivalry explains the ease with which Mauguin and Lehmann had established a warm rapport.

We have chosen as article A10 in our collection an important contribution by Grandjean in 1916. He was trying to distinguish intrinsic and extrinsic liquid crystalline effects. His article 'Orientation of anisotropic liquids on crystals' is a careful examination of the effect of crystal cleavage planes on the liquid crystals sitting on the crystal substrates. Although most of this article is devoted to detailed observations of liquid crystalline anomalies induced by defects in the crystal surface, probably the most significant observation is almost buried in one paragraph at the top of the fourth page in an article which continues for forty-nine.

Grandjean was observing drops of what he identified as focal-conic liquid crystals. These are Lehmann's flowing crystals, identified not by their degree of fluidity, but rather by their optical signature. The focal conics are the characteristic curves seen in the microscope when the flowing crystals are viewed. The focal conic liquid crystals were to be contrasted with the liquid crystals '*à noyaux*' ('with nuclei'), which is how the French school described the Schlieren texture.

The drops were attached to crystal surfaces. He found an effect which he called *phénomène des gradins*. His drops were divided into regions of apparently more or less constant height separated by narrow steps. These have passed into current liquid crystal terminology as *Grandjean terraces*. These terraces turned out to be the essential clue to the nature of the liquid crystals, for they were in fact the borders between regions in which n and $n+1$ smectic layers were to be found on a surface. We shall see how this story plays out in Section B.

Before ending this section, it is interesting briefly to discuss two further theoretical attempts at a mathematical theory of liquid crystals during this initial period. In 1916 the liquid crystal problem came to the notice of Max Born (1882–1970), professor of theoretical physics in the premier-division university of Göttingen. Born, it will be recalled, was an instrumental figure in the development of quantum mechanics during the 1920s.

We have included Bose's article on the swarm theory because of its influence, notwithstanding the fact that theory does not explain the data and is theoretically incomplete. Born's paper of 1916[36] adapted the Curie-Weiss molecular field theory of magnetism to the liquid crystal context. His basic assumption was that liquid crystalline molecules carried an electric dipole and that this quantity drives the liquid crystallinity. This theory is theoretically consistent – it *is* the Curie-Weiss theory for all intents and purposes – but unfortunately the basic assumption is wrong, because it soon turned out[37] that there were non-dipolar molecules which exhibited liquid crystalline phases. As a result, this paper has been relegated to the status of a historical footnote.

In 1917 it was the turn of Grandjean to construct a molecular field theory for liquid crystals. Because of the war he was unaware of Born's work. Perhaps for the same reason his article too remained unread and thus *ipso facto* also a historical footnote. We have included this paper elsewhere in this collection as C1, where we shall discuss it further.

The heroic period of the liquid crystalline state comes to an end around 1920. By this time a large amount of data on liquid crystals had been collected, and the number of compounds exhibiting liquid crystallinity was growing by the day. The time was ripe for the emergence of a correct picture of the molecular basis of the phenomenon, and the beginnings of a sensible description both on the molecular and macroscopic scales. This will be the topic of Section B of our collection.

References

1. Quoted by Reinitzer from an article by Gmelin in the Handbuch der Organischen Chemie **4**, 2093ff.
2. B. Raymann, Bull. Soc. Chim. (Paris) **47**, 898 (1887).
3. W. Löbisch, Ber. **5**, 513 (1872).
4. P. Planer, Ann. **118**, 25 (1861).

5. F. Reinitzer, Wiener Sitzber. **94**, 719 (1888), *ibid.* **97**, 167 (1888). These are papers presented to the Vienna Scientific Society covering more or less the same material as paper A1. The latter paper deals with the double melting properties of cholesteryl acetate and hydrocarotene benzoate.

6. An incomplete list of Lehmann's early papers includes:
   Einige Fälle von Allotropie (*Some cases of allotropy*), Zeitschrift für Kristallographie und Mineralogie **18**, 464–7 (1890);
   Ueber tropfbarflüssige Krystalle (*On drop-forming liquid crystals*), Wiedemann's Annalen für Physik und Chemie **40**, 401–23 (1890);
   Ueber krystallinischer Flüssigkeiten (*On crystalline liquids*), Wiedemann's Annalen für Physik und Chemie **41**, 525–37 (1890);
   Die Struktur krystallinischer Flüssigkeiten (*The structure of crystalline fluids*), Zeitschrift für physikalische Chemie **5**, 427–35 (1890);
   Ueber künstliche Färbung von Krystallen und amorphen Körpern (*On artistic colours in crystals and amorphous bodies*), Wiedemann's Annalen für Physik und Chemie **51**, 47–76 (1894);
   Ueber Contactbewegung und Myelinformer (*On contact motion and myelin formation*), Wiedemann's Annalen für Physik und Chemie **56**, 771–88 (1895).

7. *Struktur, System und magnetisches Verhalten flüssiger Krystalle und deren Mischbarkeit mit festen* (*Structure, system and magnetic behaviour of liquid crystals and their miscibility in solid crystals*), Annalen der Physik (series 4) **2**, 649–705 (1900).

8. O. Lehmann, *Flüssige Kristalle* (Wilhelm Engelmann, Leipzig, 1904).

9. Lehmann op. cit. p. 52.

10. L. Gattermann, *Synthese aromatischer Aldehyde* (*Synthesis of aromatic aldehydes*), Liebigs Ann. Chem. **347**, 347–86 (1906).

11. J. Perrin, *Brownian movement and molecular reality* (Taylor and Francis, London, 1910).

12. See e.g. W. Friedrich, P. Knipping and M. Laue, Sitzber. Bayer. Akad. Wiss. **303**, 363 (1912); also: W. Friedrich, *Die Geschichte der Auffindung der Röntgenstrahlinterferenzen*, Die Naturwissenschaften **10**, 363–6 (1922), Paul Knipping, *Zehn Jahre Röntgenspektroskopie*, Die Naturwissenschaften **10**, 366–9 (1922);
   W.H. Bragg and W.L. Bragg, *The reflection of X-rays by crystals*, Proc. Roy. Soc. **A88**, 428–38 (1913);
   W.L. Bragg, *The structure of some crystals as indicated by their reflection of X-rays*, Proc. Roy. Soc. A**89**, 248–77 (1913).

13. Quincke was professor of physics first in Berlin and later in Heidelberg.

14. See e.g. G. Quincke, *Ueber freiwillige Bildung von hohlen Blasen, Schaum und Myelinformen durch ölsaure Alkalien und verwandte Etrscheinungen besonders des Protoplasmas*, Wied. Ann. **53**, 593–631 (1894); G. Tammann, *Ueber die Grenzen des festen Zustandes*, Wied. Ann. **62**, 280–99 (1897).

15. See: G. Tammann, *Ueber die sogennanten flüssigen Krystalle*, Ann. Physik **4**, 524–30 (1901); O. Lehmann, *Flüssige Krystalle, Entgegnung auf die Bemerkungen des Hrn G. Tammann*, Ann. Physik **5**, 236–9 (1901);
   G. Tammann, *Ueber die sogennanten flüssigen Krystalle II*, Ann. Physik **8**, 103–8 (1902);
   O. Lehmann, *Ueber künstlichen Dichroismus bei flüssigen Krystallen und Hrn. Tammann's Ansicht*, Ann. Physik **8**, 908–23 (1902);
   O. Lehmann, *Näherungsweise Bestimmung der Doppelbrechung fester und flüssiger Kristalle*, Ann. Physik **18**, 796–807 (1905);
   G. Tammann, *über die Natur der "flüssigen Kristalle"* III, Ann. Physik **19**, 421–5 (1906).

This exchange unearthed previously neglected papers on the double-melting phenomenon, going back all the way to W. Heintz in 1849. He was quoted as having written an article in the 1849 Yearly Report on Progress in Chemistry, p. 342 in which he made just such an observation. (A more accessible reference to Heintz's work is in W. Heintz, J. prakt. Chem. **66**, 1 (1855).) Lehmann is sceptical as to whether Heintz's double-melting is the same phenomenon as occurs in liquid crystals, stating that 'allegedly all fatty acids have double melting points.' For the fatty acid case in fact, surprisingly, Lehmann is a supporter of the Tammann school of thought, noting (while citing Gibbs) that 'a liquid does not have necessarily to be a single phase. Milk, ink, colloidal gold solution and egg-white are examples of two-phase fluids'.

16. G. Coehn, *Über Flüssige Kristalle*, Z. Elektrochemie **10**, 856–7 (1904).
17. G. Bredig and N. Schukowsky, *Prüfung der nature der flüssigen Kristalle mittels Kataphorese (Test of the nature of liquid crystals using cataphoresis)*, Ber. Dt. Chem. Ges. **37**, 3419–25 (1904).
18. Georg Bredig was a leading German physical chemist in the early part of the 20th century. Born in 1868 in Glogau, he obtained his Ph.D in Leipzig in 1894 as a student of Wilhelm Ostwald. After postdoctoral periods with van't Hoff and Arrhenius, he obtained his habilitation in 1901, subsequently working in Heidelberg and Zürich. In 1911 he became full professor of physical chemistry in Karlsruhe, and became Rector there in 1921. His main scientific activities concerned catalysis and optically active compounds, decomposition processes of metal sols, and kinetics of reactions catalysed by protons. In 1933, following the Nazi rise to power in Germany, he was removed at the age of 65 from his post because of his Jewish background. He emigrated to the Netherlands in 1939, escaping to America in 1940, where even at the age of 72 he was offered a post at Princeton. This he was unable to take up as a result of ill-health. He died in New York in 1944. Schukowsky was probably his graduate student.
19. See also: R. Schenck: *Kristalline Flüssigkeiten und flüssige Kristalle* (Wilhelm Engelmann, Leipzig, 1904).
20. T. Rotarski, Ann. Phys. **4**, 528 (1901).
21. W. Kast, *Anisotrope Flüssigkeiten*, Z. Elektrochemie **45**, 184–200 (1939).
22. W. Kast, *Die Molekul-Struktur der Verbindungen mit kristallin-flüssigen (mesomorphen) Schmelzen (Molecular structure of compounds with crystalline-liquid (mesomorphic) melts)*, Angew. Chem. **67**, 592–601 (1955). Recent studies, however, suggest that in "banana"-shaped mesogens, with a complex molecular structure, different conformations in different liquid crystalline phases can in fact occur. See I. Wirth *et al.*, J. Mater. Chem. **11**, 1642–1650 (2001).
23. W. Maier and G. Englert, *Infrarotspektroskopische Untersuchungen an Substanzen mit kristallin-flüssigen Phasen (IR spectroscopic studies of compinds with crystalline liquid phases)*, Z. Physik. Chem. **12**, 123–7 (1957).
24. F. Meyer and K. Dahlem, *Azo- und Azoxybenzoäureester*, Ann. Chemie **331**, 331–46 (1903).
25. D. Vorländer, *Neue Erscheinungen beim Schmelzen und Krystallisieren*, Z. Physik. Chem. **57**, 357 (1907).
26. D. Vorländer, *Kristallinisch-flüssige Substanzen* (Stuttgart, 1908).
27. O. Lehmann, *Zur Geschichte der flüssigen Kristalle*, Annalen der Physik **25**, 852–60 (1908);
    F. Reinitzer, *Zur Geschichte der flüssigen Kristalle*, Annalen der Physik **27**, 213–24 (1908);

O. Lehmann, *Bemerkungen zu Hr. Reinitzers Mitteilung über Geschichte der flüssigen Kristalle*, Annalen der Physik **27**, 1099–1102 (1908).

28. O. Lehmann, *Die optische Anisotropieder flüssigen Kristalle*, Physikal. Zeit. **15**, 617 (1914).

29. First he wrote an article: O. Lehmann, *Scheinbar lebende Kristalle* (Apparently living crystals), Biolog. Zentralblatt **28**, 513–26 (1906); and then two years later followed it up in a longer monograph: O. Lehmann, *Die scheinbar lebenden Kristalle* (Apparently living crystals) (Verlag Schreiber, Esslingen and München, 1908). By this time Lehmann was aware of the liquid crystalline potential of certain organic molecules, in particular myelin. This had been studied by Virchow as early as 1854, and its birefringent properties had been observed by Mettenheimer in 1858. We discuss this subject in more detail in Section E.

30. The two papers in this series not reprinted here are:
    E. Bose, *Für und wider die Emulsionsnatur der kristallinischen Flüssigkeiten*, Phys. Zeit. **8**, 513–7 (1907); *Zur Theorie der anisotropen Flüssigkeiten*, Phys. Zeit. **10**, 230–44 (1909). These are not the only papers by Bose on liquid crystals. He took an active interest in the Tammann–Lehmann controversy, only rejecting the colloidal hypothesis after much thought. See also E. Bose and F. Conrat, *Über die Viskositätsanomalien beim Klärpunkt sogenannter kristalliner Flüssigkeiten*, Phys. Zeit. **9**, 169–73 (1908) and E. Bose, *Über Viskositätsanomalien anisotroper Flüssigkeiten im hydraulischen Strömungszustande (Ein Experimentalbeitrag zur Schwarmtheorie der kristallinen Flüssigkeiten)*, Phys. Zeit. **10**, 32–86 (1909).

31. J.D. van der Waals, *Over de continuiteit van den gas- en vloeistoftoestand (On the Continuity of the Liquid and Gaseous State)* [in Dutch; Ph.D thesis, University of Leiden (Netherlands), 1873]. This thesis explained, at least qualitatively, experimental results by Thomas Andrews, the successor of Michael Faraday at the Royal Institution in London. Andrews was studying the pressure–volume–temperature relations in CO_2, and presented his results to the Royal Institution in the Bakerian lecture in 1869. It was in this lecture that he introduced the term 'critical point' to describe the thermodynamic point at which liquid and vapour become identical.

32. P. Weiss, *L'hypothèse du champ moléculaire et la propriété ferromagnétique*, J. Phys. Rad **6**, 661–90 (1907); see also the German version in: *Molekulares Feld und Ferromagnetismus*, Phys. Zeit. **9**, 358–67 (1908). Note that Bose cites the German version.

33. P. Langevin, Ann. Chim. Phys. (Paris) **5**, 70 (1905).

34. G.H. Brown and W.G. Shaw, *The mesomorphic state: liquid crystals*, Chem. Rev. **57**, 1049–1157 (1957). This encyclopaedic review article played a (perhaps *the*) major role in the post-war revival of interest in liquid crystals; see Section C.

35. O. Lehmann, *Les cristaux liquids*, J. Phys. (Paris) **7**, 713–35 (1909);
    O. Lehmann, *Cristaux liquids et modèles moléculaires*, Archives des sciences physiques et naturelles (Geneva) **28**, 205–26 (1909).

36. M. Born, *Über anisotrope Flüssigkeiten. Versuch einer Theorie der flüssigen Kristalle und des elektrischen KERR-Effekts in Flüssigkeiten*, Sitzungsber. Preuss. Akad Wiss. **30**, 614–5 (1916). (This is the report of the maths and physics session of the Prussian Academy of Sciences dated 25 May 1916).

37. G. Szivessy, *Zur Bornschen Dipoltheorie der anisotropen Flüssigkeiten*, Z. Physik. **34**, 474–84 (1925).

A1

Monatshefte für Chemie (Wien) **9**, 421–41 (1888)

Beiträge zur Kenntniss des Cholesterins

von

Friedrich Reinitzer,

Assistent am k. k. pflanzenphysiologischen Institute der deutschen Universität in Prag.

Aus dem pflanzenphys. Institute des Prof. Ad. Weiss
an der k. k. deutschen Universität in Prag.

(Vorgelegt in der Sitzung am 3. Mai 1888.)

Vor etwa $1\frac{1}{2}$ Jahren theilte ich das Ergebniss einiger Untersuchungen[1] über ein in der Wurzel der Möhre vorkommendes Cholesterin mit, welches von Aug. Husemann den Namen Hydrocarotin erhalten hat. Ich führte damals aus, dass dasselbe, wenn auch nicht in der von Husemann vermutheten Art, mit dem rothen Farbstoff der Möhren, dem Carotin, in Zusammenhang zu stehen scheine und durch letzteres wieder mit dem Chlorophyllfarbstoffe. Es musste daher von Interesse sein, die nähere Natur dieses Körpers zu ergründen. Da derselbe jedoch schwierig in grösserer Menge zu beschaffen ist, anderseits aber die Cholesterine untereinander eine grosse Ähnlichkeit ihrer Eigenschaften zeigen, so beschloss ich, die diesbezüglichen Vorarbeiten erst mit dem gewöhnlichen Cholesterin vorzunehmen, welches leicht in grösserer Menge erhalten werden kann und über dessen Natur man gleichfalls noch völlig im Unklaren ist. Erst auf Grund der hiebei gesammelten Erfahrungen soll dann das ungleich kostbarere Hydrocarotin näher untersucht werden. Im Folgenden will ich einige Ergebnisse dieser Vorarbeit mittheilen.

---

[1] Sitzgsber. d. kais. Akad. d. Wissensch. Bd. 94, S. 719.

A1
Monatshefte für Chemie (Wien) **9**, 421–41 (1888)

CONTRIBUTIONS TO THE UNDERSTANDING OF CHOLESTEROL

by

Friedrich Reinitzer

Assistant at the Imperial Institute for Plant Physiology at the German
University, Prague
From the Imperial Institute for Plant Physiology of Professor Ad Weiss
at the German University, Prague
(Presented at the meeting of 3 May 1888)

About 1½ years ago, I reported the results of some studies[1] of a cholesterol occurring in the root of the carrot, which had been given the name hydrocarotene by A. Husemann. At that time I stated that this compound was related to carotene, the red pigment of carrots (although not in the manner suggested by Husemann) and thus via the carotene to the chlorophyll pigments. It was therefore of interest to investigate the nature of this substance more closely, but it is difficult to obtain in large quantities. On the other hand the cholesterols display a great mutual similarity of properties, and so I decided to undertake a preliminary study along this line first with ordinary cholesterol. This can easily be obtained in large quantities and in addition its nature remains completely unclear. Then, on the basis of the experience gained in this way, I intended to investigate more closely the far more costly hydrocarotene. In what follows I would like to communicate some of the results of this preliminary work.

The cholesterol used for the experiments described here was obtained from the factory of H. Trommsdorff and was purified there by repeated treatment with alcoholic caustic potash. It had a melting point of 147.5 °C (corr. = 148.5 °C). Normally the melting point is reported to be 145 to 146 °C. The value reported here was, however, obtained even in the case of very slow heating. The thermometer used was completely accurate and calibrated in tenths of degrees. Wislicenus and Moldenhauer have reported the melting point to be 147 °C (Annal. d. Chem., Vol. 146, p. 179).

---

[1] Sitzgsber. d. kais. Akad. d. Wissensch., Vol. 94, p. 719.

Molecular weight of cholesterol

The molecular weight of cholesterol has not yet been established with the certainty that would be absolutely necessary as a satisfactorily firm foundation for further studies of the nature of this substance. According to Gmelin (Handbuch d. org. Chemie, Vol. 4, p. 2093), the current most frequently used formula, $C_{26}H_{44}O$, was first derived by Gerhardt, on the basis that this would provide the best agreement with the derivatives. Before then, many other formulae had been proposed. Subsequently Latschinoff and Walitzky attempted to prove that the formula $C_{25}H_{42}O$ would be more likely for cholesterol. As evidence, they particularly cited facts which suggested a relationship between cholesterol and a pentaterpene. Hesse also prefers this formula, because the optical rotatory power of cholesterol is lower than that of phytosterol (Lieb. Annal., Vol. 192, p. 175). In addition, Liebermann ascribes some credibility to Latschinoff's and Walitzky's statements (Ber. d. chem. Gessell., Vol. 18, p. 1803). Recently T. Weyl[1] attempted to establish the molecular weight of cholesterol by determining the vapour density of the cholesterones and cholesterylenes derived from cholesterol. The values obtained in this way, however, are unfortunately incapable of providing complete assurance in this regard. This is because the hydrocarbons mentioned above undergo dissociation and decompose into smaller molecules. Weyl, however, tacitly assumes that they are similar and of the same size, and assigns to them the formula $C_5H_8 = \frac{1}{5}C_{25}H_{40}$. From this the formula $C_{25}H_{42}O$ for cholesterol would naturally follow. The vapour densities obtained, however, are far greater than the vapour density assignable to a compound of the formula C_5H_8. It is therefore entirely possible that the hydrocarbons derived from cholesterol have a higher molecular weight than assumed, and that, in addition, the partial molecules formed during dissociation are not of the same size. The average obtained would then exceed the calculated vapour density.

Thus, interesting and valuable as these studies are, they do not appear to me to be sufficient to decide the question of the molecular weight of cholesterol. I therefore attempted to determine its molecular weight by an exact investigation of the derivatives of cholesterol. For this purpose I used, first of all, cholesteryl benzoate, and attempted to determine the ratio in which benzoic acid and cholesterol are obtained by saponification of this material. The completely pure benzoate was obtained in solution in boiling alcohol using a reflux condenser, then decomposed with an excess of normal alkali and back-titrated with normal sulphuric acid. In this procedure, however, the determination of the end point of titration is very uncertain, and, furthermore the poor solubility of the benzoate is a problem. It was therefore attempted to determine the quantity of cholesterol obtained by saponification of the benzoate. The saponified, cooled and solidified

---

[1] Arch. f Anat. und Physiol., physiolog. Abthlg., 1886, p. 180 (Preceedings of the physiological Society at Berlin, Session of 30 October 1885), extracted in Ber. d. d. chem. Gesellsch., 1886, abstracts, p. 618.

product was filtered off, washed with aqueous alcohol and then with hot water, and then dried and weighed. The filtrate was strongly concentrated, precipitated with water, and after washing out with boiling water, also dried and weighed. It was found, however, that because cholesterol is not wetted by water, and due to the low solubility of cholesterol in aqueous solutions of alkali metal benzoate, a loss of cholesterol occurs. As a result the alkali metal benzoate was very difficult to remove completely from the cholesterol. Therefore this procedure also had to be abandoned.

I therefore proceeded to analyse a bromine derivative, and selected for this purpose the bromoacetate. The acetate was chosen because one might expect that it would change less easily during bromination than pure cholesterol, a fact which was also confirmed by experience. Moreover, the acetate and its bromide are much less soluble in alcohol than the non-acetylated cholesterol and therefore more easily purifiable. The bromination was carried out precisely in the manner reported by Wislicenus and Moldenhauer (Lieb. Annal., Vol. 146, p. 178). The dry, completely pure cholesteryl acetate was dissolved in a little dry, very pure carbon disulphide, and, while cooling with cold water, a solution of chlorine-free bromine in carbon disulphide was added until a permanent yellow colouring occurred. To bring about this condition, 4 g of bromine had to be consumed by 10 g of the acetate. During the reaction, no hydrogen bromide evolved. The liquid was evaporated at ordinary temperatures, at which time it turned dark red with partial decomposition and evolution of hydrogen bromide. The totally amorphous, yellow-coloured evaporation residue was washed with cold water and dried over sulphuric acid under reduced pressure. Both procedures had to be repeated several times in order to remove the last traces of hydrogen bromide. The product was then dissolved in as little ether as possible and precipitated with alcohol. It precipitates out as splendid crystals and can easily be freed of the red mother liquor by washing with alcohol. By repeated recrystallisation from an ether–alcohol mixture and washing with alcohol, the compound can be obtained completely colourless and with an invariable melting point. To determine the formula, two elemental analyses and two bromine determinations were performed on the substance dried over sulphuric acid under reduced pressure. To perform the latter, the compound was dissolved in ether–alcohol mixture; this solution was mixed with as much water as it could tolerate without clouding, and then decomposed with sodium amalgam. After evaporation, it was washed with boiling water and the bromine determined in the aqueous solution as silver bromide. The elemental analysis was performed with cupric oxide in a stream of oxygen in the presence of a silver wire and silver plate.

I 0.5472 g of substance yielded 0.35065 g Br Ag[1]

II 0.3249 g of substance yielded 0.7051 g CO_2 and 0.2446 g H_2O

III 0.3408 g of substance yielded 0.2173 g Br Ag

IV 0.3587 g of substance yielded 0.7790 g CO_2 and 0.2618 g H_2O

---

[1] Reinitzer transposes what would now the usual order of Ag and Br (*ed.*).

Calculated for

$C_{26}H_{43}Br_2 \cdot C_2H_3O_2$	$C_{27}H_{45}Br_2 \cdot C_2H_3O_2$
C: 58.53	C: 59.18
H: 8.03	H: 8.18
Br: 27.85	Br: 27.19
O: 5.58	O: 5.44

Found

	I	*II*	*III*	*IV*	*Average*
C	–	59.19	–	59.22	59.20
H	–	8.38	–	8.12	8.25
Br	27.26	–	27.13	–	27.19

From these volumes, it emerges unequivocally that the compound in question has the formula $C_{27}H_{45} \cdot Br_2C_2H_3O_2$. From the previously reported quantity of bromine for bromination and from the fact that no hydrogen bromide evolved at this time it follows that the substance is a bromine addition product, which also agrees with the bromine obtained by Wislicenus and Moldenhauer. Accordingly, the formula of cholesterol must read $C_{27}H_{46}O$. This result is striking because it does not agree with previous assumptions. The question now arises as to whether this can be brought into agreement with the analyses of the previously known derivatives. If one compares the results of analyses of cholesterol derivatives found by various observers with the percentage composition calculated on the assumption that cholesterol has one of the two formulae, $C_{26}H_{44}O$ or $C_{27}H_{46}O$, then one obtains the following: the analyses of cholesteryl chloride (Planer, Lieb. Annal., Vol. 118, p. 25) and of sodium cholesterylate (Lindenmeyer, Journ. f. pr. Chem., Vol. 90, p. 321) agree better with the formula with 27 carbons. The analyses of the bromides (Wislicenus and Moldenhauer, op. cit.) and of the amines (Löbisch, Ber. d. d. chem. Gesell., Vol. 5, p. 513) agree better with the formula with 26 carbons. The analyses of dinitrocholesterol, nitrocholesteryl chloride (Preis and Raymann, Ber. d. d. chem. Gesell., Vol. 12, p. 224) and of bromo-cholesteryl chloride (Raymann, Bull. de la Societe Chim. de Paris, Vol. 47, p. 898) can be interpreted in terms of both formulae. The heptachlorocholesterol of Schwendler and Meissner (Lieb. Annal., Vol. 59, p. 107) is probably not a uniform substance, since the analyses deviate from one another by up to 1 per cent, so this substance cannot be taken into consideration here.

The so-called acetic acid cholesterol, a cholesterol with crystalline acetic acid (Hoppe-Seyler, Journ. f. pr. Chemi., Vol. 90, p. 331) displays an acetic acid content which would agree better with the formula with 27 carbons than any lower one. Here also, however, the analyses deviate from one another by 1 per cent, thus significantly reducing their value for the present purpose. The studies published by Walitzky are unfortunately unavailable to me in an original form, for which reason I was unable to include his analytic results in the comparison. Finally those derivatives which contain organic radicals consisting solely of the elements occurring in cholesterol are without significance for the present question because of their high molecular weight. It therefore appears that both formulae are equally likely to be correct. Since there is no reason at the moment to doubt the correctness of the analyses or the purity of the substances, the only assumption that remains is that various homologous cholesterols occur in the animal body and were present in the compounds discussed here in more or less pure form. This assumption indeed becomes increasingly more probable upon closer examination. It is well known that E. Schulze found a second cholesterol in lanolin – isocholesterol – in addition to the ordinary cholesterol. This clearly proves that mixtures of cholesterols can occur in one and the same animal. He also points out in a comment on a study performed with Barbieri (Journ. f. pr. Chem., Vol. 25, p. 458) that the isocholesterol was probably not an isomer but rather a homologue of cholesterol. Hesse felt himself compelled for certain reasons to state that probably the cholesterol to which he gives the formula $C_{25}H_{42}O$ occurs in animal bodies mixed with phytosterol, to which he assigns the formula $C_{26}H_{44}O$ (Lieb. Annal., Vol. 192, p. 175). It also follows from the properties of the presently known cholesterols that they very probably form two homologous series, as I attempted to confirm already at the conclusion of my report on hydrocarotene and carotene (op. cit. p. 729). The discovery of two homologous cholesterols would therefore not be surprising.

As a matter of fact, the data of various observers concerning the properties of cholesterol and its derivatives also differ from one another at various points so significantly that in these cases isomerism rather than homology should be assumed. Thus, for example, the data of Walitzky concerning the behaviour of cholesterol with respect to sodium, and cholesteryl chloride with respect to alcoholic ammonia, totally contradict the data of other observers. In this case, it is worthy of notice that Walitzky performed his studies exclusively with cholesterol extracted from the brain, while most other chemists used cholesterol produced from gallstones. It would therefore not be impossible for various cholesterols to occur in the different animal organs. This is all the more probable because Schulze and Barbieri have already demonstrated a similar behaviour for a plant, i.e. *Lupinus luteus* (op. cit.). Here it was found that both in the cotyledons and in the unchanged seeds a cholesterol occurs which has greatest similarity to phytosterol or paracholesterol, while in the root and in the hypocotyl of the germinating plants, a cholesterol of a much higher melting point and optical rotatory power, the so-called caulosterol, occurs. Schulze and Barbieri are of the opinion that, because they obtained the same substance both from gallstones and from lanolin with the aid of the benzoate, and because the benzoate always has uniform properties, ordinary cholesterol is still most certainly a single substance. However, if one considers how similar homologous substances are in their properties, and how precisely their preparation using the more easily purified benzoate would have to be carried out in order to confirm that these are the same substances, then one can ascribe no very great value to this solitary observation.

31

It would now naturally require additional studies to establish whether the hypothesis advanced here corresponds to reality or whether it must be allowed to fall, and therefore, for the time being, I will refrain from making a firm claim[1]. At any rate, however, as a result, in studies with cholesterol it will be necessary to devote the greatest attention to its origin and uniformity. As regards the origin of my substance, it was obtained from gallstones. The origin of the latter are, however, unknown to me.

In the following, some previously unknown or only imprecisely known properties of several derivatives of cholesterol will be reported, in which cholesterol must be naturally assigned the formula $C_{27}H_{46}O$.

1. Cholesteryl acetate ($C_{27}H_{45} \cdot C_2H_3O_2$)

Löbisch (Ber. d. d. chem. Gesellsch., Vol. 5, p. 513) prepared this substance with the aid of acetyl chloride and reported its melting point as 92 °C. I used acetic acid anhydride for the preparation. 10 g of cholesterol were lightly boiled with 7 g of acetic acid anhydride for 1 hour under reflux, carefully washed with water and recrystallized initially from ether, then from an ether/alcohol mixture, until the substance displayed an unchanged melting point. This was found to be 114–114.4 °C (corr. 114.3–114.7 °C). Raymann, who prepared the substance some time ago in almost the same way, found 113 °C (Bull. de la Société Chim. de Paris, Vol. 47, p. 898). Mr Hofrath v. Zepharovich kindly consented to study the compound crystallographically and reported the following to me:

'Crystal system monoclinic[2]
$a:b:c = 1.8446:1:1.7283.$ $\beta = 73° 38'$

Cross-stretched narrow platelets or needles, predominantly bounded by {001}oP, {100}$\infty P\infty$, {$\bar{1}$01}$P\infty$ and {110}∞P: subordinately: {010}$\infty P\infty$, {011}$P\infty$, {111} $-P$ and {112} $-\frac{1}{2}P$. Twins along {001}.

	Calculated	Measured	Z
(001):(100) =	73° 38'	73° 43'	13
(001):($\bar{1}$01) =	50° 41.3'	50° 40'	12
(110):(100) =	60° 32'	60° 32'	10
(110):(001) =	82° 2'	82° 2'	15
(111):(100) =	57° 32.3'	57° 32'	2
(111):(110) =	25° 51.6'	25° 42'	2
(112):(111) =	16° 22.3'	16° 20'	3

[1] That is for the second, presumably new, cholesterol (*ed.*).
[2] *monosymmetrical* in the original (*ed.*).

The main extinction lies parallel to the edge between (001) and (100) and perpendicular to it: no optic axes are seen through these planes. The small dimensions of the crystals prevented a further investigation.'

During the cooling process of the molten cholesteryl acetate, a peculiar, very splendid colour phenomenon occurs before solidification (not after it, as reported by Raymann). The phenomenon can already be observed in a wide capillary tube, as is used to determine the melting point. However, it can be observed much better if the substance is melted on an object glass covered with a cover glass. When the sample is viewed in reflected light, a strong emerald green colour is seen to appear at one point. This rapidly spreads over the entire sample, then becomes blue–green, in places also deep blue, then changes to yellow–green, yellow, orange–red, and finally bright red. From the coldest places, the sample then hardens into spherocrystals which spread quite rapidly and suppress the colour phenomenon, at which point the colour simultaneously turns pale. In transmitted light, the phenomenon takes place in the supplementary colours which, are however unusually pale and scarcely perceptible. Similar colour phenomena appear to occur in several cholesterol derivatives.

Thus, Planar (op. cit.) reports that cholesteryl chloride displays a violet colour during cooling from the melt, which vanishes again upon solidifying. Raymann (op. cit.) reports similar observations on the same substance. Löbisch (op. cit.) reports that when cholesterylamine is melted it displays a bluish-violet 'fluorescence', and also mentions that the same phenomenon occurs in the case of cholesteryl chloride. I myself observed a similar phenomenon in cholesteryl benzoate (see below), and Latschinoff reports for the silver salt of cholestenic acid, which is formed by oxidation of cholesterol, that it turns steel blue when melted, which fact is probably to be explained in the same way. An accompanying phenomenon occurs in cholesteryl benzoate, to be described below, and perceptible changes are observed under the microscope at the same time as the colour phenomenon. This suggested to me that perhaps physical isomerism was present here. I therefore requested Professor O. Lehmann in Aachen, who is probably at present the person most familiar with these phenomena, to make a more detailed investigation of the acetate and benzoate along these lines. He was kind enough to perform the investigation and indeed found that trimorphism was present in both compounds. The cause of the colour phenomenon, however, has not yet been satisfactorily explained. It is only known that it is closely related to the precipitation and resolution of a substance. At the moment this substance remains a complete mystery. Whether this substance formed and disappears as a result of a physical or chemical change cannot be decided at present. In what follows the most important results from Professor Lehmann's investigations on cholesteryl acetate will be presented[1].

---

[1] He will publish them in detail later.

Modification 1. Obtained by crystallization from solvents. Monoclinic. The crystal forms are those described above, measured by Professor v. Zepharovich. When heated the crystals become cloudy before melting, breaking up into a heap of crystals of the third modification.

Modification 2. Produced by a rapid cooling of the melt in the form of sphero-crystals. If xylene is added, single crystals can be obtained. These form thin, large flakes of the monoclinic system, have a rhombic outline and an apical angle of about 63°. When heated, the crystals of this modification also become cloudy, transforming into a heap of crystals of the third modification.

Modification 3. Produced by heating the first and second modifications, as well as by slow cooling of the molten substance. They could not be obtained in definable crystals. The simplest and most regular forms were elongated rectangles with symmetrical extinction.

From these results it follows that the first modification is enantiotropic to the third, the second monotropic to the third. The first modification is most stable at ordinary temperatures, the third at higher temperatures. The second, however, was incapable under any circumstances of causing the other two to transform into their own natural state.

Professor Lehmann's study of the colour phenomenon has shown that it is produced by the precipitation of a substance whose structure resembles an aggregate of spherocrystals, as polygonal areas can be recognized, each of which displays a black cross between crossed nicols. Upon closer study, however, one sees that this substance consists of drops which acquire a jagged outline, as a result of very fine crystals which are perceptible only at strong magnifications. In other words, the substance is quite liquid, and the shape of the drops can usually be changed by moving the cover glass. If the sample is shaken in such a way that the precipitate distributes itself as finely as possible, and mixes as uniformly as possible with the remaining liquid, the brightness and beauty of the colour phenomenon is significantly enhanced. The colour-producing substance also displays a strong rotation of the plane of polarization of light. This varies with temperature and has very different intensity for the individual colours, and is rotated to the right at higher temperatures and to the left at lower temperatures. If the colour phenomenon vanishes upon further cooling and gives way to crystallization, then the precipitated substance redissolves; it is suddenly set into a peculiar motion and gradually disappears.

The nature of the colour-producing substance has not been determined to date. No impurities can be present, because the phenomenon occurs in different cholesterol derivatives and I have also already observed it in a derivative of hydrocarotene.

Cholesteryl acetate decomposes when heated above the melting point accompanied by yellow and brown colouration and evolution of pungent burnt-smelling vapours. At this point a small portion of the substance sublimates without decomposing. However, I was unable to demonstrate the splitting-off of acetic acid reported by Raymann. I also cannot confirm his statement that the acetate is decomposed by water. Even after prolonged boiling with water, I could not demonstrate decomposition. When the acetate is partially decomposed by heating, it has the peculiarity that rapid cooling brings it into a state in which the colour phenomenon mentioned above is permanently exhibited at ordinary temperatures.

2. Bromocholesteryl acetate ($C_{27}H_{43}Br_2 \cdot C_2H_3O_2$)

The synthesis of this substance has already been described earlier. It is poorly soluble in alcohol, and readily soluble in ether. It is obtained on very slow evaporation from an ether/alcohol mixture as very thin, 1–2 cm long, shiny platelets. The compound is dimorphous. The two modifications were formed under essentially the same conditions upon crystallization from ether/alcohol. The first three forms obtained by recrystallization of the same substance were triclinic;[1] the last-formed crystals were however monoclinic. Mr Hofrath v. Zepharovich was kind enough to undertake the crystallographic study and reports to me the following results:

'A. Monoclinic form $a:b:c = 1.3283:1:2.5346.$ $\beta = 82° 9'$
Orthodiagonally elongated thin platelets with predominant $\{001\}oP$, $\{100\}\infty P\infty$, $\{\bar{1}11\}P$ and subordinate $\{011\}P\infty$, $\{101\}-P\infty$, $\{\bar{1}01\}P\infty$, $\{\bar{1}12\}\frac{1}{2}P$.

	Calculated	Measured	Z
$(001):(100) =$	–	82° 9'	20
$(011):(001) =$	68° 17'	67° 17'	6
$(110):(100) =$	87° 6'	87° 5'	4
$(\bar{1}11):(001) =$	–	76° 45.6'	13
$(\bar{1}11):(\bar{1}00) =$	–	56° 43.3'	9
$(\bar{1}12):(001) =$	61° 2'	61° 10'	7
$(\bar{1}12):(\bar{1}00) =$	62° 55.5'	62° 47.5'	3

Cleavability completely along $\{001\}$. Main vibration directions parallel and perpendicular to (001:100). Plane of optic axes parallel to $\{010\}$.

B. Triclinic form[1]
The elements resemble those of A, but for the calculation a sufficient number of precise determinations is missing. The combinations of rhomboid platelets, often

---

[1] Asymmetric in original (ed.).

35

similar to the monoclinic and interpreted as {001}, {100}, {$\bar{1}$}, {11$\bar{1}$} also possess similar edge angles, as the following comparison shows:

	A	Z	Measured	Z
(001) : (100) =	82° 9′	20	81° 17′	25
($\bar{1}$11) : (001) =	76° 46′	13	77° 57′	9
($\bar{1}$11) : ($\bar{1}$00) =	56° 43′	9	56° 35′	5
($\bar{1}$12) : (001) =	61° 10′	7	60° 18′	4
($\bar{1}$12) : ($\bar{1}$00) =	62° 47.5′	3	64° 57′	4

Cleavability along {001} and {010}. On one table I found (010) : (100)=88° 47′ and (010) : (001)=96° 56′, on two others I found (010) : (001)=91° 43′ and 91° 57′. In optical behaviour, no difference is noted between the geometrically similar forms A and B.'

The substance is somewhat photosensitive. In diffuse daylight, it becomes yellow after about 3–4 weeks, later reddish-yellow to brown. At this time hydrogen bromide evolves. The discoloured substance is amorphous. In the absence of light, the compound is completely stable. The melting point is not quite the same for the two physically isomeric forms. The monoclinic form melts at 117.6 °C (corr. = 118.0 °C), the triclinic form at 115.4 °C (corr. = 115.8 °C). Upon melting, the substance usually turns weakly yellowish. After cooling, it remains glass-like and can no longer be made to crystallize. Apparently a slight decomposition occurs.

If sodium amalgam is allowed to act on the compound in ether solution, one again obtains cholesterol. However, in addition to this, yet another substance appears to have been formed. After the sodium is removed by transforming it into sodium chloride with repeated treatment with ether, a small quantity of cholesterol can be obtained by fractional crystallization in the form of colourless crystals, together with a larger quantity of a yellow, amorphous substance. The cholesterol thus obtained melted at 146.5 °C but was still not completely pure, because the molten product had a weak yellow colour. Wislicenus and Moldenhauer also obtained cholesterol, with a melting point of 147 °C (op. cit.), from bromocholes-terol using sodium amalgam.

Since the formation of a cholesterol derivative with two new hydroxyl groups is expected by reaction of an alkali with the bromoacetate, the following reaction was carried out. Aqueous caustic potash acts only very slowly and incompletely; it must therefore be used in alcoholic form. At this time a yellow substance was obtained which was poorly soluble in alcohol, readily soluble in ether, but which could not be crystallized and which therefore could not be studied further. It forms a very viscous sticky product which, after prolonged standing, becomes covered with a hard brittle pulverisable layer. If its chloroform solution is mixed with sulphuric acid, the sulphuric acid layer turns blood-red with green fluores-cence, while the chloroform layer assumes only a very pale pink-red colouring. It therefore behaves differently from cholesterol.

3. Cholesteryl benzoate ($C_{27}H_{45} \cdot C_7H_5O_2$)

This substance was first synthesized by Berthelot (Ann. de chim. [3], Vol. 56, p. 54) by heating with benzoic acid, then by Schulze (Journ. f. pr. chem. [2], Vol. 7, p. 170) by heating with benzoic acid anhydride. I used the latter procedure in a somewhat simpler version. Ten grams of anhydrous cholesterol were heated with 12 g of benzoic acid anhydride for about 1½ hours in an open flask in a sulphuric acid bath to 150–160 °C. The transformation into the ester is then almost complete and only very little remains unreacted. One also never obtains a brownish coloured melt, such as obtained by Schulze. If the heating is performed for only ½ hour to about 130–140 °C, then about 60 per cent of the cholesterol escapes reaction. The solidified melt was extracted twice with boiling methyl alcohol and the residue repeatedly recrystallized from an ether–alcohol mixture. The shape of the crystals and the solubility relationships agree precisely with the data of Schulze. The most beautiful crystals are obtained by slow evaporation of a solution in ether mixed with as much alcohol as it can tolerate without clouding.

Mr Hofrath v. Zepharovich, who was kind enough to analyse these crystals, reported the following to me concerning them:

'Crystal system tetragonal.
$a : c = 1 : 0.9045$.

Quadratic platelets with plane $\{001\}oP$, and very narrow, usually horizontal striated side faces of $\{111\}P$, $\{443\}P$, $\{221\}2P$ and $\{441\}4P$.

	Calculated	Measured	Z
$(111) : (001) =$	51° 59′	52° 25′	7
$(443) : (001) =$	59° 37′	59° 44′	17
$(443) : (4\bar{4}3) =$	75° 10.6′	75° 12′	2
$(221) : (001) =$	68° 39′	68° 43′	17
$(221) : (2\bar{2}1) =$	82° 23′	82° 24′	5
$(441) : (001) =$	78° 76.5′	78° 35′	7

The small crystals were found to be uniaxially negative in the conoscope.'

With respect to the melting point, a significant deviation from Schulze's data was noted. He found it to be 150–151 °C. However, despite continued careful purification I was able to find only 145.5 °C (corr. 146.60 °C). However, it struck me that the substance in this case melted not into a clear transparent liquid but rather always into a cloudy liquid which was only translucent. I initially considered this to be a sign of impurities, although both microscopic and crystallographic examinations of the compound revealed no signs of nonuniformity. Then, on closer examination, it was also noted that when the substance was heated to a higher temperature, the clouding suddenly vanishes. This happens at 178.5 °C (corr. 180.6 °C). At the same time I found that when one cooled the substance which had been strongly heated in this way, it exhibited colour phenomena extremely similar to those already described for the acetate. The presence of two melting points, if one may express it thus, is a remarkable phenomenon. The appearance of the colour phenomenon was primarily what made me think that physical isomerism must be present both here and in the case of the acetate. For this reason I requested Professor Lehmann in Aachen to make a closer investigation of these circumstances. The most important result of his studies of the benzoate, briefly summarized, are the following:

Cholesteryl benzoate, like the acetate, can occur in three modifications.

Modification 1. Is obtained by crystallization from solvents and forms the above-described tetragonal crystals. It melts far higher than the other two modifications. When heated the crystals remain clear, and therefore do not transform into another modification.

Modification 2. This is formed upon the solidification and rapid cooling of the molten substance. It melts a little but not much lower than the third modification and forms flat needles or narrow flakes of the rhombic system.

Modification 3. Formed by slow cooling of the molten substance, and by heating the rapidly solidified substance (Modification 2) until it almost melts. It melts a little higher than the second modification and forms thin, wide leaves with a nearly square outline and symmetrical extinction.

The three modifications are in the relationship of monotropy to one another. The colour phenomenon which occurs upon cooling of the molten substance takes place somewhat differently from that of the acetate.

When the clear-molten compound is cooled, a deep violet-blue colour appears at one point. This spreads rapidly over the entire substance and vanishes again almost equally rapidly, a uniform cloudiness appearing in its place. The substance then remains cloudy but liquid for some time. On further cooling, then the same colour phenomenon appears for the second time. As the cooling progresses further, a crystalline solidification of the substance occurs, and with it also a simultaneous disappearance of the colour phenomenon. If the molten layer of the benzoate is at least 2–3 mm thick, then in addition to the violet-blue colour, all other colours reported for the acetate also appear. The colour-producing substance here also causes the cloudiness. As in the case of the acetate, it precipitates out in drops in which crystals are found, and it dissolves again shortly before solidification. The process of precipitation and dissolution is accompanied by the colour phenomenon, but otherwise only simple clouding occurs. The colour-producing substance, as in the case of the acetate, also displays chromatic polarization, except that this is not as strong as in that case, and not so many colours occur as a result of it. The other details observable under the microscope and the explanation of the phenomenon cannot be reported at this time, because the studies in this regard have not yet been completed.

Whether the different result of Schulze for the melting point was caused by the cloudiness occurring during melting, which does indeed make observation very difficult and erroneous, or whether Schulze perhaps was studying a different cholesterol, cannot be decided from the available data.

As an aside, it might be mentioned here that now I have become aware of the colour phenomenon, I have also found it in hydrocarotene. In this case, however, only the benzoate displayed it, while the acetate is free from it. This can be used to distinguish between these two cholesterols more easily. The hydrocarotyl benzoate also displays two melting points, quite like the corresponding compound of cholesterol, a fact which I had also previously overlooked.

When boiled with water, cholesteryl benzoate does not decompose at all. It decomposes very slightly and insignificantly with aqueous caustic potash, but decomposes rather quickly with alcoholic caustic potash, especially upon the addition of ether. When heated above the melting point, it partially decomposes, at which time benzoic acid sublimes off and condenses in colder regions, sublimating in part without decomposing.

Upon stronger heating it becomes yellow, and then, after cooling, solidifies in partially glass-like fashion. In this state the colour phenomenon can be permanently preserved in it at ordinary temperatures by rapid cooling, as in the case of the acetate.

4. Sodium cholesterylate ($C_2H_{45}O \cdot Na$)

O. Lindenmeyer obtained this compound by reacting sodium with cholesterol dissolved in purified petroleum with the evolution of hydrogen (Journ. f. pr. Chem., Vol. 90, p. 321). Walitzky reports that sodium does not evolve hydrogen from anhydrous cholesterol (Beilstein, Handbuch d. org. Chem., first edition, Vol. 2, p. 1376). Although Lindenmeyer expressly states that he used completely anhydrous cholesterol and petroleum purified over sodium, it was nevertheless necessary to provide clarity on the subject by repeating the experiment. The preparation took place, with minor deviations, according to the data of Lindenmeyer. Ten grams of cholesterol, which had been completely dehydrated by drying at 100 °C, was dissolved in petroleum ether. The petroleum ether had been previously dehydrated and purified by standing and distillation over calcium chloride and then over sodium. Sodium which had been purified by remelting under petroleum and cut into paper-thin flakes under petroleum ether was then added in a small excess. Hydrogen begins to evolve immediately and the pieces of sodium become coated with a white crust, just as Lindenmeyer describes. The very fine dispersion of the sodium is very important for the greatest possible acceleration of the reaction. Otherwise at ordinary temperatures it takes place quite slowly. If on the other hand the heating is performed on a water bath using a reflux condenser, the reaction is then completed in about 2–3 days with the quantity of cholesterol reported above; this may be recognized from the fact that the sodium pieces are no longer coated with a white crust but remain completely clean. The liquid is then a thick white slurry. This slurry was filtered off, washed with a small amount of petroleum ether and then freed of sodium by dissolving in anhydrous chloroform. Dissolution takes place so rapidly that the sodium has scarcely any effect on the chloroform. When the solution is allowed to evaporate spontaneously or when it is evaporated on a water bath, one obtains the compound whose properties have already been described by Lindenmeyer. Regarding the latter one should add at most that the substance is also slightly soluble in petroleum ether.

The compound obtained was tested in order to determine that it could not be transformed in an acid by treatment with carbon dioxide.

However, when introduced into the chloroform solution, only sodium carbonate precipitated out, while pure cholesterol remained in solution.

5. Nitrocholesterol

A hot saturated solution of cholesterol in glacial acetic acid is treated with fuming nitric acid (specific gravity 1.54) at boiling heat just so long as strong evolution of red vapours occurs, and then the liquid is poured into cold water. After washing with water and drying under vacuum, a solid, odourless and tasteless nitro compound is obtained. This compound is dark-red-yellow, but yellow in the pulverised state, and it has so far resisted attempts at crystallization.

It is insoluble in water, but dissolves in aqueous ammonia and aqueous caustic alkali very readily. It then forms a dark red bitter-tasting liquid, which reacts neutrally in the saturated state, and which yields yellow or red-brown precipitates with the water-soluble salts of the alkaline earths and most metals. This substance is apparently the same as is formed in the Schiff reaction for cholesterol.

It is readily soluble in alcohol, very readily soluble in ether, chloroform, benzene and glacial acetic acid. The alkaline solution can be very easily oxidized by potassium permanganate, but I have not yet succeeded in obtaining a detectable oxidation product.

When heated on a platinum plate, the substance burns rapidly and easily but without any real deflagration. It melts at 93–94 °C with strong foaming, indicating decomposition.

Finally, permit me to express my deepest thanks to all those who supported me in carrying out the present work, including: Professor Dr Weiss for providing the materials and for relieving the burden of other work in the experiment, Professor Dr F.W. Gintl for kindly permitting the use of the facilities of his laboratory, Mr Hofrath von Zepharovich for helpfully carrying out the crystal measurements and Professor Dr O. Lehmann for carrying out the microphysical studies.

Friedrich Richard Kornelius Reinitzer was born in 1857 in Prague, at that time capital of the province of Bohemia in the Austro-Hungarian empire. He is variously described as a chemist or a botanist, and in present-day terms would probably be thought of as a biochemist. He studied at the Technical University in Prague in the period 1874–7, and worked as a teaching assistant in the Chemistry Department of the same institution from 1877 to 1882. He then switched to the Institute of Plant Physiology in the German (speaking) University in Prague, specializing in microscopy, where he was a lecturer (1882–8) and Professor (1888–95). In 1895 he was appointed to a professorship in the Technical University

in Graz in present-day Austria. He moved to the Graz Botanical Institute in 1901, where he reached the rank of Rector (1909–10).

His contributions to the liquid crystal story come from three articles written in the late 1880s on the double-melting phenomenon and from an ungentlemanly exchange in the pages of the *Annalen der Physik* in 1906–7 between himself and Lehmann, in which Lehmann's claims for absolute priority were gently rebutted. He died in 1927 two weeks before his 70th birthday.

A2
Zeitschrift für Physikalische Chemie **4**, 462–72 (1889)

Über fliessende Krystalle.

Von

O. Lehmann.

(Mit Tafel III und 3 Holzschnitten.)

Fliessende Krystalle! Ist dies nicht ein Widerspruch in sich selbst — wird der Leser der Überschrift fragen —, wie könnte denn ein starres, wohlgeordnetes System von Molekülen, als welches wir uns einen Krystall vorstellen, in ähnliche äussere und innere Bewegungszustände geraten, wie wir sie bei Flüssigkeiten als „Fliessen" bezeichnen und durch mannigfache Verschiebungen und Drehungen der ohnehin schon des Wärmezustandes halber äusserst lebhaft durcheinander wimmelnden Moleküle zu erkären pflegen?

Wäre ein Krystall wirklich ein starres Molekularaggregat, dann könnte von einem Fliessen desselben in der That ebensowenig die Rede sein als beispielsweise vom Fliessen eines Mauerwerks, das allerdings bei Einwirkung starker Kräfte in rutschende Bewegung geraten kann, welche Bewegung aber nur dann einigermassen dem Strömen einer flüssigen Masse entspricht, wenn die Fugen sich öffnen und einzelne Bausteine ausser Zusammenhang geraten und sich übereinanderschieben und durcheinanderrollen, ähnlich wie die einzelnen Körnchen einer bewegten Sandmasse.

Dass es übrigens feste, wenn auch nicht krystallisierte Körper giebt, welche ganz wie Flüssigkeiten, wenn auch unvergleichlich viel schwieriger fliessen können, ist jedem bekannt, der einmal die langsamen Veränderungen einer hohl liegenden Siegellackstange oder einer grösseren freistehenden Pechmasse beobachtet hat. Alle schmelzbaren amorphen Körper gehen kontinuierlich aus dem flüssigen in den festen Zusand über und der Punkt, bei welchem der Aggregatzustand wirklich fest wird, d. h. wo sich die ersten Anzeichen beginnender Verschiebungselastizität einstellen, ist so wenig erkennbar, dass wir häufig einen solchen Körper gerade der Fähigkeit des Fliessens halber noch flüssig nennen, wo er streng genommen bereits als fest bezeichnet werden müsste

Da in diesen Fällen schon eine sehr geringe Kraft — das eigene

A2
Zeitschrift für Physikalische Chemie **4**, 462–72 (1889)

ON FLOWING CRYSTALS

by

O. Lehmann

(With Plate III and 3 wood-engravings)

Flowing crystals! Is that not a contradiction in terms? Our image of a crystal is of a rigid well-ordered system of molecules. The reader of the title of this article might well pose the following question: 'How does such a system reach a state of motion, which, were it in a fluid, we would recognise as flow?' For flow involves external and internal states of motion, and indeed the very explanation of flow is usually in terms of repeated translations and rotations of swarms of molecules which are both thermally disordered and in rapid motion.

If a crystal really were a rigid molecular aggregate, a flowing crystal would indeed be as unlikely as flowing brickwork. However, if subject to sufficiently strong forces, even brickwork can be set into sliding motion. In a certain sense, the resulting motion corresponds to a stream of fluid mass in which the joints between the individual bricks open. The bricks then run out of control, moving over and rolling around each other in a disorderly manner, rather like single granules in a turbulent mass of sand.

As a matter of fact, there are solid – but nevertheless non-crystalline – bodies which are able to flow like liquids, although with much greater difficulty. This fact is evident to anyone who has ever observed the slow change of an unsupported stick of sealing wax or a larger free-standing mass of pitch. All fusable amorphous bodies transform from the liquid into the solid state continuously. The point at which the state of aggregation really becomes solid (i.e. where the first hints of the onset of displacement elasticity occur) is extremely difficult to recognise. Indeed, because such a material is still able to flow, we would often still regard it as fluid, even though, strictly speaking, it should already be described as solid.

As in the cases mentioned above, a very small force – the body's own weight – is easily enough to produce the phenomenon under discussion. Sometimes one might be tempted to extend the definition of the liquid state, so as to include all those sub-stances which can be made to flow by their own weight alone. However, such a definition would be completely unjustifiable. As an example, consider foundations

43

which seem to be completely rigid. But if there is a strong enough loading pressure from the brickwork above them, even they give way and begin to flow.

One must therefore admit that amorphous solid bodies indeed are able to flow. In the final analysis this causes few problems. From a theoretical point of view the disorder which is already present within the molecules of such bodies cannot be enhanced, even by the most vigorous flowing motion. However, there are extremely serious problems associated with the assumption of flow in crystalline bodies and posed by the explanation of such a possibility.

The individual elements of such an array are precisely organised with respect to each other, according to the mathematical rules of crystallography. They maintain any arrangement which they have previously adopted. They do this in the absence of adhesive or connecting joints, notwithstanding vigorous thermal vibrations. Consequently it seems completely impossible for fluid motion to disturb the whole arrangement significantly.

Thus, it used to be generally thought self-evident that a permanent deformation of crystalline bodies, e.g. bending, pressing or kneading, must necessarily be accompanied by extensive repeated destruction of their structural integrity. This would cause a large number of extremely thin cracks, splits and fissures to occur. Their small size would render them invisible or nearly so. They are not suffi- ciently extensive to destroy the whole structure, nor do they run in such a way to make it possible to pull apart the separated pieces. If one were to cut a glass pane in a zigzag fashion from *a* to *b* (cf. Fig. 1), it would not be possible to separate the two parts, although they are no longer connected to each other. Because of the spring tension of the single fragments, if many such fissures could be generated, the plate would have such a high degree of mobility that it could be easily deformed in different ways either temporarily or permanently.

Examples of deformed crystals are known for all forgeable metals, e.g. lead, tin, zinc, silver, gold, platinum, iron, copper, etc. It is known that all these metals become increasingly fragile on continued forging. This is explicable if forging increases of the number of fissures. In the same way, the perturbation caused by the cleavage process may result in an initial increase in the hardness; a starting fissure will not be as easily transmitted in the aggregate as would be the case in

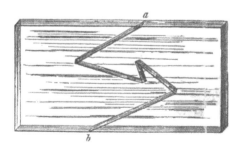

Fig. 1

the homogeneous crystal. For the same reason cane-work and sponge-like cellulose withstand the penetration of projectiles much more than the same quantity of material in a compact form, and a wire rope breaks less easily than a massive iron rod of the same weight and length.

Metals recover their initial toughness on annealing. This may be explained by the fact that at high temperatures they are changed into another enantiotropic modification. On cooling the initial state is then re-established. The fissures are closed in the course of volume changes in larger-sized unbroken crystals.

To a large extent metallic substances are deformable. Following his most interesting experiments with embossing, Tresca talks specifically about the flow of solid metals. By contrast, brittle bodies, e.g. rock crystal, can be pulverised by very careful application of pressure as soon as the elasticity limit is exceeded. Spring's results demonstrate that these observations may be attributed to the fact that fissures in metallic bodies can easily be closed under sufficient pressure by welding together their edges. However, in the case of quartz and similar hard materials, this welding capacity is very low, probably because, as a result of the rigidity of the material, the crack edges are not able to come sufficiently close to each other.

By making the above proposal, we are using a circular argument, assuming something which still has to be proved. Put another way, we suppose that all crystals, at least to a limited extent, are endowed with the capability of permanent deformation without forming cracks.

Reusch has already proposed the mobility of crystal particles in order to explain his observations concerning the deformation of rock salt. He believed this to be related to the ability of crystal particles to slide along dodecahedral faces, by which we mean those faces which truncate the cubic edges. Fig. 2 is an attempt to demonstrate the sliding of rock salt particles for a crystal pressed

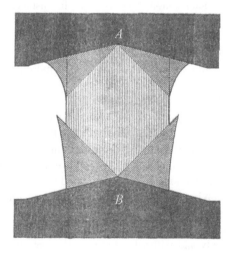

Fig. 2

between a hollow pyramidal upper punch A and an analogous lower punch B, whose initial shape is given by the dotted lines. It is assumed that the sliding only occurs for layers which are directly in contact with the punches and takes place only along the hatched lines.

Suppose such pressure were applied to rock salt and the expected sliding were indeed to take place only along the dodecahedral faces. In that case, the cleavage would have to remain undisturbed after the deformation, and the cleavage direction in the deformed particles should still be parallel to the initial cubic faces. However, Reusch himself had already observed experimentally a disturbance of the ability to cleave. Later observations, e.g. in the case of galena which can be cleaved in a similar manner, have also confirmed that the cleavage planes can be deformed to a greater or lesser extent.

From this one may conclude that real flow, without crack formation and without parallel displacement along defined gliding planes, is possible even for homogenous crystals. One may wonder how a such deformed crystal will behave, and especially how further growth will occur when we insert it in an appropriate mother liquor. Will it remain with the bent faces of the artificially imprinted form or will it gradually adopt the normal form? Perhaps it might even undergo a transition to a structureless amorphous state, which is considerably growth-retarded, since a principal difference between crystalline and amorphous bodies is the fact that the former can grow but the latter cannot, even under ideal conditions.

It has been shown by experiment that it is the first case which occurs. The crystals proceed to grow with bent faces, but not quite in the usual way, though apparently with a growth rate equal to that of normal faces. A disturbance becomes evident by the occurrence of internal tension, in such a way that each newly deposited layer can be taken as a thin flexible bent lamella which is attached to the surface but retains its own tension. Thus, by summing the effects, the resultant tends to cancel out the existing deformation. A bent crystal, for instance, shows a tendency to extend rectilinearly on growing. Not infrequently, especially in the case of very thin needles and lamellae, this process will be taking place, when, quite suddenly, the bent needle straightens up all at one go. Sometimes this occurs with such vehemence that the whole crystal bursts into two or more parts, in a certain sense nearly exploding.

Observations of this behaviour are known in abundance for microscopic crystals. It happens when the crystals are still extremely thin, hair-like or lamella-like (as in the case of so-called trichites), and which have been bent not by external forces but by internal forces (probably by action of surface tension). This behaviour can also easily be verified using artificially bent needles, e.g. those of caffeine, in which, however, the initial bend is only elastic and therefore temporary.

To date I have not been able to create a permanent deformation in larger crystals without destroying them completely. This lack of success is probably because we do not have a suitable apparatus which can perform the deformation both sufficiently slowly and without creating any shock disturbance. In order to exert pressure one could use, for instance, a rising gasometer bell jar, the rise in which will be caused by water which slowly seeps into the container. An alternative simpler method would use a bucket, into which water is poured very slowly, drop by drop, causing the weight of the bucket to increase progressively. This could work if the observation site were free from vibrations caused by passing carriages, etc.

Deformation tests of crystals involving either squeezing in a screw press (the so-called compression strength tester), or in a lever press (punch press), or by forging on an anvil, always resulted in aggregates. The initially homogenous crystals were smashed to pieces. However, the debris did not form a loose accumulation. Rather, by deforming fragments in such a way that their extremities nestled closely to the extremities of the surrounding fragments and welded together, the result was a semi-transparent, solid, connected mass. In Fig. 3 we see a microscope view of what happens when a crystal forged by this method to a thin lamella, e.g. salmiak, grows further. The small particles in contact with the solution can be observed growing further at their outer borders, in a skeleton-like fashion.

The single fragments are rather large when the deformation is weak, but the bending is too insignificant to expect any disturbance of the growth. In the case of very strong deformations the fragments are correspondingly smaller and the influence of the bending cannot be observed.

I have not so far found an approximation to the amorphous state which might have been expected from the usual ideas concerning the molecular constitution. Such an approximation might imply that a normal crystal, growing simultaneously in the same solution would gradually consume the whole material. This corresponds to the fact that amorphous bodies are without exception consumed by crystals which are brought close to them, provided they possess the property to dissolve, but they do not themselves grow in the solution. Thus the decreased content of the solution brought about by the elimination of the crystals can be increased again only by dissolving the amorphous bodies.

Crystals of the regular modification of silver iodide exhibit only a waxy consistency and can be spread with a dissecting needle on the object slide of a

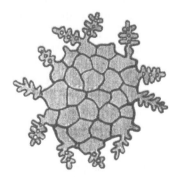

Fig. 3

microscope like hot sealing wax. Yet while they are growing, they very closely resemble thinly forged salmiak crystals between hammer and anvil. The same applies to deformed crystals of tin and lead which have been dipped as cathodes into appropriate solutions during microscopic electrolysis.

In the light of all these observations, it has not seemed possible to discover a substance whose crystals could be regarded as in a state of flow from direct observations, yet did not disintegrate and reform, but rather maintained their internal correlation under constant deformation in the same manner as do amorphous and liquid bodies. However, it seems that as a result of a recent discovery by Mr F. Reinitzer in Prague, such a substance, weakly fluid but crystalline, has indeed been detected. The nature of these crystals has not yet been fully understood, and perhaps optical illusions may be involved. Nevertheless, I have no hesitation in reporting the observations here, since so far it has proved impossible to construct an explanation of the phenomenon in terms of extremely soft crystals of a syrupy or gum-like type.

The substance in question is cholesteryl benzoate[1]. In a letter in March of last year, Mr Reinitzer, to whom I owe the substance under investigation, told me the following about the contradictory behaviour of the substance which he observed:

'If one may so express oneself, the substance exhibits two melting points. It first melts at 145.5 °C, forming a turbid but unambiguously fluid liquid. This suddenly becomes totally clear, but not until 178.5 °C. On cooling, first violet and blue colours appear, which quickly vanish, leaving the bulk turbid like milk, but fluid. On further cooling the violet and blue colours reappear, but very soon the substance solidifies forming a white crystalline mass.

When the phenomenon is observed under the microscope, the following sequence is easily detected. Eventually on cooling large star-like radial aggregates consisting of needles appear, these being the cause of the cloudiness. When the solid substance melts into a cloudy liquid, the cloudiness is not caused by crystals, but by a liquid which forms oily streaks in the melted mass and which appears bright under crossed nicols.'

These observations indeed contain many contradictions. For, on the one hand a liquid cannot melt on increasing temperature and also at the same time exhibit polarisation colours between crossed nicols. On the other hand a crystalline substance cannot be completely liquid. That a pulpy mass of crystals and liquid was not present follows from the high degree of purity of the substance under investigation; the substance came for use in the form of totally clear and well-defined crystals. In addition, at the temperatures concerned there was no possibility of chemical decomposition, and furthermore through direct visual observation in a microscope

---

[1] Both the acetate and the benzoate of hydrocarotene also exhibit the same behaviour; see also F. Reinitzer, Sitzungsber. d. Wien. Ak. **94**, (2) 719 and **97**, (1) 167, 1888. O. Lehmann, Molekularphysik **2**, 592, Anhang.

it would have been very easy to recognise clearly the edges of crystals in the liquid, especially because of the strong influence of the former on polarised light.

Reinitzer's further investigations led to still more serious complications. From them one is led to the conviction that when the material is cooled below 178.5 °C, the modification which first develops is a uniform, solid and physically isomeric modification consisting of sphero-crystals. When the system is further cooled it partially reliquifies, since the sphero-crystals 'melt (during cooling!) beginning at the periphery, and swim as solid bodies in a liquid'.

Despite all these contradictions, in my own investigations I have really been able to confirm Reinitzer's results. The impossible here really seems to become possible, but as to an explanation I was at first totally helpless.

If the substance were crystalline – as I deduced from my own observations in another context – it must be possible to prevent the crystals getting into close contact with each other as they grow by adding a small amount of a solvent. Gaps should remain filled with liquid, which would allow the edges of the single crystals to be clearly seen, and so probably to identify their polyhedral shape or at least their ability to grow. However, my attempts in this direction failed because of a lack of an appropriate solvent, until Mr Reinitzer replied to my enquiry that he had succeeded in finding such a substance which formed by itself if heated over a longer period.

And in fact I too have been able to obtain this substance by heating the sample without a cover glass over a longer period. The images which now appear allow little doubt that the mysterious modification in the temperature span of 145–178.5 °C is really a solid, crystalline, totally homogeneous enantiotropic modification.

I have not been able to recognise any polyhedral edges in the bodies separated by liquid, nor a clearly observable ability to grow. But this may be caused by the fact that the state of aggregation of these crystals, as already supposed by Mr Reinitzer, is extremely similar to the liquid state. Thus the ability to produce sharply defined forms which are totally distinct from the real liquid state, is only extremely small. Nevertheless, the other properties of the small bodies clearly point in the direction that they are crystals. Only this hypothesis can reconcile all the observed apparent contradictions, and make the total behaviour of this substance consistent with that of other similarly behaved substances.

The behaviour of the substance, contaminated in the way indicated, under the microscope between crossed nicols is now as follows.

The substance is heated until a clear melt is formed, and then cooled down slowly. Small blue-white spots appear over all the whole liquid. These grow in number and finally cover the whole space, which now appears as a cloudy-white mass. On further lowering the temperature, plates of common crystals are generated here and there. These quickly grow and eventually consume the cloudy mass; conversely they disperse into it on heating.

When the temperature is held almost constant close to 178 °C over a longer period, the plastic white mass adopts a coarser structure. Some of these coarse grains show a black cross similar to sphero-crystals, especially for samples which are not covered with a cover slip. Probably these are not multiply radial sphero-crystals in a real sense, but only more or less complicated star-like aggregates.

The most striking change of the mass occurs if it is moved by pressing on the cover slip by means of a dissecting needle. Along the lines of flow the bright spots and sphero-crystals run together, forming striations which become broader and broader on continued motion, and which finally may cover the whole visual field. However, they only appear bright if their longitudinal and transverse directions are inclined with respect to the principal directions of the nicols; otherwise they appear dark, just like crystals whose vibration directions are identical with those of the crossed nicols.

If one heats up the initial state or the striated state by only a small amount it becomes paler; conversely on cooling it becomes brighter. Depending on the thickness of the sample, instead of appearing whitish it can also exhibit various polarisation colours, like an aggregate of crystal lamellae of different thicknesses.

On heating more strongly, black dots occur here and there in the mass (cf. Plate III, Fig. 4). These are circular and become larger and larger, i.e. holes develop, filled by molten mass, so that the bright mass now forms a network.

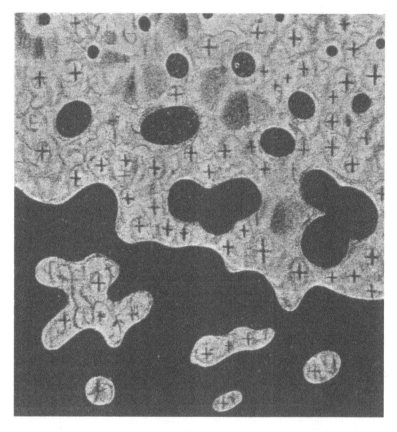

Fig. 4

Soon the network also melts as the connecting threads form an aggregate of bright spots, which would itself also eventually melt on further heating.

These isolated bright conglomerates are now most suitable for investigation. They possess razor-sharp edges and exhibit no observable surface tension effect; the result is that they can be viewed as melted crystal aggregates, and this conclusion is consistent with their polarisation properties. A closely analogous behaviour also appears during the melting of silver iodide; in this case likewise at first circular holes are formed (obviously caused by contact movement[1] in the developing melt), and finally very similar uneven rounded residues emerge while the holes are growing. However, a significant difference becomes obvious on cooling. Fragments of silver iodide rapidly grow like other crystals and enlarge to clearly observable and often multiply branched skeletons, albeit with much rounded edges and faces. The residues of our mysterious substance, however, exhibit no clear ability to grow, unless the small changes of polarisation colours are to be interpreted in that light. Rather, a dense precipitate of the aforementioned bright spots suddenly forms, which extends more and more and gradually coalesces with the larger residue (with additional movement). In this way many large isolated pieces of this curious material can be produced. The fact that there are really distinct pieces and not membranous or lamella-like species is easy to recognise in the vicinity of air bubbles. Here, as is known, a vigorous convection occurs which circulates the liquid around the air bubble from below and away from above, so that particles contained in the circulating liquid continuously roll, rotating about their axes, and so can be viewed from all sides. It will be obvious that under these circumstances they possess a certain stiffness, insofar as they generally keep their form during the movement. This stiffness, however, is extremely small, for if the whole mass is caused to flow even by a slight pressure on the slide, then the polarising particles will be totally distorted corresponding to the flow lines of the liquid. Thus the flow lines are diverted by an obstacle, even an extremely small one such as the surface of an air bubble. The mysterious crystals flow together with the fluid as if they were a part of it, except that they are endowed with a polarisation capacity. This extremely large plasticity, i.e. the near total lack of displacement elasticity, which even in the case of regular silver iodide is still clearly obvious, makes it easier to understand why the growth tendency of the crystals is so extremely small. As regards the behaviour of the substance which is not contaminated by an added oxidation product, the only difference to be seen is that it is not possible to isolate single parts of the plastic mass. The first violet colour appearing preceding the development of cloudiness is caused by the appearance of the first plastic crystals. As soon as this vanishes, the whole mass of this modification solidifies; to be more exact, it crystallises, for there is no sign of rigidity. When the mass is observed under the microscope but with a superimposed cover slip, a flow soon takes place and this is caused by the familiar crystals

[1] c.f. O. Lehmann, Molekularphysik **1**, 271 and 493 ff.

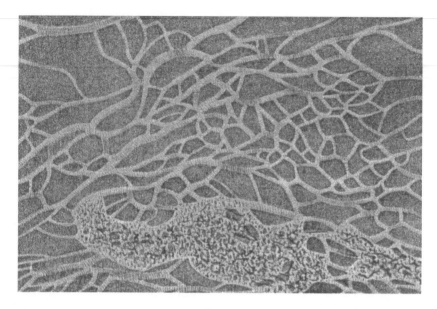

Fig. 5

developing sometimes here and sometimes there, and by the related volume change. This motion is the origin of the fact that the plastic crystals now seem to melt again. But actually they are only deformed and form an apparently homogeneous laminate. This strongly rotates the plane of polarisation of light, similar to a stack of mica lamella (more precisely: generating elliptically polarised light). They therefore appear intensely light-blue, or in different circumstances, can generate other different interference colours. Here and there, where the position of the crystals remained unchanged, they appeared bright, but run through by bright stripes (cf. Table III, Fig. 5). Because the stripes are in sharp contrast to the light-blue background, they look like bizarre 'oily streaks', generally pointing along the direction of the prevailing flow.

If the present interpretation of the observations is to be believed, a unique phenomenon is reported here for the first time. A crystalline and strongly birefringent substance has been observed which possesses such low physical strength that it cannot resist the effect of its own weight. As soon as it is not constrained by a liquid of equal density, it flows like syrup or gum; in this respect it would resemble the swimming oil ball of Plateau[1] were it to be removed from the influence of gravity.

Karlsruhe, 30 August 1889

[1] Joseph Antoine Ferdinand Plateau (1801–83) was a blind professor of physics in Ghent (Belgium). He is best known for his contribution to the molecular forces and elastic theory in liquids and thin liquid and soap films, leading to the so-called Plateau problem. This is the variational problem of minimising the surface area of a system with given volume (*ed.*).

Otto Lehmann was born in Constanz, Germany, in 1855. He was the son of a school teacher with an amateur interest in microscopy. After a peripatetic childhood he entered Strassburg (present-day Strasbourg in France, but at that time, like the rest of the disputed provinces of Alsace and Lorraine, in Germany) University in 1872, and obtained a doctorate in physical chemistry from that university in 1876. After academic posts in Mülhausen (present-day Mulhouse in France), Aachen and Dresden, he obtained in 1889 a full professorship at the University of Karlsruhe. Here he spent the rest of his life.

Lehmann was the author of an important physics textbook and an expert in polarising microscopy. Most of his career after 1889 was spent in studying liquid crystals, propagating their importance, and propagating *his* importance in their investigation. If not quite responsible for coining the term, he is responsible for advocating its wide use, even though it is now known to be a misnomer. He organised the first liquid crystal summer school in 1912, and together with Ernst Haeckel propounded the idea that liquid crystals were responsible for many of the processes of life. He was nominated, unsuccessfully, for the Nobel prize several times beginning in 1913. He died in 1922.

In the facsimile of the original article, we call the reader's attention to the reference to Plate III (Figs. 4 and 5) (which was detached from the article at the end of the journal volume), and to the woodcuttings; these were the figures which were clearly prepared specially.

1738

266. L. Gattermann und A. Ritschke: Ueber Azoxy-phenoläther.

(Eingegangen am 11. Juni.)

Die in dieser Mittheilung beschriebenen Versuche wurden in der folgenden Absicht angestellt: Bekanntlich hat Janovsky zwei isomere Azoxytoluole beschrieben, welche bei der Reduction des p-Nitrotoluols neben einander entstehen sollen. Wie der Eine von uns nachgewiesen hat, verhalten sich nun Phenoläther bei manchen Reactionen genau wie Kohlenwasserstoffe, und es lag deshalb der Gedanke nahe, zu versuchen, ob nicht vielleicht auch die den Janovsky'schen Azoxytoluolen entsprechenden Körper aus der Phenoläthergruppe ebenfalls in verschiedenen Isomeren erhalten werden könnten. Nun haben freilich Hantzsch und Werner späterhin mitgetheilt, dass die angegebenen Isomerien bei den Azoxytoluolen nicht bestehen und dementsprechend haben wir auch in unserm Falle keine Isomeren in dem gewöhnlichen Sinne beobachtet. Allein wie in der vorhergehenden Mittheilung beschrieben, zeigte sich bei der einen der von uns erhaltenen Azoxyverbindungen, dass dieselbe in einer farblosen und einer gelben Modification auftreten kann. War dies für uns ein Grund, die Azoxyderivate der Phenoläther etwas näher zu untersuchen, so wurden wir hierzu auch noch durch die folgenden Thatsachen angeregt. Diese *Körper zeigten nämlich einerseits höchst interessante physikalische* Eigenschaften, welche bislang nur bei einer einzigen Substanz beobachtet waren, und über welche am Schlusse dieser Mittheilung Näheres mitgetheilt ist. Andererseits beobachteten wir bei der Darstellung derselben auffallende Verdrängungserscheinungen unter den Alkylgruppen, welche für das Studium der Massenwirkungen nicht ohne Interesse zu sein schienen.

A3
Berichte der Deutschen Chemischen Gesellschaft **23**, 1738–50 (1890)

ON AZOXYPHENOL ETHERS

L. Gattermann and A. Ritschke

(Received 11th June)

The aim of the investigations described in this communication is as follows. It is well known that Janowsky described two isomeric azoxytoluenes, which were simultaneously produced during the reduction of *p*-nitrotoluene. One of us has demonstrated that in some reactions phenol ethers behave analogously to hydrocarbons. Therefore, the idea suggested itself to us to test whether phenol ethers could also be obtained as different isomers by analogy with Janowsky's azoxytoluenes. Hantzsch and Werner later reported, on the other hand, that the claimed isomers do not exist in the case of azoxytoluenes; moreover, in our case we also did not observe isomers in the usual sense. As reported in the previous communication, it turned out that one of the azoxy compounds that we obtained can exist as a colourless as well as a yellow modification. This was the reason for investigating the azoxy derivatives of the phenol ethers in more detail, but additionally we have been stimulated to do this by the following. Indeed these substances exhibit most interesting physical properties, which have been observed hitherto only in the case of a single compound, and about which we report more details at the end of this communication. On the other hand, during the preparation of these compounds we observed striking substitution phenomena of the alkyl groups which seem to be not without interest for a study from the point of view of mass action.

We started our experiments by the reduction of *p*-nitrophenetole, which was obtained by courtesy of the Bayer & Co. Dyeworks in Elberfeld.... (*there follow some details of the experimental procedure*)...First, long yellow needles precipitated, later followed by thin lamellae. Consequently, at the outset we supposed that two isomers had indeed been produced. As the lamellae dissolve much more easily in ethyl alcohol than the needles, we could separate both substances in this way by treating the mixtures with small amounts of alcohol at 40 °C. In this way, only the lamellae dissolved whilst the needles remained.

55

Thus, we were successful in obtaining, in a pure state, both the lamellae, which are produced only in small amounts, and the needles, which are the main reaction product. The result of the analysis of the needles was surprising: it was not consistent with the formula of an azoxyphenetole as we expected, but rather with that of an azoxyanisole ... (*we omit some quantitative details of the experiments*) ...

Thus, the ethyl group must be replaced by the methyl group by the action of methyl alcohol. If this assumption is correct the same substance must result from the reduction of *p*-nitroanisole and in fact our point of view has been confirmed experimentally. In the same way as described above we reduced nitroanisole in a solution of sodium hydroxide and methyl alcohol (1:10), and obtained an azoxy compound which was completely identical to the needles we had obtained from nitrophenetole. The chemical structure of this compound was proved beyond doubt because an exchange of the alkyl groups was not possible in this case ... (*further experimental details*) ...

Additionally, another problem has to be solved in the case of this azoxy compound. When its melting point was measured, we observed that the substance sharply transformed at 116 °C from the solid to liquid state. However, the molten substance did not form a clear, transparent liquid as usual, but it was cloudy and gave the impression of a mixture of different compounds. Despite repeated recrystallisations, the same effect was repeated each time. When the substance was heated above its melting point, we observed that the cloudy liquid suddenly became transparent very sharply at 134 °C. To begin with we found this phenomenon completely mysterious. However, the phenomenon has been clarified by a paper of Professor O. Lehmann from Karlsruhe entitled 'On flowing crystals'. In this paper very similar phenomena were described as occurring during the melting of cholesteryl benzoate. We therefore sent our crystals to Professor Lehmann, who was kind enough to inform us that our substances indeed belong to those species which he has named 'flowing crystals'. A more detailed description of these interesting physical phenomena will be given at the end of this paper.

In the light of these interesting phenomena the determination of the molecular weight seemed to be necessary. We used Raoult's method with benzene as a solvent.

The determination gave the following result:

Solvent	Substance	Depression	Molecular weight
	0.2020 g	0.3	235.8
14 g Benzene	0.4286 g	0.617	243.1

Calculated 258.

56

It follows that the substance possesses a single molecular weight.

The results of the chemical analysis of the second by-product, which results during the reduction of nitrophenetole, and which crystallises as lamellae, are in between those calculated for azoxyanisole and azoxyphenetole. Consequently, this compound is assigned as an anisole azoxyphenetole....

During melting this compound also exhibits the same behaviour as described above in the case of azoxyanisole. At 86 °C it transforms into the liquid state; the resulting cloudy liquid, however, becomes clear and transparent at 116 °C.

In order to decide if these physical and chemical anomalies were also present in the case of other homologues of the same class of compounds, we carried out reductions of other phenol ethers. For instance, we reduced *p*-nitroanisole by caustic soda solution and ethyl alcohol. The exchange of alkyl groups as described above was expected to occur and the azoxy derivative of phenetole should result.... (*More experimental details of organic chemistry*)... In this case it was also clear from the chemical analysis that the reaction product was azoxy-phenetole, and not the azoxyanisole which would normally have been expected. Consequently, also in this reaction the methyl group has been replaced by the ethyl group....

This compound melts at 134 °C, resulting in a cloudy liquid which becomes transparent above 165 °C. A determination of the molecular weight by means of the method of Raoult also resulted in a single molecular weight for this compound....

Solvent	Substance	Depression	Molecular weight
	0.1595 g	0.22	253.8
14 g Benzene	0.3547 g	0.48	258.7

Calculated 286.

Regarding the physical properties of the azoxy compounds described above, in what follows we refer to the communications which we owe to the courtesy of Professor Lehmann.

The microscopic investigation of the compounds sent to him showed that they all crystallise uniformly. The crystals of azoxyanisole belong to the monosym-metric system.... They are dichroic... Azoxyphenetole also crystallises mono-symmetrically* and is also dichroic.

* i.e. monoclinically (*ed.*).

The crystals of the third compound (anisole azoxyphenetole) are similarly also dichroic, but their boundary surfaces are so indistinct that a more detailed determination was not possible. . . .

However, the preliminary microscopic investigation seemed to me[#] to be worthwhile, in order to prove that it is also true from a crystallographic point of view that the substances are each completely uniform and different from each other in a clearly recognisable way.

Despite the considerable differences in both their chemical and their physical properties, they behave in exactly the same manner in that they transform on heating into a cloudy liquid. Furthermore, when the cloudy liquids are placed next to each other they cannot be distinguished and they are completely miscible in all proportions.

With regard to the molecular theory, especially to the theory of crystal structure, these cloudy liquids are most important. In the investigations of Reinitzer[1] it was found, for the first time, that there are substances which are definitely crystalline and strongly birefringent, but at the same time so extremely soft that it is hard to describe them as solid. The proposal of flowing crystals seemed to be a long way from the prevalent theories of crystal structure. Indications that these substances still exhibit a certain degree of displacement elasticity, which would have allowed their designation as solid, could not be obtained. Consequently, it remains open as to which state of aggregation these substances should be allocated.

The cloudy-liquid enantiotropic[2] modifications of the three substances described above remove any doubt about this open question. They lead necessarily to the conclusion that the crystal structure is not dependent on the solid state of aggregation, but that there are also drop-forming liquid crystals, which because of their optical anisotropy are comparable to common solid crystals, even those with strong birefringence. Since it follows from the investigations of H. Hertz that light is most probably not a mechanical but an electrical wave motion, objections that anisotropy of the optical elasticity can exist together with a total lack of anisotropy in mechanical elasticity seem to be irrelevant.

In order to study these liquid crystals, they must be isolated from each other. This can be achieved in the well-known way by adding solvents from which they gradually precipitate, isolated from each other, on cooling (after saturation on

---

[#] Gattermann clearly wrote the paper and here he forgot that he was speaking on Ritschke's behalf as well! This is not the only place in the paper where he is thus afflicted (*ed.*).

[1] c.f. O. Lehmann, Zeitschrift für phys. Chem. **4**, 467ff 1889. Note also that the original has Reiniker here, but this is clearly a misprint and must refer to Reinitzer (*ed.*).

[2] c.f. O. Lehmann, 'On enantiotropy' in Molecularphysik, Leipzig, W. Engelmann, Vol. I, p. 119ff.

heating). Usually I used colophony* or rigid Canada balsam as solvent. The observations were performed using a simple purpose-built hot stage microscope with a magnification of 100 to 700 times from C. Zeiss in Jena.

If there is sufficient space between microscopic slide and cover slip, each crystal precipitates as a perfect globe-shaped drop. Under normal illumination it appears just like a drop of oil in an aqueous solution except for a certain kind of stain[+] in the centre, which terminates at two diametrically opposite points of a sphere. They may be briefly termed the poles of a crystal drop.

It is obvious from the fact that the form of the drop is exactly spherical that the state of aggregation is totally fluid. If there existed any hint of displacement elasticity, the crystal could be forced by the surface tension at the crystal–liquid interface into an ellipsoidal form or into another shape similar to a sphere but not totally the same. Furthermore, the mobility, i.e. the small internal viscosity, indicates that the state of aggregation of these drops must differ widely from that of the solid state, because all known solid bodies with very small displacement elasticity (e.g. pitch, soft glass, etc.) still exhibit a very high degree of toughness.

It follows that they are really crystals, not only from the existence of the stains already mentioned above, which are always present in our drops, but never seen in normal non-birefringent liquids, but in particular from their behaviour in polarised light. By this it can be seen that they are extremely birefringent and even dichroic, very similar to the crystals of the solid modification.

A more detailed description and discussion of all the observations of these polarisation phenomena would be carrying things too far here, and will be published in another paper.[1,†] Therefore I restrict myself to a brief statement of the most important conclusions. Whereas the molecules in a solid crystal form a regular point lattice, thus for example all may have a parallel orientation, in the case of crystal drops they are ordered in rows, which connect both poles of the drop in a similar way to the curves of iron filings between the poles of a magnet. These structure lines will simply be called the line of force of the crystal drop.

---

* Sometimes known as *colophonium*. This is a pine-based resin, obtained by distilling turpentine oil and water, which has wide industrial application. It is also used a solvent and for fixing slide preparations in microscopy. The name derives from the ancient Greek city of Colophonium in Asia Minor, from where its use originates. A frequent application is to violins, in which context it is known as *rosin* (*ed.*).

[+] The original German is *Schliere*, and as far as we are aware this is the first time this term – leading directly to the present day *Schlieren texture* – appears in the liquid crystal literature (*ed.*).

[1] These communications meanwhile have been published in issue 5 of the Zeitschrift für Physikalische Chemie.

[†] The communication mentioned here does not refer to a publication of Gattermann himself but to a paper of O. Lehmann: 'The structures of liquid crystalline liquids' in Z. Physikal. Chemie **5** (issue 5), p. 427 (1890) where the structure of liquid crystalline drops was described in detail (*ed.*).

When the crystal drop is deformed, e.g. by pressing the microscopic slide and the cover slip, or by lengthening, squeezing or rolling, etc., the poles generally remain unchanged: they remain point-shaped, but they change their position on the surface of the drop. The lines of force, however, change in the same way as the curves of the iron filings, as though the poles were magnetic. If one imagines that inside the drop such a line of force between the two poles becomes directly visible, e.g. by inserting particles of dye, it becomes evident that this coloured line can never become a closed curve or acquire a node however much the drop may be twisted, pressed or stretched. However, this line can adopt the form of a spiral. The same rules also hold for real structure lines.

Still further deformation causes the poles to transform into lines. On releasing the force, however, they always contract again at once into their normal point-shaped form with the poles diametrically opposite to each other.

If a drop is divided into two or more parts, each resulting droplet acquires its two diametrically opposite poles at the moment of its formation.

If two drops flow together and merge, initially a spherical drop is formed inside which a structured interface can be observed. This is the area at which both drops, now deformed to hemispheres, join. However, this configuration is not stable, for the combined drop possesses four poles whilst it should normally have only two. And indeed, depending on how the drops have combined, one can observe inside the new drop the occurrence of a rapid or slow fluid flow, which always succeeds in making the lines of force of both drops parallel. This also causes the two poles to approach each other until they touch, at which point they suddenly vanish whilst the two remaining poles proceed to the opposite ends of a diameter.

The reason for this gradual change may be a kind of surface tension which appears at the structural interface inside the combined drop, in the region where the molecules have not yet assumed parallel orientations. The surface tension has different values at different places in the interface depending on the different orientations of the adjacent molecules. Consequently, a motion will occur which is analogous to that which occurs at the surface of water if its surface tension is reduced locally by dissolving camphor. From this there results the well-known motion of camphor granules swimming on water.

After the drops amalgamate the molecules reorient according to the mutual lines of force. I call this process copulation. If the merging drops are very small, then the copulation proceeds so quickly that it cannot be separated from the amalgamation. The same thing occurs when one of the drops is small and the other is large. The fluid may for some reason come into flow, so that frequent small drops are incident on a particular large drop. They are then immediately absorbed by the large one, and at the same time correctly oriented so that no disruption of the structure is noticeable.

If two drops are large, or if several droplets collide in a mass together, then frequently the amalgamation process remains undisturbed, but the copulation does not happen. The contact surfaces with the glass boundaries are stretched; the normal copulation process is disrupted by the surface tension against the glass and perhaps also by friction. Then peculiar complexes with many poles can occur, particularly in thin samples. The resulting set of interference colours occurring when the sample is viewed between crossed nicols is a splendid sight to behold.

If one produces the liquid crystals without any solvent, by heating the solidified melt of the pure substance between a slide and a cover slip, the formation as well as the retransformation into the solid modification is exactly the same as the mutual transition of two enantiotropic modifications. Under these circumstances of course the liquid crystal cannot adopt a drop-like form, and thus the common poles and lines of force are not formed. However, one observes, as very often in the case of allotropic transitions, that the molecules of the initial (solid) modification align the molecules of the new one (fluid in this case). Consequently, during the progress of the transition process each new molecular layer possesses the same orientation as in the crystal from which it has been formed. Thus, liquid crystals are formed with parallel oriented molecules, exhibiting uniform extinction between crossed nicols. These liquid crystals imitate by their contours exactly those of the previous solid crystals, and at first glance they can be easily confused with an aggregate of solid crystals.

If this liquid mass, which shows multicoloured patterns in polarised light, is melted, i.e. heated up to the transformation into the non-birefringent fluid modification[1], and allowed to reform on cooling, one would expect that the initial structure would be totally destroyed, and thus the newly formed mass would exhibit totally different multicoloured patterns.

Strangely this is not the case. If the heating during 'melting' was not too excessive, avoiding the fluid motion which occurs as a result of density differences, then just the same multicoloured patterns reappear which were initially present. Moreover, the melting and liquid solidification can be repeated a dozen times without any change. However, the patterns change at once on further cooling, to such an extent that the solid crystals formed cannot be oriented by the influence of the liquid crystal; if those crystals are transformed again into the liquid modification by repeated heating the patterns also change.

I know of only one explanation for this curious phenomenon, and that is that the non-birefringent melt also has a certain kind of crystalline structure, and that there is a mutual orienting effect between the molecules of both fluid modifications. Thus, it must be assumed that the non-birefringent melt crystallises[§] regularly. This conclusion, however, would lead to far-reaching consequences because there is no method to distinguish this kind of liquid possessing a regular[§]

[1] c.f. O. Lehmann, Molecularphysik I, 'On melting', p. 686.
[§] i.e. in a *cubic* structure (*ed.*).

crystal structure from other isotropic liquids. Consequently, nothing would prevent us from claiming that any liquid which can grow as a crystal possesses a crystalline structure.

Heidelberg, University Laboratory

Ludwig Gattermann was born in 1860 in Goslar, in the Lower Saxony area of North-central Germany and was the son of a baker. He studied chemistry at Leipzig and Heidelberg, and at the Gewerbe Akademie in Berlin. He took his doctorate at Göttingen and worked in Heidelberg before finally becoming Professor of Chemistry at the University of Freiburg im Breisgau in 1900. He is remembered by the Gattermann reaction (1890), the Gattermann–Koch reaction (1897), the Gattermann aldehyde synthesis (1898) and the Gattermann–Stika pyridine synthesis (1916).

Among his chemical interests were aromatic acids, but he also prepared various exotic inorganic compounds, including the pure (and highly explosive) nitrogen trichloride. He is perhaps best remembered for his textbook of practical organic chemistry, *The Practice of Organic Chemistry*, which was widely referred to (ironically or otherwise!) as *Gattermann's Cookbook*. He died in 1920.

Herr Privatdozent Dr. R. Schenck-Marburg i. H.:

ÜBER DIE NATUR DER KRISTALLINISCHEN FLÜSSIGKEITEN UND DER FLÜSSIGEN KRISTALLE.

Meine Herren! Herr Geheimrat Lehmann hat sich liebenswürdigerweise bereit finden lassen, die Gesellschaft heute Nachmittag mit seinen Untersuchungen über die flüssigen Kristalle bekannt zu machen, uns die so merkwürdigen, von ihm in ihrer Bedeutung erkannten, Phänomene durch Projektion vorzuführen.

Als Einleitung zu diesen Demonstrationen möchte ich Ihnen eine ganz kurzgedrängte Uebersicht über das Verhalten jener Flüssigkeiten, welche wir als kristallinische bezeichnen, geben, ich möchte Sie mit den Gründen bekannt machen, welche uns das Recht verleihen, sie mit einem derartig paradoxen Namen zu belegen, und des weiteren die Beziehungen zu den festen kristallisierten Stoffen erörtern.

Wir wollen direkt an unseren Gegenstand herantreten und uns mit den Erscheinungen beschäftigen, welche Reinitzer im Jahre 1888 am Cholesterylbenzoat beobachtete. Diese Substanz (bei gewöhnlicher Temperatur wunderschöne, glänzende, farblose, kristallographisch wohldefinierte Blättchen) schmilzt bei $145,5^0$ zu einem trüben Schmelzfluss von der Konsistenz des Olivenöles. Es besitzt die merkwürdige Eigenschaft, sich bei $178,5^0$ zu klären. Unter dem Polarisationsmikroskop zeigt die trübe Masse trotz ihres flüssigen Zustandes Doppelbrechung, zwischen gekreuzten Nicols bleibt das Gesichtsfeld aufgehellt. Es wird aber dunkel, sowie die Temperatur von $178,5^0$ überschritten wird, die Masse verhält sich alsdann wie eine gewöhnliche Flüssigkeit. Die optischen Verhältnisse sind von Herrn Lehmann eingehend untersucht worden.

Im Laufe der Zeit haben sich noch mehr Stoffe gefunden, welche ein ganz ähnliches Verhalten zeigen wie die Derivate des Cholesterins, aber in mancher Beziehung als Beobachtungsmaterial besser geeignet sind. Hier ist vor allen Dingen der p-Azoxybenzoësäureäthylester zu nennen, an dem Vorländer das Auftreten der Erscheinungen konstatierte. Wenn man zu dem isotropen Schmelzfluss dieser Substanz kleine Mengen fremder Stoffe mischt und das Präparat unter dem Mikroskop vorsichtig abkühlt, so beobachtet man die Ausscheidung langer, dünner Kristallnadeln, welche optisch einachsig sind, im polarisierten Lichte Dichroismus zeigen, und bestimmte Auslöschungsrichtungen besitzen.

Jeder Druck mit der Präpariernadel deformiert diese Gebilde, welche übrigens nie scharf ausgeprägte Ecken und Kanten besitzen. Die Abrundung wird verursacht durch die Wirkung der Oberflächenspannung. Ihr Einfluss zeigt sich vor allen Dingen darin, dass Nädelchen, welche

sich berühren, zu grösseren zusammenfliessen. Stossen sie unter einem Winkel aufeinander, so erfolgt zunächst Parallelrichtung und dann Vereinigung der parallelen Stücken. Die kleineren lagern sich als Wülste an die grösseren an. Noch eigentümlicher ist das Verhalten der Nädelchen, wenn sie mit einer Luftblase in Berührung kommen. Erfolgt das Aufstossen mit der Spitze, so verbreitet sich die Nadel an der Berührungsstelle unter dem Einfluss der hier kräftig wirkenden Oberflächenspannung, so dass schliesslich eine der Luftblase mit breiter Basis aufgesetzte Pyramide mit gekrümmten Seitenflächen entsteht. Die Streifung der Auslöschungsrichtung stehen stets normal auf der Luftblase Merkwürdigerweise erfolgt gar keine Beeinflussung, wenn die Nadel mit ihrer Breitseite mit der Luftblase in Berührung kommt. Man muss daraus den Schluss ziehen, dass die Oberflächenspannung an der spitzen Seite viel höhere Werte besitzt als an der breiten.

Wegen ihrer unzweifelhaften Kristallnatur und ihrer Fähigkeit, unter dem Einfluss der Oberflächenspannung zu fliessen, hat Herr Lehmann diese nadelförmigen Gebilde als „fliessende Kristalle" bezeichnet.

Bald nach der Publikation der Lehmannschen Beobachtungen fand Herr Gattermann bei einigen Abkömmlingen des p-Azoxyphenols, z. B. beim p-Azoxyanisol und p-Azoxyphenetol trübe doppeltbrechende Schmelzflüsse, welche genau wie das Cholesterylbenzoat bei einer bestimmten Temperatur klar werden.

Beobachtet man sie aber unter den gleichen Bedingungen wie die fliessenden Kristalle, so ergeben sich erhebliche Unterschiede. Aus der isotropen, mit fremden Stoffen versetzten Schmelze kommen nicht nadelförmige Gebilde heraus, sondern Tropfen von Kugelform. Diese Tröpfchen verdanken ihre Gestalt lediglich der Oberflächenspannung.

Sie sind im Gegensatz zu grösseren, aus ihnen zusammengesetzten Massen völlig klar und zeigen bei 300- bis 700 facher Vergrösserung eine ganz besondere Struktur, wie sie bei Tropfen gewöhnlicher Flüssigkeiten gar nicht vorkommt. Die scheinbare Struktur ist eine Folge davon, dass die Lichtbrechung in den verschiedenen Richtungen eine verschiedene ist. Hauptsächlich treten zwei Erscheinungen auf, welche verschiedenen Lagen der Tropfen entsprechen. Die eine, die sogen. erste Hauptlage, zeigt einen centralen dunklen Punkt, welcher von einem grauen Hof umgeben ist. In dünnen Präparaten orientieren sich die Tropfen meist in der zweiten Hauptlage, sie gleichen durch-

131*

A4
Zeitschrift für Elektrochemie **11**, 951–55 (1905)

ON CRYSTALLINE LIQUIDS
AND LIQUID CRYSTALS

R. Schenck

(Lecturer at the University of Marburg)

Good afternoon, Gentlemen. Professor Lehmann has studied the strange phenomena of liquid crystals and recognised their importance. This afternoon he has kindly consented to give a demonstration of these phenomena, making use of projection.

As an introduction to these demonstrations, I should like to give a brief overview of the behaviour of these liquids, whose properties are such that we describe them as crystalline. In this talk, I should like to explain to you why we have taken the step of labelling them so paradoxically, and then go on to discuss their relationship to solid crystalline material.

We launch into our topic immediately. We are studying the phenomena first observed by Reinitzer in cholesteryl benzoate in 1888. At room temperature this material is a shiny colourless solid with well-defined crystallographic planes. At 145.5 °C it melts, now forming a cloudy liquid with the consistency of olive oil. This liquid possesses a strange property: at 178.5 °C, the turbid liquid becomes clear.

Under the polarising microscope, notwithstanding its liquid state, this turbid material can be seen to be birefringent. When placed between crossed nicols, light transmission is not extinguished. Once above 178.5 °C, however, the sample does extinguish light, and the material behaves like a normal fluid. The optical behaviour has been studied in detail by Professor Lehmann.

Since then more materials exhibiting behaviour extremely similar to that of the cholesterol derivative have been discovered. In many respects these are more suitable as experimental materials. I mention the most significant of these substances. This is ethyl *p*-azoxybenzoate, and we owe the introduction of this material in this context to Vorländer. What happens is that a small concentration of impurity is mixed with the isotropic phase of this substance, and the sample is then cooled carefully under the microscope. One then observes the expulsion of long, thin crystallites, which are optically uniaxial, which exhibit dichroism in polarised light, and which possess definite extinction directions.

These structures are deformed whenever they come into contact with the preparation needle. They never possess sharply defined corners or edges. The rounding is caused by surface tension. The surface tension manifests itself above all when small needles in contact amalgamate into a larger ones. If the contact occurs at an angle, the needles first line up in parallel to each other, and then the parallel pieces merge. The smaller ones adhere to the larger one as a bulge. Even more characteristic is the behaviour of the small needles when they touch an air bubble. When the needle touches the air bubble at its tip, it spreads around the contact point under the influence of surface tension, which acts strongly in this situation. Eventually it takes the shape of a pyramid with a broad base and curved sides. The striations and the extinction directions are normal to the air bubble. By contrast there is no effect when a broad side of the needles comes into contact with an air bubble. From this phenomenon one concludes that the tip surface tension is much greater than that along the sides.

The needle-forming objects are evidently crystalline. They flow under the influence of surface tension, and are thus viscous. As a result of the coexistence of these two properties, Professor Lehmann has given these objects the name 'flowing crystals'.

Soon after the publication of Lehmann's observations, Dr Gattermann studied some derivatives of the p-azoxyphenols, in particular p-azoxyanisole and p-azoxyphenetole. In these materials he found turbid birefringent melts, which lose their turbidity at a definite temperature, just like cholesteryl benzoate.

However, when these materials are observed in the same set-up as the flowing crystals, considerable differences arise. When the isotropic melt is doped with an impurity, needle-forming objects no longer appear. Rather, spherical droplets appear, whose shape follows merely from surface tension effects.

These are perfectly clear, whereas by contrast aggregates of such droplets are not. Magnified 300 to 700 times they show a unique structure, unlike anything occurring in drops in normal liquids. The apparent structure is a result of the fact that the refraction of light is different in different directions. In the main two kinds of effect occur, each of which corresponds to a different drop position. The so-called 'first principal position' shows a central dark point surrounded by a grey region. However in thin samples the drops mainly orient themselves in the so-called 'second principal position'. They now resemble transparent spheres. In the centre of each transparent sphere there appears to be a lens with a different refractive index.

In contrast to drops in normal liquids, these droplets are birefringent. They exhibit dichroism in polarised light. Depending on the drop position, a cross or other figures appear when the sample is observed between crossed Nicols.

We must forego the details. I mention only that the optical phenomena can be altered by external forces in very many different ways. When the drops are rotated, the crosses turn into spiral crosses. In a magnetic field the drops take on a very definite form. They all now adopt the first principal position, if the lines of force lie vertically on the microscope stage.

If the drops are prevented from orienting themselves as a whole, effects on the molecules are felt. The magnetic effects compete with those forces which the initial crystal drop structure seeks to preserve. These forces are the so-called 'molecular directional forces'. The final result is a compromise between the magnetic and crystallographic directional forces. This compromise of course leaves an optical signature.

When several crystal drops touch each other, they flow together into a larger drop. The observations on drop amalgamation now enable the structure of larger volumes of mixing fluids to be understood. We can think of these larger volumes as large aggregates which have been deformed by pressure, thermal currents and mechanical obstacles. The images under the microscope show that the optical orientation is different in different places.

We shall define the large aggregations of liquid crystals and flowing crystals, which have been stuck together and unified, as 'crystalline liquids'.

The turbid crystalline liquids transform, at an absolutely precise temperature, into normal clear liquids which refract in the conventional way. The liquid and flowing crystals might be said to melt. The clearing point corresponds completely to a melting point. The transformation is accompanied by a volume increase. Adding impurities significantly reduces the temperature of the clearing point. From this result one can deduce that the transformation possesses a small latent heat. This conclusion is confirmed by the change of the transition point under pressure. The latent heats obtained from cryoscopic data agree with those calculated from the Thomson–Clausius equation and those given by direct calorimetric measurement.

So far, 21 substances have been found which exhibit liquid crystallinity. However, there is little doubt that this number will increase significantly in the future. Their appearance is not restricted to particular classes of chemical compounds. Liquid crystals occur in many groups of organic materials. There is even an inorganic material, so-called regular silver iodide, which one can find a justification for including amongst liquid-crystal-forming substances.

Thus the synthesis of these compounds presents little difficulty. Some may be obtained very easily indeed, such as, for example, p-azoxyanisole. These materials only need to be melted for the phenomenon of crystalline viscosity to be studied. One can then also see that they can be poured and are fluid in the way that normal fluids are. Of course the viscosity varies from substance to substance. The table below shows viscosities relative to water at 0 °C (which takes the value 100).

Material	Viscosity (water at 0 °C = 100)
cholesteryl benzoate	893–621
ethyl p-azoxy benzoate	856–472
p-diacetoxychlorostilbene	327–309
p-azoxyanisole phenetole	171–111
p-azoxyanisole	141–128
p-methoxycinnamic acid	106–91
p-azoxyphenetole	79–66

We have a complete range. The initial cases possess a viscosity of the order of olive oil. These are flowing crystals. Later cases are much less viscous than water. They rapidly follow external forces. Drops of these substances thus take on a spherical shape. Between the two there are all possible intermediate forms.

The picture adopted by Professor Lehmann and myself is that the cloudy anisotropic liquids are aggregates of birefringent flowing or liquid crystals. This point of view has not remained unchallenged. Professors Quincke and Tammann prefer to interpret these observations in terms of chemical inhomogeneity. Quincke regards crystalline liquids as mixtures in which solid crystallites are suspended. Tammann considers them to be emulsions. Both are agreed that crystalline fluids are a two-phase phenomenon.

I want now very briefly to summarise the main reasons why we are sticking to our original point of view.

Professor Tammann bases his opposition on the fact that crystalline fluids are cloudy. He believes that the experiments of Rotarski are evidence for an emulsion picture. Rotarski observed that when p-azoxyanisole is distilled, the clearing point of the distillate decreases more quickly than that of the distillation residue. But it is generally known that azoxy compounds decompose on heating. If the decomposition products are more volatile than the original material, this behaviour would immediately follow.

Professor Tammann has further observed that his samples precipitate just like emulsions. De Kock, Lehmann and I have had the opportunity of working with Tammann's original materials. We have been able to show conclusively that his samples contained a high impurity concentration, in particular of p-azoxyanisole originating during synthesis. Pure samples do not show a precipitate, and it also does not appear if the crystalline fluids are placed in a centrifuge or subjected to electrophoresis. This last experiment has been carried out by Bredig and Schukowsky, and also by Coehn. In very many cases these methods enable emulsions to be separated. In the case of crystalline liquids, however, nothing indicates the existence of emulsions.

By contrast there are a large number of facts which sharply contradict the emulsion point of view. In particular, the transition from a cloudy liquid state to an isotropic liquid is associated with a sudden change in many physical properties. I mention here the density (Fig. 330) and the viscosity (Fig. 331).

The crystalline liquid phase occurs at temperatures lower than the isotropic liquid phase. It is a strange fact that despite this, in all materials which form crystalline liquids, the viscosity is lower in the crystalline liquid than in the isotropic liquid. Equally mysteriously, at the transition to the isotropic liquid, the viscosity jumps suddenly and substantially. Emulsions also exhibit a clearing point, and the viscosity above and below the clearing point can be measured. We have made these measurements for water–phenol and alcohol–carbon disulphide mixtures. In Fig. 332, I show the relationship between flow time and temperature in these experiments. There is no sign of a discontinuity in these experiments. Furthermore, there is absolutely no theoretical reason to expect one in the case of emulsions such as these.

The fact that in turbid melts the light absorption is independent of temperature indicates that they are homogeneous. In an emulsion, however, the number of

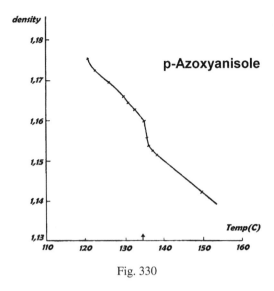

Fig. 330

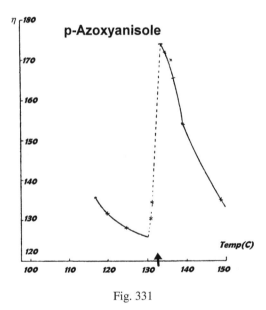

Fig. 331

droplets would have increased with decreasing temperature, causing increased absorption. The phase diagram close to the melting point was studied by de Kock. His experiments can also only be interpreted by supposing that the cloudy melts are chemically uniform.

69

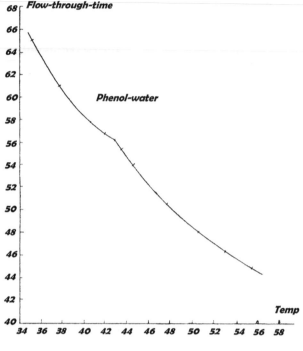

Fig. 332

One might also conceive of cases in which a reversible chemical process takes place. In this scenario a melted crystal might consist of two components which are only weakly miscible. The two components would only mix completely at a higher temperature. The clearing point would then correspond to the consolute point of the mixture. However, the experiments which I have just described really exclude this possibility.

The isotropic phase immediately above the clearing point would have to be a mixture of two components whose relative concentration changes with increasing temperature. Now we do have a quantity which will signal a uniform fluid mixture with a temperature-dependent concentration. This is the temperature coefficient of the molecular surface energy. Dr Ellenberger and I have studied this quantity. The normal value of this quantity is 2.12, but in reversible reactions deviations from this value appear[1].

---

[1] Schenck is referring to Eötvös's law: $\sigma v^{2/3} = k(T_c - T)$, where T is the temperature, T_c is the critical temperature, σ is the surface temperature, v is the molecular volume, and k is the Eötvös constant which takes the value of 2.12 for a range of normal fluids. This is a primitive version of the law of corresponding states and can of course be easily derived from density functional theory, although Schenck didn't know that. Relevant references are: R.V. Eötvös, Wied. Ann., **27**, 456 (1886); W. Ramsay and J. Shields, Phil. Trans. A **184**, 647 (1893) (*ed.*).

In p-azoxyanisole and in several other materials, our experiments always find the normal value of k just above the clearing point. We conclude with a high probability that the chemical composition is uniform and temperature-independent.

There are more reasons to believe that the cloudy liquids are homogeneous, but time forbids me to pursue this problem further. Tammann's opinion is that the crystalline liquids appear cloudy, and that therefore they must be two-phase systems. This conclusion is unjustified. An emulsion might sustain transmission through a sample between crossed Nicols and the depolarisation of polarised light. However, the strong birefringence which actually occurs would never be seen. Under the microscope one sees that in large systems, the optic axis varies from place to place. Even if the sample is transparent on the smallest scales, such media must, as Professor Lehmann has stressed, necessarily appear cloudy.

I now summarise. Objections have been raised to the hypothesis that crystalline liquids are homogeneous. These objections have been dealt with. The subject area has been exhaustively evaluated from different points of view. A broad experimental basis for Lehmann's viewpoint now exists.

I only need now to answer questions about the relationship between liquid crystals and solids. We cannot consider the liquid crystalline state in isolation. A range of experiments on different materials have shown that liquid and flowing crystals are part of a continuum of increasingly viscous materials. The flowing crystals already possess hints of a crystalline form, but in particular circumstances exhibit surface tension effects. From flowing crystals to soft crystals such as white phosphorus or camphor is but a short step. Soft crystals can be distinguished from true solids only by the degree to which they can be deformed by external forces.

We should consider liquid crystal drops as very soft crystals. Their softness makes them susceptible to surface tension effects. The special optical properties should then be explained by a surface tension deformation effect. The drop structure is a combined effect of the surface tension and the directional forces which occur in all crystals. If one could turn off the surface tension, then these bizarre materials would exhibit what might be called proper crystal structures as well as normal optical behaviour. We have no reason to consider flowing or liquid crystals as particularly special. We should be able to classify them in the normal crystallographic system. Indeed Professor Lehmann has in the past not hesitated to do this.

I have given you a rather brief overview of the field. There are other interesting questions which I have not discussed, but I must refrain from doing so because time is pressing. The demonstrations should speak more convincingly than I. I would like to hope that when you have seen the phenomena for yourselves, you will be persuaded to adopt the point of view I have advocated here.

Discussion

Professor Tammann

I am familiar with liquid crystals from personal experience. I want to discuss the central question: 'Are these materials anisotropic or isotropic? Equivalently, has birefringence in these cloudy liquids been conclusively established?'

First: the materials in question have been observed in parallel films between crossed Nicols. Professor Lehmann has described a situation in which the image is divided into segments. When the sample is rotated, their brightness changes. This seems to demonstrate birefringence, and I would not object to calling these liquids crystals. But then one observes that the segmentation and the change in brightness are not properties of the liquid itself. Rather there seems to be an anisotropic dust adsorbed onto the glass plates surrounding the sample. If the liquid is shaken violently the picture does not change. The phenomenon is thus not a property of the liquid. It can be disrupted by interfering with the adsorbed dust. Then no segmentation can be observed. I have made such observations on many occasions, and Professor Lehmann has essentially described the same thing. I conclude from this that the liquid is not itself birefringent.

A second observation is as follows. Professor Lehmann has observed droplets of liquid crystals in Canada balsam. He has seen black crosses and suchlike on their images. I have to confess that these phenomena have given me the most trouble. They do indeed seem to imply anisotropy. But very similar crosses can be seen in the images of air bubbles in Canada balsam, and air bubbles are unambiguously isotropic. Consequently I cannot concede that the so-called liquid crystals exhibit birefringence. In other respects we are certainly in agreement. The anisotropy relates only to optical properties. In all other contexts there is complete isotropy, and this even applies to growth phenomena. All liquid crystals are optically turbid media. They are thus emulsions, and contain at least two components. However, a complete analysis of this phenomenon has not yet been successfully carried out. In my own view, distillation of the liquid crystals offers the best prospect for the resolution of the problem, but this is an extremely difficult task.

We must also take into account that Dr Schenck has found discontinuous changes in certain properties as a function of temperature. However, a discontinuous change in any given property is difficult to establish beyond doubt. Finally I would like to come to the so-called flowing crystals. There are certainly such things as soft crystals. I am happy to concede flowing crystals, but *not* liquid crystals. The existence of liquid crystals is a key question when considering the lattice theory of solids. I would thus assign enormous theoretical importance to the question of the existence of liquid crystals!

> The discussion ran out of time and was closed by the chairman. As a result, a reply to Professor Tammann's points was impossible. However, the discussion reconvened in a small group for the liquid crystal demonstration. This had been arranged in the Institute of Physics as part of the afternoon

programme. Taking part in this were Messrs Lehmann, Tammann, van't Hoff, Schenck, Muller, Elbs and Bredig. Geheimrat Lehmann's main points are contained in the paper presented immediately below. At this meeting Herr Professor van't Hoff proposed the formation of a commission of experts which could examine liquid crystal problems further. This commission should include, *inter alia*, Messrs Tammann and Lehmann.

(end of the official report)

Rudolf Schenck The University of Halle plays an important role in the history of liquid crystals and it is appropriate that Rudolf Schenck was born in that city in 1870. In 1889 he entered the University in Halle to study chemistry. His Ph.D in organic chemistry under the supervision of the well-known chemist Jacob Volhard was awarded in 1894.

Remaining in Halle, his interests then turned to problems in physical chemistry, attempting to determine a relation between the heat of sublimation and the saturation vapour pressure of solids. This work was relatively unsuccessful and in any case came to an abrupt end when his apparatus was smashed by a cleaning lady in the department. Following this incident he retired to the library in a bad mood, and by lucky chance came across Gattermann's recent publication [A3] on *p*-azoxyanisole and its two liquid states. This stimulated him to study liquid crystals, and led directly to the publication translated in this volume, which supported the Lehmann view of liquid crystals against that of Tammann. This work continued at the University of Marburg, where he obtained his Habilitation degree in 1897 and was then appointed to a lectureship.

In 1906 he was appointed to a full professorship in physical chemistry at the University of Aachen, after which his interests turned to other areas of physical chemistry. In 1910 he moved to the recently founded Technical University of Breslau, where he was the first Rector. In 1916 he moved to the University of Münster in Westphalia, where he was full professor of physical chemistry and worked mainly on problems of metallurgy. He retired in 1935, and even in his seventies and eighties continued to work in the field of metallochemistry at the University of Marburg.

In later years he achieved eminence in German physical chemistry. He was President of the Deutsche *Bunsen-Gesellschaft für Physikalische Chemie* 1933–4 and 1936–41. In particular, although never a member of the National Socialist Party, he was a vociferous Nazi sympathiser and was extremely but controversially active in scientific politics.

Although his research interests had moved, he maintained a parental affection for the study of liquid crystals/crystalline fluids (unlike others he did not allow himself to be drawn into controversy over nomenclature!). As we shall see in the next section, as late as 1930–1 he contributed extensively to the general discussion on liquid crystals edited by P.P. Ewald in the review volume in *Zeitschrift für Kristallographie* [B4].

Schenck died in 1965 in Aachen. He lived to the ripe old age of 95, the only early liquid crystal pioneer to see the emergence of the beginnings of the liquid crystal display industry and the scientific maturation of the field which he influenced so profoundly.

117. D. Vorländer: Ueber krystallinisch-flüssige Substanzen.

[Mittheilung aus dem Chem. Institut der Universität Halle a. S.)

(Eingeg. am 6. Febr. 1906; mitgeth. in der Sitzung von Hrn. J. Meisenheimer.)

Das Vorkommen einer doppelbrechenden flüssigen Phase beim Schmelzen einiger Substanzen ist durch die mikroskopischen Untersuchungen von O. Lehmann[1]) und durch die physikalisch-chemischen Arbeiten von R. Schenck[2]) sichergestellt. Von den Substanzen ist zuerst das Jodsilber von O. Lehmann, die Cholesterinacyle von Reinitzer[3]) und die *p*-Azoxyphenoläther, Azoxy-Anisol und -Phenetol von Gattermann und Ritschke[4]) als anisotrop flüssig beobachtet worden. — Bei Gelegenheit einer Untersuchung der *p*-Azobenzoësäureester hatte ich mit Felix Meyer[5]) den *p*-Azoxybenzoësäureäthylester dargestellt, und nach vielen vergeblichen Versuchen, die Azo- und Azoxy-Ester in reiner Form von annähernd constantem Schmelzpunkt zu gewinnen, stellte sich heraus, dass der *p*-Azoxybenzoësäureäthylester mit zu jener kleinen Zahl von Substanzen gehört, welche in einer anisotrop flüssigen Phase aufzutreten vermögen. Der Azoxyester hat zwei scharfe Schmelzpunkte: der erste bezeichnet den Uebergang von der festen krystallinischen zu der trüben anisotrop flüssigen Form bei 114°, der zweite Schmelzpunkt die Verwandlung der trüben anisotropen Flüssigkeit in die klare isotrope Flüssigkeit bei 121°. Die Erscheinung kann schon mit blossem Auge beobachtet werden, besser mit der Lupe, und die Entstehung der anisotropen Flüssigkeit ist zwischen gekreuzten Nikols unter dem Polarisationsmikroskop deutlich zu erkennen an der Aufhellung der Flüssigkeit und der Abscheidung der eigenartigen rundlichen Gebilde mit schwarzem Kreuz, wenn man einige Krystallfragmente der Substanz völlig schmilzt und dann erkalten lässt. Die klare Flüssigkeit krystallisirt alsdann zweimal, indem die beim Schmelzen auftretenden Erscheinungen in umgekehrter Folge auftreten. Dass die trübe doppelbrechende Phase noch flüssig ist, kann man durch Schmelzen einer etwas grösseren Menge des *p*-Azoxybenzoësäureesters nachweisen. Die isotrop flüssige Phase ist dünnflüssig, verhältnissmässig leicht beweglich; dann folgt beim Erkalten die anisotrope Phase, welche sich durch Neigung des Uhrglases nur zu langsamem Fliessen bringen, mit dem Glasstab aber

---

[1]) Flüssige Krystalle, Leipzig, W. Engelmann 1904.

[2]) Kryst. Flüssigkeiten und flüssige Krystalle, Leipzig, W. Engelmann 1905.

[3]) Wiener Monatsh. 9, 455 [1888].

[4]) Diese Berichte 23, 1738 [1890].

[5]) Ann. d. Chem. 320, 122 [1902] und 326, 331 [1903].

A5
Berichte der Deutschen Chemischen Gesellschaft **39**, 803–10 (1906)

ON CRYSTALLINE-LIQUID SUBSTANCES

D. Vorländer
[Contribution from the Institute of Chemistry of the
University of Halle on Saale]
(Accepted 6 February 1906; presented at the meeting by
Mr. J. Meisenheimer)

The existence of the birefringent liquid phase of several substances on melting is proved by the microscopic investigations of O. Lehmann[1] and the physicochemical studies of R. Schenck.[2] Amongst the substances in which anisotropic liquid phases were first detected were silver iodide by O. Lehmann, the cholesterol acyls by Reinitzer[3] and the *p*-azoxyphenol ethers – azoxyanisole and azoxyphenetole – by Gattermann and Ritschke.[4] During an investigation of the *p*-azoxybenzoates in collaboration with Felix Meyer,[5] I prepared the ethyl *p*-azoxybenzoate. However after many fruitless attempts to obtain the pure azo- and azoxy-esters having nearly constant melting points, it turned out that ethyl *p*-azoxybenzoate also belongs to those small number of substances which are able to exhibit an anisotropic liquid phase. The azoxy ester exhibits two sharp melting points: the first one indicates the transition from the solid crystalline to the cloudy anisotropic liquid form at 114 °C, the second one the conversion of the cloudy anisotropic liquid into a clear isotropic liquid at 121 °C. This transition can be observed by the naked eye, but is better seen through a magnifying glass. The development of the anisotropic liquid can be clearly seen between crossed nicols under the polarising microscope. If one melts completely some crystal fragments, the liquid clears,

---

[1] Flüssige Krystalle, Leipzig, W. Engelmann **1904**.
[2] Kryst. Flüssigkeiten und flüssige Krystalle, Leipzig, W. Engelmann **1905**.
[3] Wiener Monatsh. **9**, 435 (1888).
[4] This journal **23**, 1738 (1890).
[5] Ann. d. Chem. **320**, 122 (1902) and **326**, 331 (1903).

and, on cooling, curious round droplets with a black cross separate out. The clear liquid then crystallises twice, while the transitions which occur during melting now appear in reverse sequence. By melting a somewhat larger amount of the p-azoxybenzoic ester, one can check that the cloudy birefringent phase is still liquid. The isotropic liquid phase is highly fluid and relatively easily mobile; when it is cooled, the anisotropic phase forms, which may only be caused to flow slowly by an inclination of the watch glass, or by being stirred, in the manner of syrup by a glass rod. Finally the anisotropic phase transforms on further cooling into the hard, anisotropic solid form. Both transitions are sharp and are totally different from the phenomenon of the gradual change which is observed during the melting of impure or decomposing substances, and which is well known. O. Lehmann[6] added small amounts of colophony or bromonaphthalene to the isotropic melt of our p-azoxybenzoate, and on cooling succeeded in obtaining the precipitation of long thin crystal needles of the anisotropic liquid phase. These he recorded photographically, and after highly enlarging the picture he identified crystal needles which possess all the properties of common crystals, but are soft and capable of flow. Therefore, one must acknowledge the justification for the term 'liquid or flowing crystals.'

In continuation of our investigation we draw attention to a certain parallel between the azo- and azoxy-anisoles on one hand and the azo- and azoxy-benzoic esters on the other. In both cases, namely the phenylic ethers and the benzoic esters, the crystalline-liquid phase will be created by the azoxy group, because by reduction to the azo compound the phenomenon disappears. Furthermore, the para-position of the substituents is important because ortho-azoxyanisole does not form a crystalline liquid, in contrast to the para-compounds, and also the liquid crystalline phase is absent in the case of the ortho- and meta-azoxy benzoic esters prepared by Felix Meyer and K. Dahlem.[1] In addition, in collaboration with K. Dahlem I have tested the ethyl esters and the methyl esters, but without any success. Thus the parallel with the azoxy phenyl esters is not present in this case. The methyl as well as the ethyl ether, and also the mixed methyl ethyl ether of the p-azoxybenzene, investigated by Gattermann and Rising[2] form crystalline liquids, to which we can add the n-propyl ether (1st m.p. 116 °C, 2nd m.p. 122 °C). In cooperation with C. Siebert, I prepared a series of p-azoxybenzoic esters, but a liquid crystalline substance could not be achieved neither by extending nor by branching the alkyl chain. The phenomenon remains restricted to the ethyl ester of the p-azoxybenzoic acid; at least a double melting point cannot be observed by the usual methods in the case of the other esters. Perhaps the range

[6] op. cit.
[1] Ann. d. Chem. **326**, 331 (1903).
[2] This journal **37**, 43 (1904).

of existence of the anisotropic liquid amounts to less than 1/100 or 1/1000 degrees and therefore is no longer observable.

Melting points of the *p*-azoxybenzoates
(uncorr.)

methyl	.	203 °C
n-propyl	.	103 °C
iso-propyl	.	96 °C
n-butyl	.	105 °C
iso-amyl	.	122 °C
allyl	.	88–89 °C
benzyl	.	147 °C

In the following investigations we have been guided by the idea that the formation of the crystalline-liquid state could be caused or promoted by the same atomic groups which influence also other physical properties like light refraction, colour, rotating power, etc. Firstly the ethylene group with the carbon double bond was considered. Surprisingly the experiments showed a favourable result. Together with C. Siebert and P. Hausen, I found a considerable number of crystalline-liquid substances amongst the *p*-azoxycinnamates. Not only the ethyl ester but all aliphatic esters of *p*-azoxycinnamic acid from methyl up to cetyl ester form the crystalline-liquid phase, which surpasses all examples known hitherto in respect of high viscosity, light refraction (a change due to the relative properties of benzoates and cinnamates) and by the large temperature range (more than 110 °C). The anisotropic liquid phase on melting in a test tube forms a gel which scarcely flows; however, individual droplets tend to coalesce. By insertion of the double bond, the very pale yellowish colouration of the azoxybenzoates is turned into lemon yellow. Melting the azoxycinnamates together with small amounts of colophony, one observes on cooling crystals of the anisotropic liquid phase which are similar to those found by Lehmann in the case of ethyl *p*-azoxybenzoate. The results of the more precise microscopic investigation carried out by Professor[1] Lehmann with our substances[2] were very interesting. The anisotropic liquid form of the ethyl *p*-azoxycinnamate crystallises in the form of 'polyhedral crystals with rounded edges and corners, probably quadratic prisms, sometimes hemimorphically sharpened by a steep pyramid'. 'If two of these crystals come into contact with each other they at once flow together to form a single crystal *so long as* they

---

[1] The original German uses the honorific title *Hr. Geheimrat*, which cannot really be translated, other than to emphasise the esteem in which the writer holds Lehmann. We have chosen to approximate this by including Professor in front of Lehmann's name at this point (*ed.*).

[2] Ann. d. Physik **19**, 1 and 22 (1906); Chemiker-Zeit. **30**, 1 (1906).

possess exactly the same or nearly the same orientation. Otherwise they can join as twins.'

According to preliminary observations the *p*-azoxycinnamates exhibit the melting points given below, which may be changed by further purification of the esters. Both melting points of the crystalline-liquid substances will be affected by impurities in the same way as every other melting point of an ordinary substance, which was first shown by R. Schenck.[3] The second higher melting point will be lowered most strongly by admixtures, which in this case also becomes uncertain due to the decomposition and brown coloration which occurs on heating.

Melting points of the aliphatic *p*-azoxycinnamates (uncorr.)

	1st melting point	2nd melting point
methyl	219–221 °C	254–257 °C
ethyl	141 °C	247–249 °C
n-propyl	123 °C	240–243 °C
iso-propyl	148–150 °C	184 °C
n-butyl	110–111 °C	214 °C
iso-amyl	144 °C	184–186 °C
cetyl	105 °C	139–141 °C
n-octyl	94 °C	175 °C
allyl	124 °C	234–237 °C

In contrast to the aliphatic esters, a crystalline liquid could not be observed for the benzyl ester of the *p*-azoxycinnamic acid; this ester melts in the usual way at 174–175 °C. On the other hand, both the acetic ester group and the acetophenone group also exhibit the anisotropic liquid, or rather two melting points despite the chemical analogy of benzyl $CH_2.C_6H_5$ with $CH_2.CO$;

p-azoxycinnamates

	1st m. p.	2nd m. p.
Aryl-CO.O.CH_2.COOC_2H_5[1]	146–148 °C	233–235 °C
Aryl-CO.O.CH_2.COC_6H_5[1]	229–231 °C	238 °C

Professor Lehmann, however, informed me that the benzyl ester together with ethyl *p*-azoxybenzoate forms mixed crystals or rather layered crystals: an indication that the benzyl ester of *p*-azoxycinnamic acid can also become a flowing crystal under appropriate conditions.

---

[3] Habilitation thesis, Marburg 1897.

[1] From the silver salt of the azoxycinnamic acid and bromoacetate or alternatively bromo-acetophenone.

When 2 mole of bromine is added to ethyl *p*-azoxycinnamate, the tetrabromide is formed, accompanied by decolouration, and it only exhibits a single melting point at 162 °C. Together with the colour and the carbon double bond the crystalline-liquid phase is destroyed.

Regarding the large physical and chemical activity exhibited by α-unsaturated acid esters and α-unsaturated ketones, in collaboration with R. Wilke, I also prepared and tested an α-unsaturated azoxyketone. As expected the test was positive: the azoxyketone forms a directly observable crystalline liquid, which exists within the small temperature span of 2 to 3 degrees: 1st m.p. 211 °C; 2nd m.p. 213 °C (uncorr.). In the case of *p*-azoxybenzophenone and *p*-azoxybenzonitrile a double melting point has not been found.

The fact that alkyls and acyls on their introduction into compounds sometimes act in the same way[2] led to the preparation of several other crystalline-liquid substances whose activity is obviously caused by the azine nitrogen *rather* than by the carbon double bond.

p-Acyloxybenzalazines and *p*-acetyloxycinnamic acid

Van Romburg[3] reported *p*-methoxycinnamic acid as a crystalline-liquid substance. Anisaldazine has been detected to be a crystalline-liquid by Franzen,[4] and similarly *p*-ethoxybenzalazine by Gattermann.[4] In collaboration with J.E. Hulme, I found that the hydrogen of the *p*-oxybenzalazine and the *p*-oxycinnamic acid can also be replaced by acyls in order to change substances with only one melting point into doubly melting *substances*:

p-oxybenzalazine	m.p. up to 268 °C with decomposition
p-acetoxybenzalazine	1st m.p. 185 °C, 2nd m.p. 193 °C
p-benzoylbenzalazine	1st m.p. 227 °C, 2nd m.p. up to 290 °C
p-benzosulfonyloxybenzalazine	m.p. 167 °C
p-carboxyethyloxybenzalazine	m.p. 170 °C

Amongst the derivatives of the *p*-oxybenzalazines given above, only the acetyl and benzoyl compounds exhibit a double melting point. The *o*-oxybenzal compounds, *o*-methoxybenzalazine (m.p. 143 °C) and *o*-acetoxybenzalazine (m.p. about 190 °C), which were prepared from salicylaldehydes, and the corresponding derivatives of the *m*-oxybenzaldehydes, do not exhibit double melting. The double melting also vanishes when oxygen-containing groups are introduced into the anisaldazine (azine from piperonal, m.p. 203 °C, acetyl vanillin, m.p. 158 °C,

---

[2] This journal **35**, 1683 (1902).

[3] Verhandl. d. Akad. d. Wissensch. Amsterdam 1900.

[4] This journal **37**, 3422 (1904). Cf. also R. Schenck, Krystalline Flüssigkeiten. op. cit.

veratraldehyde, m.p. 191 °C) or on substitution of OCH_3 by NH_2, $N(CH_3)_2$ and NO_2 (*p*-aminobenzalazine, m.p. about 248 °C, *p*-dimethylaminobenzalazine, m.p. 250–253 °C with decomposition, *p*-nitrobenzalazine, m.p. about 300 °C with decomposition). Only in the case of *p*-aminobenzalazine[1] did I observe under favourable conditions an anisotropic liquid – the substance decomposes on melting.

As has been known a long time, acetyl-*p*-coumaric acid does not show a sharp melting point (m.p. up to 195 °C). After recrystallisation from water, we observed the acid to melt between 200 and 205 °C (uncorr.) to a cloudy liquid. After it has been heated above the melting point and then cooled, one part of the clear liquid transforms into the anisotropic birefringent liquid (particularly observable at the border of the drop and in a thin layer), whilst another part obviously solidifies in the usual way forming the solid acid. After esterification of *p*-methoxycinnamic acid to the methyl and ethyl ester[2] the anisotropic liquid phase can no longer be detected.

When alkyloxy and acyloxy groups are added to the unsaturated groups N_2O, C:C.CO, C:N.N:C, there is a complete modification of the chemical behaviour of a large number of unsaturated compounds; when alkyloxys, or even only an alkyl and a halogen are added, there is a change in colour. The yellow *p*-oxybenzalazine is nearly bleached (pale yellow) by the introduction of acyls and alkyls. The formation of the anisotropic liquid accompanies the bleaching process. The observation of the reverse case, as mentioned above, is in accord with the experience in other fields where a definite alteration of the structure may have an effect in one direction as well as the reverse one.[3] Some less coloured azines become coloured dark yellow on heating or on recrystallisation; with acids they result in coloured addition products.[4]

In the condensation products of benzidine with aldehydes prepared by Gattermann[5] the anisotropic liquid already appears in the case of the parent substance, the benzal compound, and remains preserved in the case of many substitution products (from *p*-tolylaldehyde, anisaldehyde and so on).

The double melting crystalline-liquid substances, which gain more and more significance for physics and crystallography, generally deserve attention by chemists. A relation between the occurrence of the anisotropic liquid phase and the chemical constitution does exist. If the method of investigation can be improved it seems possible that the appearance of the double melting, hitherto

[1] Walther and Kausch, Journ. für prakt. Chem. [2] **56**, 113 (1897).
[2] Ann. d. Chem. **294**, 295.
[3] Ann. d. Chem. **341**, 5 (1905).
[4] cf. Ann. d. Chem. **341**, 45 (1905).
[5] This journal **37**, 3423 (1904).

extraordinary, may become standard for a large number of chemical compounds. To date, in addition to the oleates, 35 crystalline-liquid substances are known.

The paper concludes with a brief discussion of some details about the synthesis of *p-azoxycinnamic acid* and some other compounds discussed in the paper.

266. D. Vorländer: Einfluß der molekularen Gestalt auf den krystallinisch-flüssigen Zustand.

[Mitteilung aus dem chem. Institut der Universität Halle.]

(Eingegangen am 18. April 1907.)

Obgleich die zahlreichen krystallinisch-festen und krystallinisch-flüssigen Isomeren sich einstweilen nicht den verschiedenen Arten von Stereoisomerie und Strukturisomerie unterordnen lassen, so ist doch die molekulare Struktur neben anderen Faktoren von Bedeutung für den krystallinisch-flüssigen Zustand. Die bisher erhaltenen Resultate[1] deuten darauf hin, daß der krystallinisch-flüssige Zustand durch eine **möglichst lineare Struktur des Moleküls** hervorgerufen wird.

1. Von den Benzolderivaten haben nur die Para-Disubstitutionsprodukte eine geradlinige Struktur. Man findet dementsprechend die krystallinisch-flüssigen Substanzen bei diesen und nicht bei den *m*- und *o*-Derivaten, wo die Substituenten mit Bezug auf den Mittelpunkt des regulären Benzolsechsecks einen spitzen oder stumpfen Winkel bilden.

2. Sobald zu den beiden Para-Substituenten ein dritter Substituent hinzutritt, wird die Gerade geknickt bezw. verzweigt und der krystallinisch-flüssige Zustand damit verhindert. Die auffallendste Regelmäßigkeit dieser Art fand ich bei den Aldehydderivaten: nur die des Para-Oxybenzaldehyds sind krystallinisch-flüssig, und wenn ein zweites OR hinzukommt, so hört der krystallinisch-flüssige Zustand auf. Auch die Furfurolverbindungen sind trotz des Sauerstoffgehaltes, der starken Färbung und großen chemischen Aktivität nicht krystallinisch-flüssig. Die Beziehungen gelten in der Reihe der Azine aus Anisaldehyd, Piperonal, Veratrylaldehyd, Furfurol ebenso wie bei den Arylidenaminen, Zimtsäuren und den ungesättigten Ketonen aus denselben Aldehyden, sind also keinesfalls zufällige. Durch Methyl in Kresolderivaten und durch Einführung von Naphthalin an die Stelle des Benzols wird der krystallinisch-flüssige Zustand gehindert, doch nicht ganz aufgehoben.

3. Je mehr die Struktur nach der Länge des Moleküls ausgedehnt wird, um so günstiger ist sie für den krystallinisch-flüssigen Zustand. Die Verbindungen $RO.C_6H_4.N:N.C_6H_4.OR$ oder $RO.C_6H_4.CH:N.C_6H_4.COOH$ u. a. sind im Gegensatz zu den kürzeren $RO.C_6H_4.N:N.C_6H_5$ und $RO.C_6H_4.CH:N.C_6H_5$ krystallinisch-flüssig. Befindet sich eine der endständigen Gruppen OR in *m*- oder

---

[1] Diese Berichte **40**, 1415 [1907]; **39**, 803 [1906].

A6
Berichte der Deutschen Chemischen Gesellschaft **40**, 1970–72 (1907)

INFLUENCE OF MOLECULAR CONFIGURATION ON THE CRYSTALLINE-LIQUID STATE

D. Vorländer

[Contribution from the Institute of Chemistry of the University of Halle]
(Accepted 13 April 1907)

The numerous crystalline-solid and crystalline-liquid isomers cannot at the moment be classified in terms of different types of stereoisomerism and structural isomerism. However, it is certain that the molecular structure, in addition to other factors, is extremely important to the crystalline-liquid state. All results obtained so far[1] suggest that the crystalline-liquid state results from a molecular structure which is as linear as possible.

1. Amongst the benzene derivatives only *para* disubstitution products possess a linear structure. Accordingly one only finds crystalline-liquid substances in these cases and not with *meta*- and *ortho*-derivatives, for which the substituents form an acute or obtuse angle with respect to the centre of the regular benzene hexagon.

2. Whenever a third substituent is added to both *para*-substituents the straight line will be kinked or branched, and so the crystalline-liquid state prevented. I found the most striking regularity of this kind in the case of aldehyde derivatives: only the *para*-oxybenzaldehydes are crystalline-liquid, and if a second OR is added the crystalline-liquid state is prevented. Furthermore the fufuryl compounds are not crystalline-liquid despite the oxygen content, the strong coloration and the large chemical activity. These relationships are valid for the series of azines of anisaldehyde, heliotropin, veratryl aldehyde and fufural, in the same way as the arylidene amines, cinnamic acids and the unsaturated ketones of the same aldehydes. They are thus by no means fortuitous. As a result of the methyl group in cresol derivatives and by the

---

[1] This journal **40**, 1415 (1907); **39**, 803 (1906).

introduction of naphthalene instead of benzene, the crystalline-liquid state will be hindered but not totally excluded.

3. The more the structure is extended with respect to the length of the molecule the more the crystalline-liquid state will be favoured. The compounds $RO.C_6H_4.N:N.C_6H_4.OR$ or $RO.C_6H_4.CH:N.C_6H_4.COOH$ and others are crystalline-liquid in contrast to the shorter compounds $RO.C_6H_4.N:N.C_6H_5$ and $RO.C_6H_4.CH:N.C_6H_5$. If one of the terminal OR groups is located in the *m*- or *o*-position, the crystalline-liquid phase will be destroyed (cf. 1). These relations could be attributed also to an unsymmetrical energy distribution.[1]

4. As a result of branching of the carbon skeleton, the α-substitution products of *p*-methoxycinnamic acid[2] are not themselves crystalline-liquid, whereas the acid, according to van Romburgh,[3] exhibits an enantiotropic crystalline-liquid phase. The liquid-crystalline properties are absent in the case of α-bromo-,[4] methyl-, ethyl-, phenyl- and even for α-oxyethyl-*p*-methoxycinnamic acid (m. p. 115 °C; prepared from anisaldehyde and ethoxy ethyl acetate).

5. Comparing normal valeryl- with isovaleryl-cholesterol, F. M. Jäger[5] has found that the range of existence of the crystalline-liquid phase is very small in the case of the branched isovaleryl but large for normal valeryl. Such an effect of branching, however, is not always observable, in which case perhaps it is of less influence in comparison with the predominantly linear shape of the other molecular parts (e.g. acyloxybenzalazine).

6. No crystalline-liquid phases exist in the case of compounds with free hydroxy-phenol. Due to tautomerism or a shift and transposition of a hydrogen atom, an uncertainty exists with respect to the energy distribution, which allows the molecules to adopt any arbitrary position with respect to each other, and hence no definite direction. Elimination of the tautomeric hydrogen by alpha-substitution or acylation allows the molecules to achieve the stability necessary for the formation of the crystalline-liquid phase. Amongst all the alpha-substituents, ethyl exhibits the longest linear structure and it also possesses in comparison with other radicals a particularly favourable liquid-crystalline effect (e.g. with azoxybenzoate). However, one has scarcely any idea about the different influence of aryls and the various acyls.

The chemical investigations in this field are consistent with opinions which O. Lehmann advocated for many years.[6] According to Lehmann the origin of the

---

[1] Ann. d. Chem. **341**, 1 (1905).

[2] cf. the preceding paper (Ber. Dt. Chem. Ges. **40**, 1415 [1907]) (*ed.*).

[3] Verhandl. Akad. d. Wissenchaften, Amsterdam **1900**.

[4] By removing hydrogen bromide from methoxycinnamic acid dibromide with sodium alcoholate, John Hulme obtained two monobromided methoxycinnamic acids, one of which was crystalline-liquid.

[5] Rec. Trav. Chim. **25**, 344 (1906).

[6] Wied. Ann. d. Physik **40**, 410, 412, 422 (1890); ibid. (4) **8**, 917 (1902); **20**, 77 and **21**, 383 (1906). Kosmos **1907**, 9.

existence of the crystalline-solid and crystalline-liquid modifications does not depend on the nature of molecular aggregation, but on the constitution of the molecules themselves, and hence on their structural modifications. The birefringence of the crystalline-liquid state is caused by the anisotropy of the molecule. He compares the molecules of crystalline liquids with pieces of wire or lamellae which all orient themselves in parallel direction on shaking, whereas the molecules of amorphous isotropic liquids are spherical.

The results given above indeed seem to me to contain experimental evidence for his theory. From the linear shape of the molecules follows their parallel orientation and hence the anisotropy of the liquid. However, if the shape is jagged, as in the case of *m*- and *o*-derivatives, or branched in another way, then the molecules will rotate. They will then adopt arbitrary positions with respect to each other or become spherical in some other way, and then appear to be isotropic. The anisotropy of the molecule will be illustrated to a certain extent by the structural formula of the crystalline-liquid substances.

The chemical structure must be as uniaxial as possible. In the case of crystal drops, the uniaxiality of liquid crystals follows from the well-known investigations of Lehmann,[1] and from the appearance of pseudoisotropy, first observed by Lehmann,[1] in mixtures of superheated substances.[2] According to Lehmann, pseudoisotropy in liquid crystals is analogous to solid crystal plates, which have been cut in a plane perpendicular to their optical axis, and then viewed in the direction of the axis. The uniaxiality of crystalline-liquid rods follows from my detection of liquid crystals,[3] probably tetragonal , with straight edges and angles.

---

[1] O. Lehmann. Flüssige Kristalle, Leipzig (1904).

[2] Pseudoisotropy of pure homogeneous compounds: Vorländer, Ztscr. physik. Chem. **57**, 360 (1906).

[3] ibid., p. 363.

Daniel Vorländer was born in Eupen, near Aachen in Germany in 1867. His father owned a dyeworks but was in addition a keen amateur linguist. Daniel attended school in Dresden, where the family lived after 1871. He began university study in chemistry in Kiel in 1886, moved to Munich in 1887, and graduated in Berlin, specialising in organic chemistry, in 1890. He then obtained a graduate post at the University of Halle, obtaining his Habilitation degree in 1896. He was then successively promoted in Halle, until in 1908 he became a full professor and director of the Chemical Institute. He also saw active service in the 1914–18 war before returning to Halle in 1917. He directed the studies of a very large number of doctoral students, the most famous being the Nobel prize

winner Hermann Staudinger. Vorländer served as Vice-President of the Deutsche Akademie der Naturforscher Leopoldina, the oldest academy in the world. He retired in 1935.

In all he published 214 academic papers, on a wide variety of subjects ranging from inorganic chemistry to natural product chemistry. However, after 1906 and until the end of his career, liquid crystals were his primary interest. By the time of Vorländer's death in 1941, the 35 liquid crystalline compounds known at the start of his liquid crystal work had grown to 1050 whose properties had been catalogued, with the number observed considerably higher. A collection of his liquid crystalline samples remain stored in the Institute of Physical Chemistry in Halle.

schem Lösungspunkt Anomalien der inneren Reibung zu erwarten sind, die von ähnlicher Art sein können, wie die bei anisotropen Flüssigkeiten beobachteten Erscheinungen dieser Art. Als ich meine diesbezüglichen Überlegungen auf der XIV. Hauptversammlung der Deutschen Bunsengesellschaft vortrug, wurde in der Diskussion die Möglichkeit derart intensiver Effekte bei Emulsionen bezweifelt.

Inzwischen hat mein damaliger Diskussionsgegner, Herr Prof. Rothmund, gelegentlich einer interessanten Experimentalarbeit über die kritische Trübung[1]) Versuche über die Viskositätsanomalien von kritischen Emulsionen angestellt und in der Tat experimentell den Nachweis geführt, daß ein Verlauf der inneren Reibung, wie ich ihn vorher ohne jedes Experiment vorausgesagt hatte, wirklich beobachtet werden kann.

Nachdem nun der Verfasser inzwischen[2]) gemeinsam mit F. Conrat nachgewiesen hatte, daß auch bei den reinsten anisotropen Flüssigkeiten und unter einwandfreien Versuchsbedingungen nicht ein wirklicher Sprung, sondern eine auf ein meßbares Temperaturintervall ausgedehnte quasiunstetige Änderung der Viskosität stattfindet, ist also meine damalige theoretische Beweisführung, daß die Reibungsanomalien der anisotropen Flüssigkeiten nicht als Experimentum crucis gegen die Tammannsche Emulsionstheorie gelten könnte, nunmehr auch experimentell in jeder Hinsicht bestätigt worden. Dies ist um so mehr der Fall, als auch Herr Prof. Schenck, gegen dessen Diskussionseinwand sich die erwähnten Versuche von Bose und Conrat wandten, diese Versuche[3]) mit weiter verbesserten Hilfsmitteln wiederholt hat, aber nur den von uns (l. c.) festgelegten Kurvenverlauf bestätigen konnte.

Wenn wir (B. und C.) aber schon gelegentlich unserer gemeinsamen Veröffentlichung darauf hinwiesen, daß uns eine emulsionstheoretische Deutung des Viskositätsverlaufs anisotroper Flüssigkeiten nicht angebracht erscheine, so muß ich diesen vorsichtigen Ausspruch nunmehr dahin erweitern, daß mir nach den neuesten Befunden Vorländers[4]), insbesondere nach der Realisierung dickerer Schichten völlig klarer durchsichtiger anisotroper Flüssigkeit, jede emulsionstheoretische Deutung absolut ausgeschlossen erscheint. Das Bedürfnis nach einer Ausgestaltung der Theorie der sogenannten kristallinischen Flüssigkeiten scheint demnach dringend vorzuliegen. In dem folgenden Aufsatz mache ich den Versuch, eine früher kurz skizzierte Auffassung

1) Zeitschr. f. physik. Chem. 63, 54, 1908.
2) Diese Zeitschr. 9, 169, 1908.
3) Einer persönlichen Mitteilung zufolge.
4) Ber. d. d. chem. Ges. 41, 2033, 1908; siehe darüber auch den folgenden Artikel.

der anisotropen Flüssigkeiten etwas ausführlicher zu entwickeln.

(Eingegangen 29. September 1908.)

Zur Theorie der anisotropen Flüssigkeiten.

Von E. Bose.

Ich habe früher[1]) an dieser Stelle eine kurze Skizze einer rein kinetischen Vorstellung entworfen für die anisotropen Zustände, welche bei den sogenannten „kristallinischen" Flüssigkeiten vorliegen. Diese ging aus von dem wichtigen Befunde Vorländers, daß der anisotrope Zustand nur bei Substanzen beobachtet wird, denen eine langgestreckte Molekularkonstitution zukommt. Die Überlegungen, welche ich damals nur anhangsweise kurz an die Behandlung verschiedener anderer Fragen knüpfte, möchte ich im folgenden etwas weiter ausführen.

Boltzmann unterscheidet auf S. 20 u. folg. des 1. Bandes seiner klassischen Darstellung der Gastheorie molar ungeordnete und molekular ungeordnete Zustände. Erstere sind dadurch charakterisiert, daß innerhalb eines endlichen, eine größere Anzahl von Molekülen umfassenden Raumelementes die Geschwindigkeitskomponenten nicht mehr völlig willkürlich verteilt sind, sondern etwa eine gemeinsame translatorische Bewegung oder ein anderweitig geordneter Bewegungszustand sich über die regellose Verteilung superponiert, so daß damit eine Vorzugsrichtung in dem betrachteten Raumelement gegeben ist. Erst das Fehlen einer solchen Vorzugsrichtung charakterisiert die auch molekular ungeordneten Zustände.

Solange man die Moleln mit ausreichender Annäherung als absolut glatte, undeformierbare elastische Kugeln betrachten darf, deren Schwerpunkt im Kugelmittelpunkt liegt (z. B. für einatomige Gase), spielen irgendwelche Rotationen der Moleküle gar keine Rolle, die Koordinaten des Kugelmittelpunktes genügen völlig zur Bestimmung der Lage des einzelnen Moleküls. Dies wird sofort anders, sobald wir etwa den Schwerpunkt exzentrisch legen oder von der Kugelform abgehen; dann sind eben je nach dem Grade der Symmetrie mehr Variable zur Charakterisierung der Lage eines Moleküls notwendig. Für den von uns zu betrachtenden Fall genügt es, die Moleküle etwa als (absolut glatte, undeformierbare, elastische) verlängerte Rotationsellipsoide (mit zentralem Schwerpunkt) zu betrachten; in diesem Falle müssen zur Charakterisierung der Lage einer Molekel außer den drei Schwerpunktskoordinaten

1) Diese Zeitschr. 8, 513, 1907. Siehe die letzten Abschnitte des Artikels.

A7
Physikalische Zeitschrift **9**, 708–13 (1908)

ON THE THEORY OF THE ANISOTROPIC LIQUIDS

by

E. Bose

In an earlier paper[1] I drafted a short sketch of a purely kinetic view of the anisotropic states which lie behind the so-called crystalline fluids. It emerged from the important discovery by Vorländer that the anisotropic state is only observed in substances whose molecular shape is highly elongated. I described these ideas in that paper only briefly as an afterthought to the treatment of diverse other questions. In what follows I develop the ideas somewhat further.

In the first volume of Boltzmann's classic account of the theory of mixtures, he distinguishes molar and molecular disordered phases. The first are characterised by the property that within a finite element in space containing a large number of molecules, the components of velocity are no longer completely randomly distributed. Rather a common translational motion or another ordered state of motion is superposed on the normal distribution. This gives a preferential direction in the element of space under consideration. It is only the lack of such a preferential direction which characterises those phases which also possess molecular disorder.

It is frequently a sufficiently good approximation to consider the molecules as completely smooth undeformable elastic spheres, whose centres of gravity and physical centres coincide. An example of this would be monatomic gases. In this case molecular rotations play absolutely no role, and the coordinates of the physical centres are sufficient to define an individual molecular position. This situation immediately changes either if we displace the physical centre from the centre of gravity or if we go to non-spherical shapes. Now, depending on the degree of symmetry, more variables are required to define molecular position. For the case which we shall consider, it will suffice to consider the molecules as (completely smooth, undeformable, elastic) elongated ellipsoids (with the centre of gravity at the centre). In these cases the coordinates of the molecular centre are

---

[1] This journal **8**, 513, 1907. See the last section of the article.

insufficient to define molecular position. Two angles, defining the extension direction, must also be given.

Hitherto we have made no assumption about the mutual separation of the molecules. We have to do this in a gas, for which the mean free path is large compared to the mean separation between two molecules. For this latter case, after each collision, each molecule is well-separated from its previous neighbour. A molar disordered phase (in a system lacking external forces), as Boltzmann rightly remarked, would rapidly be transformed into a phase with molecular disorder. In liquids the mixing process is much less rapid. This leads to a very much smaller mean free path. The result is that a state with molar but not molecular disorder can be relatively long-lasting.

Now there can be a completely different way in which a molecular disorder can become a molar disorder. Let us consider, for example, a gas made of ellipsoidal molecules. Then a circulation can form, for example as a result of asymmetric heating, which as far as the motion of the centres of gravity is concerned destroys the molecular disorder. However, as far as the axis directions are concerned, complete equivalence of the axis directions may exist, so this implies molecular disorder. On the other hand we could consider the whole isotherm of a gas with an applied directional force (for example of a magnetic or electrical nature). This disturbs the molecular disorder as far as the axis directions are concerned, whereas looked at in terms of the particle motion the molecular disorder may remain completely unaffected. This last assumption, for example, is the foundation of Langevin's kinetic theory of dia- and para- magnetism.[1]

According to the kinetic view of the nature of anisotropic liquids in my previous paper, the assumption is that regarding the molecular orientation the system possesses molecular order. Otherwise, however, it assumes a state which possesses molar disorder. In this way of looking at things a finite mass of turbid anisotropic liquid would consist of swarms of molecules. In each such swarm the molecules would be able to move completely freely, but would possess a preferred orientation.[2] The turbidity might be caused by the tangle of different swarms, since the swarm boundaries, by acting as discontinuities, would affect the scattering of incident light.[3] To give a crude image of the whole collection of swarms, I might make the example of a tenuous soap foam, in which the inside of each bubble corresponds to a molecular swarm with a preferred direction, and the soap films correspond to the swarm boundaries. However, the picture will only be approximately valid for a split-second. Whereas the foam is a relatively stable structure, the swarm boundaries in the anisotropic liquid state are understood to be in constant motion.

---

[1] Ann. Chim. Phys. (8) **5**, 70, 1905.

[2] The concept of swarms with an anisotropic structure was occasionally used by H. Ambronn in an article on pleochroism on specular metallic surfaces (this journal, **8**, 665, 1907), citing its use in my previous article.

[3] The possibility of obtaining a clear anisotropic liquid would therefore coincide with the possibility of developing macroscopic swarms. These would be caused either by external conditions (temperature, etc.) or by the particular properties of the material in question.

They would probably not be seen, unless they were somehow localised by some specialised technique.[1]

This view of the structure of an anisotropic liquid is rather close to the recent successful view of ferromagnetic metals due to P. Weiß.[2] He supposes 'that the apparent isotropy of these metals arises from the random packing of elementary domains', each of which is itself anisotropic. Each domain then corresponds to a molecular swarm with a marked preferred direction, except that in the liquid state (lacking a crystal lattice), the system is very flexible as a result of the molecular mobility.

To make the swarm picture more concrete, we can construct the following picture based on the van der Waals fluid equation of state:

$$\left(p + \frac{a}{v^2}\right)(v - b) = RT$$

As is well known, the cohesive force $\frac{a}{v^2}$ serves to minimise the fluid volume. However, the magnitude of the molecular motion producing a volume increase with temperature is particularly sensitive to the magnitude of the quantity b in the expression for the molecular specific volume. The molecular motion decreases on cooling, and the overwhelming cohesive force causes a volume decrease. This continues until it is balanced by the rapidly growing internal pressure caused by the reduced average distance between the molecular centres.

I now consider a liquid consisting of rod-like molecules (extremely elongated rotation ellipsoids, with axis ratio or form factor $\frac{L}{l}$). At a sufficiently high temperature the magnitude of thermal motion will produce a phase with molecular disorder not only as regards the molecular centres but also as regards the molecular axes. Thus the liquid will be isotropic in every respect. A swarm of accidentally ordered molecules would always immediately decay again. However, let the temperature decrease. The vigorous translational and rotational molecular motions will likewise decrease. There may then come a point at which there will be a new stable phase. This phase is swarm-like regarding the axis directions, still possesses molar disorder, but no longer possesses molecular disorder. This occurs because in the swarm-like phase the strongly oriented molecule will possess favourable space-filling properties.[3] This will lead to a smaller value of b, the possibility of

---

[1] In order to make his optical observations, O. Lehmann used the trick of adding impurities such as colophonium. This would lead in our case to the break-up of the main material into very small droplets each containing an individual isolated swarm, whose boundary is well-localised by surface tension. If the swarm preferential direction is perpendicular to the microscope stage, the droplets adopt Lehmann's first principal position, and if parallel, the second principal position.

[2] This journal **9**, 358, 1908.

[3] For a crude picture of this space-filling property one should think of a box of cigars. If the cigars are ordered it is possible to fit very many more of them into the box than is possible if they are not, and they are, by contrast pointing in any direction.

a volume decrease together with an increase in the value of $\frac{a}{v^2}$. Below a certain transition temperature the cloudy liquid phase becomes stable; this phase is somewhat ordered as regards axis directions. In fact, when typical anisotropic liquids are cooled below the clearing point, one finds a quasi-sudden jump in the density.[1]

Let us consider the turbidity which begins below a certain transition temperature in such a way as to be governed by favourable space-filling properties (and thus by discontinuous change in b). Then, following my earlier work, the question is whether this change in b is really discontinuous, or just a very rapid quasi-discontinuous change. Apart from these two extreme cases, a combination of the two might also be conceivable. This would comprise a sudden jump, together with an extended change in b above a certain temperature. The three different possibilities are depicted in the sketches below as types a, b, c. The first

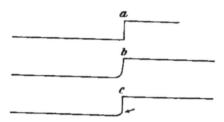

possibility of the pure sudden change should probably be rejected, particularly following the experiments of F. Conrat and the author[2] on turbid anisotropic liquids. However, it has hitherto been hard to decide with complete certainty between types b and c. Although in the opinion of the authors the experiments of Bose and Conrat are evidence for type b, they certainly seem just as understandable in terms of type c. In addition, dilatometry experiments performed by Dr Conrat at my suggestion might even point to type c as being the more probable.

So there is a change in the quantity b in the van der Waals equation which sets in below a certain temperature. Let us now consider the question as to how we could bring this change into the relevant expression. This will account for the small intermediate discontinuous factors, so long as we only require a formal description without a detailed picture.

Case a would, for example, work out as follows. In the clear liquid region, i.e. for temperatures above the clearing temperature Θ, the usual equation of state suffices:

$$\left(p + \frac{a}{v^2}\right)(v - b) = RT$$

[1] See especially R. Schenck, Crystalline liquids and liquid crystals, Leipzig 1905, p48ff.
[2] E. Bose and F. Conrat, this journal **9**, 169 (1908).

For $T < \Theta$, b should be replaced by

$$b' - \Delta,$$

where Δ is a temperature-dependent but material-independent constant. In addition, we now introduce an auxiliary variable, whose physical significance we do not consider, and amend the equation of state in the following way:

$$\left(p + \frac{a}{v^2}\right)\left(v - b + \frac{2\Delta}{\pi}\int_0^\infty \frac{\sin u}{u}\cos\frac{T}{\Theta}u\,du\right) = RT$$

The additional term $\dfrac{2\Delta}{\pi}\displaystyle\int_0^\infty \dfrac{\sin u}{u}\cos\dfrac{T}{\Theta}u\,du$ has the property that for all values of

$T > \Theta$ it disappears, whereas for values of $T < \Theta$ it yields the constant value Δ. For $T \equiv \Theta$ the expression yields the value $\frac{\Delta}{2}$, which more or less corresponds to the physically unimportant third part of the van der Waals curve.

Case b would be described, at any rate with sufficient accuracy, by the equation of state:

$$\left(p + \frac{a}{v^2}\right)\left(v - b + \frac{2\Delta}{\pi}\int_0^\infty \frac{\sin u}{u}\cos\frac{T}{\Theta}u\,du - \frac{\Delta}{\pi}\int_0^\infty \frac{u\sin(\Theta - T)u + \alpha\cos(\Theta - T)u}{\alpha^2 + u^2}du\right) = RT$$

since for values of $T \geq \Theta$ both definite integrals disappear, individually or together, while for all values of $T < \Theta$ the first additional term is equal to Δ as before, and the second becomes equal to $\Delta e^{-\alpha(\Theta - T)}$. Below the clearing temperature, b begins to decrease gradually approaching its limiting value Δ. Below the clearing temperature, the larger the constant α, the more abrupt the decrease in b. The quantity α is characteristic of the material under consideration, and at any rate to lowest order is independent of the form factor; for $\alpha \to \infty$, case b transforms into case a.

Finally, case c can be distinguished from case b in that the Δ-constants of the two additional expressions are different. The equation of state would become:

$$\left(p + \frac{a}{v^2}\right)\left(v - b + \frac{2\Delta}{\pi}\int_0^\infty \frac{\sin u}{u}\cos\frac{T}{\Theta}u\,du - \frac{\Delta'}{\pi}\int_0^\infty \frac{u\sin(\Theta - T)u + \alpha\cos(\Theta - T)u}{\alpha^2 + u^2}du\right) = RT$$

where Δ' must be less than Δ.

When the temperature is reduced from Θ the volume constant jumps from b to $b - \Delta + \Delta'$ and converges asymptotically toward the value $b' = b - \Delta$. In this case the limit $\alpha = \infty$ also transforms to case a.

As an aside I stress yet again that the equations above represent a purely formal manner of expression. Thus at best they are a description in the sense of Kirchhoff. Consequently the extension of the van der Waals equation is and remains completely inadequate from a theoretical point of view. I now consider the position of Θ in relation to the critical temperature of the clear liquid $\vartheta = \frac{8a}{27bR}$. The onset of swarm phase stability should occur at temperatures which will increase as the molecule deviates from a spherical shape, i.e. for increasing ratio of the long to the short axis L/l (which is the form factor of the stretched ellipsoid of rotation). With a growing form factor, Θ would probably approach closer and closer to the critical temperature. An increase of Θ above ϑ would be equivalent to the onset of the anisotropic phase in the gas phase (albeit a strongly compressed gas phase) and would be extremely unlikely. Given the contemporary state of chemistry, it is improbable that the form factor of any more or less rigid molecule could be larger than a relatively modest value. And extremely extended molecules, such as the higher members of the fatty hydrocarbon n-alkane series, will be subject to considerable internal motion, and we should not expect to be able to describe their behaviour even approximately by assuming rigid ellipsoids.[1]

Let us now return to our kinetic point of view of the anisotropic liquid state. Thus, inside a molecular swarm, we should consider the molecular centres as freely mobile. In particular one cannot speak of fluctuations around a definite rest position. Thus, in this respect, the material is completely consistent with the kinetic view of a fluid. There is still absolutely no question of a lattice, no matter whether rigid or slightly deformable, and thus also none of a true crystallographic symmetry. On the other hand, each individual molecular swarm is itself an anisotropic object, since the molecules in it have an evident strong average axis orientation. Along the average swarm axis, for example, the light propagation speed will be different from its value in a direction perpendicular to it. Similar considerations will apply to all vectorial properties inside a swarm. Above a certain temperature Θ it is not possible to sustain these anisotropies. A completely comparable behaviour is the sudden disappearance of the spontaneous magnetisation of ferromagnetic materials above a certain temperature. The analogy emerges especially clearly in comparison with the extremely interesting publications by Pierre Weiß on molecular fields and ferromagnetism. In ferromagnetism a molecule is subject to the molecular field acting on it, due to the total effect of the neighbouring molecules. Stable polar anisotropy is achieved below a certain limiting temperature Θ, while above Θ the stable polar anisotropy is destroyed by the vigorous thermal motion. In the anisotropic fluids a molecule feels the effect of all the neighbouring molecules through the particular type of form factor coming from

[1] Materials of this type can give rise to quite other special properties, which I intend to return to at a later opportunity.

the space-filling properties. Stable anisotropic order is thus achieved below a certain limiting temperature Θ, but this order is destroyed above Θ by the strong thermal motion.

As an illustration of the point of view I have expressed concerning the occurrence of liquid anisotropic states, this analogy seems to me to be extremely useful. However, in the ferromagnetic case a mathematical formulation is available, whereas for the anisotropic fluid case the correct expression does not yet exist.

It is now appropriate to discuss briefly the question of what conditions are most favourable for the formation of swarm phases. First, a drop in temperature and thus a reduction in both the translational and rotational molecular motion permits the individual swarms to grow and in general enhances swarm phase stability. Second, the presence of hard walls acts favourably, both because of the constraints on molecular displacement at the wall and because of the adhesion forces which occur. The adhesion forces seem to act uniformly so that the molecules point with their long axes perpendicular to the wall. Very thin layers between the microscope stage and the cover glass then appear dark between crossed nicols in parallel light. Lehmann has denoted such phases as pseudoisotropic, because a displacement of the cover glass suffices to destroy the order and permits the anisotropy to manifest itself.[1]

Now Vorländer[2] has succeeded, by distinguished and determined study, in obtaining compounds with anisotropic liquid phases at considerably lower temperatures. These phases can be investigated while still at room temperature. The conditions are here much more favourable for the formation of larger exceptionally ordered swarms with a preferred direction and only still thicker layers of the compound seem turbid, while layers of up to 3 mm in thickness can remain clear. In convergent polarised light these layers exhibit the behaviour of a uniaxial crystal splendidly. They also sometimes, in optically active molecular structures,[3] exhibit intense rotational power.

The direct mechanical demonstration of the existence of the supposed swarm aggregates would without doubt be of particular significance for the swarm theory. The question arises whether they can be produced experimentally, and by what possible means. At the outset we want to leave aside the optical properties of the turbid liquid, for since the swarm picture is a mechanical construction, a mechanical demonstration would surely be the most satisfactory.

A mechanical proof of this kind would also perhaps explain the viscosity anomalies of anisotropic liquids. Along the axis direction the molecular collision cross-sections are minimal. The internal anisotropy of a swarm will thus also exert a particular influence on the molecular displacement. The component of the

---

[1] The isolation of individual swarms is also advantageous. Lehmann achieved this by adding impurities (see note 5).

[2] Ber. d.d. Chem. Ges., **41**, 2033, 1908.

[3] For example, as a result of the existence of an asymmetric carbon atom in the molecule.

mean free path parallel to the mean swarm axis will then be considerably greater than that perpendicular to it. As a result, within each swarm one will also find an anisotropy of the internal friction, which has consequences for the viscosity measurements. This depends on the fact that the directionally ordered swarm-like states dominate the fully disordered states. This explains the smaller friction in the turbid fluid régime.

Suppose that by suitable experimental conditions the directional order of the swarm phase were to be systematically and continuously destroyed. If the friction could be studied under these conditions, then the viscosity anomalies should fail to appear, or at any rate be considerably reduced. When the capillary diameter is squeezed it seems entirely possible for the swarms to exert such an influence. In such a fine capillary the so-called hydraulic state is generated, rather than the Poiseuille state.

It is well-known that when gases and liquids flow through a tube there exist two principal flow states. These are different from each other and are linked by a broad transition region. Inside this region neither of the two states lasts very long in such a way that the two states continuously mutually supersede each other. The stable state at low flow speeds is the Poiseuille state familiar to physicists. In this state coaxial cylinders move in a smoothly ordered way with respect to each other. The velocity decreases from a maximal value on the axis according to a parabolic law and goes to zero on the outside wall of the tube. In this ordered flow state the directionally ordered state in the cloudy liquid régime can successfully be exploited.

In very high flow velocities the Poiseuille flow state is completely unstable and is replaced by the so-called hydraulic state. In this state the flow is accompanied by vigorous eddy formation. By contrast with the Poiseuille state, here a completely disordered state of motion is produced. If this disorder is able to form in such a small region that it becomes comparable to the average swarm size, then the directionally ordered state will be continually destroyed by the flow. The advantageous use of the order will then be prevented or, at the very least, considerably reduced. These questions could be attacked experimentally in the following way. In a short section of capillary (which should be thin) one first studies the dependence of through-flow-time on temperature in a slight pressure gradient. The pressure gradient is then steadily increased, until finally one is working in a purely hydraulic régime. Here also the dependence of the through-flow-time on temperature is obtained. In the Poiseuille régime one obtains the well-known discontinuous jump with a very much smaller viscosity in the anisotropic liquid state. If the discontinuity in the hydraulic state is not present, or if it disappears largely, then this represents a purely mechanical demonstration that swarm formation occurs in the cloudy liquid régime. In this process the preferred direction of least friction cannot be reached once its use by an ordered flow is no longer possible.

These experiments are currently in preparation. They will be reported later in the pages of this journal. I am grateful to the Jager Foundation for placing the means for these experiments at my disposal.

(submitted 29 September 1908)

Emil Bose was born in Bremen in 1874. He entered the University of Göttingen in 1895, graduating in 1898 in physical chemistry. In 1899 he obtained his Habilitation degree in Breslau (present-day Wroclaw in Poland, but at that time in Germany). He then returned to Göttingen as an assistant lecturer in physical chemistry under Walther Nernst, and in 1903 he moved to the Institute of Physics to work under Voigt. In 1906 he was appointed to an associate professorship in physical chemistry and electrochemistry at the University of Danzig (present day Gdansk in Poland).

Bose's interests in liquid crystals followed from a general interest in thermodynamics of mixtures. Other interests included the physics of electrolytes and hydrodynamics. He is noted not only for the theoretical work emphasised in this collection, but also for experimental work carried out both alone and with collaborators. In 1909, following a personal recommendation by Nernst, he was appointed as Professor of Experimental Physics and Director of the Institute of Physics at the National University of La Plata in Argentina. Between 1909 and 1911 he and his wife Margarethe Heibing worked tirelessly to construct a viable physics programme in La Plata. He died of typhus in La Plata at the age of 36 on 25 May 1911.

Sur les cristaux liquides de Lehmann ;

Par M. Ch. Mauguin.

L'azoxyanisol et l'azoxyphénétol ([1]), isolés par Gattermann ([2])

([2]) Gattermann. *Ber. d. d. chem. Ges.*, t. XXIII, 1890, p. 1738.

en 1890, ont fourni à Lehmann l'un des types de *cristaux liquides* les plus intéressants.

Ces corps en fondant, le premier à 116°, le second à 138°, donnent naissance à des liquides aussi mobiles ou même plus mobiles que l'eau ([1]), *deux fois plus biréfringents que la calcite.*

Une telle biréfringence, supérieure à celle de la plupart des corps cristallisés, apparaît comme tout à fait énorme quand on la compare à celle que manifestent certains liquides, siège d'un champ électrique (phénomène de Kerr) ou d'un champ magnétique (phénomène de Cotton et Mouton).

A une température plus élevée (134° pour l'azoxyanisol, 168° pour l'azoxyphénétol), ces liquides deviennent isotropes par une transformation tout à fait analogue à la fusion d'un corps cristallisé.

Au refroidissement, on retrouve les mêmes apparences dans l'ordre inverse.

L'existence d'une biréfringence aussi élevée dans des corps d'une fluidité parfaite a vivement surpris les physiciens et les cristallographes, et quelques-uns d'entre eux (non des moins éminents) n'ont pas voulu y croire; aujourd'hui même, vingt ans après ses premières expériences, Lehmann n'a pas encore réussi à convaincre tout le monde.

Pourtant, une étude attentive des phénomènes ne laisse guère de place au doute, et je me propose d'apporter ici quelques résultats nouveaux qui sont une pleine confirmation des conclusions de l'illustre savant allemand.

Les liquides biréfringents, en raison de leur fluidité même, peuvent donner naissance à des édifices très variés. Les plus simples sont des *lames d'orientation optique uniforme* qu'on

([1]) *Voir* les mesures de viscosité de Schenck (*Zeitschr. f. phys. Chem.*, t. XXVII, 1898, p. 167) et de Eichwald (*Dissertation Marburg*, 1904).

A8
Bulletin de la Société française de Minéralogie **34**, 71–117 (1911)

ON THE LIQUID CRYSTALS OF LEHMANN

by

Mr Ch. Mauguin

Azoxyanisole and azoxyphenetole,[1] isolated by Gattermann[2] in 1890, have provided Lehmann with one of the most interesting types of liquid crystal.

When these materials are melted, the first at 116 °C and the second at 138 °C, they produced liquids as fluid or even more fluid than water,[3] *and twice as birefringent as calcite.*

Such a birefringence, larger than that of most crystalline materials, appears quite huge when compared with that exhibited by certain liquids placed in an electric field (Kerr Effect) or a magnetic field (Cotton–Mouton Effect).

At a higher temperature (134 °C for azoxyanisole, 168 °C for azoxyphenetole), these liquids become isotropic by a transformation which is quite similar to the melting of a crystal.

On cooling, one obtains the same behaviour in reverse.

The existence of such a large birefringence in a perfectly fluid material greatly surprised physicists and crystallographers, some of whom (no less eminent) did not wish to believe it; even today, 20 years after the first experiments, Lehmann has not succeeded in convincing everyone.

Even so, an investigation of these phenomena leaves no room for doubt, and I propose to present here some new results which are a clear confirmation of the conclusions of the distinguished German scientist. Because of their fluidity, these birefringent liquids can give rise to very varied structures. The simplest are *optically*

Azoxyanisole   Azoxyphenetole

[2] Gattermann, *Ber.d.d.chem.Ges.*, vol XXIII, 1890, p. 1738.

[3] *See* the measurements of viscosity by Schenck (*Zeitschr. f. phys. Chem.*, vol. XXVII, 1898, p. 167) and of Eichwald (*Dissertation*, Marburg, 1904).

uniform oriented films, which one obtains on melting the material, with certain special precautions, between two clean glass surfaces. Evidently these are the most convenient films to study initially. This will be the subject of the first part of the present paper.

The results obtained can be used to understand more complex structures, the study of which appears to provide molecular physics with a large number of important facts.

In the second and third parts of this paper a very simple structure will be described, consisting of layers in a helicoidal arrangement having truly remarkable properties.[1]

Method of observation – For all of this investigation I used an ordinary (Nachet model) petrology microscope. The samples are maintained within a convenient temperature range by means of an electric current which passes through a film of carbon deposited on the sample stage, covered by a thin sheet of asbestos. This sheet, of about 1 mm thickness, a little larger than the sample microscope slide, is pierced at the centre by a circular hole to allow the light through.

This arrangement permits easy observation using convergent or parallel light. The temperature is adjusted by means of a rheostat. The only inconvenience is the need for a rather high current (5 to 10 amps).

Cleaning of the glass slides – The state of cleanliness of the glass slides used for the samples plays a big rôle in the clarity of observations. As will be shown later, it can even change completely the optical behaviour.

The method of cleaning I arrived at is as follows: the slides (sample and cover slides) which are already clean are heated for a few hours in an aqueous solution of chromic acid or in concentrated sulphuric acid, then washed meticulously in distilled water. In order to dry the slides without wiping them, which avoids dirtying them again, I place them in an apparatus in which they are immersed under reflux in a stream of ether, which is condensed in a cold trap. In this way one can wash the slides with an unlimited amount of perfectly clean liquid, without any significant consumption.

All of the experiments to be considered in what follows have been carried out with the slides treated as described (unless indicated to the contrary).

Chapter I

Liquid films with a homogeneous structure

Solid preparations – In order to have perfect homogeneous liquid films of fused azoxyphenetole,[2] it is necessary at the outset to achieve perfect preparations of

---

[1] Another structure, perhaps even more interesting, is formed in the freely floating spherical droplets for which Lehmann mentions a large number of observations in his book *Liquid Crystals* (Leipzig, 1904).

[2] All the experiments done with azoxyphenetole can be repeated with anisaldazine

CH_3O⟨ ⟩$-CH=N-N=CH$⟨ ⟩OCH_3 transition temperatures: 160 °C and 180 °C.

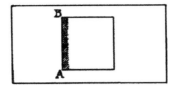

Fig. 1

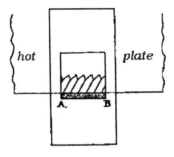

Fig. 2

the solid material. This is how they are prepared. Commercially available material, purified if necessary by crystallisation from boiling alcohol, is placed between a sample slide and a cover slide in such a way that it only occupies the edge AB of the latter (see Fig. 1).[1]

On heating, the substance melts and by capillarity spreads out regularly between the glass slides. Moving the region AB outside the heating plate (see Fig. 2), the temperature is lowered a little. Solid crystals form along the length of AB. These grow *slowly*, spreading little by little throughout the whole sample.

One obtains in this way large homogeneous crystalline domains. These are very birefringent, strongly dichroic, and appear yellow or white when illuminated by polarised light, if the vibration directions are along one or other of the extinction directions.

They are clearly biaxial. Moreover the single crystals of azoxyphenetole measured by Lehmann are monoclinic.

Liquid domains in parallel light

The solid sample heated slowly melts at 134 °C. Each solid crystal gives rise to a birefringent liquid domain, which has exactly the same topography. In the

---

[1] I avoid putting the material in the centre of the slide, because I noticed that liquid domains often showed irregularities in places where first melting occurs, due to the residues which the crystals leave on the glass slides after melting. (*see* later, surface layers).

carefully prepared samples, *all the domains*[1] *truly extinguish between nicols crossed at 90°, four times on rotating the stage, provided that the cover slide has not suffered any displacement after melting.* The extinction directions of the different liquid domains in general were the same as those of the original solid crystals from which they were formed.

If the glass slides are well-cleaned, the domains are perfect, and as clear as the previous solid domains; those people to whom I have shown them, who are familiar with microscopic observations, think that they are viewing truly crystal-line bodies. It is only by displacing the cover glass that one perceives that the substance is really liquid.

The limit of the domains can be followed throughout the thickness of the sample by varying the focus, thanks to the phenomenon of total reflection, which originates from the difference in refractive indices on both sides of the separating surface (Becke's method).

Observation with a single nicol shows dichroism (white and yellow) entirely analogous to that of solid crystals. Naturally the directions of the absorption maxima and minima are changed at the moment of melting, and follow the new extinction directions. The intensity of dichroism increases with the thickness of the sample.

If one heats more strongly to 168 °C, the birefringent liquid transforms to an isotropic liquid. The isotropic and birefringent modifications are separated by a very fine line which moves in one direction or another when the heating current is increased or decreased, tracing out in the sample at every instant the isotherm for 168 °C. The transformation resembles in every respect the reversible melting of a crystalline solid.

Finally the whole sample extinguishes between crossed nicols.

Films adhering to the glass slides – On cooling, the birefringent liquid reappears, and very remarkably, *all the previous domains reform* with the same dimensions, the same extinction directions and every smallest detail of the structures are the same as originally.

Lehmann, who first observed these effects, explained them by proposing that each solid crystal on melting leaves on the glass slides two films of oriented molecules which remain unchanged after the isotropic transition, and which influence the reforming of the domains.

All the observations I have been able to make, and which will be described in the rest of this paper, confirm absolutely this point of view.

It is moreover easy to demonstrate these surface films, as Lehmann shows,[2] by sliding the cover glass over the sample slide: then one sees *all the contours of the domains duplicate.* This process gives rise to two figures which can be superim-posed, one of which is fixed to the lower glass slide and remains immobile, while

[1] Apart from some possible irregularities in the region AB.
[2] Lehmann, *Flüssige Krystalle*, p. 58.

the other is attached to the upper glass slide and moves with it. The two surface films give exactly the impression of two *identical* petrographical preparations of granite which one can slide on top of one another.[1]

This experiment seems to me already very conclusive; here is another which is perhaps even more so.

A thin strip of paper is glued onto one of the edges of the cover glass. After melting and the formation of the birefringent domains, one lifts the cover slide around this improvised hinge. The liquid accumulates close to the paper and there is nothing to be seen by eye or under the microscope, *and all appearance of bire-fringence has disappeared*. On lowering the cover glass, the liquid returns to its original place; *all the liquid domains reform in exactly the same place and with the same extinction directions*.

I cannot see how this effect could be explained without proposing the existence of the residual surface films imagined by Lehmann, which stay on the glass slides even when the liquid has gone.

We note in passing this interesting result, in contradiction to an earlier claim by Wulff: *the films are too thin to give rise to any birefringence by themselves*.

Birefringent liquid domains in convergent light

It seemed to me that there was some interest in completing these observations by a study of the domains in convergent light, which would have the great advantage of allowing the observation of the optical properties of the material simultaneously in all directions, and at the same time to test its *homogeneity* and *anisotropy*.

In order to obtain interference fringes with such birefringent samples, it is necessary to work with monochromatic light. In white light, films of a few microns thickness already appear to be completely white.

One observes fringes of such perfection that they rival in clarity those of any solid crystal.

The fringes are identical to those which are given by thin slices of different orientations cut from a uniaxial crystal.

Some samples exhibit domains mostly parallel to the axis, others mostly oblique to the axis, with orientations that change from one domain to another. Generally the optic axis does not penetrate the field of view. However several times I observed domains rigorously perpendicular to the axis having the characteristic black cross and rings, without any sign of circular polarisation. For these I was able to determine the optical sign of the liquid as *positive*.

The number of fringes increases regularly with the thickness of the sample – For films whose thickness is of the order of one hundredth of a millimetre only one or two fringes can be observed, whereas I was able to count 12 systems of conjugated hyperbolae in a film parallel to the axis having a thickness of about 0.15 mm.

---

[1] Wulff, *Zeitschr. für Kryst.*, Vol. XLVI, p. 261.

Without doubt molten azoxyphenetole is birefringent throughout and the liquid domains have a uniform optical orientation over their full extent and thickness.

However surprising these facts appeared to certain physicists or crystallographers, they seem to be established with complete rigour by the preceding observations.

Persistence of the optical phenomena in the moving liquid – I would like now to describe other experiments which highlight an even more surprising fact: *the orientation and the optical homogeneity of the birefringent domains are not disturbed by motion which can be induced in the liquid.*

1.   In convergent light, one focuses on a domain situated towards the centre of the sample. Having immobilised the sample slide on the stage by means of two clamps, one fixes a little cork tripod, which is connected to the objective, to the cover glass with three drops of quick-setting glue. When one raises the objective by means of the focussing screw, one raises at the same time the cover glass. The liquid, moving away from the edges of the sample slide, accumulates in the centre thereby progressively increasing the thickness of the domain under observation.

The fringes stay perfectly clear despite the movement of the liquid, and they move very regularly, exactly as they would by superposition of a quartz compensator with increasing thickness. Their number increases little by little in the field of view. Having begun with one or two, one can easily obtain about twenty distinct and close-packed fringes.

Naturally when one lowers the cover glass, the fringes move in the opposite sense.

2.   The experiment is particularly instructive in the case of a film parallel to the axis. Then the fringes which one observes are hyperbolae whose axes are the two extinction directions of the film (and which are at the same time, as seen above, the directions of maximum and minimum absorption).

When one raises the cover glass, one sees the branches of the hyperbolae come together along one of the extinction directions, they merge together to form a black cross and separate in a perpendicular direction.

It is easy to be convinced, by illuminating with white light, that the direction along which the branches of the hyperbolae separate is the direction of maximum absorption.

If one relates this to the isochromatic surfaces of Bertin, it is immediately seen that the fringes must separate along the optic axis when the thickness increases. Thus:

The direction of maximum absorption (yellow colour) is the optic axis. The vibration most absorbed is the extraordinary vibration, which propagates with the smallest velocity.

Since one knows how to distinguish the ordinary and extraordinary vibration directions for these films, one can determine their optical sign; thus one verifies the result obtained with films perpendicular to the axis.

By counting the fringes which pass through the centre of the field of view, one can determine the retardation between the vibration components of the light introduced by a known increase in thickness, and thereby measure the birefringence of the liquid.

In this way I was able to cause 30 fringes ($\lambda=0.589\,\mu$m) to pass through the centre of the field of view for a change in thickness of 48 μm, read from the drum of the micrometer screw of the microscope. The result for the birefringence is a very high value (0.37), agreeing with that (0.33) which Lehmann[1] has determined using other methods.

3.   I will compare another very similar experiment performed with parallel light with the preceding experiments carried out in convergent light.

The cover glass is fixed to the objective (a weak objective) by means of small cork tripod, as before, but in such a manner that it makes an angle with the sample slide.

If one illuminates it in a suitable orientation with monochromatic light, one sees straight black fringes parallel to the dihedral angle formed by the two glass slides, analogous to those which one observes under the same conditions with a bevelled quartz compensator.

On raising and lowering the objective and cover glass to which it is fixed, more or less violent motion is induced in the liquid, which is made visible by the movement of air bubbles or floating dust particles. *In spite of this motion* the domains remain, always with the same sharp contours and always with the same clear extinctions. *The very clear fringes move regularly in one direction or another as the thickness of the domain being investigated is increased or decreased.*

All these experiments, using both convergent light and parallel light, lead us to the following conclusion, whose importance will not escape the reader:

The phenomena occur not only while the birefringent liquid is at rest, but also when it moves. Under the influence of couples from the films of solid crystals left on the glass slides, the liquid immediately forms optically homogeneous domains throughout its full extent and thickness.

Each time the liquid passes from one domain to another, by which we mean that it leaves the influence of the couples from one pair of surface films to enter that of neighbouring couples, the elements of the liquid suffer a change of orientation. This must be sufficiently rapid not to disturb the clarity of the optical phenomena observed in the course of these movements.

The changes of orientation take place both horizontally and vertically. The reason for this is that, firstly, the extinction directions vary from one domain to another, and that, secondly, the orientation of the optical axis on the glass slides differs from one domain to another.

---

[1] Lehmann, *Ann. der Phys.*, Vol. XVIII, p. 806, 1905.

These very remarkable phenomena can be identified, without contradiction, as amongst the most curious in the physics of liquid materials.

Remark on the structure of the surface films – Since under the action of the couples from some surface films, the optic axis of the liquid orients parallel to the glass slides, and under the action of other couples it orients obliquely with a variable angle, it has to be admitted that the various couples are not identical, and differ other than by simple change in orientation in the plane of the glass slides.

It is natural enough to suppose that the structure of the surface films, which originate from particular solid crystals, depends on the location of the glass attachment surfaces with respect to the crystal structure.

This is evident if one agrees with Lehmann that the molecules of the surface films are kept in the same arrangement as in the original crystal.

Be that as it may, there must exist a certain relation between the optical orientation of each of the solid crystals and the domains of birefringent liquid that correspond to them. So far I have not been able to determine the exact nature of this relation, but I do not believe it impossible to resolve this.

Birefringent liquid films perpendicular to the axis

The formation of the surface films, the role of which is predominant in all these phenomena, is influenced to the greatest extent by the state of the surface contacts with the glass, and this explains the importance of cleaning indicated at the beginning of this work.

Whereas in the case of azoxyphenetole washing with sulphuric acid, water and ether only causes an increase in the distinctness of the observations, in the case of azoxyanisole,[1] the influence of cleaning is of such importance that even the appearance of the phenomena is completely changed.

Azoxyanisole, when melted between glass slides, which have been cleaned and wiped, *but without taking any special precautions*, gives birefringent domains analogous to those of azoxyphenetole, but of poor quality, showing irregularities with streaks that appear to be traces of rubbing.

With thoroughly cleaned glass slides, one does not obtain domains at all. The molten sample, observed in parallel light, between crossed nicols, stays extinguished across its whole extent whatever the position of the stage. If one were restricted to these observations, one would think that one was dealing with an ordinary isotropic liquid.

But everything changes as soon as one looks at the sample in convergent light. Then one sees a black cross and splendid coloured rings: far from being isotropic,

---

[1] Methyloxycinnamic acid behaves just like azoxyanisole. Transition temperatures: 170 °C and 189 °C.

the liquid behaves like a slice of calcite perpendicular to the axis, but with an even higher birefringence.

Through the whole sample, the optic axis is oriented normal to the surfaces of the glass.

In monochromatic light, the number of rings increases regularly with the thickness of the sample. On raising the cover slide, one can easily obtain about fifteen rings.

The optical sign is positive. There is no evidence of optical rotation.

For these observations it is useful to provide as solid a base as possible for the microscope. The orientation of the liquid no longer has the same stability as in the azoxyphenetole films. In fact it is subject to slight vibrations which cause small intermittent deformations of the black cross and the rings, which one can avoid by making the liquid completely immobile.

If one raises the temperature, the birefringent liquid transforms, at 134 °C, to another isotropic liquid. The cross and rings disappear.

Naturally in parallel light between crossed nicols the field of view remains as well extinguished in the birefringent region as in the isotropic region, which cannot be distinguished from each other.

However, the limit of the two regions is distinctly marked by a strongly visible bright line, which moves in one direction or another following the variations in heating current.

The existence of this line is evidently due to a local alteration of the orientation of the birefringent liquid in contact with the isotropic liquid.

Chapter II

Birefringent liquid films having a helicoidal structure: Study with parallel light

All the birefringent liquid domains studied up till now exhibit a *uniform optical orientation* at all points, and have the optical characteristics of truly homogeneous crystalline plates.

One can, at the expense of these domains, create others with a more complex structure, the study of which will be the object of this new chapter.

In order to obtain homogeneous domains of liquid azoxyphenetole, a vital precaution, which I have indicated above, is to *prevent the cover glass from sliding after the sample has melted.*

If this condition is fulfilled, the two *superimposed* layers, which each solid crystal leaves on melting on the glass slides, exert a *concerted action*, which is enough to ensure the optical homogeneity of the liquid throughout the region contained between the slides.

The character of the domains is not changed when one raises the cover glass parallel to itself, which evidently does not change the relative orientation of the surface films.

However, it is very different when, as a result of the movement of the cover glass, the coincidence of the two surface films is destroyed. This can occur either when one surface film is turned with respect to the other, or if one of the surface films is displaced to be above another, so that the two films originate from different solid crystals.

Helicoidal domains in parallel light

The new domains one obtains which are contained between two non-coincident surface films in general no longer extinguish light between nicols crossed at 90°, whatever the position of the stage. They are distinctly different from the original domains and from the homogeneous crystalline films.

In order to define conveniently their optical properties, I will suppose that before any movement of the cover glass, the extinction directions have been located for each of the homogenous domains contained between the upper and lower surface films. Of these two extinction directions, one is, as is well known, a direction of maximum absorption; I will call this the first principal direction: the other direction of minimum absorption will be the second principal direction.

Having defined the system, the following are the modifications undergone by a beam of polarised white light when it propagates normal to the glass slides through one of the domains obtained after movement of the cover glass.[1]

1. If the incident vibration is linear, oriented along the first principal direction of the surface film which light passes through in order to enter the liquid, the emerging vibration is also linear, oriented along the first principal direction of the exit surface film.
   The emerging light has a yellow colour which becomes more intense in proportion to the thickness of the film traversed.
2. If the incident linear vibration is oriented along the second principal direction of the incident surface film, the process is analogous, with the emerging vibration parallel to the second principal direction of the exit surface film. But this time the light stays white.
3. Finally a vibration incident at a general azimuthal angle splits up on entering the liquid into two vibrations which behave as above. The two components propagate with different speeds giving a certain phase difference at the exit; the emerging vibration is in general elliptical.

We denote by α the angle between the same principal directions on the two overlapping surface layers. Between two nicols which have an angle of $\frac{\pi}{2} + \alpha$ between them, the liquid film contained between the surface films behaves exactly as a homogeneous domain, or a crystalline plate between two nicols crossed at

---

[1] These domains are certainly identical to the domains studied by Messrs Friedel and Grandjean (*Bull. Soc. Min.*, Vol. XXXIII, p. 192). The optical properties in parallel light are the same as those which have been described by these authors, presented here in a different manner.

90°. It extinguishes four times for one rotation of the stage, in white light and in monochromatic light.[1] Each extinction is produced when one of the principal directions of the lower surface film is parallel to the polariser, and the other principal direction of the upper film is parallel to the analyser. It stays extinguished in all positions if its thickness is such that the retardation between the two linear components defined above is an even number of wavelengths (monochromatic light). If the thickness is variable (wedged-shaped film), one sees black fringes in places where the above condition is fulfilled.

The phenomenon of dichroism manifests itself in observations with a single nicol and presents a quite remarkable effect. According to the rules given above, the colour of each domain (light or dark yellow) is determined by the relative orientation of the nicol and the surface film which faces it; the surface film situated on the other side of the sample has no effect. Therefore if one observes with a single polariser, one only sees the effect of the surface film adhering to the sample slide. The analyser, on the contrary, shows only the surface film attached to the cover slide. The two systems of overlapping surface films can have nothing in common; *thus the sample shows a totally different appearance, depending on whether it is observed with the polariser alone or with the analyser alone.* It is only by using both nicols that one can see simultaneously the contours of the surface layers adhering to both glass slides. These surface layers delineate the domains, the characteristics of which have been described above.

The simplest way to verify all these facts is the following. One focuses, with a weak objective, on a good *homogeneous* domain, which extinguishes clearly between crossed nicols. The sample slide is fixed on the stage, and the cover slide is fixed to the objective by means of the small cork tripod already used in the experiments of the preceding chapter. The objective, fixed to the microscope by means of a *clip* (not a screw), can turn about the axis of the instrument without changing the focus. It carries with it the cover slide and with it naturally, the adhering surface layer, which constrains the domain at the top, while the lower surface layer remains fixed. Thus one can make the angle between the surface layers any value between 0° and 360°.

One can also arrange a way to have several domains in the field of view of the microscope. Before fixing the cover slide to the objective, the stage is moved parallel to itself by means of the adjusting screws arranged for this purpose. The sample slide with its adhering surface layers translates, while the surface layers fixed to the cover slide remain immobile. In this way the superposition is achieved of surface layers having known orientations coming from different domains. By turning the objective (and the cover slide), one can generate any angle desired between the surface layers.

---

[1] The liquid film extinguishes between parallel nicols if the surface layers have an angle between them of 90°, the first principal direction of one being superimposed on the second principal direction of the other.

In order to make wedge-shaped films, one uses the technique described on page 81 of this article*.

Thus it is easy to verify the truth of the rules given above. I recall the following essentials of these rules:

1st   For each domain there exist two special linear vibrations. These remain linear as they propagate through the liquid structure, but during their passage they rotate by an angle equal to the angle between the surface films fixed to the glass slides.

2nd   At each instant in time these propagating-rotating vibrations are oriented along a direction of maximum or minimum absorption.

This last remark leads us naturally to think that the orientation of the liquid elements must themselves vary progressively from the lower glass side to the upper one, giving a *helicoidal structure* which connects the boundary surface films.

This structure can be to a first approximation, represented by a stack of infinitely thin birefringent (and dichroic) lamellae, in which the orientation varies progressively in the stack.

An elementary calculation, of which one will find details later, shows that such a stack has effectively the optical properties observed in films of azoxyphenetole, *provided that the product $(n - n')p$ of the birefringence of the lamellae and the helicoidal pitch is large in comparison with the wavelength of light.*

Furthermore, these results are confirmed by a more rigorous study of the propagation of light in a helicoidal medium using the theory of electromagnetism.

Helicoidal structures previously studied – These structures are not the first birefringent helicoidal structures which have been studied.

The representation as stacks of lamellae immediately suggests the idea of a comparison with the helicoidal mica stacks which Reusch used to reproduce the phenomenon of polarisation rotation by quartz and similar crystals.

In reality there are essential differences between the two types of stacking. At the outset, the stacking of Reusch is discontinuous, two consecutive lamellae making angles with each other of 45° or 60°; by contrast the angle of successive lamellae in the equivalent stacks for films of azoxyphenetole is infinitely small. But above all the greatest difference is that the incremental step of the stacks of Reusch (or more exactly the product $n - n'\ p$) is of the order of the wavelength of light. This is much smaller than the step length in the films of azoxyphenetole. It is this condition, above all, to which must be attributed the optical differences between the two types of stacking.

Bichat[1] created continuously birefringent helicoidal structures, using the Kerr Effect induced in carbon disulphide by an electric field from an appropriately twisted set of electrodes. Here the twist length is very long (17 cm), but the

---

* p. 107 of this volume
[1] Bichat, *Rev. gén. Sc.*, Vol. III, 1892, p. 248.

birefringence is extremely small, so that the product $(n-n')\,p$ is still found to be of the order of the wavelength of light. The optical properties observed compare on the whole with those of the stacks of Reusch and those of azoxyphenetole.

Mr Wallerant[1] has shown that the addition of an impurity endowed with rotatory power can induce a helicoidal twisting in a number of crystalline bodies. The structure of these is very close to that of the liquid helicoidal films, but when observed the two structures are totally different. The twisted crystals of Mr Wallerant, usually extended as rods, arrange themselves in radial clusters, which are always oriented between the glass slides in such a way that the helicoidal axis is parallel to the glass slides, and so perpendicular to the axis of the microscope. By contrast the helicoidal axis of films of azoxyphenetole is normal to the glass slides, and parallel to the axis of the microscope when the sample is on the stage. The observations on these two types of structure are mutually complementary in some respect, but cannot properly be compared.

It would be interesting to find a solid birefringent crystal sufficiently plastic to be twisted through a significant angle, and then one could try to reproduce the optical phenomena described above.

Properties of helicoidal films deduced from the Poincaré sphere representation

A birefringent film can be considered as a stack of very thin, identical, birefringent lamellae, in which the orientation in the stack varies progressively.

I will define this stack with respect to a system of three perpendicular axes $Oxyz$, with Ox and Oy parallel to the plane of the lamellae.

The directions of extinction of the lamellae situated at position z make a variable angle with Ox and Oy which I shall represent by

$$\alpha = 2\pi\frac{z}{p};$$

p is the pitch of the helix.

If dz is the thickness of each of the lamellae; two consecutive lamellae make an angle between them

$$d\alpha = \frac{2\pi}{p}dz$$

In order to find the effect of the stack on a beam of polarised light which travels normal to the lamellae, I will use a geometric representation due to Mr Poincaré, of which I have given a simple account in a recent number of the Bulletin.[2]

[1] Wallerant, *Bull. Soc. Min.*, Vol. XXX, 1907, p. 43.
[2] Mauguin, *Bull. Soc. Min.*, Vol. XXXIV, 1911, p. 6.

All the elliptic vibrations which I will have to consider will be represented on the same sphere by points similar to P: the longitude 2α is twice the angle that the vibration axes of the ellipse make with Ox and Oy; the latitude 2β is twice the ellipticity angle (Figs 3 and 4).

Each lamella transmits two linear vibrations, without alteration, along the two extinction directions. Of these two vibrations, we consider the one which propagates with the smallest velocity (corresponding to the largest refractive index). This is represented on the sphere by an equatorial point, which obviously varies from one lamella to another.

Let A1, A2,... be representative points corresponding to successive lamellae of the stack (Fig. 5). According to the basic convention given above, all these points are equidistant and separated by an angle

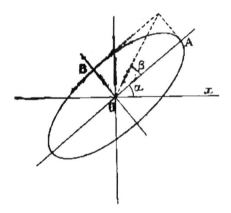

Fig. 3

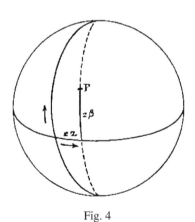

Fig. 4

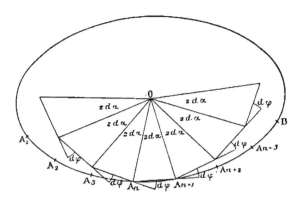

Fig. 5

$$2d\alpha = \frac{4\pi}{p}dz$$

which is twice the angle between consecutive lamellae.

From this, it follows that the effect of the lamella of rank n on an elliptical vibration which goes through it[1] is to turn the representative point of the vibration around the equatorial radius OA_n by an angle

$$d\varphi = 2\pi\frac{n - n'}{\lambda}dz$$

Then the total effect of the stack on a vibration which crosses each of the lamellae one after another translates as a set of rotations of the same amplitude $d\varphi$ about the radii OA_1, OA_2, ...

These successive moves, turning on the equatorial plane, create a regular polygonal pyramid. This has its apex at the centre of the sphere, for which the sides (faces) have the value $2d\alpha$, and for which the supplement of the dihedral angle is $d\varphi$.

It is clear that the limit of this regular pyramid, for which the faces are infinitely small, becomes a cone of revolution.

We consider the line of the regular polygon cut on the sphere by the pyramid, of which the cone is the limit (Fig. 6).

Let A_n, A_{n+1} be one of the sides of this polygon, and let I be the point where the axis of the pyramid cuts the sphere.

A_n, A_{n+1} is equal to $2d\alpha$; IA_n is the semi-angle of the apical angle ω, the dihedral angle IA_nA_{n+1} is equal to $90° - \dfrac{d\varphi}{2}$.

---

[1] See my Note on the geometrical representation of Poincare, p. 12 of the present volume.

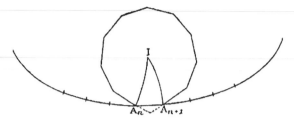

Fig. 6

Consideration of the spherical isosceles triangle IA_nA_{n+1} yields

$$\tang\omega = 2\frac{d\alpha}{d\varphi}$$

Replacing $d\alpha$ and $d\varphi$ by their values:

$$\tang\omega = \frac{2}{n-n'}\frac{\lambda}{p}$$

To summarise: in order to find the effect of a helicoidal stack on any elliptic vibration, one starts by determining the position of the equatorial points A and B (Fig. 7) which represent the linear vibrations oriented along the extinction directions of the largest refractive index of the first and last lamellae of the stack.

One supposes that the point representing the incident vibration is connected to a cone of revolution having its apex at the centre of the sphere, which is tangent to the equatorial plane along OA, and for which the magnitude of the angle 2ω is given by the formula

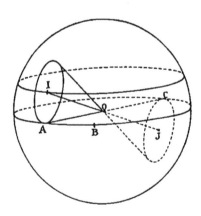

Fig. 7

$$\tan g\omega = \frac{2}{n-n'}\frac{\lambda}{p}$$

In rolling this cone on the equatorial plane from OA to OB, the representative point is translated, and its final position will determine the emergent vibration direction.

Special elliptical vibrations – We consider the representative points I and J diametrically opposite, situated on the axis of the cone. When the cone rolls on the equatorial plane, the points I and J each move on a parallel line. They are, moreover, at every instant on the meridian which passes through the end points A,C of the tangent of the cone on the equatorial plane.

If one refers to the fundamental properties of the representation of Poincaré, the geometrical relations allow us to state the following laws:

There exist two elliptical vibrations I and J of the same type, but of opposite senses, oriented in such a way that the largest axis of one coincides with the smallest axis of the other, *which propagate unchanged through the helical struc-ture*. During their propagation, these vibrations turn through an angle equal to the angle through which the lamellae themselves turn in the helicoidal structure, in such a way that at every instant their axes are oriented along the extinction direc-tions of the lamellae through which they pass.

Furthermore these two special vibrations propagate with different velocities.[1]

Explanation of the optical properties of films of azoxyanisole – The product $(n-n')p$ of the helical pitch times the birefringence is very large compared with the wavelength of light. Thus the magnitude of the cone angle defined by the equality

---

[1] The components of these vibrations along the axes of these ellipses can be written as

$$\xi_1 = a\cos 2\pi\left(\frac{t}{T} - \frac{N_1 z}{\lambda}\right), \qquad \xi_2 = b\sin 2\pi\left(\frac{t}{T} - \frac{N_2 z}{\lambda}\right),$$

$$\eta_1 = b\sin 2\pi\left(\frac{t}{T} - \frac{N_1 z}{\lambda}\right), \qquad \eta_2 = a\cos 2\pi\left(\frac{t}{T} - \frac{N_2 z}{\lambda}\right),$$

and the components along *Ox, Oy*:

$$x_1 = \xi_1\cos\frac{2\pi z}{p} - \eta_1\sin\frac{2\pi z}{p}, \qquad x_2 = \xi_2\cos\frac{2\pi z}{p} - \eta_2\sin\frac{2\pi z}{p},$$

$$y_1 = \xi_1\sin\frac{2\pi z}{p} + \eta_1\cos\frac{2\pi z}{p}, \qquad y_2 = \xi_2\sin\frac{2\pi z}{p} + \eta_2\cos\frac{2\pi z}{p},$$

One can prove without difficulty that the indices N_1 and N_2 satisfy the equation

$$N^2 - N(n+n') + nn' - \frac{\lambda^2}{p^2} = 0$$

The ellipticity angle of the special vibrations is equal to $\frac{\omega}{2} : \frac{b}{a} = \tan g\frac{\omega}{2}$ with $\tan g\frac{2}{n-n'}\frac{\lambda}{p}$.

$$\tan\omega = \frac{2}{n-n'}\cdot\frac{\lambda}{p}$$

is very small.

The same argument holds true for the ratio of the special axes of vibration, which is equal to $\tan\frac{\omega}{2}$.

This cone has a small angle. Thus, when it rolls on the equatorial plane, the points situated on its axis are clearly always very close to the tangent of the cone on the plane.

The special vibrations which they represent, quasi-linear, oriented at 90° to each other, propagate in the helicoidal medium *remaining quasi-linear*, but rotating in such a way as to be always along the extinction directions of the lamellae through which they are passing.

Any incident vibration, on entering the structure, naturally divides into the two previously identified linear components, oriented initially along the extinction directions of the first lamella. The components propagate in the stack with different velocities, turning, as has been explained, and leaving with vibration directions along the extinction directions of the last lamella, with a phase difference proportional to the thickness traversed.

If the lamellae are dichroic, it is evident that, during passage through the helicoidal stack, one of the components will always be oriented along the direction of maximum absorption.

We recover all the characteristics of the propagation of light in liquid films of azoxyphenetole.

All that remains is to check, with a particular example, that it is legitimate to assume that in these films $(n-n')p$ is large in comparison with the wavelength of light.

Let a film have a thickness of 0.05 mm: one will see later that the angle of twist cannot exceed $\frac{\pi}{2}$; the pitch then is less than 0.05 mm × 4 = 0.2 mm.

Assuming a value of 0.37 for the birefringence, 0.589 μm for the wavelength, the value for $\tan\omega = \frac{2}{n-n'}\cdot\frac{\lambda}{p}$ is close to 0.016; or the value of the ratio $\tan\frac{\omega}{2}$ of the axes of the special elliptical vibrations is close to 0.008. These vibrations are certainly close to being linear.

Electromagnetic theory of light propagation in a helicoidal medium

In this section Mauguin follows through the mathematical theory. He finds the normal modes by methods that are now standard. He finishes up with the conclusion:

> Thus the medium allows the propagation of two quasi-rectilinear vibrations, oriented at 90° to each other. These two vibrations are propagated without being changed. They rotate in the structure, and the resulting path difference is proportional to the thickness traversed.
>
> One recovers the basic conclusions of the previous section, explaining the essential optical properties of films of azoxyphenetole.

Chapter 3

Helicoidal films in convergent light

The liquid helicoidal films observed in convergent *monochromatic* light, between suitably oriented nicols, exhibit perfectly distinct fringes which are just as pretty as those coming from homogeneous films.

These fringes change according to the nature and orientation of the films which determine the liquid structure. Their study is extremely instructive, and very nicely complements the observations in parallel light described in the previous section.

This study is particularly interesting for samples contained between two surface films originating from homogeneous domains which are parallel to the axis. The structure of the liquid is then simple enough for the form of the fringes to be determined by calculation. The merit of the hypotheses can thereby be proved. It is this case which I have studied most, and on which I will expand here. First I will give the experimental facts, and then the calculations.

> Mauguin goes on to describe experiments using monochromatic convergent light. He then presents a rather detailed theory for the appearance of the fringes.
>
> He presents a particularly attractive diagram showing a set of fringes which he observed under the microscope. The fringe pattern changes as one rotates the cover-slide keeping the crossed nicols in the same position. This is shown in Fig. 8.
>
> He finishes off by remarking on the insensitivity of the domain structure to fluid motion. This is a point to which Georges Friedel would return at a later stage.

Helicoidal structure while the liquid is in motion – In the first part of this article, I have underlined the important fact that the structure of homogeneous domains is conserved, even when the liquid is in motion. Just the same can be said concerning the helicoidal domains. The optical phenomena in parallel light (extinctions, dichroism) remain the same, despite the disturbance induced in the liquid. The fringes in convergent light and the fringes in a wedge-cell retain all their clarity

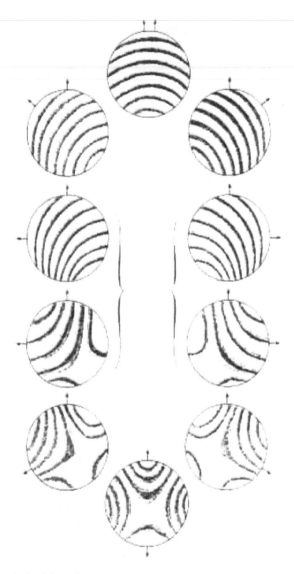

Fig. 8 Fringes obtained from the same domain of liquid azoxyphenetole, while the cover-slide
is rotated completely once.

as the cover-glass is raised or lowered. They only shift because by this means one
can vary the thickness of the liquid through which the light travels.

It is also possible to slide the cover-slide over the microscope stage. This causes
a complete change in the domain organisation. At any moment, under the action of the
films attached to the two glass surfaces, the moving fluid particles create regular
structures which exhibit all the properties which have been studied in the pages above.

Many research workers have been struck by the persistence of apparently homogeneous domains in the moving liquid. They have interpreted this as a proof that the optical anisotropy is entirely localised in immobile films attached to the glass, with the moving liquid between the films remaining isotropic. But this point of view does not sustain serious examination. Firstly, the films by themselves do not yield appreciable birefringence (see p. 78 of this article*). Secondly, the progressive increase in the number of fringes with the path length of the light shows that the helicoidal films, just like the homogeneous films (see p. 82 of this article†), exhibit a regular bulk structure.

It is certainly at first sight surprising to see that this structure is conserved in spite of the liquid motion. However, facts are facts and we must recognise them as such.

---

* p. 105 of this volume.
† p. 107 of this volume.

A9

Comptes rendus de l'Académie des Sciences **152**, 1680–83 (1911)

CRISTALLOGRAPHIE. — *Orientation des cristaux liquides par le champ magnétique*. Note de M. **Ch. Mauguin**, présentée par M. F. Wallerant.

La principale difficulté qu'on rencontre dans l'étude des liquides anisotropes de Lehmann est d'obtenir des portions de matière d'étendue notable présentant une *orientation optique uniforme* en tous leurs points. Je pense avoir trouvé dans l'action du champ magnétique un procédé qui permet de résoudre cette difficulté d'une façon satisfaisante. Avant de décrire mes expériences qui ont porté surtout sur l'azoxyphénétol et sur

A9
Comptes rendus de l'Académie des Sciences **152**, 1680–83 (1911)

ORIENTATION OF LIQUID CRYSTALS BY A MAGNETIC FIELD

Note from
Mr Ch. Mauguin
presented by Mr. F. Wallerant

The principal difficulty encountered in the study of Lehmann's anisotropic liquids is to obtain a sufficiently large region of sample having a *uniform optical alignment*. I think I have found a method, using a magnetic field, which allows the resolution of this difficulty in a satisfactory manner. Before describing my experiments carried out on azoxyanisole and azoxyphenetole, I will briefly recall what happens with these materials in the absence of an external field.

When azoxyphenetole is crystallised between two thin glass sheets and then heated to 138 °C, it gives rise to domains of homogeneous birefringent liquid. As I have shown in an earlier Note,[1] these exhibit all the optical properties of real uniaxial positive crystalline plates, whose orientations are determined by residual thin films of solid crystallites which remain on the glass sheets.

The study of azoxyanisole, carried out under the same conditions, revealed to me a very remarkable feature. After melting at 116 °C, between thin sheets of glass cleaned without special precautions, the sample exhibits domains analogous to those obtained with azoxyphenetole. By contrast, if one uses sheets of glass washed in hot sulphuric acid, distilled water and ether, the material assumes a *uniform optical orientation* across the whole sample. It looks like a thin sheet of calcite for which *the optic axis is perpendicular to the glass*. Between crossed nicols the sample extinguishes collimated light as though it were isotropic; illuminated by convergent light, the sample exhibits a black cross and the classical coloured rings for an optically positive sample. It seems that there is no region whatsoever that responds to the residual crystallites; the orientation of the liquid appears to be due to the influence of the glass alone.

---

[1] *Comptes rendus*, Vol. **151**, 1910, p. 886.

The films of liquid prepared in this way are unfortunately always very thin; I have with much trouble obtained a uniform orientation for a thickness of 0.2 mm, but I do not think it is possible to go any thicker. Furthermore, if the films of azoxyanisole are of any significant thickness, the other platelets are extremely small (1 or 2 mm). A magnetic field, as we shall see, allows us to do much better.

1. *Films perpendicular to the axis* – A film of azoxyanisole of thickness 1 or 2 mm, melted between *rigorously cleaned* glass plates, is placed in the field of an electro-magnet, perpendicular to the lines of force (position 1). The pole pieces are pierced by a longitudinal hole, which allows the sample to be illuminated with a beam of polarised light, and to be viewed with a microscope fitted with an analyser. (The beam of light is in the same direction as the lines of force.)

   The thickness of the sample is too thick for the liquid to assume a uniform structure under the action of the glass plates alone. In the absence of a magnetic field, one sees regions of different orientations separated by sharp boundaries which are constantly changing due to the motion of the fluid.

   On applying the magnetic field (of the order of 2500 Gauss*), the liquid becomes homogeneous; all the separation boundaries disappear. The sample extinguishes light between crossed nicols.[1] In convergent light the sample exhibits perpendicular to the axis the black cross and rings characteristic of a uniaxial sample.

   When the field is removed, the cross and fringes disappear and the sample resumes its original appearance. The phenomenon is perfectly clear.

   Under the action of a magnetic field, the liquid becomes equivalent to a thick uniaxial crystalline plate, for which the optic axis is parallel to the lines of force (perpendicular to the glass plates).

2. *Films parallel to the axis* – The same thick film of melted azoxyanisole is placed between two flat pole pieces, parallel this time, to the lines of force (position 2), the observations are made between two nicols appropriately oriented in a perpendicular direction (the beam of light is normal to the magnetic field).

   As in the preceding case, the sample appears homogeneous as soon as the field is applied (2500 Gauss). The sample extinguishes light when the vibration directions of the nicols are respectively parallel and perpendicular

---

* The original simply talks of *unités* – units, but they are clearly Gauss (*ed.*).

[1] The apparatus used is not sufficiently sensitive to detect if there is a small magnetic optical rotation

to the lines of force, and in all other cases the sample transmits light. In convergent monochromatic light one sees two systems of conjugate hyperbolae whose axes are oriented along the directions of extinction just defined. To summarise, *under the action of a magnetic field, the liquid assumes all the optical properties of a uniaxial crystalline plate, the optic axis of which is now directed along the lines of force (parallel to the glass plates).*

The birefringence is very high. One can be convinced of this by illuminating the sample with white light (nicols are crossed at 45° to the lines of force) and analysing the emergent beam with a Hilger prism. A spectrum is obtained with a large number of bands which clearly converge towards to the violet. If the intensity of the magnetic field is increased, the birefringence appears to increase until a field of 5000 Gauss is reached; it then remains constant for field strengths from 5000–7000 Gauss.

Naturally when the field is removed, the uniform orientation of the liquid disappears.

3. *Simultaneous orientation by the plates of glass and by the magnetic field* – The experiment above (position 2) can be repeated with a film of azoxyanisole 0.01 mm thick, a sufficiently thin film for it to be oriented by the glass plates. In the absence of a magnetic field the black cross and uniaxial rings are seen. Switching on the current of the electromagnet, the cross and rings slightly deform and move towards one or other of the edges of the field of view of the microscope along the direction of the lines of force. The film, *oriented at the same time by the glass and the magnetic field*, can be understood, at least to a first approximation, as a uniaxial film with a tilted axis.

The optic axis tilts more and more as the magnetic field strength is increased, but it is still not parallel to the lines of force in a field of 7000 Gauss. When the current is removed the optic axis returns to its original orientation.

4. *Simultaneous action of the field and residual crystallites* – A film of azoxyphenetole, which has domains differently oriented by the residual crystallites adhering to the glass plates, is placed in position 1, between the pierced pole pieces. Before applying the magnetic field, the optical retardations due to each domain are sufficiently large to give a pure white appearance. As soon as the current of the electromagnet is switched on, the colours of Newton's fringes appear which pass one after another; when the magnetic field strength is increased the retardation gets smaller and smaller, yet without ever becoming zero. *The coloured fringes are obtained equally well with a film of thickness 0.01 mm as with a film of 1 mm thickness.*

Everything that happens is as if the mass of liquid orients in the direction of the magnetic field giving the equivalent of a film perpendicular to the axis, except for a thin layer next to the glass plates, where the influence of the residual crystallites is sufficiently strong to counterbalance the effect of the magnetic field.

The phenomena described here are a little less simple than with azoxyanisole between rigorously cleaned glass plates, but they still have a simple explanation.

Charles-Victor Mauguin was born in 1878 in Provins in the Seine-et-Marne Département close to Paris. He was the son of a baker, and perhaps as a result initially did not have high academic aspirations. He trained to be a teacher in the local provincial capital at Melun and between 1897 and 1902 he worked as a teacher locally in Montereau. In that year he entered the École Normale at Saint-Cloud, subsequently passing to the École Normale Supérieure in Paris. In 1910 he submitted a thesis on organic chemistry at that institution.

His interest in liquid crystals was supported by the director of the laboratory of Mineralogy at the Sorbonne, Frédéric Wallerant, who appointed Mauguin as his assistant. He subsequently held posts at the universities of Bordeaux and Nancy, before being appointed as *Maître de conferences* by Wallerant in 1919 and succeeding to Wallerant's chair in 1933.

Apart from his liquid crystal work he is well known for his pioneering work in obtaining molecular structure by X-ray diffraction, and for inventing crystallographic group notation subsequently adapted by Hermann and adopted as a standard. His wide scientific interests stretched from astrophysics to chemistry and biology and the origin of life. He was elected to the French Academy of Sciences in 1936 and died in 1958.

127

A10
Bulletin de la Société française de Minéralogie **39**, 164–213 (1916)

L orientation des liquides anisotropes sur les cristaux ;

Par M. F. Grandjean.

Les actions auxquelles obéissent les liquides anisotropes, qui se révèlent par des structures si variées, pourront sans doute plus tard se résumer en quelques lois très simples. Bien que ces dernières ne soient pas connues, on est en droit de penser, d'après les observations très nombreuses recueillies jusqu'à ce jour, qu'elles feront jouer un rôle très important aux *actions superficielles* qui règlent l'orientation du liquide au voisinage de la surface, dans la couche capillaire, et les distingueront nettement des *actions intérieures* qui fixent l'orientation du liquide par rapport à lui-même dans toute l'étendue de sa masse. Aux premières se rattachent par exemple les plages orientées à la surface du verre; aux secondes les coniques focales, les « fils ».

Il semble que les actions superficielles donnent seules des conditions définies, impératives, tandis que les actions internes, plus souples, rendent possible un raccordement avec les orientations déterminées dans la région capillaire. Si cette manière de voir est juste, et si l'on adopte le langage mathé-

A10
Bulletin de la Société française de Minéralogie **39**, 164–213 (1916)

THE ORIENTATION OF ANISOTROPIC LIQUIDS ON CRYSTALS

by

Mr F. Grandjean

The forces to which anisotropic liquids are subject are manifest in a large variety of structures. It will no doubt be possible eventually to summarise them in a number of very simple laws. Although these are not yet known, the numerous observations accumulated up to the present give good reason to believe that a very important rôle will be played by the *surface interactions*, which determine the orientation of the liquid close to the surface and in the capillary layer. A clear distinction can be made between these and the *internal interactions*, which determine the relative orientation of the liquid throughout the bulk of the material. The former are responsible for the oriented domains at the surface of glass; the latter give rise to the focal conics and 'threads'.[1]

It seems that surface interactions give only precise necessary conditions, while the more flexible internal interactions provide a connection with the orientations fixed in the capillary region. If this way of looking at things is correct, and if one adopts the language of mathematics, it will be possible to set down the problem of the equilibrium state of anisotropic liquids thus:

Let x, y, z be the coordinates of a point in a volume containing anisotropic liquid bounded by a surface S. Let α, β, γ be the direction cosines defining at this point the orientation of the optic axis; α, β, γ are some functions of x, y, z, related to variables through a set of differential equations. These equations express the internal laws, and have an infinite number of solutions. The surface conditions give the values α_0, β_0, γ_0 of α, β, γ at each point x_0, y_0, z_0 in the surface S. They give rise to a set of boundary conditions. Of course nothing prevents *a priori* the functions α, β, γ not being completely defined by the set of differential equations and boundary conditions; in this case several structures, or even an infinite number of structures, are possible.

In any case the structure in the bulk can be considerably modified if one changes the surface conditions. For example, in the helicoidal domains of Monsieur

---

[1] *fils* in the original (*ed.*).

Mauguin, it is sufficient to turn one of the capillary layers with respect to the other for all the orientations in the interior of the liquid to be changed. Solid surfaces, in particular, usually introduce complex boundary conditions, from which result the tangled domains exhibited by azoxyphenetole and other substances from the same group, when they are melted on a glass surface without any special precautions. At least this is the case for an isotropic solid and for any surface which only appears flat and which is almost always contaminated by foreign substances. But the situation changes completely and simplifies in a remarkable way if one chooses as a surface support the cleavage plane of a crystalline material: *one almost always obtains a well-defined alignment when a drop of anisotropic liquid is placed on the cleavage planes of crystals or in the cleavage cracks.*

This has already been reported by Monsieur Mauguin in 1913 (*C.R. Acad. Sc.*, Vol. **156**, p.1246) for the case of muscovite. My observations, which were without knowing the results of this scientist, have confirmed them completely.

Below I give results for nine other minerals. These are types chosen for the greatest ease of observation, but I have seen alignment on many others, without making such a complete study as in these particular cases. These minerals are orpiment, blende, phlogopite, brucite, talc, pyrophyllite, rock salt, sylvine and leadhillite. There were five anisotropic liquids used, which were azoxyphenetole, azoxyanisole, anisaldazine, ethyl azoxybenzoate, ethyl azoxycinnamate.

The first three belong to the group of liquids showing threads or nuclei;[1] they give well-defined alignment with all the crystals studied. The two others belong to the group of focal conic liquids; they give equally well-defined alignment on orpiment, blende, talc, pyrophyllite, rock salt, sylvine and leadhillite, but nothing on phlogopite and muscovite. Brucite orients the azoxybenzoate, but not the azoxycinnamate.

Here I do not define as *alignment* an orientation of the optic axis normal to the surface, since this arrangement is not in the required correspondence with the crystal lattice, and can be observed on a surface support such as glass. The liquids of the group of azoxyphenetole do not give this orientation on the cleavage planes, or at least it occurs so rarely that I have not had occasion to observe it during my experiments. The liquids of the group of azoxybenzoate are always oriented in this way (i.e. normal to the crystal surface). The alignment of these latter liquids is accompanied by a remarkable phenomenon, which I shall call the *terrace phenomenon*,[2] and to which I shall return in a subsequent Note: a terraced or stepped drop results from the superposition of several layers parallel to the base, each layer having a constant thickness, and each having a sharp outline. Fig. 1 gives an idea of the phenomenon. It is quite like the contours of a topographical map, but the surface of each terrace or step is absolutely flat and smooth. The optic axis is perpendicular. The outline of a terrace is often marked by a line of

---

[1] *À noyaux* in the original (*ed.*).

[2] *phenomène des gradins* in the original (*ed.*).

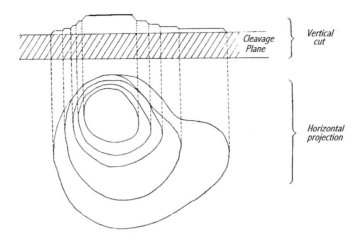

Fig. 1 Terraced droplet resting on a cleavage plane

very small focal conics, the ellipses of which are circular, or almost circular, and rest on the lower terrace. If the liquid is moved, the terraces slide over each other, changing their shape and their thickness, but the structure does not change.

When the contact between the anisotropic liquid and the crystal only takes place along a single flat face, for example if one observes a drop deposited on a cleavage plane, the drop exhibits one or more domains, in each of which the optic axis has a unique direction. The domains appear to be like uniaxial crystals. I will denote this type of structure as *parallel*, which is evidently the simplest.

If the contact takes place along two flat faces, parallel or almost parallel, for example if the liquid is viewed in a cleavage fissure, the domains obtained can still be parallel, in which case they are oriented as before; or else they have a helicoidal structure like that observed with mica by Monsieur Mauguin. Thus rock salt gives domains twisted by 90°; brucite, leadhillite and talc give domains twisted at 60°, etc. At the lower and upper boundaries, the orientations are always those of domains with a parallel structure; as a result crystals which give twisted domains have several orientations of parallel domains. Conversely, crystals like orpiment, which give a single orientation for parallel domains, never give twisted domains.

Of course, if one turns one of the cleavage planes with respect to the other, a parallel domain twists, whereas an already twisted domain twists further or untwists following the sense of rotation. These phenomena are identical to those described by Monsieur Mauguin, and I will not discuss them again in the rest of this Note. When I speak of a twisted domain, I will mean that which is spontaneously twisted in a cleavage crack whose walls have retained their relative orientation. The importance of spontaneous twist seems to me all the greater, because it is not at all exceptional.

131

Before describing the observed alignments, I will point out some precautions to be taken for their observation. I will also explain the application of the geometric procedure of H. Poincaré to helicoidal layers of constant total torsion, but variable thickness, in order to show that the twisted domains do indeed possess the theoretical properties of these layers.

In the next section of this paper Grandjean gives more details of his experimental method under the headings;

Precautions to be taken in order to observe droplets on the surface on cleavage planes.

Precautions to be taken in order to observe liquid between cleavage planes.

Then appears a section, as promised above, on the use of the Poincaré sphere to interpret the optical properties of helicoidal stacks of lamellae (as Grandjean himself has stated, this in fact repeats Mauguin's calculations which we have presented earlier in this volume in paper A8)

Application of the geometrical procedure of H. Poincaré to helicoidal layers under constant torsion, but with variable thickness (twisted domains).
Incident vibration polarised along a principal section of the lower face.
Any incident polarisation.

The next section describes how the orientations of the optic axes of drops of liquid crystal aligned on crystal surfaces are measured.

Methods used to measure the orientations

I. Domains having a parallel structure;
II. Domains observed in a cleavage crack (fissure).

The penultimate section of this paper contains the results of Grandjean's observations, given in considerable detail, and illustrated by a number of detailed sketches, of which we will reproduce one here.

Description of the orientations observed

1   *Orpiment* (cleavage plane g^1 [010])
2   *Blende* (cleavage plane b^1 [011])
3   *Phlogopite* (cleavage plane p [001])

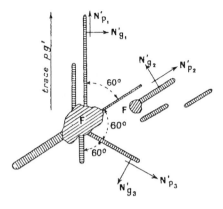

Fig. 17 *Talc* and *ethyl azoxycinnamate*. Formation of linear drops extended along g^1 and along directions at 60° to g^1.

4   *Brucite* (cleavage plane a^1 [**111**])
5   *Talc* (cleavage plane p [001])
6   *Pyrophyllite* (cleavage plane p [001])
7   *Muscovite* (cleavage plane p [001])
8   *Rock salt* (cleavage plane p [001])
9   *Sylvine* (cleavage plane p [001])
10 *Leadhillite* (cleavage plane p [001])

Conclusion

One could increase by many the number of these examples.[1] Nevertheless they are sufficient to establish a general rule:

The crystals orient the anisotropic liquids at their interface when the latter is along a perfect cleavage plane. The orientations most often obtained are simply related to the symmetry of the crystal.

I emphasise particularly the generality of this rule whenever there is a perfect cleavage plane with really smooth surfaces. I do not know of any exception for

---

[1] I have recorded a large number of other alignments but without making such a complete study as the minerals quoted here, and without making such precise measurements. I will return to this subject in a subsequent Note. Nevertheless I believe it is useful to point out even now that one can take advantage of the elongated linear drops given by azoxycinnamate during its first melting. They disclose the orientation even if the crystal is not transparent. For example, with stilbene on the cleavage plane g^{-1}(010) one obtains beautiful linear extensions (of the liquid drop), as perfect as on orpiment, directed along the pg^{-1} edge. Azoxybenzoate also behaves on stilbene as it does on orpiment.

the three liquids tried. Azoxyphenetole, azoxyanisole and anisaldazine, that is to say for thread-like liquids or liquids with nuclei. For focal conic liquids the orientation is usually very good, but occasionally poor or non-existent. In all cases there can be various degrees of alignment. For example the alignment of azoxyphenetole is more perfect on orpiment and rock salt than on muscovite. With liquids of this group the alignment strengths emerge above all through the influence, more or less strong, of perturbations such as those provided by solid crystals of the same material, or by dust particles, or by deformations of the liquid surface.

The focal conic liquids then differ from the thread-like liquids in that they do not always give defined orientations parallel to the cleavage plane. They are distinguished also by their striated (streaked) domains, by their terraced drops, and by their tendency to orient with their optic axis normal to the surface of the walls. Nevertheless, in spite of the very clearly identified different characteristics, the two groups of liquids are very similar in their alignments which are frequently the same. In the examples quoted, azoxybenzoate always aligns exactly as azoxyphenetole, unless it does not align at all.

The orientation of the anisotropic liquid phase on solid crystals, already established by all observers, must be a particular case of the law above. It cannot be an orientation of the bulk liquid in relation to the bulk solid, but must depend on the boundary face of the solid crystal.

These phenomena do not in any way possess the randomness which occurs when one crystal is oriented on top of another of a different type. In that case a coincidence is required, involving a simple numerical relation between some parameters and a simple geometrical relationship between the lattice planes. Here the alignment conditions are of a different nature altogether; they are fulfilled more often than not, although there cannot possibly be a physical or chemical relationship between all the crystals studied and all the anisotropic liquids tried. This is why I attribute great importance to this general character. One is forced to accept that a solid crystalline body bounded by a lattice plane exerts exterior forces, all over the plane and throughout the thickness of the capillary region, which tend, among other things, to align foreign molecules along defined directions. This alignment is not ordinarily visible, because the capillary zone is too thin for the phenomena that occur there to be directly visible to us, but it can become accessible to observation in certain cases. For example with anisotropic liquids: because these liquids orient by themselves.* If a very thin film is oriented, all the rest orients parallel to this layer, provided that there is no other obstacle.

One can denote the collection of forces exerted by the crystal, supposedly limited to a lattice plane, on external molecules as the contact field. The field depends on the nature of these molecules. Amongst the effects it causes, the simplest are those of orientation.

* Not a sentence in the original either! (*ed.*)

One could say that the observed orientations in the presence of the field are the positions of stable equilibrium of the molecule, which corresponds to the minima of the energy. When there is only a single orientation, this is because there is only a single minimum. If there are several orientations equally spread out, it can be concluded that the minima are equal. This will be the case for orientations symmetric with respect to an element of symmetry of the crystal. But two orientations can behave very differently when they are not symmetric. For example it has been shown that the linear elongations of azoxycinnamate adopt very different forms adjacent to pyrophyllite, according to whether they are perpendicular to the elongation axis of the crystal or at $61°^{†}$ to it. The same occurs on rock salt, according to whether they are at 45: or parallel to the side of the square. It is probable that in this case the minima are very different.

[†] A misprint or printing problem? What is so special about 61° as opposed to 60°? (*ed.*)

François Grandjean was born in Lyon in 1882, but was educated in Paris from the age of 10. Already recognised as a brilliant student in high school, he entered the École Polytechnique in 1902 and the École des Mines in 1905. His first academic post was at the École des Mines in St Étienne, where he was assistant to Georges Friedel. Subsequently he transferred to the École des Mines in Paris, where he was first professor of palaeontology and then of mineralogy. In 1937 he was elected to the French Academy of Sciences, occupying the chair previously occupied by Pasteur. During the immediate prewar period he was also elected to the Academy of Agriculture and became Inspector-General of Mines and Director of the French Geological mapping Service.

Grandjean's first scientific contributions were made in the area of liquid crystals. He started to work in this area in 1908. Between 1910 and 1921 he published 20 papers in this field, in which he was able to put his talents in microscopy to good use in elaborating liquid crystalline properties close to surfaces and liquid crystal textures, as well as making some important theoretical advances. Paper C1 is the first successful molecular field theory of liquid crystalline behaviour, but was not widely read, and thus not influential. We owe its rediscovery to Y. Bouligand.

After 1921, failing to receive funding for continuing his liquid crystalline work, Grandjean turned to mineralogy, palaeontology, geology, cartography, and most notably to acarology. This is the study of those minute arachnids (of the order *Acarida*) known as mites, and as such was a perfect subject for microscopy. In 1941 he retired from his professional work to devote himself to acarology. Grandjean is recognised as a giant in the field and is credited with having created the modern subject. All in all Grandjean published 241 papers in this area between 1928 and his death, in Geneva aged 93, in 1975.

Section B

THE INTER-WAR PERIOD: ANISOTROPIC LIQUIDS OR MESOMORPHIC PHASES?

THE INTER-WAR PERIOD:
ANISOTROPIC LIQUIDS OR
MESOMORPHIC PHASES?

Introduction

As we have seen, the newly emergent field of liquid crystals, crystalline liquids, anisotropic liquids – in the absence of convincing physical pictures, there could not yet be a consensus on terminology – had been started in the German-speaking world and leaked out to France following Lehmann's successful visits to Geneva and Paris in 1909. There was a continuing vibrant interchange of ideas, samples and even visits between Germany and France in the years leading up to the First World War. The Great War, however, temporarily interrupted free scientific interchange, although not by as much as was the case during the Second World War. Thus the important theoretical papers of Born and Grandjean did not seem to penetrate the front line. Although most of the French researchers in the liquid crystal field were allowed to continue their work uninterrupted, their German colleagues were unable to escape war duties. In 1914 Lehmann, at 59, was too old for military service, but Vorländer, already 47, dropped his scientific work in order to become a battery commander on the Eastern and Western Fronts.

With the end of the war, serious research could begin again. The big puzzle remained as to the nature of liquid crystals. Let us summarise the situation as it appeared in 1920. Some substances which were liquid-like from a hydrodynamic point of view (i.e. they flowed), nevertheless seemed crystalline from a crystallographic and optic point of view. Roughly speaking, in the context of *viscosity*, the anomalous materials could be divided into two classes, which had been denoted by Lehmann as flowing (or slimy liquid) crystals on the one hand (with high viscosity), and drop-like liquid crystals on the other (with lower viscosity), with the term 'liquid crystals' serving as an overriding classification. The crystallinity stared out at the observer from the stage of the polarising microscope. Nevertheless, because of their flow properties, some workers had been unwilling to accept these materials as 'crystals', and described them as anisotropic or crystalline liquids. A few stragglers, of whom Tammann remained the most vociferous, continued to reject the observations, still insisting as late as 1922 that the so-called liquid crystals were simply colloidal mixtures of some sort.

Under the polarising microscope, a whole slew of multi-coloured patterns had been observed. The flowing crystals in general seemed to look different from the liquid crystals. The liquid crystals exhibited the Schlieren texture, with bands of bright and dark crossing at special points, named *Kernpunkten* (hard points) in German, with the whole samples labelled by Georges Friedel as *liquides à noyaux* (liquids with cores) in French. Sometimes, when observed from a different aspect, these liquids also displayed long lines threading through them, in which case Friedel names them *liquides à fils* (liquids with threads), and sometimes the threads could be seen running into the cores. Contrast this with the flowing crystals, which exhibited under the microscope peculiar patterns resembling slices of cones, identified by mathematicians as *focal conics*, which Friedel had named *liquides à coniques*. These substances also had been found by Grandjean to exhibit some layering phenomena in drops on surfaces.

Finally there was a class of liquid crystals which rotated the plane of polarisation of light. These materials exhibited some aspects of both the flowing and the ordinary liquid crystals, and the rotatory power could be tuned by mixing materials.

The stage was set for a systematic attempt at classifying liquid crystal properties, and this was the task set for himself by Georges Friedel, at that time newly translated from his Mining School in St Étienne to Strasbourg in the newly reconquered Alsace. Friedel, as an Alsatian by birth, was keen to be a pioneer in this regard, not least because he was able to reoccupy his ancestral family home, from which he had been banished in 1911 following a dispute with what he regarded as the occupying German authorities. In January 1922 he sent a student to Karlsruhe to bring back some samples from Lehmann, and on the 23rd of that month, before even the student had returned, he sent a polite note to Lehmann to thank him for his cooperation.

Otto Lehmann died, unexpectedly, on 17 June 1922. Friedel's review article was published in the *Annales de Physique* in November 1922. We reprint translated extracts as article B1 in our collection. As we shall see below, it was perhaps as well, given the tone of this article, that Lehmann died before Friedel's article appeared. This paper is probably the most influential article on liquid crystals which has appeared, before or since. Even to refer to it as an article fails to do it justice, for in fact it is 201 pages long, and had it not appeared in a review journal, would more merit the appellation monograph than simple review.

Friedel's review marks the coming of age of liquid crystal science; with it, studies of liquid crystals pass from a collection of disparate observations to a unified body of knowledge. So influential has it been that it has acted as a screen between the early years and more modern approaches to the subject. One reason that it has acted as a screen is that it contains remarkably few references; much of the reference to earlier work is implicit rather than explicit, and Friedel is not as generous as perhaps he should have been in recognising the contributions of others. The result is that it has been difficult to follow the subject back directly from this article. Be that as it may, in later years mere citation of this work has served as a sign that a writer is knowledgeable about the roots of the field. For several years, in fact, following the renascence of interest in liquid crystals in the 1950s and 1960s which we come to in Section C of this collection, the Friedel review was *the* most widely cited paper in the French language scientific literature.

The Friedel paper is unfortunately too long to reprint in its entirety. We nevertheless recommend serious students of liquid crystals and the history of science to follow up the extracts we have chosen to make the effort to read the whole paper, preferably in the original. There are *samizdat* translations (one for example from the US army in the 1960s) which suffice for basic understanding, but which fail to do justice to Friedel's magnificent use of language. As translators we have endeavoured here to maintain the spirit, as well as the meaning, of Friedel's tome.

As an aside, it is worth remarking that Friedel's life was not always easy. He lost his mother when he was still a very small child. He carried the burden of the success of the family Friedel (described in more detail in his biography after the article), and following the Franco-Prussian war he was at least alienated from his family seat in Alsace. As a professional he spent his life in the provinces, in St Étienne rather than in Paris, the result of which was that despite his impressive family pedigree, he was an oppositional figure, not fully part of the scientific establishment. He was never, for example, elected to the Académie des Sciences. Whether as cause or result, he seems to have nursed a kind of private bitterness, in that he did not suffer fools gladly. The article demonstrates this in its vivid language, as we shall see; there is extra evidence from parenthetical written remarks in his archives, and his close colleague Grandjean hints as much in his extremely affectionate 1933 obituary.

We also draw the reader's attention to the reprinted title page of article B1. The article is often wrongly cited. Often the incorrect citation is simply the result of the enthusiasm of the naïve, for of this article more than others it can truly be said (at least nowadays!) that it is often cited but seldom read. In addition, notwithstanding appearances, there was no M.G. Friedel, only a Monsieur Georges Friedel. *Caveat lector*!

We now pass to a brief discussion of the article itself. The very title of the article – *The mesomorphic states of matter* – indicates the change of paradigm that Friedel was advocating. Not for Friedel a gentle opening summarising the present state of play, followed by mild suggestions for improvement. Friedel launches straight into his subject matter, guns blazing, metaphorically speaking. In the very first paragraph he tells us that liquid crystals were wrongly named by Lehmann, and this incorrect terminology had hindered the development of the subject. He, Friedel, will introduce new more appropriate terminology! Furthermore, not only did Lehmann err in denoting these substances as liquid crystals, but all the German scientists were confused...He continues in like vein.

As can be seen in the translated text, he reserves particular scorn for Lehmann's explanation of the mechanism for liquid crystal formation. Lehmann had posited a molecular directional force (*molekulare Richtkraft*) or structural force (*Gestaltungskraft*). Friedel dismisses this is as a *singulière logomachie* – mere mysticism, accusing Lehmann and colleagues of invoking a mysterious German divinity! It is the lack of reference to this exaggerated rhetoric which convinces this writer that the article has been cited by many who have not read it. The natural embarrassment of the scientist when faced with ostentatious display of emotion must be surely balanced, at least in this case, by a sense of exhilaration, as one watches the penmanship of the master cutting up lesser opponents.

The crucial thesis of Friedel was that the important feature of the so-called liquid crystals was not their degree of fluidity, but rather their molecular structure. With the help of his daughter Marie, a classicist, he invented the terms *nematic* (from the Greek word νημα = nema = thread), *smectic* (Greek σμηγμα = smegma = soap) and *cholesteric* (for many of the chiral materials, including Reinitzer's original liquid crystals were cholesterol derivatives). The ensemble was neither some peculiar crystal, nor yet a peculiar liquid, nor even a system combining the properties of both, which as we shall see below, became a point of disagreement with other workers. Rather (*and the Germans had missed this*!) one was considering a completely new state of matter, which he denoted *mesomorphic*, from the Greek words μεσος (mesos = intermediate or middle) and μορφη (morphe = shape or form). The state was mesomorphic because it was intermediate between the solid and the liquid phase. In fact Friedel did not use the term *phase*, but employed rather *stase*. The *stase* terminology, as we shall see, excited not a little controversy of its own, and was used as late as 1960 by Luzzati and colleagues in paper E5, but has not entered the canon. Friedel felt it necessary to introduce new terminology of his own here, because of the usage of phase in the context of *solid, liquid* or *gas*. Only the classification of phase transitions introduced in the 1930s by the Ehrenfests allowed the identification of states of matter, phases and *stases*.

The nematics could be identified with the drop-forming liquid crystals, his own *liquides à fils*, a term which he now dropped. He discusses at great length what is meant by a crystal, and the relationship between microscopic periodicity and the existence of macroscopic facets. This is a prelude to noting the *lack* of evidence for periodicity in *liquides à fils*. Thus nematics were liquid-like phases in which rod-like molecules were aligned. In fact, this much had been realised by Grandjean in his Forgotten Theory Paper (C1) of 1917. Beyond this Friedel also discusses the structure of what we now recognise as defects, as well as their origin, and makes some remarks as to how defects combine. He seems also to have understood that the nematic turbidity was not an essential property, but rather the consequence of fluctuation, for he notes that a magnetic field can quench the fluctuations and restore the transparency of the material.

Turning to the smectic phases, he was the first to realise, albeit using circumstantial evidence, that these were layered, and he went so far as to suggest that this hypothesis could be tested using X-ray diffraction. What he denoted as the cholesteric phase had hitherto been confusing, possessing some features of *liquides à fils* and some of *liquides à coniques*. Friedel realised that this was

142

a chiral nematic phase, and consequently exhibited some features of layering. There is also a sophisticated discussion of the expected optics of cholesterics, and how this was consistent with the observations, as well as a description of helix inversion in a mixture, passing through the infinite pitch state which was functionally equivalent to a nematic.

Although his article is wide-ranging, Friedel does not claim to have solved all problems, and with the benefit of hindsight, we do find some, though not many, substantive errors. Thus he seems not have realised that nematics could transmit torques, nor that there were several smectic phases. In this latter case he is inclined to dismiss observations of others hinting of what we now call polymesomorphism. He is aware that a set of confocal ellipses and hyperbolae are the focal lines of a family of parallel surfaces. Indeed he uses this, together with his observed focal conics, as a piece of evidence in building up his picture of smectics as layered materials. Nevertheless, he claims in this article to be mystified by the 'peculiar tendency of the smectic liquid to form Dupin cyclides around confocal conics'. It seems to have been the particular existence of the Dupin cyclides which puzzled him, for his general level of understanding seems to have been deep.

To sum up, the impact of this article was profound, and it indeed led to change in world-view and language in the description of liquid crystal science. Meso-morphic phases, nematics, smectics and cholesterics were soon to grace the pages of physics and chemistry journals around the world. Only one battle did Friedel lose comprehensively, and that concerned the term 'liquid crystal' itself. Intellectually his battle was won; liquid crystals were not crystals at all, but peculiar liquids with some hint of solid properties. But the epithet was vivid, and the terminology already widespread. Georges Friedel's righteous hostility to Germans and liquid crystals was insufficient to shift scientific usage of the term. Liquid crystals were here to stay.

As we have seen in Section A, the key experiments confirming the existence of solid crystal lattices were the X-ray diffraction experiments of the Braggs and of von Laue and colleagues. Friedel's article had pointed out that direct confirmation of the existence of smectic layers required an analogous experiment. And indeed, even as his article was appearing, his son Edmond was already collaborating with Maurice de Broglie in an X-ray experiment designed to do just this. The resulting paper was published on 12 March, 1923 in the *Comptes rendus de l'Académie des Sciences*, and we include this as article B2 in our collection. This article, unsurprisingly, uses Georges Friedel's mesomorphic terminology as though it were standard (which it soon would be!).

We may recall that X-ray scattering from a single crystal gives rise to spots in scattering directions related to the lattice parameters of the crystal. A powdered solid redistributes these spots into circles. Likewise, a disordered smectic would be expected to yield a circular diffraction pattern, with a simpler distribution of radii than in a solid. The authors remark that Hückel[1] had carried out the analogous experiment for nematics but had failed to discern any evidence of periodicity. Here, by contrast, the smectic signature rings out loud and clear.

From now on the pace of liquid crystal research quickens. The choice of papers in our collection is now for the first time influenced by language; for the first time there is an accessible relevant literature in English. In a number of cases we (like others) have chosen the easy route, by which we mean the English language version of a paper when equivalents have appeared slightly earlier in German or Russian. It is not quite historically fair, but we have tried nevertheless to give credit where credit is due.

The narrative is further complicated by the fact that there are several contemporaneous subplots. One subplot is the antagonistic development of Bose's swarm theory, in particular by Ornstein and Kast, and of the competing distortion theory by Oseen and later Zocher. Another subplot concerns the discovery by Frederiks* in the Soviet Union of a threshold effect for the alignment of liquid crystals in thin cells subject to a magnetic field. At the same time the long-lasting rivalry of the French and German schools over terminology and intellectual priority continues to simmer. And finally, the realisation of the existence of a scientific field with reams of open questions draws a number of major figures to organise major symposia to explore the important issues.

Paper B3 is the first paper in English by the ultimately tragic figure of Vsevolod Konstantinovich Frederiks, and in it his eponymous effect is described. Let us first remind the reader of the physics of the Frederiks effect. Fields (either electric or magnetic) align liquid crystals. Often they align parallel to the field (*positive* dielectric or diamagnetic anisotropy). So do properly prepared surfaces. Let us suppose competing effects in a thin cell. The field tends to align the liquid crystal (let us say) perpendicular to the walls, and the surfaces are prepared so as to align the molecules along the same (arbitrary) easy axis in the plane of the cell walls. As the field is increased, to begin with nothing happens, but then at some *threshold* field, the molecules in the liquid crystal begin rapidly to realign, and at only slightly higher fields than threshold the system is almost entirely realigned according to the whim of the field.

The existence of a threshold is central to the operation of modern display devices, and it is this fact which forces this effect into the early part of any elementary liquid crystal text. But by itself the effect has little obvious technological significance, although it is of considerable interest as an exemplar of the effects of competing bulk and surface fields. Nowadays the imposed field is

---

* The literature contains a number of spellings of this gentleman's name. In Russian he was Всеволод Константинович **Фредерикс** and there is no ambiguity. The transliteration to the German literature in 1930s used Fréedericksz (sometimes without the accent) – and seems to have been the original spelling of the Swedish family name. Even so, it seems peculiar given the actual pronunciation. It was also Frederiks's own spelling of his name during the time he spent outside Russia before 1919, and we must presumably give him the benefit of the doubt. Even in the English language literature of the period this transliteration carried the day, but the more-or-less phonetic transliteration which we have used is more usual nowadays. We revert to the early spelling when referring specifically to events which occurred at that time.

usually an alternating electric field (constant voltages interact with low ionic concentrations to complicate the physics), because the coupling of electric fields with liquid crystal orientations is much stronger than the analogous magnetic effect. And nowadays too, there is much more experience in the surface alignment procedure, a problem we shall return to in papers B11 and B12. But in the 1920s the complications of applying an electric field overwhelmed experimentalists, and bespoke surface preparation was almost non-existent. Thus it was magnetic fields that were the experimental probe of choice.

We may recall that in 1926 Frederiks was working in Leningrad in the then seven-year-old Soviet Union, and was in the process of building up a research group. Frederiks's initial junior co-worker in the liquid crystal field was Alexandra Nikolaevna Repiova,[#] and Frederiks was able to use his German contacts to obtain anisotropic liquid samples from Daniel Vorländer in Halle.

Their original paradigm was Born's theory of the dipolar origin of anisotropic liquids. They reported some initial results at the Congress of the Russian Physical Society, held in Moscow in December 1926. A longer paper was published in the physics section of the Journal of the Russian Society of Physics and Chemistry the following year, entitled 'On the problem of the nature of the anisotropic liquid state of matter'.[2] Here we find for the first time, amongst other results, what was to become the Frederiks effect. It is interesting to note that Frederiks never refers to liquid crystals, nor even crystalline liquids, despite his Halle contacts. Already in his 1927 Russian paper, he is regularly using the term 'nematic'. As was the custom, a substantively similar paper was published in German in the *Zeitschrift für Physik*.[3]

The experimental set-up involves liquid crystal trapped between a lens and a flat plate, subject to a magnetic field in the plane of the flat plate which tend to realign the liquid crystal molecules away from the direction normal to the flat plate. The sample is placed between crossed nicols. The basic result is that outside a critical radius, fringes are observed; this corresponds to the region in which the field succeeds in competing with the surfaces. Inside the critical radius, the sample looks black. This is where the surface overwhelms the field.

Thus the critical field, found in a more modern context by scanning through field strengths at constant thickness, was originally observed in inverse fashion as a *critical film thickness* at constant field, and thus clearly visible in a sample of variable thickness. Frederiks and Repiova did change the field as well, and estimated that the critical thickness decreased with field according to a power law of $z \propto H^{-1.75}$. Further work by van Wyk[4] in Utrecht, a student of Ornstein, published in 1929, found similar behaviour but with a different power law, now with $z \propto H^{-0.93}$.

This work was sufficiently noteworthy to earn Frederiks an invitation to the 55th annual meeting of the American Electrochemical Society, held in Toronto in

---

[#] c.f. our previous note on the transliteration of Russian names. This is Репиёва in Russian, and we have transliterated phonetically what is Repiewa in the contemporary literature.

May 1929. One imagines that it was the German rather than the Russian version which had attracted the attention of the American electrochemists. By now Repiova's place in this project had been taken by a new graduate student, Valentina Vasilevna Zolina, who carried out further experiments in an attic room just beneath the roof of the Ioffe Institute, increasing the maximum magnetic fields eventually up to 25 kG. The political situation was such that it was impossible for Frederiks to obtain permission to leave the Soviet Union (from his biography we see that things were to get worse), but nevertheless he was able in March to submit his manuscript to be read at the meeting. It is this paper which we include as article B3.

The interpretation of these results was more problematic. Frederiks was initially inclined to regard them as confirmation of the swarm theory; we shall return to this in article B6. Soon after they appeared, an alternative explanation in terms of distortion theory was suggested by Zocher and coworkers (article B7). Eventually around 1934, following a good deal of debate in the liquid crystal community, Frederiks transferred his allegiance from the former to the latter school of thought.

The increase in interest in anisotropic fluids led the eminent crystallographer P.P. Ewald to organise on the subject what we might now call a virtual symposium. The papers were circulated and recirculated, thus including an extensive discussion section, during 1929, and the final result published in the *Zeitschrift für Kristallographie* in 1931. Were it not for the shift in scientific *lingua franca* from German to English in the period between then and now, we should surely have reproduced several papers from this volume. As it is, we nevertheless include extracts from the General Discussion section, as well as reproducing the contents page, in paper B4.

This General Discussion section is extremely long, lasting from page 269 to page 347, and clearly we are in a position to translate only a few highlights. An introductory survey is provided by Rudolf Schenck. It is many years since Schenck worked in the field, but he retains an affection for liquid crystals. We left Schenck in Karlsruhe in June 1905. He had just presented a talk which comprehensively undercut the foundations of Gustav Tammann's emulsion picture of liquid crystals, and subject to the consequent wrath of Tammann himself. We may recall that van't Hoff has intervened from the chair to call for a commission of experts to examine liquid crystal questions and to resolve the raging controversies.

In his introduction, Schenck recalls the drama of the Karlsruhe meeting and further tells us how the commission fared. Sadly, as we might ourselves have predicted, the answer is not so well. Schenck seems to have been the secretary to this committee, for he prepared reports annually over the period 1906–8 for the committee chair, based on work occurring both in Germany and abroad. The final meeting of the committee was set to take place at the 1909 annual meeting of the *Deutsche Bunsen-Gesellschaft*, to be held in Aachen. However, apparently some misunderstanding broke out between van't Hoff and Lehmann, for their correspondence broke down and the final meeting never took place. Van't Hoff promptly resigned as chair, and the work of the commission came to a premature end. What was to be done with all the accumulated work? At the suggestion of Johannes Stark, Schenck collected his reports together and published them (not from our standpoint using an obvious journal!) in the 1909 edition of the *Jahrbuch der Radioaktivität und Elektronik*.[5]

In the General Discussion proper, an important issue concerned the nature of Lehmann's *molekularen Richtkräfte*, the forces which Georges Friedel had so contemptuously dismissed. Chemists, such as Vorländer, interpreted these in terms of the chemical forces which lead to compound formation, whereas physicists sought an electromagnetic origin. There was extensive somewhat heated debate about swarm theory, including debate between Zocher (*contra*) and Ornstein (*pro*). Friedel's use of the term *stase* also comes under scrutiny. Finally the debate about terminology, overlain as it was with emotional significance, continued to drag on.

We can reasonably assume that Otto Lehmann himself would not have been best pleased with the Friedel classification. Vorländer and Lehmann themselves were at times bitter rivals, as the following footnote to a 1923[26] paper shows:

> I cannot agree with O. Lehmann's interpretation of the so-called liquid crystal characteristics of ammonium oleate hydrate and similar soaps. These materials, when melted with water, are said to yield liquid crystals. What Lehmann really observed, photographed and then described, at least in part, were probably suspensions of swollen soft birefringent phases in water. In my experience, liquid crystalline phases of fatty acid salts are only formed at high temperatures, in the molten phase of the (incompletely decomposed) anhydrous salts. Lehmann repeatedly referred to his discovery, mainly in order to establish his priority with respect to Reinitzer.

Luckily, by this time Lehmann had already been dead for two years and Vorländer was not called upon to defend his remarks against an irate Lehmann (and we know that Lehmann was no shrinking violet!). Vorländer was also offended by Friedel's new 'mesomorphic' classification. Vorländer had his own classification,[6] dating all the way back to 1907. In this classification, he used the term *pl-phases* for the higher melting, low viscous liquid crystalline phases, derived from the first and last letter of the German word of Gattermann's *p*-azoxyaniso*l*. The lower melting liquid crystalline phase he called the *Bz-phases*, derived from the letters *B* and *z* of the German word *Benzoesäure* (benzoic acid) because he observed these phase in the case of benzoate esters. He saw no need for new nomenclature, as is witnessed in the following sarcastic extract:[7]

> If anybody beginning to work in this field wants to introduce new terms – please, do it with great enjoyment.

Vorländer's contribution to the Ewald symposium was a forceful restatement of his own position, both academically and as a pioneer in the field. By the time he had received copies of the papers and it was time to contribute to the Discussion, his blood-pressure was reaching seriously elevated levels, and what his contribution lacked in formal rigour, it gained in richness of language. His words were clearly carefully chosen, and add much to our historical understanding of what it felt like to be a liquid crystal scientist during this period. Unfortunately their emotional content meant that they had less of a persuasive effect on his colleagues. Thus, in spite of Vorländer's high prestige in the field, the terms *pl-* and *bz-* were relatively rapidly replaced – within a decade or so – by Friedel's new terminology. Nevertheless, we do find Vorländer's terminology employed by his own former student Conrad Weygand in 1939,[8] and even as late as 1955 in a review article by Wilhelm Kast.[9]

His criticisms of the swarm model favoured by Ornstein were matched only by his excoriation of the Friedels, father and son, for their new crystalline liquid nomenclature. Not only did he feel that the data were not well described, but he was also seriously offended (with good reason) by the omission of any reference to his valuable work. We can get a feeling for Vorländer's strong feelings from the following short extracts

> There are experiments for recognising correct and incorrect names for scientific phenomena. A name is correct when it is based on fact and experimental observation, but incorrect if it has at its core theories or hypotheses....

> Mesomorphism, mesophases, or intermediate phases for crystalline-liquid phases, are *incorrect* names because there is not a single fact or phenomenon to prove that the liquid crystals stand between solid crystals and amorphous melts. Give me a *single* property of crystalline liquids that indicates their intermediate position as crucial to the phenomenon. *I know of no such property*! Everywhere, even in X-ray diffraction: sharp, discontinuous change; no intermediate position. In contrast, one could cite more than a *dozen* properties that prove that on the one hand the liquid crystals behave just like solid crystals, and on the other, just like amorphous oils. In other words, hermaphroditism! At best, one could introduce, instead of crystalline liquids, the term crystal-like, crystalloid liquids, or liquid crystalloids, but that does not constitute essential progress.

> I reject the words mesomorphic states and mesophases as a designation of crystalline-liquid properties or phases. I consider the expressions mesomorphism and mesophases, as well as the words nematic and smectic, *completely misguided*, **even if the theory which led to the words at some later time turns out to be correct**.

The italics are ours, but the drift of his remarks is clear! Notwithstanding these problems, always faced by practitioners in a rapidly developing interdisciplinary

148

field, it is important to reiterate the debt which the liquid crystal community continues to owe to Daniel Vorländer, not only for his work during the pioneer era, but also for his continuing work in the inter-war period. Many of his intuitions did not bear fruit during his lifetime. For example, in paper in 1932,[10] apparently in contradiction to much of his earlier work on molecular shape, he extended his conception to systems with *bent* molecular shape, e.g. derivatives of resorcinol (*m*-disubstitution) and catechol (o-disubstitution),[11,§] writing:

> From these facts we may derive some essential results concerning the structure of crystalline liquids. From the crystalline-liquid properties of the *angled* catechol and resorcinol derivatives, one is forced to conclude that, so long as the lateral sides of the angles are sufficiently linearly efficient for liquid crystallinity to occur – as in the present case (caused by) the long *p*-phenobenzoyl groups – the bending by itself in no way completely prevents the crystalline-liquid properties.

By the time of his retirement in 1937, more than 2000 liquid crystalline compounds had been synthesised in his Institute, some of which were used two decades later by Horst Sackmann to work out his system of polymorph smectics, as we shall see in paper C5. A collection of Vorländer's compounds in sealed glass tubes and kept in attractive cigar boxes can still be inspected in the Institute of Physical Chemistry of the Martin-Luther-University in Halle (see Fig. B1).

Other themes of the Symposium were no less controversial. The dominant theoretical framework was the swarm theory due to Ornstein and collaborators. We last met the swarm theory in an incomplete state as constructed by the unfortunate Emil Bose in 1908. We include extracts from the debate in the Discussion section as to whether Ornstein's swarm theory is the same as that of Bose. Ornstein drew a distinction between his swarms and Bose's, averring that, unlike in Bose's theory, there is no element of emulsion included in his swarm picture. As the memory of the argument between Tammann and Lehmann was still alive, and the emulsion picture of liquid crystals known to be Officially Incorrect, it was important to establish theoretical intellectual credentials by denying any emulsion content.

However, Ornstein's denials attracted some scepticism from a number of workers, most notably K. Herrmann, who insisted that the two swarm theories *were* the same, that there had therefore to be some degree of isotropy between the swarms, and that anyway the picture was not supported by X-ray data. Behind the swarm theory lay a belief that there had to be what we would call a mean-field

---

§ Vorländer was not to clarify the reason for the liquid crystal behaviour of his bent molecules. However, very recently workers in the Halle group descended in a direct intellectual line from Vorländer found evidence that actually he had been the first to prepare a *banana phase (B$_5$)* in his resorcine derivatives.[11]

Fig. B1 One of the cigar boxes in which Daniel Vorländer used to keep his collection of crystalline liquids. See Plate II (courtesy of G Pelzl).

theory which would provide a more solid foundation for the swarms, and that this theory would be the analogue of the Langevin theory of magnetism. We now know, of course, that that theory had already been constructed by Grandjean (article C1), that it had been missed by the liquid crystal community, but that unfortunately it provides no further justification for the swarms.

Zocher was equally unimpressed by the swarm theory, but his objection was more theoretical, for he and Oseen had an alternative theoretical viewpoint, which we shall discuss in more detail when we look at the 1933 Faraday Symposium. There was a vigorous exchange between Ornstein and Zocher on the swarm theory, with Ornstein providing a list of experimental data justifying the swarm picture, and Zocher retorting, roughly speaking, that consistency and justification were different matters. Zocher's problem was that his alternative picture was unable to provide quantitative calculations, a point used to effect in Ornstein's rhetoric. Interestingly Oseen, who might have been in a position to do so, tended to confine his interventions to technical matters, and avoided explicit debate on grand theoretical questions.

The contemporary liquid crystal practitioner should not be too disdainful of this ultimately unsuccessful theory. Ornstein was driven to the swarm theory from the success of his theory with Fritz Zernike on the nature of the critical point. He was one of the most successful theoretical physicists of the twentieth century. He had success galore elsewhere to compensate his failure in this problem. What he did not have was the accumulated experience of the behaviour of systems with slowly varying order parameters to warn him off this approach; indeed he did not yet have the concept of order parameter itself.

There was also considerable debate on the various terms associated with what we now call liquid crystal phases. The Friedels were using the terms *stase*, *phase* and *texture*.[12] What was the difference? Without a detailed knowledge of the internal structure of the liquid crystals (or whatever one might call them), it was difficult to answer these questions fully. Could one draw a distinction? Zocher was unconvinced that a stase and a phase were distinct objects, but Ewald was more persuaded. The Gibbs Phase Rule, which gives information as to the number of possible coexisting phases was invoked by Zocher to render less plausible the idea of a stase. Ewald, by contrast, was seeking some intuitions from geometry as well as from thermodynamics. With the benefit of hindsight we can see that symmetry is indeed very important, and this was a fruitful avenue to follow, although the resolution of the problems lay simply in understanding general problems of statistical mechanics better.

Oseen noted that the molecular organisation in the region of apparent singularities would be different from that elsewhere in the system. The Friedels agree, but don't know what it would be like.[13] In the same way, the Friedels are willing to consider the possibility of the existence of more than just the smectic and nematic phases, but are unconvinced by the experimental evidence. Vorländer has seen more than two phases, and has promised the Friedels to send them a sample, but for some unknown reason, despite the intervention of Ewald as interlocutor, it never arrives.

After the Discussion, Ewald once again sends out all the manuscripts so that some final remarks may be made. At this stage the Friedels realise that they have never seen the intemperate contribution from Vorländer, and have not yet an opportunity to respond. Ewald clearly was a wise man, for by the time the offending article is finally transmitted, their passions have died down, and barely taking the time for a gesture of contempt, they slide it by and on to the next subject. What is striking from the Friedels' final contribution is the articulate plea they make for more experiments, drawing attention to controversy on theoretical questions which they regard as unimportant, and which can in any case as yet only be inadequately resolved on the basis of current data.

Up to 1930, most work in liquid crystals had been carried out in Germany and France, with some contributions from groups in Russia, Sweden and the Netherlands. However, Ewald's virtual meeting was to have an unanticipated influence on broadening interest in the subject. The proceedings of that meeting, as we have seen in paper B4 above, were published, principally in German, but with contributions in French from the Francophone contributors in the *Zeitschrift für Kristallographie*. Not in English, however, although, as it happened, the English crystallographer Sir William Bragg[+] was an Editorial Board member of the journal.

Perhaps struck by the fact that Anglo-Saxon contributions to liquid crystals had been less than crucial, Sir William resolved not to be left behind. The best way to catch up is to arrange a *meeting*, to which the leaders in the field can be invited. So it was that in May 1933 Sir William Bragg (1862–1942), Director of the Royal Institution, and John Desmond Bernal (1900–71), then a rising young star of the crystallographic world, organised a Discussion Meeting of the Faraday Society[§] in London on *Liquid Crystals and Anisotropic Melts*.

All the major figures in the liquid crystal world were invited (although at least one died before he was able to come, and others, for a variety of reasons, did not show up). The list of invited speakers for the Faraday Discussion was almost a replica of the contributors to the Ewald virtual meeting, though for some reason Ewald himself did not attend. The topics of the papers presented (this time in many cases in person) at the London meeting similarly echo the pages from the *Zeitschrift für Kristallographie* published only two years previously.

Repetition notwithstanding, the record of that Faraday Discussion is the first major document in English on liquid crystals. As we have already emphasised, the peculiar status of English, as the modern *lingua franca* of the scientific world, has conferred on the papers read at the 1933 London meeting a greater historical importance than might otherwise have been the case. The organisers of the meeting went to great trouble to ensure that papers originally written in other languages were translated so as to be accessible to the Anglophone audience. The meeting was fittingly held at the Royal Institution in Albermarle Street in Central London, the academic home of those ancient sages Sir Humphrey Davy (1778–1829) and Michael Faraday (1791–1867).

One hundred and fifty members and visitors attended that meeting, which is a rather impressive number, given the relative novelty of the field. Among eminent foreign visitors at the meeting, the proceedings record, were Professors Leonard Ornstein of Utrecht and Hans Zocher of Prague, and Professor Dr Rudolf Schenck, now of Berlin.[*] Schenck was a particularly important visitor, not only because of his seminal contribution to the field (in which he was no longer an active participant) but because he was the current President of the *Deutsche Bunsen-Gesellschaft*. Not present, except in spirit – for papers were presented on their behalf – were the distinguished theorist Professor Carl Wilhelm Oseen of the University of Uppsala in Sweden, and the equally

[+] This is the elder Bragg, W.H., to be distinguished from the younger (his son), W.L., later Sir Lawrence (1890–1971).

[§] The Faraday Society much later became the Physical Chemistry and Chemical Physics section of the Royal Society of Chemistry.

[*] The proceedings record this, but were wrong. Schenck was never a professor in Berlin, and at this point worked at the University of Münster.

distinguished experimental physicist Professor V.K. Freedericksz of the Physico-Technical Institute of Leningrad, who had once again been refused an exit visa by the Soviet authorities.

Dominating the meeting (and, reading between the lines, also the Organising Committee) was the ebullient figure of Bernal. An Irish-born polymath, Bernal was to become a dominant figure in twentieth century crystallography, known as well for his left-wing views, his womanising and his contributions to the history of science. So learned was Bernal, that he was known, at least to his co-workers and acolytes, as 'Sage'. Bernal's own interests were so wide that he was never able to win the Nobel prize in his own right, but several of his students did win the Nobel prize for ground-breaking work in the resolution of the structure of molecules important in molecular biology. In the discussion it is always Bernal who is first on his feet with some point or other.

The first paper in the Discussion proceedings is by Oseen, and we include this in our collection as article B5. In fact in these discussions the papers are circulated beforehand and taken as read at the meeting, with perhaps five minutes of introduction from the author. As a result Oseen's physical absence at the meeting would not have been a major inconvenience. In 1933 Oseen had been working on anisotropic fluids for 12 years, and had already published 20 papers and he had also written a monograph[14] on the subject published in 1929. All in all he was to publish 26 papers on the topic.

His first three papers were entitled *Versuch einer kinetischen Theorie der anisotropen Flüssigkeiten* – Essay on a kinetic theory of crystalline fluids – and apart from this one, all the others were entitled, in one language or another, 'Contributions to theory of anisotropic fluids'. In 1934 he switched from German to French as a protest against the rise of Nazism in Germany. Paper B5 is his only paper in English, and judging from the published proceedings (in which translators were acknowledged on the title page), he nevertheless wrote it himself with presumably only minor editorial help. So by May 1933 his ideas were somewhat mature. It is worth retreading our steps back to 1923 to follow some of the early steps in Oseen's thinking.

Much, though not all, of Oseen's work was in theoretical hydrodynamics, and it is within this perspective that he approached the theory of anisotropic fluids. His first efforts were directed towards a hydrostatic theory, and we quote from the introduction to his third article in 1923:[15]

The first question which strikes a theoretician in this field is the origin of the two forms in which anisotropic liquid drops appear: the sphere-forming "liquid crystals", and the "flowing" crystals which also appear crystalline. Lehmann gave the answer to this question, which was that in liquid crystals the structural force is insufficient to overcome the surface tension. That this explanation is valid is hardly open to doubt. The task of the theoretician is to define this concept of structural force, and connect it to the usual molecular forces....

153

There follow almost 40 pages of almost impenetrable algebra, before Oseen emerges again with something we can recognise . . . .

The number of independent constants is therefore four.

Difficult as the working is, we are already seeing the four elastic constants of what has come to be called the continuum theory of liquid crystals, which are now traditionally labelled K_{11}, K_{22}, K_{33} and K_{24}. But let us compare Oseen's Swedish tone to that exuded almost simultaneously by Friedel in Strasbourg. Whereas the fierce Friedel castigated the structural force in forceful terms as essentially meaningless, the milder Oseen merely noted that some further progress was necessary in order to connect this idea with more normal physics!

By 1933 his formulation of the elastic energy had settled down, although it is not yet quite in its modern form. The detailed form of the terms is probably equivalent but differently stated, and the conventions are of course different from today. We may note also that the paper is focussed in a rather modern way: in successive sections Oseen deals with statics, defect structure, optics, dynamics and finally smectics. The modern eye is drawn to the use of the term 'aeolotropic' (from the Greek αιολος = 'changeful'), which is now no longer used in a liquid crystal context. There are speculations about the molecular origin of the 'cholesterine-nematic' (cholesteric) phase.

In the dynamics section he presents new work by his Ph.D student Adolf Anzelius (1894–1979). That work was not entirely successful, at least partly because it does not use the time derivative of the director as a dynamical variable in the theory. This is to some extent the result of a lack of careful experiments in the literature at the time to direct the theorists. The deficiency is made up later by Mięsowicz (articles B9 and B10) and by Tsvetkov[16] from the Leningrad school, whom we shall come across again in article C2. Anzelius's theory would nevertheless be the starting point for the successful 1968 Ericksen-Leslie theory, which we shall come across later as article C6.

What is interesting is that, despite his technical brilliance, Oseen seems to have had some difficulty with the basic physics of anisotropic liquids (we can perhaps sympathise!). He seems, for example, to have believed that anisotropic fluids no longer obeyed Newton's laws of motion. From the perspective of 2002, what is missing is a well-defined scale separation between continuum theories on the one hand and molecular theories on the other.

Article B6, entitled 'New arguments for the swarm theory of liquid crystals' is by Ornstein and Kast. This article is included because it represents the eventually unsuccessful swarm theory at the height of its influence. The swarm theory, we underline, was a brave attempt to formalise physical intuition into a viable calculational scheme. Ornstein had developed the swarm theory beyond the rather limited picture of Bose, and his swarms were much more concrete than those of Bose. He really believed that inside liquid crystals there existed macroscopic entities which could be aligned by magnetic fields, and which could be regarded as 'physical' – as opposed to the usual chemical – molecules. In B6 he and Kast marshal a whole slew of experimental evidence in support of the swarm theory.

Article B7, by Hans Zocher, entitled *The effect of a magnetic field on the nematic state*, is concerned with the explanation of the Frederiks threshold experiments. To some extent this article recapitulates a paper which appeared in 1929 in the *Zeitschrift für physikalische Chemie*.[17] Note the use of the term *nematic*, despite Zocher's German background. In ten years, Friedel's terminology had achieved clear victory. Apart from Oseen, Zocher was the other main protagonist in this period for the continuum theory (called the 'distortion theory' by Zocher). Whereas Oseen somewhat regally simply ignored the swarm theorists, Zocher was more inclined to confront it head-on, as we find in the very first paragraph of his paper:

.... It is extremely important to decide which is the correct (theory), not only for the particular problem dealt with in this paper, but for the whole general question of the structure of these phases ....

There is also experimental work by his colleague Eisenschimmel (also present at the conference with his wife), who is acknowledged underneath the title, but nevertheless not elevated to full co-authorship.

Zocher discusses various different possible Frederiks geometries, and refers back to the van Wyk experiment. Van Wyk had used the distortion theory to examine the spatially dependent behaviour of the optic axes. Zocher shows by way of example that this same theory will indeed give rise to threshold behaviour. He addresses experimental work, including that which Kast[18] previously adduced in favour of the swarm theory. Although the distortion theory apparently does not reproduce Kast's result in the low field régime, Zocher has a host of excuses – probably, with the benefit of hindsight, more-or-less correct – concerned with implicit assumptions about the experimental set-up, which provide himself with a satisfactory excuse. There are hints of a discussion of what we would now call anchoring (and which he calls variable surface tension), and a remark that with plausible anchoring energies there would be negligible differences in (in modern language) pretilt.

Eisenschimmel's experimental section makes a number of innovative contributions. He first notes that by choosing suitable material it is possible to align the optic axes perpendicular to the magnetic lines of force, whereas in previous experiments it had always been found to be parallel. He then carries out a somewhat qualitative experiment demonstrating the Frederiks effect with parallel aligned in-plane easy axes, positive diamagnetic anisotropy, and an in-plane magnetic field. There is a final experiment essentially showing that when the surfaces are antagonistic (in this case, one easy axis is in-plane and one normal to the surface) and there is a perpendicular in-plane field, the critical field is substantially increased. Although the geometry is not the same as for the twisted nematic phase, the physics is, and this paper provides a link between the work of Mauguin in articles A8 and

A9 (which are cited by Zocher) and the twisted nematic device itself, in particular Leslie's article D5.

The importance of Zocher's article lies in the progress that had been made since the Ewald symposium. In 1930–1, Zocher had been fending off demands by Ornstein not merely to criticise the swarm theory but actually to provide some viable alternative. The rhetoric is unconvincing, even if Zocher himself was sure of his ground. But the Frederiks experiment provided a specific procedure to calculate the elastic constants. Now the distortion theory was up and running, and the writing was on the wall for the swarm theory. As we have seen, soon the Frederiks group also was to begin to conceptualise not in terms of Ornstein's swarms but in terms of the Oseen–Zocher elastic constants.

Before passing onto extracts from the General Discussion section (article B8), it is of some interest to survey briefly other highlights of the Symposium. As an echo of the 1905 meeting in Karlsruhe in which Lehmann had given demonstrations (article A4), the introduction records that:

> After Professor Oseen's Introductory Paper had been taken as read and discussed, very beautiful demonstrations were given by Professor Vorländer, Professor Van Iterson and Dr. A.S.C. Lawrence. Further experimental demonstrations were given on Tuesday by Professor van Iterson and Mr. Bernal.

The proceedings include fully four separate articles by the by-now-veteran Vorländer, including his demonstrations. There are two articles on liquid crystal viscosity, one by Wolfgang Ostwald (son of the Nobel prize winner), and one by Herzog and Kudar. At this stage the role played by symmetry and geometry in defining exactly what was meant by viscosity in a liquid crystal was as yet unclear. We also note a contribution, entitled *Lyotropic Mesomorphism*, by A.S.C. Lawrence (at that time in Cambridge, but later to be a professor in Sheffield). We return to lyotropic substances in more detail in Part E, but the following extract from the first paragraph summarises the analogies between thermotropics and lyotropics as well as any textbook:

> It is interesting to recall that Lehmann's liquid crystals of ammonium oleate, which formed the foundation of the subject under discussion, were deposited from solution; and also that the amount of water present profoundly modified their form. Since that time, however, most of the substances studied have been in the form of anisotropic melts of single compounds. The action of heat and of a solvent are not dissimilar insofar as they both loosen the directive forces holding the molecules in their normal crystal lattice. For a mesoform to appear, it is necessary for these forces to persist in either one or two dimensions after loosening of the third. But even when this occurs and a mesoform exists it is unlikely that heat and all solvents will bring this about equally well.

156

The key word here is 'loosening'. The word *lyotropic* comes from the Greek verb λυειν, to loosen,[19] known to generations of classical scholars because it is the *only* regular verb in Ancient Greek!

Finally we note an article by Bernal and Crowfoot, who conclude:

> ....the mesophases, far from being an anomalous manifestation, take their place in a regular procession from the disorder of the ideal liquid to the regularity of the ideal crystal....

The young Dorothy Crowfoot, later Hodgkin (1910–94), would later be celebrated for determining the atomic structure of Vitamin B_{12}, an achievement for which she was awarded the 1964 Nobel Prize in Chemistry.

The General Discussion ranged widely. The transcript of the Discussion is not a true and accurate representation, of course. There are inserted contributions, and Discussants are able to edit their words so as to make it appear that what they said is more intelligent than it appears on the day. But, by contrast with the Ewald Symposium, underlying this transcript there is here the record of a human occasion. Is this why the debate seems to have taken a more sedate path than its Ewald counterpart? Or is it simply that Anglo-Saxons are more measured in their expression of emotion? Or perhaps the absence (presumably due to illness, for he was to die in December of the same year) of a Georges Friedel who might have provoked Vorländer?

Ornstein reiterates again and again his belief in the swarm theory, emphasising the manner in which surfaces affect the local orientation in their neighbourhood. He stresses that swarms are necessary to understand the turbidity of liquid crystals – this seems to underlie his real belief in the theory – but Zocher says no, statistical fluctuations and static deformations are quite sufficient. Zocher has evidently had more time to think about his debate with the swarm theorists. He now is able to point to internal contradictions: different calculations of the swarm size from experimental data give very different answers. He finds an articulate ally in Bernal, who yet again points out that although the swarm theory does explain the experiments of Kast and Ornstein, it does not necessarily *compel* this explanation.

Hermann suggests the existence of a tilted smectic (in general agreement with Vorländer's point that there are many unexplored liquid crystalline phases), but Bernal is extremely sceptical. Mention is made of the existence of the Dupin Cyclides in the smectics – a problem left open in Friedel's review – and Sir William Bragg presents an impromptu explanation, which, it was felt, was sufficiently important to codify *ex post facto*; a short paper about this thus appears in the proceedings.[20]

In papers B9 and B10 we return to the viscosity of anisotropic fluids, which have been addressed, albeit imperfectly, in the Faraday Symposium. It had long been known that the dielectric response of a liquid crystal depended on liquid crystal orientation; this is the essence of what is understood by an *anisotropic* fluid when observed in an optical experiment. As far back as 1913, Neufeld had attempted to find a magnetic field dependence of the viscosity, but he reported a negative result.[21] The problem was revived again by Marian Mięsowicz in 1934, a student of Professor Mieczysław Jeżewski in the Mining Academy in Cracow, Poland.

Much of the early French progress, we may recall, had been made by Friedel and Grandjean, working in a School of Mines. Despite the implication of its rather practical name, *L'École de Mines* was a centre for the fundamental study of geology, geophysics and more generally earth science. As such the classification of minerals and crystal types fell within its legitimate sphere. Liquid crystals were thus a natural extension. No doubt the long-term aim of governments in funding such institutions involved a practical pay-off, eventually. The Poles followed a similar train of thought, though whether the Polish Government could visualise liquid crystal displays as the long-term technological result is doubtful. Be that as it may, Jeżewski had been making dielectric studies of liquid crystals. Mięsowicz returned to the viscosity problem with apparatus considerably improved compared to that available to Neufeld in 1913.

In contrast to Neufeld, Mięsowicz obtained positive results. Paper B9, published in *Nature* in 1935, reported proof of principle. A fuller set of results was presented in a paper in German in the Bulletin of the Polish Academy of Sciences in 1936. However, this journal is not widely read and so a paper for *Nature* was in preparation when the war intervened. Paper B10, published after the war, reports these detailed results. It is said that at the outbreak of war, Mięsowicz's apparatus was summarily wrecked by the invading German forces. In fact, Mięsowicz never returned to serious study in the liquid crystal field (after the war he became a nuclear physicist). Luckily the results were not completely lost.

The important point about Mięsowicz's results is that by applying a magnetic field, he was able to anchor the nematic director in a specific orientation with respect to a flow field. He is then able to determine the viscosity when the director is parallel to the flow, or alternatively parallel to the velocity gradient, or out-of-plane. This makes three viscosities altogether. Nowadays they are known as the *Mięsowicz* viscosities. The highest viscosity (the second of the three) is between four and seven times the lowest (the first of the three). Averaging viscosities is a very dangerous thing to do!

This section is concluded with two papers with the same title – *On the orientation of liquid crystals by rubbed surfaces* – by Pierre Chatelain, which appeared during the war. For many years gentle rubbing was the method of choice in order to create a surface with suitable liquid crystal alignment properties. In the technological era this has been a very important imperative. In this problem the skill of the engineers has by far outpaced the understanding of the scientist. It is probably fair to say that even today a detailed and reliable model for rubbing-induced surface anchoring is lacking. Only recently have other surface orientation methods begun to compete with the noble art of rubbing.

Chatelain did not invent rubbing. Already in 1927, Zocher and Coper[22] had found that rubbing a glass surface oriented a nematic phase, and Eisenschimmel

158

had employed just this approach in his Frederiks-like experiment presented in article B7. However, in the articles we have translated, Chatelain is the first to investigate systematically rubbing effects. B11 is a short note in *Comptes rendus*, submitted by Charles Mauguin, who by this time was an eminent crystallographer and holder of a chair at the *Laboratoire de Minéralogie* at the Sorbonne in Paris. The follow-up B12 is a longer account in the *Bulletin de la Société Française de Minéralogie* (Mauguin's journal of choice), and from this article we have chosen some extracts.

Although even after these papers, rubbing remained as much an art as a science, the art was now considerably more educated and experienced. We may note how Chatelain systematically changed the amount of rubbing, the rubbing material, the nature of the slides, the mesogenic material; everything indeed that could in principle be altered. From a contemporary perspective, the degree of concentration seems all the more wonderful when we reflect on the difficult political situation in France caused by the Second World War. We may speculate that perhaps only by restricting his interest to mundane and repetitive scientific tasks was Chatelain able to divert his attention from these difficulties.[23]

With Chatelain's articles we conclude Section B. Liquid crystal science then began a somewhat somnolent phase which lasted almost 20 years. The war and its attendant instability surely played a major role in generating this sleepy period. Frederiks was *de facto* a victim of Stalin, and Ornstein similarly a victim of Hitler. As we have seen, the foundations of liquid crystal science were laid in Germany and France, and in both cases, recovery from the war was not easy. By the end of the war Vorländer had died of old age, and his younger colleague Weygand[24] had fallen in battle, a victim of German desperation in the last days of the war. Oseen also was dead, and Zocher,[25] regarded as politically unreliable by all sides, eventually emigrated to Brazil where he made a successful new career. Finally the technological promise of liquid crystals was not yet obvious, so that many workers in the field of anisotropic fluids (on all sides) had been diverted to more pressing war work.

It was not till the late 1950s before progress was resumed. We shall return to discuss this in Section C.

References

1. E. Hückel, *Zerstreuung von Röntgenstrahlen durch anisotrope Flüssikeiten* (*Scattering of X-rays through anisotropic fluids*) Phys. Z. **22**, 561–3 (1922). Hückel's main contribution to physical chemistry is through his molecular orbital theory. He is also known for the Debye–Hückel theory of electrolyte screening in colloids. This paper is his only contribution to liquid crystal science, and is often, perhaps unjustly, neglected because of its negative result. Hückel's interest in colloids may well have been inspired by his contacts in Göttingen with Richard Zsigmondy (see article E1), whose daughter Annemarie later became his wife.

2. A. Repiova and V. Frederiks, *к вопросу о природу анизотропно-жидкогососояния вещества* (*On the problem of the nature of the anisotropic state of matter*), Zh.R.F.Kh.O **59**, 183–200 (1927).

3. V. Freedericksz and A. Repiewa, *Theoretisches und Experimentelles zur Frage nach der Natur der anisotropen Flüssigkeiten* (*Theory and experiment concerning the question of the nature of anisotropic fluids*), Phys. Zeit. **42**, 532–46 (1927).

4. A. van Wyk, *Die orientierenden Einflüsse von Magnetfeld, Wand und gegenseitiger Wechselwirkung auf die Schwärme des flüssig-kristallinischen* p-*Azoxyanisols* (*The orientational influence of a magnetic field, a wall and mutual interactions on swarms in the liquid crystal p-azoxyanisole*), Ann. Phys. **3**, 879–933 (1929).

5. R. Schenck, *Bericht über die neueren Untersuchungen der kristallinischen Flüssigkeiten* (*Report on recent studies of cryalline liquids*), Jahrbuch der Radioaktivität und Elektronik **6**, 572–639 (1909).

6. D. Vorländer, *Neue Erscheinungen beim Schmelzen und Kristallisieren* (*New observations of melting and crystallisation*), Z. phys. Chem. **57**, 357–64 (1907).

7. D. Vorländer, E. Daümer, W. Selke and W. Zeh, *Die einsachige Aufrichtung von festen weichen Kristallmassen und von kristallinen Flüssigkeiten*, Z. phys. Chem. **129**, 435–74 (1927). The remark cited is on p. 468.

8. C. Weygand (Discussion remark on an article by W. Kast on *Anisotropic fluids*), Z. Elektrochemie **45**, 200–2 (1939).

9. W. Kast, *Die Molekul-Struktur der Verbindungen mit kristallin-flüssigen (mesomorphen) Schmelzen* (*Molecular structure of compounds with crystalline-liquid [mesomorphic] melts*), Angew. Chem. **67**, 592–601 (1955).

10. D. Vorländer and A. Apel, *Die Richtung der Kohlenstoffvalenzen in Benzolab-kömmlingen II*, Ber. Dtsch. Chem. Ges. **65**, 1101–09 (1932).

11. G. Pelzl, I. Wirth and W. Weissflog, *The first 'banana phase' found on an original Vorländer substance*, Liq. Cryst. **28**, 969–72 (2001).

12. Note, however, that in the ground-breaking 1922 article B1, Georges Friedel tended to use *structure* to describe a specified set of defects, when later workers would have used *texture*. Later, however, when X-ray crystallographic investigation of these effects took off, the term *structure* was reserved for X-ray crystallographic patterns.

13. This is of particular interest to one of the present authors, who studied just this problem (without realising its historical provenance!), calculating the local order close to a defect line: N. Schopohl and T.J. Sluckin, *Defect core structure in nematic liquid crystals*, Phys. Rev. Lett. **59**, 2582–5 (1987). However, this was not a problem which exercised Georges Friedel greatly, and we may speculate that this is because he realised the relevant length scales were not amenable to analysis under the microscope. He understood perfectly well that at a defect the optical axis directions converged.

14. C.W. Oseen *Die anisotropen Flüssigkeiten. Tatsachen und Theorien* (*Anisotropic liquids: experiments and theories*) (Fortschritte der Chemie, Physik und physikalischer Chemie, Vol. 20, no. 2, Berlin 1929).

15. C.W. Oseen, *Versuch einer kinetischen Theorie der kristallinischen Flüssigkeiten, III. Abhandlung* (*Essay on a kinetic theory of crystalline fluids, part 3*), Kungl. Svenska Vetenskapakademiens Handligar. **63**, no. 12 (1923).

16. See e.g. G.M. Mikhailov and V.N. Tsvetkov, *Влияние электрического поля на скоорсть течения анизотропно-жидкого р-азоксианизола вкапиляре*; (*Influence of an electric field on the speed of flow of the anisotropic liquid p-anoxyanisole in a capillary*), Zh.E.T.F. **9**, 208–14 (1939).

17. H. Zocher and V. Birstein, *Beitrage zur Kenntnis der Mesophasen* (Zwischenaggregatzustände) *V: Über die Beeinflüssung durch das elektrische und magnetische Feld.* Z. phys. Chem. **142**, 186–94 (1929). Note that already by 1929 Zocher has switched to using the term 'mesophase', although he does still feel the obligation to explain himself; *Zwischenaggregatzustände* literally means 'intermediate aggregation state'.

18. W. Kast, *Dielektrische Untersuchungen an der anisotropen Schmelze des para-Azoxyanisol* (*Dielectric studies of the anisotropic melt of para-azoxyanisole*), Annalen der Physik **83**, 391–417 (1927).

19. The principal meaning of λυειν is 'to loosen', and it is clearly to this meaning that Lawrence is making allusion in the extract just cited. However, a derived meaning of the word is 'to dissolve = *lösen*', or 'to break into constituent parts'. It is with this meaning that the root 'lyo' first enters the physical chemistry literature, for example in 'lyophobic' = 'fearful of being dissolved' = 'water repellent'. See H.G. Liddell and R. Scott, *A Greek-English Lexicon* (The Clarendon Press, eighth edition, Oxford, 1897). We have not been able to identify the first time the term 'lyotropic' was used in the liquid crystal literature, but it seems likely that this term too was an invention of Marie Friedel's.

20. The mathematics of the Bragg paper, although somewhat heavy going for a contemporary theoretical physicist, contained relatively standard nineteenth century undergraduate-level geometry. This paper is not universally felt to have added greatly to the understanding of the Dupin cyclide problem already achieved by Friedel. Note also that as Bragg co-organised the meeting, the decision to include this exposition in the proceedings might not be thought of as entirely dispassionate.

21. M.W. Neufeld, *Über den Einfluß eines Magnetfeldes auf die Ausflußgeschwindigkeit anisotroper Flüssigkeiten aus Kapillaren* (*On the influence of a magnetic field on the flow velocity of an anisotropic fluid through capillaries*), Phys. Z. **14**, 646–50 (1913) cited in M. Mięsowicz, Mol. Cryst. Liq. Cryst. **97**, 1–11 (1983). *Liquid Crystals in my memories and now – the role of anisotropic viscosity in liquid crystals research.*

22. H. Zocher and K. Coper, *Über die Erzeugung der Anisotropie von Oberflächen* (*On the production of surface anisotropy*) Z. phys. Chem. **132**, 295–302 (1928).

23. The present writers do not know anything about Chatelain's politics. Chatelain worked at the University of Montpelier, which was in the so-called 'unoccupied' (Vichy) part of France in the period 1940–42. The role of the middle classes in Vichy France, and the extent to which they collaborated with Hitler, is a subject of much contemporary historical debate.

24. Hans Kelker, in his fascinating article *History of liquid crystals*, Mol. Cryst. Liq. Cryst. **21**, 1–48 (1973), reports that Weygand 'died as a storm-trooper in the last days of the war'. Another conjecture is that Weygand died by his own hand. Indeed he was an active member of the Nazi party and published a dubious book entitled "Deutsche Chemie".

25. For reasons more fully explained in his biography. As an ethnic German he was super-numerary in post-war Czechoslovakia, despite family and other reasons for opposition to Nazism.

26. D. Vorländer, Die Erforschung der Molecularen Gestalt mit Hilfe der Kristallmischen Flüssigkeiten, Z. Phys. Chem. **105**, 211 (1923).

B1
Annales de Physique **18**, 273–474 (1922)

LES ÉTATS MÉSOMORPHES
DE LA MATIÈRE

Par M. G. FRIEDEL

INTRODUCTION

Je désignerai sous ce nom les états particuliers que présentent les corps signalés par Lehmann à partir de 1889 sous les noms de *cristaux liquides* ou *fluides cristallins*. Sur la foi de ces dénominations, très malheureuses mais sans cesse répétées depuis trente ans, beaucoup de gens s'imaginent que les corps si curieux sur lesquels Lehmann a eu le grand mérite d'attirer l'attention, mais qu'il a eu le tort de mal nommer, ne sont autre chose que des substances cristallisées, différant simplement de celles qui étaient antérieurement connues par leur degré plus ou moins grand de fluidité. En fait, il s'agit de tout autre chose, et de quelque chose d'infiniment plus intéressant que ne seraient de simples cristaux plus ou moins fluides.

Mettons à part l'iodure d'argent cubique, que Lehmann a joint à tort aux corps que nous nous proposons d'étudier ici, et qui est tout simplement cristallisé, sans rapport aucun avec les autres « cristaux liquides ». Il n'est d'ailleurs nullement liquide, mais seulement plastique, comme beaucoup d'autres corps cristallisés. Sous cette réserve, les corps de Lehmann ne sont ni des cristaux ni des substances cristallisées. En les nommant cristaux ou en leur appliquant l'épithète de cristallisé on ne donne aucune idée de leurs proprié-

B1
Annales de Physique **18**, 273–474 (1922)

THE MESOMORPHIC STATES
OF MATTER

by

Mr G. Friedel

Introduction

I use the term mesomorphic to designate those states of matter observed by Lehmann in the years following 1889, and for which he invented the terms *liquid crystal* and *crystalline fluid*. Lehmann had the great merit of drawing attention to these materials, but he erred greatly in naming them. The unfortunate names have been repeated again and again over the last 30 years. As a result many people suppose that these substances are merely crystalline materials, albeit rather more fluid than those hitherto known. The exact opposite is the case. Indeed, these materials are infinitely more interesting than they would be if they were simply crystals exhibiting some unexpected degree of fluidity.

Let us put to one side the cubic silver iodide which Lehmann incorrectly regarded as one of the substances which we propose to discuss here. This is simply a crystal, without any connection with the other 'liquid crystals'. This material is, by the way, not liquid at all, but merely plastic, like many other crystalline materials. Subject to that reservation, Lehmann's materials are neither crystals, nor crystallised substances. By calling them crystals, or by applying to them the epithet crystallised, one gives no hint of their properties. These properties are entirely different from those of substances which usually carry this designation. On the other hand, the terms 'liquid' or 'fluid' must not be taken to carry their usual meaning in this context. The majority of Lehmann's materials, it is true, are quite fluid indeed, certainly as fluid as water. But there are some whose structure does not differ in any way from the others, but which are nevertheless obviously solid. It is sufficient to use as an example ordinary soap in cake form. It is known, besides, that physics is unable to fix a natural boundary, a discontinuity between the solid and the liquid state. What characterises Lehmann's materials is not their state, which is more or less fluid. It is their special structures. There are only a small number of structural types and they are always the same. Everything that

163

follows will, I hope, show that the Lehmann materials constitute two completely new forms of matter. These are separated, without exception, from the crystalline and amorphous forms by discontinuities, in the same way that these two forms are separated from each other by a discontinuity. These two new forms themselves possess some common properties. It will be necessary from time to time to concatenate them under a single banner, but they are distinctly different from each other, and can be present in the same body. They are separated from each other, without exception, by a discontinuity. Before Lehmann's time it was not known that the two very general crystalline and amorphous forms of matter corresponded to two types of molecular structure. Nowadays, in admittedly a relatively small number of materials of a very special chemical type, two more are known. These correspond to special types of structure, neither crystalline nor amorphous.

However distasteful one finds the coining of new terms, it should be seen that it is absolutely necessary in these circumstances to find an adjective which unambiguously describes all these forms. This will be the analogue of the adjectives crystallised (or crystalline) and amorphous, which are used to describe the forms, substances, phases and so on which were already known.

The terms 'birefringent liquids' and 'anisotropic liquids' are generally used by those people who have not accepted the identification of the Lehmann materials with crystals. They are inadequate because some of these materials are not liquids. In addition, the birefringence, and in general the anisotropy, although remarkable, are not what is most special about these materials. These properties are also associated with crystalline substances, and can also occur in amorphous substances either subject to mechanical forces, or subject to a magnetic or electric field, or even permanently when they are solid. What is required is a language which absolutely eliminates all these possibilities.

Without exception, the regions of stability of the two new types, in any given material, come between the crystalline region of stability at low temperatures and the amorphous region of stability at higher temperatures. Similarly, their molecular structures form an intermediate stage between perfectly ordered periodic crystalline structure and the completely disordered amorphous structure. For these two reasons the term *mesomorphic* seems to be appropriate. No doubt it would have been preferable to have found a term like 'crystallised', linked to a very ancient usage, completely devoid in itself of any meaning which could make it one day inappropriate in the light of new facts. Nevertheless, the term seems vague enough to avoid any embarrassment.

On the other hand, and for the same reasons, we will have to be able to describe the two mesomorphic forms separately. The distinction between them was recognised by Lehmann almost right at the beginning of his discovery. He called the first 'Fliessende Krystalle', or 'Schleimig flüssige Krystalle' (flowing crystals, or slimy liquid crystals), and the other 'Flüssige Krystalle', or 'Tropfbar flüssige Krystalle' (liquid crystals, or drop-forming liquid crystals), thus attributing an exaggerated importance to the latter's normal degree of fluidity.[1] On the other hand, although he did notice a difference between the two types, he was sure that they were two variations of one and the same thing. This he wanted to call a 'crystal', even though he realised that it did not possess a periodic structure. He was inventor of the term 'liquid crystal', and was aware of the striking effect that this association of words could have, and indeed unfortunately did have, on many people. He did not, however, know how to escape from this awkward concept, and too often described the properties of the two types in a confused way.

Actually, as we shall see, there is a strict distinction between the two types. What constitutes the essential difference between them is not the degree of fluidity but the structures. Having recognised this, Grandjean and G. Friedel called materials of the first type 'conic section liquids', because of the role played in their structure by groups of focal conics. In a similar way, they called those of the second type 'thread-like liquids' (*liquides à fils*) or 'liquids with nuclei' (*liquides à noyaux*), so as to emphasise the crucial structural features. I am no longer going to use this terminology because, as I have said, the word liquid should not play any role. Here again, what are required are two adjectives without too exact a meaning, whose only purpose is that of unambiguous description. I am going to use the term *smectics* (σμηγμα, soap) to describe the forms, bodies, phases, etc. of the first type (this includes Lehmann's Fliessende Kr. and Schleimig flüssige Kr. and also conic section liquids). The motivation for this terminology is that soaps at ordinary temperatures belong to this group, and that ammonium and potassium oleates were in particular the first materials of this type to attract attention. I am going to use the term *nematic* (νημα, thread) to describe the forms, bodies, phases, etc. of the second type (Lehmann's Flüssige Kr., Tropfbar flüssige Kr., as well as thread-like liquids) because of the linear discontinuities, which are twisted like threads, and which are one of their most prominent characteristics.

It is not without hesitation that I propose new terminology. For the example of 'liquid crystals' gives us the opportunity to observe the influence of a word badly chosen. Lehmann did not understand clearly how the substances he was studying would take their places in the hierarchy of material phases. He did however perceive, albeit in a rather confused fashion, that here indeed was something entirely new. We must be grateful to him for trying to convince his contemporaries of that fact. He was misled by the inadequate definitions of crystalline materials which were current at that time, and thus erred in classifying his liquids as crystals. But much more serious were the errors of the numerous scientific opponents whom he encountered, especially in Germany. These opponents failed to see anything in the new materials other than crystal suspensions in an amorphous liquid

---

[1] It should be added that Lehmann's 'Fliessende Kr.' include cholesteric materials, which are, as we shall see, very different from the others. In this respect the classification by degree of fluidity is misleading.

(Quincke, Wulff), or emulsions of two liquids (Tammann, Nernst). They were not aware that these hypotheses were in no way able to explain the characteristic structural properties. They did not even try to provide a demonstration which would show, above all, that a suspension or an emulsion could exhibit such structures. Lehmann was very definitely closer to the truth than the authors we have just named. If he had not invoked any principle, but merely called his materials 'birefringent liquids', or some other description which only involved establishing the facts, no harm would have come of it. By bad luck, he made them 'liquid crystals'. This term was soon accepted more or less everywhere either by scientists who had not had the opportunity to see for themselves, or by those who had only vague ideas about crystals and crystallised matter. Now old traditions, which were justified once upon a time, have removed from the physicist the study of crystallography, by which we mean the physics of the large majority of solid bodies. It has been handed over to the mineralogist. Physicists who hear about liquid crystals have in general taken it as read that this subject belongs in the domain of crystallography, and thus of mineralogy. But the study of these materials, which are all organic and almost all artificial, has no link with the natural field of mineralogy. So in this way, with very few exceptions, perhaps more specially in our country, the study of these materials and their mesomorphic forms, which transcend the limits of crystallography, remains entrusted to naturalists, and almost unknown to physicists. I want to be able to show them that they have had in front of their noses, over the last 33 years, a field of study that is scarcely penetrated by exploration and extraordinarily rich in surprises. At each step one can find unresolved problems. The mineralogists' methods have been able to elucidate these problems in part, and these methods will still be able for a long time yet to uncover something new. However, these problems will only really be conquered for science using the tools of the physicist.

Either directly or through the good offices of Messrs Mauguin, Gaubert, Weick, Jaeger and Lehmann, I have been able to get hold of about forty substances in mesomorphic phases. I have then been able to carry out a general review of their properties using what seems to be a sufficiently large set of examples to enable general conclusions to be drawn. It is these studies on which I propose to report here. In trying to impose some order in this topic, I shall not hesitate to emphasise the numerous questions which remain, and which require further research.

History

I shall only recount the most important facts.

We are indebted to German chemists for the discovery of substances in the mesomorphic phases. Starting in 1889, Lehmann drew attention for the first time to the peculiar properties of cholesteryl benzoate. F. Reinitzer, who had prepared this material, had himself recognised in some way two melting points. When it is cold, the material is crystalline and solid. At 145.5 °C the crystals melt forming a turbid and viscous liquid (which actually is a nematic liquid of the cholesteric type).

Then, at 178.5 °C this turbid liquid suddenly switches into an ordinary liquid state, which is now completely clear and flows easily. Both of these transformations are completely reversible, and under the microscope one sees a perfectly discontinuous boundary between the two phases. When Lehmann looked at the intermediate turbid phase between crossed nicols, he recognised that it was birefringent and thus anisotropic. He was misled by poor definitions of what was meant by a crystal. And so, from that moment, believing that he could see a crystalline material in this phase, he regrettably combined its properties with the completely different properties of silver iodide. In the cubic form that it takes below 146 °C, this material is a weak plastic solid. However, all of its properties are those of a crystal and none are those of a mesomorphic substance, with which it has nothing in common.

Soon after this first example, it was Lehmann again who recognised birefringence and the main structural properties of the mesomorphic phase in para-azoxyphenetole and para-azoxyanisole, which were nematic substances prepared by Gattermann. He then found more or less analogous characteristics in potassium and ammonium oleate, which were the first smectic substances reported. It was he too who established the properties of the ethyl para-azoxybenzoate and of the ethyl para-azoxycinnamate prepared by Vorlænder, which were the first non-soap smectic materials, and then of lecithin , also a smectic.[1]

Since that time, many other examples of mesomorphic materials have been discovered. One should mention particularly Vorlænder[2] for discovering a large number of substances which can exist in the new phase, and for investigating their principal chemical properties. Above all he showed that all these substances possess molecules which were remarkably long, linear, and unbifurcated, and in which the bisubstituted benzene rings only allowed a para- position, and forbidding ortho- or meta- isomers. His principal work (1908) already brought the number of 'crystalline fluid' compounds to some 200 or 250, all organic compounds with very complicated molecules. However Vorlænder still included silver iodide amongst them, regarding it as 'the only inorganic compound'. This shows to what extent the true nature of these compounds continued to be poorly understood in Germany, even 20 years after the first discovery. Be that as it may, thanks to Vorlænder's measurements it was possible from that time on to prepare a large quantity of new mesomorphic substances more or less at will.

Physical studies advanced more slowly. German physicists appear to have had two main concerns. One of these was to counter the idea of 'liquid crystals', and substitute it with hypotheses which were even less able to explain the facts than Lehmann's own. The other involved struggling against these hypotheses and in favour of the idea of a liquid crystal. To begin with Quincke, Tammann and Rotarski doubted the purity of the substances studied by Lehmann. Quincke and then

[1] O. Lehmann, *Flüssige Kristalle*, Leipzig, 1904; *Die Neue Welt der Flüssigen Kr.*, *1911*.

[2] D. Vorlænder, *Kristallinisch-flüssge Substanzen*, Stuttgart 1908.

Wulff imagined a suspension of solid crystals in an amorphous liquid. Tammann and Nernst favoured an emulsion of two liquids. All these ideas, suggested in a very rough way by the characteristic turbidity of the mesomorphic phases (which furthermore was not uniform) did not hold up a minute once the structures were examined under a microscope. It does not matter whether one looks from far off or from close in; no impure amorphous liquid, no crystalline suspension in a liquid, and certainly no emulsion has ever shown anything at all resembling these absolutely constant and completely characteristic structures. Nevertheless, even without going into the microscopical study of the structures, it is as well to dispose of these objections. One can show conclusively that Lehmann's materials are, or at least can be in principle, completely pure. They are phases as well defined as the crystalline and amorphous phases, whose properties remain unchanged by any process of purification. A suspension or an emulsion simply would not be acceptable. This proof was provided by the researches of Schenck and his students.[1]

Schenck came out in favour of the Lehmann 'liquid crystal'. But this is just language. He does not examine the hypothesis in which there is a new phase of nature. Rather he supposes, *a priori*, that there are only two possibilities. The one associates known crystalline and amorphous phases in a mixture, suspension or emulsion. The other proposes a homogeneous phase, which he calls, without examining it further, a liquid crystal. And the first group of hypotheses he is able to rule out.

From the German work we should add a certain number of measurements which were sometimes fruitful but often premature, in that the thing to be measured was not sufficiently well understood. One of the most interesting of these was the work of Stumpf[2] on the rotatory power of amyl cyanobenzalamino-cinnamate.

When it comes to the detailed study of the mesomorphic phases whose independence Schenck had established, the German authors seem to be little concerned. They are little concerned with their structures, little concerned with their connections with crystalline and amorphous phases, and little concerned by the need to establish order in this jungle of new facts. An exception should be made for Lehmann, whose large number of publications contain many observations with the microscope. However, these are confused, and are rendered still more unclear by the theoretical considerations which accompany them.

It has been in France above all that the main steps in understanding the principal concepts have been made, although understanding is still far from complete. We should recall particularly the study by Mauguin[3] of nematic twisted films, using the experiments to provide a complete explanation of the optical properties of these liquids. These properties were so bizarre that to begin with

[1] R. Schenck, *Kritallinische Flüssigkeiten und flüssige Kristalle*, Leipzig, 1905.
[2] F. Stumpf, *Doctoral thesis, Gottingen*, 1911.
[3] *Bull. Soc. Minéralogie*, 1911.

G. Friedel and Mauguin had described them without finding the interpretation.[1] We also recall Mauguin's work on the effect of a magnetic field on nematic liquids,[2] that of G. Friedel and Grandjean on focal conic structure in smectic materials,[3] and Mauguin's work on the effect of Brownian motion in nematic liquids earlier seen by Friedel and Grandjean. Mauguin and subsequently Grandjean[4] examined the orientation of mesomorphic materials on crystals. Grandjean also discovered terraced drops in smectic materials,[5] rectilinear flow of the smectic liquid over crystals which oriented it,[5] and the equally spaced planes which characterise nematic materials of the cholesteric variety,[6] which remain extremely puzzling. Out of this research has come the idea that there are two new mesomorphic forms of matter. These are completely distinct from the crystalline and amorphous forms, and possess properties which absolutely cannot be found in either of the two types of matter previously known.

Crystalline materials

> In this section he goes on to review what is meant by crystalline phases, going back into the history of the subject, to the ideas of Mallard and Haüy. Towards the end of this section his frustration is clearly getting the better of him . . .

. . . Lehmann understood the point, albeit in a confused way, that he could save the word 'liquid crystal' only if he discovered in his materials discontinuous properties. So he searched either for polyhedral crystal shapes with planar faces, or for twinning phenomena. He thought he had found them, but we shall see that it was only an illusion. In despair he was forced to recognise the lack of a periodic structure. All that remained was to find a definition of a crystal which neither referred to discontinuous surface properties governed by Haüy's Law, nor to the periodic structure which, depending on your point of view, is either its cause or its manifestation. Thus we obtain the bizarre definition: 'A crystal is an anisotropic body which has a "molekulare Richtkraft" and which has as a result the property of growth.' The 'Richtkraft' or 'Gestallungskraft' is a mysterious force which is supposed to make the vertices stick out, the edges straight and the faces planar. It is supposed to counteract the surface tension which causes surfaces to be round. In fact, it is only a nebulous new Germanic Divinity, which acts against

[1] Bull. Soc. Min., 1910.
[2] C. R., 1911. I, p. 1680.
[3] Bull. Soc. Min, 1910.
[4] Bull. Soc. Min. 1916 and 1917.
[5] Bull. Soc. Min., 1916.
[6] C.R., 1921, I. CLXXII.

itself in an extraordinary collection of meaningless words,* and, as we shall see, lacks any basis other than that of an inexact view of the role of surface tension. As far as the growth capacity is concerned, it would force us to consider both the drop of water growing in vapour, and the vapour bubble growing in water, as crystals. The idea of growth capacity is obviously completely incapable of helping us to place a boundary between those materials which are crystalline and all those other materials which are not.

> Friedel goes on to consider what is meant by an amorphous material, drawing a distinction between the concepts 'amorphous' and 'isotropic', but pointing out that if an amorphous body is not isotropic, then this is due to some external conditions extant while it was being formed. Most amorphous bodies, he avers, are fluid and all materials do exhibit a high temperature fluid amorphous phase.

Mesomorphic materials

Mesomorphic materials, like crystalline materials, are always spontaneously anisotropic, whatever the circumstances under which they have been formed. As in the case of non-cubic crystals, their anisotropy manifests itself through a birefringence which can be extremely strong. By contrast, they resemble amorphous matter in that the envelope of planes with discontinuous vector properties, which is a signature of crystalline matter and which is governed by Haüy's Law, does not exist.

All mesomorphic phases can be observed under essentially the same circumstances, and we shall shortly describe these. There are some exceptions to this rule; these are some smectic materials, such as soaps, lecithin and protagon, which are mesomorphic at room temperatures and above. It seems that the crystalline forms of these materials are not known, but they may be stable at very low temperatures. Other than that the conditions for observing mesomorphic behaviour are as follows:

At room temperature, the material is crystalline. When the temperature is increased one sees a reversible and discontinuous transformation at a well-defined characteristic temperature T_1. The temperature T_1 depends on the type of material and is a function of pressure, and is exactly analogous to polymorphic transformations or melting. As in these latter cases, the microscope reveals a well-defined boundary between the crystalline and mesomorphic phases, lacking any transition region. Although this boundary is extremely clear, and separates two phases which look very different, it has nevertheless sometimes been able

* *Une singulière logomachie (ed.).*

170

to pass unnoticed, leading one to believe that the change from the crystalline to the mesomorphic phase is continuous.

This has been the case for a number of smectic materials. In these cases the crystallisation occurs when the mesomorphic phase is frozen and the crystalline phase orients itself more or less on top of the mesomorphic phase and 'pseudomorphises' its structure. Often this is very approximate and does not hide the transformation at all. Examples include ethyl azoxybenzoate – ethyl azoxycinnamate mixtures, where the phenomenon was first noticed, and several other cases which we shall mention later. However, this pseudomorphism can sometimes be so close that the crystalline phase faithfully reproduces the conic structure characteristic of the smectic phase, right down to the most minute microscopic detail. The most notable example of this occurs in cholesteryl stearate. Almost as perfect an example is the case of cholesteryl myristate. Here the boundary between the crystalline and the smectic phases, although completely well-defined, is so difficult to observe and separates two phases which look so similar that it was missed by Lehmann, who thought he had seen an example of a continuous transformation from 'liquid crystal' to solid crystal. In fact this boundary is always present. There is always a perfect discontinuity between the mesomorphic and the crystalline phases.

The mesomorphic phase is normally either a liquid, which sometimes can be extremely fluid (particularly if it is nematic), or a viscous material (particularly if it is smectic). This is what has drawn attention to the mesomorphic phase in the first place. But the degree of fluidity varies with temperature. When the mesomorphic state continues to exist right down to room temperature, either by supercooling, e.g. ethyl anisalaminocinnamate, or in equilibrium e.g. soaps or protagon, it can sometimes reach a state that one could only describe as solid. Like amorphous materials, mesomorphic materials are normally liquid or pasty, but can exceptionally be solid. The transformation T_1 thus normally looks like the melting of a crystalline solid.

On the other hand, to the naked eye the mesomorphic phase is usually turbid. Looked at between nicols, under the microscope, it is birefringent. The turbidity disappears and gives way to complete transparency. This sometimes occurs spontaneously and sometimes only when certain precautions are taken. Under the microscope the material appears either optically homogeneous or made up of a small number of optically homogeneous regions. The cloudiness, which gave rise to all the different objections by Lehmann's opponents, and led some to believe in an emulsion or a suspension, is completely explained by the birefringence. Added to this are the rapid changes in optic axis from one point to the next, which most often occur when no precautions are taken and the structure is complex. As Lehmann clearly realised, the mesomorphic phase is turbid for roughly the same reason as a closely packed collection of small birefringent crystals is turbid. The crystallites are individually clear but oriented in all directions, in such a way that a light ray cannot go through it without undergoing reflections and refractions as it passes from one crystallite to another. In fact this analogy, although true in principle, holds only approximately. In mesomorphic materials, it is true that optical discontinuities exist and that indeed they govern the turbidity by causing light rays to scatter. But these discontinuities do not occur on separating surfaces, as they would in the crystal case. Rather they are linear, both in smectic and in nematic materials. The scattering of the light rays which causes the turbidity occurs when the light rays pass close to these discontinuity lines.

A single substance may exhibit two mesomorphic phases, one smectic and one nematic. In this case, and there are no exceptions, the smectic phase occurs at lower temperatures and the nematic at higher temperatures. At T_1, the smectic phase replaces the crystalline. At another higher temperature T_2 a discontinuous reversible transition occurs and the nematic phase appears. This is always separated from the smectic phase by a perfectly well-defined surface with no transition region. Finally at an even higher temperature T_3 (or T_2 if there is only one mesomorphic phase) the birefringent mesomorphic phase gives way to an isotropic and amorphous liquid. The transition is always reversible, sudden and discontinuous, with a well-defined boundary between the phases and no transition region.

Lehmann and other scientists who accepted his viewpoint were influenced by the notion of 'liquid crystals'. They thought they had seen cases of three and four mesomorphic phases. In their view mesomorphic and crystalline phases were identical in nature. By taking this point of view they necessarily had to consider the transformations T_1 and T_2 as polymorphic transformations which were completely analogous to similar transitions frequently seen in these materials below T_1. Thus, by analogy, one might sometimes expect to observe a number of such transformations above T_1 and be tempted to discover them. In fact even if one accepts that the mesomorphic form is completely distinct from the crystalline, at first sight it does not seem impossible for a single material to exhibit several smectic or nematic phases. And it could be that it will be like that. But my observation of the facts leads me to believe that it is not. In the reported cases of triple mesomorphic phases which I have been able to observe myself – see below for ethyl anisalaminocinnamate (Vorländer and Wilke), cholesteryl laureate (Jaeger), cholesteryl myristate (Jaeger) – I have always been able to establish that a structural change within a single phase had been confused with a distinct phase. The same error has led to a belief in mesomorphic phases in several cases when only one exists (see below Stumpf's amyl cyanobenzalaminocinnamate and Gaubert's anisalamidoazotoluene). As I have not been able to obtain samples of any of the other substances reported as exhibiting two 'crystalline liquid' phases, I cannot be completely sure. However, in my opinion it is very likely that there is only one nematic and one smectic phase in each material. In this the mesomorphic phases resemble the amorphous phase rather than crystals, which can often exhibit many different forms. The least that one can say is that statements about multiple mesomorphic phases should be accepted only with great hesitation. The existence of more than two of these phases, one smectic and the other nematic, should only be accepted if an absolute proof can be provided, and at the moment there is no such proof.

In summary, and subject to this last reservation, the series of forms which a single substance goes through as temperature is increased is as follows:

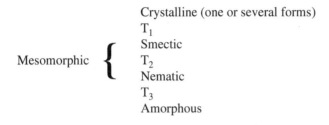

Crystalline (one or several forms)

Mesomorphic $\left\{ \begin{array}{l} T_1 \\ \text{Smectic} \\ T_2 \\ \text{Nematic} \end{array} \right.$

T_3

Amorphous

However, one or other of these forms may be absent, and the mesomorphic forms in particular only exist in a certain number of chemically special organic substances.

First mesomorphic type: smectic materials

As a typical smectic material one can take the mesomorphic phase of ethyl para-azoxybenzoate, or ethyl para-azoxycinnamate, or better still octyl para-azoxycinnamate, or even ethyl anisalaminocinnamate. They are all very stable and easy to study. The structures are identical for all smectic materials, whether pure or not. The study of mixtures is very instructive in this respect. Let us consider, for example, a mixture of a smectic material such as ethyl para-azoxybenzoate and a nematic material such as para-azoxyphenetole. These substances mix completely in the amorphous state; in variable proportions, depending on the temperature, in the mesomorphic state; and only finally separate in the crystalline state. Depending on the temperature and the relative concentrations, the mixture can be: (a) a completely homogeneous smectic phase, *containing all the material*, and completely indistinguishable from a pure smectic material; or (b) a completely homogeneous nematic phase, *containing all the material*, and completely indistinguishable from a pure nematic material; or (c) two coexisting phases, one smectic and the other nematic, which are completely distinct and which are separated by a surface of discontinuity. We recall that when we mix ordinary crystals and amorphous liquids, the result is either crystalline or amorphous matter, and not something in between. The two types are separated by an absolute discontinuity. In the same way, when we mix mesomorphic materials, the result is crystalline, or smectic, or nematic, or amorphous, but with no transition region, and nothing in-between. And furthermore we recall that it is impossible to distinguish one amorphous liquid from another qualitatively or structurally, regardless of whether it is pure or a mixture. In the same way a smectic material (or a nematic material) exhibits exactly the same structures and properties in all cases, regardless of degree of purity. We shall return to this very important point later in this article.

The smectic state is thus one of the general classes of matter, on the same level as the crystalline and amorphous states. What then are its characteristics?

It can, depending on the circumstance, present several structures:[1]

1. *Homogeneous structure* – Let us crystallise ethyl azoxybenzoate by melting it and then cooling it again between two glass plates. It is then easy to arrange things so as to obtain large uniform crystalline regions. First one must first melt the sample on a microscope hot stage. Then one must retrace one's steps slowly towards lower temperatures. It is important to initiate the crystallisation process in a corner by putting the sample in contact with a trace of crystal. This avoids supercooling. The crystallisation can thus take place as slowly as is desired, and beautiful elongated homogeneous crystalline regions are obtained.

One then increases the temperature slowly just past the temperature T_1. At this temperature the crystals transform suddenly into a perfectly clear smectic material. Under the microscope the large homogeneous birefringent regions can still be seen. The boundaries of these regions are the same as for the previous crystalline regions. In the case of ethyl azoxybenzoate the birefringence is very strong, and about twice that of Iceland spar. It can be examined either in a parallel light beam, or in converging light. However it is done, the regions exhibit all the

---

[1] It will be useful here to specify the meaning of the words structure and phase, which can lead to ambiguity, and which some authors use without defining them sufficiently. When we talk of *structures*, we mean different ways of organising a particular homogeneous element, which is small but still made up of a large number of molecules, and whose *molecular structure* is constant. As long as this molecular structure does not change discontinuously, the properties of this small homogeneous element also do not change discontinuously, and the transformation points remain the same whatever the *structure*. There can be a structural change without a phase change. The thing that remains the same if there are no *structural* modifications, but alters when the *molecular structure* changes or when there is a phase transformation, is this small homogeneous element. In the crystalline phase, we call this the crystal, and in the mesomorphic phases it should be given a name. Lehmann wrongly described it as a 'molecule' when he described the organisation of elements in the complex structures exhibited by mesomorphic materials. This was also what Stumpf described as an 'elementary crystal' while describing amyl cyanobenzalaminocinnamate. Stumpf had already expressed his belief in the existence of these 'elementary crystals' in the first phase. He then asked himself whether in the second phase they were the same, or whether rather they constituted a different 'modification'. If we translate this question into the language used here, this amounts firstly to affirming the existence of two phases, and then to asking: Is one of these phases a simple structural modification of the other, or is it rather really a distinct phase. It was a question that Stumpf, despite his belief in the existence of two phases, never really answers. We shall resolve it further on in this article in favour of a structural modification, but the answer is not *a priori* given by stating that there are two phases.

It seems that in most cases it is the misuse of the term 'phase change' which has led to a belief in the existence of multiple mesomorphic phases. A phase change has been confused with a mere structural change, not involving modification of the molecular structure nor consequently of the physical properties of the small homogeneous element.

properties of a uniaxially positive birefringent material whose optical axis is oriented at some arbitrary angle with respect to the glass.

Is each region oriented consistently with the orientation of the crystal which has just been melted? It seems very likely that this is the case. Azoxybenzoate crystals are biaxial, and the plane of the optical axes is normal to the long axes of the regions. The optical axis of the smectic region always seems to be in the same plane, and not far from the acute bisector of the crystal. I will limit myself to indicating that there would be significant interest in measuring this orientation with more precision. It does not seem to have been studied, and is directly linked to the problem of the relative orientation of the smectic and crystalline phases. We shall return to this when we discuss the conic structure.

When one melts large crystalline regions one obtains large smectic regions which appear homogeneous. However, in fact they never are, except right in the centre of the region. One can examine the contact surfaces between these regions under magnification. When one does so, one realises that the surfaces are not real surfaces of discontinuity, as they had been in the crystals. Rather the optical orientation of one region changes gradually into that of its neighbour, passing through a zone in which there is a complex structure. In this zone there appears a very large number of small black lines. These are the focal groups about which we shall have more to say later. We shall come back to this point after we have studied the conic structures.

Without exception, the homogenous regions in all smectic materials are *uniaxial*, whether they have been obtained by melting crystalline regions or in some other way. If a lateral force is exerted on the cover-slide during an observation of a viscous material in convergent light, then the black cross comes apart and the material becomes biaxial. But this is only a deformation, and is similar to what happens when a uniaxial crystal is subject to the same conditions. By themselves, left to their own devices, all smectic materials are uniaxial.

In addition, all smectic materials, without exception, are optically *positive*.

Homogeneous regions of any size can occur. They often occur spontaneously in another way, reported long ago by Lehmann. Smectic materials possess the property that they orient themselves with their optical axis perpendicular to the preparation surface. Lehmann called this 'spontaneous homeotropy' and 'forced homeotropy', and attributed it to the action of the glass. Sometimes this occurs by itself, as for example in the smectic phase of cholesteryl caprylate. In other materials it only occurs when one presses on the cover-slide with small movements, and the degree to which it occurs differs greatly from one substance to another. Good examples are: octyl para-azoxycinnamate, ammonium oleate, and the smectic phases of cholesteryl pelargonate, cholesteryl caprate, cholesteryl laurate and cholesteryl myristate. Lehmann called the regions which he obtained in this fashion 'pseudo-isotropic', because between nicols they extinguished parallel light, and if one was not careful, it was easy to confuse them with isotropic regions. A special name for these regions would be pointless because it would run the risk of confusion with a special state of matter. Such errors already occur in this field and we would not wish to compound them. We shall simply call these regions *normal regions*.

Lehmann's statement that homeotropy is the result of the effect of the glass has generally been accepted uncritically. I have shown, however, that this is mistaken. An ethyl azoxybenzoate drop freely suspended over a hole punched through any kind of support – for example glass or mica – immediately and spontaneously orients itself with its optical axis normal to the surface. Furthermore the surface becomes completely flat and resembles Grandjean's terraced drops, which I shall shortly come to. This shows conclusively that the glass plays no role in the orientation, which is rather the consequence of a surface film in the liquid itself. The glass or other solid support, far from causing the homeotropy, in fact acts against it by a kind of friction. This friction fixes the surface film and impedes its motion. During 'forced homeotropy', the small motions to which the glass is subject disrupt the surface film. Lehmann believed that spontaneous homeotropy was evidence of stronger forces by the glass plates. In fact it is not so, but on the contrary the result of an imperfect contact between the liquid and the glass. When the action of the glass reaches a maximum, the homeotropy is simply prevented because the glass holds the surface film in place. We shall find the same situation in nematic liquids.

In fact homeotropy is not a phenomenon confined to smectic materials. It is at least as noticeable in nematic materials. It even occurs in certain soft materials which appear crystalline, such as beeswax and ozocerite. But here one is dealing with materials whose structure is so delicate and confused that it is still difficult to say whether they are smectic, or as seems more likely, crystalline.

If one heats a drop of smectic material on a glass plate into the isotropic state with the upper surface remaining in contact with air, and then cools it again until the smectic phase reappears, homeotropy is still observed. In this case, what appears in the centre of the drop is a focal conic structure, as we shall see. But if the drop is very flat, this structure does not reach the edges of the drop. One then sees a border region of variable size in which the orientation is normal. More precisely, the optical axis is almost certainly normal to the flat surface in contact with the glass, but deviates near the free surface, and thus towards the edges of the drop, to a direction perpendicular to the free surface. Between nicols the boundary region exhibits, not complete extinction, but some degree of birefringence, and a black cross can be seen in the drop (photograph 3).

When the supporting surface is rigorously clean, it can happen that the substance does not come into contact with it. These conditions favour the occurrence of the peculiar 'terraced drops', reported for the first time by Grandjean. Glass is extremely difficult to clean sufficiently well, and so the phenomenon occurs only rarely on glass. But it occurs all the time if one uses a support consisting of a freshly cleaved crystal which does not orient the material under investigation. Muscovite, for example, is very suitable for ethyl azoxybenzoate or ethyl azoxycinnamate. To avoid close contact, the following steps are necessary. First take a small quantity of crystalline azoxybenzoate, or azoxycinnamate, or better still

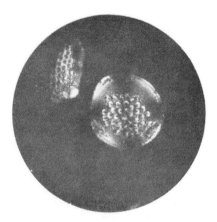

G. Friedel, photograph 3.

prepare a mixture of these two materials. Put it on to cold mica. Heat it to a temperature between T_1 and T_2, but do not allow it to reach T_2. If melting to the isotropic phase occurs, the contact is re-established, and from then on only a complex conic structure will be observed. But if one stays below T_2, the terraced drops can be seen to form and to spread. These same drops form, more easily and without any special precaution, if the support is eliminated. This is done by suspending the drop over a hole in any support whatever. In this case it does not matter whether one starts with the crystal or with the isotropic liquid. This provides evidence that it really is the absence of any close contact with the solid support which governs the terraced drop structure, the most perfect homeotropic form (photographs 4 and 5).

Terraced drops possess smooth and planar free surfaces. More precisely, they possess free surfaces consisting of flat steps exactly parallel to each other. These reflect light simultaneously, like the surfaces of a cleaved crystal. Each of these steps is bounded at the edge by a contour. Along the contour the slope increases dramatically, as it links one step of the ladder with the next. A chain of tiny identical and contiguous focal groups, which become smaller as the height of the step decreases, cling to the abrupt boundary of the step (photograph 5). In this region where the surface suddenly acquires a strong curvature, they reveal a complex conic structure whose details have not so far been studied. When we describe conic structures and particularly when we describe the rod-like 'bâtonnet' structure, we shall again find the same evidence of a relation between the surface curvature and the structure in its neighbourhood.

We may note here that many observations of mesomorphic structures require strong magnification. This is particularly true when one examines the chains of conics sitting on the terrace edges. However, in this respect one is very quickly limited by the proximity of the objective. This fact is difficult to reconcile with

177

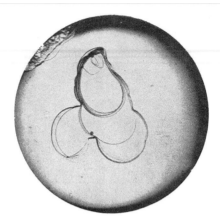

G. Friedel; photograph 4.

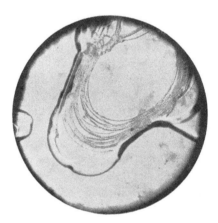

G. Friedel; photograph 5.

the necessity of heating the sample and maintaining it at a constant and sometimes high temperature. If two objectives could be used, each with a weak magnification, the distance to the objective could then be increased at will. A
microscope based on this principle would be very important indeed in the study of
mesomorphic materials. For the moment, all that one can say for sure is that
chains of focal groups obviously exist along the steps, but it is impossible to be
more precise about their detailed organisation.

The steps are of very different heights. If the drop is pressed by moving it, the
steps are so easily displaced that they seem to slide, one on top of the other, without any change in thickness. The material evidently exhibits an enormous fluidity
in the directions of the plane of the steps, even though it is extremely viscous in

all other directions. Furthermore, apart from in the narrow zones on the edges of the steps, the drop behaves as if it is homogeneous and uniaxial, with its axis normal to the surface plane.

There is some resemblance between the terraced drops and the soap films studied by many physicists, and in particular by J. Perrin and P.V. Wells. These films are formed from a solution of a smectic material. They appear, at least to begin with, to be constructed from amorphous material, that is to say, from an ordinary liquid. Does evaporation make the smectic properties reappear in sufficiently thin films? Or is it not rather that the properties of the smectic substance are preserved in very thin films, but the extra water destroys them in the bulk? This last interpretation would be consistent with what we have observed in nematic materials. In this case we sometimes observe that a surface film retains its mesomorphic state well outside the bulk stability region, i.e. at temperatures where the bulk is crystalline or amorphous. Whichever of the explanations is true, there is little doubt that the terraced drops and Perrin's films are very similar, in that they both consist of steps of constant thickness. Soap films are probably not the simple ordinary amorphous liquid films that most people seem to believe they are. Rather they probably consist of uniaxial smectic material with an optical axis perpendicular to the film. However, because soap birefringence is weak, it seems difficult to check this directly.

In Grandjean's drops the step thicknesses are not the same. However, by analogy with Perrin's results, they should be multiples of the same very small thickness, of the order of a molecular length. In impure potassium oleate, mixed with water and other impurities, Wells obtained 44 Å. There we find the first hint of the molecular structure of smectic materials. We are using Perrin's results, while at the same time supposing that in smectic materials the molecules are distributed in equidistant planes. More generally, the molecules would be distributed on parallel surfaces, which are planar when the structure is homogeneous. We shall see that there are many other facts which point towards the same hypothesis.

In smectic materials the plane perpendicular to the optical axis plays a very special role. A first example of this can be seen in the terraced drops. The free liquid surface spontaneously organises itself so as to be parallel to this plane, whose direction must therefore correspond to a minimum of the capillary constant. It is very likely that this is a discontinuous minimum. By this I mean that the capillary constant in directions very close to this plane must take values which are suddenly much larger. The plane normal to the optical axis is thus analogous to the planes with discontinuous properties in a crystal, i.e. the lattice planes. But whereas in a crystal there is a whole family of such planes, governed by Haüy's Law, here there is only one. When considering drop motion, this same plane is exactly analogous to a crystal glide plane. The planes in this direction slide very easily over each other, whereas the resistance to relative displacement is very much larger in all other directions. If we exaggerate a little, we could say that the smectic material is fluid in the direction perpendicular to the axis, and solid in all other directions. We shall see that studies of the rod-like 'bâtonnets' and of rectilinear flow confirm this conclusion. We are led, then, to regard the plane normal to the axis as playing a role in the molecular structure of smectic materials which is the analogue of that played by the dense lattice planes in crystals.

2. *Focal conic structures* – Smectic materials are usually not very fluid. Their consistency in general is somewhat like a soft soap. But they become more fluid with rising temperature. As a consequence the homogeneous regions obtained between two glass plates by melting the crystal eventually become unstable. When the temperature approaches T_2 what is most commonly seen is that the homogeneous regions are destroyed, and invaded by a new structure. This is the focal conic structure. This same structure can be seen at lower temperatures if one nudges the cover-slide, destroying the homogeneous regions without, however, going so far as to induce homeotropy.[1] But the structure in that case is fine-scale and unclear. It is also obtained in a material subject to homeotropy by bringing the temperature above T_2 as though passing into the isotropic liquid, and then returning to the interval T_1 T_2, while at the same time avoiding any movement of the cover-slide. Under these conditions some substances which are very susceptible to normal orientation exhibit homeotropic regions. But it is rare even then not to find a few conic structure regions. Usually, however, the conic structure is easily obtained everywhere. When it is cooled from the isotropic liquid, the smectic phase manifests itself through the bâtonnets which we shall discuss later. These bâtonnets, which themselves possess conic structures, gradually increase in size. They then merge into a continuous body, itself possessing the same structure. The same thing occurs when the smectic material is obtained by cooling or evaporating a solution (oleates). In summary, the conic structure affects the whole sample whenever it does not form homogeneous regions. Furthermore, as we have stated, it can be seen locally even when the homogeneous domains are present. Then we see it: (a) along contact surfaces between regions formed by melting crystals; and (b) on the step edges in the terraced drops. So, we see the focal conic structure whenever we do not see the homogeneous structure.

The material exhibiting the conic structure is immobile. Under the microscope we do not see the continuous and often violent movement of suspended dust which is seen in nematic materials and amorphous liquids. However, in most cases the material is quite fluid. A very weak mechanical force is enough to modify the structure. This can be verified by touching the cover-slide or by putting a thin glass thread into the sample. The sample then rearranges itself and the structure is modified. Immediately afterwards, however, the immobility returns.

Let us consider the well-developed conic structure. One can obtain this, for example, in ethyl azoxybenzoate by bringing the temperature very close to isotropic melting, and then cooling just a little (photograph 6). Under the microscope one

---

[1] The best way of avoiding homeotropy is to use glass supports, with the glass roughened by dilute hydrofluoric acid.

G. Friedel; photograph 6.

can see the whole sample criss-crossed by a large number of fine black lines. These are easily as visible in natural light as they are between nicols (photographs 7 and 8). At first these lines appear to be stationary threads made from some real material. They always take the form of conics, by which I mean ellipses and branches of hyperbolae. They are also always grouped in twos, with an ellipse linked to a branch of a hyperbola by a focality relation. By this we mean that planes of the two conics are perpendicular, that they have the same centre, that the major axis of the ellipse coincides with the transverse axis of the hyperbola, and that the vertex of one of the curves situated on the same side of the centre coincides with the focus of the other. Only one of the two branches of the hyperbola ever appears. In this case near the vertex the line is very black and pronounced. However,

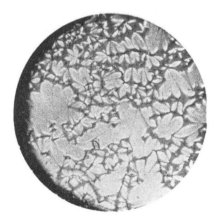

G. Friedel; photograph 7.

181

G. Friedel; photograph 8.

as one goes further away from it in either direction, the line becomes symmetrically thinner and thinner, and finally sufficiently far away it can no longer be seen. Similarly, the ellipse is strongly pronounced close to the vertex coinciding with the focus of the real branch of the hyperbola, but on the opposite side it becomes much thinner. These variations are directly related to the eccentricity. When the ellipse is a circle, the line is uniform. In this case the hyperbola is a straight line, or nearly so, perpendicular to the plane of the ellipse. It is very pronounced when it crosses this plane, but increasingly less so on either side. When the ellipse is very eccentric, it is very black towards the vertex on the side of the hyperbola, but becomes difficult to see towards the other vertex. In this case, the asymptotes of the hyperbola make a very small angle between them and both almost coincide with its transverse axis.

The appearance of such a *focal group* varies with orientation. If, as often happens, one of the two curves lies in the surface of a sample viewed directly from above, then the appearance is as shown in Fig. 1 (I) if an ellipse, and Fig. 1 (II) if a hyperbola. In the latter case the curve perpendicular to the glass is then reduced to one of its two halves. Fig. 1 (III) represents the focal group situated in the bulk and viewed from some random angle of incidence.

In all cases, from whatever angle they are being observed, in projection the two curves are perpendicular to each other at their intersection. This is sufficient, using a well-known geometrical theorem, to establish the focality relation between the two conics. The elliptic and hyperbolic shapes of the two curves, as well as the focality relation, have furthermore been verified in a number of other ways which we shall not go into here. The results are unambiguous given the degree of accuracy of the measurements that were possible.

It is possible to try to deform one of these singular curves, such as the ellipse. This can be done, for example, by introducing a thin glass thread into the sample.

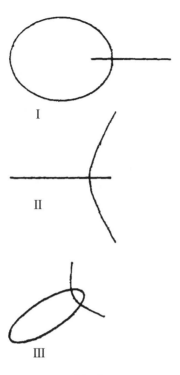

Fig. 1

If the material is very thick and almost solid, then the result is only a local and irregular deformation. In the majority of cases the material is more fluid. If it is sufficiently so, then the ellipse deforms but it remains an ellipse. It may contract, or dilate, or change its eccentricity; the curve deforms everywhere and not just close to the point which has been touched. At the same time all the deformation in the ellipse causes a corresponding deformation in the hyperbola, which follows the motion of the ellipse while always remaining its focal partner.

The ellipse may be complete or it may be interrupted. The hyperbola by necessity must stop somewhere. When the ellipse closes, the focal group is said to be *complete*; otherwise it is *fragmented*. We know that each of the confocal conics is the locus of the vertices of the cones of revolution passing through the other. Take the two points where the hyperbola associated with a complete group stop. From these two points draw all the straight lines supported by the ellipse. The two cones of revolution which are defined in this way bound a region of space known as the focal group *domain* (Fig. 2). Similarly, for a fragmented group, the domain of the focal group is bounded by these two cones and by the two other cones of revolution supported by the hyperbola, whose vertices are the extremities of the fragment of the ellipse (Fig. 3).

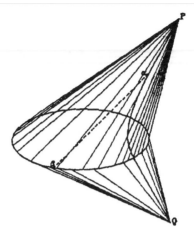

Fig. 2

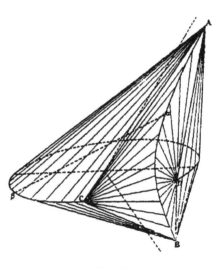

Fig. 3

Even though the relative positions of the focal groups are very varied, they are nevertheless not distributed at random. They interact through completely deterministic laws. The simplest case we obtain can be called the *polygonal* structure. In general this occurs in a rather thin sample, when the fluidity is rather large. An example is when ethyl azoxybenzoate, very slightly doped with colophony, is heated between glass plates treated with hydrofluoric acid. The temperature is

184

brought close to T_2 until one observes the sample start to melt. The sample is then cooled again some degrees while lightly shaking the cover-slide. The hydrofluoric acid treatment is necessary to avoid homeotropy. The field of view in this case is divided, both on the bottom face of the sample and on the top, into closed polygons forming two networks R and R'. These networks are very different and at first sight seem to have nothing to do with each other (Fig. 4; photographs 9 and 10). The sides of these polygons have the same appearance as the conics. They are straight or slightly curved and correspond to each other in pairs on the two faces

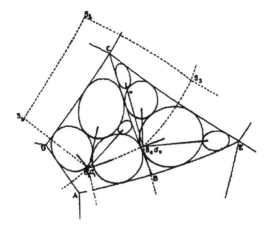

Fig. 4

G. Friedel; photograph 9.

185

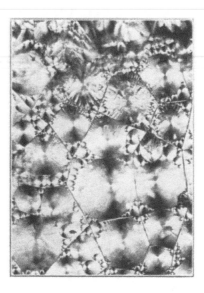

G. Friedel; photograph 10.

of the sample. Each element of R matches an element of R', which is itself exactly perpendicular at the point where their projections cross. These polygon sides are thus very likely to be focal conic groups themselves. However, their weak curvature does not usually permit this to be verified directly. A polygon can be made up of any number of sides. There can even be only two, although in this case they must necessarily be curved. Each polygon is filled with ellipses which are tangent to each other and to the sides of the polygon, and which are stuck to the surface. Occasionally one finds an empty polygon. If one observes the sample from directly above, associated with each ellipse is its corresponding hyperbolic branch, which starting at one of the foci, dives into the sample and is projected onto the plane of the ellipse along a straight line coinciding with its major axis. By focussing we can follow this hyperbola, increasingly weak but still very visible, until it hits the opposite face. In this way we find that all the hyperbolae which correspond to ellipses contained in the same polygon P in the network R converge at one of the vertices of the network R'. Conversely all the ellipses in the same polygon P in R possess major axes which pass through a single point, and this point is the projection of a vertex of R'. And those sides of the network R' which are perpendicular to the sides of the polygon P of the face R converge at precisely this vertex in the plane R'. The number of vertices in the network R' and the number of polygons in the network R is thus equal, and *vice versa*. Starting from the network R and the sample thickness, one can very simply reconstruct the positions and sizes in the network R', and the results are in complete agreement with experiment.

Friedel continues his discussion of focal conics, adding some particular examples, and then goes on to discuss further the pseudomorphism which he has mentioned earlier....

We know that straight lines which rest on two focal conics are at every point perpendicular to surfaces known as *Dupin cyclides*. Cyclides corresponding to the same focal group form a family of parallel surfaces. In these parallel surfaces the perpendicular distance between any two surfaces is constant. The Dupin cyclide is shaped like a ring which is cut by the plane of the hyperbola along two circles of different diameter whose centres are the vertices of the ellipse, i.e. also the foci of the hyperbola. It is also cut by the plane of the ellipse along two circles. One of these circles is of no concern to us because it is outside the focal region, and the centre of the other is the vertex of the hyperbola, or equivalently the focus of the ellipse (Fig. 5). A special case is a torus, which occurs when the ellipse reduces to a circle and the hyperbola to a straight line.

In one of the cases of orientation mentioned above we sometimes observe a focal group whose hyperbola is attached to one of the surfaces of the sample.

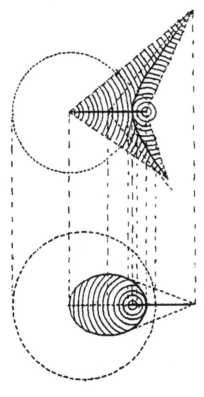

Fig. 5

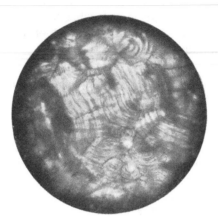

G. Friedel; photograph 15.

As a consequence the ellipse reduces to a half-ellipse perpendicular to this surface. At the instant of crystallisation we see crystalline lamellae trace two series of circles around the vertices of the ellipse, with complete circles restricted to the focal region. In the same way, a series of circles is traced around the vertex of the hyperbola when the ellipse is attached to the surface (photograph 15). These circles are the traces of the Dupin cyclides. At each point the crystalline lamellae are directed parallel to one cyclide and perpendicular to the optical axis of the smectic material. In addition, at each point of the focal domain, this optical axis is directed along the straight line which, passing through this point, rests on the two conics. In other words, the plane perpendicular to the optical axis is at each point tangent to the Dupin cyclide passing through this point, and is defined by the two focal conics. The focal domain is thus a region in which the surface perpendicular to the optical axis of the smectic material curves inward in the form of a Dupin cyclide, rather than growing mainly along a plane, as in stepped drops or more generally in a homogeneous structure. The focal conics are winding axes for the structure, and around them the optical axes diverge.

This explains to us what the conics look like and why they are so visible. The curves can be seen because they constitute the loci of optical discontinuity, the lines near which the optical orientation changes very rapidly. As a result, a light ray passing near one of these lines encounters optical orientations which change over very short distances, and thus is scattered.

Grandjean's experiment on this subject is described, and the conclusion is reached that the only structural discontinuities in smectics are focal conics.

To what does the peculiar tendency of the smectic liquid to wind Dupin cyclides around focal conics correspond? The answer cannot yet be given. This is one of the many questions which are still open in the field of mesomorphic materials. But it is possible at least to make two remarks which may help to provide a physical cause for this geometry.

First, the fact that the Dupin cyclides corresponding to a single group of focal conics are *parallel* surfaces must play an important role. It certainly agrees with what Perrin's observations have led us to suppose for the molecular structure of smectic materials. This is that the molecules arrange themselves in parallel surfaces which are equally separated along the optical axis.

Next, it can be shown geometrically that the Dupin cyclide is the only surface all of whose normals rest on two curves. If, all normals of a surface S hit two curves, or, which amounts to the same thing, if the two sheets of the focal surface reduce to two curves, then the surface S is a Dupin cyclide and the two curves are focal conics. Equivalently, within the small regions which are the focal domains, the optical axes of the smectic substance rest on two lines, which are the only discontinuities in the structure, and which are necessarily focal conics. We shall see that in nematic materials the optical axes diverge along a single line, which can take any arbitrary shape. Why the optical axes rest on two lines in a smectic material, but on only one in a nematic material, is not yet clear. But it seems a property which is less strange than the simple existence of conics and cyclides. It will no doubt be possible to find a physical cause, probably related to the molecular structure of the equidistant surfaces.

Friedel goes on to discuss some more exotic aspects of the focal conics, and shows a picture of the fan-shaped texture.

Oily Streaks – When we destroy the conic structure by shaking the cover-glass we obtain, more-or-less easily depending on the substance, a perpendicular orientation through homeotropy. But it is unusual to obtain by this means a single entirely homogeneous region all in one go. In general there remain regions where very fine focal groups form what Lehmann called 'oily streaks'. These are birefringent bands with negative elongation and transverse streaks. In them we find, if we magnify sufficiently, chains of focal groups whose hyperbolae are perpendicular to the elongation of the band. Ethyl azoxybenzoate possesses little homeotropic tendency between glass slides and when cooled from the isotropic liquid provides beautiful structures with large focal groups. However, when the cover-glass is shaken, it changes to structures which demonstrate the nature of the oily streaks well. These latter make it possible to follow all the transitions between the large bands of transverse fan-shaped zones, where the conics are often still very distinct, and the narrow bands where the focal groups are too thin

189

for them to be observed in detail. These bands have all the properties of oily streaks. In the majority of smectic materials the oily streaks are relatively fine. We can only guess at the transverse chains of focal groups of which they are formed, but not really observe them with any precision. These curious forms are still little known, and are similar to the chains of focal groups which stand at the step edges in the stepped drops.

Bâtonnets – The smectic phase can appear within the amorphous phase, or inside a cooled nematic phase, or form from a supersaturated amorphous solution. In such a case it does not appear in spherical droplets, as the nematic phase does, but in small bodies with special shapes. These appear at first sight to be crystals, and were initially described by Lehmann as such.

These are the peculiar small 'rodlets' (bâtonnets) (Fig. 9) which are in general cylindrically symmetric, and which in special cases can actually be just cylinders with rounded ends, and which can be extinguished between crossed nicols. This is particularly the case when they form in a rather viscous liquid, in such a way that they cannot move freely and coalesce. One often encounters this type in protagon when it is cooled from the isotropic phase.

Very much more often, they are decorated by bulges which give them the shape of banisters, and each of these bulges is made of a necklace of small pearls, of small completely identical projections, corresponding to a garland of focal domains. When two bâtonnets meet each other, they combine just like two drops of an ordinary liquid. But before joining together, if the angle between their respective axes is not too great, they orient each other, and the smaller is then swallowed by the larger. It is evident that the fluidity is not the same in the directions parallel and perpendicular to the rods. The small bâtonnet stays close to where the bâtonnets first touched, and hardly flows, if at all, in a longitudinal direction. It simply produces a localised bulge in the longer of the two bâtonnets. The added mass then distributes itself perfectly symmetrically around the axis. The majority of the bulges form in this way. Nevertheless, even without collisions like this, the bâtonnets usually spontaneously exhibit at least one centre bulge. These bulging

Fig . 9

bâtonnets are by far the most frequent, do extinguish between nicols, and possess an inhomogeneous structure.

The narrative continues with further discussion of the bâtonnets. Friedel is then led to discuss surface tension, in order to understand the shapes of the bâtonnets. He continues...

...Consequently, in an anisotropic liquid whose molecules are oriented parallel to each other but whose positions are randomly distributed just as in an amorphous liquid, we can expect the capillary constant to vary little or not all as a function of direction. Despite the anisotropy, the drops will be spherical or nearly so. We shall see that this is what happens in nematic liquids. By contrast, we find in smectic materials strong variation of the capillary constant with direction, manifesting itself by strong departures from a spherical shape for the drops. And so we can plausibly conclude that we must seek the cause not in the mutual orientation of the molecules, but rather in their spatial organisation. This strongly supports what the stepped drops have suggested to us. A smectic material, unlike a nematic material, but like a crystal, possesses in its molecular structure something beyond the fact that the molecules are parallel to each other. This structure obeys some law of spatial distribution of the points homologous to the molecules. We are thus almost necessarily led to the idea of the distribution of the molecules within parallel equidistant surfaces, which are flat in a homogeneous structure.

The constant uniaxiality of all smectic materials, and their symmetry of revolution around the normal to the structural plane, forces us to admit that within each of these planes the molecules are distributed at random, and rotated in all sorts of ways. However, they do at the same time possess one direction in common, and this is oriented perpendicular to the plane. If one looks along the direction of the optical axis, such a material has to some extent an amorphous molecular structure. However, when viewed from a perpendicular direction, it exhibits one unique direction, that of the equidistant planes, analogous to the directions of the lattice planes in a crystal. In this way, smectic materials do indeed constitute a stepping stone between crystalline and amorphous materials.

However, we should point out a difficulty encountered by this hypothesis, albeit a minor one. The plane perpendicular to the axis, which is the only one to be compared with the many crystal lattice planes, should exhibit a minimum for the capillary constant. Indeed it is this which shows up in the stepped drops. On the other hand, the bâtonnets, rather than being flattened parallel to this plane, as one should apparently expect them to be, by contrast extend in a direction perpendicular to this. Those which are homogeneous are always cylindrical and extended parallel to the optical axis. This means that the capillary constant is minimal for the planes (all identical, because of the cylindrical symmetry) parallel to the optical axis, that is to say, for the directions at right angles to those for which the stepped drops define a minimum. There is no contradiction, for there is no reason why there should not be two relative minima of the capillary constant. But if the molecular structure which we are giving to the smectic material really explains the minimum for the plane perpendicular to the axis, then it will not explain the minima for the planes parallel to it. In order to explain this, we shall probably have to make use of the extremely elongated linear shape which is characteristic of the molecules in mesomorphic materials and particularly in smectic materials. If the molecules whose homologous points occupy the parallel planes were approximately spherical, there is no doubt that because these planes correspond to a density maximum, the perpendicular directions would be characterised by a density minimum and thus by a maximum in the capillary constant. But if the molecules, being linear and very long, are arranged so that their long axes are normal to the plane, it could be that in the direction normal to the plane the atomic elements provide a density maximum, even though the equal distances of the *homologous* points are maximal in this direction.

Subject to this reservation, all the facts which we have described so far agree with the structural hypothesis which we have proposed. Mauguin was the first to observe parallel orientation of mesomorphic materials by crystals, and Grandjean made many observations on this subject. Grandjean's observations all lead also to the same conclusion.

There follows a discussion of the orienting effect of crystal surfaces, both in nematics and in smectics, and some mention of flow properties of smectics at surfaces. Finally Friedel addresses himself to the question of a direct experiment to confirm his layer hypothesis . . .

We will notice that the measurements of Perrin and Wells on soaps are sufficient in principle to confirm, more or less definitively, the structural hypothesis that so many facts suggest. But there is without doubt another means of finding evidence for the equidistant planes in smectic materials, and to measure their separation in pure materials which are better defined than those which Perrin studied. This involves using X-rays. One could study the stepped drops, for example using drops suspended over a hole and which in principle are the same as soap films. One would then examine photographically the reflections of different orders at the same wavelength from the plane of the homogeneous drop. In view of the measurements by Perrin and Wells, in order to do this one would have to use rather long wavelengths. The same conclusion follows when we consider the extreme length of the molecules. The interplane separation is probably of the order of several tens of Ångstroms. Thus, in order to obtain a large enough angle of reflection one would have to use X-rays with a rather large wavelength. It would be easier, however, to avoid the problem of maintaining a large enough stepped drop for a long time, and to use the method of Debye on a smectic material in a conic structure. In this case the equidistant surfaces would be oriented in all possible directions. A number of attempts to use X-rays to study the molecular structure of mesomorphic materials have been made. However, so far all experiments have used nematic (or cholesteric) materials. The complete absence of any X-ray diffraction in these materials shows only that for these materials there is no periodic atomic arrangement. It does not seem to have been realised that smectic materials might behave differently. As a result of all these considerations, I think it extremely likely that such an experiment would be successful. The experiment in these materials would involve the reflection of X-rays from a single plane normal to the optical axis.

I now want to complete this summary of the properties of smectic materials and the contrast which they make with nematic materials. One should again record the fact that the strongest magnetic fields that one can produce have no effect on even the most mobile smectic liquids, for example ethyl azoxybenzoate. By contrast, the same fields strongly align true nematic liquids (Mauguin). On the other hand, nematic materials of the cholesteric type, in which we have found strong analogies with smectic materials, appear to remain unaffected (at least in the fields of some thousands of Gauss which it has been possible to produce).

Electric fields (Yngve Björnstahl[#]), which orient true nematic materials, have no aligning effect on smectic materials.

It does not seem that these two differences between the two types of mesomorphic material can be the result of a greater or lesser viscosity. However, there is much to do to understand these phenomena, for experiments have only been carried out in a small number of substances.

We also note the absence in smectic materials of Brownian motion of the dark regions observed between nicols. This is observed in nematic materials as soon as they are slightly mobile. The phenomenon is closely linked to the viscosity, and disappears, even in nematic materials, for slightly larger viscosities.

At the end of this paper a list of materials in the smectic phase which I have been able to observe can be found. In addition to pure substances, one should add a large number of mixtures, which are as useful as pure materials in the physical study of the smectic form of matter. Usually these are mixtures of two smectic materials, or between one smectic material and another material which, when pure, does not possess a smectic phase. But there also cases in which a mixture of two materials exhibits a smectic phase, even though neither pure material possesses a smectic phase. This is the case for mixtures of azoxyphenetole or azoxyanisole (both with only one mesomorphic, nematic, phase) with amyl cyanobenzalaminocinnamate

[#] Yngve Björnstahl (1880–1942) was professor of experimental physics at the University of Uppsala (*ed.*)

193

(a material with a single cholesteric-nematic mesomorphic phase). These mixtures exhibit, like many pure materials, a nematic phase (of the cholesteric type) at higher temperatures, and at lower temperatures, a positive smectic phase of the ordinary type not shown by the two materials separately.

Second mesomorphic type: nematic materials

We need to distinguish two different types of nematic materials. Between them, as we shall see, one can find continuous transitions. I shall call the first the *true nematic type* (thread-like liquids and liquids with nuclei, positive nematic liquids), and the second the *cholesteric type* (liquids with Grandjean planes, negative nematic liquids).

True nematic liquids

The nematic phase, whether true or cholesteric, is always, just like the smectic phase, separated from the crystalline and amorphous phases by an absolute discontinuity. When the same material exhibits both a nematic and a smectic phase, these phases are likewise separated by an absolute discontinuity. Some authors, such as Vorlænder, believe in the existence of continuous change between the two types. However, this is a theoretical belief, rather than one based on observation of the facts, since the properties of these alleged intermediate forms have never been described. By contrast, in my observations I have always found a completely sharp boundary between the nematic and the smectic phases. On each side of this boundary the properties are very different, they are well defined and always the same, and there is nothing which remotely resembles a continuous transition.

It is also always the case, that when the two phases do exist, the smectic phase always occurs at lower temperatures and the nematic phase (whether true or cholesteric) at higher temperatures. When these phases exist in equilibrium, the smectic phase appears at temperature T_1 when the crystal melts. The nematic phase then appears at T_2 at an abrupt and reversible transformation, and subsequently this nematic phase transforms at a temperature T_3 into an amorphous liquid at a final discontinuity. Thus, considering the classification of the stability domains on a temperature scale, the smectic phase comes between the crystalline phases and the nematic phase (of either variety), and the nematic phase between smectic and the amorphous phases. I know of no exception to this rule, which appears to be completely general.

The smectic phase presents some analogies with crystals. By contrast, we shall see that, in complete agreement with the findings of the last paragraph, the properties of the nematic phase are much further away from those of a crystal. The smectic possesses many properties which have given reason to believe that the smectic phase has a molecular structure involving a degree of periodicity, and to suppose a single direction of 'lattice' planes similar to lattice planes in a crystal. All these properties disappear simultaneously in the true nematic material and we shall first of all study this situation. It is at first sight disturbing that these properties partly reappear in the cholesteric type, and we shall later see how this happens.

1. Homogeneous structure – ...

Friedel discusses some aspects of experiments on homogeneous structures (i.e. films with the director uniformly aligned in the plane of the sample). He then goes on to discuss the physics of what eventually turns out to be the twisted nematic cell.... Note, however, how strongly (and wrongly, as it would eventually turn out) Friedel guards against a mechanical torsion effect, as opposed to merely an optical effect, between two slides twisted with respect to each other.

The same holds true if, beginning with large homogeneous uniaxial regions obtained by melting a crystal, one rotates one of the slides, the cover-glass, for example, by an angle α with respect to the other. Those parts of the same homogeneous zone which remain superposed by this action possess all the properties of the respective zones. Suppose the nicols to be initially crossed at a right angle. Then, for perfect extinction of the new zone in white light to occur, one must also turn the nicol analyser with respect to the polariser through an angle α.

G. Friedel and Grandjean have described these properties in detail. However, they were victims of the illusion that the optical properties seemed localised in the surface films. They were thus not able to find an interpretation of their experiments, which was given completely by Mauguin.

Mauguin has shown that these zones are simply *twisted*. There are extra-ordinary stable films attached to the two glass slides. These films each have their own optical orientation and they each determine the optical orientation of the nearby bulk. In these homogeneous zones, the two films, each aligned by the crystal which preceded them, have the same orientation. The same holds true for the liquid which comes between them. In twisted zones, the main parts of the two films make some angle between them. The bulk liquid between them links these two orientations by a helicoidal twist whose pitch is constant over the whole sample. To be exact, the word twist should not be taken in its mechanical sense. One realises, for example, that a twist of 360° is equivalent to no twist at all. If the optical axis of the initial homogeneous zone is parallel to the slide, then in the same way a twist of 180° is the same as a zero twist. Furthermore, the twist always takes the shortest route. Thus, when the angle α between the two films goes beyond 90°, one can see (in monochromatic convergent light) a twist which, for example, had been right-handed for $\alpha < 90°$, suddenly become left-handed for $\alpha > 90°$, or the other way round. All this occurs suddenly and instantaneously. In addition, we should take care not to confuse this peculiar type of twist with mechanical twist. It is sufficient to recall that, despite the stability of each optical zone, and of the twist which then occurs, the liquid is, or at least could be without changing anything, in continual motion. The liquid molecules which make up the bulk between the films attached to the glass slides are constantly being renewed. It is extremely peculiar, but abundantly clear, that nevertheless, when these molecules enter a particular zone, they immediately take on the orientation defined either by the films themselves or by the helicoidal connection rule.

Mauguin has shown that the optical properties of the twisted samples are completely explained as follows. One describes an incident vibration by its representative point P on the Poincaré sphere. The effect of going through a layer twisted through an angle α is to roll the point P around the cone of revolution on the equator of the sphere. This motion is such that the generatrix tangent to the equator, which is initially parallel to the principal positive section of the first surface film, traverses an angle α in the sense of the twist, and so eventually coincides with the representative azimuth of the principal positive section of the second surface film. The vertex angle of the cone is given by:

$$tg\omega = \frac{\lambda\alpha}{\pi e(v_1 - v_2)}$$

where λ is the wavelength, e is the thickness of the layer, and v_1, v_2 are the indices of vibrations parallel to the two principal sections in the homogeneous sample.

There follows a discussion of the persistence of films, how to prepare slides, and Friedel notes the phenomenon of optical flashing characteristic of homeotropic textures which occurs in nematics, but not in smectics. The next heading deals with structures *à fils* and *à noyaux*. He discusses 4-brush and 2-brush defects, with indices ±1/2 and ±1, noting also the sum rule for the combination of defects. A short section deals with the negative results of X-ray scattering experiments on nematics, and then he goes on to discuss Mauguin's experiments on the orientation of nematics in magnetic fields, noting the connection of this phenomenon with the Cotton–Mouton effect. We take up the story again as he is summing up his conclusions so far . . .

Molecular structure of nematic substances – I will briefly recall here the properties of smectic materials which agree with the idea of a distribution of molecules on equidistant parallel surfaces. I shall then compare these with the properties of nematic materials.

1. Smectic materials are immobile, in the sense that if the structure stays the same there is no bulk displacement of matter. All displacements are accompanied by structural change, save only slippage perpendicular to the optical axis. This agrees well with the idea of a molecular structure consisting of well-defined spatial arrangement of the molecules, in which the plane perpendicular to the optical axis plays a special role.

    By contrast, in nematic materials we have seen that the liquid is often very fluid, even more so than the amorphous phase, and that very violent motion of the liquid changes the structure not a jot.

2. Inhomogeneous smectic materials arrange themselves in focal domains, organised by a family of Dupin cyclides. These are *parallel* surfaces which are normal to the optical axis. This suggests that there exist in the molecular structure, in one or another form, parallel surfaces perpendicular to the optical axis.

    In nematic substances there is nothing remotely like this.

3. The stepped drops in smectic materials show that there is a unique planar direction with discontinuous properties. This plane is precisely the one perpendicular to the optical axis, and the one which in inhomogeneous materials bends itself into a Dupin cyclide.

    In nematic substances stepped drops do not occur.

4. Rectilinear flows in smectic materials on crystalline cleavages also contribute to show that plane normal to the optical axis is a plane with discontinuous properties.

    Rectilinear flows are not observed in nematic materials.

5. The bâtonnets formed by smectic materials, whose shapes are far from being spherical, show that there are strong variations in the capillary constant as a function of direction. The study of crystals teaches us that these variations are related to the spatial organisation of the molecular elements, and little, or not at all, with their nature or their orientation. From this we derive another reason to believe in a certain degree of regularity in the spatial organisation of molecules in smectic materials.

    In nematic materials the drops are spherical and the capillary constant hardly varies with direction. The spatial organisation of the molecules thus seems likely to be similar to that in an amorphous material.

6. The orientation of smectic materials by crystal surfaces is selective and is thus subject to some conditions. The study of crystals teaches us that these conditions are related to the periodicity in the spatial organisation of the molecular elements. Thus there is a certain degree of periodicity in the molecular structure of smectic materials. However, this is less complete than that which exists in the crystalline state.

Nothing similar seems to exist in nematic materials. The molecules in this case align freely on top of those of the crystal without being hindered by the necessary matching between the two periodicities.

To summarise, there are many reasons leading to the conclusion that the molecules in smectic materials are organised in parallel equidistant planes. All these reasons disappear at the same time when we go from smectic to nematic materials. This partial periodicity, which still makes smectic materials resemble crystals, thus probably does not exist in nematic materials. The molecules in nematic materials are probably distributed at random, just as in amorphous materials. But on the other hand, nematic materials are spontaneously birefringent, and often very strongly so. We therefore must attribute this anisotropy to a common molecular alignment, caused by a field determined by their mutual interaction. The constant uniaxiality leads one to suppose that the molecules rotate in all possible ways, because it is not very likely that these very different molecules are themselves cylindrically symmetric. In this respect the same holds true for smectic materials. The whole system has a cylindrical symmetry around the common direction, which is that of the optical axis.

I cannot hold myself back from noticing how much the existence of a common direction for all the molecules is consistent with Vorländer's* observation concerning the chemical formula of mesomorphic materials. This holds true for both mesomorphic forms. This formula was such that the chemists imagined that the molecule is always linear, has no bifurcations of any importance, and is remarkably long. It is probably rectilinear in the mesomorphic form, and has, so to say, the shape of a needle. The collection of molecules with this shape would then behave rather like a packet of needles. Indeed, one can hardly imagine how a collection of such molecules could be organised with an ordinary density, unless they were parallel, or very nearly so, to each other. This is especially true in a very mobile liquid like the usual nematic material. It is true that at higher temperatures the same material can be found in an amorphous state. But it is always possible to imagine that when the material melts to the amorphous form, the molecules lose their rigidity while remaining linear. They would then become able to bend, and become entangled with their neighbours. The viscosity measurements of Schenck and Eichwald agree well with this idea for true nematics, whose interior friction increases suddenly when going from the nematic to the amorphous states. According to these authors, the same is not true for smectic materials (ethyl azoxybenzoate). But we should recognise that viscosity measurements using flow for smectic liquids only yield mean values whose meaning is not well defined. Smectic liquids possess an interior friction which depends

* For some reason, Friedel has reverted to the orthodox spelling of Vorländer's name, having used æ for ä earlier! (*ed.*).

strongly on direction, and in particular exhibits a discontinuous minimum for a direction in the plane.

The molecular structures to which we have been led for smectic and nematic materials are in extremely good agreement with the order in which we place the four types of matter on the temperature scale. At low temperatures we find the crystallised type in which the periodic spatial organisation of the elements is perfectly well defined, and in which thermal fluctuation plays but a superficial role. At higher temperatures the smectic type still exhibits some degree of periodic structure. However, what remains of the perfectly periodic crystal structure is the parallelism of a common direction for the molecules and their distribution over equidistant surfaces. Thermal fluctuation has broken the other bonds. At still higher temperatures the nematic phase shows us the thermal fluctuations which have now destroyed all periodicity. All regular distribution of the homologous points of the molecules has gone. All that remains is the parallelism of the common direction of the molecules. Even this only occurs on average, and in detail it is obscured by visible fluctuational motion. Finally at still higher temperatures, all order disappears into detail; all that remains are average properties. This is the amorphous phase.

It is notable that this invariant order can be found when these metamorphoses are caused, not by varying the temperature, but for some other reason. Consider the notable case of 10-bromphenathro-6-sulphonic acid. This material can be crystallised at room temperature. It dissolves in an excess of water, yielding an ordinary amorphous liquid solution. But if one adds only a small amount of water to the crystalline material, a thick paste is obtained. This paste is smectic; adding more water yields a turbid much more mobile liquid, and this is a nematic material. More water still finally returns the amorphous liquid. These transformations can easily be followed under the microscope.

The narrative continues with a detailed experimental discussion of the phase progression for this lyotropic material. Friedel then passes on to a discussion of cholesteric materials

---

Nematic materials of the cholesteric type

Friedel first deals with the conic structures (analogous to the smectics). We take up the story again when he turns to the signature of cholesteric planes

---

2. *Planar structure* – The *planar* structure of cholesteric materials possesses very special properties. In part these have been known for a long time. The special properties make the cholesteric phase a rather unique type of matter. The

first case is when it consists of uniform zones, which only exist in the perfectly regular planar structure, and are aligned perpendicular to the optical axis. These zones, when observed in natural light, reflect bright colours which vary according to the angle of incidence. Between nicols, they exhibit colours which do not change when the hot stage is rotated. *In part* they are the result of what is in general a very strong rotatory power. In part, however, as we shall see, they have another cause entirely. They are very regular, and when it is possible to examine them closely, it can be seen that they consist of layers of equal thickness bounded by surfaces of discontinuity (Grandjean planes). The three properties – coloured appearance, rotatory power and the presence of the Grandjean planes – are intimately linked, and do not occur separately. There are numerous further details, some still puzzling, and we shall try to use these to elucidate the question of the structure of the planar zones as best we can. So far this structure is badly understood.

The second case is when the planar structure is not perfectly regular. Now it is disturbed by 'oily streaks', which are completely analogous to those we have met in smectic materials. If we move the cover-glass and destroy the small conical zones, we see these zones seem to deform. They flow, giving rise to oily streaks. Between the oily streaks, uniform regions appear, and these possess the regular planar structure. If the sample is placed between a planar and a convex glass slide, or in a wedge cell, the motion of the support does not establish a regular planar structure in the thinnest part of the sample. Rather, the oily streaks are expelled towards regions where the cell is thicker. There they remain, surrounding the planar structure regions. Only in very thin films of the order of a hundredth of a millimetre or less can the oily streaks be eliminated entirely so as to create a completely regular planar structure. As for the oily streaks, as in smectic materials they are badly understood. Just as in smectic materials, chains of small equal birefringent figures appear. These are surely chains of focal groups, but the precise details are indistinct.

Optical sign – We have seen from the conical structure that, whereas all smectic and true nematic materials possess a positive optical sign, all cholesteric materials have a negative optical sign. The oily streaks confirm this. The transverse striation, homologue to the fan-shaped zones, is negative, and the elongation positive. This is the reverse of what occurs in the smectic oily streaks. The sign of the planar zones observed in convergent light is always negative, again confirming the negative optical sign.

We shall see that there is good reason to suppose that this striking sign anomaly in cholesterics materials is misleading. It is misleading in the sense that the optical sign of a small element of a cholesteric material is just as positive as it is in any other mesomorphic material. The negative sign can be understood in terms of the very strong twist around a direction normal to the elementary positive optical axis. This twist causes the whole system to be negatively uniaxial. We shall see that the positive sign of the element can be directly observed under certain special conditions.

Scattered colours – When the planar zones are illuminated by white light, the scattered light consists of bright colours, whose appearance is reminiscent of peacock feathers, or of the opal of Lippmann's photographs. When a cholesteric material is melted between glass slides, there are no scattered colours when the structure consists of small conical zones, but the colour appears immediately if the cover-glass is lightly shaken. The colour is a function essentially of the angle of incidence, or more precisely of the two angles made between the normal to the sample and incident ray and between the normal to the sample and the reflected ray. The wavelength is a maximum when the two rays are normal to the sample, and diminishes as the angles increase. It reaches a minimum when they are both at grazing incidence. It diminishes momentarily when the cover-glass is touched or displaced, and then returns rather quickly to its original value. Thus it evidently depends on the structure. If one ensures that the structure is completely regular, then the colour is uniform and appears like a spectral colour. In fact, it is not monochromatic, but is characterised by a strong intensity maximum at a well-defined wavelength, whose colour is only slightly different from the reflected colour, which is actually complex. Without precautions, the colour is usually inhomogeneous. Under the microscope it appears to be made up of small differently tinted spots.

Once a regular structure, and hence a rather uniform set of colours, has been obtained, it is easy to establish that the scattered colour comes entirely from *circularly polarised* light (Giesel). Depending on the material, the coloured light is either completely extinguished by a left-circular analyser, and thus made up of right-circular vibrations,[1] or *vice versa*. In this way we can distinguish two types of cholesteric material. We label those materials which reflect right-circularly coloured light as *dextrorotatory* (*d*), and those materials which reflect left-circularly coloured light as *levorotatory* (*l*). The properties of these two types, particularly with respect to their structures, are mirror-images of each other.

This circular polarisation of the reflected light continues to hold true even if the structure is no longer perfectly regular. But in this case, the circular analyser only extinguishes the light completely for film thicknesses which are not too large. One can observe these properties particularly easily in cholesteryl benzoate-cholesteryl acetate and cholesteryl benzoate-cholesteryl caprate mixtures. These mixtures in fact remain in a supercooled cholesteric state right down to room temperatures. They can, in suitable relative concentrations, remain in this

---

[1] That which allows *right*-circular vibrations to pass, I am calling a right circular analyser. *Right* here means that the representative point, for the observer receiving the ray, turns in the same direction as the hands of a clock.

state for hours, and sometimes for days and even months. The best relative concentrations in the first case are close to 55 per cent benzoate to 45 per cent acetate. If one melts this mixture in a thin film between two glass slides, and then cools it again, one finds first a fleeting violet tint. This is not the result of a special phase, but simply the appearance of the cholesteric phase in a cloud of extremely fine droplets immersed in the isotropic liquid. Then, once the transformation has occurred, one sees the gradual appearance of a violet tint. To begin with this is very dark, and only visible at normal incidence and in normal light, whereas at oblique angles the tint goes black. Little by little this tint approaches a clear violet, and then successively blue, green, yellow and red at normal incidence and in normal light. Grazing incidence always yields a tint with a lower wavelength. Cooling further to $0°$, the red itself disappears at normal incidence, and all we see is black. Now the colour is only visible at oblique angles. Thus we go through the whole of the visible spectrum, from the ultraviolet to the infra-red. If the cover-glass is pushed down or moved, the tint revives and returns towards the violet. However, it does so unequally from point to point, and varying *as a function of the orientation given by the movement*. The material is now very viscous because of the cold, and only returns to its original colouring very slowly. By using the circular analyser, it is found that the colour only comes from left-circular vibrations, and thus we must be dealing with a levorotatory substance.

In one respect Friedel was mistaken. He saw a "first fleeting violet tint" at the onset of the cholesteric phase, and attributed it not to a new phase, but to a cholesteric-isotropic emulsion effect. We now know that this effect was in fact due to a special phase – the blue phase – in which there are bright colours over a restricted temperature range close to the onset of cholesteric behaviour. We omit much of the further treatment of cholesterics (some a repetition of Mauguin's work).

Among interesting phenomena pointed out by Friedel is the inversion of the pitch as one changes the concentration of a mixture, although the critical concentration varies as one changes the temperature. Thus at fixed concentration one can change from l-type to d-type cholesterics simply by changing the temperature.

We return to Friedel right at the end of his discussion of cholesterics...

The relation between cholesteric and other mesomorphic materials – Some cholesteric mixtures are able to pass through a perfectly characterised nematic state in the liquid at the θ-point from a dextrorotatory to a levorotatory structure. This implies unambiguously that cholesteric materials belong to the nematic type. They are only a special form of nematic material, related to two types of structure which both may occur in the same material. One is the planar structure and this remains puzzling. But we know that it possesses holoaxial structure, with neither a plane nor a centre of symmetry, and is able to sustain two enantiomorphic forms which constitute the l- and the d- types. This structure, which consists of right-handed or left-handed twists imposing a rotatory power, always has the same sense in a particular material at a defined temperature (apart from delay phenomena which can occur at the transformation). The structure is completely consistent with a lack of holoaxial symmetry in the molecules. It is certainly possible to observe twists in true nematics, such as azoxyphenetole, for which the molecules are not asymmetric. But these twists take any arbitrary value or sense. The planar structure and the twists with a determined sense and magnitude which it implies are necessarily linked to a lack of molecular symmetry. This is not the direct cause of the rotatory power, but it does determine it indirectly, because of the structure to which it gives rise. Cholesteric materials thus appear to be the form taken by nematic materials manufactured from molecules endowed with holoaxial asymmetry.

The other structure is that with negative conical zones. In this case one normally sees neither Grandjean planes, nor rotatory power, nor yet selective reflection. These smectic analogies seem initially to contradict the classification of cholesteric materials within the nematic type. We need to try to resolve this apparent contradiction.

The passage from a cholesteric type to a true nematic by a continuous transition is not only observed in those mixtures of d- and l-materials where inversion easily occurs. In the first case, in cholesteric materials with a planar structure, it is frequent to observe extremely thin films which sometimes extend by capillarity over the support and whose thickness is very small by comparison with the thickness of the planes. One also observes in the first planar interval, near the contact with the support, zones whose nematic nature is clear. We have seen an example of this in the zones with nuclei (*plages à noyaux*) exhibited by very thin films of a mixture of cholesteryl benzoate and cyanobenzalaminocinnamate. But the same observation could be made in many other cases. In the same way we can consider a mixture of more or less equal proportions of cyanobenzalaminocinnamate and azoxyphenetole. This can exhibit a very nice planar structure. At the glass slide contact, it also exhibits, in the first planar interval, zones marking the trace of solid crystals, completely identical to those made by pure azoxyphenetole. These persist even above the melting temperature to the isotropic phase. Nothing like this occurs in smectic materials.

Furthermore, it is easy to follow the gradual transformation of a cholesteric material into a nematic material. One observes a mixture of a cholesteric and a true nematic material and increases the nematic proportion. The transformation can be seen to be completely continuous. A mixture of cholesteryl benzoate (l-cholesteric) and azoxyphenetole, which are miscible in all proportions in the mesomorphic state, is very suitable for this observation. The mixture is completely homogeneous and behaves like a pure substance. There is only a single mesomorphic phase, and this is a 1-cholesteric type. We have already referred to this above, in connection with the extreme sensitivity of its structure to cover-glass motion. The structure with small conical zones is very stable in a stationary sample, but immediately gives way to a planar structure as soon as one touches the slide, even ever so lightly. This enables us at the same time to assure ourselves again that the transition between cholesteric and nematic types is completely continuous, and to understand better the negative conic structure of cholesteric materials. The proportions indicated above for the mixtures are very rough, and were usually evaluated without weighing small amounts of material. As a result they only give an idea of the order of magnitude.

Let us consider mixtures rich in cholesteryl benzoate, for example 5 parts of benzoate to 1 of azoxyphenetole. When this mixture is cooled from the isotropic liquid, one obtains the structure with small grey zones. Then, if one touches the cover-glass, the planar structure forms immediately. The planes can be seen, for example in a mica slit. They are closely packed, but dark in colour, and become much sharper at lower temperature. The reflected colour is blue. The same holds true, with little change in the reflected wavelength and the distance between the planes, so long as the benzoate proportion does not go below about 1:1. Beyond that, the planes begin to rapidly pull apart. The result is that the reflected wavelength is shifted towards the red. It is already in the extreme red for a proportion of 2:3 (by weight), and beyond that moves into the infra-red. Even while the planes are moving further and further apart, the planes remain completely sharp up to 1:5 and 1:10 mixtures. In a mica slit these mixtures exhibit the most beautiful types of planar structure that it is possible to see, and very sharply. However in thin drops, the mobile fringes and striations resembling commas can be seen with less regularity. At proportions of the order of 1:20, the edges of the planes, now very far apart from each other, begin to take on the appearance of the threads of a nematic material. At the same time independent threads appear between the planes. Their ends are frequently attached to these edges; alternatively they may go from one side of a plane to the other. However, these threads do not seem different from very fine oily streaks, and coalesce in the same manner. In 1:40 mixtures the liquid has become extremely mobile. The planes are however still very pretty, and their spacing is still growing. They can still be seen as clearly, but now very far apart, at proportions 1:100 and 1:200. By this stage one can see only 2 or 3 planes in a sample, whereas in a 1:5 mixture, for example, more than 50 planes could be seen. The edges of the planes can no longer be distinguished from threads, save by their arrangement in steps parallel to the lines of equal thickness. The liquid has now become as fluid as water. Nevertheless, if one presses on the mica slide to expel the liquid, and then lets it return, the edges of the planes come back, still unbroken and still occupying the same positions. The thick oily streaks have disappeared. Only very fine streaks remain, which are indistinguishable from pure azoxyphenetole threads. If one crystallises the sample and then returns to the mesomorphic phase, the uniform zones which mark the traces of the crystals, and which are identical to those in pure azoxyphenetole, remain attached to the surface of the support. In some places one also can find zones with nuclei. The same phenomena, and particularly the persistence of the edges of the planes, can be followed right up to mixtures of the order of 1:400. Apart from the existence of these planes, the liquid has now assumed the qualities of a true nematic liquid. Beyond this, it cannot be distinguished from pure azoxyphenetole.

For a number of different reasons it would be very interesting to study quantitatively the variations in the properties of these mixtures. But what we need to remember for the moment is the evidence of a gradual change from the cholesteric type to the true nematic type. There is no discontinuity. There is also the fact that nearly pure azoxyphenetole with less than 1 per cent of cholesteryl benzoate already displays the characteristic properties of the cholesteric material.

Let us now start again with the same mixtures. Instead of looking at the planar form, we study the form with small conical textures, making sure not to ruin them by moving the cover-glass.

For the pure benzoate or mixtures rich in benzoate, the small grey zones which form between glass slides on cooling are indistinct. They become more distinct and larger as the proportion of azoxyphenetole is increased. Starting with proportions of about 1:1, they become very pretty. Their structure is now clearly conical, with fan-shaped textures and even very distinct focal groups. Between nicols, one can see very nice birefringence colours, with a negative sign. There is no longer any trace of rotatory power. Between a planar and a convex slide there are no rings of rotatory power shifting with a continuous movement as one rotates the nicols. There are only birefringence rings which, when one of the nicols is rotated through 45°, disappear but do not shift, and reappear in intermediate positions.

If one touches the cover-glass the regular planar structure reforms in the thinner part of the sample, with reflected colours and rotatory power. In the thicker regions, oily streaks develop. These are large, and seem to be made up of chains of conics, whose hyperbolae are perpendicular to the elongation. These oily streaks are the result of a deformation of the initial oily streak structure. But if one only touches the cover-glass very lightly and delicately, it is easily possible to see how they are arranged with respect to the fan-shaped textures. One finds that they form perpendicular to the hyperbolae in the fan-shaped textures and that their transverse striation is completely the same as these hyperbolae.

When the isotropic liquid is cooled, one can observe the edge of the mesomorphic phase. It is seen to be uneven and jagged, with tapering bulges which run into each other, while curving round and standing up. This is exactly the same behaviour as that of the surface of a smectic material when it is formed in the isotropic liquid by colliding *bâtonnets*. The majority of the pointed bulges are connected to the bulk, but occasionally isolated drops can be seen. These drops have exactly the same shape as the smectic *bâtonnets*. They coalesce and are absorbed by the bulk just in the same way as are *bâtonnets* in smectic materials.

205

The same observations can be made in mixtures richer in azoxyphenetole. By the time one reaches proportions of 1:10, the stability of the conical texture is significantly reduced, and the liquid is becoming increasingly mobile. Although the conical structure is still quite stable for low thickness samples, in thick films simply moving the liquid often switches the texture to a structure with oily streaks surrounding regions with rotatory power. However, by contrast, in thin films the conical structures have become more distinct and larger. The isolated smectic *bâtonnets* tend to disappear, and can only be seen as conical bulges of the surface of contact with the isotropic liquid. These bulges are frequently elongated and almost cylindrical. But now something new appears which does not occur in smectic materials. When one looks under rather strong magnification, one sees a system of very fine lines. These are exactly equidistant, but were too closely spaced in the previous systems to be visible at practical magnifications. But here they appear clearly. They are normal to the hyperbolic streaks of the fan-shaped textures, and thus are normal at each point to the negative optical axis of the liquid. We recall that one has but to touch the cover-glass, and the optical axis will align perpendicular to the slide, at least in thin samples, and a system of equidistant Grandjean planes will form parallel to the slide, i.e. perpendicular to the negative axis. So it must be that in the conical textures the equidistant surfaces are identical to the Grandjean planes. They are too finely spaced to be visible under the weak magnifications available in normal small textures in cholesteric liquids. But here they appear because, as we have discovered from the study of the planar structure, the distance between the Grandjean surfaces has become much larger.

It appears that what we are seeing are Grandjean's surface textures; Lehmann observed these in drops consisting of azoxyphenetole doped with colophony, and described them as 'Schraffierung'.[1]

Now the proportion of cholesteryl benzoate is further reduced, towards 1:20. The mobility of the liquid is much increased. This means that, when the isotropic liquid is cooled, it is no longer possible to sustain the distinct conical and fan-shaped textures. But these textures form nevertheless, and continue to exist in places where the liquid is relatively immobile. If the cooling is done rapidly, it is even still possible to obtain a conical structure over the whole of a thin sample. The immobile textures exhibit Grandjean surfaces, still more distinct and further apart than in the previous case. Over almost all of the sample, this texture is almost immediately destroyed by the liquid motion, but traces of the equidistant surfaces remain on the surface film attached to the glass slide.

In this way, when the distance between the equidistant Grandjean planes becomes large enough, it is possible to distinguish these surfaces in the conical structure. In the planar structure, these surfaces are planar and normal to the negative optical axis. This is just as in the case of the stepped drops in smectic

[1] Literally: 'shading' (*ed.*).

materials. Here there are equidistant surfaces separated by molecular distances, whose existence we have been obliged to accept, and which are planar but perpendicular to the positive axis. In the cholesteric conical structure, the surfaces wind themselves into Dupin cyclides, everywhere perpendicular to the negative axis. The same holds true for the conical structure in the smectic materials. There one can also find equidistant surfaces, in this case only molecular distances apart, and they also are wound into Dupin cyclides.

By this means it is possible to understand the appearance of one and the same structure in both cholesteric and smectic materials. This structure is the conical structure, and in each type of matter its existence is linked to the existence of parallel equidistant surfaces. Why it is that in any material which contains equidistant surfaces these surfaces take the shape of a family of Dupin cyclides is not known yet. But the two features do seem to be related. In cholesteric materials, however, these surfaces play a very different role than in smectic materials. In smectic materials, the distances between neighbouring surfaces are of the order of a molecular length. This amounts to some 40 to 50 Å at the very most for very long soap molecules, according to Perrin's measurements. The selectively reflected wavelengths, at least for obliquely incident light, are almost certainly in the X-ray region (but the experiment remains to be done, however). In cholesteric materials, the distances between the surfaces are at another order of magnitude altogether. Examples are some 2000 Å in amyl cyanobenzalaminocinnamate according to Grandjean's measurements, and 80,000 Å and above in the other cases we have just discussed. The selectively reflected wavelengths are no longer in the X–ray region, but now are in the visible spectrum, and even frequently in the far infra-red. In smectic materials, the equidistant surfaces belong to the molecular structure. However, in cholesteric materials they belong on a much larger-scale. They are the result of the mutual interaction of homogeneous elements, each of which contains large numbers of molecules and possesses a nematic structure without any periodicity. This is confirmed by the fact that there is no X-ray scattering from cholesteric materials, just as there is none from nematic materials.

The study of the planar structure has taught us that these homogenous elements of cholesteric material are optically positive. This is exactly like all other meso-morphic materials. They assemble themselves into a structure the details of which are not yet understood. However, it does involve strong twists around a direction normal to the positive optical axis, thus giving rise to a complete system with a negative uniaxiality. It is this complete complex structure which displays, at equal distances, structural discontinuities normal to the negative axis, and hence parallel to the elementary positive axis. These are the Grandjean surfaces. It is completely different in smectic materials. But nevertheless, these equidistant surfaces, whose nature is so different in the two cases, give rise to the same winding into Dupin cyclides, with converging optical axes towards lines which take the same focal conic shapes. And so it seems highly probable that this 'conical' arrangement is related purely and simply to the existence, within the structure, of parallel and equidistant surfaces, absolutely regardless of the role of the surfaces within the structure. The small negative textures are reminiscent of the smectic phase. But, in view of the considerations above, this is as consistent with an absolute and continuing distinction between the cholesteric and the true positive smectic forms as it is with the complete continuity which, by contrast, exists between the cholesteric and nematic forms.

We still need to understand why the reflected colours and the rotatory power disappear completely in the conical structure in cholesteric materials. If in this conical structure the Grandjean planes exist, as one in fact sees that they do, and if they are simply curved around into Dupin cyclides, it would seem at first sight that one should see reflected colours and a rotatory power. They might be blurred by the rapid orientational variations, but they should not be completely suppressed. This is a difficulty which needs to be noted.

I do not want to disregard the imperfections and gaps in the present study. No doubt the conclusions can be improved. Nevertheless, I believe the conclusions that I have reached combine the known facts into a form which expresses actual relations between different types of mesomorphic materials. Consequently they form a suitable basis for classification, and a good starting point for new research. In the course of this presentation we have been able to see how imperfect and limited our understanding remains, and how much there is still to do.

Classification of the substances investigated

I think it will be useful to list below the classification of the substances in the mesomorphic phase which I have had the opportunity to observe. I also give a very brief indication of the relevant principal properties, and at the same time their chemical structures. As far as the last point is concerned, the different conventions adopted by different authors for the same substance can frequently lead to confusion.[1]

We should recall that pure materials are not the only materials which can be profitably studied in the mesomorphic field. The same properties can be just as well found in a large number of mesomorphic mixtures, and occasionally even better. Frequently there is complete miscibility both in the isotropic and in the mesomorphic state, and separation only occurs when the substances crystallise. Very exceptionally it is impossible to form the mixture at all, or the components only combine in limited proportions. But all mesomorphic phases which have been obtained, whether pure or mixtures, have qualitatively the same physical

[1] By relying on labelling L. Royer and myself (C.R. **174**, 1523, 1922) distinguished cholesteryl cinnamate and cholesteryl cinnamylate samples coming from different sources, which did not have exactly the same properties but which were both cinnamates, and also, following Lehmann's labelling convention, called paramethylphene methylol what turned out actually to be cholesteryl paratyolic ether.

properties and the same structures as those of pure materials. They all belong to one of the types we have described, and which is given in the table below. One never obtains any intermediate form between a crystalline material, a smectic material, a nematic material and an amorphous material. There is never a smooth transition from one to another of these four constant types, which are separated by sharp discontinuities. This observation gives a truly special importance to the following classification:

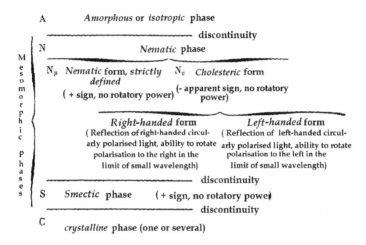

The regions of stability on the temperature scale correspond to temperatures which decrease from A to C.

Of course, a phase can exist in a pseudo-stable equilibrium outside its range of stability. It may well be the case that such a phase has no stability domain, but still exists, sometimes fleetingly, and sometimes for even longer periods. Many of the substances we shall mention do not exhibit an equilibrium state of any mesomorphic material between the crystalline and amorphous states, whether on heating or on cooling. However, one does appear if one rapidly cools an isotropic liquid, super-cooling in such a way that the crystallisation velocity becomes very slow, and occasionally disappears. In this latter case, one can observe the mesomorphic phase at leisure, even though it is only in pseudo-equilibrium. But in all these cases, whether there be equilibrium or not, the order in which the four phases appears stays the same, so long as the temperature change is monotonic. For example, a given substance will pass from A to C, depending on the circum-stances, either directly, or through an interpolating phase S. But one never observes the phases to appear except in the order ASC. Delay phenomena are particularly important in crystallisation. They are much less important in AN, AS and NS transformations.

The review is completed by a list of 42 compounds, divided into types of phase diagram. For each compound, Friedel gives a brief summary of the phase diagram as a function of temperature, a sketch of its chemical structure, the values of the temperatures at which the phase transitions occur, the kind of textures which are observed, and gives the original worker or workers who first found the relevant 'liquid crystal'. This was perhaps the first 'Handbook of Liquid Crystals'

Georges Friedel was born in 1865 in Mulhouse (Alsace, France), and came from a family with long-standing roots in Alsace. He was the son of the celebrated chemist and crystallographer Charles Friedel (1832–99), known for the Friedel–Crafts reaction, and as a consequence spent his childhood in his father's apartment at the School of Mines in Paris. Charles Friedel's grandfather Georges Duvernoy (1777–1855) was a naturalist of renown, collaborator of Cuvier, and also member of the French Academy of Sciences. As we see in the next paper, the Friedel scientific line continued after Georges.

Friedel entered the École Polytechnique in 1885, graduating in 1887 to join the School of Mines in a junior position. During the period 1891–4 he was a mining engineer within the School of Mines at Moulins. In 1894 became professor in the St Étienne branch of the School of Mines, becoming director in 1907. He was elected corresponding member of the Academy of Sciences in 1917. In 1918, following the logic of his family history, he became professor of Geological Sciences in Strasbourg, which had just been recovered from Germany following the end of the First World War. He retired early as a result of ill-health in 1930 and died in 1933.

Friedel worked in several different areas, beginning with an early collaboration with his father which involved chemical synthesis. After this he worked on the effects of water in zeolites. Apart from his liquid crystal research, he is best known for his work on so-called Bravais lattices. He established, even without the X-ray confirmation by von Laue and the Braggs, that optical observations of crystals were best explained in terms of regular atomic structures which had been categorised by Auguste Bravais (1811–63).

ÉLECTRO-OPTIQUE. — *La diffraction des rayons X par les corps smectiques* (¹).
Note de MM. M. de Broglie et E. Friedel, présentée par M. Brillouin.

Les travaux de G. Friedel sur les corps méromorphes ont mis en évidence qu'il existe, pour la matière, entre l'état amorphe et l'état cristallisé, deux états intermédiaires possibles : l'état nématique, dans lequel les molécules seraient distribuées au hasard mais auraient toutes une direction commune (liquides à fils et à noyaux); l'état smectique, dans lequel les molécules, ayant une direction commune, seraient en outre réparties par surfaces parallèles équidistantes (liquides à coniques). Les corps cholestériques (liquides à plans de Grandjean), qui paraissent former un troisième groupe, ne sont que des corps nématiques doués de torsion (²).

Si ces idées sont exactes, ni les corps nématiques ni les corps cholestériques ne doivent diffracter nettement les rayons X. C'est ce qui a été constaté par Huckel (³). Au contraire, les corps smectiques possédant une répartition périodique des molécules, doivent agir sur les rayons X comme ferait un des systèmes de plans réticulaires parallèles d'un cristal. En particulier, une portion de matière smectique qui présente toutes les orientations possibles de ces surfaces parallèles, doit diffracter un faisceau monochromatique et donner sur une plaque normale au rayon direct un spectre formé d'anneaux concentriques.

Tel est le résultat que nous avons cherché à vérifier.

Il fallait, selon toute probabilité, employer des rayons de grande longueur d'onde : la formule chimique des corps smectiques comporte toujours une longue chaîne rectiligne, et tout amenait à considérer ces chaînes comme orientées parallèlement à l'axe optique, donc suivant la normale aux plans privilégiés; la distance entre points homologues de deux couches successives, la période, devait donc être au moins égale à la longueur de ces chaînes et considérable par rapport aux distances interatomiques.

Nous disposions même d'une mesure probable de cette période pour certains corps smectiques, les oléates : les mesures de Perrin et de Wells (⁴)

(¹) Nous emploierons dans cette Note, pour désigner les liquides anisotropes, les expressions proposées par G. Friedel, *Ann. de Phys.*, novembre-décembre 1922.

(²) G. Friedel, *Comptes rendus*, t. 176, 1923, p. 475.

(³) Huckel, *Phys. Zeitsch.*, t. 22, 1921, p. 561.

(⁴) P.-V. Wells, *Annales de Physique*, p. 69.

B2

Comptes rendus de l'Académie des Sciences **176**, 738–40 (1923)

DIFFRACTION OF X-RAYS BY SMECTIC SUBSTANCES[1]

Note from Messrs

M. de Broglie and E. Friedel

presented by Mr Brillouin

The work of G. Friedel on mesomorphic* substances has shown that in matter there exist two possible intermediate states between the amorphous state and the crystalline state. These are: the nematic state in which the molecules would be distributed randomly but all having the same direction (*liquides à fils* and *à noyaux*); and the smectic state, in which the molecules have the same direction, and are in addition distributed on equidistant parallel surfaces (conic liquids). Cholesteric substances (Grandjean plane liquids), which appear to form a third type, are only nematic substances with a twist.[2]

If these ideas are true, neither nematic nor cholesteric substances should give a sharp X-ray diffraction pattern. This has been noted by Huckel.[3] By contrast, because smectic materials have a periodic distribution of molecules, they must act on X-rays in the same way as the system of parallel lattice planes of a crystal. In particular, when a sample of smectic material possesses all possible orientations of the parallel surfaces, it should diffract a monochromatic beam and give rise to an image of concentric rings on a plate normal to the direct beam.

This is the result we have sought to verify.

It was almost certain that it would be necessary to use long wavelength radiation. The chemical structure of the smectic substances always includes a long linear chain, and everything suggests that these chains are oriented parallel to the optic axis, and hence normal to the special planes. The distance between equivalent

---

[1] In order to describe anisotropic liquids, we will use in this Note the notation proposed by G. Friedel, *Ann. de Phys.*, November-December 1922.

* The original French has *meromorphes*, probably a misprint.

[2] G. Friedel, *Comptes rendus*, **Vol. 176**, 1923, p. 475.

[3] Huckel, *Phys. Zeitsch.*, **Vol. 22**, 1921, p. 561.

points in two successive layers, the period, must then be equal at least to the length of these chains, and large in comparison with interatomic distances.

We were even in a position to estimate the probable dimension of this period for certain smectic materials, the oleates. The measurements of Perrin and Wells[4] on thin films obtained with a mixture of potassium oleate, glycerine and water have established that these films are formed from identical elementary layers with thickness of around 42 to 44 Ångstroms. It seems that soapy water has, in thin films, the smectic structure possessed by the anhydrous oleates. One can thus anticipate that with the oleates one will obtain diffraction rings corresponding to a repeat distance of the order of 40 Å.

We used a tube with a copper target electrode, which gave effectively monochromatic radiation (rays at K_α 1.541 and 1.537 Å) and allowing us to work in the open air.

Our first experiments were carried out on ammonium, sodium and potassium oleates. Several negatives were obtained, notably with cold sodium stearate. They confirmed the anticipated results.

In the middle, the first ring is seen, almost as black as the central spot which corresponds to the direct beam. Then two rings are visible with diameters twice and three times the first, with decreasing intensity, corresponding to the second and third order reflections. The rings of large diameter, which are less visible, correspond to atom–atom diffractions within the molecule.

The measurement of the diameters of the inside rings, in particular the third which could be measured to around 2 parts in 100, gives a value of 43.5 Å for the repeat distance, in good agreement with the measurements of Wells.

Similar results were obtained for potassium and ammonium oleate.

We are continuing this work in order to check if the same results are seen with other smectic substances, and to establish, particularly by using the large diameter rings, the atomic structure of these substances.

Conclusions – 1.   As predicted by G. Friedel, smectic materials do indeed consist of equidistant molecular layers.

2.   The thin soap films, for which J. Perrin discovered his law of multiple spacings, are not films of amorphous material, but of smectic material. This should also be borne in mind when the capillary properties discovered on soap films are extrapolated to amorphous liquids.

3.   Combined with the measurements of Wells, our experiment constitutes without doubt the first direct measurement of the wavelengths of X-rays, which starts from the wavelength of light, without knowing either Avogadro's Number or Planck's Constant. Neither our measurements, nor those of Wells (which depend on the refractive index assumed for the layer), are yet precise enough to give anything more than the order of magnitude of the quantities already accepted.

[4] P.-V. Wells, *Annales de Physique*, p. 69.

4. The direct measurement of the wavelength of very soft X-rays using diffraction is known to have been limited by the impossibility of obtaining crystals which have lattice planes with a large repeat distance, and which are sufficiently dense to give significant diffraction. The use of smectic materials will perhaps allow the achievement of the direct measurement of the longest wavelengths (theoretically up to about 80 Å with the oleates).

Louis-César-Victor-Maurice, 6th Duke de Broglie (usually known as Maurice) was born in Paris in 1875. The de Broglies are a French noble family of Piedmontese origin who settled in France in the seventeenth century. The family boasts numerous distinguished political figures, including two marshals of France and a number of French ambassadors to London. Maurice was the elder brother of the perhaps even more distinguished physicist Louis

Victor de Broglie who introduced the idea of particle-wave duality and paved the way for the development of quantum mechanics.

He graduated from the Naval School in 1895, and then served as a naval officer until 1904. At that time he became interested in physics. As he was independently wealthy, he was able to develop his own private laboratory in his country house. He was awarded a doctorate in 1908 as a result of his researches into atomic structure. He continued research with collaborators in his own laboratory until 1942, interrupted only by military service in the First World War (1914–19). He then became a professor at the Collège de France (1942–6).

His main speciality was in X-ray studies. He introduced the so-called rotating crystal method for determining crystalline structures, and in 1921 discovered the nuclear photoelectric effect. The paper included in this volume is, however, his only direct contribution to the study of liquid crystals. He was elected to the French Academy of Sciences in 1924, to the Académie Française in 1934, and to the Royal Society of London as a foreign member in 1946. He died in 1960.

Edmond Friedel was born in 1895 and was the son of Georges Friedel. He was decorated for his service during the 1914–18 war, before entering the École Polytechnique in 1919. In 1920 he entered the Mining Section of the Civil Service and was sent to Strasbourg. During this period, apart from assisting his father's liquid crystal research, he drew a geological map of the Sarre district and lectured at the newly created *École de Pétrole*. After his father's death he was promoted in the Mining Service, working until 1937 in Béthune in northern France, before being transferred to the School of Mines in Paris, firstly as Sub-director, and then, after 1944, as Director. He was also decorated for his war activities during the period 1939–40.

In Paris, he taught crystallography, but his activities were mainly oriented towards geology. For many years before and after the war he advised the Moroccan Administration for Mines, and he also created the French Office for Geological and Mineral Research, which continues its activities to this day. He retired in 1965 and died in 1972.

ЖУРНАЛ

РУССКОГО

ФИЗИКО-ХИМИЧЕСКОГО

ОБЩЕСТВА

ПРИ ЛЕНИНГРАДСКОМ УНИВЕРСИТЕТЕ

ЧАСТЬ ФИЗИЧЕСКАЯ

Ответственный редактор А. Ф. Иоффе

ТОМ LIX—1927

ГЛАВНОЕ УПРАВЛЕНИЕ НАУЧНЫМИ УЧРЕЖДЕНИЯМИ (ГЛАВНАУКА)

ГОСУДАРСТВЕННОЕ ИЗДАТЕЛЬСТВО
МОСКВА 1927 ЛЕНИНГРАД

Front cover of volume 59 of the Journal of the Russian Physical Chemistry Society, in which the original Russian version of the Frederiks paper B3 appeared.

К ВОПРОСУ О ПРИРОДЕ АНИЗОТРОПНО-ЖИДКОГО СОСТОЯНИЯ ВЕЩЕСТВА. [1]

А. Репьева и В. Фредерикс.

1. Дана теория опытов К а с т а, основанная на анизотропии диамагнитных свойств анизотропных жидкостей.

2. С помощью простых оптических наблюдений показано, что внутренняя ориентирующая сила согласно теории Б о р н а по всей вероятности дипольного происхождения.

3. Указан способ измерения внешней ориентирующей силы, основанной на анизотропии диамагнитных свойств вещества.

4. Установлено, что, вопреки общераспространенному мнению, магнитное поле действует на все виды анизотропных жидкостей.

1. Теория анизотропного состояния вещества, данная в 1916 г. Б о р н о м [2], построена на том же начале, как и теория магнетизма Л а н ж е в э н а. Роль внутреннего магнитного поля играет в ней поле электрических диполей, существование которых предполагается в каждой отдельной молекуле вещества. Как и в теории магнетизма, внутреннее «дипольное» поле возможно не при всех температурах.

Пусть T — абсолютная температура, p — момент диполя, P — момент единицы объема, k — постоянная Б о л ь ц м а н а, $\mathfrak{N}$ — число молекул (или диполей) в единице объема, пусть кроме того a — сокращенное обозначение для $\frac{pP}{kT}$ и θ для $\frac{\mathfrak{N} p^2}{9 k}$.

Внутреннее поле возможно только при таких температурах, при которых уравнение для a

$$\frac{T}{3\theta} a = \operatorname{cotg hyp} a - \frac{1}{a}$$

имеет вещественные корни. Это будет иметь место, если $T < \theta$. θ является, таким образом, критической точкой, своего рода точкой К ю р и.

По Б о р н у, эта точка должна приблизительно совпадать с точкой перехода из анизотропно-жидкого в аморфно-жидкое состояние.

[1] Доложено на пятом съезде русских физиков в Москве 20/XII. 1926 г.
[2] M. B o r n. Ber. d. Berl Ak. d. Wiss. **30**, 614 1916.

The first page of the original Russian paper by A. Repiova and V. Frederiks entitled "On the nature of the anisotropic liquid state of matter".

следовательно темные кольца, идя от центра, соответствуют значениям $\delta = 0, \lambda, 2\lambda, \ldots m\lambda \ldots (m + k)\lambda$.

При $H > 0$ (рис. 4) имеем до z первоначальное направление осей, а, следовательно, и расположение колец, дальше — новое.

Наклон осей, начиная от z, изменяется различно для различной толщины слоя, а именно так, что с возрастанием толщины слоя δ убывает. В этой области убывание разности хода вследствие увеличения угла наклона оптической оси покрывает приращение разности хода вследствие возрастания толщины слоя. Здесь темные кольца соответствуют, начиная от z, значениям $\delta = (m - 1)\lambda, (m - 2)\lambda, \ldots (m - k_1)\lambda$.

С возрастанием напряжения поля изменение наклона осей увеличивается, а δ убывает.

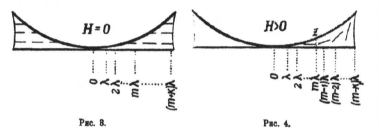

Рис. 3.   Рис. 4.

При ослаблении магнитного поля это явление идет в обратном порядке, а с исчезновением поля препарат восстанавливает свой первоначальный вид.

II. Нормальный слой.

a) Нормальный слой в продольном магнитном поле (рис. 2) остается без изменений, так как оптические оси в нем уже расположены параллельно направлению поля.

b) В поперечном поле (рис. 1) названный слой проявляет следующее: с возникновением магнитного поля на краях препарата появляются очень резкие, тонкие кольца, которые по мере возрастания силы поля тоже стягиваются к центру (фот. 1, 2, 3, табл. 1), указывая на появление двойного лучепреломления и возрастание его с усилением поля.

В этом случае, как это видно на снимках (центральные темные круги), для каждого H, тоже существует предельная толщина слоя z_n, до которой сохраняется первоначальное, нормальное положение оптических осей.

При $H = 0$ оптические оси расположены как показано на рис. 5, двойное лучепреломление отсутствует, весь препарат сплошь темный.

При $H > 0$ расположение осей следующее (рис. 6): от центра до z осталось нормальное, дальше — изменилось.

This page describes the original Frederiks effect experiment.

B3

Transactions of the American Electrochemical Society **55**, 85–96 (1929)

A paper presented at the Fifty-fifth General Meeting of the American Electrochemical Society, held at Toronto, Canada, May 27, 1929, Mr. Floyd T. Taylor in the Chair.

ON THE USE OF A MAGNETIC FIELD IN THE MEASUREMENT OF THE FORCES TENDING TO ORIENT AN ANISOTROPIC LIQUID IN A THIN HOMOGENEOUS LAYER[1]

By V. Freedericksz[2] and V. Zolina.[2]

ABSTRACT.

Forces arising in a thin layer of the surface of an anisotropic liquid, and to which the orientation of the drops in the form of a homogeneous layer is due, may be equilibrated by the effect of a magnetic field. This circumstance may permit quantitative as well as qualitative deductions as regards the character and the strength of these forces. Up to the present we succeeded in establishing first, that these forces do not depend on the kind of material on which the anisotropic liquid is placed; secondly, that they are not subject to great variations when the temperature conditions are changed, and, finally, that they decrease with the distance of the orientated drops from the surface of the liquid. The experiments and measurements have been completed with *para*-asoxyphenetol and anisaldazine. Both of these substances have given similar results.

It is known that an anisotropic liquid in a thin layer, for instance when held between a microscope slide and a covering glass (Fig. 1), represents a layer similar in many respects to a layer of a solid homogeneous crystal.

If the optical axis in such a layer is perpendicular to its surface (Fig. 1b) then the layer of the liquid does not differ at all from a crystalline plate; but if the axis is parallel to its surface the principal sections of the layer may possibly be differently orientated in the various portions of the layer (Fig. 1a).

[1] Manuscript received March 8, 1929.

[2] National Physical and Technical Laboratories, Leningrad, Russia.

Opinions differ as to the factors determining the orientation of a liquid. Undoubtedly we have here to deal with substances whose molecules are greatly extended in one direction. Some of the existing theories, among which the theory of Ornstein seems to be the most acceptable, involve the idea that individual molecules unite into separate droplets, fully homogeneous as to their structure. The questions regarding the controlling forces

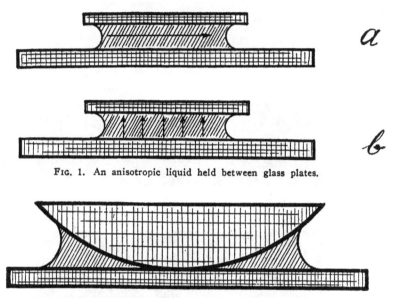

FIG. 1. An anisotropic liquid held between glass plates.

FIG. 2. An anisotropic liquid between a plane and a convex glass.

of such a grouping and, in particular, the question on the compatibility of Born's dipol theory will not be discussed here.

An anisotropic liquid, representing a collection of orientated drops, should be generally considered turbid, since owing to temperature fluctuations the disposition of the drops can not be supposed to be of any regularity or symmetry. However, the introduction of collateral forces (such as act from the surface of the liquid) may bring about conditions owing to which all the individual drops would be orientated in the same direction.

These collateral forces should be strong enough to withstand the disturbing effects of temperature fluctuation. It is evident that in a layer of an anisotropic liquid the action of such forces should be observable. The remarkable experiments of Langmuir render it unnecessary to consider individual orientated drops, the dimensions of which might be reduced to those of a molecule. If we are at present referring to drops of linear dimensions having a value of 10^{-5} cm., it is because we wish to take into account some observations on diffused light in an anisotropic liquid, and

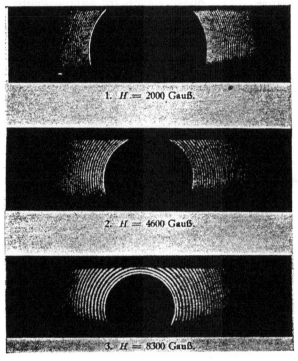

FIG. 3. Effect of magnetic field on a drop of para-acetoxyben-zalazine. (The drop is mounted as shown in Fig. 2 and photograph taken perpendicularly.)

observations on the changes of the dielectrical character of the drops in a magnetic field. Compare the experiments of Kast[3] and Jesewski.[4]

In order to measure the force orienting some individual drop, it is evidently necessary that another measurable force should be introduced, and from the equilibrium of these two forces an inference be drawn as to the character of the former. In our case the auxiliary measurable force is that of a magnetic field.

[3] W. Kast, Ann. d. Physik., 18, 377 (1922).
[4] K. Jesewski, Zeit. f. Physik., 52, 268 (1928).

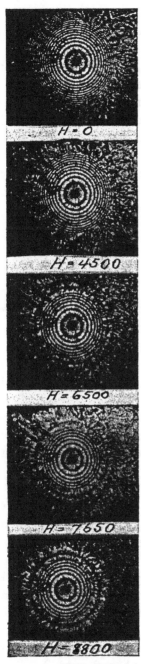

FIG. 4. Effect of magnetic field on a drop of para-azoxyphenetole. Optical axis parallel to the glass plate.

FIG. 5. Effect of magnetic field on a drop of anisaldazine. Optical axis parallel to the glass plate.

As has been shown in a previous paper[5] the specific behavior of the magnetic field may be ascribed to the anisotropy of the diamagnetic permeability.

If v be the volume of the drop and μ_{11} and μ_1 the magnetic permeability in a direction which is parallel or perpendicular to the optical axis of the drop, H the value of the magnetic field, a the angle between the optical axis and the magnetic field, then the moment of a pair of forces acting on the drop will be:
$2(\mu_{11} - \mu_1)vH^2 \cos a \sin a$.

If the dimensions of the drop be reduced to that of a molecule, the only change which this formula will undergo will be that the two first factors must be given a molecular interpretation.

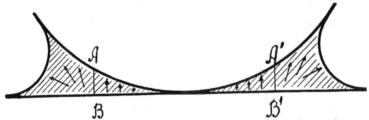

Fig. 6. Effect of a magnetic field on a drop of anisotropic liquid.

The moment of the force controlling the orientation of the drop and acting from the surface of the anisotropic layer depends in a different manner on the angle a.

If the optical axis forms with the normal to the surface an angle β, and if in the state of equilibrium $\beta = 0$, then the moment of a pair of forces, arising from a deviation from the state of equilibrium, will attain its maximum value D when $\beta = \dfrac{\pi}{2}$.

If, for simplicity's sake, we admit that this moment is equal to D sin β, where D is an empirical coefficient,[6] the admitted simplification will not have any influence on the results, as we shall see further on.

If in the state of equilibrium the optical axis runs parallel to the surface, then the moment of a pair of forces appearing at the deviation from the state of equilibrium may be represented as D cos β.

If in the first and in the second case we produce a magnetic field perpendicular to the optical axis, then $\beta = \pi/2 - a$. Let

[5] V. Freedericksz and A. Repiewa, Zeit. f. Physik., 42, 532 (1927).
[6] This coefficient might be interpreted on the basis of Born's dipols theory.

us further admit that there is no need that any other forces be taken into consideration; then in the first case, the condition of equilibrium will be expressed by

$$2(\mu_{11} - \mu_1)vH^2 \cos a \sin a = D \cos a$$

and in the second case, by

$$2(\mu_{11} - \mu_1)vH^2 \cos a \sin a = D \sin a$$

We see that each time two possibilities are offered. For instance, in the first case

$$\text{either } a = \frac{\pi}{2}$$

$$\text{or } \sin a = \frac{D}{2(\mu_{11} - \mu_1)vH^2}$$

If $D > 2(\mu_{11} - \mu_1)vH^2$ then there exists only one solution, namely $a = \pi/2$. If $D < 2(\mu_{11} - \mu_1)vH^2$ then only the second solution will correspond to the stable equilibrium. Both solutions coincide when $D = 2(\mu_{11} - \mu_1)vH^2$.

Analogous inferences may be drawn with regard to the second case. Thus, we see that the magnetic field may disturb the orientation of an anisotropic layer, but only in case it exceeds a certain value.

The orientating pair of forces acting on a certain drop proves dependent upon the distance from the surface at which the drop is located.

Thus far, the questions regarding the mechanism of translation of this force from one drop to another as well as those concerning the causes of their becoming weaker, with the increase of the distance from the surface, are left undiscussed.

If the anisotropic liquid is placed between a plane and a convex glass (Fig. 2) then the orientated drops located in the middle of the layer will be differently affected by the orientating force. In Fig. 3, 4 and 5, are given reproductions of *para*-acetoxybenzalazine, *para*-asoxyphenetole and anisaldazine taken in the form of such a plane-concave anisotropic layer and placed in different magnetic fields, the strength of which is given in the figures. In the case of *p*-acetoxybenzalazine the optical axis is directed perpendicularly to the glass, in the cases of *p*-asoxyphenetole and anisaldazine, it runs parallel to it. The photographed layers are

placed between two crossed Nicol prisms; this gives on the photographic plate a series of dark and light concentric rings at the
points where the optical axis is not normally directed to the glass;
these rings represent curves of equal difference of paths of

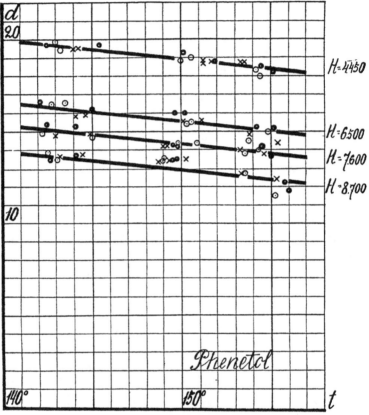

FIG. 7. Orienting force is independent of the kind of glass used.
Drop of phenetol.
t = temperature ° C.; d = diameter of the circle in arbitrary units.
Glasses of three different densities (3.8, 3.0 and 2.5).

ordinary and extraordinary rays in the plate. It is quite distinctly
seen that the effect of the magnetic field extends only up to a
certain circle of a definite radius. The larger the magnetic field,
the smaller is this radius.

This phenomenon previously[7] studied by us in respect to
p-acetoxybenzalazine (Fig. 3) could have given rise to a suspicion that the action of the magnetic field is unperceivable in the

[7] V. Freederickz and A. Repiewa, loc. cit.

central portion of the preparation, because of a feeble intensity of double refraction in this middle portion of it. But the observations made with the two other substances mentioned above relieve us from any doubt as regards the fact that the action

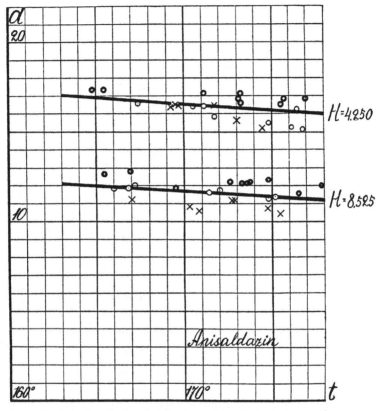

FIG. 8. Orienting force is independent of the kind of glass used.
Drop of anisaldazine.

t = temperature ° C.; d = diameter of the circle in arbitrary units.
Glasses of three different densities (3.8, 3.0 and 2.5).

of the magnetic field is confined to definite limits. The curves of equal differences of paths in the central portion of the field (Fig. 4 and 5) are identical and do not change with any strength of the magnetic field.

Let the points A B in Fig. 6 be the limits of action of the magnetic field. We may admit that throughout the whole distance from A to B there is not a single drop for which the value of D would exceed that of $2(\mu_{11} - \mu_1)vH^2$, and that for the drops

lying in the vicinity of A B, and which are on the left of this
line, the force D is less than $2(\mu_{11} - \mu_1)vH^2$.

It is natural to assume that the effect of the magnetic field,
acting on the line A B, commences with the drops remotest from
the limits of the liquid, that is, with the drop disposed in the
middle of the section A B. If this be correct the value of D thus
determined must correspond to the distance from the surface

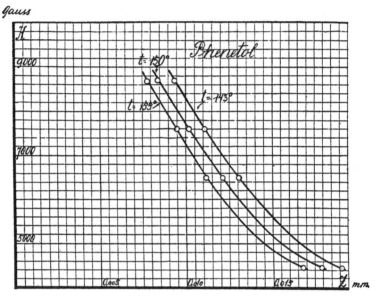

Fig. 9. The dependence of D on the thickness of the layer of
anisotropic liquid (phenetole).
Z = thickness of liquid layer in mm.

A B/2. A molecular interpretation of the force may be given
only after a complete theory of the anisotropic state of a liquid is
worked out. As long as this theory does not exist, we are confined to make experimental attempts to find the factors controlling the force D.

At present we have investigated the dependence of this force
on the thickness of the layer, as well as on the temperature and
on the kind of glass holding the anisotropic liquid. Systematic
observations have been made with p-asoxyphenetole and anisaldazine. The results of these are represented in Fig. 7 and 8, in
which the abscissas represent the temperature, and the ordinates
the diameter of a circle d in arbitrary units, limited to that por-

tion of the layer which is affected by the magnetic field. o-o-o denotes the results obtained with a glass, the density of which was equal to 3.8; x-x-x, those obtained for a glass with density = 3.0; o-o-o, those obtained for a glass with density = 2.5

The observations have been made with four values of H, shown in Fig. 7 and 8.

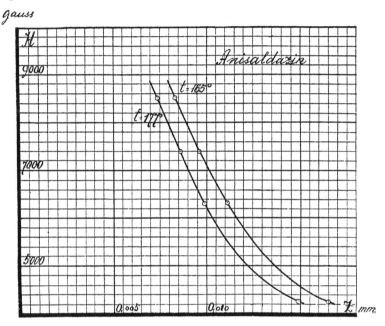

FIG 10. The dependence of D on the thickness of the layer of anisotropic liquid (anisaldazine).

Z = thickness of liquid layer in mm.

We see that within the limits of errors of observations the orienting force does not depend on the kind of the glass. We may add to this that it also does not depend on the method applied for cleaning the glasses, if the cleaning is done carefully.

This leads to a very important inference, namely, that *the value of* D *depends only on the anisotropic liquid itself.*

The dependence of D on the thickness of the layer is seen in Fig. 9 and 10, in which the magnetic fields are ordinates and the corresponding thickness of the layer at different temperatures indicated on the figures are abscissas.

For phenetole as well as for anisaldazine we find for magnetic fields, the strengths of which are within the limits of 4,000 and 9,000 gauss, that

$$D \sim \frac{1}{Z^{20}}$$

where Z is the thickness of the layer. We see that the dependence on the thickness of the layer is almost identical for both substances. In order to ascertain how much D increases when the values of Z become smaller, stronger fields than those which we used would be necessary. Supplementary experiments with very high values of H are in progress.[8]

As to the effect of temperature, we see in Fig. 7 and 8 that the temperature does not produce a great effect, although for phenetole it is sufficiently perceptible.

The absolute value of the force D may be computed if the factor $(\mu_{11} - \mu_1)v$ is known; the approximate value of this factor may be obtained from the experiments of Kast and Jesewski, but the usefulness of their numerical results in the present case depends on the influence of the force D. If we compute the factor from the results of their experiments, leaving the force D out of consideration, the value of the factor $(\mu_{11} - \mu_1)v$ will be 10^{-20}. This gives for D a value of the order 10^{-13}.

But the computation of absolute values of these forces will be of definite interest only when we have at our disposal a fully satisfactory theory of the anisotropic liquid state. Undoubtedly the conditions of temperature have to be taken into consideration when computing these forces, as well as the forces to which the spontaneous orientation of molecules inside each drop is due.

Presently systematic observation of all the forces acting by means of holding the molecules in equilibrium with impressed magnetic fields, will, we hope, lead us to results which will prove satisfactory both quantitatively as well as qualitatively.

DISCUSSION.

V. FREEDERICKSZ (*Communicated*): Recently we have made a number of new experiments using stronger fields up to 25260

[8] See author's "communicated" discussion below.

gauss. We were thus enabled to determine the interdependence of D and Z with reasonable accuracy. In Fig. 11 the ordinate represents log H, and the abscissa log Z. (Z is here expressed in microns.) The heavy, straight line, inclined at an angle of

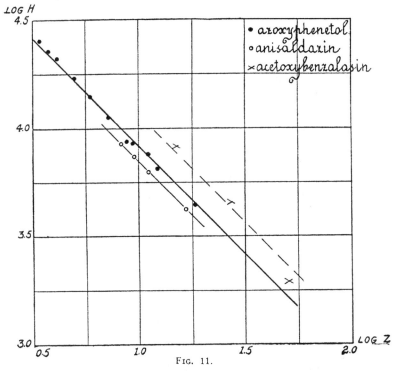

FIG. 11.

45° to the axis, closely represents the results of our measurements. This indicates that *D is equal to the inverse square of the distance Z from the surface with an accuracy of ± 1 to 2 per cent.* As is seen from the figure this interesting law holds true for the two recently examined substances, azoxyphenetole and anisaldazine. The measurements on acetoxybenzalazine are not to be taken into account, as they were made in 1927 with insufficient accuracy; they are included in Fig. 11 to indicate that they might be interpreted by the same law. This law is *independent of the temperature.*

Vsevolod Konstantinovich Frederiks (sometimes spelt Freédericksz) was born in Warsaw, at that time part of Russia, in 1885. His family was aristocratic, and traced its roots to Swedish emigrants in the early 18th century. Frederiks spent the greater part of his youth in Nizhny Novgorod. He was sent abroad to University in Geneva in 1903 and graduated in physics in 1907. His first scientific work was conducted in Geneva in low temperature physics. After 1911 he worked in Göttingen with Waldemar Voigt on piezoelectricity. In 1918 he returned to Russia, working first in Moscow, and then from 1919 in the State Optical Institute in Petrograd (St Petersburg), as well as lecturing at the university.

He was catholic in his intellectual interest, working in metal physics, on relativity and cosmology (with A.A. Friedman), and even publishing papers on the new quantum mechanics. In the late twenties and thirties his main interests lay in anisotropic fluids, in which after an early flirtation with swarm theory, he was a firm adherent of the elasticity theory of Zocher and Oseen. In 1931 he was the first to observe periodic domains induced by hydrodynamic motion.

In 1927 Frederiks married Maria Dmitrevna Shostakovich, elder sister of the well-known composer. His aristocratic background, while conferring many advantages, made him an obvious target during Stalin's purges in the 1930s. He was arrested in 1936, and accused of preparing terrorist acts against the Soviet regime and Stalin in particular. He was tortured and like many in the same situation was persuaded to plead guilty to absurd accusations. Despite the pleadings of his numerous high-placed friends, he was sentenced to ten years in prison and in 1937 deported to the gulag. While in prison he caught tuberculosis. He was released in 1943 and died of pneumonia on 6 January 1944 in hospital in Gorky (Nizhny Novgorod).

Frederiks was subsequently rehabilitated in the post-1956 Krushchev thaw. This was of course little consolation to Frederiks himself, but early enough to be of great comfort to his family and colleagues and to rescue his scientific reputation.

B4

Zeitschrift für Kristallographie **79**, 269–347 (1931)

Inhaltsverzeichnis des 79. Bandes.

IV   Inhaltsverzeichnis des neunundsiebzigsten Bandes.

Zeitschrift für Kristallographie **79**, 269–347 (1931)
First page of General Discussion

Vorwort
zu der Diskussion über die vorstehenden Aufsätze.

Die Aufsätze, die in der ersten Hälfte 1930 verfaßt worden waren, wurden in Satz und Korrektur gegeben und in der ersten Juliwoche 1930 an die Verfasser und eine Reihe weiterer Forscher auf dem Gebiet der flüssigen Kristalle versandt. Bis Ende August war eine erste Reihe von Diskussionsbemerkungen eingelaufen und in der Folgezeit setzte eine — teils direkte, teils durch den Herausgeber vermittelte — weitere Korrespondenz ein. Es war schwierig, diese ganz systematisch zu gestalten, und es ist ebenso schwer, sie in die zur Veröffentlichung geeignetste Form zu bringen. Immerhin läßt sie sich in eine Reihe von Zuschriften allgemeiner Art und in solche Zuschriften gliedern, die an bestimmte Aufsätze anknüpfen. Die Diskussion wurde Anfang 1931 ebenfalls in Fahnenabzügen an die beteiligten Herren versandt, woraufhin nochmals Gelegenheit zu kurzen Schlußbemerkungen gegeben wurde, die durch die Überschrift: ›Schlußdiskussion‹ bezeichnet worden sind. Die beiden letzten der voranstehenden Aufsätze (Wo. Ostwald und Fréedericksz-Zolina) wurden als Diskussionsbemerkungen eingesandt und sind demnach nicht, wie die anderen Aufsätze, im Juli 1930 verschickt worden. Ihr Umfang ließ es geraten erscheinen, sie aus der Diskussion herauszunehmen; aber die Diskussion konnte nicht an sie, wie an die eigentlichen Aufsätze, anknüpfen.

Allgemeine Diskussionsbemerkungen.

Rudolf Schenck (Münster):
Richtkräfte und Schwarmbildung.

Mit Interesse und Freude muß man das Unternehmen des Herrn Herausgebers der Zeitschrift für Kristallographie begrüßen, den Lesern einen Überblick über den heutigen Stand unserer Erkenntnis von den flüssigen Kristallen zu geben, indem er ein ganzes Heft der Zeitschrift für diesen Gegenstand zur Verfügung stellte und die Forscher, welche sich experimentell oder theoretisch mit der Materie befaßt haben, zu Berichten über ihre Arbeiten oder zur Äußerung ihrer Meinungen veranlaßte.

Schon im Jahre 1905 hatte die Deutsche Bunsen-Gesellschaft den Versuch gemacht, durch Anregung zu einer ähnlichen Zusammenarbeit eine Klärung der eigenartigen Phänomene herbeizuführen, nachdem O. Lehmann auf der Karlsruher Jahresversammlung einen größeren Kreis der Teilnehmer mit den Erscheinungen der flüssigen Kristalle bekannt gemacht und ein heftiges Für und Wider hervorgerufen hatte. Dieses gab den Anstoß zur Einsetzung eines Ausschusses unter dem Vorsitz von I. H. van 't Hoff, welcher die experimentellen Ergebnisse durch eigene Kenntnisnahme auf ihre Zuverlässigkeit prüfen und aus ihnen die weiteren Folgerungen ziehen sollte. Man wäre auf diesem Wege sicher zum erstrebten Ziele gelangt, wenn die Kommissionsmitglieder zu einer richtigen Prüfung des Beobachtungsmateriales hätten zusammengeführt werden können. Der Schreiber dieser Zeilen hat während der Jahre 1906 bis 1908, um die notwendigen Unterlagen für die Arbeit des Ausschusses

B4
Zeitschrift für Kristallographie **79**, 269–347 (1931)

GENERAL DISCUSSION ON LIQUID CRYSTALS

Foreword to the discussion on the articles*

The articles were written in the first half of 1930, and were then prepared in proof form and corrected. In the first week of July 1930 they were sent to the authors and also to a set of further researchers in the liquid crystal field. The first set of Discussion Remarks was submitted in the period up till the end of August. Subsequently there was further correspondence, sometimes directly, and sometimes through the intermediary of the editor. It was very difficult to construct a systematic whole, and just as difficult to bring it into a form suitable for publication. At any rate it can be divided into a set of general communications, and into communications which can be attached to definite articles. At the beginning of 1931 the Discussion was sent in proof form to the participants, in order to give short final remarks, which have been included under the title 'Concluding Discussion'. The final two articles included (those of Wo. Ostwald and Fréedericksz-Zolina) were submitted as Discussion Remarks. Thus, unlike the other articles, they were not submitted in June 1930. Their size justified publication outside the specific Discussion section, but unlike the other articles, there is no reference to them in the Discussion.

General discussion remarks

The introductory contribution was from Rudolf Schenck (Münster). Schenck makes some historical remarks which we refer to further in the Introduction to Section B, and follows them by some general questions concerning the nature of liquid crystalline behaviour.

* written by Ewald (*ed.*).

D. Vorländer (Halle a. S)

On the crystalline liquids

SWARMS AND CLUSTERS[#]

... Bose took my experimental results, and my theory of molecular rods and their clustering, as the foundation of an emulsion hypothesis of crystalline fluids. The Bose hypothesis states that crystalline fluids are not uniform, but consist only of isolated variable swarms of parallel-oriented rod-like molecules. Between them can be found a variable number of identically constructed but disordered vagabond-like molecules – thus a mixture of anisotropic and amorphous material....

The experiments undertaken by Bose and his school to justify the swarm hypothesis failed. Ornstein has contributed experimental material in support of his swarm hypothesis, but I now think I can conclude that Ornstein's swarms and Bose's are not the same thing. I think that it is far from Ornstein's purpose to use his swarms to interpret the nature of crystalline fluids in terms of emulsions or colloids. If Ornstein were to replace the word 'swarm' by 'cluster', i.e. parallel-oriented molecules, then his experimental results would be in much closer accord with my work and my chemical theory....

With reference to this, Ewald noted at this point in the Discussion that it might really only be a question of taste whether one referred to swarms or clusters, inasmuch that a precise definition was required of the meaning of the word swarm. On this point, however, I profoundly disagree. The distinction between swarm and cluster is of an absolutely fundamental nature:

1. The swarm as an incorrect picture....
2. Experimental demonstration of anisotropic molecular swarms or chains....
   (in a model by Kundt based on Bose's model) One obtains anisotropic fluids, but not crystalline fluids with their stationary birefringence. It is known, however, that the expression 'anisotropic fluids', which Oseen uses to describe crystalline fluids, is not correct. The anisotropic fluid in my model is only anisotropic just so long as the external compulsion remains. It loses its anisotropy immediately the compulsion is removed....
3. Further experimental proofs for the crystallinity of fluids and against the swarm hypothesis....

Those who seek swarms in crystalline fluids, and by doing so seek a symbol of disorder, are not able to understand the powerful ordering effects which are manifest in liquid crystals....

There are experiments which enable correct and incorrect names for scientific phenomena to be recognised. A name is correct when it is based on fact and experimental observation, but incorrect if it has at its core theories or hypotheses.

[#] *Bündeln* in the original (*ed.*).

Mesomorphism, mesophases, or intermediate phases for crystalline-liquid phases, are *incorrect* names because there is not a single fact or phenomenon to prove that the liquid crystals stand between solid crystals and amorphous melts. Give me a *single* property of crystalline liquids that indicates their intermediate position as crucial to the phenomenon. *I know of no such property!* Everywhere, even in X-ray diffraction: sharp, discontinuous change; no intermediate position. In contrast, one could cite more than a *dozen* properties that prove that on the one hand the liquid crystals behave just like solid crystals, and on the other, just like amorphous oils. In other words, hermaphroditism! At best, one could introduce, instead of crystalline liquids, the term crystal-like, crystalloid liquids, or liquid crystalloids, but that does not constitute essential progress.

I reject the words mesomorphic states and mesophases to denote crystalline-liquid properties or phases. I consider the expressions mesomorphism and mesophases, as well as the words nematic and smectic, to be completely misguided, even if at some later time it should turn out that the theory which led to the words is correct....

The conclusion from these facts that 'since liquid crystals do not yield an X-ray pattern resembling solid crystals, liquid crystals are not crystals' is false....

If in the future each researcher were to want to introduce new names on the basis of any old opinion, then unholy confusion would be brought into this beautiful field of work. If that had been the case, three-dimensional crystallography would already have been under strain some decades ago, and countless beginners would have been misled.

If one looks up in a register of chemical journals, one finds at least 20–30 meso-compounds, by which one understands preparatory substances. In addition chemistry possesses one or two meso-positions. I will thus never make out of my crystalline liquid substances mesogens, nor mesophases out of crystalline liquid phases. Again, I hope that no chemist will be found taking up the Friedel nomenclature....

SCHLIEREN AND BÂTONNETS

In 1922 Friedel wanted cr. fl. forms to be recognised in a precisely dual classification, comprising nematic and smectic forms. As long ago as 1902–3 I distinguished these forms as the drop- or stain-forming and the bâtonnet-forming crystalline liquids. At that time, in joint work with Felix Meyer on *p*-azoxybenzoic acid ethyl ester, I was the first to see the appearance of each such spot (without cover glass) and bâtonnet (with cover glass). These appear to be of a completely different variety than the cr. fl. melts of Gattermann.... Thus it was that in 1922 Friedel managed to arrive exactly where I had been in 1907. I dispute Friedel's right to endow these principal forms with new names and to describe these different types of phenomena without any hint of my work, as though Friedel had been the first to recognise the fundamental nature of the phenomenon....

241

Those colleagues who use the expressions nematic and smectic probably often do not actually know what they are talking about. Most of the thinner forms should be nematic and the more viscous smectic – a completely inadequate partition classification of cr. fluids. The two cr. fl. phases of my p-ethoxy-benzalamino-α-methyl cinnamic ethyl ester should be: (I) nematic, and (II) smectic. However, in reality, their apparent crystal structures hardly differ. (I) does not resemble at all p-azoxyanisole, with its drops and stains. Similarly, (II) is unlike azoxybenzoic acid ethyl ester, with its spots and bâtonnets. The differentiation of the two phases of the ethoxy ester into nematic and smectic makes no sense....

Halle a. S. January 1931.

Discussion of the Friedel paper

G. Foëx (Strasbourg)

On the mechanism of the passage from the nematic to the isotropic stase. Concerning the very precise distinction between the two stases established by G. and E. Friedel.

...In ferromagnetics the magnetic material $(t < \vartheta)$ only differs from the non-magnetic material $(t > \vartheta)$ by the orientation of the magnetic axes. The two types of material do not constitute two distinct stases. Physical properties such as the density, electric resistance and so on change in a continuous way when one passes through the Curie point by varying the temperature. Only derived quantities (coefficient of dilatation, rate of change of resistance with temperature, etc.) undergo a discontinuity.

The properties of the nematic stase when taken together imply that it differs from the isotropic stase above all and perhaps only by molecular orientation. It is then natural to wonder whether the temperature T_3 is not the nematic analogue of a Curie point – i.e. of a temperature at which the orientation ceases to exist, with a progressive increase of the orientation below this temperature....

G. and E. Friedel (Strasbourg)

Response to Oseen's remarks

1.   Oseen notes that in a mesomorphic substance the molecular distribution in a uniform texture would not be the same as in the immediate neighbourhood of linear singularities (*fils* or *coniques*), which are regions where the molecules cannot maintain their quasi-parallelism. He would like us to give the term 'structure' a more precise definition which includes both cases. If we are not to do this, looking at the problem mathematically, one might say that we were attributing to each

242

stase, smectic or nematic, two, or even an infinity, of structures. To this we reply that, in distinguishing 'structure' and 'texture' we have not claimed to do mathematics, and that we leave to mathematicians the problem of imagining perfectly general definitions sufficiently precise for their calculational needs.... From our point of view, this is how to resolve the question posed by Oseen. We have never observed, in a mathematical sense, linear singularities. We have observed *fils* and *coniques* which are seen with a definite thickness. As to what happens in the immediate neighbourhood of the lines which the mathematicians uses to replace the *fils* and *coniques*, we have no idea....

...3   On the subject of the possible coexistence of other mesomorphic stases, we have never said that we consider the existence of other stases to be impossible. We have only affirmed that the mesomorphic substances that we are familiar with so far can be classed within the two categories of smectic and nematic. If it is true that other types of stase have recently been observed, this would be a very important discovery. In any event, this is a question of fact and not of theory....

H. Zocher (Berlin-Dahlem)

Discussion remarks to G. and E. Friedel

It is in general not usual to consider a colloidal solution as a unique phase in the sense of the Phase Rule. Rather we consider it as a two-phase system consisting of a dispersant and a dispersed phase.... The original rule of Gibbs is certainly not to be used in this case. For example, water vapour above a dispersed phase differs according to the degree of dispersity. In any case, each phase is characterised by the molecular order, and it seems to me that the word stase clarifies nothing new other than that which we have hitherto known as a phase. In addition, further on in the article, the word stase is required in exactly the same sense as phase in the original sense of the Phase Rule....

G. and E. Friedel (Strasbourg)

Observations for H. Zocher on his recent note

We are absolutely in agreement with Zocher that in general the two notions of phase and stase apply to the same objects. But it suffices that this be not always the case. For example, there is at least one case when two distinct phases in the sense of the Phase Rule only constitute a single stase. This concerns a liquid and its vapour. But mainly, if we consider it useful to preserve the two terms phase and stase, it is because they represent different points of view. The first is concerned with ideas of equilibrium and variation of equilibrium states, whereas the second focuses on the notion of molecular structures and their irreducible discontinuous transformations....

P.P. Ewald (Stuttgart)

Notes on the concepts 'stase', 'texture' and 'phase'

The Friedels distinguish molecular order, through the word 'stase', from the multifarious states of order into which the same homogeneous molecular volume can be built, which are 'textures'. Using these two words should above all achieve a clean division between different meanings of the words 'order' and 'structure'.

In typical cases the distinction is clear.... However, the boundary between the concepts of texture is not sharp.... In liquid crystals the indistinct boundary between the concepts of stase and texture can lead to difficulties, e.g. in cholesterol-like materials. Let us assume that for a state with Grandjean planes, the molecular order is distinguished from that of a nematic material by a spiral deformation. Then the question of whether the deformed structure should be seen as a new stase, or as a deformation (or texture) of another stase depends on the magnitude and regularity of the deformation. This decision has a direct consequence on the controversial problem concerning the number of fundamentally different liquid crystal states in the same material.

...(Ewald then attempts to make a distinction between 'geometric' and 'thermodynamic' types of stase.)

I am therefore not able to agree with Zocher's discussion note that the contents of the concepts 'phase' and 'stase' are identical. This holds only if one omits the fine-scale study of the medium which contains just the features which are important in the colloidal and mesomorphic state.

(The Friedels reply, that after mature reflection, they do not think it possible to include any geometric considerations in the definition of 'stase'.)

Discussion of the Ornstein paper

K. Herrmann (Charlottenburg)

On swarm theory

...That Oseen also makes no distinction between Bose's picture and that of Voigt is based on his own words: 'Obviously the hypothesis enunciated here is in all fundamental points identical with that already published by Bose in 1907'....

I thus believe that one is justified in supposing that according to the swarm theory partly ordered material (inside the swarms) and disordered material (between the swarms) coexist in the anisotropic temperature régime.

Precisely this idea seems to me to be ruled out by the X-ray data.

The experimental fact is that, above the clearing point, the magnetic field has no effect on the typical liquid (this is also clear from observations made with the microscope). The halo (amorphous ring) in the X-ray picture remains unaltered.

Below the clearing point, however, a whole set of halos appears. If amorphous material still existed in this temperature régime, then the halo would become stronger in two quadrants and weaker in the two others, but it would not disappear. We now have some data in which a weak remnant of the halos can be seen in the liberated quadrants, but we also have other data – and this is the decisive point – in which one can find not the slightest hint of the original halo in any quadrant.

This must be seen as a proof against this version of the swarm hypothesis, since a trace of the halo would always be observable in the pattern coming from an emulsion . . . .

L.S. Ornstein (Utrecht)

Answer to K. Herrmann

In my opinion, the foundation of the swarm hypothesis is:

(a) Two molecules of a liquid-crystal-forming substance exert mutual forces on each other thorough parallel positioning, requiring: 1. exceptional axes, 2. special atomic groups within the molecules.
(b) These effects work only through directly neighbouring molecules or above surface layers which are some molecules thick.
(c) The tendency for parallel-positioning for larger groups can only be understood in terms of indirect effects. This means that (a) directly affects (b), (b) directly affects (c), and so (a) indirectly affects (c).

The centres of mass can be spatially completely disordered – moreover, surface order is also possible.

Swarms are regions in which indirect forces exist.

The phenomenon of 'directional swarm formation' is possible in all liquids whose molecules lack spherical symmetry. The degree of parallel positioning and the radius of the swarm will be different for different sized molecules.

The boundary of the different regions will not be sharp. The transition will, however, be vanishingly small by comparison with the swarm size. In a mathematical treatment it is possible – as always in such cases – to make the transition discontinuous. Here I am rejecting Bose's foam theory. There is no question of an emulsion – the states are real phases (even when they consist of swarms) . . . .

The picture of parallel positioning of the exceptional molecular axes probably corresponds to the 'nematic' state. . . . In 'smectic' (materials) the situation is more complicated . . . .

H. Zocher (Dahlem)

Discussion notes to L.S. Ornstein

To prove the swarm hypothesis it is without doubt primarily important to show that the other possible explanations of the phenomena under consideration are incorrect. Such explanations have been given by, amongst others, Friedel and myself. Secondly, one must also show that all other phenomena will be explained by it, or equivalently, be in accord with it. Regarding this last point, I might make use of the abundance of material described most of all by Friedel....

This explanation uses swarms which are in principle without boundaries, and which possess tiny oscillations of small groups of molecules. At what moment does this explanation become implausible?...

L.S. Ornstein (Utrecht)

Reply to H. Zocher

The swarm hypothesis gives a precise quantitative description of the following phenomena:

1. Extinction wavelength-dependence (Riwlin).
2. Dielectric constant as a function of E and H (Kast, Jeżewski).
3. Thermal effect in magnetic field (Moll, Ornstein).
4. Crystalline interference as a function of magnetic field and wavelength (van Wyk).
5. Extinction in a magnetic field (Moll, Ornstein).
6. Conductivity as a function of magnetic field.
7. It can give a simple quantitative explanation of X-ray interference in a magnetic field.
8. Gives a prediction, yet to be checked that the magnetic permeability is a function of magnetic field.

By contrast, the considerations of Friedel and Zocher have produced only qualitative results. They have not brought this field closer to the ideal of a mathematical description of nature....

The swarm hypothesis does not mean that a liquid crystal is to be considered as a disperse system....

I do admit the possibility of interaction between the swarms, but for the most part I have neglected it in order not to complicate the mathematical treatment. However, my student van Wyk has found phenomena which he has explained using just such an interaction.

Utrecht, 15 September 1930.

H. Zocher (Dahlem)

To Ornstein

Most of the above eight points supporting the swarm hypothesis (No's 2,5,6,7,8) concern the same phenomenon, namely the average orientation in different field strengths, and should not therefore be considered as independent confirmations. The explanations in the remaining points (No.s 1,3 and 4) in principle have nothing to do with the swarm hypothesis. Consequently they could equally be repeated by those who take an opposing point of view. The apparent accord of the field-orientation curve with the swarm hypothesis is absolutely no proof. In these observations there is a hidden fundamental assumption that the material consists of independent thermally vibrating colloidal parts....

The quantitative treatment of the orientation-field dependence using our point of view is of course not impossible. At the moment, however, the fundamental elastic properties are unknown. As a result it is difficult in practice, and considerably harder than finding the long-sought translation of the Langevin theory. The Mesophase world still offers such an abundance of qualitative effects, that one can put off the quantitative analysis of any particular case until more is known qualitatively....

Concluding Discussion

. . .

G. and E. Friedel (Strasbourg)

Professor Vorländer's note 'Discussion on the articles by G. and E. Friedel etc.', situated in the chapter 'General Discussion Notes', was not communicated to us during the course of the Discussion. In fact, perhaps it is just as well that this violent diatribe against our ideas was thus shaded from all controversy. It self-destructs, all on its own.

Professor Vorländer thinks that he sees three, four or more 'crystalline fluid' stases in some substances. Professor Vorländer had allowed us to hope for the communication of a sample of one of these many substances. Despite a friendly intervention on the part of Professor Ewald, we still have no sample of this type. We still believe that very probably this is a case of erroneous observation. Perhaps this is a result of material impurity, or perhaps confusion between simple texture change and true change of stase. It is more and more likely that a single material never offers more than one smectic and one nematic stase....

To conclude, let us permit ourselves but one observation. One can see the position occupied by theory, and the tiny degree of importance that, in general unconsciously, seems to be devoted to establishing the facts exactly. In view of this, it would seem that in the domain of mesomorphic materials the facts are so well studied that there is nothing else to do other than let the imagination roll on their theoretical interpretation. In reality it is not at all like that. There remains in this field a rich harvest of facts to be observed by extremely simple methods. It is only the effect of a regrettable and widely

247

spread fashion that, especially in recent years, theories, essentially ephemeral and hence secondary mind games, have come to reduce the facts to the status of humble servant . . . .

There follows a final comment from Vorländer, who seems to realise that perhaps his previous contribution might have been too forceful. He makes a plea to his colleagues to concentrate on the facts, and to draw up a dictionary between the various terminologies currently in use. He finishes by inviting colleagues to visit him in Halle so that he might be able to *demonstrate* to them experimentally the error of their ways! The mantle then passes to P.P. Ewald to wind up the proceedings.

P.P. Ewald (Stuttgart)

This Discussion now draws to a close, not because the participating researchers have reached agreement on fundamental matters, but simply as a result of external constraints. The editor's purpose was to identify differences on established experimental questions and in the interpretation of the phenomena, and then by this means to influence current opinions. This goal seems only to have been achieved to a rather modest extent. Nevertheless this document can be presented to the public in good conscience. It contains new theoretical and experimental results. In addition, it demonstrates to all those interested in this well-studied field a cross-section of the views of contemporary researchers. The names of these researchers cannot even be mentioned once without, on one page or another, annoying an indignant opposition. Nevertheless, future generations, with clearer insight in the fundamentals of the phenomena, will with good cause count them as the pioneers in the field. The editor is convinced that the present work will accelerate research progress by clarifying and stimulating ideas.

This volume has taken a long time to prepare. The collaborators involved have had substantial other demands on their time and still have not flagged. The editor's only remaining duty is express his own thanks, and those of the public, to them for their work.

Liquid Crystals and Anisotropic Melts
A General Discussion held by the Faraday Society April 1933

Contents.

CONTENTS

LIQUID CRYSTALS AND ANISOTROPIC MELTS.

A GENERAL DISCUSSION.

THE FIFTY-EIGHTH GENERAL DISCUSSION organised by the FARADAY SOCIETY was devoted to " LIQUID CRYSTALS AND ANISO- TROPIC MELTS."

On the 24th and 25th April, 1933, a meeting for the discussion of the above subject was held, by the courtesy of the Managers, in the Lecture Theatre of the Royal Institution. About 150 members and visitors were present. The President of the Society, Dr. N. V. Sidgwick, F.R.S., occupied the Chair throughout the meeting.

At the opening session Sir William Bragg, O.M., F.R.S., on behalf of the Managers of the Royal Institution, welcomed the Society and ex-pressed their wishes for the success of the Society's deliberations.
The President then referred briefly to the lamented death of Geheimrat Professor F. Rinne, whose paper would be discussed later at the meeting. Those present stood in silence for some moments in memory of the Geheimrat.

The President then introduced the overseas members and guests individually to the Society and called upon them to rise in their places whereupon they were welcomed with acclamation by the Society. Those so welcomed were : Professor Dr. R. Schenck (*Berlin*), President of the *Deutsche Bunsen-Gesellschaft*, Professor D. M. Bose (*Calcutta*), Dr. W. Eisenschimmel and Frau Eisenschimmel (*Prag*), Professor K. Herrmann (*Berlin-Charlottenburg*), Professor R. O. Herzog (*Berlin*), Professor G. van Iterson (*Delft*), Dr. W. Kast (*Freiburg*), Professor L. S. Ornstein (Utrecht), Professor F. Paneth (*Königsberg*), Professor Dr. D. Vorländer and Dr. H. O. Vorländer (*Halle a. Salle*), Professor H. Zocher and Frau Zocher (*Prag*).

The Secretary then read messages of regret at their inability to be present from Professor C. W. Oseen, Professor V. Freedericksz, Professor W. Ostwald, Professor G. Foëx, Professor G. W. Stewart, Professor J. J. Trillat, Professor H. Mark, and from the immediate Past President, Sir Robert Mond.

After Professor Oseen's Introductory Paper had been taken as read and discussed, very beautiful experimental demonstrations were given by Professor Vorländer, Professor van Iterson and Dr. A. S. C. Lawrence. Further experimental demonstrations were given on Tuesday by Professor van Iterson and Mr. Bernal.

The social headquarters of the meeting were, by kind invitation of the Executive Committee, at the Chemical Club, Whitehall Court, the Society's lady guests being entertained at Whitehall Court.

Members of Council and English contributors of papers met the overseas guests informally for luncheon at Whitehall Court before the meeting commenced, and the Society entertained its overseas guests at dinner at Kettners' Restaurant on Monday evening.

At the conclusion of the meeting votes of thanks were accorded to the overseas guests for their presence and assistance in the success of the meeting; to Sir William Bragg and the Managers of the Royal Institution for their hospitality, and to Mr. Green and his assistants and other officers of the Royal Institution for their hospitality during the meeting; to the Committee of the Chemical Club for their hospitality; to the contributors of papers; to the organising committee (the President, Mr. Bernal and Mr. Rawlins), and to the translators of papers. In conclusion Professor Vorländer expressed the thanks of the overseas guests to the President and the Society.

The contributions, which had all been circulated in advance, were taken as read, the authors each devoting a few minutes to indicating the lines upon which they hoped their contributions would be discussed. The contributions will be found in the succeeding pages. The printed arrangement of the papers does not exactly follow the order in which they were presented at the meeting, since it was found that the General Discussion tended to classify certain of the papers in a different order. The General Discussion is reported at the end of the volume (p. 1060 onwards) in the order in which it was found convenient at the meeting.

B5
Transactions of the Faraday Society **29**, 883–900 (1933)

THE THEORY OF LIQUID CRYSTALS.

By C. W. Oseen.

Received 22nd November, 1932.

In my monograph " *Die anisotropen Flüssigkeiten, Tatsachen und Theorien* " Fortschritte der Chemie, physikalischen Chemie und Physik. Bd. 20, 1929, I have among other things given a review of the results to which my theory had led by that time. The object of this article is to give a review of the results arrived at later on. In this connection I shall also enter upon some questions still unsolved.

1. On the Forces that give Rise to Liquid Crystals.

As far as I know, the physicists who have paid attention to liquid crystals have all taken it for granted that the forces which cause molecules to combine so as to form a liquid crystal, are not the chemical valence forces, but belong to the large group of molecular forces. This assumption has been verified by Vorländer,[1] whose statements are based upon the examination of a very large amount of chemical material. Now, as is well known, Debye and others have shown that the molecular forces may to a large extent be interpreted as electrostatic forces between the different nuclei and electrons that constitute molecules. Under these circumstances the question arises whether the forces between the molecules of a liquid crystal can also be supposed to be of an electrostatic nature. In my opinion the answer to this question must be negative.

Hitherto it has been possible in the theory of liquid crystals to ascribe to molecules rotational symmetry about an axis. For the determination of the positions and orientations of two molecules with regard to a definite co-ordinate system, then, ten quantities are needed. These may be, for either molecule, the three co-ordinates of the centre of gravity and the two angles that fix the direction of the axis of symmetry. Now there are ∞^6 equivalent co-ordinate systems. From this it follows that the positions of the two molecules in relation to each other must be determined by four quantities. These quantities may be chosen as the distance between the centres of gravity (r_{12}), the cosine of the angle that the axis of symmetry of the first molecule, when drawn in a given direction, forms with the line joining the centres of gravity, drawn from molecule 1 to molecule 2, the cosine of the angle that the axis of symmetry of the second molecule, when drawn in the same direction, forms with the line of connection just mentioned, drawn from molecule 2 to molecule 1, and the cosine of the angle that the two axes of symmetry form with each other. If by $\mathbf{r}_{12}$ is meant the vector that is directed from the centre of gravity of molecule 1 to that of molecule 2, and if by $\mathbf{L}^{(1)}$ (or $\mathbf{L}^{(2)}$) is

[1] D. Vorländer, *Chemie der kristallinen Flüssigkeiten.* Z. *Kristallographie,* **79,** 64.

meant a vector of unit length which has the same direction as the axis of symmetry of the molecule 1 (or 2), then these quantities will be :

$$r_{12}, \quad \mathbf{L}^{(1)} \cdot \frac{\mathbf{r}_{12}}{r_{12}}, \quad \mathbf{L}^{(2)} \cdot \frac{\mathbf{r}_{21}}{r_{21}}, \quad \mathbf{L}^{(1)}\mathbf{L}^{(2)} . \qquad . \qquad . \quad (1)$$

$$(\mathbf{r}_{21} = - \mathbf{r}_{12}).$$

It is at once obvious, however, that these four quantities do not yield a single-valued determination of the positions of the two molecules relative to each other, but only a two-valued. The angle between the two planes which cut one another along the line joining the two centres of gravity and which contain each one of the two axes of symmetry, is determined as to its magnitude, but not with regard to its sign. For a single-valued determination of the positions of the two molecules relative to one another a further quantity is necessary, and this may be chosen to be :

$$\mathbf{L}^{(1)} \times \mathbf{L}^{(2)} . \frac{\mathbf{r}_{12}}{r_{12}} \qquad . \qquad . \qquad . \qquad (2)$$

The square of this quantity, but not the quantity itself can be rationally expressed in the quantities (1). Accordingly it is only the sign of this quantity that introduces anything new.

From this it will be seen that there are two different kinds of liquid crystals. On the one hand there is a group characterised by the fact that the potential energy of two molecules may be expressed as a uniform function of the four quantities (1). On the other hand there is a group in the case of which such a description of the potential energy of two molecules is not possible, whereas this energy may be expressed as a uniform function of the four quantities (1) and of the quantity (2). It is one of the most important results of the theory, that the last-mentioned group of liquid crystals is identical with the group of cholesterine-nematic substances. From this must be inferred that the molecular forces that give rise to liquid crystals, are of such nature that the potential energy of two molecules may depend on the sign of the quantity (2) ; in other words, of such nature that it may occur that the potential energy of two molecules is not invariant for a reflection at a plane in space. This requirement excludes electrostatic forces. Electrostatic energy, which depends only on the distances between the charges, cannot change when the whole system is reflected at a plane, since at this reflection all charges and all distances remain unchanged.

If the force between two molecules in a liquid crystal of the cholesterine-nematic type cannot be regarded as purely electrostatic, then it will be an obvious conclusion that the non-electrostatic part of the potential energy of two molecules is of a magnetic nature. This theory is corroborated by the fact that the magnetic force actually has a directing effect upon anisotropic molecules. From the point of view of atomic theory, too, there are reasons to believe that two molecules can, among other things, also exert magnetic forces on each other. If a molecule has an axis of symmetry, the total moment of momentum of the molecule, which is composed of the mechanical moment of momentum and the spin-moment of the electrons, must have a component along the axis of symmetry, which for every state with a fixed energy value has also a fixed value. In the case of liquid crystals only those states are at present taken into consideration where the moment of momentum coincides with the axis of symmetry. Evidently two molecules constructed in this way will exert magnetic forces on each other. On closer examination, however, it becomes evident that even in this way we do not obtain a satisfactory interpretation of the molecular forces that appear in liquid crystals. A moment of momentum along the axis of symmetry may, so far as the physical effects are concerned, to a first approximation be substituted by a magnetic moment in the direction of the axis of symmetry. According to the interpretation sketched above the cholesterine-nematic substances must necessarily be paramagnetic. But, so far as we know, this is not the case. Besides, even in that case we should not be any nearer the solution of the problem. For magneto-static energy, like electrostatic, is invariant for reflection at a plane.

We have seen that, according to the theory, cholesterine-nematic substances are characterised by the fact that the potential energy of two molecules is not invariant for reflection. This agrees well with Vorlander's proposition that cholesterine-nematic substances are always so constructed that the molecule contains at least one asymmetric carbon atom.[2] Now, as is well known, the asymmetric carbon atom is characterised by optical rotary power. There naturally arises the question whether this effect does not suggest where the solution of the problem is to be found. Now the theories of optical rotary power that were started at the same time by M. Born and myself, both show that the essential fact here is the difference of phase that will exist between the different vibrators of the active molecule. It may be asked, then, whether the finite velocity with which the force is transmitted does not effect the interaction between two molecules. If we imagine two molecules which contain electrons rotating about axes in them, the interaction between them may, to a first approximation, be described by means of electric and magnetic dipoles, quadrupoles, and so on. But in an exact description the finite velocity of propagation of the force must also be taken into consideration. Here it is possible to understand how, in the case of cholesterine-nematic substances, the potential energy of two molecules can undergo a change on reflection at a plane. If this is correct, the solution of the problem of liquid crystals is to be found in a chapter of atomic physics that is still unwritten.

2. On Singularities of Structure in Nematic Substances.

If we suppose that the force between two molecules decreases rapidly enough as the distance increases, and if further we suppose that the density of the substance considered is constant, we shall find that the laws holding for the orientation of the axes of the molecules may be summed up in the requirement that an expression of the form

$$\int \rho^2 \{ K_1 \mathbf{L} \operatorname{rot} \mathbf{L} + K_{11} (\mathbf{L} \operatorname{rot} \mathbf{L})^2 + K_{22} (\operatorname{div} \mathbf{L})^2 \\ + K_{33} ((\mathbf{L}\nabla)\mathbf{L})^2 + 2K_{12} \operatorname{div} \mathbf{L} \cdot \mathbf{L} \operatorname{rot} \mathbf{L} \} d\omega,$$

(where $d\omega = dx_1 \, dx_2 \, dx_3$),

will have an extreme value. In the case of nematic substances $K_{12} = 0$, $K_{22} = K_{33}$. In this case we obtain a solution :

$$\mathbf{L}_1 = \cos \phi, \ \mathbf{L}_2 = \sin \phi, \ \mathbf{L}_3 = 0,$$

$$\phi = \frac{K_1 x_3}{2(K_{11} - K_{33})} + f(x_1, \ x_2),$$

$$\frac{\partial^2 f}{\partial x_1{}^2} + \frac{\partial^2 f}{\partial x_2{}^2} = 0.^3$$

[2] Loc. cit., p. 84.
[3] Cf. my above-mentioned monograph, Die anisotropen Flüssigkeiten, p. 49.

If we put: $f = \pm \psi +$ constant, $\pm \frac{1}{2}\psi +$ constant, $\psi = \tan^{-1}\frac{x_2}{x_1}$, we obtain structures with a single line $x_1 = x_2 = 0$, which must optically show the phenomena described by Lehmann under the names of whole *Kernpunkte* $(+ \psi)$, whole *Konvergenzpunkte* $(- \psi)$, half *Kernpunkte* $(+ \frac{1}{2}\psi)$, half *Konvergenzpunkte* $(- \frac{1}{2}\psi)$. But as pointed out in the before-mentioned monograph, the theory also meets with difficulties. The energy of such a structure would be infinitely great, and, besides, it seems difficult to see why structures of the type:

$$f = k \log r$$

should not also be found.

We obtain a solution of these difficulties, if we abandon the assumption that the density is constant. Obviously the theory thus obtained will be far more complicated than before, and the formulæ are indeed so long that they cannot be reproduced here. For the new theory it is a matter of vital importance that the thermodynamic function which is to take a minimum value is the free energy. The entropy of an æolotropic substance is not known empirically. According to the theory every state of an element characterised by the distribution of the axes of the molecules in the different directions in space has its entropy. What is here to be taken into consideration, is, however, the most probable distribution and the entropy corresponding to this. Per unit of mass this entropy may be expressed as a function of the density and the absolute temperature $F(\rho, T)$. Hence we may write:

$$S = \int F(\rho, T)\rho d\omega.$$

If we develop this theory by the methods of mathematical physics, we shall obtain a system of three differential equations for the determination of the direction of the axis and of the density. If we specially want to examine plane structures, *i.e.*, structures in which, for instance, $L_3 = 0$, and L_1 and L_2 only depend on x_1, x_2, the equations are reduced to two. In order that these equations may represent a nematic substance it is necessary to impose certain conditions on the coefficients. If we do so and put $L_1 = \cos \phi$, $L_2 = \sin \phi$, we obtain for the determination of ϕ and ρ the equations:

$$\Delta \phi - \frac{1}{\rho}\left(\frac{\partial \phi}{\partial x_1}\frac{\partial \rho}{\partial x_1} + \frac{\partial \phi}{\partial x_2}\frac{\partial \rho}{\partial x_2}\right) = 0,$$

$$C_2^{(0)}\Delta\rho + 2C^{(0)}\rho + 2K_{22}\rho\left[\left(\frac{\partial \phi}{\partial x_1}\right)^2 + \left(\frac{\partial \phi}{\partial x_2}\right)^2\right] = \beta + 2m^2 T\left(F(\rho, T) + \rho\frac{\partial F}{\partial \rho}\right).$$

Obviously these equations may be satisfied by putting

$$\phi = k\psi, \quad \psi = \tan^{-1}\frac{x_2}{x_1}, \quad \rho = f(r) = f(\sqrt{x_1^2 + x_2^2}).$$

For the determination of the function f we obtain an ordinary differential equation of the second degree. For this differential equation $r = 0$ is a singular point. By means of successive approximations we can find a solution of this equation which fulfils, for $r = 0$, the conditions $f = 0$, $f' = 0$. Our system of equations has, however, no solutions of the form $\phi = k \log r$, $\rho = f(\psi)$. The two above-mentioned difficulties inherent in the theory of the singularities of structure of nematic substances, are thus removed.

A singular line in a nematic substance, the direction of which coincides with that of view, gives rise to a *Kernpunkt* or *Konvergenzpunkt* and so forth. If its direction is perpendicular to the direction of view, it gives rise to other phenomena. It appears, then, that we have to distinguish two cases. It may occur that the image of the singular line is on both sides surrounded by a half-shade, owing to the fact that one of the two rays in the aeolotropic substance, to which the incident ray has given rise, undergoes a deviation in the neighbourhood of the singular line, which prevents it from penetrating into the microscope. It may, however, also occur that there appears no such half-shade. These facts give rise to the question whether the difference between singular lines with a half-shade and such lines without a half-shade has any connection with the distinction between whole *Kernpunkte*, whole *Konvergenzpunkte*, half *Kernpunkte*, half *Konvergenzpunkte*. In order to solve this problem the author has made a geometrical-optical examination of the progress of the rays in the vicinity of a singular line. The result of this examination is that in all these cases the extraordinary ray undergoes a deviation in the vicinity of the singular line, but that this deviation is generally small, if the singular line is of the type " whole *Kernpunkt*" ($\phi = \psi +$ Konst.), whereas it is always great, if the singular line belongs to any of the other types. Though this result is to be received with a certain reserve, as the methods of geometrical optics are not applicable when a ray undergoes sharp deviations, yet it seems probable that a singular line without a half-shade is always of the type " whole *Kernpunkt*" ($\phi = \psi +$ Konst.), and that a singular line with a half-shade is in general of another type than this.

3. On the Iridescent Structures of Cholesterine-nematic Substances.

A substance, for which $K_{12} = 0$, $K_1 \neq 0$ has a tendency to appear in a structure that may be characterised as helicoidally-twisted. Mathematically such a structure may be described by means of the simple formulæ :

$$L_1 = \cos \alpha x_3, \ L_2 = \sin \alpha x_3, \ L_3 = 0.$$

Here **L** is as usual a vector of unit length which indicates the direction of the molecular axis. In my monograph and in the discussion arranged by Professor Ewald and published in *Z. Kristallographie*, I have given some information on the remarkable optical qualities of such structures. Here I shall give a more detailed account.

If the dielectric coefficient is regarded as a tensor with the components

$$\epsilon_{jk} = (\epsilon_3 - \epsilon_1)L_jL_k + \delta_{jk}\,\epsilon_1(\delta_{jk} = 0, \ if\, j \neq k, = 1, \ if\, j = k)$$

Maxwell's equations takes the form :

$$\text{rot } \mathbf{H} = \frac{\epsilon_3 - \epsilon_1}{c}\Big(\mathbf{L}\frac{\partial \mathbf{E}}{\partial t}\Big)\mathbf{L} + \frac{\epsilon_1}{c}\frac{\partial \mathbf{E}}{\partial t}, \quad \text{rot } \mathbf{E} = -\frac{1}{c}\frac{\partial \mathbf{H}}{\partial t}.$$

We consider first the solutions of these equations that correspond to waves propagated in the direction of the x_3 axis. It appears that these solutions find their simplest expressions if, in addition to the fixed x_1, x_2 axes, we introduce a moving (x_1', x_2') system so chosen that the x_1' axis has at every point the same direction as the vector **L** and that, consequently, the x_2' axis is perpendicular to **L** and to the x_3 axis. We put :

$$\frac{1}{2}\Big[\alpha^2 + \frac{\omega^2}{2c^2}(\epsilon_1 + \epsilon_3)\Big] = a, \quad \frac{1}{4}\Big[\alpha^2 - \frac{\omega^2}{2c^2}(\epsilon_1 + \epsilon_3)\Big]^2 - \frac{\omega^4}{16c^4}(\epsilon_1 - \epsilon_3)^2 = b,$$

$$\Omega_1 = \sqrt{a + \sqrt{b}} + \sqrt{a - \sqrt{b}}, \quad \Omega_2 = \sqrt{a + \sqrt{b}} - \sqrt{a - \sqrt{b}},$$

259

and prescribe that, if $b > 0$, in which case Ω_1 and Ω_2 are real, the roots shall be taken with their positive values, and, if $b < 0$, in which case Ω_1 is real and Ω_2 purely imaginary, the signs shall be so chosen that Ω_2 takes the form $- ki$ $(k \geq 0)$, whereas the real part of Ω_1 is still positive. If then we put:

$$-\Omega_1 = \Omega^{(1)}, \quad -\Omega_2 = \Omega^{(2)}, \quad \Omega_1 = \Omega^{(3)}, \quad \Omega^2 = \Omega^{(4)}$$

we obtain the following expressions for the components of the strengths of the electric and the magnetic fields in the moving system just mentioned.

$$E_1' = \frac{1}{2}\left[(\Omega^{(j)} + \alpha)^2 - \frac{\omega^2 \epsilon_1}{c^2}\right] e^{i(\omega t\, +\, \Omega^{(j)} x_3)},$$

$$H_1' = \frac{i}{2}\left[\frac{c}{\omega}(\Omega^{(j)} + \alpha)^2 (\Omega^{(j)} - \alpha) + \frac{\omega}{c}(\epsilon_1 \alpha - \epsilon_3 \Omega^{(j)})\right] e^{i(\omega t\, +\, \Omega^{(j)} x_3)},$$

$$E_2' = \frac{i}{2}\left[(\Omega^{(j)} + \alpha)^2 - \frac{\omega^2 \epsilon_3}{c^2}\right] e^{i(\omega t\, +\, \Omega^{(j)} x_3)},$$

$$H_2' = -\frac{1}{2}\left[\frac{c}{\omega}(\Omega^{(j)} + \alpha)^2 (\Omega^{(j)} - \alpha) + \frac{\omega}{c}(\epsilon_3 \alpha - \epsilon_1 \Omega^{(j)})\right] e^{i(\omega t\, +\, \Omega^{(j)} x_3)}.$$

A discussion of these expressions already gives valuable information on the strange optical qualities of helicoidally-twisted strata. If $b > 0$, which occurs, if ω lies outside the interval $c\alpha/\sqrt{\epsilon_1}$, $c\alpha/\sqrt{\epsilon_3}$, all $\Omega^{(j)}$, as observed are real. Our formulæ show that in this case there are two elliptically polarised waves in either direction. Their velocities of propagation are ω/Ω_1 and ω/Ω_2. During the propagation the ellipse of vibration rotates thus that its principal axes always fall along the x_1' and x_2' axes. If, on the contrary, $b < 0$, that is, if ω lies within the interval $c\alpha/\sqrt{\epsilon_1}$, $c\alpha/\sqrt{\epsilon_3}$, Ω_2 and consequently $\Omega^{(2)}$ and $\Omega^{(4)}$ are purely imaginary. The corresponding waves are, referred to the moving co-ordinate system, to be taken as standing waves. These waves, too, are elliptically polarised so that the principal axes of the ellipse of vibration fall along the x_1' and x_2' axes. The amplitude increases, for the wave [4], exponentially with x_3. For the wave [2] the amplitude decreases exponentially with x_3. Within the interval $c\alpha/\sqrt{\epsilon_1}$, $c\alpha\sqrt{\epsilon_3}$ there will generally be a region where $kd \gg 1$, d being the thickness of the æolotropic stratum. This region is called the interior part of the interval. In this part of the spectrum there is then a wave the upper amplitude of which is enormous compared to the lower, and another wave the lower amplitude of which is enormous compared to the upper. The above-mentioned propositions are of fundamental importance for the study of the reflection of a wave in a cholesterine-nematic stratum of iridescent structure, and of its passage through this stratum. Let us imagine that a wave that comes from the region $x_3 < 0$, meets the æolotropic substance in the plane $x_3 = 0$ and leaves it in the plane $x_3 = d$. The requirement that in the plane $x_3 = d$ there shall be no incident radiation, gives rise to two linear connections between the vector components $E_1(d)$, $E_2(d)$, $H_1(d)$, $H_2(d)$. Let us now specially consider a wave that may in the interior of the substance be represented by the formulæ:

$$E_1' = \sum_{j = 1,\, 2,\, 4} A_j \frac{1}{2}\left[(\Omega^{(j)} + \alpha)^2 - \frac{\omega^2 \epsilon_1}{c^2}\right] e^{i(\omega t\, +\, \Omega^{(j)} x_3)},$$

$$E_2' = \sum_{j = 1,\, 2,\, 4]} \frac{1}{2}\left[(\Omega^{(j)} + \alpha)^2 - \frac{\omega^2 \epsilon_3}{c^2}\right] e^{i(\omega t\, +\, \Omega^{(j)} x_3)}.$$

The conditions just mentioned, applicable to $x_3 = d$, make it possible to express A_1 and A_4 in the forms $A_2 e^{-kd} F_1(\alpha, \omega, C, \epsilon_1, \epsilon_3)$ and $A_2 e^{-kd} F_2(\alpha, \omega, C, \epsilon_1, \epsilon_3)$, respectively. Hence it will be seen that A_1 and A_4 belong, roughly speaking, to the same order of magnitude as $A_2 e^{-kd}$. In the interior of the interval $c\alpha/\sqrt{\epsilon_1}$, $c\alpha/\sqrt{\epsilon_3}$ they are infinitesimal compared to A_2. Hence in the vicinity of the plane $x_3 = 0$ we obtain with very great accuracy :

$$E_1' = A_2 \frac{I}{2}\left[(\alpha - \Omega_2)^2 - \frac{\omega^2 \epsilon_1}{c^2} \right] e^{i\omega t - kx_3}.$$

From these expressions for E_1', E_2', H_1', H_2' we can now determine the corresponding expressions $E_1^{(i)}$, $E_2^{(i)}$, $H_1^{(i)}$ $H_2^{(i)}$ for the incident wave, as well as the components $E_1^{(r)}$, $E_2^{(r)}$, $H_1^{(r)}$, $H_2^{(r)}$ of the reflected wave. In both cases we obtain comparatively simple expressions showing that both waves are elliptically polarised. We can also determine the components, $E_1^{(g)} \ldots$ of the wave coming from the plane $x_3 = d$. It appears, as may also be concluded from what is said above, that the amplitude of this wave bears to that of the incident wave a proportion of the order of magnitude $e^{-kd} : I$. The incident wave thus undergoes a total reflection, when it meets the æolotropic stratum.

It is interesting to examine more strictly the expressions for the incident and the reflected wave. We refer them, of course, to the fixed (x_1, x_2, x_3) system. In order to obtain the simplest possible formulæ we put :

$$E_1^{(i)} + iE_2^{(i)} = E_1^{*(i)}, \quad E_1^{(i)} - iE_2^{(i)} = E_2^{*(i)},$$

and in the same way :

$$H_1^{(i)} + iH_2^{(i)} = H_1^{*(i)}, \quad H_1^{(i)} - iH_2^{(i)} = H_2^{*(i)}.$$

We have in these denominations :

$$E_1^{*(i)} = -iH_1^{*(i)} = \frac{\omega}{4c}(\epsilon_3 - \epsilon_1)\left(\frac{\omega}{c} - \alpha + \Omega_2\right)A_2 e^{i\omega(t - x_3/c)},$$

$$E_2^{*(i)} = iH_2^{*(i)} = \frac{c}{2\omega}\left[(\alpha - \Omega_2)^2 - \frac{\omega^2}{2c^2}(\epsilon_1 + \epsilon_3)\right]\left(\frac{\omega}{c} + \alpha + \Omega_2\right)A_2 e^{i\omega(t - x_3/c)},$$

$$E_1^{*(r)} = iH_1^{*(r)} = \frac{\omega}{4c}(\epsilon_3 - \epsilon_1)\left(\frac{\omega}{c} + \alpha - \Omega_2\right)A_2 e^{i\omega(t - x_3/c)},$$

$$E_2^{*(r)} = -iH_2^{*(r)} = \frac{c}{2\omega}\left[(\alpha - \Omega_2)^2 - \frac{\omega^2}{2c^2}(\epsilon_1 + \epsilon_3)\right]\left(\frac{\omega}{c} - \alpha - \Omega_2\right)A_2 e^{i\omega(t - x_3/c)}.$$

The quantities $\Omega^{(j)}$ $(j = $ I, 2, 3, 4$)$ are roots of the equation :
$$\Omega^4 - 4a\Omega^2 + 4b = 0.$$

This equation may also be written :

$$\left[(\alpha - \Omega)^2 - \frac{\omega^2}{2c^2}(\epsilon_1 + \epsilon_3)\right]\left[(\alpha + \Omega)^2 - \frac{\omega^2}{2c^2}(\epsilon_1 + \epsilon_3)\right] = \frac{\omega^4}{4c^4}(\epsilon_3 - \epsilon_1)^2.$$

If Ω_2 is purely imaginary, it follows that

$$\left| \frac{\frac{\omega^2}{2c^2}(\epsilon_3 - \epsilon_1)}{(\alpha + \Omega_2)^2 - \frac{\omega^2}{2c^2}(\epsilon_1 + \epsilon_3)} \right| = I.$$

There is then such a number t_0 that:

$$\frac{\omega^2}{2c^2}(\epsilon_3 - \epsilon_1) = \left[(\alpha + \Omega_2)^2 - \frac{\omega^2}{2c^2}(\epsilon_1 + \epsilon_3)\right]e^{-it_0}$$

$$= \left[(\alpha - \Omega_2)^2 - \frac{\omega^2}{2c^2}(\epsilon_1 + \epsilon_3)\right]e^{+it_0}.$$

On account of this our expressions for $E_1^{\bullet(r)}$ and $E_2^{\bullet(r)}$ may be written:

$$E_1^{\bullet(r)} = \frac{c}{2\omega}\left[(\alpha + \Omega_2)^2 - \frac{\omega^2}{2c^2}(\epsilon_1 + \epsilon_3)\right]\left(\frac{\omega}{c} + \alpha - \Omega_2\right)A_2 e^{i\omega(t - t_0 + x_3/c)},$$

$$E_2^{\bullet(r)} = \frac{\omega}{4c}(\epsilon_3 - \epsilon_1)\left(\frac{\omega}{c} - \alpha - \Omega_2\right)A_2 e^{i\omega(t - t_0 + x_3/c)}.$$

If we write $E_1^{(i)}$ and $E_2^{(i)}$:

$$E_1^{(i)} = R\{ae^{i\omega(t - x_3/c) + i\alpha}\} = a\cos\{\omega(t - x_3/c) + \alpha\},$$

$$E_2^{(i)} = R\{ibe^{i\omega(t - x_3/c) + i\beta}\} = -b\sin\{\omega(t - x_3/c) + \beta\},$$

$E_1^{(r)}$ and $E_2^{(r)}$ will take the form:

$$E_1^{(r)} = R\{ae^{i\omega(t - t_0 + x_3/c) - i\alpha}\} = a\cos\{\omega(t - t_0 + x_3/c) - \alpha\},$$

$$E_2^{(r)} = R\{-ibe^{i\omega(t - t_0 + x_3/c) - i\beta}\} = b\sin\{\omega(t - t_0 + x_3/c) - \beta\}.$$

The formulæ for $E_1^{(r)}$ and $E_2^{(r)}$ will be transformed into the formulæ for $E_1^{(i)}$, $E_2^{(i)}$, if $t_0 - t$ is substituted for t. This proves that the ellipses of vibration of the incident and the reflected wave have exactly the same form and the same size. With regard to the different directions of propagation it also shows that in both cases the vibration, regarded by an observer towards whom the wave advances, takes place in the same direction. There is then no difference between the two waves but the opposite directions of propagation and a difference of phase.

We have found that there exists a wave of light, elliptically polarised in a given manner, which is subjected to a total reflection in the helicoidally twisted structure, at which reflection the reflected wave has, but for a difference of phase, exactly the same character as the incident wave. In order to master the problem of reflection we must also take another wave into consideration. We obtain it by examining a wave which in the æolotropic substance is determined by the formulæ:

$$E_1' = \sum_{j=1,3,4} B_j \frac{I}{2}\left[(\Omega^{(j)} + \alpha)^2 - \frac{\omega^2\epsilon_1}{c^2}\right]e^{i(\omega t + \Omega^{(j)}x_3)},$$

$$E_2' = \sum_{j=1,3,4} B_j \frac{I}{2}\left[(\Omega^{(j)} + \alpha)^2 - \frac{\omega^2\epsilon_3}{c^2}\right]e^{i(\omega t + \Omega^{(j)}x_3)}.$$

In this case I shall only mention the result of the computation. We put:

$$-I + \frac{c^2}{\omega^2}\frac{\Omega_1^2 - \Omega_2^2}{\alpha(\epsilon_1 + \epsilon_3 - 2)}\left(\frac{\omega}{c} + \alpha + \Omega_2\right) = M_1,$$

$$\frac{\Omega_2}{\Omega_1} + \frac{c^2}{\omega^2}\frac{\Omega_1^2 - \Omega_2^2}{\alpha\Omega_1(\epsilon_1 + \epsilon_3 - 2)}\left(\frac{\omega}{c} + \alpha + \Omega_2\right)\left(\frac{\omega}{c} + \alpha\right) = M_2,$$

and:

$$M_1 \cos(d\Omega_1) + iM_2 \sin(d\Omega_1) = N_1,$$
$$M_2 \cos(d\Omega_1) + iM_1 \sin(d\Omega_1) = N_2.$$

With this notation we have :

$$E_1^{*(i)} = \frac{\omega}{4c}(\epsilon_3 - \epsilon_1)\left\{\left(\frac{\omega}{c} + \alpha\right)N_1 - \Omega_1 N_2\right\}B_4 e^{i\omega(t - x_3/c)},$$

$$E_2^{*(i)} = \frac{c}{2\omega}\left\{\left[\left(\frac{\omega}{c} + \alpha\right)\left(\alpha^2 - \frac{\omega^2}{2c^2}(\epsilon_1 + \epsilon_3)\right) + \left(\frac{\omega}{c} - \alpha\right)\Omega_1^2\right]N_1 \right.$$
$$\left. + \Omega_1\left[\frac{2\alpha\omega}{c} - \alpha^2 + \Omega_1^2 - \frac{\omega^2}{2c^2}(\epsilon_1 + \epsilon_3)\right]N_2\right\}B_4 c^{i\omega(t - x_3/c)},$$

$$E_1^{*(r)} = \frac{\omega}{4c}(\epsilon_3 - \epsilon_1)\left\{\left(\frac{\omega}{c} - \alpha\right)N_1 + \Omega_1 N_2\right\}B_4 e^{i\omega(t + x_3/c)},$$

$$E_2^{*(r)} = \frac{c}{2\omega}\left\{\left[\left(\frac{\omega}{c} - \alpha\right)\left(\alpha^2 - \frac{\omega^2}{2c^2}(\epsilon_1 + \epsilon_3)\right) + \left(\frac{\omega}{c} + \alpha\right)\Omega_1^2\right]N_1 \right.$$
$$\left. + \Omega_1\left[\frac{2\alpha\omega}{c} + \alpha^2 - \Omega_1^2 + \frac{\omega^2}{2c^2}(\epsilon_1 + \epsilon_3)\right]N_2\right\}B_4 e^{i\omega(t + x_3/c)}.$$

For the wave that emerges from the plane $x_3 = d$ we finally obtain :

$$E_1^{*(g)} = \frac{\epsilon_3 - \epsilon_1}{2\alpha(\epsilon_1 + \epsilon_3 - 2)}(\Omega_1^2 - \Omega_2^2)\left(\frac{\omega}{c} + \alpha + \Omega_2\right)B_4 e^{i\omega(t - x_3/c) + i\alpha d},$$

$$E_2^{*(g)} = -(\Omega_1^2 - \Omega_2^2)\left\{1 + \frac{1}{2\alpha}\left(\frac{\omega}{c} + \alpha + \Omega_2\right)\right.$$
$$\left. - \frac{c^2}{\omega^2\alpha(\epsilon_1 + \epsilon_3 - 2)}\left(\frac{\omega}{c} + \alpha + \Omega_2\right)\left[\left(\frac{\omega}{c} + \alpha\right)^2 - \Omega_1^2\right]\right\}B_4 e^{i\omega(t - x_3/c) - i\alpha d}.$$

From these formulæ it will be seen that if a plane wave, advancing in the direction of the normal, meets a helicoidally-twisted structure, the intensity of the reflected wave as well as that of the passing one must in general be a periodic function of $\Omega_1 \cdot d$. Now, if the two bounding surfaces of the æolotropic substance are not exactly parallel, but form a small angle, then both in the reflected light and in that which has passed there will appear dark stripes at constant distances from one another, all parallel to the line of intersection of the bounding surfaces. If, however by means of an elliptical polariser we cause the incident ray to be of the first type considered above, these dark stripes will not appear. The conclusions here drawn from the theory agree perfectly with the observations that have been made with relation to the so-called stripes of Grandjean. It cannot very well be assumed that this agreement is due to a mere chance. There will then be no reasons why the so-called planes of Grandjean should really exist.

We have mentioned above some consequences of the theory which agree with known facts. I shall now point out some consequences, the correctness of which is questionable. According to the statements of our experimentalists, the iridescent structures of cholesterine-nematic substances have an optical rotary power which in certain cases is very great. Now the above formulæ show that in the interior of the region $c\alpha/\sqrt{\epsilon_1}$, $c\alpha/\sqrt{\epsilon_3}$, there is no optical rotary power in the usual sense of the word. Every wave which, having passed through the æolotropic stratum, emerges from the plane $\chi_3 = d$, must be obtainable through a linear combination of those emergent waves that correspond to the two types of elliptically polarised waves in the structure. We know, however, that the former type, which gives rise to a total reflection, does not give any perceptible wave emergent from the plane $\chi_3 = d$. The ellipse of

263

vibration of the wave that emerges from this plane has then exactly the same form, and the principal axes of this ellipse have exactly the same directions, whatever the incident wave may be. Hence a rotary power in the usual sense of the word is out of the question. As we see, the argument stated here is based on the existence of an elliptically polarised wave that on incidence in the substance is subjected to a total reflection. Now this consequence of the theory agrees with facts. Then it seems probable that the conclusion that has here been drawn from this consequence, also agrees with facts. According to Friedel the optical rotary power actually changes its sign for a certain wave-length, taking then an infinitely large value. But in the place of the spectrum where this infinite rotary power was supposed to be found, no rotation at all can be shown, because, according to Friedel, the emergent wave is circularly polarised. If we leave Friedel's theory of the infinitely great rotary power out of the question and keep to what can be directly observed, the theory seems to agree with facts on this point too.

We shall now leave the region $c\alpha/\sqrt{\epsilon_1}$, $c\alpha/\sqrt{\epsilon_3}$ and consider the parts of the spectrum for which ω falls outside this interval. The result of the theory is that in these regions of the spectrum there is no rotary power in the usual sense of the word. It is true that an incident, plane-polarised wave that passes through the stratum, is changed into an elliptically-polarised wave whose greater axis of vibration forms an angle with that of the incident wave, but this angle does not depend linearly on the thickness of the stratum, but in a far more complicated way. If, as the experimentalists state, the iridescent structure of a cholesterine-nematic substance really has an optical rotary power of the usual kind, then, on this point, the theory in its present state disagrees with facts. It must be added, however, that our experimentalists seem to have taken it for granted that, when a rotation of the plane of polarisation is perceptible, there will also be a rotary power in the usual sense of the word. A careful examination of phenomena of this kind is at present one of the most pressing needs with reference to liquid crystals.

An account has been given of the laws which hold for the perpendicular incidence of a wave in the iridescent structure of a cholesterine-nematic substance. But it is also very interesting to study the reflection and refraction of a non-perpendicular wave. It is found empirically that, in the case of a non-perpendicular wave also, there is a region in the spectrum in which a component of the wave undergoes a total reflection. As the incidence of the wave becomes oblique instead of perpendicular, the region of total reflection is displaced towards the violet part of the spectrum. It is this fact, above all, which the theory has to explain. The investigation shows that an incident wave, for which the components of the electromagnetic field are proportional to :

$$e^{i\omega\left(t - \frac{\beta_1\chi_1 + \beta_2\chi_2 + \beta_3\chi_3}{c}\right)} \quad \text{(where } \beta_1{}^2 + \beta_2{}^2 + \beta_3{}^2 = 1)$$

gives rise to a wave of the type :

$$E_j = e^{i\omega\left(t - \frac{\beta_1\chi_1 + \beta_2\chi_2}{c}\right)} e^{\frac{ih\omega\chi_3}{c}} F_j(\chi_3),$$

(where $j = 1, 2, 3$).

The functions $F_j(x_3)$ are periodic with the period $2\pi/\alpha$. They satisfy the conditions :

$$F_j\left(x_3 + \frac{\pi}{\alpha}\right) = -F_j(x_3).$$

h is a transcendent function of c, ω, α, E_1, E_3 and $\beta^2 = \beta_1{}^2 + \beta_2{}^2$. In order that a total reflection may occur, it is necessary that h should have a complex value. This is, as we know, the case for $\beta = 0$ if ω falls in the interval $c\alpha/\sqrt{\epsilon_1}$, $c\alpha/\sqrt{\epsilon_3}$, i.e., if the wave-length lies within the limits :

$$\frac{2\pi}{\alpha}\sqrt{\epsilon_1} \quad \text{and} \quad \frac{2\pi}{\alpha}\sqrt{\epsilon_3}.$$

Now, if $\beta^2 > 0$ but so small that β^4 may be neglected compared with unity these limits will be displaced to :

$$\frac{2\pi}{\alpha}\sqrt{\epsilon_1} - \beta^2 \quad \text{and} \quad \frac{2\pi}{\alpha}\sqrt{\epsilon_3 - \beta^2 \frac{\epsilon_1 + \epsilon_3}{2\epsilon_1}}.$$

We see that the theory explains why the displacement takes place towards the side of the short waves. Besides the region of total reflection, which remains for the case of $\beta = 0$, there is, however, an infinite number of other such regions. They all lie on the violet side of the region first found. Now, if we pass from perpendicular incidence to oblique, it will not only occur that the region of reflection already present is displaced towards the violet part of the spectrum, but new colours will appear on the violet side of this region.

One of the most remarkable properties known as to the iridescent structures of cholesterine-nematic substances, is the fact that their colour depends both on the angle that the incident light forms with the normal and on that which the reflected light forms with this normal. By this it seems that one of the oldest optical laws which exists, namely that according to which the angle of reflection is determined by the angle of incidence and equal to this, does not hold for these structures. The theory developed does not explain this fact. A theory of it may, however, be given, which is based upon the properties of helicoidally-twisted strata. We only have to remember that no body has an absolutely plane surface and that, if two plane surfaces existed, it would never be possible to adjust them so as to be perfectly parallel to each other. From this it follows that the theory concerning the iridescent strata of choles-terine-nematic substances must also take into account the case where a bounding surface divides the structure obliquely, so that the prepara-tion takes the form of a wedge. In this case other phenomena will appear. If, for the sake of simplicity, we assume the stratum to be infinitely thick, the most important result of the theory can be expressed as follows : A plane polarised wave incident on the plane bounding surface of a heli-coidally twisted stratum, the normal of which makes an angle with the axis of the stratum, gives rise to a reflected wave, which is composed of an infinite number of plane waves with different directions of propagation. The relation between the angle of incidence and that of reflection is there-fore not single-valued. An infinite number of angles of reflection corre-spond to a given angle of incidence.

4. On the Motion of Æolotropic Liquids.

It has been known for some long time that many liquids do not move in accordance with the Navier-Stokes equations of motion. In 1925 Kruyt summed up the investigations then made and arrived at the result that disperse-systems with globular particles follow the law given by Poiseuille which holds for a liquid passing through a tube, whereas deviations from this law appear in the case of systems with non-spherical particles. Further material since 1925 has confirmed the correctness of Kruyt's conclusion. Under these circumstances Dr. A. Anzelius [4] set himself the task of formulating laws of motion valid for a liquid with oblong molecules. We shall give below a survey of the theory of Anzelius and the results to which it has led hitherto.

Anzelius assumes that in a liquid consisting of oblong molecules with symmetry of rotation, the direction of the axis of a molecule will vary continuously from point to point, even if the liquid moves, so that at a certain moment there will be a certain direction of axes for every element of volume. Under these circumstances we can no longer suppose that Navier's and Stoke's assumptions concerning the tensor of viscosity

$$\tau_{jk} = - p\delta_{jk} + \mu\left(\frac{\partial u_j}{\partial x_k} + \frac{\partial u_k}{\partial x_j}\right)$$

holds good. Instead, we must assume that this tensor will depend also on the direction of the axis of the molecule, i.e., on $\mathbf{L}$. On the presumption that a dissipation-function exists, Anzelius determines the form of this function. His results may be summed up in the statement that the dissipation function, F, must have the form:

$$A_1 \sum_{\substack{j,k=\\1,2,3}} D_{jk}{}^2 + A_2 \sum_k \left(\sum_j L_j D_{jk}\right)^2 + A_3 \left(\sum_{j,k} L_j L_k D_{jk}\right)^2$$
$$+ 2A_4 \sum_{j,k} L_j L_k D_{jk} \sum_m D_{mm} + A_5 \left(\sum_m D_{mm}\right)^2,$$

where
$$D_{jk} = \frac{1}{2}\left(\frac{\partial u_j}{\partial x_k} + \frac{\partial u_k}{\partial x_j}\right),$$

and $A_1 \ldots A_5$ are constants. From F we obtain the tensor of viscosity by the formulæ:

$$\tau_{jk} = - p\delta_{jk} + \frac{1}{2}\frac{\partial F}{\partial D_{jk}}.$$

Now the question arises, how the directions of the axes of the molecules are to be determined. Anzelius assumes that between the molecules there are forces in action that have rotary moments about the centres of gravity. The analytical expression for these forces he takes from the present author's theory of liquid crystals. But he assumes further that the motion of the liquid has a directive effect on the molecules. This, he supposes, is due to the fact that on account of the motion the impacts of the molecules give rise to a rotary moment. Taking the conditions of symmetry for his basis he finds for this moment, $\mathbf{M}$, an expression, which may be written in the form:

$$\mathbf{M} = \{(C_1 - C_2)(\mathbf{L}D) + \tfrac{1}{2}(C_1 + C_2)(\mathbf{L} \times \text{rot } \mathbf{u})$$
$$+ (C_{11} + C_{12})(\mathbf{L} \times (\mathbf{L}D)) - \tfrac{1}{2}(C_{11} - C_{12})\text{rot } \mathbf{u}\} \times \mathbf{L}.$$

C_1, C_2, C_{11}, C_{12} are constants. $(\mathbf{L}D)$ is the vector whose components are $L_j D_{j1}$, $L_j D_{j2}$, $L_j D_{j3}$. Here, a term in which the same index appears twice, will, with regard to this index, be summed up over the values 1, 2, 3.

Finally I should mention the boundary condition which Anzelius introduces into his theory. He assumes that, in a flowing liquid, asymmetrically constructed molecules arrange themselves perpendicularly to the wall.

[4] A. Anzelius, *Über die Bewegung der anisotropen Flüssigkeiten.* The Annual of the University of Uppsala, 1931.

Basing our computations on these assumptions we may now proceed to calculate the motion of a liquid in given cases. As appears from what is said above, one of the most interesting cases is that of a liquid flowing through a tube, because in this case a liquid with oblong molecules does not give the same result as a liquid with spherical molecules. Anzelius summarises the results of his calculations on this case as follows :

1. The theoretical relation between the pressure gradient and the amount of liquid passing per unit of time agrees with the experimental relation, if $A_3 < 0$.

2. When A_3 is negative, the distribution of velocity over the transverse section of the tube is different from that found in the case of isotropic substances, in that the velocity in the middle of the tube is comparatively greater. Ostwald supposes, though he cannot refer to any theory or observations, that in flowing æolotropic liquids an equalisation of the velocities will take place in the transverse section, analogous to that which occurs in turbulent isotropic liquids. The result arrived at does not agree with Ostwald's assumption.

3. The orientation of the molecules depends on the sign of the pressure gradient, that is, on the direction of the flow. This agrees with an observation by Professor R. Fåhreus, Uppsala, according to which blood that is absorbed by a capillary, seems to change its colour, when the direction of the flow is reversed.

4. As the velocity of the flow increases, the molecules tend to take a direction which is parallel to the axis of the tube. This phenomenon is analogous to the rotation of the molecules in electric and magnetic fields, which may, if the field strength is sufficiently high, cause the destruction of the edge stratum.

Anzelius has further investigated the behaviour of an æolotropic liquid in the case of Couette's arrangement, that is, when the liquid is enclosed between two coaxial cylinders, the inner of which is stationary, whereas the outer one rotates with constant velocity round the common axis. If the inner radius is r_1 and the outer one r_2 if $(r_2 - r_1)/r_1 = E$, and if ω is the angular velocity of the outer cylinder, and M_d is the rotary moment acting per unit of length on the inner cylinder, he finds :

$$M_d = \alpha\omega + \beta\omega^3$$

where :

$$\alpha = \frac{2\pi A_1 r_1 r_2}{2E - E^2 + E^3}, \quad \beta = \frac{\pi A_3 C_2{}^2 r_1{}^3 r_2{}^3}{60 K_{22}{}^2}E \text{ for asymmetrical molecules,}$$

$$\text{or} = \frac{\pi A_3 C_1{}^2 r_1{}^3 r_2{}^3}{60 K_{22}{}^2}E \text{ for symmetrical molecules.}$$

To this must be added an expression for the largest angle that the axis of a molecule forms with the radius vector from the axis of the cylinder :

$$\lambda_{\text{max.}} = \frac{C_2}{8K_{22}}r_1 r_2 \omega E.$$

We see that by determining ω, M_d, $\lambda_{\text{max.}}$ we shall also be able to determine three constants A_1, A_3 and C_2/K_{22}. K_{22} is a constant taken from the author's theory of liquid crystals.

Anzelius has further investigated the manner in which, in Couette's case, the stability of the motion is affected by the æolotropy of the molecules. In order to make the complicate calculations possible he had to assume that the axis of a molecule adjusts itself radially everywhere.

The investigation is of interest, because in this case, too, one must assume $A_3 < 0$.

I hope that the above review of the theory of Anzelius has shown the great interest it deserves. Attention must, however, be called to a difficulty inherent in this theory. During the motion of a liquid the moment **M** has to do work. It is not clear how this work stands with regard to the principle of energy.

On account of these doubts as to the theory of Anzelius the author has developed another theory of the motion of an æolotropic liquid. According to this theory the state of an element is characterised by a vector **L**, whose components are the mean values of the unit vectors that indicate the directions of the axes of the molecules belonging to the element. Accordingly the direction of the vector **L** indicates the mean directions of the axes of these molecules. Its length indicates the æolotropy of the element. For the flow of the liquid and for the changes of the vector **L** the following laws can be formulated :

$$\rho \frac{du_j}{dt} = X_j - \frac{\partial p}{\partial x_j} - \rho \frac{\partial}{\partial x_j} \int F(x, x', \mathbf{L}, \mathbf{L}') dm' + \sum_k \frac{\partial \tau_{jk}}{\partial x_k},$$

$$\rho d^2 \frac{d^2 L_j}{dt^2} = \Lambda_j + \rho A \frac{L_j}{a} - \rho \frac{\partial}{\partial L_j} \int F(x, x', \mathbf{L}, \mathbf{L}') dm' - \sigma_j.$$

Here X_j $(j = 1, 2, 3)$ are the components of the external force that acts on the liquid, Λ_j are the components of a generalised external force, for instance the magnetic. A is a quantity characteristic of the theory and called the pressure of orientation. $Fdm \, dm'$ is the potential energy of the element dm in the neighbourhood of the point x and of the element dm' in the neighbourhood of the point X'. In order to determine the tensor of viscosity τ_{jk} and the corresponding vector σ we assume that they depend on a dissipation function $G(D_{jk}, dL_j/dt ; \mathbf{L}, \mathbf{rot} \, \mathbf{u})$ so that

$$\tau_{jk} = \frac{\partial G\left(D_{jk}, \dfrac{dL_j}{dt} ; \mathbf{L}, \mathbf{rot} \, \mathbf{u}\right)}{\partial D_{jk}},$$

$$\sigma_j = \frac{\partial G}{\partial \dfrac{dL_j}{dt}}.$$

The heat produced per unit of volume and unit of time is then $2G$.

The two equations given above do not suffice to determine the motion. To them must be added the condition of continuity :

$$\rho \frac{dv}{dt} = \sum_j \frac{\partial u_j}{\partial x_j},$$

and the equation of the conduction of heat :

$$\rho T \frac{d}{dt} \frac{\partial \zeta}{\partial T} + 2G = \mathbf{div} \, \mathbf{j}.$$

Here $v = 1/\rho$ is the specific volume, ζ the free energy, and $\mathbf{j}$ the current of heat. Further the equations of the state of the liquid must be added :

$$p = -\frac{\partial \zeta}{\partial v}, \quad A = -\frac{\partial \zeta}{\partial A}.$$

The theory expressed in these formulæ agrees with the principle of energy.

According to the opinions generally accepted it may be assumed that the dissipation function G is a homogeneous, quadratic function of the nine quantities D_{jk}, dL_j/dt the coefficients may be functions of L_j and **rot u.** A natural hypothesis enables us to determine G more closely. It is a plausible assumption that all forces of friction vanish, when the liquid moves like a solid body. This assumption is equivalent to the assumption that G is a homogeneous, quadratic function of the eight quantities D_{jk}, $\mathbf{L}\dfrac{d\mathbf{L}}{dt}$, $\dfrac{d\mathbf{L}}{dt}$ **rot u.** The coefficients may as before depend on $\mathbf{L}$ and **rot u.**

The theory sketched above has not yet been used for the calculation of given problems of motion. I have mentioned it here, because it seems to me to throw light upon one of Lehmann's most remarkable observations on liquid crystals. He found that in certain cases a substance, spread out between two glass surfaces, would be put into motion, when influenced by a flow of heat coming from below, during which motion the different drops of liquid seemed to be in violent rotation. Further investigations convinced Lehmann that in this case it was not the drop itself, but the structure, that moved. According to my opinion the observed motion was due to the fact, that the molecules rotated with the same speed round vertical axes drawn through their centres of gravity. At this rotation $\mathbf{u} = 0$, $\mathbf{L}\dfrac{d\mathbf{L}}{dt} = 0$. Consequently all forces of viscosity vanish. The " violent speed " is hereby explained.

5. On the Forms of Liquid Crystals.

When Lehmann introduced the term " liquid crystals," he did it not only on account of the double refraction found in certain liquids, but also on account of the form that drops of certain substances, namely those which are now usually called the smectic ones, will take under favourable conditions. That this form is not spherical is the first important fact in this connection that a theory of liquid crystals has to explain. Lehmann's further investigations showed that smectic drops can assume a multitude of different forms, some of which are characterised by considerable grace and symmetry. To find these forms theoretically is the final object of the theory in this connection.

In my contribution to the discussion on liquid crystals arranged by Professor Ewald and published in Z. *Kristallographie*, I formulated a mathematical problem, the solution of which was calculated to throw light upon the above questions. Roughly speaking, it stands to the exact thermodynamic treatment of them as Laplace's theory of capillarity stands to Van der Waals' theory. The problem was formulated like this : It is required to determine a closed surface, S, a region outside it, U, and a vector function $\mathbf{L}$ which fulfils the condition $L^2 = 1$, in the interior of S so that the expression :

$$\frac{\rho^{(a)2}}{2m^{(a)2}}\int_K \{K_{11}(\mathbf{L}\ \mathbf{rot}\ \mathbf{L})^2 \times K_{33}((\mathbf{L}\Delta)\mathbf{L})^2\}d\omega$$

$$+ \frac{1}{2}\int_S \left\{ \frac{\rho^{(a)2}}{m^{(a)2}}f_1(L_n) + \frac{2\rho^{(a)}\rho^{(i)}}{m^{(a)}m^{(i)}}f_2(L_n) + \frac{\rho^{(i)2}}{m^{(i)2}}G_2 \right\}dS,$$

where $\rho^{(a)}$, $\rho^{(i)}$, $m^{(a)}$, $m^{(i)}$, K_{11}, K_{33}, G_2

269

are constants and K is the region enclosed by S, takes the smallest value that is compatible with the conditions :

$$\rho^{(a)} \int_K d\omega = Nm^{(a)}, \quad \rho^{(i)} \int_U d\omega = Mm ;$$

N and M being here constant, positive numbers.

The above problem has been subjected to an investigation by Mr. N. E. Larsson. The following is for the most part based upon this investigation.

If we want to solve a problem in the calculus of variations, we may begin by formulating the necessary conditions to be fulfilled by the solution, which conditions follow from the requirement that a small variation of the wanted functions must leave unchanged the expression that is to take the minimum value. In the problem under consideration we can firstly vary the directions of the axes of the molecules, or in other words, the vector $\mathbf{L}$, and secondly the form of the crystal, in other words that of the closed surface S. Through the former variation we obtain the laws holding for the orientation of the molecules in a liquid crystal. These laws, which must of course agree with the laws ealier formulated by the author, have the form :

$$[(K_{11} - K_{22})\{2\ \mathbf{L}\ \text{rot}\ \mathbf{L} \cdot \text{rot}\ \mathbf{L} - \mathbf{L} \times \mathbf{grad}\ (\mathbf{L}\ \text{rot}\ \mathbf{L})\}$$
$$+ K_{33}\ \text{rot rot}\ \mathbf{L}] \times \mathbf{L} = 0 \qquad (1)$$

Further we obtain a condition that must be fulfilled at the bounding surface S. If, for the sake of simplicity, we put :

$$\frac{m^{(a)2}}{\rho^{(a)2}}\left\{ \frac{\rho^{(a)2}}{m^{(a)2}} f_1(L_n) + \frac{2\rho^{(a)}\rho^{(i)}}{m^{(a)}m^{(i)}} f_2(L_n) + \frac{\rho^{(i)2}}{m^{(i)2}} G_2 \right\} = f(L_n),$$

this condition may be written :

$$(K_{11} - K_{33})L_n \cdot \mathbf{L}\ \text{rot}\ \mathbf{L} \cdot \mathbf{L} - K_{11}\ \mathbf{L}\ \text{rot}\ \mathbf{L} \cdot \mathbf{n}$$
$$+ K_{33}\ L_n \cdot \text{rot}\ \mathbf{L} - \tfrac{1}{2} f'(L_n) \cdot \mathbf{n} \times \mathbf{L} = 0 \qquad . \quad (2)$$

The variation of the form of the crystal gives a new condition which the surface S must fulfil :

$$K_{11}(\mathbf{L}\ \text{rot}\ \mathbf{L})^2 + K_{33}((\mathbf{L}\Delta)\mathbf{L})^2 + f'(L_n)\mathbf{div}\ \mathbf{L} + \sin\psi\ \frac{df'(L_n)}{dt}$$

$$+ \{f(L_n) - f'(L_n) \cdot L_n\}\left(\frac{1}{R_1} + \frac{1}{R_2} \right) = \text{constant} = K \qquad . \quad (3)$$

Here ψ is the angle between the vector $\mathbf{L}$ and the normal $\mathbf{n}$ drawn outwards. df'/dt is the derivative of the function f' in the direction given by the projection of the vector $\mathbf{L}$ on the tangent plane of the surface. R_1 and R_2 are the radii of curvature of the surface, taken as positive, if the corresponding centre of curvature lies within the surface S.

As the problem is difficult, it will be allowable to begin by investigating simple cases.

It appears from Lehmann's photographs that one structure, which is at least approximately realised in many smectic drops, is so shaped that the axes of the molecules are parallel, and the form of the drop is determined by the fact that its bounding surfaces are a right circular cylinder and two planes perpendicular to this. The direction of the axes of the molecules coincides with the direction of the generatrices of the cylinder. Now the question arises whether this structure agrees with the theory developed.

It is at once obvious that the law holding for the structure of the interior of a liquid crystal, is fulfilled. The first boundary condition, too, is fulfilled, if we make the plausible assumption that $f'(0) = 0$. The second boundary condition is fulfilled by the cylindrical boundary surface. But it is not fulfilled with regard to the plane end surfaces, because the left-hand member of the equation (3) has the value 0, whereas the right-hand member must have the value K. From this it follows that our assumptions concerning this structure cannot all be correct. We see also where we shall have to make a correction. The assumption that the non-cylindrical bounding surfaces are plane, cannot be absolutely correct. From this follows that in the vicinity of these surfaces $\mathbf{L}$ cannot be a constant. If the end surfaces have a different form, for instance spherical, it will not be difficult to satisfy the equation (1) and the first boundary condition. This may be done thus, that the axes of the molecules are assumed approximately to fulfil the condition $\mathbf{rot}\ \mathbf{L} = 0$ and to arrange themselves perpendicularly to the end surfaces. But the question still remains to be answered why these end surfaces are but slightly curved. We see, however, from our formula (3) that an explanation is afforded at once. If we apply this formula to the cylindrical bounding surface, we shall obtain $f(0)$ as coefficient of the medium curvature. If, on the contrary, we apply it to the end surfaces, we shall obtain $f(1)$. We only have to assume $f(1) \gg f(0)$ and the approximately plane form of the end surfaces will be explained. It should be noticed that the assumption $f(1) \gg f(0)$ is also necessary for the explanation of the prolate form that is characteristic of the smectic drops considered. The fact that the molecules at the greatest part of the bounding surface adjust themselves so as to be parallel to this surface, proves that $f(1) \gg f(0)$. The theory thus traces the prolate form and the approximately plane end surfaces to a common origin.

Much work will yet be required before the strange and beautiful forms that smectic drops will often take are explained. But the framing of a theory of these forms has begun.

Carl Wilhelm Oseen was born in Lund, in Southern Sweden, in 1879. He entered the University in Lund in 1896, graduating with a master's degree in mathematics and mechanics in 1900. He spent the academic year 1900–1 at Göttingen, where the leading mathematicians were David Hilbert and Felix Klein. In 1903 he was awarded the Dr. phil. degree for a dissertation on contact transformations. At this point he returned to Lund, where he worked as a researcher and lecturer until he was appointed to a chair in mechanics and mathematical physics at the prestigious University of Uppsala in 1909. He remained in Uppsala until 1933. In 1933 he resigned his chair in Uppsalla to become Director of the new Nobel Institute of theoretical physics in Stockholm.

Oseen's main contributions are in hydrodynamics and the theory of liquid crystals. His best known contribution to general hydrodynamics comes from his calculation of the drag force on a sphere in flow. By a clever limiting process he was able to evaluate a leading order correction to the classical Stokes formula which turns out to be independent of the viscosity. He also wrote more than 25 papers in liquid crystals, as well as making contributions to optics, relativity, quantum theory (unsuccessfully!), quantum electrodynamics, and even the history and philosophy of science. All in all he published 155 scientific papers, almost all by himself, in a period when many workers were more parsimonious about publication. He died in 1944.

B6
Transactions of the Faraday Society **29**, 930–44 (1933)

NEW ARGUMENTS FOR THE SWARM THEORY OF LIQUID CRYSTALS.

By L. S. Ornstein and W. Kast (*Utrecht*).

Received in German on 20th January, 1933, and translated by
Helen D. Megaw.

Before dealing with some new measurements on liquid crystal melts of *p*-azoxyanisole that have furnished new arguments for the swarm theory, we shall give as introduction a short summary of the main features of the theory and the most important evidence in its favour.

I.

The spontaneous anistropy which characterises liquid crystals must be due to an arrangement of the molecules. X-ray photographs have now shown that no lattice arrangement exists, and this has recently been confirmed by one of us (Kast), who, using strictly monochromatic Cu-radiation selected from the resolved spectrum, obtained photographs (not yet published) which showed that, except for the rather smaller intensity throughout, no difference exists between the crystalline liquid phase and the amorphous liquid phase. M. and G. Friedel[1] have now introduced the conception of mesomorphous structure. Vorländer and Bose had already worked on the assumption that the molecules have a tendency to set parallel, and this parallel-setting, in which each molecule must be displaced in the direction of its axis, and rotated about that axis, by random amounts, here appears as the lowest order of the mesomorphic structure, the nematic structure.

It is not implied, however,—and here we come to the formulation of the Swarm Theory—that the structure stretches uniformly over the whole macroscopic preparation. A large number of experimentally observed facts suggest rather that it exists only in smaller agglomerations or groups, comprising about 100,000 molecules, which groups lie with their principal directions irregularly disposed, and have interactions much weaker than those between molecules in the same group. Any further arrangement is due to external forces. We might speak of a kind of polycrystalline structure, and we can in fact treat these groups as independent micro-crystals, in cases where the processes concerned are of short duration compared with the individual existences of the groups. But, as is shown by the kinetic considerations we discuss later, the length of their individual existences is limited by diffusion-like processes, and so we prefer to call these groups " swarms."

The first exact mathematical treatment of the Swarm Theory was given by Ornstein and Zernike[2] for systems in the neighbourhood of the critical point. They showed that the correlation of the density deviations in the neighbourhood of the critical point became very large,

[1] M. and G. Friedel, *Ann. Physique*, **18**, 273, 1922.
[2] L. S. Ornstein and F. Zernike, *Physik. Z.*, **19**, 134, 1918.

273

and hence they succeeded in giving a formula for the opalescence up to the critical point, which, in contrast to the Einstein-Smoluchowski formula, does not become infinite at the critical point. In detail, the correlation occurs as follows. The relationship between the density of neighbouring volume elements, being dependent according to Boltzmann's principle on the potential energy of the molecular forces, like that is *directly* effective only for very small distances, of the order of molecular dimensions. For example, if at a point P_1 a known density deviation is given, then the most probable density (or density fluctuation) is only influenced by this over quite a short distance. But, in addition to the direct effect, there is an indirect effect. For example, if the point P_2 is directly influenced by P_1, and similarly P_3 by P_2, then P_3 is also influenced indirectly by P_1, since the density at P_2 depends on that at P_1. Therefore the density at P_3 is indirectly dependent on the density at P_1, and similarly also with a point P_4 which is directly influenced by P_3. The further one goes from P_1 the weaker, naturally, become these indirect probability effects ; and as Ornstein and Zernike have shown, one can define a mean effective radius of indirect influence. The ranges over which this indirect correlation exists are the swarms of density-fluctuations.

If we now apply these considerations to liquid crystals, the dense packing of the molecules in the liquid makes it possible to neglect density-fluctuations. We now consider, instead, fluctuations of direction. We assume that the molecules, by reason of their elongated form (and if necessary also their permanent dipole moment) exert on one another rotation moments which tend to set them parallel. Further, two molecules only exercise direct mutual influences on their positions when they are quite close together ; but through the indirect influence we get larger fields in which the directions of the molecules are nearly the same, and these are the swarms of direction-fluctuations. We shall not discuss the orienting forces in detail, but we may say that the hypothesis for the formation of such swarms presupposes an un-symmetrical shape of the molecules. Monatomic liquids, or those with spherically symmetrical molecules, can never form such swarms. On the other hand, from co-operation of the rotation-moments of dipoles, which have a much larger effective radius, mathematical difficulties as to convergence arise, if, indeed, a screening effect does not occur. The fact that the mutual attractive or repulsive forces of the dipoles are effective only over small distances tends to make the molecules link on end to end rather than lie side by side. The swarm will therefore have an anistropic shape whose long axis coincides with the direction of the dipole.[3]

A liquid which, because of the structure of its molecules, has a tendency to swarm-formation, will therefore consist of swarms which

[3] One of us, elsewhere (L. S. Ornstein, *Z. Krist.*, **79**, 117, 1931), starting from the potential energy of two molecules which exert a rotation-moment on one another, has given a mathematical treatment of the direction-swarms. It is better to put for the value of the potential energy

$$f(x' - x, y' - y, z' - z) \{\cos \phi \cos \phi' + \sin \phi \sin \phi'(\cos \psi \cos \psi' + \sin \psi \sin \psi')$$
$$- 3(\cos \phi \,.\, l + \sin \phi \sin \psi \,.\, m + \sin \phi \cos \psi \,.\, n$$
$$+ \cos \phi' \,.\, l + \sin \phi' \sin \psi' \,.\, m + \sin \phi' \cos \psi' \,.\, n)\}.$$

The symbols have the same meaning as they have there ; l, m and n are the direction-cosines of the line of junction. The conclusion reached earlier, however, is not altered in any way.

are of approximately equal size. The swarm-direction (or direction of the molecules in the swarm) will, however, be different from one swarm to another, and transition ranges will occur between the separate swarms. If now at a given instant the distribution of molecules between the swarms is known, it will only alter very slowly, because that involves a sort of diffusion process. Hence for many purposes a swarm may be considered as an independent elementary particle, *e.g.*, one may even speak of the Brownian movement of the swarm as a whole.

2.

As important evidence for this idea of swarms in liquid crystals, we may call to mind a series of measurements made in the Utrecht Institute.

The marked turbidity characteristic of liquid crystals makes it plausible to assume that their elementary ranges are disordered. But, more than this, the agreement of the measurements of transparency[4] with the swarm theory is a quantitative one. Ornstein and Zernike[5] have calculated the light scattered by a medium where arbitrary gradients of refractive index occur, and have shown that the transparency depends on the product $\omega^2 d$, where d is the thickness of the layer and ω^2 the mean square of the scattering angle for unit length of the path traversed, a quantity which is proportional to the square of the double refraction. Experimentally, Miss Riwlin found, varying the wave-length of the light and the thickness, that for the same transparency the values of $(n_1 - n_2)^2 d$ were constant to within less than 1 per cent. We are therefore concerned with an aggregate of irregularly arranged doubly-refracting regions.

In many cases where one works with thinner layers one sees no turbidity, but larger regions which extinguish uniformly under the polarising microscope. We shall now show that these larger ordered regions only come into existence under the action of external forces. It is characteristic of liquid crystals that, for example, very small fields suffice for magnetic orientation ; and similarly very small electric double layers or capillary forces will suffice to introduce at the edge of the preparation a far-reaching arrangement of the swarms, which extends by correlation for a few hundredths of a millimetre depth into the preparation. Miss Riwlin's measurements give a good example of this also. When the preparation was obtained by melting the solid, it always acted as if its thickness were 0·04 mm. less than if it had been obtained by cooling from the clear liquid. This shows that when it is obtained from the solid, the orientation of the solid crystals with respect to the walls gives rise to a layer 0·02 mm. thick on each side of the preparation which lets nearly all the light through and is therefore regularly ordered.

But even when the preparation is obtained from the liquid a similar arrangement is set up from the outside inwards. This is shown by the extinction experiments of Moll and Ornstein[6] in a magnetic field. In a strong field perpendicular to the surface of the bulb the transparency, and therefore the arrangement of the swarms, is greater than in the absence of the field. But after the removal of the field it is actually

[4] R. Riwlin, *Dissertation*, Utrecht, 1923.

[5] L. S. Ornstein and F. Zernike, *Proc. Acad. Amsterdam*, **21**, 115, 1917.

[6] W. I. H. Moll and L. S. Ornstein, *Versl. Acad. Amsterdam*, **25**, 682, and 112, 1916-17.

much less than in the initial state. Hence it follows, in the first place, that we are not dealing with a structural relationship over the whole layer; because in that case, after a temporary deformation of the structure by the magnetic field, the old structure, and therefore the old transparency, would reappear. Secondly, it follows that a certain arrangement must have existed in the initial state. We then understand the photometer curves of the transparency given in Fig. 1 to mean that the magnetic field destroys the arrangement with axes parallel to the wall but random azimuth, which predominates in the original state (A) (lower portion of the curve), and changes it to an arrangement with vertical axes (B), which is more complete in strong fields, but less in weak ones, than the original state. When the field is removed, the arrangement of the swarms breaks up completely; a very opaque state (C) is then obtained, which, however, is not stable, but generally goes over after some hours into the initial state arranged from the outside inwards. That we are really concerned here with an arrangement with axes parallel to the wall is shown by measurements with a parallel magnetic field. Here (B) the field only slightly improves the arrangement of A, by re-orienting the azimuths, and this better arrangement is to a large extent preserved after the removal of the field (C), and is stable. In this experiment, too, the arrangement from the outside occurs to a greater extent when the preparation is obtained from the solid. In this case one obtains in strong fields the same curve as with the preparation from the liquid in weak fields.

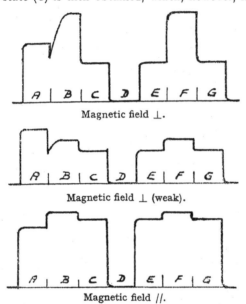

Magnetic field ⊥.

Magnetic field ⊥ (weak).

Magnetic field //.

FIG. 1.—Light-Transmission of Liquid-crystalline *p*-Azoxyanisol.

In these thin layers at the edge, and in these only, we have now in fact an extensive structural relationship; in thin microscopic preparations it may frequently extend throughout the whole preparation. This is shown by the experiment of Moll and Ornstein,[7] which demonstrates the heat effect of magnetic orientation. This succeeds only if a silver plate is hung in the melt to act as a thermo-element, and the field applied perpendicular to it. It is then found that after the application of the field a slow heating of the plate occurs, and after its removal a slow cooling. From the fact that a parallel field alone has no effect, but when applied in conjunction with a vertical field it accelerates the cooling, it is apparent that adiabatic temperature changes from the

[7] W. I. H. Moll and L. S. Ornstein, *Versl. Acad. Amsterdam*, **26**, 1442, 1918.

deformation of the outside layers are involved. Clearly the outside layers manifest themselves in the same way in the peculiar effect of the magnetic field strength on the optic axial figure, which has been observed by van Wijk.[8] The optic figure is not steadily improved by increasing field-strengths ; instead, it is very prominent for certain field-strengths and disappears for others. This may be explained by considering quantitatively the arrangement of the swarms at the outside. In the interior all the swarm axes lie parallel to the field, but at the outside they lie at an angle depending on the interaction of the surface and the magnetic field. Hence with increasing field strength the path difference in the outside layers alters, and an optic figure only occurs when it is equal to an integral number of wave-lengths. The expression obtained for

the " optical thickness " of the edge layer is $\delta = \dfrac{K}{H^a}$, where K is a constant

depending on the material and treatment of the bulb ; the index α, in accordance with the hypothesis of magnetic rotation-moments of diamagnetically anisotropic particles, is nearly equal to 1. The value of the thickness δ for 1200 gauss is 0·01 mm., for 10,000 gauss 0·002 mm.

All these experiments, then, show that invariably the molecules of the liquid are arranged only into individual swarms, and that any more extensive arrangement has its origin in external forces, whether these are produced by arbitrarily applied fields, or arise by themselves at the surface of contact between the melt and solid bodies, particularly glass. We now turn to experiments that give information as to the size of the swarms. Up to the present we could only make the qualitative statement that the reason why weak fields have so large an arranging effect is that they attack not single molecules but the whole swarm. We now know from the experiments of Foëx and Royer[9] that the molecules are diamagnetically anisotropic with $\mu_1 - \mu_2 \approx 10^{-6}$, and from the dielectric measurements of one of us[10] in a magnetic field that they are also dielectrically anisotropic with $\epsilon_1 - \epsilon_2 = -0·1$ (the subscript 1 applies to the direction along the axis). From the calculation of the dielectric constant as a function of the magnetic field strength, the constants of the magnetic rotation moment of the swarm are found :

$$(\mu_1 - \mu_2)v \approx 10^{-21}, \text{ hence } v \approx 10^{-15} \text{ cm.}^3.$$

In this, the assumption is made that on account of the thickness of the layer used only the opposing effect of the heat motion enters the calculation.[11] The figures thus arrived at indicate that in such a swarm about 10^6 molecules are held in parallel positions. A further estimate of the size is given by orientation experiments in an electric field. The dielectric measurements of Jeżewski[12] had already suggested the probable existence of a permanent dipole along the axis of the swarm, and the orientation of the swarms parallel to the lines of electric force was proved directly by one of us,[13] by means of X-ray photographs. These X-ray pictures show, in an electric as well as in a magnetic field,[14] a splitting of the broad interference ring into two arcs whose centre of gravity

[8] A. van Wijk, *Ann. Physik* (5), **3**, 879, 1929.
[9] G. Foëx and L. Royer, *C.R.*, **180**, 1912, 1925.
[10] W. Kast, *Ann. Physik* (4), **73**, 145, 1924.
[11] V. Fréedericksz and A. Repiewa, *Z. Physik*, **42**, 532, 1927.
[12] M. Jeżewski, *Z. Physik*, **51**, 159, 1928, and **52**, 878, 1929.
[13] W. Kast, *Z. Physik*, **71**, 39, 1931 and **76**, 19, 1932.
[14] W. Kast, *Ann. Physik* (4), **83**, 418, 1927.

lies in the direction perpendicular to the field. This leads inevitably to the assumption of a dipole moment in the swarm-direction, since the dielectric anisotropy must itself bring about a setting of the molecules across the lines of electric force. Thus we are here concerned with a case similar to that of molecules with negative Kerr constants, where the direction of the greatest polarisability (*i.e.*, of the induced dipole moment) and of the permanent dipole moment are at right angles. In consequence two rotation-moments appear in the electric field, one turning across proportional to $(\epsilon_2 - \epsilon_1)vF^2$ and one along the length, proportional to pF. Since the X-ray picture for a field-strength of 7200 volts/cm. still shows perfect parallel formation, we obtain the inequality

$$pF \gg (\epsilon_2 - \epsilon_1)vF^2 \; ;$$

and for $F = 24$ e.s., $v = 10^{-15}$ cm.^{3}, and $(\epsilon_2 - \epsilon_1) = 0 \cdot 1$; $p \gg 10^{-14}$ e.s. This is more than 10^4 molecular dipole moments, whose magnitude [15] is fixed as $2 \cdot 3 \times 10^{-18}$ e.s., and the consequence is, as is required by the swarm theory, that the molecular dipoles must lie in the swarm so that they add.

3.

The new investigations follow on from the separation of the two electric rotation moments which was carried through by one of us [16] with alternating fields of increasing frequency.

In the experiments so far the X-ray picture was obtained in alternating fields of different frequency. Up to about 25,000 oscillations per second the picture did not change ; above that, the fibre character of the picture became more marked. Hence the time of relaxation of the particles is $\frac{1}{25000}$ seconds. Above that they no longer follow instantaneously. At about 300,000 oscillations per second, the splitting-up has completely vanished. This marks a second critical time, that which the particles need to turn through 180° under the influence of the field ; here the phase difference between the alternating field and the movement of the dipoles has the value 180°. At yet higher frequencies, the dipoles no longer reach the position parallel to the field during one reversal of the field, and so only oscillate about their equilibrium positions, which are independent of the frequency and in which the induced dipole lies in the direction of the field. Correspondingly, the diagram again shows a fibre character, but now with the fibre direction perpendicular to the field. The measurements therefore give a relaxation time of the order of 10^{-5} sec. as against 10^{-11} sec. for ordinary molecules. It is not yet possible, however, to enter into further quantitative considerations, because it is not easy to see at a glance the nature of the frictional mechanism, which cannot, indeed, without unduly forcing it, be visualised as Stokesian friction. There are, however, particular cases where the relaxation time is found to be very low for normal viscosities throughout (about 135 referred to water at 0° C. as 100), so that to obtain times which can be measured with the oscillation apparatus it is necessary to raise the viscosity artificially by adding oil at low temperatures. The importance of this experiment does not, however, merely lie in its determination of at least the order of magnitude of the swarms ; it is rather that the introduction of a friction dispersion makes probable the separate existence of swarms and the uniformity

[15] J. Errera, *Physik. Z.*, **29**, 426, 1928.
[16] W. Kast, *Z. Physik*, **71**, 39, 1931.

of their size. It appeared, therefore, desirable to obtain further confirmation of these results by a more direct method.

For this purpose, measurements were made of the dielectric loss in a condenser filled with liquid crystals of p-azoxyanisol over the frequency range 5000 to 3,000,000. The dielectric loss plotted against the angular frequency ω gave a hyperbola, so long as only a pure conduction loss was involved, on account of the relation $\tan \delta = \dfrac{1}{\omega RC}$; $\omega \tan \delta$ is constant. This is the same whether polar or non-polar molecules are concerned, so long as the relaxation frequency is not exceeded. If that happens, the phase-difference introduced between the motion of the dipole and the alternating field gives rise to an additional loss. We do not attempt to give any further account of the experimental details, except to say that the relaxation frequency must show itself by the commencement of an increased loss, and the second critical frequency referred to above by the maximum of this additional loss. Now the value of the relaxation frequency

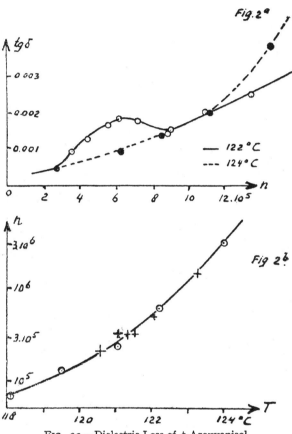

FIG. 2a.—Dielectric Loss of p-Azoxyanisol.
FIG. 2b.—Critical Frequencies of p-Azoxyanisol.

depends on the magnitude of the forces which resist turning, and therefore in some way on the size of the particles. But the size of the swarms decreases with rise of temperature, as determined from measurements of their double refraction and their diamagnetic and dielectric anisotropy ; hence we can arrange the measurements so that we either maintain a fixed temperature and look for the relaxation frequency by varying the frequency, or else apply a fixed frequency and vary the temperature.

We have carried out measurements by both methods, and found maxima of dielectric loss that agree in the temperature range between

118° and 124° C. for frequencies from 70,000 to 3,000,000. Fig. 2*a* shows two loss curves for varying frequency at temperatures 122° and 124° C. The distinct general increase of loss with increasing frequency occurs also in experiments with the condenser empty, and may therefore be subtracted. The occurrence of a frictional dispersion for frequencies of the order of 10^5 is thereby confirmed as expected. But the very surprising thing is the extraordinarily large variation with temperature (as shown in Fig. 2*b*) of the frequency at which the friction dispersion occurs. It confirms the conclusion already mentioned, that we may not visualise it as a Stokesian friction. In that case, from this large variation with temperature, having regard to the normal small alteration of viscosity, we should have to postulate a very large alteration of the size of the swarms, such as is not even approximately consistent with other experiments. For example, it appears from the measurements of Miss Riwlin [4] that the transparency increases slightly with increasing temperature, although the double refraction decreases. From the formula

$$D = C(n_1 - n_2)^2 d$$

it can thus be seen that the constant C, *i.e.*, the number of differently oriented layers which the light traverses on its way through the preparation, increases rather more rapidly than the square of the double refraction diminishes; but it changes at most by a factor of 3 over the whole range in which liquid-crystalline melts exist. It is therefore impossible that the linear dimensions of the swarms should change more than this; and the same estimate is reached from the values of $\mu_1 - \mu_2$ and $\epsilon_1 - \epsilon_2$ derived from the measurements of dielectric constant in a magnetic field.[17] Hence we see that the explanation of the large variation with temperature must lie in a friction mechanism of such a kind that the swarms, in turning, knock against one another

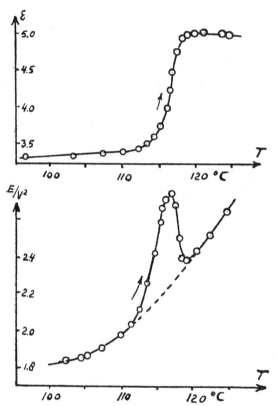

FIG. 3.—Upper Diagram: Dielectric Constant of *p*-Azoxyanisol. Lower Diagram: Dielectric Loss of *p*-Azoxyanisol.

[17] L. S. Ornstein, *Z. Krist.*, **79**, 331, 1931.

and are deformed. Then the forces resisting turning involve the elastic properties of the swarm, which may vary rapidly with temperature.

From the measurements with varying temperature we found an interesting new effect. After the loss-maximum expected for a given frequency had been passed through somewhere in the liquid-crystal range, there appeared a second maximum, just before crystallisation occurred, which was very nearly in the same place for all the frequencies used. Fig. 3 shows this effect, as it occurred for uniformly increasing temperature at a frequency of 10^6. The upper curve gives the dielectric constant for the same temperature-scale, and shows that the anomalies occur just at the melting-point. For these phenomena the only possible explanation is that here there occur again swarms of the same size as those which had been present in the ranges where liquid crystals exist, or, that here temporary swarms occur, of the same size as appear again later in the liquid-crystal melt. We think that those molecules, which in the melt do not belong to swarms but lie in the spaces between them, become associated in the immediate neighbourhood of the melting-point with swarms of quickly altering size. We shall deal later with the importance of this phenomenon for the melting-point; but here we are only interested in the conclusion, that, as shown by the magnitude of the effect, the spaces between the liquid-crystal swarms are of considerable size, about the same order of magnitude as the swarms themselves. This, however, raises no difficulties in our calculation. For example, Hermann [18] has concluded that those molecules which do not lie in the swarms must, in a magnetic field of any considerable strength, cause the appearance of a faint complete ring in the X-ray diagram, but this is not so. For if all the swarms were oriented parallel, then the spaces between would lose their identity. The parallel arrangement would spread out from the swarms over what were the spaces between, and it would perhaps be interesting to follow how the breaking-up of the general arrangement in swarms happens when the field is removed. And this even obviates the difficulty that one might perceive in the absence of anomalies of dielectric constant corresponding in a normal way to friction dispersion. Thus for one swarm there occur in the layers between, which are of the same order of magnitude, the same number of 10^5 separate molecules which do not show frictional dispersion at this place. Moreover, one must consider that, on account of the occurrence of saturation phenomena for the orientation of the dipoles during one alternation, the dielectric constant has no longer a constant value.

4.

The effect here described, in which shortly before crystallisation even those molecules which do not lie in swarms in the liquid-crystal melt proceed to form swarms of rapidly increasing size, must not be confined to the liquid-crystal melt.

The same temporary increase of the dielectric constant which Errera [15] has observed at the melting-point of p-azoxyanisol and which we may connect with this effect is also indicated in ordinary dipole liquids, e.g., in water. In both cases it is limited to very low frequencies. Errera finds it with 680 oscillations per second, but not with 300,000; and the dielectric constant of ice at the melting-point for frequencies

---

[18] K. Hermann, Z. Krist., **79**, 309, 1931.

of the order of 10^6 is stated to be about 2, but for frequencies of the order of 100 to be from 80 to 90.

I. TABLE—DIELECTRIC CONSTANT OF p-AZOXYANISOL FOR FREQUENCY 680.

151° C.	5·2	117° C.	5·4
145	5·2	116·6	12·6
136	5·3	116·4	11·5
122·8	5·3	115·8	7·2
119	5·4	111·2	3·8

TABLE. II—DIELECTRIC CONSTANT OF ICE NEAR THE MELTING-POINT.

— 2° C.	Frequency 40-80	93·9	P. Thomas, *Physic. Rev.*, **31**, 278, 1910.
0	100	78	J. Dewar and J. A. Fleming, *Proc. Roy.*
0	10^7	2	*Soc. London*, **61**, 2, and 316, 1897.

We therefore extended our measurements of dielectric loss to ordinary dipole liquids, and next investigated benzophenone (C_6H_5—C—C_6H_5).

$$\overset{\|}{O}$$

Benzophenone has a very strong dipole moment, as follows from the large increase of dielectric constant (from 3 to 12) at the melting-point. The measurements were difficult because of this large change of dielectric constant, and also because with decreasing temperatures there is much super-cooling. The expected effect was, however, indicated. To control the conductivity, because of its big influence on the loss, we proceeded to make provisional measurements of conductivity. The arrangements were such that a potential of 6 volts was applied every time until the galvanometer reversed again; this ballistic deflection was then read. From this it appeared that the conductivity followed a peculiar course at the melting-point, which we could show to be compatible with the temporary existence of swarms. The conductivity at the melting-point increased, first slowly, then more steeply. Thus, when the temperature-time curve shows the end

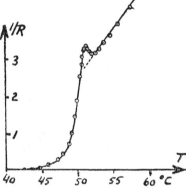

FIG. 4.—Upper Diagram: Conductivity of Benzophenone (6 volts).
Lower Diagram: Conductivity of p-Azoxyanisol (6 volts).

of the melting process, it still does not bend down again at once, but increases steeply for a short time, goes through a pronounced maximum, and finally passes over into a straight line which corresponds to the normal temperature gradient of conductivity (Fig. 4). Here, of course, we are concerned with conductivity by foreign ions, and the maximum therefore indicates a temporary ionic mobility which is abnormally large. In this connection, we remember that in liquid-crystalline p-azoxyanisol [19] a corresponding rise in conductivity can be produced by magnetic or electric orientation of the swarms, and we therefore conclude that in benzophenone in quite a small temperature range above the melting-point there exist swarms which have so large an electric moment that they can be appreciably arranged by 6 volts in $\frac{1}{2}$ mm. or 120 volts per cm.

As a control we have performed the same experiment on p-azoxyanisol, for which we find experimentally, over the whole range in which it is an anisotropic liquid, the same course of the conductivity as for benzophenone at its melting-point (Fig. 4). We have now evaluated these curves; we find the difference of conductivity relative to the straight line of the normal liquid range produced backwards, and this shows that the curves for p-azoxyanisol and benzophenone can be brought into coincidence by merely adjusting the temperature scale. If we now add the rotation moment to the dipole moment of the swarm, and write σ_1 for the conductivity in the direction of the swarm axis and σ_2 for that in both directions perpendicular to it, we can calculate the increase in conductivity by Boltzmann's hypothesis. We obtain the same formula that was previously derived by one of us [20] for the change in conductivity in a magnetic field, using the assumption of a permanent magnetic moment of the swarm. In consequence of the conductivity anisotropy of the swarms, a rotation moment will of course occur, which gives rise to an apolar setting of the swarms in the direction of the electric field. But, just as was the case for the rotation moment due to the dielectric anisotropy, this can be neglected in comparison with the rotation moment from the strong permanent dipole of the swarms. We thus obtain :—

$$\Delta\sigma = \frac{2}{3}(\sigma_1 - \sigma_2)\left\{1 - \frac{3\coth C}{b} + \frac{3}{b^2}\right\}$$

where

$$b = \frac{pf}{kT}.$$

We now treat $(\sigma_1 - \sigma_2)$ as approximately independent of temperature; then only the expression in the bracket depends on the temperature, because the dipole moment is connected with the size of the swarm, which depends on the temperature. The absolute temperature, which appears directly in b, only changes from $391°$ to $407°$ and can therefore be left out of consideration. By comparison of the calculated curve $\Delta\sigma = \text{const. } f(b)$ with the experimental curve $\Delta\sigma = g(T)$, we now obtain b and hence also the dipole moment p as a function of the temperature (Fig. 5). This result is in excellent agreement with our earlier estimate $p \gg 10^{-14}$ e.s.; because we find in the neighbourhood of the melting-point 200×10^{-14} and in the neighbourhood of the clearing-point

[19] Th. Svedberg, *Ann. Physik* (4), **44**, 1121, 1914; and W. Kast, *Ann. Physik* (4), **73**, 145, 1924.
[20] L. S. Ornstein, *Ann. Physik* (4), **74**, 445, 1924.

10×10^{-14}. A more accurate estimate of the extreme value is not possible, since the experiments were done with steadily increasing temperature and therefore there was a temperature fall in the melt. From the known size of the molecular dipole moments, 2×10^{-18} e.s., we deduce that the number of molecules contained in a swarm is from 100×10^4 down to 5×10^4; and the same values apply also for benzophenone, where, however, they cover a very small temperature range immediately above the melting-point. Corresponding to the alteration, by a factor of 20, of the number of molecules in a swarm, the linear dimensions of the swarm will alter by a factor of about 3, a value which agrees very well with the estimate given above.

It must now, of course, be shown that liquids not containing dipoles do not give this effect; because even if the dipole moment were not essential for swarm formation, its presence is the condition which makes possible the orientation of swarms in weak fields. This control raises a new and interesting difficulty. A molecule such as benzene has, of course, no dipole moment as a whole; but it has six local moments which in general compensate externally. But now if, in solution, or perhaps even at the melting-point, the molecules come particularly closely together, then the local dipoles can be effective. Thus, Isnardi[21] states that there is an anomalous increase in the dielectric constant of liquid

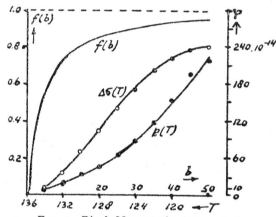

FIG. 5.—Dipole Moment of p-Azoxyanisol.

benzene shortly before the melting-point; and we are indebted to Professor Holst of the Philips-Eindhoven Company for the information that Dr. v. Arkel has made similar observations in benzene solutions. It is very characteristic of this effect that it disappears when the hydrogen atoms of benzene are substituted by F or Cl atoms. These large atoms prevent a sufficiently close approach of the molecules. On account of this difficulty we have not yet been able to make a valid control experiment with substances not containing dipoles. Naphthalene ($C_{10}H_{10}$) shows the conduction anomaly similarly, though to a still smaller degree than benzophenone, but this is understandable from what has just been said about benzene. On the other hand, experiments with carbon tetrabromide have proved impracticable at present because of the readiness with which it vaporises even from the solid state.

5.

We can summarise these considerations about the melting-point by saying that apparently every liquid which has not spherically sym-

[21] H. Isnardi, Z. Physik, 9, 173, 1922.

metrical molecules forms swarms, in the immediate neighbourhood of the melting-point, of the same kind as are stable for liquid crystals over a larger temperature range. We now make some observations upon the clearing-point.

The objection is frequently urged against the swarm theory of liquid crystals that it cannot account for a sharp point of transition from ordinary liquids. This is due to a misunderstanding of the Phase Theory. The Phase Theory always gives sharp transitions when, for two modifications with different internal potential energies, at a given temperature and pressure, the thermodynamic potentials per unit mass are equal. These values of the temperature and pressure define the critical point at which the two phases co-exist. Here, fluctuation phenomena have, as a whole, no effect. In a kinetic treatment in which density deviations or deviations from ideal lattice structure do have an effect, their influence ceases as soon as one turns to the derivation of equilibrium conditions. This is due to the fact that, kinetically expressed, thermodynamics is a science of mean values, and in considering mean values fluctuation phenomena are eliminated. One can hardly speak of the influence of fluctuations on the (macroscopic) pressure, and it is no more legitimate to speak of their influence on the entropy or thermodynamic potential.

To decide, therefore, the existence of a sharp transition point in the transformation liquid crystal ⇌ amorphous liquid, it is only necessary to determine whether it is possible to construct two different liquid modifications with different internal potential energies. Now it is easy to think of two possibilities; the first as described above, with swarms of molecules oriented parallel, and the second where the needle-shaped molecules stick confusedly in all directions. Both the energy of shape and the dipole energy would be different, and they would also have different specific volumes, and therefore different internal energies of the molecular attractive forces (van der Waals' forces). From our knowledge of liquids derived from X-ray photographs, particularly those of Stewart, we know, however, that swarms in our sense of the word exist even in ordinary liquids. In one of his latest papers Stewart [22] has summed up the properties of the so-called cybotactic groups, which include a few hundred or thousand molecules, and possess temporary individual existences, badly defined limits, an optimum size, internal regularity, and anisotropic extension, all properties which we attribute to our swarms. It is only the number of molecules which is smaller. It is very interesting that in this respect cybotactic groups form a continuous series with nematic groups at the clearing-point. Thus we have molecular structures which are broken up into swarms in the amorphous liquid phase as well as in the liquid-crystal phase. Then the clearing-point means for the swarms the transition from one structure to another with higher internal potential energy. In particular, it is possible to say of these structures that in the nematic group the anisotropy of the single molecules is considerably magnified, but in the cybotactic group it is to a large extent cancelled out. If we now think of the molecules in nematic swarms as parallel, and with their dipole moments parallel, then we can check this assumption if we think of the molecules, either singly or in small groups, as lying antiparallel or crossed, still, of course, without any arrangement of their centres of gravity.

[22] G. W. Stewart, *Physic. Rev.* (2), **37**, 9, 1931.

We shall sum up the points considered in a number of statements that may perhaps serve as a basis for discussion :—

(1) Liquid crystals are aggregates of sub-microscopic homogeneous ranges, the so-called swarms, in which the molecules all lie parallel.

(2) The structure is to be visualised as of a polycrystalline nature ; the swarms have an individual existence of considerable duration.

(3) The swarms lie perfectly irregularly ; their axes are distributed at random.

(4) Any further arrangement of the swarms into larger homogeneous ranges is caused by external forces.

(5) At the surface of contact of the melt with solid bodies there occur forces, due to electric double layers and capillarity, which give rise to an arrangement of the swarms up to a few hundredths of a millimetre deep in the layer. The special phenomena in thin layers have therefore no bearing on the nature of liquid crystals.

(6) The rotation moments of electric and magnetic fields are large because they act, not on single molecules, but on the whole swarm. Quantitative experiments on orientation give the number of molecules in a swarm as of the order of 10^5.

(7) The swarms are doubly refracting, diamagnetically and dielectrically anisotropic. The axis of the swarm is the direction of least diamagnetism and least electric polarisability.

(8) The swarms possess, in the direction of their axis, a permanent electric moment of about 10^5 times the value of the molecular moment. Consequently saturation effects occur even in weak electric fields.

(9) From the effect of the orientation of the swarms on the mobility of foreign ions in the melt, the size of the swarms can be calculated over the whole range of existence of liquid crystals of p-azoxyanisol ; they are found to include from 10×10^5 to 0.5×10^5 molecules.

(10) From X-ray photographs in an alternating electric field, as well as from measurements of dielectric loss, it is seen that the swarms show the phenomenon of frictional dispersion for frequencies of the order of 10^5 to 10^6.

(11) The critical frequency varies extremely rapidly with temperature, much more than can be explained by the decrease in size of the swarms. This indicates that a resistance to the turning of the swarms arises from the way they knock against one another and are deformed, the magnitude of which depends on the elastic properties of the swarms.

(12) The swarm formation comes about, in the first place, because of the unsymmetrical shape of the molecules. The dipole moment only comes into consideration secondarily, but must act in such a way that the swarms have an elongated shape.

(13) Several liquid modifications with sharp inversion points can exist if there are different liquid molecular structures with different internal potential energy. It is, however, immaterial whether these structures are homogeneous over the whole volume or are broken up into swarms.

(14) The nematic swarm and the cybotactic swarm represent two aspects of such structures. In the first the molecules lie so that their anisotropies are magnified ; in the second, which must have a high internal potential energy, so that the molecular anisotropies are to a large extent cancelled out.

Leonard Salomon Ornstein was born in Nijmwegen in the Netherlands in 1880. He studied mathematics and physics at the University of Leiden between 1898 and 1908. In that year he graduated with a doctoral degree having written a thesis on statistical mechanics under the supervision of the Nobel prize winner (1902) Hendrick Antoon Lorentz. From 1909–15 he worked as a lecturer in theoretical physics at the University of Gronigen. In Gronigen Ornstein collaborated with the future Nobel prize winner (1953) Frits Zernike, creating the classical  theory of critical opalescence which bears their names.

In 1915 Ornstein moved to the University of Utrecht as professor of theoretical physics, transferring to be Director of the Physical Institute in 1921. His most influential work from this period was his collaboration with George Eugene Uhlenbeck *On the theory of Brownian motion* in 1930. This work has had a profound impact on the kinetic theory of stochastic processes in a whole variety of fields, including in recent years, the theory of stock market fluctuations. In the 1930s Ornstein turned to other work, including nuclear physics and biophysics, the latter work being funded by the Rockefeller Foundation. After the German invasion of the Netherlands in 1940, this funding was cut off and moreover, Ornstein, as a Jew, was expelled from his own laboratory. Following this, Ornstein unsurprisingly became depressed, withdrew from scientific life, and shortly thereafter died in 1941 at the relatively young age of 60.

Wilhelm Kast, another native of Halle in the liquid crystal story, was born in 1898. After finishing his secondary education in 1917, he served in the German army until the end of the First World War. He then studied physics at the University of Halle from 1919 to 1922, and during this period he attended Daniel Vorländer's chemistry lectures. His Ph.D. work on the dielectric aniso-tropy of liquid crystals was supervised by Otto Lehmann's former assistant Gustav Mie and submitted in 1922. Following Mie's appointment as full Professor of Physics in Freiburg in the same year, Kast also began work in Freiburg. After his Habilita-tion degree in physics Kast worked as a lecturer in liquid crystals in Mie's institute until 1937. During

1932–3 he was awarded a Rockefeller fellowship to work as a guest scientist with Ornstein in Utrecht on the swarm theory of liquid crystals. In 1937 Kast returned to Halle as full Professor of Physics.

In addition to his continuing research in liquid crystals, he performed X-ray studies on the structure of synthetic fibres in collaboration with his doctoral student Wilhelm Maier. After the end of the war in 1945, many scientists from the universities of Halle, Leipzig, and Jena were evacuated on the orders of the US Army. Among those evacuated was Kast, who was transported to the Frankfurt am Main area.

After 1947 he was employed by the Bayer Company in Leverkusen to work on synthetic fibres at its factory in Dormagen. Here he was able to continue his X-ray structural investigations. In 1954 he was awarded the titles of Honorary Professor by the University of Cologne in 1954 and by University of Freiburg in 1957. In Freiburg his principal collaborator was Hermann Staudinger. He was appointed as full Professor Emeritus in Freiburg in 1959. He lived to see the vigorous progress in liquid crystal research in the sixties but because of his deteriorating health he could no longer be active in research. He died in 1980.

B7
Transactions of the Faraday Society **29**, 945–57 (1933)

THE EFFECT OF A MAGNETIC FIELD ON THE NEMATIC STATE.

By H. Zocher (*Prague*).

(*Experimental Section in collaboration with W. Eisenschimmel.*)

Received in German 21st *February*, 1933, *translated by H. D. Megaw.*

Two fundamentally different hypotheses have hitherto been used to account for the changes in nematic systems in a magnetic field; these are the Swarm Theory and the Distortion Theory (*Verbiegungstheorie*). It is extremely important to decide which is the correct one, not only for the particular problem dealt with in this paper, but for the whole general question of the structure of these phases.

The experimental evidence for the swarm theory depends essentially on the changes of dielectric constant in magnetic fields, to which Ornstein [1] applies Langevin's theory of ferromagnetism, while Fréedericksz [2] uses Gans's theory of diamagnetism. A quantitative comparison with the distortion theory has not hitherto been possible, as the necessary mathematical treatment was not available. This is given in what follows.

A theory of distortion in a magnetic field was given some years ago by the author.[3] In that a case was considered which involved no changes in the dielectric constant. We shall consider it briefly. The following assumptions are made. (1) The whole substance tends to take up a position such that the axial direction at every point is the same. (2) Any force acting so as to disturb this state where the directions are uniform causes a distortion in which the direction changes continuously until a restoring force of an elastic nature holds the applied force in equilibrium. (3) At surfaces of solid bodies (*e.g.*, glass or metal) the positions at first assumed are almost unchangeable.

Let us now consider a case in which the substance is placed between a fixed plate and one which can be rotated; and assume that at both plates it orientates itself with the principal axes parallel. Then when no forces act, the movable plate sets with its direction of axis parallel to that of the fixed plate. A torque D acting on the movable plate causes a rotation which increases with the turning-moment. For small distortions, the angle ϕ between the azimuths at the upper and lower plates will be proportional to the turning-moment, *i.e.*, it will obey Hooke's Law. The elastic resistance is directly proportional to the area q of the layer, and inversely proportional to its thickness z. This gives the relation

$$D = q\frac{\phi}{z}k_t. \qquad \qquad . \qquad . \qquad . \qquad . \qquad (1)$$

The quantity k_t corresponds to a kind of torsion modulus, and has the dimensions of a force; hence it is expressed in dynes. It can be seen that there is an analogy with the deformation of ordinary solid bodies,

[1] L. S. Ornstein, *Ann. Physik*, **74**, 445, 1924.
[2] V. Fréedericksz and A. Repiewa, *Z. Physik*, **42**, 532, 1927.
[3] H. Zocher, *Physik. Z.*, **28**, 790, 1927.

though the deformation is actually of a different nature. It consists of a pure torsion without sheer, which in a solid cylinder only occurs in a line along the axis.

If a magnetic field is now allowed to act on an undeformed layer perpendicular to the axial direction, a state will be arrived at in which the elastic forces are in equilibrium with the magnetic forces. The energy of a layer of thickness dx in a field H is given by

$$dE = q\frac{H^2}{2}(\kappa_1 \cos^2 \phi + \kappa_2 \sin^2 \phi)dx,$$

where κ_1 and κ_2 are the susceptibilities parallel and perpendicular respectively to the axis of the molecule, and ϕ the angle between the axis and the direction of the field. The increase in the turning-moment in a direction from the movable to the fixed plate is, for each layer,

$$dD = q\frac{H^2}{2}(\kappa_1 - \kappa_2) \sin 2\phi dx.$$

But, from equation (1),

$$D = qk_t\frac{d\phi}{dx}, \text{ and } \frac{dD}{dx} = qk_t\frac{d^2\phi}{dx^2},$$

hence

$$k_t\frac{d^2\phi}{dx^2} = \frac{H^2}{2}(\kappa_1 - \kappa_2) \sin 2\phi,$$

$$\therefore \frac{d\phi}{dx} = H\sqrt{\frac{\kappa_1 - \kappa_2}{k_t}}\sqrt{\sin^2 \phi - \sin^2 \phi_0}, \qquad . \qquad . \quad (2)$$

where ϕ_0 is the angle between the direction of the field and the axis at the movable surface. For large thicknesses, $\phi_0 = 0$, and therefore

$$\frac{d\phi}{dx} = H\sqrt{\frac{\kappa_1 - \kappa_2}{k_t}} \sin \phi.$$

By integrating (2), a relation can be obtained connecting x, the distance from the movable plate, with the axial azimuth :

$$xH\sqrt{\frac{\kappa_1 - \kappa_2}{k_t}} = \int_{\phi_0}^{\phi} \frac{d\phi}{\sqrt{\sin^2 \phi - \sin^2 \phi_0}} \qquad . \qquad . \quad (3)$$

This integral can easily be brought to Legendre's normal form of an elliptic integral of the first kind :

$$xH\sqrt{\frac{\kappa_1 - \kappa_2}{k_t}} = \int_{\psi}^{\frac{\pi}{2}} \frac{d\psi}{\sqrt{1 - \cos^2 \phi_0 \sin^2 \psi}},$$

where

$$\sin \psi = \frac{\cos \phi}{\cos \phi_0}.$$

We can now imagine a second layer with a second fixed plate so placed above the movable plate that it is a mirror image of the first ; it must then necessarily be in equilibrium. Finally, if we make the movable plate infinitely thin, we obtain the case of a layer between two fixed plates. If z is now the total thickness of the layer,

$$\frac{z}{2}H\sqrt{\frac{\kappa_1 - \kappa_2}{k_t}} = \int_0^{\frac{\pi}{2}} \frac{d\psi}{\sqrt{1 - \cos^2 \phi_0 \sin^2 \psi}} \qquad . \qquad . \quad (4)$$

for the case, that $\phi = \pi/2$ at the plates.

The experiment considered here, which we shall call Case III, had not previously been performed, but it will be dealt with in the experimental part of the work. The other two possibilities for the position of the layer in a magnetic field, which will next be discussed, were first investigated in detail by Mauguin,[4] who described qualitatively the optical effect. More recently, van Wyk[5] and Fréedericksz[6] have made optical measurements, Jezewski[7] and Kast[8] dielectric measurements. The first possibility is that the position of the substance at the surface is the same as in Case III, but that the field is applied perpendicular to the plate (Case I). The second is that the substance has its axes perpendicular to the plate (so-called uniaxial orientation), and that the field is applied parallel to the plate at any arbitrary azimuth (Case II). In both these cases, distortion must occur, such that the axis of rotation of the distortion no longer coincides with the x-axis but is perpendicular to it. The angle made by the axis of the molecules in each layer with the axis of x, at a distance $x = 0$ (i.e., in the middle of the layer), approaches 0° in Case I, 90° in Case II, as the field strength increases. The axial lines, i.e., the lines which give the direction of the axis at any point, will be as shown in Figs. 1a and b. The configuration

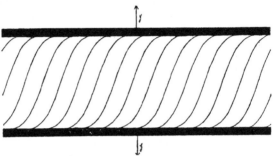

FIG. 1a.—Case I. Axial lines in a magnetic field. $\phi_0 = 20°$.

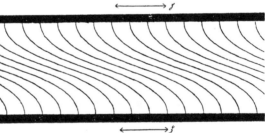

FIG. 1b.—Case II. Axial lines in a magnetic field. $\phi_0 = 20°$.

will always be given by plane curves if the setting assumed to occur at the surface is homogeneous.

Both these cases are complicated by the fact that the angle made by the axial direction with the x-direction varies, so that even for constant $\frac{d\phi}{dx}$ it is not certain whether the torque producing the distortions is the same. The case when the principal axis coincides with the x-direction was previously called by the author " lengthwise distortion " (Längsbiegung), that in which the two directions are at right angles, " fan formation " (Auffächerung) or " cross distortion " (Querbiegung). The

[4] Ch. Mauguin, Comptes Rendus, **152**, 1680, 1911.
[5] A. van Wyk, Ann. Physik, **3**, 879, 1929.
[6] V. Fréedericksz and V. Zolina, Z. Krist., **79**, 255, 1931.
[7] M. Jezewski, Z. Physik, **40**, 153, 1926.
[8] W. Kast, Z. Krist., **79**, 146, 1931.

moduli k_1 and k_2 corresponding to these may be very different, the behaviour of smectic substances results from this. In these, k_1 is so much greater that the lengthwise distortion never occurs, and the axial lines are always straight lines. Hence follows the principle of the deformation of the smectic state, i.e., the conical structures which have been established by the beautiful work of G. Friedel. The law is obeyed perfectly by this whole range of complicated and remarkably interesting phenomena.

In nematic substances, on the other hand, the two moduli may be very nearly equal. Experiment shows that this may be true to a first approximation. Van Wyk[5] has worked out a theory for Case I that is almost identical with equation (2) given above. In his optical experiments, the great thickness of the layer used enabled him to put $\phi_0 = 0$. The observations are in good agreement with the theory. Further, Fréedericksz[6] has shown that his optical observations agree very well, both for Case I and Case II, with the theory which he adopted from van Wyk. In this, he does not assume that $\phi_0 = 0$, but applies the calculation to thin layers by the introduction of the elliptic function.

The difference between van Wyk's treatment and that given above will not be considered here. A practical application of Fréedericksz's measurements will, however, be made. It appears that for any given field strength there is a certain thickness of layer above which distortion first occurs. For less than this thickness, the distortion is zero, that is, any distortion would require a bigger expenditure of mechanical energy than would be available from the magnetic energy released by orientation in the field. The phenomenon is closely analogous to the behaviour of a loaded column, which only bends sideways at a fixed "breaking stress."

In equation (4) this implies that $\phi_0 = \dfrac{\pi}{2}$, and hence

$$\frac{z}{2}H\sqrt{\frac{\kappa_1 - \kappa_2}{k}} = \int_0^{\frac{\pi}{2}} d\psi = \frac{\pi}{2}.$$

Fréedericksz showed that the product of z_0 and H_0 as given by this relation is constant; for p-azoxyanisole below 120° it has the value 8·4. From this,

$$z_0 H_0 = \frac{\pi}{\sqrt{\dfrac{\kappa_1 - \kappa_2}{k}}} = 8·4,$$

and therefore

$$\sqrt{\frac{\kappa_1 - \kappa_2}{k}} = ·37.$$

From the measurements of Foex and Royer, $\kappa_1 - x_2$ is about $0·15 \times 10^{-6}$. Thus k must be $1·0 \times 10^{-6}$ dynes.

For these very small magnitudes of the elasticity, it is obvious that, as Mauguin[4] has observed, it is not possible to obtain thicker layers in an unbent state without magnetic field. It is only possible to do so up to a thickness of 0·2 mm. Then, according to the above, a turning-moment of 10^{-4} ergs per cm.² suffices to give the middle of the layer an angular distortion of unity. A torque of this magnitude might easily be introduced by differences in density caused by small temperature changes. Since the friction depends on the direction. heat convection can give distortions. The thicker the layer, the greater the difficulty with which a homogeneous orientation is achieved.

It is of interest to consider the highest torque per unit area of surface that can be achieved with the strongest available fields. This, of course, occurs at the outside of a thick layer, where it has the value

$$k\frac{d\phi}{dx} = H\sqrt{k(\kappa_1 - \kappa_2)}.$$

For 25,000 Gauss it gives, in the example just mentioned, 0·0094 ergs per cm.² The dimensions are those of a surface tension, the magnitude with which it competes. The orientation of the phase at the fixed plate is determined by the difference of surface tensions in the different positions. When the axes are parallel to the wall, the surface tension (σ_1) may be quite different from that when they are perpendicular (σ_2), since the two surfaces actually possess quite different structures. The difference may be of the order of magnitude of 10 ergs per cm.², which is 1000 times greater than the greatest attainable torque. This explains the fact observed by van Wyk and by Fréedericksz, that the angle at the surface against glass is independent of the field strength. If the surface tension is given by the relation

$$\sigma = \sigma_1 \sin^2 \phi + \sigma_2 \cos^2 \phi,$$

where ϕ is the angle between the axis and the normal to the plate, then for $\sigma_1 - \sigma_2 = 10$ ergs per cm.², the change of angle would only be about two minutes.

The effect of the field on the capacity of a condenser filled with a nematic substance will next be considered. It will be assumed that we are dealing with Case I, and that the normal to the condenser plates makes an angle α with the lines of magnetic force. The axial lines will then only lie in a plane if the direction of the axes at the surface lies in the same plane as the lines of electric and magnetic force. This case will be dealt with first in what follows. Let the angle between the direction of the axes and the magnetic field be ϕ, that between the axes and the lines of electric force $\chi = \phi + \alpha$. The equation is the same as before (equation 4), except that ϕ at the solid boundary has the value $\frac{\pi}{2} - \alpha$ instead of $\frac{\pi}{2}$. Then

$$\frac{z}{2}H\sqrt{\frac{\kappa_1 - \kappa_2}{k}} = \int_{\phi_0}^{\frac{\pi}{2} - \alpha} \frac{d\phi}{\sqrt{\sin^2 \phi - \sin^2 \phi_0}}$$

$$= \int_{\psi_0}^{\frac{\pi}{2}} \frac{d\psi}{\sqrt{1 - \cos^2 \phi_0 \sin^2 \psi}}$$

$$= F(\cos \phi_0, \pi/2) - F(\cos \phi_0, \psi),$$

where

$$\sin \psi_0 = \frac{\sin \alpha}{\cos \phi_0}.$$

The numerical values of the two elliptic integrals of the first class, F, may be found from the tables.[9] The dielectric constant ϵ of the nematic layer in the field varies continuously from place to place in the condenser along the x-direction :

$$\epsilon = \epsilon_1 \cos^2 \chi + \epsilon_2 \sin^2 \chi,$$

[9] E.g. E. Jahnke and F. Emde, *Funktionstafeln*, Teubner, Berlin, 1928; L. Kiepert, *Integralrechnung*, Hanover, 1910.

where ϵ_1 and ϵ_2 are the dielectric constants parallel and perpendicular to the axis respectively. If the change of capacity with field strength is attributed to an alteration of "the" dielectric constant, the quantity so obtained is an average value ϵ'. This can be calculated from the "air distance" of the condenser plates, i.e., the distance between the plates for which the capacity in air (or more accurately in vacuo) would be the same: The "air thickness" of the nematic layer is

$$\frac{z}{\epsilon'} = 2\int_0^{\frac{z}{2}} \frac{dx}{\epsilon_1 \cos^2 \chi + \epsilon_2 \sin^2 \chi}.$$

Let $\epsilon_2 - \epsilon_1 = \Delta$.
Then for zero field strength, the "air distance" is z/ϵ_2, and for infinite

field strength, $\dfrac{z}{\epsilon_2 - \Delta \cos^2 \alpha}$. The difference

$$\frac{z}{\epsilon_2 - \Delta \cos^2 \alpha} - \frac{z}{\epsilon'}$$

can be put equal to

$$(\epsilon_2 - \epsilon' - \Delta \cos^2 \alpha) z / \epsilon^2$$

if Δ is small compared with ϵ_1 and ϵ_2. From the measurements of Jezewski[7] in the example taken above, $\epsilon_2 = 5\cdot31$, $\epsilon_1 = 5\cdot15$, so that the above gives a good approximation. Hence

$$\epsilon' - \epsilon_2 + \Delta \cos^2 \alpha = \frac{2\epsilon^2}{z}\int_0^{\frac{z}{2}} \left(\frac{1}{\epsilon_2 - \Delta \cos^2 \alpha} - \frac{1}{\epsilon_2 - \Delta \cos^2 \chi} \right) dx$$

$$= \frac{2\Delta}{z}\int_0^{\frac{z}{2}} (\cos^2 \alpha - \cos^2 \chi) dx.$$

Putting (eq. 2)

$$dx = \frac{d\phi}{H\sqrt{\frac{\kappa_1 - \kappa_2}{k}}(\sin^2 \phi - \sin^2 \phi_0)},$$

and $\chi = \phi + \alpha$,
we obtain finally

$$\epsilon_2 - \epsilon' = \Delta \Bigg[\cos^2 \alpha$$

$$- \frac{\{E(\cos \phi_0, \pi/2) - E(\cos \phi_0, \psi_0)\} \cos 2\alpha + \cos \phi_0 \cos \psi_0 \sin 2\alpha}{F(\cos \phi_0, \pi/2) - F(\cos \phi_0, \psi_0)} \Bigg] \quad (5)$$

Here E is the elliptic integral of the second kind,

$$E(\cos \phi_0, \pi/2) - E(\cos \phi_0, \psi_0) = \int_{\psi_0}^{\frac{\pi}{2}} \sqrt{1 - \cos^2 \phi_0 \sin^2 \psi}\, d\psi.$$

It may be found in the tables.

This equation will be briefly discussed. Fig. 2 represents graphically the relation between $\dfrac{\epsilon_2 - \epsilon'}{\Delta}$ and $\dfrac{2}{zH\sqrt{\frac{\kappa_1 - \kappa_2}{k}}}$ for several values of α.

When $\alpha = 0$, the quotient in the bracket becomes $\dfrac{E}{K}$; this expression was derived by Fréedericksz for the double refraction in Case I, and $1 - \dfrac{E}{K}$

in Case II. This can easily be understood, because the double refraction, like the difference of the dielectric constants, is proportional to $\sin^2 \chi$. The axial lines in Figs. $1a$ and $1b$ have been drawn for Case I and Case II when $\alpha = 0$ and $\phi = 20°$. It is interesting in this formula that for $\alpha \neq 0$ the discontinuous character of the function disappears; ϕ_0 reaches the value $\dfrac{\pi}{2} - \alpha$ at $z = 0$. Even the thinnest layer has a distortion, which, however, is very small for values of z below z_0. The effect is therefore not essentially different for values of α of a few degrees; the function here shows a strong curvature. (See the curve for $\alpha = 5°$ in Fig. 2.) For large values of H the equation becomes

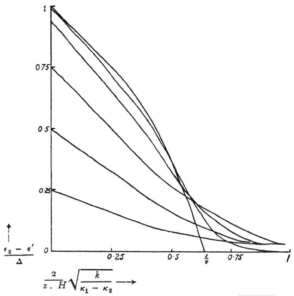

FIG. 2.—Graphs of $\dfrac{\epsilon_2 - \epsilon'}{\Delta}$ plotted against $\dfrac{2}{zH}\sqrt{\dfrac{k}{\kappa_1 - \kappa_2}}$ for $\alpha = 0°, 5°,$ arc $\sin \frac{1}{4}, 30°, 45°$ and $60°$.

$$\epsilon_2 - \epsilon' = \Delta\left(\cos^2\alpha - 2\,\frac{\cos 2\alpha + \sin \alpha}{zH\sqrt{\dfrac{\kappa_1 - \kappa_2}{k}}}\right).$$

If the values of $\epsilon_2 - \epsilon'$ are plotted against the reciprocal of the field strength, then at high field strengths straight lines are obtained which cut off an intercept on the y-axis proportional to $\cos^2 \alpha$. Their slope is proportional to $(\cos 2\alpha + \sin \alpha)$, which has a maximum at $\sin \alpha = 0.25$ ($\alpha = 14° 29'$ approx.). For $\alpha = 30°$ the line is parallel to that for $\alpha = 0$, and becomes continuously less steep as α increases. The curve for $\alpha = 0$ is everywhere concave to the x-axis, but the other curves for small field strengths show a convex curvature towards it which is predominant above $\alpha = 30°$. It is also of interest that the effect of the field strength in the neighbourhood of

$$H_0 = \frac{\pi}{z\sqrt{\dfrac{\kappa_1 - \kappa_2}{k}}}.$$

at first increases with increasing α, and later decreases.

The measurements of Kast [10] and Jezewski (*loc. cit.*) may be adduced in support of the theory. In Fig. 3, curve (1) shows Kast's measurements I on *p*-azoxyanisole at $119°$ C. plotted against $\frac{I}{H}$, and curves (2) and (3) similar measurements for *p*-azoxyphenetole at $138°$ C. and $153°$ C. respectively. For large field strengths they are very nearly straight lines. It is not possible to calculate the curve *a priori*, as the thickness of the layer in the condenser is not given. But we can assume the value of $\sqrt{\frac{\kappa_1 - \kappa_2}{k}}$ from the measurements of Fréedericksz,[6] and calculate the thickness from this:

$$z = \frac{2}{H\sqrt{\frac{\kappa_1 - \kappa_2}{k}}\left(1 - \frac{\epsilon_2 - \epsilon'}{\Delta}\right)}.$$

From the value for *p*-azoxyanisole at $119°$ C., for which

$$\sqrt{\frac{\kappa_1 - \kappa_2}{k}} = \frac{\pi}{8\cdot4},$$

z is found to be approximately $0\cdot05$ mm.; from *p*-azoxyphenetole at $153°$ C.,

$$\sqrt{\frac{\kappa_1 - \kappa_2}{k}} = \frac{\pi}{8\cdot06},$$

and hence z is $0\cdot03$ mm. Curves (4) and (5) give the measurements of Jezewski for *p*-azoxyphenetole at $143°$ C. and *p*-azoxyanisole at $122°$ C. The curves appear to be irregular. The thickness z being given as $0\cdot07$ cm., the value of $\sqrt{\frac{\kappa_1 - \kappa_2}{k}}$ calculated from the approximate course of curve (4) is about half as big as that found from Fréedericksz's values of $z_0 H_0$, while from curve (5) it is about equal to it.

For small field strengths, however, Kast's measurements deviate considerably from the theoretical curve. His curves are convex instead of concave to the axis. Apart from the fact that the theory involves several doubtful assumptions, this discrepancy is not surprising. In the first place, the measurements are made with a condenser surrounded with nickel wire, so that the field strength, and field direction, must differ appreciably from that in the absence of nickel. Secondly, the theory only refers to that part of the condenser where the axis of the nematic substance is parallel to the condenser plate. But near the edge this is certainly not true. Further, dust particles (particles of the mica used for insulation) may cause deformation in the absence of a field. It has been shown that the effect is greatest when the angle between the axis and the magnetic field has an intermediate value, and the field strength is small.

The measurements II of Kast [11] for varying angle α will next be compared with the theory. Unfortunately Kast gives no complete table of his numerical results, so that the points in question have had to be taken from Figs. 3 and 4 of Kast's paper. Table I gives the change of acoustic

[10] L. S. Ornstein, *Ann. Physik*, **74**, 445, 1924.
[11] W. Kast, *Ann. Physik*, **83**, 391, 1927.

interference frequency thus obtained; in this, the values taken from Kast's Fig. 3 for $\alpha = 0$ have been decreased by 1 per cent. to correct for the temperature difference of 0·5° C. All the measurements lie in a range where $\epsilon_2 - \epsilon'$ decreases linearly with the reciprocal of the field strength. In Fig. 4 are shown the corresponding straight lines calculated for the actual distance between the plates of 0·75 mm. It is clear that the measured results agree well with those calculated. This is the more noteworthy, that only one parameter Δ was to be derived from the measure-

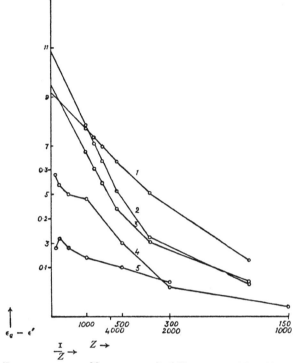

Fig. 3.—1, 2, 3. Measurements I of Kast. $\epsilon_2 - \epsilon'$ in arbitrary units (right-hand numbers along y-axis), H in gauss (upper numbers along x-axis).
4, 5. Measurements of Jezewski. $\epsilon_2 - \epsilon'$ in absolute units (left-hand numbers along y-axis), H in gauss (lower numbers along x-axis).

ments; and incidentally, the straight lines here obtained could have been arrived at beforehand, using only the optical measurements of Fréedericksz. The swarm theory introduces a second parameter,

$$x = \frac{\kappa_1 - \kappa_2}{kT}v,$$

TABLE I.—CHANGE OF ACOUSTIC INTERFERENCE FREQUENCIES ACCORDING TO KAST, II.

α \ H	$\dfrac{2900\ f.\ 0°}{2250}$	725.	375.	225.
0°	537	494	—	377
25°	432	391	357	262
35°	348	318	289	184
45°	276	251	—	—
60°	165	124	—	68
70°	96	—	72	51

297

besides that mentioned above ; it is to be noted that this second parameter is about twenty times bigger when calculated from measurements
II than when calculated from measurements I. It is easy to see that
this depends on the layer thickness, which is twenty times as great in II.
A systematic deviation is only shown by the values for $\alpha = 70°$, which
are greater than were to be expected. This is evidently connected with
the fact that a change of acoustic frequency is observed even for $\alpha = 90°$,
which is not easy to account for. It would seem reasonable to subtract
this value for $\alpha = 90°$ from all the other values, when good agreement
is obtained even for $\alpha = 70°$. According to Jezewski, the values for the
greatest field strengths are represented very accurately by the formula

$$\epsilon = \epsilon_1 \cos^2 \alpha + \epsilon_2 \sin^2 \alpha.$$

The way in which the effect depends on temperature is of particular
interest. According to Ornstein, the change in dielectric constant can
be written

$$f(H) \cdot g(T).$$

This can only be correct to a first approximation ; only, in fact, if the
variation of Δ with the tempera-
ture is appreciable, while that of
$\frac{\kappa_1 - \kappa_2}{k}$ is negligible. According

to the experimental results of
Fréedericksz, $z_0 H_0$ decreases with
rise of temperature, and therefore
$\frac{\kappa_1 - \kappa_2}{k}$ increases with rise of tem-

perature. This means that for
increasing temperature k decreases
more rapidly than $\kappa_1 - \kappa_2$. Hence
$\frac{\epsilon_2 - \epsilon'}{\Delta}$ is reached with smaller

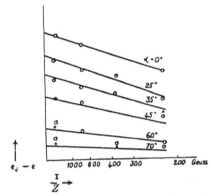

Fig. 4.—Measurements II of Kast.
Graph of $\epsilon_2 - \epsilon'$ plotted against $1/\zeta$.
The straight lines represent the
calculated values.

field strengths as the temperature
rises. $\kappa_1 - \kappa_2$ should to a first
approximation decrease in the
same way with rising temperature
as do Δ and the double refraction $n_1 - n_2$. Even at the temperature of
transition into the amorphous phase, the various kinds of anisotropy
possess values different from zero. From the variation of k with
temperature follows, according to the Second Heat Theorem, that a
change of deformation with change of temperature is connected (cooling associated with deformation, heating with release from deforming
forces). The very small size of the effect (order of magnitude $10^{-5°}$)
makes proof scarcely possible. Moll and Ornstein [12] observed such an
effect, but it was of the opposite sign. For Case III no change of temperature was observed, though the order of magnitude of that expected
was the same. The observation that was made must be due to some
other effect.

[12] L. S. Ornstein, Z. Krist., 79, 90, 1931.

Experimental Section.

Our experiments on the magnetic behaviour of nematic layers relate to two different problems. Till now no exception had been found to the rule that optic axes tend to set parallel to the lines of magnetic force ; in other words, that the susceptibility along the axis is greater (the diamagnetism smaller) than at right angles to it :

$$\kappa_1 - \kappa_2 > 0.$$

We now tried to see if substances of quite different constitution from those hitherto examined might not furnish examples of the opposite behaviour. With this in view, we investigated a number of aqueous nematic systems. The experiments showed that nematic concentrated solutions of bromo-phenanthrene sulphonic acid and chlorophenanthrene sulphonic acid, which showed large effects in a magnetic field, both set with their principal axes perpendicular to the lines of force. Since both these systems have negative double refraction, the vibration corresponding to the greater refractive index here lies parallel to the lines of force, just as in the substances previously examined. But the law, even as thus stated, is not generally valid. Thus a concentrated solution of salvarsan is definitely affected by a magnetic field, though to a smaller degree, and sets with the direction of its axis perpendicular to the lines of force. Its double refraction is positive, and hence the vibration direction of greater refractive index is perpendicular to the lines of force. On the other hand, the phase which occurs in a moderately concentrated aqueous solution of potassium laurate seems to show a tendency to set with its axis parallel to the lines of force. Its double refraction is negative, like that of the halogen phenanthrene sulphonic acids, so that in this, as in salvarsan, the vibration direction of greater refractive index is perpendicular to the lines of force.

These experiments show that the rule that the axes tend to set parallel to the lines of force does not always hold.

The rest of our work involves the actual performance of the experiments described as Case III in the theoretical part of the paper. They are purely qualitative, and only give a preliminary account of the orientation. It was necessary for these experiments to obtain preparations with the greatest possible uniformity of orientation of their axes parallel to the surface (which was of glass). For this purpose, we made use of the effect on the orientation of the nematic phase produced by rubbing the glass surface.[13] If a glass plate is rubbed in a given direction, then the nematic phase in contact with it sets with its symmetry axis parallel to the direction of rubbing. The substances to be investigated, p-azoxy-anisole and anisaldazine, were placed between a plane glass plate and a plano-convex lens, the adjoining surfaces of which were rubbed in the same direction by means of a rotating leather pad. (Layers between a lens and a plate have been used by Lehmann and also by Fréedericksz.) The orientation is almost perfect if the force used in polishing is sufficient. If the direction of rubbing on the plate and the lens are accurately the same, the substance appears dark between crossed nicols when the direction of one nicol is parallel to the direction of rubbing. It is in thicker layers that the well-known self-contained distortions about the lines of discontinuity mostly occur. The preparation was placed in a perforated copper block, electrically heated

13 H. Zocher and K. Coper, Z. physik. Chem., 132, 295, 1928.

above and below the magnetic field, in such a position that the direction of rubbing made only a very small angle with the normal to the lines of force. Observations were made with a polarising microscope. If the polarising prism was set in the extinction position for zero field strength, then when the field was put on the thicker layers became brighter. Near the point of contact of lens and plate, out to a fixed radius from the centre, no effect occurred. (See photograph,* Fig. 5.) This phenomenon is completely analogous to that observed by Fréedericksz for Cases I and II. The values of $z_0 H_0$ are of the same order of magnitude as his, and hence k_t is also of the same order of magnitude as k_1 and k_2. In the part where the effect occurs, the optical behaviour of the layer is of a rather complicated nature. If white light is used, the thicker layers show high order white. The actual tint, however, is not the same as for zero field, but is brownish. This was to be expected. Mauguin [14] showed that for a system in which one surface is rotated relative to the other, a vibration which is initially parallel to the first layer only changes its direction, and emerges parallel to the second layer, so long as the rotation effect is small compared with the double refraction. Hence in the above case, since the first and second layers are parallel, no change would be observed if the ratio of the rotation effect to the double refraction were small. Differences are only observed when this condition is not fulfilled. But for red light the double refraction is very much smaller than for light of shorter wave-length. Therefore the effect must be much more marked in the long-wave part of the spectrum than in the short-wave region, and hence the emergent light must have a brownish tint.

Extinction cannot be obtained by turning the nicols, as in the case of an unrotated layer. If the nicols are respectively parallel and perpendicular to the direction of rubbing, and if a gypsum plate of first order red is introduced between the specimen and the analyser, at 45° to the vibration direction of the nicols, a change appears in the red interference colour which is superimposed on the rather dull high order white. The axial lines of the layer between the two plates must, of course, lie in the same pair of quadrants between the field direction and its normal as does the direction of rubbing. With the gypsum of first order red, a blue addition colour is then obtained if the vibration direction of greater refractive index of the gypsum lies in the other pair of quadrants; if it lies in the same pair of quadrants as the axial lines, a yellow subtraction colour results. The same effect could be brought about without the application of the field, by turning the preparation through a small angle in the pair of quadrants containing the axial lines, and at the same time introducing between the preparation and the gypsum a thin flake of a substance of weak double refraction, with its vibration direction of greater refractive index at about 45° in the second pair of quadrants. The layer rotated in a non-uniform manner by the field therefore produces on the plane-polarised light from the polariser an effect which is to a first approximation the same as that of a layer of high order white on which is superimposed a weakly doubly-refracting layer of constant thickness and different orientation.

The observations in monochromatic light agree with these results. In sodium light, in the absence of a magnetic field, the field of the microscope is dark; but when a magnetic field is applied, it shows a large

---

[14] Ch. Mauguin, *Bull. Soc. Min.*, **34,** 1911.   * Facing page 927.

number of interference rings in the parts where the thickness is great. These cannot be made to disappear by turning the polariser. But if a thin flake of mica (for example, a quarter-wave plate) is introduced between analyser and specimen, then by turning the flake and the analyser a position can always be reached in which the fringes vanish in a region covering a large number of rings (from 5 to 10). In this region there is uniform illumination which varies as the polariser is rotated. The vibration direction of the mica with the greater refractive index is here only turned through a small angle from the direction of rubbing in the quadrants containing the axial lines. The thicker the layer and the stronger the magnetic field, the greater must this angle be. For a given field-strength, the interference fringes in a fixed zone can be made to disappear without turning the analyser, by turning the mica only. Then zones nearer the centre, where the layers are thinner, require the analyser to be turned in the same direction as the mica, those further from the centre in the opposite direction. This shows that for increasing thickness of the layer, as for increasing magnetic field, the double refraction of the imaginary flake of constant thickness must be increased. No appreciable displacement of the interference rings depending on the field strength was observed, which means that the double refraction of the layer of variable thickness (high order white) did not alter appreciably. The azimuth of this can be determined by turning the polariser into the position where the uniform illumination is of minimum intensity. The greater the field strength and the thickness of the layer, the more does it differ from the direction of rubbing. On removal of the mica, the fringes in this position reappear, though they are of small intensity. This implies that the model assumed does not correspond accurately to the properties of the layer.

More accurate observation shows that the position of the axis at the surface is not absolutely independent of the field strength. The difference between the surface tensions when the axes are parallel and perpendicular to the direction of rubbing depends on the anisotropy of the glass surface, and will not necessarily be of the same order of magnitude as the difference previously referred to between σ_1 and σ_2.

Finally, another experimental possibility should be mentioned which does not correspond to any of the three Cases I, II, or III, and which we have been able to realise in practice. In this, the assumption is no longer made that the orientation at the two surfaces is the same. For example, if one of the glass surfaces is treated with acid, p-azoxyanisole tends to set with its axis perpendicular to the surface. Hence in this case the axes at one surface are parallel to the direction of rubbing, and at the other they are perpendicular to the surface, and so a distortion occurs when no field is present. This will be constant throughout the layer,

$$\frac{d\phi}{dx} = \frac{\pi/2}{z}.$$

Calculation shows that the phase-difference due to the double refraction is half as great as when the axes at both surfaces are parallel to the direction of rubbing. The experiments confirm this result, at least to a first approximation. A magnetic field acting parallel to the surface then brings about a large decrease in the diameter of the interference rings, that is, an increase in the path difference, which becomes almost double its previous value.

301

Hans Ernst Werner Zocher was born in 1893 in Bad Liebenstein in Thuringia in the central part of Germany. Between 1912 and 1914 he attended Leipzig and Jena universities, studying mathematics and physical sciences. He saw two weeks active service in the First World War, but was seriously wounded in the face and spent two years in hospital. After being released from hospital he returned to his studies, this time at Berlin University, where he graduated in 1920. In 1921 he became assistant to H. Freundlich at the Kaiser-Wilhelm Institute für physikalische Chemie und Electrochemie. He began work on mesophases while in Berlin, and also worked during this period on chemiluminescence and photospectroscopy. In 1931 he was appointed as extraordinary professor at the German Technische Hochschule in Prague. Here he became director of the Institute for Physical Chemistry and Electrochemistry, and in 1936 also director of the photographic laboratory. He was appointed to a full professorship in 1937.

During the Second World War, his situation in Prague was difficult, because his wife was Jewish, and he had some record of opposition to Nazism. He lost his job at the university and became technical advisor to the Society for Chemical and Metallurgical Production. In 1946 he emigrated to Brazil. He worked in the Laboratorio de Producaõ Mineral in Rio de Janeiro and subsequently for the Brazilian National Research Council.

In his early work in Berlin Zocher discovered the orienting effects of rubbing on dyestuffs. His main effort was in optics. He applied polarisation optics to many photochemical problems, to molecular structure in a variety of biophysical problems, as well as most notably in liquid crystals. He is best known for theoretical work in liquid crystals discussed in this volume. He became a member of the Brazilian Academy of Sciences in 1949, was awarded an honorary doctorate by the University of Brazil in 1963, and in 1965 was awarded the Einstein Prize by the Brazilian Academy of Science. He was also honoured for his lifetime contributions to liquid crystals; he received a citation at the third international liquid crystal conference in Berlin in 1970, but unfortunately died in October 1969 before he was able to receive the citation in person. This brief biography is based on the citation by J.F. Dreyer published in the proceedings of that conference.

302

B8

Transactions of the Faraday Society **29**, 1060–85 (1933)

LIQUID CRYSTALS AND ANISOTROPIC MELTS.

A GENERAL DISCUSSION

Mr. Bernal (*Cambridge*) said : The theory of the forces giving rise to liquid crystals in section I. of Professor Oseen's Paper would seem to involve serious physical difficulties. If the force between molecules in a cholesteric liquid crystal depends on $L_1 \times L_2 . r_{12}$ so that it is not invariant for reflection in a plane then, as Professor Oseen points out, the laws of physics would have to be altered, or some quite unknown type of force invoked. It seems to me that this argument rests on a concealed error in the premises. By assuming in the first place that the molecules possess rotational symmetry he tacitly assumes that they possess internal symmetry of the second sort (*i.e.*, planes or centres of symmetry). Now this is not the case for cholesteric molecules which are optically active and exist in enantiomorphic forms. Bearing this in mind the fact that for instance a system of right-handed molecules mutually arranged in a right-handed screw should have different energy from one where right-handed molecules are arranged in a left-handed screw is not only not surprising but physically necessary so that the necessity for invoking new principles disappears

In section 3 a very beautiful and elaborate theory is given for the behaviour of light in a spirally twisted medium. The question arises as to whether this theory which formally accounts for the best known properties of cholesteric substances is based upon a physically acceptable model. In particular the planes of Grandjean visible as fringes in a wedge-shaped preparation of an iridescent cholesteric substance are explained as an optical illusion. Before accepting this view it would be worth while to see whether the same phenomena could not be explained on the basis of some real physical periodicity of the order of 1000 to 10,000 A. Friedel from exhaustive microscopic observations points to three main arguments in favour of the physical reality of Grandjean's planes.

(1) The complex pattern of *franges mobiles, virgules*, etc., that appear between the Grandjean planes

(2) The nematic nature of the first Grandjean layer.

(3) The changing of the edges of the planes into nematic lines on " detorsion " by adding another cholesteric substance of opposite sign.

The whole phenomenon strongly suggests schiller structure in minerals

303

and that due to regular twinning in potassium chlorate. The regularity may be due to "swarms" or "secondary structure" as in Zwicky's theory of solids. It would be very useful to find some critical experiment that could distinguish between these two apparently very different explanations of the same phenomena. The difficulty of finding such an experiment may however be an indication that here again as in the case of quantum physics we are trying to set up a fictitious difference between unobservables.

Dr. Lawrence (*Cambridge*) (*communicated*): Professor Oseen's remarks on the flow of ælotropic liquids cover what are at present two distinct problems; the flow of liquids containing anisodimensional molecules of normal size; and, secondly, the flow of those containing giant polymer-molecules or colloid micelles. Concerning the first, flow orientation requires very large rates of shear and departures from Newtonian flow have not been found. In the second class such departures are general, although confused in some cases by other phenomena such as elasticity. The factor determining the rate of shear required to orient any given particle depends, of course, upon its random kinetic motion. The distinction between the two problems is real insofar as anomalous flow requires, even for the most anisodimensional particles, molecular weights of at least 20,000; so that there is a big gap between them. In this discussion, the distinction is of peculiar interest since it makes Ostwald's demonstration of the anomalous flow of nematic melts evidence for Ornstein's Swarm theory.

Professor Oseen, in reply, wrote: The theory of the movements of anisotropic liquids is still in its infancy. The movements that lead to the forms of equilibrium of liquid crystals, cannot obey the law of Newton. Moreover, as to the theory of anisotropic liquids of small molecular weight, a theory of the motion is needed.

Professor Zocher (*communicated*): The energy and any other scalar quantity of a system cannot be changed, if the *whole* system is reflected at a plane. A change is only possible by a partial reflection, *i.e.*, by reflection of the arrangement of the molecules without reflecting the molecules itself, if the molecules are not identical with their mirror images. Neither an ordinary vector nor an axial vector can represent the symmetry of a molecule of the cholesteric phase, only a combination of both representing the axis of symmetry of the molecule. Bearing this in mind, we find no difficulty in the assumption that the forces between the molecules are of an electrical nature. The magnetic moments of paramagnetic molecules cannot bring about orientation, because these magnetic moments are not attached to the molecules.

It is true that Oseen's theory does not yet explain all such observations as *franges mobiles, virgules*, etc., but it is not known what these observations indicate, and therefore it is scarcely possible to evaluate them as strong arguments against this theory. (At this point it may be emphasised that the supporting surfaces, *e.g.*, mica, may be of great influence on such observations.) On the other hand, this theory explains several very important and striking phenomena, which are not explained by Grandjean and Friedel. I will mention only the following three facts for perpendicularly incident light:—

(1) Only circularly polarised light of one sense of rotation is reflected.

(2) The sign of rotation is inverted by reflection, for instance right-handed light is reflected as right-handed.

(3) Only one order of interference colours is observed.

In discussing Professor Vorländer's papers (page 899).

Professor Ornstein (*Utrecht*) said : It would be well to bear in mind the work of C. Hermann on the liquid crystal,[1] which shows that there are many possibilities of orientation of molecules, to which different phases can be attributed. He thought that Professor Vorländer ought to recognise this work.

Mr. Bernal (*Cambridge*) said : **1.** Professor Vorländer's conception of mixed dimensionality is difficult to grasp in its mathematical aspect, but it plainly contains a physical principle of critical importance to the formation of liquid crystal phases. In this connection I should like to insist on the preponderating influence of spatial anisotropy of molecules. Optical, magnetic and even electrical anisotropy are far less important, because, for crystals that form anisotropic melts, the intermolecular forces are chiefly Van der Waals' attraction, the centres of which are spread pretty uniformly along the molecules. As regards the predominating influence of linear over planar forms it should be noticed that the aromatic ring is far less a plane structure than its chemical formula would suggest. If we allow a radius of action of $1 \cdot 8$A. for the C—H groups derived from X-ray measurements, the benzene molecule appears as an oblate spheroid of polar semi-axis $2 \cdot 2$A. and equatorial radius $3 \cdot 2$A.; this is in marked contrast to its high optical and magnetic anisotropy. Under these circumstances it is not surprising that even simple *para* substituents lead to a molecule in which length rather than flatness is the predominating dimension.

2. The observed supracrystallinity of *p*-azoxybenzoic acid is in perfect accord with modern views as to the strength of the binding between carboxy groups due to the mobile H bond. The dicarboxylic acids, where the carboxyl groups are placed so as to be far enough apart and to be capable of linking together through hydrogen bonds in indefinitely long chains are particularly well crystallised and insoluble. A large number of organic substances supposed to be amorphous or polymerised undoubtedly must owe this appearance to extended hydrogen bonding.

3. In his remarks on the polymorphism of liquid crystals Professor Vorländer puts forward much evidence for the existence in certain cases of at least four intermediate stages between solid and isotropic liquid where Friedel maintains that only two (Nematic and Smectic) exist. I have examined some of the substances in question and am firmly convinced of the reality of the transformations observed (for details see the paper by Bernal and Crowfoot). I feel inclined to differ as to their interpretation. In every case the two highest temperature forms correspond to the classical description of Nematic and Smectic states. Whether the lower temperature forms, when they exist, are also liquid crystalline is more open to doubt, as here K. Herrmann's work shows that some lattice regularities exist and we may have to do with crystalline states admitting of molecular movement. I think the difference of these views is almost purely one of nomenclature and that, if this were recognised, most of the differences between the theses of Friedel and those of Vorländer would disappear.

[1] *Z. Krist.*, Tl. Kr. heft 1.

In the Discussion of this paper and that of Professor Zocher (page 945).

Professor Ornstein (*Utrecht*) said : One of the most striking properties of liquid crystals is their turbidity. It is necessary to give an explanation of this if we are to accept the idea that the liquid crystal forms an aggregate of homogeneous anisotropic regions. By reason of the double refraction the light is scattered. The effect of the clustering tendency shown clearly by liquid crystals is a quite general phenomenon (compare the Barkhausen effect in monocrystalline iron which shows the same phenomenon). Zocher, in his theory, neglects the action of Brownian disturbing forces, and his equation gives no better a description of the facts than Okuba's and Honda's non-static heat considerations of the theory of fero-magnetism.

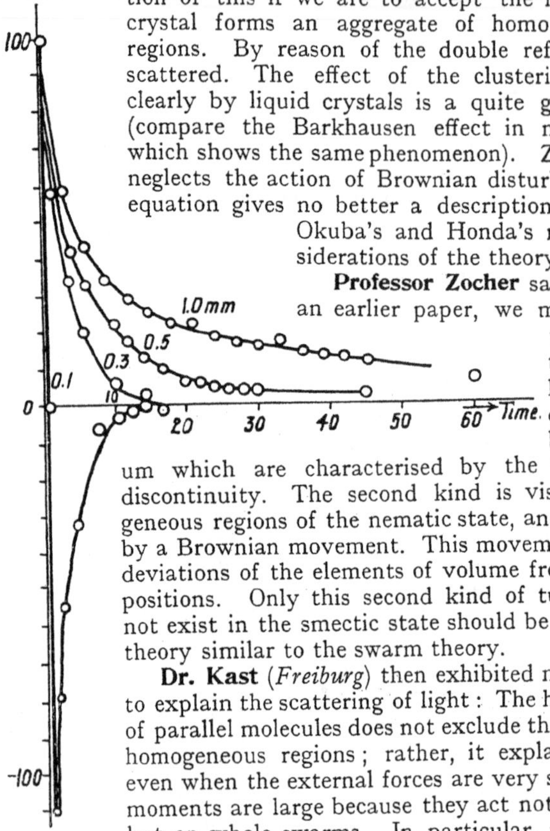

Professor Zocher said : As I showed in an earlier paper, we must distinguish between two kinds of turbidity. The first kind results from the deformations in the bi-refringent medium which are characterised by the well-known lines of discontinuity. The second kind is visible also in homogeneous regions of the nematic state, and is probably caused by a Brownian movement. This movement consists in small deviations of the elements of volume from their equilibrium positions. Only this second kind of turbidity, which does not exist in the smectic state should be treated by a kinetic theory similar to the swarm theory.

Dr. Kast (*Freiburg*) then exhibited new curves as follows to explain the scattering of light : The hypothesis of swarms of parallel molecules does not exclude the possibility of larger homogeneous regions ; rather, it explains their occurrence even when the external forces are very small. The turning-moments are large because they act not on single molecules, but on whole swarms. In particular, reference may again be made to the occurrence of such uniform regions immediately in contact with the limiting surfaces of glass or metal.

The swarm theory, however, postulates an essential difference between the surface of the preparation and its interior. This is clearly shown by some new measurements on the scattering of light by layers of different thicknesses of p-azoxyanisole in the anistropic-liquid state, illuminated with sodium light. It is known that the transparency in the direction of the field is increased in the magnetically oriented state. We have recently recorded the disappearance of this increased transparency on

Fig. 3.

306

the removal of the iron-free magnetic field, using an apparatus with a small time-lag (a selenium-silver photo-element, used with a Moll galvanometer).   In the diagram, the increase in transparency for a field of 1650 gauss was put equal to 100 for each bulb.   It is clear that the type of recovery of the original transparency is completely different for the different thicknesses.   In particular, it can be seen that up to a distance of more than 1/100 mm. from the outside the increase disappears within 1 second (bulb 0·1 mm.), and is succeeded by a large temporary increase in turbidity ; whereas for thicker bulbs (0·5 and 1 mm.) it is some minutes before the increase in transparency has disappeared.   This is in excellent agreement with our hypothesis that the preparation at the surface is affected by external forces which bring about parallel setting, while in the interior it is only the temperature fluctuations of direction that can eventually destroy the magnetic orientation.

Professor Zocher (*Prag*) said : The enormous difference between the layers of 0·5 and 1 mm. proves, that the influence of the walls reaches over a distance of at least several tenths of a millimetre.   The same fact is well known from the optical observations of Mauguin, and seems to be incompatible with the statement of the swarm theory, that the diameter of one swarm is 10^{-5} cm.

Dr. Kast (*Freiburg*) said : The structure of the parallel molecules is the determining point for their liquid crystalline character ; clusters are found in many other cases, *i.e.*, isotropic liquids as well as solids.   The significance of the melting phenomenon is the transition of one phase with one structure into another phase with another structure (see Nos. 16-17 of the Summary).

Dr. N. K. Adam (*London*) said : How sharply does Professor Ornstein think that the swarms are defined at their boundaries ?   In other words, how thick, approximately, is the transitional layer between one swarm and another, or the surrounding liquid ?   The phase boundary between a liquid and a gas is quite thin of the order, 10^{-7} cm. ; are these swarms to be considered as separate " phases " ?

Professor Ornstein (*Utrecht*), in reply, said : The swarms are differently oriented parts of the same phase.   In the optical considerations on turbidity a sudden transition between the separate parts is postulated, and the calculations on this basis are in accordance with the facts. However, it is conceivable that transition layers exist which can be of larger dimensions that those between liquid and gaseous phase under normal circumstances, but are of the order of those which are present near the critical point where the clustering tendency as has been shown by Zernike and Ornstein, is of great importance.

Mr. Bernal (*Cambridge*) said : Professor Ornstein and Dr. Kast have collected in this paper a large number of results, both experimental and theoretical, and have shown that all of these are in accordance with their swarm hypothesis of liquid crystals.   What they have not in my opinion succeeded in proving is that the existence of swarms follows necessarily from the observations or that another hypothesis might not as well account for the facts.   Of the phenomena discussed some (the orienting effects of surfaces, the slow re-establishment of equilibrium on the removal of disturbing fields) find a common explanation in any theory which is based on a mutual tendency to parallelism between molecules.   There remain two important arguments for the swarm theory, one based on Riwlin's measurements of extinction in nematic *para*-azoxyanisol, and Kast's measurements on the orientation of nematic *p*-azoxyanisol in electrical fields. As to the first, the swarm theory here leads to the correct formula for the extinction varying as $(d - 0.04)(n_1 - n_2)^2$ where d is the thickness and n_1 and r_2 the refractive indices. However, extinction would arise without swarms, merely from the continuous variation of direction of molecules from place to place. This must necessarily be a function of the birefringence, and would probably depend on the second power, as the size of the swarm does not, in Ornstein's theory, explicitly enter into the index. Similarly, on any theory the effective thickness of the layer would be bound to be less than the real thickness, because of the more perfect orientation near the walls. The agreement with theory is not, therefore, conclusive for the existence of swarms.

The evidence from electric orientation gives more definite information, but is not so simple to interpret. In the first place, it is unfortunate that so much elaborate physical measurement has been made on the classical liquid crystal p-azoxyanisol because its real chemical constitution is still in doubt and what is known of it is difficult to reconcile with the physical data. These are essentially that :—

(1) The dipole moment is 2.3×10^{-18}, measured in solution.

(2) For ordinary frequencies the dielectric constants are $\epsilon_1 = 5.4$, $\epsilon_2 = 5.5$, $\epsilon_2 - \epsilon_1 = 0.1$, where 1 and 2 refer to orientations with the molecular axes parallel and perpendicular to the electric field.

(3) In fields of low strength and relatively low frequency, the axis of the molecule lies parallel to the field.

Combining (1) (2) and (3), Ornstein assumes that the dipole moment of the molecule lies along its long axis, but that the polarisibility of the molecule is greater perpendicular to the long axis.

The chemical formula of p-azoxyanisol can be written

(a)

(b)

In (a), where the O atom is held by a semi-polar bond to one N atom, the dipole axis is highly inclined to the molecular axis ; in (b) where the O atom is shared in some unspecifiable way with both N atoms, the dipole moment is perpendicular to the molecular axis. No way of formulating p-azoxyanisol has been suggested that will make the dipole moment lie along the molecular axis. If we are to reconcile this fact with the electrical observations, the most plausible assumption is that the molecules in the liquid crystal state are rotating. This is not unreasonable, as already in the crystal their benzene planes are nearly at right angles.

If we accept formula (a) this would give a component of the dipole moment along the rotation axis, and would also explain the larger polarisibility at right angles to the axis where the effect of the perpendicular component of the dipole moment overcompensates the greater longitudinal polarisibility of the static molecule. This assumption would reduce the effective dipole moment of the molecule to $1 \cdot 1 \times 10^{-18}$ e.s. or less, but otherwise would fit with Ornstein's assumptions. If the molecules do not rotate, on the other hand, it is quite plain that with either formula (a) or (b) the dipole moment does not orient the molecules, even for low fields and some other mechanism, streaming phenomena, electrostriction, etc., must be invoked.

This possibility is not to be excluded as the experiments Freedericksz and Zolina show. The first estimate of swarm size given by Ornstein and Kast is based on the existence of a swarm dipole moment $p \gg 10^{-14}$ e.s. But, apart from the uncertainties mentioned above, this swarm dipole moment may be considered as a simple measure of the directional coherence of neighbouring molecules whereby the electrical turning moments of external fields on individual molecules are compensated over a far greater range than for normal liquids.

More significant are the extremely important experimental facts, brought out by Kast, that for fields of greater frequency than a certain critical value, the molecular axes set perpendicular instead of parallel to the lines of force and that, near this frequency, there is a considerable dielectric loss due to the reorientation. The order of magnitude of this critical frequency, $10^5 \rightarrow 10^6$ per sec., shows conclusively that there is involved something of much greater magnitude than a simple molecular orientation. But the further investigation of the variation of critical frequency with temperature shows that the simple swarm theory is not adequate to account for the phenomena, for this frequency changes nearly a hundredfold within a temperature change of $6°$ C., while the size of the swarm decreases, at most, to a third. The knocking together and deformation of the swarm is not a happy hypothesis to explain this discrepancy, for if the swarms are so deformable as to be completely broken up in times of 10^{-8} secs., much shorter than their presumed normal " life," why need they be postulated in the first place ?

We may conclude that the best defence of the swarm theory would seem to be at present as follows. A somewhat discontinuous aggregation of molecules is to be expected in liquid crystals by analogy with similar textures occurring in other states of matter (i.e., Zwicky texture in solids, association swarms at the critical point, etc.). It is also a useful model for imagining the structure of a liquid crystal in bulk (as against its microscopical examination in thin layers). But up to the present we have neither theoretical knowledge to prove the necessity of swarms from the properties of molecules, nor the experimental observations to deduce it from the study of liquid crystals themselves.

More experimental work needs to be done, particularly with other substances than p-azoxyanisol, to separate the essential from the accidental influences of magnetic, electric and optical anisotropy.

Finally, while agreeing with Professor Ornstein that the swarm theory has been an extremely fruitful working hypothesis, as is amply shown by his own work and that of his collaborators, it would be a grave mistake to assume further that any other theory such as the deformation theory of Zocker has not as much right to justify itself.

With regard to Professor Zocher's paper, I would like to call attention to the physical meaning of two of the constants, k_1, k_2, referring to lengthwise or cross distortion between molecules. Zocher states that when

$k_1 \gg k_2$ a smectic substance results with straight axial lines. This, it seems to me, is confusing cause and effect. The fundamental physical characteristic of smectic states is the existence of regularly spaced planes of molecules which is due to a minimum of potential energy when two molecules lie not only parallel but with ends aligned $=\!\!=$ rather than $\underline{\quad=}$. These planes can be distorted only into Dupin's cyclides, as Sir William Bragg has pointed out, the normals to which form a system of straight lines along which the molecules must lie, hence arises the straightness of the axial lines and the fact that $k_1 \gg k_2$.

As regards the very interesting observations on orientation anomalies in nematic substances, it would seem that these are more apparent than real. Although for the greater number of substances the axes for greatest refractivity and susceptibility are the same, there is no theoretical reason why this should be the case. In molecular substances both are vectorially additive properties of homopolar bonds, but the constants for the bond types do not necessarily follow in the same order, and optical refractivities, as they depend on electrical deformability, are much more affected by ions and local dipoles in neighbouring molecules. The exceptional cases of potassium laurate and salvarsan are probably to be explained in this way. In the first case, the bi-refringence of aliphatic chains is small, and might be reversed by the action of water molecules or K^+ ions ; in the second, effect of the arsenic atom must be large on the optical properties, but small on the shape of the molecule.

Professor Zocher (*communicated*) wrote : It seems that Dr. Bernal has not understood me, in consequence, possibly, of a mistake of translation. I wrote in my paper, that the *behaviour* of the smectic phases results from the fact, that $k_1 \gg k_2$. In general, it is possible to derive the mechanical properties of the phases from their molecular structure, *i.e.*, that in the smectic phases $k_1 \gg k_2$, and from the mechanical properties we can derive the arrangement of the optical axis in a deformed region. This is what Oseen did, and Sir William Bragg's derivation can be considered from the same point of view.

I think also that there is no unconditional reason why the axes of greatest refractivity and susceptibility should be the same. What we wished to prove first was, whether the axis of symmetry is always the axis of greatest susceptibility. This might be expected, because the axis of symmetry is regarded as the length of the molecule, *i.e.*, the direction of the greatest number of chemical bonds. In another paper I examined whether the axis of symmetry is always that of the greatest refractivity. It could be shown, that the greater polarisibility (for light) of the length of the molecule can be overcompensated by the action of those parts of the molecule which have a stronger polarisibility perpendicularly to it. This negative component of double refraction may be due to ionisation, since it is magnified by heating or by addition of water. Salvarsan has only positive double refraction....

Professor Lowry (*Cambridge*) said that he was very glad that Professor Iterson had mentioned the " confocal conics " which Friedel had observed in smectic media. The parallel orientation of molecules, which had formed the principal topic of discussion up to this point, was simple and easy to understand, but was less fundamental and important than the remarkable figures which were observed when this regular orientation was destroyed, and it would be of great interest if an explanation could be given of the arrangement of the molecules which gives rise to these figures.

Mr. Rawlins and **Dr. Lawrence** (*communicated*) : The texture of the focal conic structure varies with the conditions and with the nature of the substance. It is clear that the form of any surface of a family will depend upon its value of k, but the actual shape observed is the particular member k_{max}. of the family, where k_{max}. is the value of k for the outermost surface. The texture under different conditions will also vary with the values of the constants a, b, c, and it scarcely seems likely that these will all have the same temperature coefficient : thus, any one substance may show a different texture at different temperatures.

For molecules of different shapes, the texture may vary : for small values of k, where the strain is greatest (as Sir William Bragg has pointed out, page 1058), molecular configuration may be expected to exert its greatest effect upon the packing; here the need for maintaining dk constant (as the geometry requires) results in the most marked differences of arrangement of molecules of different shapes. This effect must show itself in the relative values of a, b and c. These phenomena can be observed by the traces.

Professor Herrmann (*Charlottenburg*) said : With reference to the orientation of molecules of p-azoxyanisole in an electric field, found by Dr. Kast as well as myself and my collaborators,[4] we now think it due to another cause. We attribute the orientation in a steady field to a secondary effect, the convection currents which are always present (*loc. cit.*)...

In discussing Professor Herrmann's paper (page 972) :

Professor Zocher (*Prag*) said : Since the smectic phases of thallium soaps are optically uniaxial, it is improbable that the molecules are straight-tilted chains. I would believe that the chains are not quite straight, perhaps partly spirals so that their length standing parallel to the optical axis is shorter.

Dr. Kast (*Freiburg*) said : In the X-ray diagrams of the nematic phase the ring means the random orientation of the swarm axes (texture). The effect of the magnetic field is to bring about an orientation of the swarm axes.

Mr. Bernal (*Cambridge*) said : The point raised in this paper is of the greatest interest in its theoretical implications. It is very difficult to conceive how in a truly smectic substance molecules can exist inclined to the plane normals. In each plane of such a substance the molecules must be supposed to have every degree of freedom except perpendicular to the plane. Now if the molecules are inclined but haphazard in azimuth they cannot be close packed as they would be when normal and would comprise a much larger area in each layer in an arrangement very much like that of an expanded liquid monomolecular film.

It is difficult to imagine that such an arrangement would be stable,

---

[4] K. Herrmann, A. H. Krummacher, and K. May, *Z. Physik*, **73**, 419, 1933.

besides this would imply a much larger change of volume on melting than is observed in thallium stearate. If we except the experimental proof of inclined molecules it therefore follows that the state cannot be truly smectic but must belong to one of the intermediate states already suggested by K. Herrmann in his other papers and treated theoretically by C. Hermann. In this case it would still remain to determine to which of these states thallium stearate belonged. Each layer with inclined molecules must be biaxial and the optical uniaxial character could only be due to the complete lack of azimuthal parallelism between planes.

An observation I have made since reading Dr. Herrmann's paper makes it still possible, though not likely, that the change from solid to liquid crystal is not so great as he has found. Thallium stearate is certainly smectic at 140° when the high temperature photographs are taken, but it crystallises in feebly biaxial plates (apparently rhombic) in which the optic axial angle changes continuously with the temperature in a way similar to that of gypsum. It becomes uniaxial a few degrees below the melting-point, then the axes cross and it becomes increasingly biaxial.

At about 100° there is an abrupt change and another form with smaller bi-refringences but the same orientation appears. This is the state stable at room temperatures and for which K. Herrmann found the spacing 42·05A. Now the change in optical properties suggests forcibly that not only is there a change of slope of the chains in this high temperature form leading to a shorter spacing but that this change is reversed in the stability range of the high temperature form and that the chains are nearly perpendicular at the melting-point. It would be very interesting to check these observations by X-rays and thus find by direct measurement the slope angle in the smectic state.

In discussion of Dr. Müller's contribution (page 990) :

Professor Zocher said : As to the swarm theory, it is impossible to find out the diameter of the swarm. Professor Ornstein calculates from experiments in a magnetic field the volume of a single swarm to 10^{-15} cm. or the diameter thereof to 10^{-5} cm. An aggregate of swarms of this size orientated at random gives no double refraction as a whole. On the other hand, from measurements of turbidity he calculates to 10^{-3} cm. the thickness of a layer influenced by a surface. It is in fact possible to get homogeneous layers of the thickness of $2 \cdot 10^{-2}$ cm., and therefore the thickness of a single swarm should be of the same size. It is evident that such large swarms cannot exhibit Brownian movements, which will cause distribution at random.

According to Professor Ornstein's opinion, it is not necessary that the substance should be broken up into swarms, and it seems the experiments show that the nematic phases do not break up into swarms. To the smectic phases the swarm theory cannot be applied because a magnetic field has no effect on them under analogous circumstances.

Therefore I think that consideration of the Brownian movement, which is the main content of the swarm theory, cannot be used for the explanation of the essential properties of mesophases. It finds application only to the before-mentioned turbidity of the second kind, *i.e.*, the turbidity of homogeneous regions in nematic phases.

With regard to the contribution of Mr. Bernal and Miss Crowfoot (page 1032) :

Professor Ornstein (*Utrecht*) said : The effect of the wall of one group of molecules is propagated through the liquid crystal to a depth of the order of a swarm, *i.e.*, about 0·01 mm. The experiments of v. Wyk show that the layers at the wall also orient one another—interaction of swarms. The fact that there are swarms implies that on large complexes of molecules more force is exerted by the field. The experiments of the late Miss Riwlin show that a liquid crystalline aggregate consists of homogeneous layers of fixed double refraction. The extinction can be calculated as a function of $d\omega^2$, ω being the numerical value of the double refraction, d the thickness of the layer. The extinction of all nematic substances can be described by one formula which applies to all thicknesses of the preparation, and all wave-lengths. There can be no doubt that the swarm theory is a heuristic working hypothesis which has brought to our knowledge a great number of facts without requiring any new and special hypotheses. The chief problem now is to show from the molecular theory the necessity of the existence of swarms, in other words, to find from the molecular image forces of interaction between the molecules, which, together with heat motions, determine the mean cluster or swarm radius.

B9
Nature **136**, 261 (1936)

INFLUENCE OF A MAGNETIC FIELD ON THE VISCOSITY OF PARA-AZOXYANISOL

It is known that a magnetic field has an influence on the orientation of the molecules in anisotropic liquids[1]. As it seems probable that the coefficient of viscosity of these substances will depend on whether, and in which direction, the molecules are oriented, a magnetic field may be expected to influence the value of the viscosity coefficient. The first experiments on this subject were made by M.W. Neufeld[2], who investigated the influence of a magnetic field on the velocity of flow of anisotropic liquids (*p*-azoxyanisol and anizaldazin) through a capillary tube. But, as is well known, in a layer of anisotropic liquid so thin as that in the capillaries used by Neufeld (diameter 0.09 mm), the action of the field may be impeded by the directive action of the walls. Consequently, Neufeld's experiments cannot be regarded as definitive.

In order to examine the question further, I applied a method in which thicker layers of liquid were in motion, and the surface of friction was a plane, so that it was possible to introduce a magnetic field perpendicular to this plane. This method is somewhat similar to that used by Quincke[3] for measurements of the influence of an electric field on the viscosity of liquids.

On one side of the beam of an analytical balance, a glass plate, 48 mm × 24 mm × 0.8 mm, was hung in a vertical plane. The plate was immersed in the liquid under investigation, the liquid being in a vessel having two parallel walls 6 mm apart. The vessel was placed in a thermostat between the poles of an electromagnet. Thus the plate could oscillate vertically in its own plane. The period of the oscillations was about 5 sec; the maximum amplitude was 0.5 cm. The damping of such oscillations depends upon the viscosity of the liquid in which the plate is immersed. So far, the influence of the density of the liquid on the

---

[1] See, for example, M. Mięsowicz und M. Jeżewski, *Phys. Z.*, **36**, 107: 1935. M. Trautz und E. Fröschel, *Ann. Phys.*, **22**, 223; 1935.

[2] M.W. Neufeld, *Phys. Z.*, **14**, 645; 1913.

[3] G. Quincke, *Wied. Ann.*, **62**, 1; 1897.

damping has been neglected. The apparatus was calibrated with sugar solutions, and then the measurements with *p*-azoxyanisol were made. Its temperature was 125 °C (anisotropic-liquid phase). The damping of the oscillations of the plate was measured without and with the magnetic field (2,400 Gauss) perpendicular to the plate. The experiments have shown a rather unexpectedly great influence of the magnetic field on damping. From the values of the logarithmic decrements, the viscosity coefficients were determined, and it has been found that the viscosity in the magnetic field is about 3.5 times greater than without the field. This effect disappears completely after the transition to the isotropic-liquid phase (at a temperature of 135 °C).

The experiments are being continued in order to examine how far this effect depends on temperature as well as on the direction of magnetic field. Other substances in their anisotropic-liquid phase are now also under investigation, and the results will be published shortly.

I wish to express my best thanks to Prof. M. Jeżewski for his helpful interest in these experiments.

M. MIĘSOWICZ

Physical Laboratory,
Mining Academy,
Cracow.
June 19.

B10
Nature **158**, 27 (1946)

THE THREE COEFFICIENTS OF VISCOSITY OF ANISOTROPIC LIQUIDS

Before and during the War, investigations were reported on the viscosity of anisotropic liquids[1,2,3,4]. As is well known, the flow of an anisotropic liquid influences the orientation of the molecules. On the other hand, the value of the viscosity coefficient depends on this orientation. Therefore this coefficient is a function of the velocity-gradient, and the usual definition of the viscosity coefficient for these liquids loses its significance. If under the influence of any factor the molecules of the liquid should be orientated in one direction and the motion is unable to change this orientation, then we have the viscosity coefficient in the ordinary sense. But in this case we have to deal with the anisotropy of the viscosity, and in case of a liquid of the type of p-azoxyanisol we have three principal viscosity coefficients belonging to the three directions of orientation; these are: (1) direction of the flow; (2) direction of the velocity gradient; (3) perpendicular to both these directions. Having given the molecules an orientation by means of a magnetic field in such circumstances that the flow did not change this orientation, I obtained the following values for the three principal viscosity coefficients for p-azoxyanisol and p-azoxyphenetol[2].

Substance and temperature	Molecules parallel to the direction of the flow, η_1	Molecules parallel to the gradient of velocity, η_2	Molecules perpendicular to the direction of flow and to the velocity gradient, η_3
p-Azoxyanisol 122 °C	0.024 ± 0.0005	0.092 ± 0.004	0.034 ± 0.003
p-Azoxyphenetol 144.4 °C	0.013 ± 0.0005	0.083 ± 0.004	0.025 ± 0.003

[1] Międowicz, M., *Nature*, **136**, 261 (1935).
[2] Międowicz, M., *Bull. Acad. Pol.*, A, 228 (1936).
[3] Zwetkoff, W.N. and Michajlow, G.M., *Acta Phisicochim.* URSS., **8**, 77 (1938).
[4] Becherer, G. and Kast, W., *Ann. Phys.*, **41**, 355 (1942).

These results throw light on those obtained by the other investigators. The results of the older investigators (Eichwald[5] and Dickenschied[6]) obtained by the method of flow through capillary tubes are in agreement with my results. Evidently, in both cases we were dealing with the orientation of molecules parallel to the direction of flow. Zwetkoff and Michajlow[3], using the method of flow through a tube with rectangular cross-section, by application of the strongest available magnetic field and with the smallest possible velocity of flow, obtained values about 80 per cent of my value, η_2. From the dependence of the results on the intensity of the magnetic field, it is clear that these investigators did not reach the state of constant orientation of molecules, and that the flow of the liquid changed this orientation. The results obtained by these authors for the different values of velocity and for different intensities of magnetic field lie between η_1 (orientation parallel to the flow) and η_2 (parallel to the velocity gradient) $\eta_1 \leq \eta_{\text{Zwetkoff}} < \eta_2$.

The measurements recently published by Becherer and Kast[4] were not carried out with constant orientation of the molecules. They were, however, orientated (at least in the layers where the phenomenon of viscosity chiefly takes place) in the planes of friction, but without a definite angle in this plane. The value obtained by these investigators is therefore not one of the three principal coefficients in the sense given by me. Clearly, $\eta_1 < \eta_{\text{Kast}} < \eta_3$, because in my measurements of η_1 and η_3 we have also an orientation parallel to the plane of friction, but once parallel and then perpendicular to the direction of the flow.

Hence the coefficient η_1 was measured by other workers as well as by me. The method of flow through capillaries gives usually the result corresponding to an orientation of molecules in the direction of flow; the other values given by different investigators do not correspond to constant orientation of molecules.

M. MIĘSOWICZ

Physical Laboratory,
Mining Academy,
Cracow.
May 10.

[5] Landoldt-Börnstein, I., Erg. Bd. (1927).
[6] Landoldt-Börnstein, I. Bd. (1923).

Marian Mięsowicz was born in Lwów (then in Poland, but now Lviv, Ukraine) in 1907. He studied physics at the Jagiellonian University in Cracow. His interest in liquid crystals was aroused by Professor Mieczysław Jeżewski of the Mining Academy in Cracow. Although some of his important liquid crystal research was not published until after the Second World War, in fact Mięsowicz's work in liquid crystals was confined to the period 1934–8. The Second World War dislocated completely Mięsowicz's work and after the war he changed field completely. He became interested in cosmic rays, and was later appointed professor in the Institute of Nuclear Physics in Cracow.

He retained an affection for the liquid crystal field, however, and participated in the international liquid crystal conferences in Bordeaux in 1978 and Bangalore in 1982. He died in 1992.

319

B11
Comptes rendus de l'Académie des Sciences **213**, 875–76 (1941)

CRISTALLOGRAPHIE. — *Sur l'orientation des cristaux liquides par les surfaces frottées; étude expérimentale.* Note de M. Pierre Chatelain, présentée par M. Charles Mauguin.

Soit une préparation de liquide anisotrope obtenue de la manière suivante : de l'azoxyanisol, dans sa phase nématique, est introduit par capillarité entre un porte-objet et un couvre-objet, placés sur la platine chauffante d'un microscope polarisant; ces deux lames ont été préalablement frottées sur du papier suivant une direction D bien déterminée, les surfaces frottées étant celles qui sont au contact du liquide, les directions D coïncidant. Observée en lumière parallèle, entre nicols croisés, une telle préparation rétablit la lumière; mais contrairement à ce qui se produit avec des lames non frottées, elle s'éteint dans toute son étendue pour quatre positions de la platine, comme le ferait une lame monocristalline, la vibration lente étant dirigée suivant D. Si les lames n'ont pas été nettoyées spécialement, ni frottées un grand nombre de fois, la préparation, observée en lumière monochromatique avec le montage de lumière convergente, montre les figures caractérisant un uniaxe, l'axe optique ayant une inclinaison variable d'une région à l'autre, les régions parallèles à l'axe étant les plus nombreuses.

Action de la durée du frottement. — Pour obtenir une préparation s'éteignant uniformément suivant D, il suffit de frotter la lame 3 ou 4 fois sur du papier (trajet 3o^{cm} environ). Si la durée du frottement est accrue, lames frottées 5o fois par exemple, le nombre de plages non parallèles à l'axe diminue notablement; en frottant la lame 100 fois, la préparation obtenue est équivalente à une lame monocristalline, dont l'axe optique serait parallèle à la direction D du frottement. Comme l'axe optique d'une préparation orientée d'azoxyanisol dans sa phase nématique est confondu avec la position moyenne de la direction d'allongement des molécules, il est possible d'énoncer les résultats suivants :

Le frottement exerce deux actions, la première, immédiate, oriente les molécules d'azoxyanisol de façon que leur allongement soit dans le plan défini par la normale aux lames frottées et la direction D du frottement; la deuxième, plus lente à se faire sentir, progressive et un peu irrégulière, impose à cette direction d'allongement une orientation moyenne de plus en plus fréquemment confondue avec D.

Action du nettoyage. — Si les lames ont été flambées ou nettoyées

B11
Comptes rendus de l'Académie des Sciences **213**, 875–76 (1941)

ON THE ORIENTATION OF LIQUID CRYSTALS BY RUBBED SURFACES: EXPERIMENTAL STUDY

Note from
Mr Pierre Chatelain
presented by Mr Charles Mauguin

Consider a preparation of anisotropic liquid obtained as follows. Some azoxyanisole, in its nematic phase, is introduced by capillarity between a sample slide and cover slip. It is then placed on the heated stage of a polarising microscope. The two glass slides have been previously rubbed on some paper along a well-defined direction D, with the rubbed surfaces in contact with the liquid and the directions D coincident. When it is observed in parallel light between crossed nicols, such a preparation transmits light. However, by contrast to the situation occurring when the plates are not rubbed, this sample extinguishes light across its entire extent for four stage positions, just like a single crystal sample, with the slow vibration direction directed along D. If the slides have been neither specially cleaned nor rubbed a large number of times, the sample exhibits figures characteristic of a uniaxial sample when observed with monochromatic convergent light. In this case the optic axis has a direction which varies from one part of the sample to another, with the domains parallel to the axis (D) the most numerous.

Effect of the amount of rubbing – In order to obtain a sample extinguishing uniformly along D, it is sufficient to rub the slide 3 or 4 times on some paper (over a distance of about 30 cm). If the amount of rubbing is increased – slides rubbed 50 times for example – the number of domains not parallel to the axis decreases considerably. If the slide is rubbed 100 times, the resulting sample is equivalent to a monocrystalline sheet whose optic axis would be parallel to the rubbing direction D. As the optic axis of a sample of oriented azoxyanisole in the nematic phase can be identified with the mean direction of the molecular long axis, it is possible to state the following results:

The rubbing exerts two effects. The first effect, which occurs immediately, orients the azoxyanisole molecules such that their alignment is in the plane defined by the normal to the rubbed slides and the direction of rubbing D. The second effect, which takes effect much more slowly, progressively and with a little irregularity imposes on this direction of alignment a mean orientation, which is increasingly coincident with D.

Effect of cleaning – If the slides have been flamed or chemically cleaned before being rubbed, the results are the same, except that it is sufficient to rub the slides 50 times in order to have remarkably well-aligned samples. It is very difficult to remove the effect of rubbing, particularly if the slides have been flamed before being rubbed. Only very powerful chemical cleaning can remove from a slide all memory of previous rubbing and even prolonged sterilisation in a flame is not always sufficient.

Effect of the substances on which the slides are rubbed – Paper has been replaced by a variety of substances such as silk, woollen cloth, rubber, leather, lead or glass. The results remain effectively the same. The differences observed result from the fact that the slides can be rubbed on some substances more easily than on others.

Effect of the nature of the slides rubbed – The glass slides have been replaced by mica[1] or quartz slides. The experiment carried out with mica is particularly important. If the slides are freshly cleaved, the sample has a helicoidal structure, with the molecules at the bottom surface oriented at 30° to the right of the projection of g^1 on the cleavage plane, and those at the upper surface at 30° to the left of this line. If, however, the mica slides are rubbed along the g^1 direction, the resulting sample is equivalent to a monocrystalline plate whose axis would be directed along that line; in other words the rubbed mica behaves like glass. The same phenomenon occurs with quartz. There is every good reason to think that this result is general; the effect of rubbing is the same, whatever the nature of the surface rubbed.

Variation with the nature of the melted material – The tests were carried out on azoxyanisole, azoxyphenetole, anisaldazine and *p*-anisal-amino-ethylcinnamate. The effects observed permit the following results to be stated. The effect of rubbing is effective only on the nematic phase and not on the smectic phase. For the nematic phase, it is all the more effective when the material examined has a greater tendency to give rise to adhering layers.

Length of retention of the property – Slides rubbed more than a year ago, and kept only in a bell jar, have retained the ability to align liquid crystals almost

---

[1] Ch. Mauguin, *Comptes rendus* **156**, 1913, p. 1246.

intact, for slides which had been flamed or chemically cleaned. There was a tendency for a reduction in the extent of the domains parallel to the axis, and a tendency for the formation of irregularities for non-wettable slides.

(read on 15 December 1941)

B12

Bulletin de la Société française de Minéralogie **66**, 105–30 (1944)

SUR L'ORIENTATION DES CRISTAUX LIQUIDES PAR LES SURFACES FROTTÉES

Par Pierre Chatelain.

J'ai signalé dans ma thèse (1) qu'une préparation d'orientation uniforme dans toute son étendue, de para-azoxyanisol ou de para-azoxyphénétol à l'état de liquide nématique, peut être obtenue de la manière suivante : le couvre-objet et le porte-objet en verre, entre lesquels le corps sera fondu, lavés avec de l'éther ou de l'alcool, sont frottés sur du papier dans une direction bien déterminée ; les faces frottées ayant été mises en regard et les directions de frottement coïncidant, la substance est fondue ; pendant toute la durée de la phase nématique, cette préparation vue au microscope polarisant, en lumière parallèle, s'éteint dans toute son étendue quatre fois pour un tour de la platine, le grand indice étant dirigé suivant la ligne neutre parallèle à la direction du frottement.

On sait depuis longtemps que les parois ont une action sur les couches de liquides nématiques situées dans leur voisinage, cette action, en général assez désordonnée, est une cause d'erreur importante dans toutes les études entreprises pour mesurer les grandeurs caractéristiques de ces liquides et leurs variations en fonction des divers facteurs : température, champ électrique ou magnétique. Le phénomène indiqué ci-dessus, qui permet d'ordonner à volonté cette action des parois, apparaît comme particulièrement intéressant ; c'est pourquoi j'en ai entrepris une étude plus approfondie.

324

B12
Bulletin de la Société française de Minéralogie **66**, 105–30 (1944)

THE ORIENTATION OF LIQUID CRYSTALS BY RUBBED SURFACES

Pierre Chatelain

I have reported in my thesis[1] that a sample of para-azoxyanisole, which is uniformly aligned over its whole extent can be obtained in the following manner. The top and bottom glass plates, between which the sample will be melted, are washed with ether or alcohol, and are then rubbed on paper in a well-defined direction. The rubbed faces are then placed opposite one another with their rubbing directions coincident, and the sample between them is melted. When viewed in a polarising microscope with parallel collimated light, this sample extinguishes the light four times for each complete rotation of the microscope stage across the entire sample area, with the largest refractive index along the direction of rubbing.

It has been known for a long time that the walls have an effect on films of nematic liquids in their neighbourhood. This complicating effect is an important source of error in all the studies carried out to measure the characteristic quantities of these liquids and their variation as functions of various factors, such as temperature and electric or magnetic fields. The phenomenon mentioned above, which allows the effect of surfaces to be prescribed at will, seems to be particularly interesting. For this reason I have carried out a very detailed study.

The aim of this work therefore is to determine the precise conditions required to obtain as perfect samples as possible, and to explore all aspects of the phenomenon, identifying its origin as well as exploring the theoretical and practical consequences.

I. Experimental study of the phenomenon

Because the phenomenon under consideration is a specifically surface phenomenon, above all one must specify the state of the surface in the best way possible. Out of many different methods, I have focussed on that which is most usually used: the wetting ability, and a related effect, that of misting. This last effect permits the state of the surface to be characterised as soon as it appears to be wettable.[2] On the other hand, it is difficult to make identical solid surfaces. The results have therefore been reported after repeating the experiments a number of times and recording the effects which occur most frequently.

Method of obtaining the samples

In the first series of experiments, the surfaces were cleaned with a rag soaked in ether or alcohol, and then wiped. This gives the plates a clean appearance, but they are not wettable and a drop of water slides on the glass. The plates must therefore be covered with a layer of impurities, probably grease from the fingers, which was dissolved and transferred to the glass by the rag. The plates are then rubbed on paper, usually typewriter paper; the sheet was placed on a stack of filter papers, permitting a degree of flexibility to the set-up. A well-defined rubbing direction is obtained by using a back-and-forth movement of the glass plate, with one of its edges pressed against a metal ruler placed on the paper. The distance travelled by the glass plate for one to and fro motion is about 30 cm. I will call 'a singly-rubbed plate' one which has been rubbed with a single pass to and fro. This process is more difficult when rubbing a cover-slip, because of its fragility. When observation in convergent light is not necessary, it is better to replace the cover-slip with a piece of microscope slide.

The two plates are then placed on the hot-stage of a polarising microscope. The substance is introduced between the slides in the molten state by capillarity. The melting occurs on the supporting lower slide, outside the region occupied by the cover slide. This precaution is taken both to avoid traces of solid left from the preparation and to eliminate, by a kind of filtration, dust particles which disturb the organisation of the liquid.

Observation of the samples in the polarising microscope

In parallel white light, the sample extinguishes over its entirety, four times for each rotation of the microscope stage, with the slow vibration direction along the direction of rubbing D. I had deduced from this observation that the long axes of the molecules of azoxyanisole, under these conditions, align parallel to D, and that this film of liquid crystal was equivalent from an optical point of view to a crystalline plate cut parallel to its optic axis, the latter being coincident with D. A more complete study with parallel monochromatic light, and then with convergent light, has shown that this result is not always exact.

A sample in which the glass plates had been rubbed only ten times was illuminated with yellow sodium light. The sample extinguished light well, but away from the extinction positions it no longer appeared as homogeneous as with white light. Numerous very fine lines of equal retardation appeared, indicating that there could be a significant variation from one point to another. These variations are too large to be explained by variations of thickness or temperature. Rather they are due to variations of orientation from one point to another. This is confirmed by observation with convergent monochromatic light. At the edge of the domains, most of which are parallel to the axis, are found some domains where the optic axis, that is to say the alignment direction of the molecules, is inclined to the plane of the glass. This variable inclination can be as much as 30°.

This is reminiscent of a fact reported by Mauguin,[3] who oriented para-azoxyanisole by making it crystallise slowly and then heating it afresh. He obtained homogeneous domains, defined by the boundaries of the now-melted crystals, usually parallel to the crystallite axes, but sometimes inclined to them. There is an important difference here; with a rubbed glass plate, the projection of the optic axis onto the plane of the glass plate has a fixed direction when one passes from one domain to another. Thus the extinction axes are the same for all domains. That is not the case for the samples of Mauguin.

If the temperature of the sample is raised above the isotropic clearing point and then lowered again, the effect of rubbing is as strong as it was before. In the same way the adhering films of Mauguin remain after such an operation. If the liquid crystallises after a fresh melting, the effect of rubbing reappears, but slightly perturbed, as we shall see later.

The next section of this paper (pp. 109–117) reports the results of a series of experimental studies. These results had been previously reported, at least in summary, in Chatelain's paper, reprinted above as article B11. The translation continues with p. 118 of the original paper.

II.   On the probable mechanism of this orientation effect

Having carried out this experimental study, it remains to consider in what possible way the effect of a rubbed surface on the molecules of nematic substances can be imagined.

Electrical charging by rubbing has been eliminated by putting the rubbed plates under X-rays. This radiation ionises the air, and the ions created in this way neutralise any free charges to be found on the glass plates. Plates treated in this way retain the property of orienting liquid crystals. Charging by rubbing is not then the cause. This was almost certain from the outset, since such charging would create a field normal to the plates, and this would have the effect of orienting the molecules normal to the surfaces of the glass.[5]

It seems likely that rubbing has the effect of orienting electric dipoles along the direction D. These dipoles give rise to an electric field in the neighbourhood of the wall which is approximately parallel to the rubbing direction D. The molecules next to the wall[5] are oriented parallel to this field, and the orientation in the bulk occurs subsequently, step-by-step, through molecular interactions. This interpretation immediately explains why rubbing has no orienting effect on substances in the smectic phase, since the electric field has no significant effect on this phase.

It remains to clarify the nature of these dipoles. Two hypotheses are presented. The dipoles could either be the result of impurity molecules on the surfaces, or be due to ion pairs pulled out of the supporting surfaces. Either of these could be oriented along the direction D by rubbing.

Because of the general nature of the phenomenon, the first hypothesis appears to be the most probable. Let us examine the consequences. According to all reports, the impurities found at the surface of solid materials consist of organic molecules, particularly fats, associated with molecules of water. The molecules of the fatty material are extended, and their associated electric dipole is directed along the extension direction of the molecules. Rubbing would have the effect of flattening the molecules onto the surface, their direction of alignment being parallel to D. The molecules of fatty material under usual conditions arrange themselves normal to the surface on which they are deposited, and give rise to a film which is isotropic in the plane of the surface. Thus no special direction is defined in this plane. On the other hand *rubbing will define a new type of molecular layer*, a layer in which the molecules are parallel to the surface and largely aligned along a definite direction. This is without doubt the first time that such a type of molecular layer has been demonstrated on solid surfaces. Langmuir proposes the existence of such an arrangement on the surface of water, which he calls gaseous layers.[6]

This hypothesis allows one to understand the independence of the effect of rubbing on the substance rubbed and the rubbing material. At the same time it seems likely that the effect of rubbing makes itself felt slowly. It would first result in the long axes of the molecules of fatty material being aligned parallel to the plane defined by the normal of the surface and the direction D. Then, slowly, the direction of the long axes of the molecules would become parallel to D. In the case of non-wettable surfaces, the very great thickness of impurities prevents, by its very elasticity, a perfect orientation of molecules parallel to D. As a result, if the surface of the slide has not been rubbed for a long enough time, these samples exhibit domains whose optic axis is more or less tilted. This also explains the progressive reduction in the effect of rubbing with time, as the fatty molecules gradually straighten up.

For those slides which have been flamed, the layer of impurities is very slight. According to the experiments of Devaux, it is of the order of a monomolecular layer. Under these conditions, the molecules affected by the rubbing are in direct contact with the glass. Thus the rubbing effect is very rapid, and more perfect, and above all the fatty molecules will be fixed much more strongly in their new position. It is thus only with extreme difficulty that the film loses all memory of rubbing, and so one obtains the longest-lasting alignment effect.

Slides cleaned chemically are perfectly clean, but it appears likely that they become contaminated at the time of rubbing. Only a very small amount of fatty material is needed. Despite all precautions, the vapour pressure of the grease found on the assistant's fingers is sufficient to cover the glass slides with an oriented layer of fatty molecules. Flamed slides appear to give slightly more perfect samples that those cleaned chemically. This may be a consequence of a result reported by A. Marcellin,[7] which is that the molecules adhere to glass much more easily when the glass has been dehydrated by intense heat.

This hypothesis provides a ready interpretation of the experimental results. It will remain a hypothesis so long as one is unable to rub glass slides without dirtying them. Such an experiment will never be possible because it is known that when two surfaces are very clean, they cannot slide on one another.

In all our experiments, in which our slides have been cleaned in this manner, after rubbing, the mist which settles on the slides is grey, and droplets are aligned in such a way as to reveal the direction D. Different authors[8,9] have studied this effect on mist through the phenomenon of breath figures. They appear to think that the observations can be interpreted through the presence of impurities on the slides studied, but they have not carried out precise experiments. The hypothesis of a superficial modification of the supporting surface structure by rubbing is not definitely ruled out. A study using electrons would enable an unequivocal choice to be made between the two hypotheses. Given the state of development of this technique, such a study appears to be very difficult.

Nevertheless, as a result of the analogy presented by the phenomenon of orientation of nematic liquids by rubbed surfaces with the phenomena of misting, images of breath, and in an more general way, with all the phenomena demonstrated since the discovery of photographic technique of Daguerre,[2,10] it is possible to conclude that it has the same origin, and that like them, it demonstrated the anisotropic properties of surfaces.

III.   Some consequences of this phenomenon

The next section of this paper (pp. 121–128) compares the results of alignment by rubbing with the much earlier observations of Mauguin and Grandjean on the orientation of nematics on crystal faces. Grandjean's classic paper on this subject is included in this collection as article A10. In the interests of brevity, we have not included this section of the paper in the translation reproduced here. We conclude the translation with the final section of the paper, from p. 128.

Practical consequences of this phenomenon

This study provides a very practical way to determine *a priori* the effect of the necessary boundary walls of all receptacles in which liquid crystals must be contained.

I have been able to observe samples of more than 1 mm thickness oriented parallel to the axis by the sole effect of rubbed walls. It must be possible to exceed this thickness, but it is difficult to check optically the good alignment. This is because the depolarised scattered light rapidly becomes more intense than the transmitted light, which causes the extinction and fringes in convergent light to slowly disappear.

Thanks to this method, it is easy to create helicoidal films of great extent, which have a predetermined twist. It is sufficient to turn the cover slide in such a way that the direction of rubbing on it makes the desired angle with the direction of rubbing on the support slide. In this way I have been able to observe that the principal vibrations transmitted by a helicoidal film with a twist close to 90° can be considered to be rectilinear up to a thickness corresponding to the third order of coloured fringes. For less than this thickness the principal vibrations become more and more elliptical. In white light the sample is strongly coloured, and whatever the orientation of the rubbing direction with respect to the symmetry planes of the nicols and whatever the angle of the nicols, the light is still transmitted.

It seems that it would be interesting to return to the measurements of Kast[5] of the dielectric anisotropy and those of Foex[12] on the permeability, since both these authors were hampered in their experiments by not knowing the effects of the walls. Azoxyanisole in particular should lend itself to such experiments, since it is readily possible to obtain perfect samples in which the average direction of the molecules is either perpendicular to the walls (with chemically cleaned slides), or parallel to one direction determined by the wall (rubbed slides).

This effect must also allow the problem of measuring the indices of these substances to be approached using the prism method. However, because of scattered light, the thickness cannot be large. The angle of the prism with liquid will necessarily be small, of the order of 10°, and the precision will not be much better than I obtained using Newton's rings, but the measurements will certainly be faster to perform.

Finally these samples should permit the study of scattering from incident polarised light.

In summary, this study specifies in a complete way the conditions necessary to obtain good samples of nematic liquids oriented by rubbing the support surfaces. Even if from a theoretical point of view the conclusions I have reached are not absolutely certain, as is almost always the case with phenomena occurring at the surfaces of solids, the straightforward practical consequences identified show that this phenomenon should allow important steps to be made towards a deeper understanding of the nematic phase.

References

1. P. Chatelain. *Bull. Soc. franç. Minéral.*, 1937, **60**, 300.
2. R. Merigoux. Thesis, Paris, 1938, 76.
3. Ch. Mauguin. *Bull. Soc. franç. Minéral.*, 1911, **34**, 3.
4. Ch. Mauguin. *C. R.*, 1913, **156**, 1246.
5. Kast. *Ztsch. F. Phys.* 1931, **71**, 39.
6. D. Dervchian. *Soc. de Chim. Phys.* Oral communication at meeting of 23 March 1938.

7. A. Marcelin. Surface solutions, two dimensional fluids and monomolecular layering. Conference reports.
8. Lord Rayleigh. Collected scientific papers, Vol. VII.
9. T.J. Baker. Phil. Mag. 1922, **44**, 752.
10. Bouasse. Vision and reproduction of shapes and colours, part III, chapter V.
11. F. Grandjean. *Bull. Soc. franç. Minéral.*, 1916, **39**, 206.
12. Foex. Trans. of the Faraday Soc. 1933, **29**, 958.

Pierre Chatelain was born in 1907 in the spa town of Bagnères de Bigorre in the Hautes Pyrénées province of Southern France, the son of teachers. He received his secondary education in Toulouse and his degree level education at the École Normale Supérieure in Paris, graduating in physical sciences in 1933. He was a graduate student in the Optics laboratory at the same institution, which at the time was directed by Charles Mauguin, obtaining his doctoral degree in 1937.

His subsequent career was spent at the University of Montpelier, where he eventually became professor of mineralogy and crystallography. Apart from the paper reproduced in this volume on the effect of rubbing on the orientation of liquid crystals at surfaces, his main research effort was devoted to X-ray diffraction of nematic and cholesteric liquid crystals. Chatelain is known not only for his work, but also for his fortitude in the face of adversity; he suffered from polio as a child and for the whole of his life walked with crutches. He retired in 1972 and died in 1982.

Section C

THE MODERN PHYSICAL PICTURE

THE MODERN PHYSICAL PICTURE

Introduction

Paper C1 in our collection of modern papers is by François Grandjean. This paper is not so modern, dating as it does back to 1917. It outlines the first successful attempt at what we now call a molecular field theory for nematic liquid crystals. Grandjean is quite explicit in attempting to follow the ideas of Langevin and Weiss, who had created a molecular field theory of magnetism. In Section A we have seen how Max Born[1] attempted to do the same thing, and developed a theory of a ferroelectric liquid, which was rejected on experimental grounds. In essence, Grandjean developed the theory we use today, albeit using a slightly different notation. However, sadly, in fact this paper played no role in the subsequent development of the subject, because it remained unread and almost uncited. We postpone discussion of the reason for this, so that we can examine paper C1 together with paper C3, to which it bears an uncanny resemblance. At that stage we shall also attempt to compare and contrast their relative fates.

Apart from Grandjean's Forgotten Paper, Tsvetkov's paper C2 represents the first step towards a successful molecular theory of nematic behaviour. The primary reason for including this paper in our collection is that this paper exhibits, for the very first time in the liquid crystal literature,[2] the nematic order parameter

$\bar{P}_2 = \left\langle \frac{1}{2}(3\cos^2\vartheta - 1) \right\rangle$ (denoted by Tsvetkov as S), where ϑ is the angle between

the molecule and the optical axis, and the brackets represent averaging over all molecules. At this stage, he is not yet using the term *order parameter*. If he has to use a word at all, he uses *Ordnungsgrad* (degree of order), but mostly, he uses simply '*S*-value'. But over and above that, Tsvetkov shows us a fine example of physical reasoning at its very best. Just like both Grandjean (C1) and Maier and Saupe (C3), he was seeking to describe the theory of a cooperative phenomenon. Indeed, he says so explicitly in his introduction, citing a number of workers in the then relatively undeveloped field of statistical mechanics. But finding himself unable to provide an *a priori* argument, rather than give up, he combines experiment and general physical theory in order to extract the maximal amount of information.

When this paper was written, Tsvetkov was a mere 31 years of age, but already had a long and impressive scientific record, having inherited Frederiks's group in

Leningrad after Frederiks's arrest and deportation in 1936. The paper was submitted in April 1941, just two months before the German invasion of the Soviet Union, and was published in 1942, at the height of the German siege of Leningrad. Nonetheless the paper is written in German, the premier scientific language of the age, so as to be accessible to the international scientific community. There is perhaps an irony that in order to enable it to be accessible to the present-day community, we have had to provide it in an English-language version.

His argument goes through a number of steps. He is able to use some experimental results on *solid para*-azoxyanisole[3] to determine the molecular electric and magnetic susceptibilities of a *para*-azoxyanisole molecule, using relatively standard arguments from the electrodynamics of continuous media. These in turn can be used to relate the values of the extraordinary and ordinary refractive indices to his order parameter. As he remarks, S is related to *macroscopic* properties of the anisotropic fluid. But now he makes an imaginative leap. He is also able to measure the (not yet Frank!) elastic constant as a function of temperature, and he notices that it is proportional to S^2. And furthermore he himself had already made a calculation, some three years previously, based on equipartition of energy, showing that the *low temperature* angular fluctuation is inversely proportional to the elastic constant, but dependent also on a cut-off wave number of the order of the inverse intermolecular distance.

So now we have the framework for a self-consistent argument. The order parameter depends on the angular fluctuation. The fluctuation depends on the temperature and the elastic constant. And the elastic constant depends on the order parameter. The result is a curve of $S(T)$, with the temperature scale determined by the molecular size. There is a maximum temperature T_k for which the nematic phase is possible, with $S(T_k)$ taking the universal value of 0.67. Tsvetkov is well aware that this is an incomplete theory, and that the true theory would predict two free energy branches with the phase transition occurring when the free energies are equal. For this reason he is not too disappointed that the experimentally measured S at the phase transition is not 0.67, but closer to 0.4. But still, the very fact of a discontinuous jump is encouraging, and in the absence of a full free energy calculation, he identifies T_k as the phase transition temperature. Of course, T_k is not itself predicted without knowing the intermolecular distance, but its experimental value (in this case 408 °K) is sufficient to specify this one remaining parameter.

At the time the real physical content was the $S(T)$ curve with its van der Waals loop, and its unstable retrogressive branch. Later studies, as we shall see, were able to provide a much more convincing curve, albeit with the same qualitative structure, with the added benefit of a lower limit of stability to the isotropic phase. We can admire the ingenious argument, which gives good qualitative and indeed semi-quantitative agreement with experiment, despite lacking the important fundamental self-consistency of a Curie-like theory. What remains is the importance of the order parameter S, rather than something to do with $\langle \cos\vartheta \rangle$, as occurs in the theories of Born.[1]

336

In the immediate aftermath of the war, it took some time before the pre-war interest in liquid crystals was to be rekindled. By the late 1950s a renaissance of interest in mesomorphic phenomena began to flower. Two indicators of the quickening of the pace of liquid crystal research were the appearance in 1957 of a major review article[4] by Glenn Brown and his student Wilfrid Shaw in *Chemical Reviews*, and the subsequent appearance, in 1962, of a liquid crystal monograph,[5] entitled 'Molecular Structure and the Properties of Liquid Crystals', authored by George Gray from the University of Hull in England.

We shall return to Gray's work, which had begun in the late 1940s when few researchers felt that liquid crystals had any further surprises in store, later in this book. It was to be extremely influential, not only for the development of the subject as a whole, but also, as we shall see in Section D, to the commercialisation of the science of liquid crystals which was to take place in the 1970s. But in 1962 this was all in the future.

Let us now return to Brown and Shaw's review. Coming, as it did, after a fallow period in liquid crystal research, it contains an extraordinarily rich bibliography, going back right to the early days of liquid crystals. Also included is a comprehensive survey of the chemistry and physics, not only of liquid crystals as we currently understand them, but also of the rich menagerie of lyotropic phases. As is appropriate for an article in *Chemical Reviews*, Brown and Shaw were chemists, from the University of Cincinnati in Ohio, United States. It is perhaps surprising that only some 20 pages of the article is devoted to the detailed molecular structures of known liquid crystals. The article still contains elements of scientific interest, but some parts are now of more obvious historical interest, most notably the theoretical section entitled 'Hypotheses as to the Nature of the Nematic Structure'. We quote some brief extracts which show the rather surprising consensus of the time.

A theoretical interpretation of the mesomorphic state has intrigued many physicists and chemists since its discovery in 1888. There can be found in the literature exchanges of arguments for and against two hypotheses that evolved out of the data that were interpreted. Even today there is no one theoretical interpretation of the mesomorphic structure that completely explains all of the experimental data. However, of the two hypotheses that have been proposed, namely the swarm hypothesis and the distortion hypothesis, the swarm hypothesis is the more widely accepted.

First, brief mention will be made of the distortion hypothesis as it was proposed by Zocher ... the hypothesis has its limitations when one attempts to interpret the properties of light extinction and wall effects. The reader is referred to (an article by) Zocher in which he argues that the two hypotheses do not have the same physical significance.

337

The London Faraday symposium, in which Zocher had argued so forcefully against the swarm hypothesis, had been 24 years earlier than this. The Leningrad school, as we have seen from Tsvetkov's 1942 paper, had long since abandoned the swarms. It was perhaps a mark of the paucity of recent research in this area, and the lack of real debate, that Brown and Shaw were able to write off the distortion point of view in just one paragraph. But added to this is the fact that Oseen's and Zocher's work was essentially macroscopic. Their point of view was directed to an audience of physicists, ready to think about problems at the relevant necessary length scale. But it was quite unappealing to a chemist, who seeks explanation in terms of molecular structure.

But what made the dumbbell-shaped molecules line up? Daniel Vörlander had provided an appealing picture, but no theoretical physicist, with the possible exception of Grandjean whose work had been forgotten, had been able to begin to transform Vorländer's intuitions into tractable statistical mechanical calculations. And certainly neither Oseen nor Zocher had the remotest idea. All Brown and Shaw probably noticed was Zocher's increasingly irritable tone, and they used normal social cues to deny credibility to their author. So it was no wonder that they did not really take Zocher seriously.

Shaw's subsequent career was spent in the chemical industry. Brown remained an academic, later moving from Cincinnati to Kent State University, where he founded the famous KSU Liquid Crystal Institute. Kent is an unprepossessing middle American small town some 50 miles south of Cleveland. Apart from the Liquid Crystal Institute, it is otherwise unknown, apart from a brief interlude of notoriety in the late 1960s, when State Troopers managed to shoot dead a number of demonstrating students on the campus. In addition to their scientific contributions, Brown and the Liquid Crystal Institute were to play an important organisational role in the transformation of liquid crystal science from a niche interest to a major contributor to modern science and technology.

Article C3 by Maier and Saupe begins to remedy the deficiencies in the pictures of Oseen and Zocher. By contrast with Oseen and Zocher, Maier and Saupe concentrate on the molecular forces, and the degree of order, rather than its orientation and variability. Here finally the community was introduced to a viable molecular field theory of the nematic state. The article in our collection is just the summary, and the ideas were subsequently developed in two rather longer papers in 1959 and 1960.[6]

The molecular field theory requires a number of crucial inputs. The first is a intermolecular interaction. The second is an order parameter. The third is an averaging procedure, enabling a feedback process enabling spontaneous order to occur at sufficiently low temperature. The order parameter Maier and Saupe have acquired – directly, for they cite him – from Tsvetkov. The intermolecular potential energy is a problem. In the summary paper and the first longer paper, they concentrate on the dispersion forces, that is, the induced dipole–dipole interaction between the molecules. To obtain this they go through a rather long and involved quantum mechanical perturbation calculation. But they believe it is important:

> In an earlier communication (*article C3; ed.*) we have pointed out some relationships between the molecular structure of liquid-crystalline substances and the clearing point of their nematic phase. In our opinion, these relationships indicate that the dispersion forces between the optically strongly anisotropic molecules are the determining factor for the existence of liquid crystalline states of order. We have therefore reconsidered the attempt made by Born....

By contrast the importance of the linearity of the molecule, first noticed as we have seen by Vorländer in 1907 and much repeated subsequently, is minimised:

MODEL AND ASSUMPTIONS

3...The effect of the repulsion forces ("the shape of the molecule") is neglected for the time being.

Having calculated the effective interaction, Maier and Saupe are then able to parameterise its effect on a single molecule using a single parameter with the unit of energy. And now, they use the classical mean field procedure to determine conditions on that parameter to yield a critical temperature, in units of that energy, at which the order parameter S can maintain a spontaneous non-zero value. Furthermore, it turns out that the phase transition is first-order (like the experimental nematic-isotropic phase transition), and that there is a clever graphical construction to determine the self-consistent value of S to be determined. The theory is mathematically elegant, contains a first order isotropic-nematic phase transition, with an order parameter tending to unity in the limit of low temperature. So there is strong presumptive evidence of its applicability. Molecular shape is unimportant, and dispersion forces govern the nematic phase. But there is the small matter, which they point out themselves in paper II, that:

> ...a nematic phase could in principle arise even in the case of spherical molecules, so long as their optical anisotropy were sufficiently strong. This statement appears to be a complete contradiction of the very well-known fact that all substances with liquid crystalline phases are characterised by a strictly elongated shape of their molecules. But in fact there is no discrepancy because the required optical anisotropy can only be found in elongated molecules....

So now Maier and Saupe reinterpret their crucial parameter in terms of an effective steric interaction; their agreement with experiment is maintained but they have shifted the ground on which they base their theory. Paper II also includes a discussion of the density change at the transition, which requires a little nimble theoretical footwork (since the original theory is essentially a constant volume theory), but seems to give reasonable results.

We now know that this reinterpreted theory is actually rather good. More modern 'density-functional' theories of the transition are only reiterating Maier–Saupe in a rather more formal way. The temperature dependence of the order parameter is in remarkably good agreement with experiment, but is subject to the one fitting parameter. If they had included such an analysis in one of their papers they would have been all the more convinced that the theory was correct, even if the foundation of the theory remained slightly shaky. The description of the phase transition region turns out to be not so good. This is because there are strong order parameter fluctuations in the neighbourhood of the transition, but an analysis of this process was well beyond the statistical mechanics of the period.

Now let us return to the Forgotten Theory of Grandjean (C1).[7] The physics is identical to that of Maier and Saupe. Even Maier and Saupe's graphical construction for the evaluation of the order parameter is present. Indeed in some sense Grandjean's treatment is more modern. He does not worry about the details of the intermolecular interaction. This is just the approach of a present-day theoretical physicist (who might recruit a chemist colleague to help with molecular detail!) Grandjean is only interested in the symmetry, and that suffices, given just one parameter. His certainly is the long-sought-after Langevin-like theory which Ornstein was looking for, and which Zocher mocked, in the Ewald symposium discussion in 1930–1. But notwithstanding the resemblance between the two approaches, there is no question of plagiarism. Maier and Saupe would not have struggled manfully towards a resolution of their problems had they known the answer already. Their papers would have been phrased entirely differently. Grandjean's work really was forgotten. The question is why?

The immediate reason is that the paper was published during the Great War in France, while much of the relevant work was going on in Germany. So the obvious people (Born, Zocher, etc.) would not have read it. By the time they were in a position to read it, time had passed by, and there was no database, as there is today, to draw it to their attention. Secondly, Georges Friedel's review paper in 1922 (B1) has been so influential that it has tended to relegate prior scientific work to the status of history. Now we know that Georges Friedel held Grandjean in very high regard (some would say, only Grandjean!). Friedel would not have downgraded Grandjean's work gratuitously, and Friedel must have been aware of this paper. But Friedel's long 1922 paper does not cite it. As we have seen, it does not cite very much! Even those papers it does cite, it does so inadequately. A modern referee would send it back and require a better literature search. But that was then, and now is now. There was no such referee. Copious literature tells us Georges Friedel was at the very least impatient with the theoretical arguments between swarmists and distortionists. His plea was for more experimental work. The theory could sort itself out later. So his omission of the Grandjean theory from the 1922 paper was no accident. He felt that the time was not right.

As we have seen in article B5, the major liquid crystal theorist of the 1920s was undoubtedly Oseen. He summarised his work in his 1929 review article,[8] which lays great store in the Born[1] theory, as well as in its experimental refutation,[9] chiefly because of its contrast with his own theory, which he sees as being more successful. The Oseen theory, as we have seen, is a macroscopic theory describing spatial structure, for example within a layer, rather than a thermodynamic theory predicting equations of state. Grandjean's work is summarily dismissed in the same footnote which cites Born's paper:

> A similar theory was developed at the beginning of the following year by Grandjean....

Had Oseen not really read Grandjean's paper? Or simply not realised (because he was concentrating on other things) that changing from $\langle\cos\vartheta\rangle$ to $\langle\cos2\vartheta\rangle$ is of fundamental significance? Is the fact that the reference only occurs in the footnote a hasty response to a referee (perhaps Georges Friedel?) rather than a genuine reaction to Grandjean's work? It is difficult to know. Whatever the case, the fate of Grandjean's work was determined by this "double whammy", ignored as it was by the principal experimental review, and dismissed by the principal theoretical review.

There are probably other factors too. Grandjean's paper was written before the era of the order parameter. His use of $\langle\cos2\vartheta\rangle$ rather than Tsvetkov's

$$\left\langle\frac{1}{2}(3\cos^2\vartheta - 1)\right\rangle$$ may also have played a role. They do indeed have same physical

content, but somehow not the same emotional strength; the latter is zero in the isotropic phase, whereas the former is not. Thus the scientific community was in some sense not *ready* to receive a theory of this nature; nowadays such theories are two-a-penny, and would be regarded as the standard approach to a given new problem. So the casual reader would not happen upon Grandjean's theory and immediately seize upon it as *the* answer to long-posed problems. Finally, no doubt, there is Grandjean himself. By the time the question might have become important, Grandjean had abandoned this field, and had become a biologist. His studies of mites and ticks were to make him famous and probably consumed all his passions. If the liquid crystal people were not reading his paper, Grandjean was unconcerned. More fools them!

We pass now from studies on a molecular scale back to the continuum régime. Sir Charles Frank (he was not yet Sir Charles at that time, of course) was originally a chemist, like Brown and Shaw, but unlike them he spent most of his career as a theoretical physicist. His article C4 is often cited as the origin of the so-called 'Frank' elastic constants. It is a misleading citation, as we have already seen from Tsvetkov, for reasons that we shall shortly relate and that should anyway be familiar to those readers who have followed our argument carefully. This comment is not meant in any way to demean Sir Charles's high standing as a scientist, and Sir Charles himself was always quick to deny any elevated status in the development of the theory of liquid crystals.

The article we include in this collection appears in the *Discussions of the Faraday Society*. Many readers will be familiar with these Discussions. We have met already the London 1933 Faraday Discussion on *Liquid Crystals and anisotropic Melts*. This was organised in major part by J.D. Bernal, and the 1958 meeting was meant quite openly as a reprise of the successful 1933 meeting. Bernal is again the major figure in this Discussion, which rather than in London, took place at the redbrick and thus perhaps more plebeian University of Leeds. We have included the title pages to give the reader some flavour of the breadth of the material under discussion, as well as some indication of the personnel involved. The meeting was entitled *Configurations and Interactions of Macromolecules and Liquid Crystals*, but only the first few papers were devoted to liquid crystals as we now understand the term. The main focus was rather on the newly emergent field of polymers; the macromolecular element was introduced because of the growing realisation of their importance and rôle in a host of physical, chemical, and indeed biological processes.

The preamble to the written Discussion records the attendance of 38 foreign visitors, of whom we might note in particular Prof. Dr W. Maier of Freiburg im Breisgau, whose article C3 we have just met. Another contributor to the liquid crystal discussion was the young Dr G.W. Gray of the University of Hull, who communicated his discovery of what has come to be called the odd–even effect, and pointed out that some apparent hysteresis at the N^*-I phase boundary was a misinterpretation of complex data involving surface effects. Professor Maier contrasted his molecular point of view to Frank's continuum approach. His goal was to publicise his molecular field theory, which at that stage he believed (as we have seen, wrongly) supported the crucial role of dispersion forces in mesophase behaviour. As the Maier–Saupe papers had thus far only appeared in German, a personal appearance by Maier at a major international meeting would justifiably be thought to be important for the acceptance (as we would now say) of the Maier–Saupe paradigm.

Other contributors to the macromolecular part of the proceedings whose names would later be well-known included the late Andrew Keller of Bristol University (known for his work on polymers), Aaron Klug (later Sir Aaron, Nobel prize winner and President of the Royal Society) of Birkbeck College (known for his X-ray work on biological macromolecules), and Michael E. Fisher of King's College (later of Cornell and Maryland) (known for his work on the theory of critical phenomena). A co-author of Klug's, indeed the first author of a paper on the structure of RNA, was (the late) Rosalind E. Franklin. Franklin, who had only recently died from cancer, was well-respected during her lifetime, and achieved massive posthumous fame as the 'Dark Lady' in J.D. Watson's popular account of his discovery of the structure of DNA.[10] The rather contemptuous and cavalier manner in which she was thought to have been treated by the DNA pioneers has built her into a feminist icon, and has led to two comprehensive biographies.[11]

Frank's paper is not so much an original piece of work, as a digest and revitalisation of calculations principally due to Oseen more than a quarter of a century earlier. He starts in characteristic mode:

> One of the principal purposes of this paper is to urge the revival of experimental interest in its subject. After the society's successful Discussion on liquid crystals in 1933....

The paper cites only five references. Reference four is by Frank himself (from 1951) on dislocations, in which subject he was at the time perhaps the leading theorist in the world. Reference two is the classic review by Georges Friedel in 1922 (our B1), and references one and three are to the articles by Oseen (B5) and Zocher (B7) in the 1933 Discussion.

His paper, using the notation that has now become standard and following an argument that is almost standard, introduces the 'Frank' elastic constants. He draws attention to the role of each constant in describing bend, splay and twist, and he is the first who explicitly uses this terminology. He discusses the K_{24} saddle-splay constant, pointing out its unimportance in bulk problems, because of the disappearance of the K_{24} term in the relevant Euler–Lagrange equation. He explains how the chiral term enters in the description of the cholesteric phase, and hints at the existence of the flexoelectric effect (a point taken up by R.B. Meyer[12] eleven years later). He discusses briefly how the theory might be extended to describe the smectic phase. Finally he goes on to discuss defect lines. It is an impressive paper, but even Sir Charles himself would probably have stated that it was a recap in more modern language of points made by Oseen and Zocher in the late 1920s and early 1930s. In fact careful readers will be able to compare the articles of Oseen (B5) and Frank (C4). They may find an uncomfortable overlap. Sir Charles would not have been embarrassed, given his essentially propagandist role.

But in this propagandist role he was remarkably successful. So successful that (like Georges Friedel before him) he partially excised that part of the history of the subject which preceded him. The elastic theory is due to Oseen, with perhaps a major assisting role played by Zocher. Zocher is important because he seems to have seen more clearly the distinction between macroscopic and molecular-level theories. We may recall him referring to 'the long sought after Langevin-like theory...' in his debate with Ornstein and the swarm theorists. Certainly all the major theorists in the late 1930s (there were few, but nevertheless, all of them) accepted the elastic theory. We have seen Tsvetkov in article C2 using the concept of an elastic constant – which he is able to evaluate quantitatively – completely unselfconsciously. So how did Oseen and Zocher's elastic constants come to be Frank's as well?

343

The most plausible scenario is that readers of Frank's review read the science carefully, but his historical acknowledgements less so, especially since they may not have had ready access to journals dating back to 1933. Pierre-Gilles de Gennes (see article C7) was careful to refer to the Frank–Oseen theory, but other workers, including his colleagues from Orsay, were less so. We find a reference from the Orsay Liquid Crystal group in *Physical Review Letters* from 1969[13] simply to the 'Frank' elastic moduli. The mathematical group of Ericksen and Leslie, whom we shall meet again in article C6, were usually scrupulous in assigning credit to Oseen. Occasionally in haste they omitted references to Oseen.[14] But those who read their articles were not always so fastidious. So if mathematicians and experimental physicists were both crediting Oseen's theory to Frank, it is hardly surprising if this view passed into the canon.

There is a (perhaps apocryphal) story concerning an occasion at which Sir Charles was due to receive an honour for his contributions to the theory of liquid crystals. The orator was oleaginous but not terribly well-informed. He referred to Sir Charles's many important papers in the field of liquid crystals. Sir Charles, who did not suffer fools gladly, smartly interrupted him. 'Only one, actually', he is reported to have said, in his typically no-nonsense, matter of fact, and rather *mezzoforte* style.

Which brings us to the other major contribution of article C4: the invention of the term *disinclination*. Frank draws attention to the analogy with dislocations in solids. Let us recall that a dis*location* is a line in a crystal on which the *location* of the lattice points is not well defined. If one takes a path around it, one seems to go sensibly from lattice point to lattice point, but by the time one arrives back at the starting point one has gone through the wrong number of steps. The degree of mismatch (which has both a number and a direction) is known in material science as the *Burgers vector* (another Frank invention).

Now other workers, most notably Oseen and Georges Friedel, had of course noticed that physics in the neighbourhood of the anomalous lines of singularity would be different. But it is Frank who recognises the essential topological role of defect or singularity lines. Indeed his section on this subject begins with the pithy statement:

> The nematic state is named for the apparent threads seen within the fluid under the microscope.... They are line singularities such that the cardinal direction of the preferred axis changes by a multiple of π on a circuit taken round one of the lines.

In modern language, Frank's essential point is that these observations, and the characteristic patterns seen using crossed polarisers, are the signature of *topological defects* in the fluid. Furthermore, these topological defects are themselves the signature of the type of order parameter in the system. This governs his choice of language. Along the circuit the director changes sensibly locally, but by the time one has returned to the initial point, it seems to have

344

rotated through a multiple of π. This rotation is not locally noticeable, but globally enforces a line along which the *inclination* of the liquid crystal director is not well defined. Hence dis*inclination*.

The subsequent story becomes, as might be expected for a nematic liquid crystal, somewhat murky. The influence of Frank's paper took some time to have an effect, as can be seen to by examining its citation count. It is not the most cited of Frank's papers, for his influence on the science of defects in solids was by far greater than his influence on liquid crystals. But for all that, it is all the same a much-cited paper. Over the period 1980–97, the paper was cited 497 times, and the citations continue unceasingly. In the period 1975–9, it was cited 160 times, and a further 241 times in the period 1970–4. As befits a classic paper, citations have increased with time, and indeed we find only 38 citations in the period 1965–9, and *none at all* in 1963 or 1964.

So the paper was only beginning to worm its way into the consciousness of liquid crystal scientists in the late 1960s, when interest in the subject began to explode. In 1968, de Gennes and Jacques Friedel (yes! *another* Friedel, this time the son of Edmond and grandson of Georges) were thinking about defect lines which closed in on themselves in loops.[15] The paper is in French and uses an inexact French analogy: *disclination*. But the terminology of the Orsay group seems to have been somewhat plastic. In different papers around this time workers associated with this group use disinclination (in English),[16] *disinclinaison* (in French)[17] or *désinclinaison* (in the very neighbouring paper).[18] However, for the English translation of both abstracts the sub-editor at *Les Editions de Physique* by now preferred 'disclination'. Someone, somewhere, was disinclined to be disinclined.[19] *Disclination* it became and disclination it has remained.

From the theory of Frank we pass back to experimental observation. In article C5[20] we find unveiled for the first time the now familiar classification of the smectic phases into smectics A,B,C....[21] In 1959 we find merely smectics A,B and C, and even then the story only begins slowly; the others follow only considerably later. Let us recall the historical background from our discussion in earlier sections. Daniel Vorländer was convinced at an early stage that he had observed several modifications of what he could not bring himself to call the smectic phase. Georges Friedel, in his elegant 1922 classification (article B1), was likewise convinced that there was but one smectic phase. The difference of opinion was vociferous and sometimes heated. As we have seen ourselves, in the discussion following the Ewald virtual symposium (article B4), The Friedels, father and son, did leave the intellectual door slightly ajar. *They* had not observed different modifications; *they* were persuaded on balance that all reports of different modifications were simply error by inexperienced microscopists; samples of the offending materials had mysteriously not been communicated to *them*. But still, in the end, this was a matter for experiment and not for invective.

However, in fact, it was not only a matter for experiment, at least not if the experiment is confined to the microscope stage and cover-slide. We will also recall the debate about the distinction between *stase*, and *phase*, and *modification*, and *texture*. How are we to recognise that something is indeed *significantly* different? In order to be sure about these differences, it would be necessary simultaneously to include in the experimentalist's toolbox, at the very least, microscopy, X-ray crystallography and calorimetry, together with a good understanding from other were sources of the relevant molecular structure. By the time of the 1933 Faraday meeting, Karl Herrmann, of Berlin-Charlottenburg, was convinced that there were indeed significant differences. In an article entitled 'Inclinations of molecules in some crystalline fluid substances', he notes that[22]

> ... The distance between the planes, calculated by means of Bragg's formula, agrees in order of magnitude with the lengths which are necessitated by the molecules of the substances in question.
>
> The question arises whether it is *essential* (editor's emphasis) for G. Friedel's idea that the molecules should be placed *perpendicularly* (author's emphasis) to these smectic planes. It is stated by de Broglie and Friedel (article B2) that "... . the distance between homologous points of two successive layers *i.e.* the period, must then be at least equal to the length of these chains, and considerable with respect to the atomic distances". ... the present author and his collaborators have for a long time held the same point of view.

Now, however, he wants to recant. Herrmann was investigating the smectic phase of thallium oleate and thallium stearate. From X-ray investigations of the solid crystal structure he is able to calculate the chain lengths of the relevant fatty acids. And from X-ray studies of the smectic phases he was able to determine the distance between the layers. First he notices that actually this distance is *greater* than the molecular chain length. But this cannot be! So maybe a *double* molecular layer is more plausible? But now the interlayer distance is too short! Herrmann's solution, albeit reached indirectly, is that the chains must be tilted, and he calculates the tilt to be about 47° with respect to the normal to smectic planes.

What Georges Friedel would have made of this we do not know. The Faraday meeting was held in April 1933, and Friedel was already chronically ill. We do not know whether he was invited to the Faraday symposium and turned down the invitation, or whether Bernal, uncharacteristically allowing discretion to be the better part of valour, preferred on balance to invite only Vorländer in order to avoid an early start to the Second World War. Whatever the situation, on 10 December Georges Friedel died, and Edmond Friedel had other intellectual fish to fry. It seems possible that they might have argued that, put in modern language, the layers were *interdigitated*, that is that the ends of the molecules in adjacent layers overlapped. This argument was certainly used by some later authors, and of course to some extent it is valid. Or perhaps Georges Friedel would have beaten an honourable retreat and admitted that here was indeed evidence of more than one smectic phase.

346

Be that as it may, evidence slowly accumulated of further smectic modifications. First Vorländer,[23] in one of his last papers, and then his former student Conrad Weygand,[24] observed more than one smectic phase. Weygand, stubbornly resisting the Friedel nomenclature and holding to Vorländer's, discovered a substance with what he called one *Pl*- and three *Bz*-phases. The three *Bz*-phases were labelled *Bz-I*, *Bz-II*, and *Bz-III*. In the 1950s George Gray[25] and colleagues in Hull made long and detailed studies of the connection between phase behaviour and thermodynamic properties on the one hand, and molecular structure on the other. By examining a homologous series, they found globally three smectic phases, with individual compounds exhibiting two of these as well as a nematic phase. They too had a labelling scheme: *Sm-I, Sm-II*, etc. Recording all this work, Brown and Shaw's review article reports several instances of 'smectic polymorphism'. The scene was set, therefore, for a definitive resolution of this problem.

Horst Sackmann's approach (Heinrich Arnold is his postgraduate student) was somewhat different from that of Herrmann. They were in Halle, and luckily had access to Vorländer's incomparable collection of mesogenic compounds. Returning to the tools of the specifically *physical* chemist, they made a classic microscopical study of smectic mixtures. The method is simply of categorisation; they wanted to distinguish between different types of liquid crystal simply on the basis of what they look like. Recognising the essence of the smectic textures from the Friedel classification, they observed optically that there were in fact different smectic textures. Classifying them, they found, not two, but three, distinct types. One, the usual texture, which they labelled as smectic A, is the familiar smectic. There were two others, which they labelled as smectic B and C. Modifications B and C seemed to contain some extra features. The paper reports that they both contain some brush-like light and dark features, which reminded them of the Schlieren texture so characteristic of nematics. And in the phase they have labelled as the smectic C, the fan-shaped textures and polygonal textures which occur in the other smectic modifications appear to be 'broken'.

The important point was they looked at *mixtures*. They changed the mixture concentrations. There is never a smooth passage between what they now identified as essentially different smectic classes. Whenever there was a shift from one class to another, there was a miscibility gap. This is strong evidence that the three classes have a different structure at the molecular level. The smectic A is clearly the original Friedel smectic. But what of smectics B and C? For various reasons, they understood that smectics B possessed a local lattice-like order within the layers. But as far as smectic C is concerned, they were mystified. In their own words:

> Phases of type C should possess a common third type of structure,
> of which at present there is no indication of its nature.

But no mention, not even a hint, of a tilted phase.

Paper C5 is in German. In the early 1960s the first signs of what we now call 'globalisation' were emerging. One of these signs is the emergence of English as a common scientific standard. Outside the German-speaking community the article elicited little response. As we see in our collection, the late 1950s was a fertile time for fruitful developments in the liquid crystal field, but time was necessary for them to recombine to make a more coherent whole. In 1965, Glenn Brown organised the first international liquid crystal conference at Kent State University in Ohio. Among those he invited was Horst Sackmann from the German Democratic Republic – far away communist East Germany. Unlike many from the eastern block Sackmann was able to travel to this international conference.[26] And in Kent, in collaboration with his student Dietrich Demus, Sackmann presented an English version of his ground-breaking work.[27] Here there is no mention of the possible crystalline structure of smectic B, still less of tilt in smectic C. All that Sackmann can bring himself to say is:

> Because of the general connection between double refraction and structure of matter our present texture system points to a structural connection between the phases with the same characteristics of texture, especially because there is a coincidence of texture characteristics of the liquid crystal phases with their miscibility characteristics. But an explanation of the given texture characteristics on structural backgrounds is needed urgently. Today no explanation between texture and structure of phases is known beyond that of Friedel.

And following this article, Sackmann's classification finally did excite a response, most specifically from Alfred Saupe, by now relocated at the Mecca of liquid crystal science at Kent State University. In 1968, Brown organised a second international conference, and invited Saupe to give a plenary address. In an overarching review, entitled 'On molecular structure and physical properties of thermotropic liquid crystals', Saupe discusses textures in nematic, cholesteric, and what he now specifically calls smectic A liquid crystals, before passing onto what he calls 'tilted smectic liquids'. He draws attention to Herrmann's work on thallium stearate and thallium oleate, and makes the comment that

> The tilted smectic liquid has some interesting similarities with a nematic liquid which are not obvious at first sight.

Saupe points out (in more modern language) that the tilt would involve a broken symmetry, and that the tilted smectic liquid would therefore be able to sustain twist, just like a nematic. Not only that, but there may be defects in the director projection onto the smectic planes (he does not say this, but leaves it understood by the reader) and these defect lines will leave their signature in a Schlieren texture. Making reference not only to the paper by Sackmann and Demus in the 1965 international conference, but also to observations he has made in collaboration with his KSU colleagues Arora and James L. Fergason (whom we will come across again in a more technological context in Section D), he continues:

Indeed such Schlieren textures can be observed with some smectic liquids.

But *still* no identification of the tilted phase with the smectic C, although there is hint enough here that privately Saupe had convinced himself of their identity. Interestingly enough at the same conference there were papers by I.G. Chistyakov and colleagues from Ivanovo in the Soviet Union, investigating polymorphism in smectic liquid crystals.[28, 29] Chistyakov, who uses both X-ray diffraction and differential calorimetry, does identify the smectic C phase with a tilted smectic phase. However, his tilted phase is herringbone-like, with the direction of tilt alternating in opposite directions in neighbouring rows, which is rather different to the structure imagined by Herrmann, although given Herrmann's indirect method of analysis, presumably not in principle in contradiction with it.

By now, of course, Saupe and colleagues had the bit between their teeth, and in the period between the 1968 Kent State conference and the third international conference held in Berlin in August 1970, there was relatively rapid progress. In a paper published with Arora and Fergason in 1970, but presumably submitted in mid-1969,[30] he reports:

An additional lower temperature liquid crystalline phase is observed with chloro-substituted compounds for *n*-alkoxy chain lengths of C_8 and longer and with the methyl substituted compounds for chain lengths of C_{10} and longer. This lower temperature phase can show a threaded texture very similar to the nematic phase. Furthermore, both phases can be uniformly oriented by surface action and can assume a twisted structure. We suppose that the lower temperature phase has a layered structure in which the molecules are inclined to the layers, and that it corresponds to a smectic liquid crystal classified by Sackmann and Demus as smectic C.

The paper is an interesting example of the way in which various different strands of indirect evidence, none of which is by itself conclusive, are built up to provide an eventually overwhelmingly persuasive picture. It starts off modestly, simply talking about 'nematic II', as opposed to the traditional 'nematic I', but notes that nematic II occurs at lower temperatures than nematic I.

349

The anomalous phase can twist, having been oriented by the surface, and it exhibits the Schlieren texture, and both of these properties should be characteristic only of the nematic phase. But the phase resembles a smectic in that it exhibits a focal conic texture. But the twisting has also been observed by some French authors in an unpublished lecture at the 1968 Kent State meeting, so this observation is not some fluke or error. The phase transition temperature between the two 'nematic' phases changes with the molecular chain length in a manner typical of nematic–smectic phase transitions, and this also is suggestive of a smectic phase. Some unpublished X-ray pictures by their Kent State colleagues A. de Vries and G.H. Brown (of which more later) are also consistent with a smectic identification. And then in some similar systems, in which X-ray studies have been carried out, the layer length is some 7 per cent less than molecular length, which could be due to interpenetration of the layers, but could also be due to tilt....

The delay between submission and publication unfortunately detracts from the drama of the story. Already in the summer of 1969, at the 8th International Congress of Crystallography in New York State, de Vries had presented X-ray studies on the BACP liquid crystal systems. In the published long abstract of this paper[31] (see Fig. C1), we find that the inner ring associated with the nematic has broken up into four spots, and this, according to Vries, is consistent with a tilt of 45° and the smectic C phase. At roughly the same time two further more careful optical studies from the Kent state group[32,33] had finally confirmed the identification. These papers were justifiably published in *Physical Review Letters*, the most highly prestigious journal for new results in physics.

The first of these, entitled 'Biaxial liquid crystals', was submitted on 1 December 1969 and already published on 23 February 1970. It focuses on the fact that, for symmetry reasons, a tilted liquid crystal would not only be birefringent, but in fact have different refractive indices in three principal directions. Making conoscopic observations (i.e. using convergent light) on the liquid crystal DOBCP (bis (4-*n*-decyloxybenzal)-2-chloro-1-4-phenylenediamine), they find that the black cross which is characteristic of the nematic phase splits into two crescent-like objects, which is consistent with a tilt angle of about 45°.

The second paper, submitted in May and published in September 1970, already takes the identification of the smectic C and the tilted smectic phases as read. They find that the liquid crystal terephthal-bis (4-*n*-butylaniline) – TBBA for short! – has calorimetric signatures of a whole range of phase transitions. With decreasing temperature they are able to identify the isotropic and the nematic phase, then successively smectic A, C and B phases, and finally a crystalline phase. Not only that, but optically they find that it is consistent to suppose that the tilt angle in the smectic C phases starts from zero, and increases rapidly with decreasing temperature, consistent with the signature of a continuous phase transition, although the data do not absolutely rule out a first order transition.

Taken together, this set of papers finally laid the ghost of Georges Friedel to rest. His scepticism of the existence of more than one smectic modification was

XIII-28. X-RAY PHOTOGRAPHIC STUDIES OF LIQUID CRYSTALS.
By *Adriaan de Vries*, Liquid Crystal Institute, Kent State University, Kent,
Ohio 44240, USA.

X-ray photographs have been made from samples contained in thin walled glas capilleries of 0.5 mm diameter. The samples have been studied in their solid, mesomorphic and liquid phases. Most materials showed in the liquid phase two diffuse rings, one corresponding to interatomic distances of about 5 Å (the interatomic distances between atoms of neighbouring molecules), the other corresponding to distances equal to the lengths of the molecules. This last ring (the inner ring) can be explained as the diffraction effect from the gaps in the electron density found at the ends of the rod-like molecules. Many nematic phases show this same pattern, only the rings often are split up into two crescents because of preferred orientation. For most bis-(4'-*n*-alkoxybenzal)-2-chloro-1,4-phenylenediamenes (BACP's), however, and for 4,4'-di-*n*-heptyloxy-azoxybenzene (DHA), the inner ring is much more intense in the nematic phase than in the liquid, and sometimes it is split up into four spots (Fig. 1a). This indicates that the molecules in these nematic phases are not just parallel, as usually is assumed, but that a substantial part of the molecules is arranged in groups (cybotactic groups); in each group the centers of the molecules lie in a fairly well defined plane which makes an angle of about 45° with the direction of the long axes of the molecules. For those BACP's for which

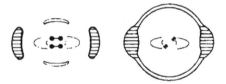

Fig. 1a Cybotactic nematic. Fig. 1b Tilted smectic.

the nematic phase is stable over a sufficiently long temperature range, the photographs show that the number of molecules arranged in cybotactic groups, and the rigidity of these groups, decreases as the temperature increases. The highest degree of organization was shown by the BACP with 8 carbon atoms in each alkoxy chain, at temperatures not more than 20° above the crystal-nematic point (for transition temperatures of the BACP's see Arora, Fergason and Saupe, Mol. Cryst., 1969, to be published).

The BACP's with 9-18 C-atoms in the alkoxy chains, and DHA, have a smectic as well as a nematic phase. The smectic phases could be identified from the photographs (Fig. 1b) as tilted smectic phases: the molecules do not stand perpendicular to the smectic planes, but at an angle (about 45°). The tilted smectic phase of DHA has been classified by Sackmann and Demus (Mol. Cryst., 2, 81, 1968) as smectic C.

Fig. C1 Abstract of the crucial X-ray crystallographic study establishing that molecules in a smectic C phase are tilted with respect to the layer normal; reset from the original.

misplaced. His opponents (as Lehmann would so lovingly have put it) were not fooling themselves; the different smectic modifications they had seen were not simply flights of imagination brought on by wishful thinking, as Friedel had thought. Just as there are many crystalline phases, so there are also many meso-morphic phases. Nevertheless, to establish this fact required some considerable time, and several different strands of evidence. There was X-ray evidence, combined with calorimetry, combined with a better understanding and more experience of what was meant by a phase transition. But the primary signature is that the textures themselves were subtly differentiated. Friedel, with all his observational care, either had not noticed, or more likely had not been exposed to the errant smectics. It is this insight, due to Sackmann, which justifies the choice in our collection of his pioneering article, rather than one of the several other articles, ground-breaking though they were, which together made up the unfolding story of the discovery of the smectic C phase.

In article C6, by F.M. Leslie, we see the first full expression of the contemporary continuum theory of nematic liquid crystals. The theory has come to be known as the Ericksen–Leslie theory, despite the fact that Ericksen's name does not appear as a co-author in this ground-breaking paper.[34] Although we include here only Leslie's authoritative version of the theory, the intellectual strands leading to this paper come from a number of sources. The primary source for a continuum theory is, as we have seen in paper B5, the extensive body of work developed by Oseen. In paper B5, Oseen presented some preparatory work by his graduate student Anzelius on the dynamics of nematic liquids. Here he refers to the 'tensor of viscosity' entering the dissipation function. Anzelius never published his thesis work; one of his examiners drily commented that 'the dissertation should have technical importance for lubrication problems for semi-dry fabrication'.[35] There was no further work on the dynamics of liquid crystals *per se* until J.L. Ericksen (1924–) returned to the problem in the late 1950s at Johns Hopkins University in Baltimore, Maryland, USA.

In fact Ericksen, who was approaching the problem from the point of view of theoretical mechanics, interested himself in the foundations of a whole set of theories of anisotropic fluids. F.C. Frank had confined his interest to static problems, and as such he was able merely to consider a free energy minimisation problem. But for Ericksen, the static problem was simply a special case of the full *dynamical* problem, and for this a free energy functional was by necessity insufficient. In his own words:[36]

> In studies of the static behaviour of liquid crystals, it is common to ignore or to treat in an obscure way the stress tensor and other representations of force. Our treatment emphasises these. From the viewpoint of general continuum mechanics, the theory seems interesting. It is one of the few mechanical theories which involve an asymmetric stress tensor. The fact that it is consistent with known principles of mechanics puts it in a still more exclusive class. As is discussed by TRUESDELL & TOUPIN,

general principles of mechanics do not require the stress tensor to be symmetric. Asymmetry is associated with the presence of what they call (body) couples and couple stresses. In principle and, to some degree of satisfaction, in practice, one can calculate these for some physically realizable situations....

There are two important points here. One is his emphasis on the stress tensor rather than the free energy density. For Ericksen static equilibrium occurs when forces balance, whereas for Frank it occurs when the free energy is minimised. Both statements are true, but in the context of developing a complete dynamical theory, the former is more fruitful. The second point is equally crucial. *The stress tensor does not have to be symmetric.* In conventional theories an element of fluid cannot sustain a body couple. Not so for anisotropic fluids, and what can be more anisotropic than a liquid crystal? Given this starting point a full dynamical theory was but a question of time, though the complications in writing down the full invariances of the liquid crystal problem still presented significant difficulties. In the years that followed, Ericksen returned to the problem on several further occasions.[37]

At the same time, Leslie was beginning to work on the continuum mechanics of complex fluids in Newcastle, UK. The applied mathematics group (Leslie is the only self-proclaimed mathematician with a paper in this collection, but compensates for his solitude with two entries!) was directed by Albert Green (1912–99), who had wide interests in this area. The classic Navier–Stokes equations were perceived to be inadequate to describe a variety of systems in which there might be internal degrees of freedom. The study of such systems goes back to the idea of a generalised continuum, proposed by E. and F. Cosserat in 1907.[38] In two papers in 1964 and 1965, Green and co-workers[39] discussed the continuum mechanics of systems with local multipolar displacements. The simplest multipole case occurs when there is a local vector, and this Green and colleagues labelled a *director*.

Green suggested to Leslie that the director case might be worth pursuing in greater detail, and the result of that study was published[40] in 1966 under the suitably modest title of *Some constitutive equations for anisotropic fluids*. At this stage, the motivation for this programme was still essentially rheological and theoretical. Leslie does compare his work to the parallel work of Ericksen, noting some similarities and some differences (e.g. his solutions do not seem to exhibit plug flow). But when it came to liquid crystals, Leslie was less sure:

The theory described in this paper allows a fluid to have a preferred direction at each point.... Materials called liquid crystals possess such preferred directions. However, since the present theory does not reduce to Frank's hydrostatic theory of liquid crystals, it would appear that it is inadequate to describe these materials.

353

In fact Leslie was only persuaded to add even these comments (at the end of the introduction) by a referee who was perhaps rather more empirically oriented than Leslie was himself at that stage. The reason that there is a discrepancy between Leslie's 1966 theory and Frank's hydrostatics, as Leslie himself was to find out shortly afterwards, is that no elastic effects have been built into the theory.

At that point Leslie and Ericksen's paths crossed physically, for Green had arranged for Leslie to spend the academic year 1966–7 at Johns Hopkins. After some discussions with Ericksen about the form the full theory of liquid crystals should take, Leslie spent the whole of the next month in the library. Ericksen gave him a copy of Anzelius's doctoral thesis on the dynamical theory of nematics. The library owned a comprehensive German dictionary. With thesis in one hand and dictionary in the other, progress was slow. The final form of the theory, which we reproduce in article C6, was published in 1968. The theory builds on Leslie's 1966 paper, but elastic contributions are added to the stress tensor in addition to the dissipative terms already included. The result is a theory which now *does* reduce to the Frank theory in the hydrostatic limit.

It is worth summarising the essential points of the Leslie–Ericksen theory. The motion of the director (which had existed in the liquid crystal literature since Oseen and Zocher, but only now acquired a canonical name) is coupled with the usual fluid motion. The fluid motion is governed by Newton's law, but the recipe for the stress tensor now includes some *elastic* terms, which derive essentially from the fact that the free energy is not minimised, as well as some *dissipative* terms. The dissipative terms generalise the single viscosity of simple fluids to a group of six quantities, and involve not only the familiar rate of strain tensor, but also the symmetry-breaking effects of the director, and a suitably covariant rate of change of the director. In fact, as we have seen in articles B9 and B10, the existence of more than one viscosity was already known, though *exactly* how many viscosities would be required remained unknown. Included in the formulation is also an equation of motion for the director. This is purely dissipative – there is no local moment of inertia – and the relaxation viscosity is related through a dissipation theorem to a combination of the other viscosities.

Attractive as the Ericksen-Leslie theory was as an intellectual edifice, the true test remained one of experiment. We have seen two major attempts at a liquid crystal theory – the swarm theory and Born's dipole theory – fail at this hurdle. The natural testing ground for the Johns Hopkins group, given their fluid mechanics background, was viscometry. The long history of observations of liquid crystalline viscosity – going all the way back to Schenck in 1898 – demonstrated clearly that liquid crystals were non-Newtonian fluids. A Newtonian fluid has certain constraints forced upon it. But once *non*-Newtonian, almost anything was possible. Ericksen's hunch was that the non-Newtonian behaviour of liquid crystals was the result of competition between the elastic effect of orienting surfaces and the viscous effect of orienting flows. Another suggestion at the time was that the non-Newtonian behaviour was due to so-called 'fractional orientation', in which stronger flows would orient the molecules more perfectly (which turned out not to be the case here, but is in fact true for lyotropic liquid crystals in strong flow fields).

At Ericksen's suggestion, Ray Atkin, a postdoctoral worker from England, examined the properties of Poiseuille flow – flow in a capillary with circular cross-section. There is a nineteenth-century formula for the fluid flux Q of a standard Newtonian fluid. One obtains $Q = \dfrac{\pi P R^4}{8\eta}$, where R is the capillary radius, P the pressure gradient, and η the fluid viscosity. This can be inverted so as to provide a formula for the fluid viscosity η in terms of the measurable quantities q, P and R. Now we have $\eta = \dfrac{\pi P R^4}{8Q}$. In other words, if we change Q,R,P, we still get the same value for η.

What happens when the fluid is non-Newtonian? One can define an effective viscosity empirically using Poiseuille's formula: $\eta_{\mathrm{eff}} = \dfrac{\pi P R^4}{8Q}$. Somewhat surprisingly, some regularity remains. For a classical isotropic viscoelastic fluid, it was known that η_{eff} was a function of $\dfrac{Q}{R^3}$. But for a nematic liquid crystal, by contrast, Atkins found that η_{eff} was a function of $\dfrac{Q}{R}$. So here was an unambiguous test.[41]

This result, though published only in 1970, was ready in 1968 before the 2nd international liquid crystal conference at Kent State University. And to this conference, unbeknownst to Atkin, Ericksen or Leslie, came the experimentalist Frederickson and his student Fisher, who had performed just the viscometric experiment envisaged by Ericksen, in which they changed the size of the capillary, and also the nature of the orienting condition at the boundaries. The effective viscosity was emphatically *not* a function of $\dfrac{Q}{R^3}$, for each capillary and boundary condition yielded a different curve for $\eta_{eff}\!\left(\dfrac{Q}{R^3}\right)$. However, encouraged by Atkins, Fisher and Frederickson rapidly replotted their data, finding that indeed when η_{eff} is plotted as a function of $\dfrac{Q}{R}$, the curves from different capillaries collapse onto one plot (or strictly speaking two, for the plots do depend on the boundary conditions).[42] This scaling result gave strong reinforcement to the

355

Ericksen–Leslie theory, and in the years that followed, it became clear that it was possible to interpret essentially all macroscopic dynamical data on nematic liquid crystals in terms of this theory. Indeed using similar principles, Leslie was later able to extend the theory not only to discuss cholesteric materials[43] (a relatively straightforward matter) but also smectic materials.[44]

The story was completed by Parodi,[45] who used a thermodynamic argument to reduce the number of independent viscosities from six to five. This might appear a trivial improvement, but in fact it is rather important, for it permits the viscosities to be completely determined using four viscometric experiments and one relaxation experiment. We mention finally the work of Martin, Parodi and Pershan[46] who used a different more physics-based point of view to derive more or less equivalent equations for nematodynamics close to equilibrium. What is important, to the physicist if not to the applied mathematician, is what variables are required to be included in a hydrodynamic theory. This question can only be answered within the scope of a kinetic theory in which the order parameter correlations are well understood. By contrast the rational mechanician or applied mathematician is likely to suppose the static equilibrium properties of the important hydrodynamical variables to be known (by observation or revelation), and proceed from that point.

We have now examined both the current standard microscopic and macroscopic theories of the nematic state. In the final paper C7 in this section we turn to the intermediate-scale Landau-type theory developed by Pierre-Gilles de Gennes. With this theory, liquid crystal science came of age, in that it was now possible to use the full panoply of tools available to the theoretical physicist to analyse liquid crystal phenomena.

To put this paper in perspective, let us return to the work of Georges Friedel (paper B1). We have discussed Friedel's use of the term *stase*. In hindsight we regard the distinction between *stase* and phase as a reflection of the undeveloped understanding of the concepts of phase and phase transition in the 1920s. Van der Waals in the 1870s had shown semi-quantitatively how cooperative effects lead to a distinction between low density gases and high density liquids, and how as pressure increases a sudden *phase transition*, involving a jump in density, can occur between them. Similarly Curie[47] and Weiss[48] had shown near the turn of the twentieth century that ferromagnetism was the result of cooperative interaction between the magnetic moments of individual molecules (at that stage of unknown provenance). At the *Curie point* there was a sudden onset of spontaneous magnetism as temperature was reduced, but no sudden *jump* in mean magnetic moment. This concept spurred liquid crystal theorists, who expected a nematic Curie point. However, when the nematic Curie–Weiss theory was discovered – independently by Grandjean (C1) and Maier and Saupe (C3) – the so-called Curie point did in fact involve a sudden jump in degree of order. What explains the difference between magnetism and anisotropic liquid order? Where are the systems analogous and where are they not?

The contemporary view of phase transitions can be traced back to a number of sources. The standard classification into transitions which are 'first-order', 'second-order' and so on, is due to Paul Ehrenfest,[49,50,51] working in Leiden in the Netherlands in 1933. The idea is simply that in an nth order phase transition there is a discontinuity in the nth derivative of the free energy. Up to that time the term 'phase transition' was restricted to those transitions which we would now label as first-order. Ehrenfest's theory was motivated by the discovery by careful calorimetric observations by W.H. Keesom[52] and his daughter A.P. Keesom[53] of the bizarre change in behaviour of Helium close to 2.2K. On the basis of Keesom's observations, Ehrenfest concluded that the specific heat reached a maximum just below the sudden change in properties, and then fell back at the 'transition' to a more normal value. The term λ-point for this transition is due to Ehrenfest, and reflected the shape of the specific heat in the transition region. Ehrenfest was then able to show that a 'second-order' transition would exhibit a specific heat rather like that observed. He then proceeded to develop the theory further, in particular developing an analogue of the Clausius–Clapeyron relation for the pressure-dependence of the transition temperature.

A few years later, Lev Landau, working in Kharkov in the present-day Ukraine, presented an alternative theory of phase transitions.[54] The mythology records that here we find the free energy expanded in terms of an order parameter, with the coefficient of the squared term in the order parameter linearly dependent on temperature. In fact, though there is mention of representations of symmetry groups of distribution functions, the words 'order parameter', first introduced by Felix Bloch in his habilitation thesis in 1932, do not appear. There is nevertheless strong overlap between the deep content of the Landau and Ehrenfest theories. Landau makes the point that a change in symmetry of a phase *necessarily* involves some kind of transition, because symmetry can either be present or absent. As a result a solid–liquid critical point analogous to the liquid–vapour critical point *cannot* occur, because of the necessity of symmetry change at a liquid–solid transition. He is further clear that the presence or absence of a cubic term in what Landau describes as η (and which we recognise as the order parameter) depends on the symmetry change at the transition. Furthermore if there is a cubic term the transition is a phase transition properly defined, whereas if there is not, then the transition is like a Curie point.

From our standpoint, it is interesting that as long ago as 1937 Landau was already thinking about liquid crystals. The final part of his second paper on phase transitions is devoted to liquid crystals alone. He asserts that:

> One often finds the opinion that liquid crystals represent bodies in which the molecules are arranged in "chains", oriented in one direction,

i.e. bodies in which ρ is a function of one variable. However, it has been shown in section 1 that such bodies cannot exist.

Instead we can imagine liquid crystals as bodies in which the molecules, or more precisely their centres of mass, are distributed completely randomly, as in ordinary liquids. Anisotropy of the liquid crystal is caused by the equal orientation of its molecules. For instance, if the molecules have an elongated shape, then they can all be arranged with their axes in one direction.

He goes further to explain that the crucial quantity to investigate is the intermolecular distribution function ρ_{12}, but in a remarkably prescient comment, notes that:

... Therefore the possible symmetry groups of ρ_{12}, i.e. of liquid crystals, are not the 230 space groups, but point groups. Of course, the number of these groups is not limited to 32 as in solid crystals; the symmetry of liquid crystals should be classified in the same way as the symmetry of molecules. In symmetry axes of any order (and not only of the second third, fourth and sixth orders) are possible. In particular liquid crystals are possible with total axial symmetry. It is experimentally known that liquid crystals are uniaxial. It would be very interesting to establish whether they possess total axial symmetry, or simply have axes of higher than second order.

In principle, liquid crystals with cubic symmetry are possible. Such crystals are impossible to distinguish from ordinary crystals in their optical properties. It is possible that liquid He II is such a crystal ....

And thus is his attention diverted towards the nature of superfluid helium. In fact he is sceptical that He II is a liquid crystal, but that was the important problem of the day. We can only speculate what progress he might have made on the subject of more orthodox liquid crystals, had the mood taken him.

As it is we had to wait another 30 years for further reflections on liquid crystals in this vein. In the meantime statistical mechanics synthesised the Ehrenfest and Landau pictures, realising along the way that the picture was inadequate. In 1944 Onsager's exact solution of the two-dimensional Ising model showed that neither picture was complete, because the specific heat diverged at what should have been the Curie point, and in 1955 more careful experiments at the Helium λ-point established that even in the archetypal case, the specific heat actually diverged and so the Ehrenfest theory did not apply. Out of these crises came the modern theory of critical phenomena.

In article C7 de Gennes finally carries through the programme so tantalisingly alluded to by Landau in 1937. He was ideally placed to do this. He had a solid

358

background in the theories of magnetism and superconductivity, as well as the backing of an experimental team – the Orsay Liquid Crystal group. It is de Gennes who realises that the true order parameter is not Tsvetkov's S, but rather that this is a surrogate for the true order parameter, which is a traceless symmetric tensor. In the ideal case, this will be characterised by only one number (S), but strictly speaking it is the tensor on which we should concentrate, for this governs the behaviour of tensor susceptibilities. These susceptibilities mark the liquid crystal as birefringent with an optical axis. Given this order parameter, it is not difficult to show that there is a third order invariant, and thus that the isotropic–nematic phase transition should be first order. And from here, he explores a large number of consequences of the theory – susceptibilities, coherence lengths, and the nematic–isotropic interface. He then extends the theory to the time-dependent domain, as well as to cholesterics. In this article de Gennes concentrates on the easy cases, but it is de Gennes who makes them easy. It is not unfair to say that this paper started a whole industry, and the full consequences of what came to be known as the *Landau–de Gennes* theory are still being worked through even 30 years later.

De Gennes's participation in the liquid crystal community lasted perhaps seven years. After 1975 his interests moved elsewhere – to macromolecular systems, and later still to other aspects of what came to be called *Soft Condensed Matter*. He left a subject transformed, in which his paradigm (building on that of Landau) was the first port of call for the puzzled experimentalist. His reward, later, partly for his work in the liquid crystal community, but partly elsewhere, was the Nobel prize for Physics in 1991. From the point of view of fundamental science, after de Gennes the paradigm was set, and the practice of science was, in the sense of Thomas Kuhn, normal rather than extraordinary. This is thus a suitable stage at which to move from the foundation to application, which we shall address in the next section.

References

1. M. Born. *Über anisotrope Flüssgkeiten. Versuch einer Theorie der flüssigen Kristalle und des Elektrischen KERR-Effekts in Flüssigkeiten*; Sitzungsber. Preuss. Akad Wiss. **30**, 614 (1916).
2. Tsvetkov's article C2 is the first paper in which the orientational order parameter $\frac{1}{2}\langle 3\cos^2\vartheta - 1\rangle$ appears in the liquid crystal literature as properly understood, and is the route through which it found general acceptance. However, it does not appear to be the very first time that it is used. The earliest use of the orientational order parameter that we have been able to find is in a paper on gels by P.H. Hermans and P. Platzek, *Beiträge zur Kenntnis des Deformationsmechanismus und der Feinstruktur der Hydratzellulose. IX Über die theoretische Beziehung zwischen Quellungsanisotropie und Eigendoppelbrechung orientierter Fäden* (*Contributions to the understanding of the deformation mechanism and fine structure in hydrated cellulose part IX: On the*

theoretical relationship between anisotropic expansion and birefringence in oriented threads) Kolloid-Zeitschrift **88**, 68–72 (1939).

3. G. Foëx and L. Royer. *Le diamagnétisme des substances nématiques*, Comptes rendus **180**, 1912–13 (1925). See also: J. Phys. (France) **10**, 426 (1929).
4. G.H. Brown and W.G. Shaw. *The mesomorphic state*, Chemical Reviews **57**, 1049–1157 (1957).
5. G.W. Gray, *Molecular Structure and the Properties of Liquid Crystals* (Academic Press, London and New York 1962).
6. W. Maier and A. Saupe. *Eine einfache molekulare Theorie des nematischen kristallinflüssigen Phase. Teil I*, Z. Naturforschung **14a**, 882–9 (1959); Teil II, *ibid* **15a**, 287–92 (1960). For translations see: P.E. Cladis and P. Palffy-Muhoray (eds) *Dynamics and Defects in Liquid Crystals: a Festschrift in honour of Alfred Saupe* (Gordon and Breach, London) pp. 389–440.
7. We are grateful to Yves Bouligand for drawing our attention to Grandjean's paper.
8. C.W. Oseen. *Die anisotropen Flüssigkeiten. Tatsachen und Theorien* (Fortschritte der Chemie, Physik und physikalischer Chemie, Vol. 20, no. 2, Berlin 1929).
9. G. Szivessy. *Zur Bornschen Dipoltheorie der anisotropen Flussigkeiten*, Z. Physik **34**, 474–84 (1925).
10. J.D. Watson. *The double helix: a personal account of the discovery of the structure of DNA* (Weidenfeld and Nicholson, London 1968).
11. Ann Sayre. *Rosalind Franklin and DNA* (WW Norton & Co., New York 1975); Brenda Maddox, *Rosalind Franklin: the dark lady of DNA* (Harper Collins, London 2002).
12. R.B. Meyer. *Flexoelectric effects in liquid crystals*, Phys. Rev. Lett. **22**, 918–21 (1969).
13. Orsay Liquid Crystal Group. *Quasi-elastic Rayleigh scattering in nematic liquid crystals*; Phys. Rev. Lett. **22**, 1361–3 (1969).
14. J.L. Ericksen, *Inequalities in liquid crystal theory*, Phys. Fluids **9**, 1205–7 (1966). We are grateful to Professor Ericksen for correspondence on this subject. He points out that he has referred to the static continuum theory as the 'Oseen–Zocher–Frank' theory, but that elsewhere, while clearly attributing the theory to Oseen, he did suggest that Frank's article might be a good place to read about it.
15. P.G. de Gennes and J. Friedel. *Boucles de disclination dans les cristaux liquides*, Comptes rendus de l'Académie des Sciences **B 268**, 257–9 (1969).
16. C. Caroli and E. Dubois-Violette. *Energy of a disinclination line in an anisotropic cholesteric liquid crystal*, Solid State Comm. **7**, 799–802 (1969).
17. M. Kléman and J. Friedel. *Lignes de dislocations dans les cholésteriques*, J de Physique **30**, **C4**, 43–53 (1969).
18. Groupe experimental d'Etudes des cristaux liquids. *Lignes doubles de désinclination dans les cristaux liquids à grand pas*, J. de Physique **30**, **C4**, 38–42 (1969).
19. We are grateful to Maurice Kléman for this *bon mot*, which he reports second hand, but of uncertain provenance.
20. The article in our collection is actually part 4 of a long collection of papers. Curiously, part 4 appeared before parts 1,2 and 3! These appeared in: (1) Z. phys. Chem. (Leipzig) **213**, 137 (1960), (2) *ibid.* **213**, 145 (1960), (3) *ibid.* **213**, 262 (1960).
21. Many, but not all, readers will be aware that the current classification scheme continues well beyond smectic C, at least to smectic K, and in some dialects,

as far as O. The great part of the crucial work establishing the classification, and the relationship between the phases, lies outside the period covered by this volume. Only smectics A and C involve no periodicity at all in the smectic layer plane. The other phases involve a combination of local in-plane crystalline structure, and in-plane molecular orientation with respect to the local crystalline axes. For a pedestrian semi-popular description of the physics of smectic phases up to smectic I, see, for example, T.J. Sluckin, *Liquid crystals: physics and technology*, Contemporary Physics **41**, 37–56 (2000). An important point to note is that many phases originally identified as smectic (and thus fluid) were later reclassified (particularly by George Gray and collaborators in the 1980s) as crystalline, albeit with a crystalline order much better established in the direction normal to the 'smectic' planes.

22. K. Herrmann. *Inclinations of molecules in some crystalline fluid substances*, Trans. Faraday Soc. **29**, 972–6 (1933).

23. D. Vorländer, R. Wilke, H. Hempel, U. Haberland and J. Fischer. *Die kristallin-flussigen und festen Formen des Anisal-p-amino-zimmtsaureäthyl esters $C_{19}H_{19}O_3N$*; Z. Kristallographie A**97**, 485 (1937).

24. C. Weygand, *Über formständige, isolierte kristallin-flüssige Bildungen. 4 Beitrag zur chemischen Morphologie der Flüssigkeiten*; Z. phys. Chem B**53**, 75–84 (1943).

25. See in particular G.W. Gray, Brynmor Jones and F. Marson, *Mesomorphism and chemical constitution, part VIII: The effects of 3' substituents on the mesomorphism of the 4'n-alkoxydiphenyl-4-carboxylic acids and their alkyl esters*. J. Chem. Soc. (London) 393–401 (1957).

26. Sackmann's freedom to travel is no signal of his political conformity. He had a long record of vocal, if private, opposition to the East German communist régime. Rather it was his elevated scientific status, as vice-president of the prestigious *Deutsche Akademie der Naturforscher Leopoldina*, which conferred on Sackmann some immunity from the persecution to which a lesser mortal would surely have been subject.

27. H. Sackmann and D. Demus. *The polymorphism of liquid crystals*, Mol. Cryst. **2**, 81–102 (1966).
Note the year's delay between the conference and the publication of the proceedings!

28. I.G. Chistyakov and W.M. Chaikowski. *The structure of p-azoxybenenes in magnetic fields*, Mol. Cryst. Liq. Cryst. **7**, 269–78 (1969).

29. I.G. Chistyakov, L. Schabischev, B.I. Jarenov and L.A. Gusakova. *The polymorphism of the smectic liquid crystal*; Mol. Cryst. Liq. Cryst. **7**, 279–84 (1969).

30. S.L. Arora, J.L. Fergason and A. Saupe. *Two liquid crystal phases with nematic morphology in laterally substituted phenylenediamine derivatives*, Mol. Cryst. Liq. Cryst. **10**, 243–57 (1970).

31. A. de Vries. *X-ray photographic studies of liquid crystals*, Acta Crystallographica A**25**, s135 (1969).

32. T.R. Taylor, J.L. Fergason and S.L. Arora. *Biaxial liquid crystals*, Phys. Rev. Lett. **24**, 359–62 (1970).

33. T.R. Taylor, S.L. Arora and J.L. Fergason. *Temperature-dependent tilt angle in the smectic C phase of a liquid crystal*, Phys. Rev. Lett. **25**, 722–5.

34. As the subsequent story indicates, it is with good reason that the theory has taken the names of both Ericksen and Leslie. Ericksen's strict policy was to permit his postdoctoral

workers and graduate students to publish under their own names alone. He reports surprise, shared by few others in the field, at finding his name appended to the theory. The theory is sometimes known as *nematodynamics*, or as the *ELP* theory (the *P* is Parodi; see ref. [45] below).

35. T. Carlsson and F.M. Leslie. *The development of theory for flow and dynamic effects for nematic liquid crystals*, Liquid Crystals **26**, 1267–80 (1999).

36. J.L. Ericksen. *Hydrostatic theory of liquid crystals*, Arch. Ration. Mech. Anal. **9**, 371–8 (1961).

37. see e.g. : J.L. Ericksen. *Anisotropic fluids*, Arch. Ration. Mech. Anal. **9**, 231–7 (1961); Arch. Ration. Mech. Anal. **10**, 189–96 (1962).

38. see P. Chadwick. *Albert Edward Green (1912–99)*, Biog. Mem. Roy. Soc. **47**, 257–78 (2001).

39. A.E. Green and R.S. Rivlin. *Multipolar continuum mechanics*, Arch. Rat. Mech. Analysis **17**, 113–47 (1964); A.E. Green, P.M. Naghdi and R.S. Rivlin. *Directors and multipolar displacements in continuum* mechanics, Int. J. Eng. Sci. **2**, 611–20 (1965).

40. F.M. Leslie. *Some constitutive equations for anisotropic fluids*, Q.J. Mech. Appl. Math., **19**, 357–370 (1966).

41. R.J. Atkin. *Poiseuille flow of liquid crystals of nematic type*, Arch. Rat. Mech Analysis **38**, 224 (1970).

42. J. Fisher and A.G. Frederickson. *Interfacial effects on the viscosity of a nematic mesophase*, Mol. Cryst. Liq. Cryst. **8**, 267 (1969).

43. F.M. Leslie. *Continuum theory of cholesteric liquid crystals*, Mol. Cryst. Liq. Cryst. **7**, 407–20 (1969).

44. F.M. Leslie, I.W. Stewart and M. Nakagawa. *A continuum theory for smectic C liquid crystals*, Mol. Cryst. Liq. Cryst. **198**, 443–54 (1991).

45. O. Parodi. *Stress tensor for a nematic liquid crystal*, J. Physique **31**, 581–4 (1970).

46. P.C. Martin, O.Parodi and P.S. Pershan. *Unified hydrodynamic theory for crystals, liquid crystals and normal fluids*, Phys. Rev. **A6**, 2401–20 (1971).

47. P. Curie. *Propriétés magnétiques des corps à diverses temperatures*, Annales de chimie et de physique (7th series) **5**, 289–405 (1895).

48. P. Weiss. *L'hypothèse du champ moléculaire et la propriété ferromagnétique*, J. Phys. Rad **6**, 661–90 (1907).

49. P. Ehrenfest. *Phasenumwandlungen im üblichen und erweiterten Sinn, classifiziert nach dem entsprechenden Singularitäten des thermodynamischen Potentiales*, Verhandlingen der Koninklijke Akademie van Wettenschappen (Amsterdam) **36**, 153–7 (1933).

50. See G. Jaeger. *The Ehrenfest classification of phase transitions: introduction and evolution*, Arch. Hist. Exact. Sci. **53**, 51–81 (1998) for a fuller discussion of the historical background to the theory of phase transitions.

51. It is not strictly relevant to the story, but shortly thereafter Ehrenfest, who had long suffered from depression, committed suicide.

52. Keesom was a low temperature physicist. In this role he collaborated with Daniel Vorländer in a study of solid nitrogen: W.H. Keesom and D. Vorländer, *Über den kristallisierten Stickstoff*, Ber. Dt. chem. Ges. **59**, 2088 (1926).

53. W.H. Keesom and A.P. Keesom. *On the anomaly in the specific heat of helium*, Verhandlingen der Koninklijke Akademie van Wettenschappen (Amsterdam) **35**, 736 (1932).

54. L.D. Landau. *К теории фазовых переходов*; Ж.Е.Т.Ф. (paper 1) **7**, 19 (1937); (paper 2) **7**, 627 (1937). English translation in: *Collected works of L.D. Landau* (Pergamon, Oxford, 1965) pp. 193–216. For an assessment of Landau's career and impact, see e.g. A. Kojevnikov. *Lev Landau: physicist and revolutionary*, Physics World, **15**, No. 6 pp. 35–9 (2002).

C1

Comptes rendus de l'Académie des Sciences **164**, 280–83 (1917)

CRISTALLOGRAPHIE PHYSIQUE. — *Sur l'application de la théorie du magnétisme aux liquides anisotropes.* Note (¹) de **M. F. GRANDJEAN**, présentée par M. L. De Launay.

Pour adapter la belle théorie de Langevin et P. Weiss aux liquides anisotropes, il m'a paru nécessaire de remplacer la relation de proportionnalité entre le champ intérieur et l'aimantation par une autre relation, équivalente, qui soit susceptible d'être généralisée. On y parvient en admettant qu'à chaque instant les molécules du corps qui tombent en direction dans un petit angle solide $d\omega$, parallèle à Δ_1, créent un champ magnétique constant de direction Δ_1. Ces molécules changent continuellement, mais leur nombre dN est fixe. En négligeant la fluctuation due aux changements de position des molécules, on admettra que le champ est proportionnel à dN, de sorte que si l'on appelle du_m l'énergie potentielle de ce champ, pour une molécule M de moment magnétique μ, écartée de Δ_1 d'un angle α, on a

$$du_m = -\mu \cos\alpha \, dh = -\mu b \cos\alpha \frac{dN}{N};$$

N est le nombre total des molécules; dh représente le champ magnétique intérieur dû aux dN molécules.

Pour avoir l'énergie totale intérieure u_m, il suffit d'intégrer dans tout l'espace. La valeur de $\frac{dN}{N}$ est donnée par la relation de Maxwell-Boltzmann

$$\frac{dN}{N} = C e^{-\frac{u}{rT}} d\omega,$$

(¹) Séance du 22 janvier 1917.

C1
Comptes rendus de l'Académie des Sciences **164**, 280–83 (1917)

ON THE APPLICATION OF THE THEORY OF MAGNETISM TO ANISOTROPIC FLUIDS[1]

Note from
Mr F. Grandjean
presented by Mr L. De Launey

In order to adapt the elegant theory of Langevin and P. Weiss to anisotropic liquids, it seemed to me necessary to replace the equation of proportionality between the field and the magnetisation by another equivalent equation, which could be generalised. I propose that at each instant the molecules of the substance which are in an increment of solid angle $d\omega$, parallel to Δ_1, give rise to a constant magnetic field in the direction Δ_1. These molecules are changing constantly, but their number dN is fixed. Neglecting the fluctuation due to changes in the position of the molecules, I will assume that the field is proportional to dN, such that if one denotes du_m the potential energy of this field, for one molecule M with a magnetic moment of μ, making an angle of α with Δ_1, one has

$$du_m = -\mu\cos\alpha \, dh = -\mu b \cos\alpha \frac{d\mathrm{N}}{\mathrm{N}}$$

N is the total number of molecules, dh represents the internal magnetic field due to dN molecules.

In order to obtain the total energy u_m, one integrates over all space. The value of $\frac{d\mathrm{N}}{\mathrm{N}}$ is given by the Maxwell–Boltzmann relation

$$\frac{d\mathrm{N}}{\mathrm{N}} = C e^{-\frac{u}{r\mathrm{T}}} d\omega$$

---

[1] Meeting of 22 January 1917.

F. GRANDJEAN

u being the total energy, which is the sum of u_m and the external energy u_e; T is the absolute temperature; C is a constant. Δ is the direction of the internal field, and γ and V are respectively the angles of M and Δ_1 made with Δ. In choosing as integration variables V and the azimuth of Δ_1 as a ratio to Δ, one notes that the integration can be carried out for the azimuthal angle on which u does not depend, and finally one has, removing C by normalisation,

$$u_m = -\mu b \cos\gamma \ \frac{\int_0^\alpha e^{-\frac{u}{rT}} \sin V \cos V \, dV}{\int_0^\alpha e^{-\frac{u}{rT}} \sin V \, dV}$$

Now the quotient of the two integrals appearing on the right-hand side represents precisely the value of the magnetisation divided by $N\mu$. The result of this is therefore that the internal field is proportional to the magnetisation.

For anisotropic liquids I will follow exactly the same calculation process, starting from the expression,

$$du_m = -v \cos 2\alpha \cdot di = -vq \cos 2\alpha \cdot \frac{dN}{N}$$

di will be the internal field due to the dN molecules, v is a coefficient which plays the same role as the magnetic moment μ. The proportionality to $\cos 2\alpha$ is pure speculation, but it is the simplest which is appropriate for anisotropic liquids.

For these liquids, the internal field and the molecular axis are both directions having a centre of symmetry; $\alpha=0$ and $\alpha=\pi$ must be two identical positions of stable equilibrium.

Carrying out the calculation in the same way, one finds

$$u_m = -\frac{1}{3}vq(S \cos 2\gamma + T)$$

where S and T are quantities independent of γ. By analogy with magnetic fields we will call the quantity i the molecular field

$$i = \frac{1}{2}qS$$

where q is the intensity i_0 the field would have if the N molecules were parallel to D. One obtains, replacing S by its value (obtained by normalisation),

$$I = \frac{i}{i_0} = \frac{1}{2} \ \frac{\int_0^{\pi/2} e^{-\frac{u}{rT}} \sin V (3 \cos^2 V - 1) \, dV}{\int_0^{\pi/2} e^{-\frac{u}{rT}} \sin V \, dV}$$

366

If there is no external field, putting

$$\alpha = \frac{vi}{r\mathrm{T}}, \quad \beta = \frac{r}{vi_0}$$

this becomes

(1) $$I = \frac{1}{2} \frac{\displaystyle\int_0^{\pi/2} e^{-\alpha\cos 2\mathrm{V}} \sin \mathrm{V}(3\cos^2 \mathrm{V} - 1)d\mathrm{V}}{\displaystyle\int_0^{\pi/2} e^{-\alpha\cos 2\mathrm{V}} \sin \mathrm{V}\,d\mathrm{V}} = \frac{\displaystyle\sum_0^\infty \frac{2^{m+1} m\alpha^m}{(2m+1)(2m+3)m!}}{\displaystyle\sum_0^\infty \frac{2^m \alpha^m}{(2m+1)m!}}$$

(2) $$I = \beta \mathrm{T} \alpha$$

Taking α as the abscissa and I as the ordinate, equation (2) represents a straight line of slope $\beta\mathrm{T}$ and equation (1) represents a curve which is shown in the figure given here. From the intersection of the line and the curve, at each temperature

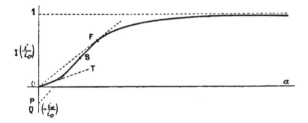

one obtains the value of I. F is the contact point of the tangent from O to the curve. Between O and F the equilibrium is unstable; it is stable beyond F. For example, if one lowers the temperature progressively, the field disappears suddenly at a temperature Θ defined in such a way that $\beta\Theta$ is equal to the slope of the tangent OF. This is the point of isotropic melting.

If there is an external field i_e superimposed on the internal field I, and express- ing it in the same way as a function of the disorienting angle γ, the same curve remains, but the slope $\beta\mathrm{T}$ rotates about the point Q on the ordinate axis. Following the position of this point as a ratio of that where the tangent of inflexion SP cuts the ordinate axis, several cases can be produced. I will not pursue this discussion here, which in any case is very simple. I will simply remark that the form of the curve explains the persistence of the contact layers with the solid material beyond the point Θ. At these contacts, the solid produces an effective molecular field in the capillary region and this plays the role of an external field. The anisotropy persists in this region even above Θ, when the rest of the liquid is already iso- tropic. Because these layers have not ceased to exist, the rest of the liquid orients on these during cooling. Thus when the temperature is reduced, the layers reform exactly at the same places with the same orientation and the same contours.

367

Über die Molekülanordnung in der anisotrop-flüssigen Phase

Von *W. Zwetkoff*

Die vorliegende Untersuchung ist ein Beitrag zur Erforschung der Existenzbedingungen des anisotrop-flüssigen Zustands und der Umwandlung der anisotropen Flüssigkeit in eine amorphe. Dabei wird die anisotrope Flüssigkeit als ein System von Molekülen mit Achsensymmetrie und einer Ordnung der Molekülachsen über grosse Entfernungen betrachtet. Die Untersuchung der gegenseitigen Abhängigkeit des Ordnungsgrades und des Elastizitätsmoduls der anisotrop-flüssigen Substanz ermöglicht eine Erklärung ihrer Existenz und die sprunghafte Umwandlung der anisotrop-flüssigen Phase in die isotrope; ausserdem wird eine Schätzung des Wärmewertes der Umwandlung und der Differenz der Wärmekapazität der Substanz in den anisotropen und isotropen Phasen möglich.

1. Einleitung

Eine charakteristische Eigenschaft der Substanzen im anisotrop-flüssigen Zustand ist die parallele Orientierung ihrer stäbchenförmigen Moleküle, die beim Übergang der Flüssigkeit in die amorphe Phase verschwindet.

Nach der Theorie von Ornstein und Kast[1, 2, 3] findet diese parallele Orientierung nur in den Schwärmen, d. h. in den Gebieten statt, deren lineare Abmessungen der Grössenordnung einer Lichtwellenlänge entsprechen (übrigens wird in einigen Fällen für die Abmessung des Schwarmes ein bedeutend höherer Wert angegeben[4]), während die gegenseitige Orientierung verschiedener Schwärme als vollkommen regellos vorausgesetzt wird. Man setzt ebenfalls voraus, dass die Schwärme nicht aufeinander einwirken, bzw. dass wenigstens ihre Wechselwir-

---

[1] Ornstein a. Kast, Trans. Farad. Soc., 29, 900 (1933).
[2] Ornstein, Koll. Z., 69, 137 (1934).
[3] Kast, Z. Elektrochem., 45, 184 (1939).
[4] Ornstein u. Braaf, Kolloidchem. Beih., 44, 427 (1936).

C2
Acta Physicochimica U.R.S.S. **15**, 132–47 (1942)

ON MOLECULAR ORDER IN THE ANISOTROPIC LIQUID PHASE

by

W. Zwetkoff

The present study is a contribution to the investigation of the conditions under which the anisotropic fluid phase exists and of the transition of the anisotropic liquid into its amorphous counterpart. We consider the anisotropic fluid as a system of axially symmetric molecules, and with the molecular axes displaying order over long distances. Studying the mutual dependence of the degree of order and the elasticity modulus explains the existence of the anisotropic fluid phase and also its sudden transformation into an isotropic fluid. In addition, it now becomes possible to estimate the latent heat of the transformation and the difference in heat capacity between the anisotropic and isotropic phases.

1. Introduction

A characteristic property of substances in the anisotropic fluid phase is the parallel orientation of their rod-shaped molecules. This disappears when the fluid transforms into the amorphous phase.

According to the theory of Ornstein and Kast,[1,2,3] this parallel orientation can only be found within swarms, that is, in a region whose linear dimensions correspond to the order of magnitude of the wavelength of light – though in fact in some cases the swarms are assigned distinctly larger dimensions.[4] At the same time the mutual orientation of the different swarms is assumed to be completely random. One also assumes that the swarms do not affect each other, i.e. that their interaction energy is negligible by comparison with their thermal energy.

---

[1] Ornstein and Kast, Trans. Farad. Soc., **29**, 900 (1933).

[2] Ornstein, Koll. Z., **169**, 137 (1934).

[3] Kast, Z. Elektrochem., **45**, 184 (1939).

[4] Ornstein and Braat, Kolloidchem. Beih., **44**, 427 (1936).

By contrast, according to Zocher's continuum theory,[5] the parallelism of the molecular axes can in principle spread over an unlimited large volume. This is, however, impossible in practice, because random currents and impurities in the fluid have a disorienting effect, due to the very small orientational interaction between the molecules in the anisotropic fluid. This orientational interaction is characterised macroscopically by the modulus of elasticity A, which is numerically equal to the turning moment per unit area and unit deformation.

The present writer[6] has measured the value of A for the three different types of deformation (see further below) of the anisotropic liquid azoxyanisole.

Diverse observed phenomena in anisotropic fluids, such as, for example, light scattering[4,7] or relaxation phenomena in rotating fields[8,9] lead to the conclusion that large molecular swarms are present in the material. These swarms participate wholly in the rotational Brownian motion, and, for example, rotate under the influence of external fields.

However, both the existence of elastic forces in the anisotropic fluid medium permitting it to remain homogeneous in layers of thickness up to 1 mm,[10] and a whole set of other experiments[11] prove unambiguously that the interaction energy of neighbouring molecular groups is not at all small in comparison with their thermal energy.

The study of the rotational Brownian motion of particles in anisotropic fluids leads to the conclusion that this motion does not consist of free rotation.[12] Rather it consists of rotational vibrations around the mean axial direction of all the oriented particles.

Thus we can consider the anisotropic fluid phase as a system in which a long-range molecular orientational interaction is indicated. Consequently there should also be a long-range order with regard to the alignment of the molecular axes.[13]

The concept of long-range order proposed by Bragg and Williams[14] has proved very fruitful in the study of 'cooperative' phenomena. Several authors have been able to use it to explain the phenomenon of melting.[15-19]

[5] Zocher, Physik. Z., **28**, 790 (1927); Trans. Faraday. Soc., **29**, 945 (1933); Ann. Phys., **31**, 570 (1938).

[6] Zwetkoff, Acta Physicochimica URSS, **6**, 865 (1937).

[7] Zwetkoff, Acta Physicochimica URSS, **9**, 111 (1938).

[8] Zwetkoff, Acta Physicochimica URSS, **10**, 555 (1939); **11**, 537 (1939).

[9] Zwetkoff, Bull. Acad. Sc. URSS, in press.

[10] Zwetkoff and Ungar, Z. Physik, **110**, 529 (1938).

[11] Marinin and Zwetkoff, Acta Physichimica URSS, **11**, 837 (1939); **13**, 219 (1940).

[12] Zwetkoff, Acta Physicochimica URSS, **11**, 97 (1939).

[13] This point of view was also proposed by J. Frenkel, Bull. Acad. Sci. URSS, physics series, No. 3, 307 (1937).

[14] Bragg and Williams, Proc. Roy. Soc., **145**, 699 (1934).

[15] Frank, Proc. Roy. Soc., **170**, 182 (1939). (*In the original, Zvetkov or the journal misspell Frank as Franck.*)

[16] Bresler, Acta Physicochimica URSS, **10**, 491 (1939).

[17] Lennard-Jones and Devonshire, Proc. Roy. Soc., **169**, 317 (1939); **170**, 464 (1939).

[18] Cernuchi and Eyring, J. Chem. Phys., **7**, 547 (1939).

[19] Wannier, J. Chem. Phys., **7**, 810 (1939).

The sudden transformation of the anisotropic phase into the isotropic phase has been explained recently by Frank[20] using ideas from chain reaction theory.

In the present work we seek to establish a quantitative relationship between the degree of long-range order in the anisotropic liquid phase and its physical properties. We also use this point of view to consider the transformation from the anisotropic to the amorphous liquid.

2. Magnetic and optical properties of the material as a function of the degree of order

The degree of order reaches a maximum value in the solid state and becomes zero (as far as long-range order is concerned) at the transition of the substance to the amorphous liquid state. In the anisotropic liquid phase the order takes an intermediate value, which goes up as the temperature is decreased.

The quantitative measure of the long-range order can be expressed using a method analogous to that used by Frenkel in his theory of 'orientational melting'.

The symmetry axis (the axis of the largest magnetic and electrical polarisability) of a macroscopic homogeneous anisotropic fluid material is taken to coincide with the direction of the Z-axis of a fixed XYZ coordinate system. Obviously the molecular axes fluctuate around the mean direction Z as a consequence of the thermal motion, making an angle ϑ with respect to it. In this way, the order of the structure is expressed using a definite quantity S, which is a function of the mean value of ϑ.

We put:

$$S = \frac{1}{2}(\overline{3\cos^2\vartheta - 1}) \tag{1}$$

Thus complete order $(\overline{\cos^2\vartheta} = 1)$ is characterised by the value $S=1$, whereas in complete disorder $\left(\overline{\cos^2\vartheta} = \frac{1}{3}\right)$, $S=0$. The quantity S can be expressed by using those macroscopic properties in the anisotropic fluid which depend on the magnitude of the molecular order. Amongst these properties are, for example, the birefringence and the diamagnetic anisotropy of the material.

Tsvetkov then goes on to explain how, using results from the solid state of azoxyanisole, he is able to relate the order parameter S to measurements of the diamagnetic anisotropy and birefringence data. The results are given in Table 1, and in Figs 1 and 2, which we show, and in which he presents values for $S(T)$ derived in two different ways, as well as $A(T)$, where A is the elastic modulus.

---

[20] Frank, Physik. Z., **39**, 530 (1938).

Table 1

$t(C)$	n_e	n_ϑ	ρ	R_e	R_o	$\frac{1}{3}(R_e - 2R_e)$	$\Delta R = R_e - R_\vartheta$	$S = \dfrac{\Delta R}{\Delta R_0}$
160		1.636	1.126		0.318	0.318	0	0
137		1.651	1.145		0.319	0.319	0	0
131	1.822	1.580	1.156	0.376	0.288	0.317	0.088	0.50
127	1.853	1.569	1.161	0.385	0.282	0.317	0.103	0.58
120	1.882	1.565	1.168	0.394	0.279	0.317	0.115	0.65
112	1.900	1.562	1.175	0.397	0.276	0.316	0.121	0.69

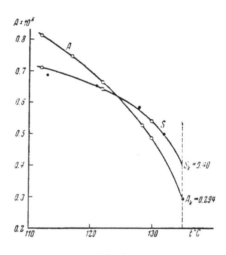

Fig. 1

Table 2

$t\,^{\circ}C$	$S = \dfrac{\Delta_x}{\Delta_0}$	$S = \dfrac{\Delta R}{\Delta R_0}$
112	0.71	
113		0.69
117	0.68	
121		0.65
122	0.64	
128		0.58
128.5	0.57	
130	0.54	
132		0.50
135	0.40	

We continue the translation as Tsvetkov begins to develop theoretical ideas about the feedback loop which maintains orientational order.

4. Magnitude of the long-range order and ordering energy

The experimentally determined temperature dependence of the quantity S can be interpreted if one supposes the order to be a 'cooperative' phenomenon. In order to do this we introduce the 'ordering energy' V, i.e. the energy necessary to transfer a molecule from the ordered to the disordered phase. For this one can obviously use the elasticity modulus which already has been mentioned. In fact:

$$V = \frac{1}{2} a \overline{\vartheta_{max}^2}$$
(12)

where $\overline{\vartheta_{max}^2}$ is the mean square of departure angle of the molecular axis with respect to the molecular axis direction in the completely ordered state.

$\overline{\vartheta_{max}^2}$ is the size of this quantity for complete disorder. Since for chaotically oriented axes $\overline{\sin^2 \vartheta} = \frac{2}{3}$, we can put $\overline{\vartheta_{max}^2} = 0.9$ approximately.

The value a is the coefficient of an elastic couple which permits the molecular axis in the direction of the minimal potential energy to remain. As shown elsewhere,[12] on can consider the departure of the particle axes from the mean direction as an elementary deformation of the material, yielding the dependence:

$$a = 3A\pi r$$
(13)

where r is the mean particle radius.
From this we have

$$V = 1.35 A\pi r$$
(14)

The degree of long-range order S is a function of the ordering energy V and decreases when the latter is removed. It is not difficult to determine the functional dependence $S(V)$ if one puts:

$$S = 1 - \frac{3}{2} \overline{\sin^2 \vartheta} \approx 1 - \frac{3}{2} \overline{\vartheta^2}$$
(15)

and if one expresses $\overline{\vartheta^2}$ using eq. (15), derived in ref.[12]:

$$\overline{\vartheta^2} = \frac{kT}{a} = \frac{kT}{3Ar}$$

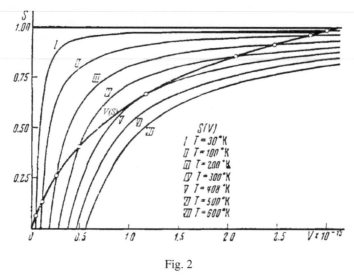

Fig. 2

In this case, substituting back from (14), we find:

$$S(V) = 1 - 0.675\frac{kT}{V} \tag{16}$$

In Fig. 2 we show the dependences $S(V)$ given by eq. (16) for several temperatures (the curves are marked with Roman numerals).

We see in the measurements that as V tends to infinity, S tends to unity.

On the other hand the ordering energy must change with increasing degree of order S through the interaction force between the molecules. Thus the relation $V = V(S)$ must hold. For the case they studied, Bragg and Williams[14] assume a linear dependence of V on S. By contrast, Ketelaar[26] determined experimentally an exponential dependence for $Ag_2 Hg J_4$- crystal.

For the anisotropic fluid azoxyanisole, the nature of the function obviously can be determined experimentally, since (see Table 2) the values of S and A are known experimentally at different temperatures.

We use the values A_1, A_2, A_3 corresponding to the three types of deformation in azoxyanisole, which we have studied ourselves.[6] We have calculated the mean value of the moduli $A = \frac{1}{3}(A_1 + A_2 + A_3)$ at different temperatures. The results are displayed in the second column of Table 3, and are repeated next to the curves $S = S(t)$ in Fig. 1.

A comparison of the values of A and S enables the following relationship to be formulated:

[26] Ketelaar, Trans. Farad. Soc., **34**, 874 (1938).

Table 3

$t\,^{\circ}C$	$A \times 10^7\,dyn$	S measured	S calculated	$V \times 10^{14}$
99	9.75	0.76	0.760	15.40
112	6.10	0.71	0.700	12.80
117	7.45	0.68	0.680	11.77
132	6.65	0.64	0.635	10.50
128.5	5.28	0.57	0.564	8.36
130	4.88	0.54	0.540	7.72
135	2.94	0.40	0.401	4.65

$$A = A_0\,(e^{\beta S} - 1) \tag{17}$$

which holds very well with $\beta = 2.52$ and $A_0 = 1.68 \times 10^{-7}$ dyn. This is illustrated in Table 3, in which values of S are given, both from measured experimental measurements and from calculations using eq. (17).

It is of course true that the relationship (17) can only be checked over a relatively restricted range of temperatures of less than 40° (99–135 °C). Whether it is also valid for lower temperatures is a question which cannot be determined without further work.

Using eqs (14) and (17), we now obtain:

$$V = V_0\,(e^{\beta S} - 1) \tag{18}$$

The values of V calculated from eq. (18) are likewise presented in Table 3, in which the value $V = V_0 = 2.65 \times 10^{-14}$ is assumed. This was determined using eq. (14) and by substituting the values A_0 and $r = 3.73 \times 10^{-8}$ cm.

The anisotropic fluid state is derived by solving simultaneously eqs (16) and (18). It is thus represented by the crossing points of the curves $S(V)$ and $V(S)$. In a certain temperature régime the curves have two crossing points (for example, curves *III* and *IV*). However, it is not hard to show that only states with larger values of S correspond to a stable equilibrium point. At the temperature T_k the curves touch at one point. At temperatures higher than this the curves $S(V)$ and $V(S)$ possess no common point, and consequently the anisotropic liquid state is impossible here.

5. Anisotropic liquid-amorphous liquid phase transition

From the simultaneous solution of eqs (16) and (18) we obtain:

$$S = 1 - 0.675 \frac{kT}{V_0(e^{\beta S} - 1)} \tag{19}$$

375

or

$$S = 1 - \frac{kT}{2\pi A_0 r(e^{\beta S} - 1)}. \tag{20}$$

The curves derived from eq. (19) are shown in Fig. 3.

We set the anisotropic liquid–amorphous liquid phase transition temperature to be the maximum temperature T_k at which long-range order is possible. Strictly speaking, this requires comparison of the free energies of the two phases. However, our qualitative idea is not significantly impaired by the approximation we have made.

As the curves in Fig. 3 show, the anisotropic liquid phase transforms into the amorphous liquid phase at the end-value of S. Consequently the transition is accompanied by a sudden change in the degree of long-range order. Hence the anisotropic liquid–amorphous liquid phase transition is first order, in which is accompanied not only by a change in heat capacity but also by a sudden jump in volume with a latent heat.

In fact an unambiguous density decrease at the anisotropic fluid–amorphous fluid phase transition was already demonstrated in the experiments by Schenck.[24] Using that author's experiments on the reduction of the clearing temperature after adding impurities, as well as knowing the effect of pressure on the transition point (see the experiment of Hullet[27]), the order of magnitude of the latent heat can be estimated.

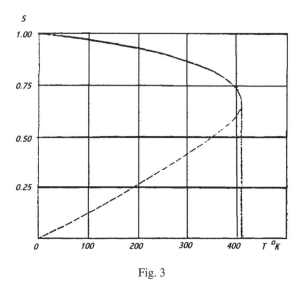

Fig. 3

[24] Schenck, 'Kristallinische Flüssigkeiten', Leipzig, 1905.
[27] Hullet, Z. phys. Chem., 28, 629 (1899).

Recently Kreutzer and Kast[28] have carried a direct calorimetric measurement both of the latent heat and also of the heat capacity difference between the anisotropic and isotropic phases. The latent heat per gram for azoxyanisole was found to be 1.79 cal, and the heat capacity difference $C_{anis.} - C_{is.}$ taken at constant pressure is 0.04 cal/degree.

As the curve in Fig. 3 shows, at the anisotropic–amorphous transition, the degree of long-range order changes abruptly from $S_k = 0.67$ to zero. On the other hand the experimental value, as can be seen in Fig. 1, is $S_k = 0.4$. The source of the disagreement can evidently be found in two factors. First, one of the fundamental equations on which our theory relies, namely eq. (17), though doubtlessly valid for the temperature range of 40° studied (99–135 °C), cannot be generalised to temperatures below 100 °C. Second, one can only be sure that the other fundamental equation, eq. (15), is valid for small fluctuations of the molecule axes, and in general this is not the case close to T_k. Nevertheless the main thrust of this discussion is confirmed by calculating the mean molecular radius from eq. (20). If we substitute in this equation $S_k = 0.67$, $T_k = 408°$ and $A_0 = 1.68 \times 10^{-7}$ dyn, we obtain:

$$r = 3.73 \times 10^{-8} \, \text{cm}$$

while the value of the mean radius of azoxyanisole obtained from the relation $\frac{4}{3}\pi r^3 N = \frac{M}{\rho}$ ($N = 6.06 \times 10^{23}$, M – the molecular weight) is equal to 4.45×10^{-8} cm.

If the imprecision of this calculational procedure is taken into account, then the agreement may be regarded as satisfactory.

Using the same considerations, it is possible to calculate the latent heat Q associated with the anisotropic–isotropic phase transitions.

In fact, according to Table 3, the sudden energy jump per molecule behaves as:

$$U = V_{S = S_k} - V_{S = 0} = 4.65 \times 10^{-14} \, \text{erg}$$

The number of molecules per gram of material is $n = \dfrac{1}{\frac{4}{3}\pi r^3 \rho}$, from which the

specific latent heat is:

$$Q = nU = \frac{3U}{4\pi r^3 \rho} \tag{21}$$

Substituting the values $U = 4.65 \times 10^{-14}$ erg, $\rho = 1.15$ and $Q = 1.78$ cal $= 7.45 \times 10^{-7}$ erg, yields:

[28] Kreutzer and Kast, Naturwiss., **25**, 233 (1937); Ann. Physik, **33**, 192 (1938).

$$r = 5 \times 10^{-8} \, cm$$

whose order of magnitude agrees with the values given.

Finally, the theory presented here explains why the heat capacity in the anisotropic fluid material is greater than it is in the amorphous phase.

The natural suspicion is that this excess specific heat accounts for the work associated with disorientation. If this is case, the heat capacity difference $C_{anis.} - C_{is.}$ must agree in order of magnitude with the temperature coefficient of the ordering energy.

$$C_{anis.} - C_{is.} = -n \frac{\Delta V}{\Delta T} = \frac{3}{4\pi r^3 \rho} \frac{\Delta V}{\Delta T} \tag{22}$$

From this, we obtain, using eq. (14):

$$r^2 = -\frac{3 \times 1.35}{4 \rho (C_{anis.} - C_{is.})} \frac{\Delta A}{\Delta T} \tag{23}$$

We calculate the mean value $\frac{\Delta A}{\Delta T}$ using the information in Table 3, in which $\Delta T = 36$ and $A = (2.4 - 9.7) \times 10^{-7}$ is assumed. The value $C_{anis.} - C_{is.}$ we take equal to 0.04 cal/deg . g $= 1.67 \times 10^6$ erg/deg . g.

Substituting this value in (23), we obtain

$$r = 9.6 \times 10^{-8} \, cm$$

whose order of magnitude agrees with the other information.

Physics Institute,
Leningrad State University

Submitted
22 April 1941

Victor Nikolaevich Tsvetkov was born in St Petersburg in 1910. He graduated from from Leningrad Pedagogical Institute in 1931, enrolling as a Ph.D student of V.K. Frederiks in Physics at Leningrad State University. He received his Ph.D in 1935, and after the arrest and deportation of Frederiks, he became the leading worker in liquid crystal science in Russia. He received his Doctor of Science degree in 1940 and was elected as a member of the Russian Academy of Sciences in 1968.

Tsvetkov's early work in liquid crystals was concerned with the orientational effects of electric and magnetic fields. By using a rotating magnetic field he was the first to determine the rotational viscosity and diamagnetic anisotropy of liquid crystals. He also worked on dynamical light scattering. The paper included in this volume came out of Tsvetov's experimental work in this area.

In his later work Tsvetkov turned to polymer physics, in which area he was a pioneer in the Soviet Union. He occupied the first chair in the Soviet Union in polymer and liquid crystal physics. He led a large group whose work ranged widely in the field: from flow birefringence, through optical effects, to equilibrium and non-equilibrium statistical mechanics. Latterly he also organised the Laboratory of Molecular Hydrodynamics and Optics at Institute of Macromolecular Compounds of the Russian Academy of Sciences. He continued to work and direct his laboratory right up to death from a heart attack in 1999 at the age of 89.

Eine einfache molekulare Theorie des nematischen kristallinflüssigen Zustandes

Von W. Maier und A. Saupe

Physikalisch-Chemisches und Physikalisches Institut der Universität Freiburg i. Br.

(Z. Naturforschg. 13 a, 564—566 [1958] ; eingegangen am 20. März 1958)

Nematische Phasen verdanken ihre Anisotropie einzig und allein der mehr oder weniger vollständigen Parallelorientierung der Längsachsen ihrer gestreckt geformten Moleküle [1]. Ihr Ordnungszustand läßt sich dementsprechend durch eine einzige Größe, den „Ordnungsgrad" $S = 1 - \frac{3}{2} \overline{\sin^2 \Theta}$ kennzeichnen; mit Θ ist hierbei der Winkel zwischen der Moleküllängsachse und der symmetrieachse der Flüssigkeit, mit $\overline{\sin^2 \Theta}$ der über sämtliche Moleküle gemittelte Wert von $\sin^2 \Theta$ bezeichnet. Bei idealer nematischer Ordnung ist $S = 1$, für die isotrope flüssige Phase gilt $S = 0$. Die Werte, die an nematischen Flüssigkeiten tatsächlich beobachtet wurden, liegen zwischen [2] $S = 0{,}2$ und $S = 0{,}8$. S nimmt mit steigender Temperatur ab und geht am Klärpunkt sprunghaft auf 0. Von einer Theorie nematischer Flüssigkeiten wird man in erster Linie verlangen, daß sie die Existenz dieser Fernordnung, die Größenordnung des Ordnungsgrades und dessen Temperaturabhängigkeit erklären kann.

Es liegt nahe — wie dies Born [3] und Zwetkoff [4] auch schon getan haben —, eine einfache statistische Theorie dieser Flüssigkeiten analog zur Weissschen Theorie des Ferromagnetismus zu entwickeln. Man nimmt also an, daß die das Verhalten eines einzelnen Moleküls bestimmenden zwischenmolekularen Kräfte durch ein Inneres Feld ersetzt werden können, welches allein vom Ordnungsgrad S und von der Dichte bzw. dem Molvolumen V der Flüssigkeit abhängt. Der uns hier allein interessierende winkelabhängige Anteil der potentiellen Energie eines beliebigen, durch den Index 1 gekennzeichneten Moleküls läßt sich also durch

$$D_1 \equiv D_1(\Theta_1, \overline{\sin^2 \Theta}, V)$$

beschreiben. Die Boltzmann-Statistik ergibt für das zeitliche Mittel von $\sin^2 \Theta_1$ die Beziehung

$$\overline{\sin^2 \Theta_1} = \frac{\int_0^{\pi/2} \sin^3 \Theta_1 \exp\{-D_1/k\,T\} \, d\Theta_1}{\int_0^{\pi/2} \sin \Theta_1 \exp\{-D_1/k\,T\} \, d\Theta_1} . \qquad (1)$$

Im Gleichgewicht muß $\overline{\sin^2 \Theta_1}$ gleich dem über sämtliche Moleküle gemittelten Wert $\overline{\sin^2 \Theta}$ sein:

$$\overline{\sin^2 \Theta_1} = \overline{\sin^2 \Theta} . \qquad (2)$$

Mit diesen beiden Gleichungen ist das statistische Problem im wesentlichen gelöst, denn sie gestatten uns, sobald D_1 bekannt ist, $\overline{\sin^2 \Theta}$ in Abhängigkeit von der Temperatur zu berechnen.

Born [3] hat für D_1 die potentielle Energie des permanenten elektrischen Dipolmoments im elektrischen Inneren Feld der Dipolflüssigkeit angenommen und keine voll befriedigenden Ergebnisse erhalten. Es erscheint dies heute auch nicht mehr verwunderlich, nachdem man weiß, daß die elektrischen Dipolkräfte keinen wesentlichen Anteil an der Existenz nematischer Phasen haben. Zwetkoff machte keine Annahmen über die Natur der zwischenmolekularen Kräfte, sondern ging von den am Azoxyanisol gemessenen S-Werten aus und leitete aus ihnen eine empirische D_1-Funktion ab. Mit dieser konnte er einige wesentliche Eigenschaften des kristallinflüssigen Zustands verständlich machen, womit die Brauchbarkeit des Weissschen Verfahrens nachgewiesen war.

Aus den Zusammenhängen zwischen Molekülstruktur und Klärpunktstemperatur scheint uns nun hervorzugehen, daß die Dispersionskräfte zwischen den stark anisotropen Molekülen die entscheidende Ursache für die Bildung der nematischen Phase sind [5]. Mit einigen vereinfachenden Annahmen haben wir durch eine quantenmechanische Störungsrechnung zweiter Ordnung folgenden Ausdruck für eine allein durch die Dispersions-

[1] Zusammenfassende Darstellungen: W. Kast, Angew. Chem. 67, 592 [1955]. — P. Chatelain, Bull. Soc. Franç. Minéralog. Cristallogr. 77, 323 [1954].

[2] W. Maier u. G. Englert, Z. Phys. Chem., N. F. 12, 123 [1957]. — P. Chatelain, Bull. Soc. Franç. Minéralog. Cristallogr. 78, 262 [1955].

[3] M. Born, S.B. Kgl. Preuß. Akad. Wiss., Phys.-Math. Kl. 1916, 614 und Ann. Phys., Lpz. 55, 221 [1918].

[4] V. Zwetkoff, Acta Physicochim. URSS 16, 132 [1942].

[5] W. Maier u. A. Saupe, Z. Naturforschg. 12 a, 668 [1957].

C3
Zeitschrift Naturforschung **13a**, 564–66 (1958)

A SIMPLE MOLECULAR THEORY OF THE NEMATIC LIQUID-CRYSTALLINE STATE

by

W. Maier and A. Saupe

Physical Chemistry and Physics Institute of the University of Freiburg i. Br.

(submitted 20 March 1958)

The anisotropy of nematic phases is due only to the more or less complete parallel orientation of the longitudinal axes of the elongated molecules in such phases.[1] Consequently, the state of order can be defined by a unique quantity, the 'degree of order' $S = 1 - \frac{3}{2}\overline{\sin^2\vartheta}$; where ϑ is the angle between the longitudinal axis of the molecules and the axis of symmetry of the liquid, and $\overline{\sin^2\vartheta}$ is the mean value of $\sin^2\vartheta$ taken over all molecules. For an ideal nematic order, $S = 1$; for the isotropic liquid phase $S = 0$. The values actually observed for nematic liquids vary between $S = 0.2$ and $S = 0.8$.[2] S decreases with increasing temperature and abruptly becomes equal to 0 at the clearing point. A theory of nematic liquids must above all provide an explanation for the existence of this long-range order, as well as for the order of magnitude of the degree of order, and for its temperature dependence.

It is desirable to develop a simple, statistical theory of these fluids similar to the Weiss theory of ferromagnetism. This has already been done by BORN[3] and ZWETKOFF.[4] Thus, the intermolecular forces determining the behaviour

[1] Reviews: W. KAST, Angew. Chem. **67**, 592 (1955); P. CHATELAIN, Bull. Soc. Franc. Mineral. Crist. **77**, 323 (1954).

[2] W. MAIER and G. ENGLERT, Z. Phys. Chem. (New Series) **12**, 123 (1957); P. CHATELAIN, Bull. Soc. Franc. Mineral. Crist. **78**, 262 (1955).

[3] M. BORN, Sitzungsber. Kgl. Preuss. Akad. Wiss. Phys.-Math. Kl. **1916**, 614 and Ann. Physik **55**, 221 (1918).

[4] V. ZWETKOFF, Acta Physicochim. URSS **16**, 132 (1942).

of an individual molecule are assumed to be replaceable by an internal field which depends only on the degree of order S and on the density (i.e. the molar volume V) of the liquid. Only the part of the potential energy of any molecule which depends on angle is of interest to us. This potential energy associated with a particular molecule designated by index 1 can be expressed by

$$D_1 = D_1\left(\vartheta_1, \overline{\sin^2\vartheta}, V\right)$$

Boltzmann statistics yields for the time-average of $\overline{\sin^2\vartheta_1}$ the relationship:

$$\overline{\sin^2\vartheta_1} = \frac{\displaystyle\int_0^{\pi/2} \sin^3\vartheta_1 \exp\{-D_1/kT\}\, d\vartheta_1}{\displaystyle\int_0^{\pi/2} \sin\vartheta_1 \exp\{-D_1/kT\}\, d\vartheta_1} \tag{1}$$

At equilibrium, $\overline{\sin^2\vartheta_1}$ must be equal to the mean value of $\overline{\sin^2\vartheta}$ taken over all molecules:

$$\overline{\sin^2\vartheta_1} = \overline{\sin^2\vartheta} \tag{2}$$

The statistical problem is essentially solved by these two equations because one can calculate $\overline{\sin^2\vartheta}$ as a function of the temperature from these equations if D_1 is known.

BORN[3] assumed D_1 to be the potential energy of the permanent electrical dipole moment in the electrical internal field of the dipolar liquid but he did not obtain completely satisfactory results. Today, this is not surprising because we know that the electrical dipole forces do not substantially contribute to the existence of nematic phases. ZWETKOFF did not make any assumptions about the nature of the intermolecular forces. Rather he proceeded from the measured S-values for azoxyanisole and derived from them an empirical D_1 function. He was able to explain some essential properties of the liquid crystalline state by means of this function, and in this way the usefulness of the WEISS method was confirmed.

It seems to us that the relationship between molecular structure and clearing point temperature indicates that the dispersion forces between the strongly anisotropic molecules are the decisive factors in the formation of the nematic phase.[5] On the basis of some simplifying assumptions we obtained the following expression for an energy D_1 determined only by dispersion forces from second-order perturbation calculation:

---

[5] W. MAIER and A. SAUPE, Z. Naturforsch. **12a**, 668 (1957).

$$D_1 = -\frac{A}{V^2}\left(1 - \frac{3}{2}\overline{\sin^2\vartheta_1}\right)\left(1 - \frac{3}{2}\overline{\sin^2\vartheta}\right) \tag{3}$$

A is a characteristic constant for each substance, and is independent of the temperature and of the volume.

When this energy function is used to calculate $\overline{\sin^2\vartheta_1}$ according to (1) as a function of $\overline{\sin^2\vartheta}$ for various values of the parameter A/TV^2, the curves shown in Fig. 1 are obtained. The intersections with the broken straight line $\overline{\sin^2\vartheta_1} = \overline{\sin^2\vartheta}$ yield the equilibrium values possible for $\overline{\sin^2\vartheta}$. It can be seen that in general several such solutions are obtained. From these, only those which correspond to a stable or at least to a metastable equilibrium are of physical significance. For these solutions the following condition holds:

$$\frac{d\left(\overline{\sin^2\vartheta_1}\right)}{d\left(\overline{\sin^2\vartheta}\right)} < 1$$

For instance, in the case of curve 3, only solutions P_1 and P_3 correspond to a possible equilibrium state. In fact, P_1 represents a nematic order ($S>0$) and P_3 represents the isotropic liquid phase ($S=0$). Which of the two equilibria represents the stable modification cannot yet be decided. Passing on to higher A/TV^2 values, (i.e. to lower temperatures) curve 2 is approached. The latter corresponds to the limiting case in which the slope of the curve at P_3 has risen to 1. The isotropic liquid phase can no longer exist at this temperature, whereas

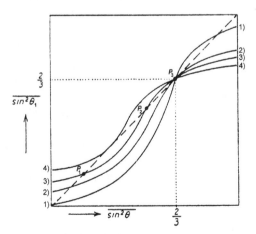

Fig. 1 Qualitative graph of $\overline{\sin^2\vartheta}$ for 4 different values of the parameters A/TV^2. TV^2 grows monotonically with the numbering.

the nematic phase has assumed a correspondingly higher degree of order. When the temperature is decreased to a value below T_2 (curve 1), a possible solution with $\overline{\sin^2 \vartheta_1} > 2/3$ occurs; this would correspond to a different type of order. However, it can be shown that this state can never be realized. When one proceeds from curve 3 to higher temperatures, another limiting case is reached with curve 4: there no longer exists a solution with $S > 0$ and the nematic phase can no longer exist.

The isotropic-liquid–nematic transition point thus lies between the two limiting cases (2) and (4). To determine the transition point, the free enthalpy difference between both phases must be calculated. This can be done without difficulty. We have introduced a constant m to take care of the mutual steric hindrance of the molecules. Among other things, this constant enters into the expression for the molar energy of the nematic order:

$$U_{order} = \frac{1}{2} \frac{N_L}{m} \frac{A}{V^2} S^2$$

The theory now contains two unknown constants. They can be determined from the transition temperature and the volume jump at the clearing point. For azoxyanisole: $A = 1.292 . 10^{-8}\,\mathrm{erg\,cm^6}$ and $m = 1.6$. With these two values S can now be calculated over the entire temperature region of the nematic phase of azoxyanisole.

The result of the calculation is shown by the broken curve in Fig. 2. The experimental S values presently available are given for comparison.* It is seen that the uncertainty of the experimental values is greater than their discrepancy with respect to the theoretical curve and that the theory correctly yielded the absolute magnitude as well as the temperature dependence of S.

The possibilities of the theory are not yet exhausted with the calculation of the degree of order. The following quantities can also be calculated from it: the heat of transition at the clearing point, the specific heat, the compressibility, the absorption of ultrasonic waves in the nematic phase, and the temperature dependence of the elastic constants of the nematic liquid. Finally, it is also possible to interpret the anomalous COTTON–MOUTON effect observed in the isotropic liquid phase.

We have calculated these quantities for azoxyanisole, and, wherever these were available, we have compared them with the experimental values. In all cases rather good agreement was found. It may thus be said that the theory is able to

---

* Recently, H. LIPPMANN and K.H. WEBER of the Institute of Physics at the University of Leipzig have determined the degree of order using proton resonance methods. According to a private communication from these authors, their values lie approximately 7 per cent higher than those obtained from diamagnetic susceptibility measurements.

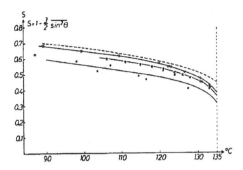

Fig. 2 4,4'-di-methoxy-azoxybenzene. Degree of order of the nematic phase determined from: ∇ diamagnetic susceptibility, + refractive index for $\lambda = 456\mu$, $\times$ refractive index for $\lambda = 589\mu$, $\lozenge$ UV-absorption, ◆ IR absorption, S theoretical curve.

explain both the existence of a nematic phase and also the majority of its properties. The theory must of course be applied to other substances before a final decision can be made. A detailed publication is being prepared.

These studies were carried out at the Institute of Physics at the University of Freiburg. We express our gratitude to Prof. Dr W. GENTNER, Director of the Institute, for his kind support. Furthermore, we are very grateful for the material and financial support received from the Deutsche Forschungsgemeinschaft [German Research Association].

Wilhelm Maier was born in 1913 in Villingen in the Black Forest region of Germany. From 1932 to 1937 he studied physics at the University of Freiburg. Together with his doctoral supervisor Wilhelm Kast, he then transferred to Halle, where Kast had been appointed to a professorship in physics. He was awarded Ph.D in physics from the University of Halle in 1937. He continued his collaboration with Kast, and was awarded his habilitation degree in the field of liquid crystals in 1943.

During the Second World War, in addition to his duties at the university, he worked with the optical firm Carl Zeiss Jena as part of his compulsory war work. After the end of the war he was one of those, together with Kast, who was directed by the U.S. Army to leave Halle. In 1946 he returned to the University of Freiburg as lecturer and senior scientist at the Institute of Physics. He was promoted to a titular professorship in 1952, and the next year he moved to the Institute of Physical Chemistry directed by the well-known spectroscopist Reinhard Mecke. In 1958, in collaboration with his graduate student Alfred Saupe, he developed the molecular-statistical theory of nematics included in this collection. In 1962 he was offered professorships by the universities of Kiel and Darmstadt, but remained in Freiburg, having been promoted to a full professorship in physics.

Maier died in Italy at the age of only 50 on 25 April 1964. He had been invited to deliver a lecture at the University of Pisa. After the lecture he had planned to rest for some days at the sea coast, where he met with a tragic lethal accident, which was never fully solved.

Alfred Saupe was born in Baden-weiler in the Black Forest region of Germany in 1925. At the time that he should have entered university the Second World War was raging, and Saupe was drafted into the air force. He was captured by the British and spent the period 1945–8 as a prisoner of war in England. He finally entered the Albert-Ludwigs University in Freiburg to study physics in 1949, graduating in 1955 for work on the UV spectroscopy of *p*-azoxyanisole. From there he continued work in Freiburg under the direction of Wilhelm Maier, both as an experimentalist and a theorist in liquid crystal problems. It was during this early period that Saupe carried out the work reprinted in this collection. In 1958 he was awarded a doctoral degree for this work.

Saupe continued to work in Freiburg in a number of positions until 1968, having received his Habilitation degree in 1967. In 1968 he accepted an invitation from Glenn Brown to join the Liquid Crystal Institute at Kent State University in Ohio. He retired from Kent State University in 1992, and returned to Germany as director of the liquid crystal group at the University of Halle. He retired from Halle in 1996.

Saupe has been one of the leading researcher in liquid crystals in the second half of the twentieth century. His interests have ranged from continuum theory and molecular theory, through optics, defects, and various different liquid crystal spectroscopies, through devices to lyotropic and amphiphilic systems.

A GENERAL DISCUSSION

ON

CONFIGURATIONS AND INTERACTIONS OF MACROMOLECULES AND LIQUID CRYSTALS

A GENERAL DISCUSSION on Configurations and Interactions of Macromolecules and Liquid Crystals was held in the Department of Chemistry, University of Leeds (by kind permission of the Vice-Chancellor) on the 15th, 16th and 17th April, 1958. The President, Sir Harry Melville, K.C.B., F.R.S., was in the Chair and over 200 members and visitors were present.

Among the distinguished overseas members and guests welcomed by the President were the following:

Dr. Acloque (France), Dr. K. Altenburg (Germany), Dr. H. Benninga (Netherlands), Prof. Dr. S. E. Bresler (Leningrad, U.S.S.R.), Mr. J. Bussink (Netherlands), Prof. D. G. Dervichian (France), Dr. E. W. Fischer (Germany), Mr. K. G. Götz (Germany), Mr. Y. M. de Haan (Netherlands), Dr. F. Halverson (Stamford, U.S.A.), Dr. B. Hargitay (Belgium), Dr. K. Heckmann (Göttingen), Dr. J. Hijmans (Netherlands), Dr. M. Joly (France), Mr. R. Koningsveld (Netherlands), Kr. V. A. Kropatshev (Leningrad, U.S.S.R.), Dr. M. Kryszewski (Poland), Dr. H. Lippmann (Germany), Prof. V. Luzzati (France), Prof. Dr. W. Maier (Germany), Dr. M. B. Matthews (Chicago, U.S.A.), Dr. H. Morawetz (New York, U.S.A.), Mr. H. Mustacchi (France), Dr. F. Patat (Germany), Prof. S. A. Rice (Chicago, U.S.A.) Mr. C. Ruscher (Germany), Dr. H. Schuller (Germany), Mr. A. Skoulios (France), Dr. H. Tompa (Belgium), Dr. J. Trommel (Netherlands), Mr. G. A. Voetelink (Oklahoma, U.S.A.), Dr. de Vries (France), Mr. R. A. Vroom (Netherlands), Prof. D. F. Waugh (Cambridge, U.S.A.), Dr. Karl-Heinz Weber (Germany), Dr. R. Westrik (Netherlands), Dr. J. Willems (Germany), Dr. Wippler (France).

C4

Discussions of the Faraday Society **25**, 19–28 (1958)

I. LIQUID CRYSTALS

ON THE THEORY OF LIQUID CRYSTALS

By F. C. Frank

H. H. Wills Physics Laboratory, University of Bristol

Received 19th February, 1958

A general theory of curvature-elasticity in the molecularly uniaxial liquid crystals, similar to that of Oseen, is established on a revised basis. There are certain significant differences: in particular one of his coefficients is shown to be zero in the classical liquid crystals. Another, which he did not recognize, does not interfere with the determination of the three principal coefficients. The way is therefore open for exact experimental determination of these coefficients, giving unusually direct information regarding the mutual orienting effect of molecules.

---

1. INTRODUCTION

One of the principal purposes of this paper is to urge the revival of experimental interest in its subject. After the Society's successful Discussion on liquid crystals in 1933, too many people, perhaps, drew the conclusion that the major puzzles were eliminated, and too few the equally valid conclusion that quantitative experimental work on liquid crystals offers powerfully direct information about molecular interactions in condensed phases.

The first paper in the 1933 Discussion was one by Oseen,[1] offering a general structural theory of the classical liquid crystals, i.e. the three types, smectic, nematic and cholesteric, recognized by Friedel[2] (1922). In the present paper Oseen's theory (with slight modification) is refounded on a securer basis. As with Oseen, this is a theory of the molecularly uniaxial liquid crystals, that is to say, those in which the long-range order governs the orientation of only one molecular axis. This certainly embraces the classical types, though in the smectic class one translational degree of freedom is also crystalline. Other types of liquid crystal may exist, but are at least relatively rare: presumably because ordering of one kind promotes ordering of another—it is already exceptional for *one* orientational degree of freedom to crystallize without simultaneous crystallization of the translational degrees of freedom. Fluidity (in the sense that no shear stress can persist in the absence of flow) is in principle compatible with biaxial orientational order, with or without translational order in one dimension. It is very unlikely that translational order in two dimensions, and not in the third, can occur as an equilibrium situation. The existence of a three-dimensional lattice is not compatible with true fluidity. A dilute solution with lattice order, which appeared to be fluid, would not be considered a liquid crystal from the present viewpoint, but rather a solid with a very low plastic yield stress.

The Oseen theory embraces smectic mesophases, but is not really required for this case. The interpretation of the equilibrium structures assumed by smectic substances under a particular system of external influences may be carried out by essentially geometric arguments alone. The structures are conditioned by the existence of layers of uniform thickness, which may be freely curved, but in ways which do not require a breach of the layering in regions of greater extension than lines. These conditions automatically require the layers to be Dupin cyclides and the singular lines to be focal conics. Nothing, essentially, has been added to the account of these given by Friedel, and not much appears to be needed, though a few minor features (the scalloped edges of Grandjean terraces, and scalloped frills of battonets) have not been fully interpreted.

The case is quite different for the nematic and cholesteric liquid crystals. This is particularly clear for the former. In a thin film, say, of a nematic substance, particular orientations are imposed at the surfaces, depending on the nature or prior treatment of the materials at these surfaces; if the imposed orientations are not parallel some curved transition from one orientation to the other is required. Curvature may also be introduced when, say, the orienting effect of a magnetic field conflicts with orientations imposed by surface contacts. Something analogous to elasticity theory is required to define the equilibrium form of such curvatures. It is, however, essentially different from the elasticity theory of a solid. In the latter theory, when we calculate equilibrium curvatures in bending, we treat the material as having undergone homogeneous strains in small elements: restoring forces are considered to oppose the change of distance between neighbouring points in the material. In a liquid, there are no permanent forces opposing the change of distance between points: in a bent liquid crystal, we must look for restoring torques which directly oppose the curvature. We may refer to these as torque-stresses, and assume an equivalent of Hooke's law, making them proportional to the curvature-strains, appropriately defined, when these are sufficiently small. It is an equivalent procedure to assume that the free-energy density is a quadratic function of the curvature-strains, in which the analogues of elastic moduli appear as coefficients: this is the procedure we shall actually adopt.

Oseen likewise proceeded by setting up an expression for energy density, in terms of chosen measures of curvature. However, he based his argument on the postulate that the energy is expressible as a sum of energies between molecules taken in pairs. This is analogous to the way in which Cauchy set up the theory of elasticity for solids, and in that case it is known that the theory predicted fewer independent elastic constants than actually exist, and we may anticipate a similar consequence with Oseen's theory.

It is worth remarking that the controversial conflict between the " swarm theory " and the " continuum theory " of liquid crystals is illusory. The swarm theory was a particular hypothetical and approximative approach to the statistical mechanical problem of interpreting properties which can be well defined in terms of a continuum theory. This point is seen less clearly from Oseen's point of departure than from that of the present paper.

2. BASIC THEORY

We first require to define the components of curvature. Let L be a unit vector representing the direction of the preferred orientation in the neighbourhood of any point. The sign of this vector is without physical significance, at least in most cases. If so, it must be chosen arbitrarily at some point and defined by continuity from that point throughout the region in which L varies slowly with position. In multiply-connected regions it may be necessary to introduce arbitrary surfaces of mathematical discontinuity, where this sign changes without any physical discontinuity. At any point we introduce a local system of Cartesian co-ordinates, x, y, z, with z parallel to L at the origin, x chosen arbitrarily perpendicular to z, and y perpendicular to x so that x, y, z form a right-handed system. Referred to these axes, the six components of curvature at this point are (see fig. 1):

$$\left. \begin{array}{ll} \text{" splay ":} & s_1 = \partial L_x/\partial x, \quad s_2 = \partial L_y/\partial y; \\ \text{" twist ":} & t_1 = -\partial L_y/\partial x, \quad t_2 = \partial L_x/\partial y; \\ \text{" bend ":} & b_1 = \partial L_x/\partial z, \quad b_2 = \partial L_y/\partial z. \end{array} \right\} \tag{1}$$

Then, putting

$$
\left.
\begin{aligned}
L_x &= a_1 x + a_2 y + a_3 z + 0(r^2), \\
L_y &= a_4 x + a_5 y + a_6 z + 0(r^2), \\
L_z &= 1 + 0(r^2), \ (r^2 = x^2 + y^2 + z^1),
\end{aligned}
\right\}
\tag{2}
$$

we have

$$
s_1 = a_1, \ t_2 = a_2, \ b_1 = a_3, \ -t_1 = a_4, \ s_2 = a_5, \ b_2 = a_6.
\tag{3}
$$

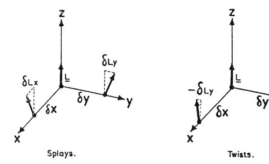

Splays.   Twists.

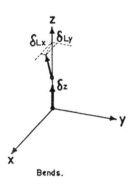

Bends.

FIG. 1.

We postulate that the free energy G of a liquid crystal specimen in a particular configuration, relative to its energy in the state of uniform orientation, is expressible as the volume integral of a free-energy density g which is a quadratic function of the six differential coefficients which measure the curvature:

$$
G = \int_v g\,d\tau,
\tag{4}
$$

$$
g = k_i a_i + \tfrac{1}{2} k_{ij} a_i a_j, \ (i,j = 1 \ldots 6, \ k_{ij} = k_{ji}),
\tag{5}
$$

where summation over repeated suffixes is implied.

In so far as there was arbitrariness in our choice of the local co-ordinate system (x, y, z), we require that when we replace this by another equally permissible one (x', y', z') in which we have new curvature components a'_i, g shall be the same function as before of these curvature components:

$$
g = k_i a'_i + \tfrac{1}{2} k_{ij} a'_i a'_j.
\tag{6}
$$

This requirement will impose restrictions on the moduli, k_i, k_{ij}.

The choice of the x-direction was arbitrary, apart from the requirement that it should be normal to z, which is parallel to the physically significant direction $\mathbf{L}$:

391

hence any rotation of the co-ordinate system around z is a permissible one. Putting $z' = y$, $y' = -x$, $z' = z$, gives us the equations*:

$$\left.\begin{aligned}
&k_1 = k_5, \ k_2 = -\ k_4, \ k_3 = k_6 = 0, \\
&k_{11} = k_{55}, \ k_{22} = k_{44}, \ k_{33} = k_{66}, \\
&k_{12} = -k_{45}, \ k_{14} = -k_{25}, \\
&k_{13} = k_{16} = k_{23} = k_{26} = k_{34} = k_{35} = k_{36} = k_{46} = k_{56} = 0
\end{aligned}\right\} \tag{7}$$

The working is omitted here: the simpler examples of eqn. (13-17) below exhibit the principle.

A rotation of 45° gives a further equation

$$k_{11} - k_{15} - k_{22} - k_{24} = 0, \tag{8}$$

and rotation by another arbitrary angle just one more

$$k_{12} + k_{14} = 0, \tag{9}$$

which with those obtained previously gives

$$k_{12} = -\ k_{14} = k_{25} = -\ k_{45}. \tag{10}$$

Thus, of the six hypothetical moduli k_i two are zero and only two are independent:

$$k_i = (k_1 \ k_2 \ 0 \ -k_2 \ k_1 \ 0), \tag{11}$$

while of the thirty-six k_{ij} eighteen are zero and only five are independent:

$$k_{ij} = \left\{\begin{matrix}
k_{11} & k_{12} & 0 & -k_{12} & (k_{11} - k_{22} - k_{24}) & 0 \\
k_{12} & k_{22} & 0 & k_{24} & k_{12} & 0 \\
0 & 0 & k_{33} & 0 & 0 & 0 \\
-k_{12} & k_{24} & 0 & k_{22} & -k_{12} & 0 \\
(k_{11} - k_{22} - k_{24}) & k_{12} & 0 & -k_{12} & k_{11} & 0 \\
0 & 0 & 0 & 0 & 0 & k_{33}
\end{matrix}\right\} \tag{12}$$

If the molecules are non-polar with respect to the preferentially oriented axis, or, if polar, are distributed with equal likelihood in both directions, the choice of sign of $\mathbf{L}$ is arbitrary. It is a significant convention in our definition of curvature components that z is positive in the positive direction of $\mathbf{L}$: and if z changes sign, one of x and y should change sign also to retain right-handed co-ordinates. Hence a permissible transformation *in the absence of physical polarity* is $\mathbf{L}' = -\mathbf{L}$, $x' = x$, $y' = -y$, $z' = -z$. This gives us

$$\left.\begin{aligned}
L'_{x'} &= -a_1 x' + a_2 y' + a_3 z' + 0(r^2), \\
L'_{y'} &= \quad a_4 x' - a_5 y' - a_6 z' + 0(r^2).
\end{aligned}\right\} \tag{13}$$

Since, compared with (2), the coefficients with indices 1, 5 and 6 have changed sign, the required invariance of (6) gives us the equations

$$k_1 = 0, \ k_5 = 0, \ k_6 = 0, \tag{14}$$

and (from the second order terms in which only one factor has changed sign):

$$k_{12} = k_{13} = k_{14} = k_{25} = k_{26} = k_{35} = k_{36} = k_{45} = k_{46} = 0. \tag{15}$$

Some of this information is already contained in eqn. (7-10). The effect upon (11) and (12) is that k_1 and k_{12} vanish.

There is a further element of arbitrariness in our insistence on right-handed co-ordinates, unless the molecules are enantiomorphic, or enantiomorphically

392

arranged. Empirically, it appears that enantiomorphy does not occur in liquid crystals unless the molecules are themselves distinguishable from their mirror images, and that it also vanishes in racemic mixtures. *In the absence of enantiomorphy*, a permissible transformation is $x' = x$, $y' = -y$, $z' = z$, giving

$$\left. \begin{array}{l} L_{x'} = a_1 x' - a_2 y' + a_3 z' + 0(r^2), \\ L_{y'} = -a_4 x' + a_5 y' - a_6 z' + 0(r^2), \end{array} \right\} \tag{16}$$

whence, by the same argument as before (omitting the redundant information),

$$k_2 = 0 \text{ and } k_{12} = 0. \tag{17}$$

Hence, while (15) with (11) and (12) expresses the most general dependence of free energy density on curvature in molecularly uniaxial liquid crystals, k_1 vanishes in the absence of polarity, k_2 vanishes in the absence of enantiomorphy, and k_{12} vanishes unless both polarity and enantiomorphy occur together.

The general expression for energy density in terms of the notation of eqn. (1) is

$$g = k_1(s_1 + s_2) + k_2(t_1 + t_2) + \tfrac{1}{2}k_{11}(s_1 + s_2)^2 + \tfrac{1}{2}k_{22}(t_1 + t_2)^2 + \tfrac{1}{2}k_{33}(b_1{}^2 + b_2{}^2)$$
$$+ k_{12}(s_1 + s_2)(t_1 + t_2) - (k_{22} + k_{24})(s_1 s_2 + t_1 t_2). \tag{18}$$

By introducing

$$s_0 = -k_1/k_{11}, \quad t_0 = -k_2/k_{22}, \tag{19}$$

and

$$g' = g + \tfrac{1}{2}k_{11}s_0{}^2 + \tfrac{1}{2}k_{22}t_0{}^2, \tag{20}$$

i.e. by adopting a new and (in the general case) lower zero for the free-energy density, corresponding not to the state of uniform orientation but to that with the optimum degree of splay and twist, we obtain the more compact expression

$$g' = \tfrac{1}{2}k_{11}(s_1 + s_2 - s_0)^2 + \tfrac{1}{2}k_{22}(t_1 + t_2 - t_0)^2 + \tfrac{1}{2}k_{33}(b_1{}^2 + b_2{}^2)$$
$$+ k_{12}(s_1 + s_2)(t_1 + t_2) - (k_{22} + k_{24})(s_1 s_2 + t_1 t_2). \tag{21}$$

An alternative form of this expression is given as eqn. (25) below.

2.2 COMPARISON WITH OSEEN'S THEORY

According to Oseen, the energy is expressed by

$$\frac{1}{2m^2} \int\int \rho_1 \rho_2 Q(\xi_1, \xi_2) \mathrm{d}\tau_1 \mathrm{d}\tau_2$$

$$+ \frac{1}{2m^2} \int \rho_2 \{ K_1 \mathbf{L} \cdot \nabla \times \mathbf{L} + K_{11}(\mathbf{L} \cdot \nabla \times \mathbf{L})^2 + K_{22}(\nabla \cdot \mathbf{L})^2$$

$$+ K_{33}((\mathbf{L} \cdot \nabla)\mathbf{L})^2 + 2K_{12}(\nabla \cdot \mathbf{L})(\mathbf{L} \cdot \nabla \times \mathbf{L}) \} \mathrm{d}\tau. \tag{22}$$

We are not concerned with the first integral, which is not related to the dependence of energy on curvature (and which plays only a minor role in Oseen's theory). It is the integrand of the second integral which should be compared with our free-energy density g. Noting that

$$\begin{array}{lll} \mathbf{L} \cdot \nabla \times \mathbf{L} = \partial L_y/\partial x - \partial L_x/\partial y & = -(t_1 + t_2), \\ \nabla \cdot \mathbf{L} = \partial L_x/\partial x + \partial L_y/\partial y & = (s_1 + s_2), & (23) \\ ((\mathbf{L} \cdot \nabla)\mathbf{L})^2 = (\partial L_x/\partial z)^2 + (\partial L_y/\partial z)^2 & = (b_1{}^2 + b_2{}^2), \end{array}$$

we see that with

$$\begin{array}{ll} -(\rho^2/2m^2)K_1 = k_2, \ (\rho^2/m^2)K_{11} = k_{22}, \ (\rho^2/m^2)K_{22} = k_{11}, \\ -(\rho^2/m^2)K_{12} = k_{12}, \ (\rho^2/m^2)K_{33} = k_{33}, \end{array} \tag{24}$$

Oseen's expression is equivalent to (18), except that the latter contains the additional terms,

$$k_1(s_1 + s_2) - (k_{22} + k_{24})(s_1 s_2 + t_1 t_2).$$

This accords with the anticipation that Oseen was in danger of missing terms by adopting a Cauchy-like approach to the problem.

The first omission is not very important, since it is virtually certain that there is no physical polarity along the direction **L** in any of the normal liquid crystal substances which were discussed by Oseen; and k_1 is then zero. The second omission is of more general significance: but $(s_1s_2 + t_1t_2)$ relates to an essentially three-dimensional kind of curvature. It occurs in pure form (with $(s_1 + s_2)$ and $(t_1 + t_2)$ equal to zero) in what we may call " saddle-splay ", when the preferred directions **L** are normal to a saddle-surface; and then contributes a positive term to the energy if $(k_{22} + k_{24})$ is positive. It is zero if **L** is either constant in a plane or parallel to a plane. It may be disregarded in all the simpler configurations which would be employed for the determination of moduli other than k_{24}, provided only that $(k_{22} + k_{24})$ is non-negative.

The most gratifying result of the comparison is to notice that there is one term in Oseen's expression which can be omitted, namely the last, since K_{12} is always zero under the conditions which justify omitting the term $k_1(s_1 + s_2)$. For many purposes we actually have a simpler result than Oseen's. In detailed application, he in fact assumed $K_{12} = 0$, supposing this to be an approximation.

We may conveniently use relations (23) to cast eqn. (21) into co-ordinate-free notation:

$$g' = \tfrac{1}{2}k_{11}(\nabla \cdot \mathbf{L} - s_0)^2 + \tfrac{1}{2}k_{22}(\mathbf{L} \cdot \nabla \times \mathbf{L} + t_0)^2 + \tfrac{1}{2}k_{33}((\mathbf{L} \cdot \nabla)\mathbf{L})^2$$
$$- k_{12}(\nabla \cdot \mathbf{L})(\mathbf{L} \cdot \nabla \times \mathbf{L}) - \tfrac{1}{2}(k_{22} + k_{24})((\nabla \cdot \mathbf{L})^2$$
$$+ (\nabla \times \mathbf{L})^2 - \nabla \mathbf{L} : \nabla \mathbf{L}), \tag{25}$$

where

$$\nabla \mathbf{L} : \Delta \mathbf{L} = \left(\frac{\partial L_x}{\partial x}\right)^2 + \left(\frac{\partial L_y}{\partial y}\right)^2 + \left(\frac{\partial L_z}{\partial z}\right)^2 + \left(\frac{\partial L_y}{\partial x}\right)^2 + \left(\frac{\partial L_y}{\partial y}\right)^2 + \left(\frac{\partial L_y}{\partial z}\right)^2$$
$$+ \left(\frac{\partial L_z}{\partial x}\right)^2 + \left(\frac{\partial L_z}{\partial y}\right)^2 + \left(\frac{\partial L_z}{\partial z}\right)^2,$$

or $\quad (\nabla \cdot \mathbf{L})^2 + (\nabla \times \mathbf{L})^2 - (\nabla \mathbf{L} : \Delta \mathbf{L})$

$$= 2\left\{\frac{\partial L_x}{\partial x}\frac{\partial L_y}{\partial y} + \frac{\partial L_y}{\partial y}\frac{\partial L_z}{\partial z} + \frac{\partial L_z}{\partial z}\frac{\partial L_x}{\partial x} - \frac{\partial L_y}{\partial x}\frac{\partial L_x}{\partial y} - \frac{\partial L_z}{\partial y}\frac{\partial L_y}{\partial z} - \frac{\partial L_z}{\partial z}\frac{\partial L_z}{\partial x}\right\},$$

in a fully arbitrary system of co-ordinates.

3. Particular cases

3.1. The smectic state

According to Oseen, the smectic state corresponds to the vanishing of all the moduli except k_{22} and k_{33} (our notation). The (free) energy is then minimized when

$$\mathbf{L} \cdot \nabla \times \mathbf{L} = 0, \quad (\mathbf{L} \cdot \nabla)\mathbf{L} = 0. \tag{26}$$

The second of these equations states that a line following the preferred direction of molecular axes is straight: the first, that the family of such straight lines is normal to a family of parallel surfaces (defining the surface parallel to a given curved surface as the envelope of spheres of uniform radius centred at all points on the given surface). Hence he formally predicts the geometry explicable by molecular layering without apparently appealing to the existence of layers: their real existence he explains separately by use of his Q integral. The present writer considers this a perverse approach. We need to explain why k_{22} and k_{33} are so large that the other moduli are negligible, and can do so from the existence of the layering. There is no real need to employ the theory of curvature strains to interpret smectic structures, until one requires to deal with small departures from the geometrically interpretable structures, which will be permitted if the other moduli are merely small, instead of vanishing, compared with k_{22} and k_{33}.

Before leaving the subject of the smectic state, we may remark that to explain deformations in which the area of individual molecular layers does not remain constant, it is necessary to invoke dislocations of these layers. It is likely that these dislocations are usually combined with the focal conic singularity lines, being then essentially screw dislocations.

3.2. THE NEMATIC STATE

The nematic state is characterized by $s_0 = 0$, $t_0 = 0$, $k_{12} = 0$, corresponding to the absence both of polarity and enantiomorphy. There are four non-vanishing moduli, k_{11}, k_{22}, k_{33} and k_{24}. The last has no effect in " planar " structures. Hence Zocher's [3] three-constant theory is justified (his k_1, k_2, k_t are the same as k_{11}, k_{33}, k_{22}). Formerly, this appeared to be only an approximate theory, neglecting k_{12} which appeared in the theory of Oseen. The simplest way to measure the moduli is to impose body torques by imposing magnetic fields. k_{22} can be determined straightforwardly; k_{11} and k_{33} are more difficult to separate from each other. There is evidence that they are about equal. Using the experimental data of Fréederickz and of Foëx and Royer, Zocher shows that if they are equal the value for p-azoxyanisole is 1.0×10^{-6} dynes. Information about the relative magnitude of these moduli is obtainable from the detailed geometry of the " disinclination " structures described in § 4. The fact that there is not another unknown in k_{12} ought to encourage a complete experimental determination of the moduli for some nematic substances.

3.3. THE CHOLESTERIC STATE

Enantiomorphy, either in the molecules of the liquid-crystal-forming substance, or in added solutes, converts the nematic into the cholesteric state. $k_2 \neq 0$, and the state of lowest free energy has a finite twist, $t_0 = k_2/k_{22}$. In the absence of other curvature components, there is only one structure of uniform twist, in which L is uniform in each of a family of parallel planes, and twists uniformly about the normal to these planes. The torsion has a full pitch of $2\pi/t_0$: but since L and $-$ L are physically indistinguishable, the physical period of repetition is π/t_0. When, as is usually the case, this is of the order of magnitude of a wavelength of light or greater, it can be measured with precision by optical methods. Some additional information should be obtainable by perturbing this structure with a magnetic field. Uniform twist can also exist throughout a volume when the repetition surfaces are not planes, but curved surfaces : for example, spheres, though in this case there has to be at least one singular radius on which the uniformity breaks down. Since the cholesteric substances, like the smectic substances (though for entirely different reasons) give rise to structures containing families of equidistant curved surfaces, their structures show considerable geometric similarity to those of the smectic substances. This was appreciated by Friedel, who also realized that it was misleading, and that the cholesteric phases are in reality thermodynamically equivalent to nematic phases.

3.4. THE CASE $k_1 \neq 0$

If $k_1 \neq 0$, the state of lowest free-energy density has a finite splay, $s_0 = k_1/k_{11}$. This can only exist when the molecules are distinguishable end from end, and there is polarity along L in their preferred orientation. Then, almost inevitably, the molecules have an electric dipole moment, and therefore, unless the material is an electric conductor, the condition $\nabla . \mathbf{L} \neq 0$ implies $\nabla . \mathbf{P} \neq 0$ (where $\mathbf{P}$ is the electric polarization), so that finite splay produces a space-charge. As a second consideration, it is not geometrically possible to have uniform splay in a three-dimensionally extended region. The simple cases of uniform splay are those in which L is radial in a thin spherical shell of radius $2/s_0$, or a thin cylindrical shell of radius $1/s_0$. These considerations relate this hypothetical polar class of liquid crystals to the substances which produce " myelin figures ": but since the only allowed structures for this class have a high surface-to-volume ratio, a theory of their configurations which does not pay explicit attention to interfacial tensions will be seriously incomplete.

4. " DISINCLINATIONS "

The nematic state is named for the apparent threads seen within the fluid under the microscope. Their nature was appreciated by Lehmann and by Friedel. In thin films they may be seen end on, crossing the specimen from slide to cover-slip, and their nature deduced from observation in polarized light. In this position they were named *Kerne* and *Konvergenzpunkte* by Lehmann, positive and negative nuclei (*noyaux*) by Friedel. They are line singularities such that the cardinal direction of the preferred axis changes by a multiple of π on a circuit taken round one of the lines. They thus provide examples of the configurations excluded (under the name of " Moebius crystals ") from consideration in the establishment of a general definition of crystal dislocations.[4] In analogy with dislocations, they might be named " disclinations ". It is the motion of these disclination lines which provides one of the mechanisms for change of configuration of a nematic specimen under an orienting influence, in the same way as motion of a domain boundary performs this function for a ferromagnetic substance. It is the lack of a crystal lattice which allows the discontinuities of orientation to have a line topology, instead of a topology of surfaces dividing the material into domains, in the present case. Disclination lines occur in cholesteric as well as in nematic liquid crystals.

The actual configuration around disclination lines was calculated by Oseen for the case $k_{11} = k_{33}$, $k_{12} = 0$, the latter assumption actually being exact. Then if L is parallel to a plane and ϕ is the azimuth of L in this plane, in which x_1, x_2 are Cartesian co-ordinates, the free energy is minimized in the absence of body torques when

$$\partial^2\phi/\partial x^2 + \partial^2\phi/\partial y^2 = 0. \tag{27}$$

The solutions of this equation representing disclinations are

$$\phi = \tfrac{1}{2}n\psi + \phi_0, \tan \psi = x_2/x_1, \tag{28}$$

n being an integer. These configurations are sketched in fig. 2. Changing ϕ_0, merely rotates the figure in all cases except $n = +2$, for which three examples are shown. Of these, the first or the third will be stable, for a nematic substance according as k_{33} or k_{11} is the larger. In the other cases, non-equality of k_{11} and k_{33} should not make drastic changes, but only changes of curvature in patterns of the same topology. Thus, for $n = -2$ or -1, if k_{33} is larger than k_{11} the bends will be sharpened, and conversely. This would be observed as a non-uniform rotation of the extinction arms with uniform rotation of polarizer and analyzer. The ratio of k_{11} and k_{33} can thus be determined from a simple optical experiment.

In cholesteric substances ϕ_0 is not constant, but a linear function of the co-ordinate x_3, normal to the x_1, x_2 plane: except for the case $n = 2$, with $k_{11} \neq k_{33}$, which should show a periodic departure from linearity near the core, from which the relative magnitudes of k_{22} and $(k_{11} - k_{33})$ could be deduced, though the observations would not be simple.

Oseen was puzzled at the non-occurrence of configurations corresponding to

$$\phi = c \ln r + \text{const.}, \ r = (x_1^2 + x_2^2)^{\frac{1}{2}}. \tag{29}$$

The reason is obvious: unlike n in (28), c is not restricted to integral values, and can relax continuously to zero. Alternatively stated, this configuration requires an impressed torque at the core for its maintenance.

At the time Friedel and Oseen wrote their papers, the values of the disinclination strength n which had been observed were $-2, -1, 1, 2$. Since then the case $n = 4$ has been observed by Robinson [5] in the radial singularity of a cholesteric " spherulite". The non-occurrence of high values of $|n|$ is explained by the fact that the energy is proportional to n^2. The fact that higher values than one occur indicates a relatively high energy in the disorderly core of the disinclination line, which must be as large as its field energy so that it becomes profitable for a pair of disinclinations to share the same core.

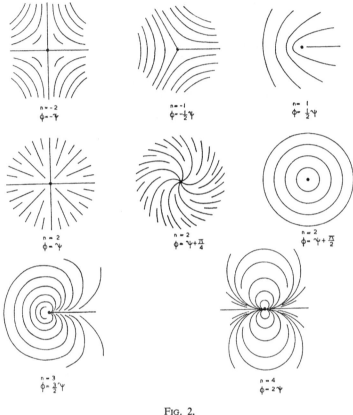

FIG. 2.

5. k_{24}

Let us leave aside the question of how to determine k_{24}, necessarily involving the observation of three-dimensional curvatures, until we have better information about the moduli of plane curvature, k_{11}, k_{22} and k_{33}.

6. RELATIONSHIP TO THE ORDINARY ELASTIC CONSTANTS

Let us take note that the molecular interactions giving rise to liquid crystal properties must also be present in solids. This indicates that conventional elastic theory is incomplete: the direct curvature-strain moduli should also be included. This is true, but does not seriously invalidate the accepted theory of elasticity. Consider the bending of a beam, of thickness $2a$, to a radius R. Then the stored free energy according to ordinary elasticity theory is $g_E = Ea^2/24R^2$, where E

397

is Young's modulus. The stored free energy arising from a curvature modulus k (corresponding to k_{11} or k_{33} according to the molecular orientation in the beam) would be $g_k = k/2R^2$. Their ratio is

$$g_k/g_E = 12k/Ea^2.$$

Taking k as 10^{-6} dyne, and E as 10^{10} dyne/cm^2, this ratio is unity when a is about $3\cdot5 \times 10^{-8}$ cm, and negligible for beams of thickness as large as a micron. Thus, on the visible scale, the curvature-elastic constants are always negligible compared with the ordinary elastic constants, unless the latter are zero.

[1] Oseen, *Trans. Faraday Soc.*, 1933, **29**, 883.
[2] Friedel, *Anales Physique*, 1922, **18**, 273.
[3] Zocher, *Trans. Faraday Soc.*, 1933, **29**, 945.
[4] Frank, *Phil. Mag.*, 1951, **42**, 809.
[5] Robinson, this Discussion.

*Editors' note: $x' = y$ ($z' = y$ is clearly a misprint).

Frederick Charles Frank (usually known as Charles), was born in 1911 in Durban, South Africa. His parents were English and his father a military man, and at the age of 11 weeks Frank moved to England, where apart from short periods abroad he spent the rest of his life. He entered Lincoln College, Oxford in 1929 as a student of chemistry, graduating in 1932. He then continued in Oxford, receiving his doctorate in 1938. This was followed by a postdoctoral period in the laboratory of Peter Debye at the Kaiser-Wilhelm Institut für Physik in Berlin. A short period doing colloid science in Cambridge was followed by war service. Frank worked in intelligence in the Air Ministry with R.V. Jones.

His particular skill was in deciphering photographs of sensitive installations, the story of which is told in Jones's book *My Secret War*.

After the war Frank accepted an invitation from Sir Nevill Mott (1977 Nobel prize winner in Physics) to join the Physics Department at the University of Bristol. He was promoted to research fellow in 1948, reader in 1951, professor in 1954, and directed the laboratory between 1969 and his retirement in 1976. His influence attracted many other fine researchers to Bristol.

Frank's most significant contributions are to the theory of solid dislocations. He also worked in crystal growth, geophysics, nuclear physics and polymer physics. Despite the significance of his contribution to the theory of liquid crystals, he was the author of very few papers in this subject, the first having been written in German while in Debye's laboratory in 1938.

Frank received many honours arising from his academic work, apart from receiving the O.B.E. in 1946 in recognition of his war work. He was elected to the Royal Society of London in 1954, was its vice-president in 1969, and received the Copley Medal of the Royal Society in 1994. He was knighted in 1977 and was also elected as a foreign associate both of the U.S. National Academy of Engineering (1980), and of the U.S. National Academy of Sciences (1987). He remained scientifically active up to his death in 1998.

C5
Zeitschrift für Elektrochemie **11**, 1171–77 (1959)

Isomorphiebeziehungen zwischen kristallin-flüssigen Phasen
4. Mitteilung: Mischbarkeit in binären Systemen mit mehreren smektischen Phasen

Von *HEINRICH ARNOLD* und *HORST SACKMANN*

Aus dem Institut für Physikalische Chemie der Universität Halle/Saale

(Eingegangen am 13. Juli 1959)

Mit Hilfe des Heiztischmikroskopes wurden 14 binäre Systeme, davon 10 ausschließlich nach der Kontakt-methode, untersucht. Die smektischen Modifikationen der Komponenten konnten auf Grund ihres gegen-seitigen Mischbarkeitsverhaltens in drei Gruppen zusammengefaßt werden. Die Resultate deuten darauf hin, daß diese Gruppen drei verschiedene Strukturtypen kristalliner Flüssigkeiten repräsentieren.

1. Einleitung

In Mitteilung I[1] [die Mitteilungen werden im folgenden mit (I, II, III) bezeichnet] wurde eine Zu-sammenfassung der bisherigen Untersuchungsergebnisse über die Mischbarkeit von Stoffen mit kristallin-flüssigen (kr.-fl.) Phasen gegeben und der Zusammenhang mit der Frage der Polymorphie im kr.-fl. Zustand aufgezeigt.

In der vorliegenden Arbeit werden Resultate von Untersuchungen über die Mischbarkeit smektischer Phasen wiedergegeben. Bereits aus der Möglichkeit des

Auftretens mehrerer smektischer Modifikationen bei ein und derselben Substanz ist zu schließen, daß die unter dieser Bezeichnung zusammengefaßten Phasen ver-schiedene Strukturen (molekulare Ordnungszustände) besitzen können. Es kann erwartet werden, daß sich derartige Strukturunterschiede im Mischbarkeitsverhal-ten bemerkbar machen. Die folgenden Versuche gehen nun der Frage nach, ob und in welchem Umfange dies geschieht[2].

[1] Mitt. I–III, Z. physik. Chem. (Leipzig) (im Druck).

[2] Eine ausführliche Darstellung ist H. Arnold, Disser-tation Halle (S.) 1959 zu entnehmen.

C5
Zeitschrift für Elektrochemie **11**, 1171–77 (1959)

ISOMORPHIC RELATIONSHIPS BETWEEN CRYSTALLINE-LIQUID PHASES

Part 4: Miscibility in binary systems with several smectic phases

by
Heinrich Arnold and Horst Sackmann
From the Institute of Physical Chemistry, University of Halle/Saale
(submitted 13 July 1959)

Using a hot stage microscope, 14 binary systems have been investigated, of which 10 systems have been investigated only by the contact method. Because of their mutual miscibility the smectic modifications could be classified into three groups. The results suggest that these groups represent three different structural types of crystalline liquids.

1. Introduction

In part I[1] [in the following the parts are named I, II, III] a summary is given of previous results concerning the miscibility of substances with crystalline-liquid (cr.-l.) phases and the question of polymorphism in liquid crystals.

In this paper results about the miscibility of smectic phases are given. From the possibility that one and the same substance exhibit several smectic modifications, one can conclude that phases grouped according to different classifications can exhibit different structures (molecular state of order). It can be expected that those structural differences should become obvious from miscibility properties. The following experiments answer the question if and to what extent this takes place.[2]

---

[1] Part I–III, Z. physik. Chem. (Leipzig) (in press) [*Editors note: Vol.* **213** *(1960), I: p. 127; II: p. 145; III: p. 262*].

[2] For details cf. H. Arnold, Ph. D Thesis, Halle 1959.

It should be mentioned here that smectic modifications of different substances being completely miscible will be classified to a particular group and labelled according to a simple notation (A, B or C).

The investigations were carried out using a hot stage microscope as described in Part II.

Substances

The preparation of 4,4′-dialkoxyazobenzenes is given in Part II; that of 4,4′-diethylazoxybenzoate in Part III. Vorländer mentioned only one cr.-l. phase in the case of dialkylazoxycinnamates (Table 1). However, our own microscopic investigations indicated a readily reproducible transition within the cr.-l. range for higher homologues (column 3) having a sharp transition onset with decreasing temperature.

In contrast to the compounds previously mentioned, the substances given in Tables 1 and 2 show a decrease in the transition temperatures after long heating, caused by decomposition.

Using a rapid procedure, however, reproducible data ($\pm 0.2\,°C$) could be obtained except with diethylazoxycinnamate. The transition of the smectic into the isotropic liquid in the case of dialkylazoxycinnamates took place over a noticeable temperature range (Table 1, column 2), and this was probably caused by impurities which

Table 1 4,4-Azoxy-di-n-alkylcinnamate*

Alkyl group	Temperature (°C)				
	1		2		3
	Vorl.	this	Vorl.	this	this
Ethyl	141	140	~264	~260–265	–
Hexyl	97	96.5	189	190.5–191	158.5
Heptyl	93	93	182	183–184	161.5
Decyl	98	98.5	164	166–166.5	160
Dodecyl	100	102.5	156	157.5	154.5

Column 1 = melting point
Column 2 = transition smectic A/isotropic liquid
Column 3 = transition smectic C/smectic A
Vorl. = after Vorländer;[3] this = this work

*The substances given in Tables 1 and 2 were given to us by courtesy of Prof. Langenbeck from the collection of the Institute of Organic Chemistry of the University of Halle. They date back to Vorländer's time.

[3] D. Vorländer, Z. phys. Chem. Abt. A *126*, 449 (1927).

Table 2 4-Alkoxybenzal-4-amino-ethylcinnamate*

Alkoxy group	Temperature (°C)							
	1		2		3		4	
	Vorl.	this	Vorl.	this	Vorl.	this	Vorl.	this
Methoxy	I: 83–85	81	90–92	92	118	118.5	139	139.5
	II: 106	108						
Ethoxy	76–78	77	112	116.5	153	157	157	158.5

Column 1 = melting point
Column 2 = transition smectic B/smectic A
Column 3 = transition smectic A/nematic
Column 4 = transition nematic/isotropic-liquid
Vorl. = after Vorländer;[4,5] this = this work

could not be eliminated by recrystallisation. This also occurred during the transition of the smectic into the nematic phase of the alkoxybenzalaminoethylcinnamates (Table 2) over a temperature range of typically 1–2 °C (and in their mixtures). Often, some degrees below the transition very small nematic areas become visible. In Table 2, column 3 (this), as well as in the phase diagram (Fig. 6) the temperatures given are those at which the last part of smectic phase disappeared on increasing temperature.

The methoxybenzalaminoethylcinnamate exhibits two different solid modifications named 'I' and 'II' by Vorländer. Modification 'I' could be obtained regularly after keeping the supercooled crystalline liquid at 63–66 °C for several hours, whilst the modification 'II' formed on cooling without any special precautions.

3. Binary systems of didodecyloxy-azobenzene, diethylazoxybenzoate, and dihexylazoxycinnamate

The cr.-l. phase of didodecyloxy-azobenzene together with non-nematic cr.-l. phases of the other dialkoxyazoxybenzenes were assigned to the smectic or diethylazoxybenzoate type by their discoverers Weygand and Gabler, without further comment. They were investigated microscopically by us, and as far as we could find (Part III) there were no noticeable differences between them. However, we did observe in the microscope a considerable difference between these phase modifications with respect to the smectic modification of the diethylazoxybenzoate.

[4] D. Vorländer, R. Wilke, U. Hempel, U. Haberland and J. Fischer, Z. Kristallogr., Mineralog., Petrogr., Abt. A 97, 485 (1937).
[5] D. Vorländer (together with R. Wilke, U. Haberland, K. Thinius, H. Hempel and J. Fischer), Ber. dtsch. chem. Ges., 71, 501 (1938).

Fig. 1 4,4'-Azoxy-di-n-hexylcinnamate 90×(a) Smectic phase A; (b) Smectic phase C.

Quite similar differences occur in the case of both cr.-l. modifications of dihexyl-azoxycinnamate and its higher homologues (Table 1). The cr.-l. modification of this compound which is stable above the transition temperature was labelled a smectic A phase, whilst that existing below this temperature has been designated as a smectic C phase.

The A phase, similar to the smectic diethylazoxybenzoate, is shown in Fig, 1a. This texture is formed from the isotropic liquid between the cover slip and object slide. We used polarised light with the oscillation directions of polariser and analyser parallel to the picture edges. We refer to Friedel[6] concerning the detailed description and structural interpretation of this 'fan texture'. This name originates from the picture of the material, which is divided into several areas with lines diverging like a fan. The extinction direction coincides with the direction of the lines running parallel to the picture edges, so they appear dark. Within the 'fan' the extinction direction always changes continuously.

In Fig. 1b the same substance is shown in the same detail, but after transformation into the smectic C modification. The fanlike areas are now transformed into domains sharply divided by boundaries, at which the extinction direction changed discontinuously. There is no longer any regularity for the extinction directions which is characteristic of smectic dialkoxyazobenzenes.

Further macroscopic texture investigations also indicated similarities between the high temperature smectic modification (A) of dihexylazoxycinnamate and the smectic diethylazoxybenzoate on one hand, and the low temperature smectic modification (C) of dihexylazoxycinnamate and the smectic dialkoxyazobenzenes on the other.[2]

Now by means of the three binary phase diagrams, the miscibilities of the smectic phases of these three compounds will be determined, and it will be seen if they correspond with the isomorphic relationships obtained from the microscopic textures.

[6] G. Friedel, Ann. Physique *18*, 274 (1922); G. and E. Friedel, Z. Kristallogr., Mineralog., Petrogr. Abt. A *79*, 1 (1931).

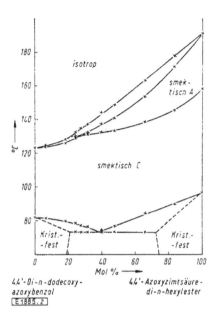

Fig. 2 Phase diagram.

(a) The system didodecyloxyazobenzene and dihexylazoxycinnamate

As shown in Fig. 2, the smectic C modification of dihexylazoxycinnamate is completely miscible with the cr.-l. phase of didodecyloxyazobenzene. The latter was also designated as a smectic C phase together with the mixtures. As already mentioned in Part I it was also found that in this case the textures of the homogeneous cr.-l. mixtures cannot be differentiated microscopically from those of the related pure phases. No heterogeneous region was observed for the transition between the smectic A and C phase mixtures. The same occured for several transitions between liquid phase mixtures in systems described in Part II. In both cases one can suppose that the heterogeneous region is extremely small and thus could not be detected. The three-phase coexistence at 130 °C of a smectic A phase mixture with a smectic C phase mixture and an isotropic-liquid mixture could be observed easily, especially by the contact method.

(b) The system diethylazoxybenzoate and dihexylazoxycinnamate

The smectic phase of diethyl-azoxybenzoate is completely miscible with the smectic A modification of the dihexylazoxycinnamate (Fig. 3). Therefore it was also identified as a smectic A phase. Near its minimum, the curve of the onset transition into the isotropic liquid of the smectic A phase mixture closely approaches the primary crystallisation curve of the diethylazoxybenzoate (Fig. 3). A miscibility gap, however, could be excluded by the investigation of

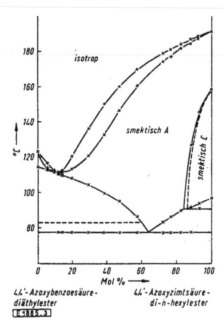

Fig. 3 Phase diagram.

the supercooled smectic mixtures. There is some indication of a heterogeneous area for the transition between the smectic A and C phase mixtures. However, this cannot be taken as certain. The enantiotropic transition of solid diethyl-azoxybenzoate between 81 and 85 °C also became obvious in the mixtures, and is shown by a dashed line; the same holds for Fig. 4.

(c) The system diethylazoxybenzoate and didodecyloxyazobenzene

The phase diagram is given in Fig. 4. In addition to regions of both smectic A and C, starting from the pure compounds, two further intermediate cr.-l. phases occur ('nematic' and 'smectic A'' in Fig. 4). One of them is typically nematic because of its microscopic appearance. According to Part II, section 5, one can expect a latent nematic phase for didodecyloxyazobenzene, for which a transition into the isotropic liquid would take place just below the 'smectic/isotropic' transition temperature. Furthermore the occurrence of a stable nematic phase in the mixtures can be understood by assuming that the addition of diethylazoxybenzoate causes a smaller decrease of the latent smectic/nematic transition point than the 'smectic/isotropic' transition. The nematic phase could be seen as a 'stabilised intermediate phase'. Incidentally we found for the first time a heterogeneous region of detectable width for the transition between the nematic and isotropic-liquid mixture phases.

406

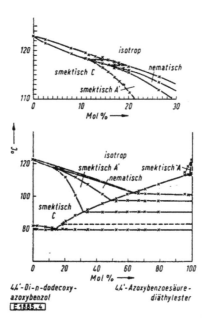

Fig. 4 Phase diagram (The region between 110–125 °C and 0–30 per cent is enlarged above).

No heterogeneous region could be detected for the transition between the second intermediate cr.-l. phase (smectic A′) and either the nematic intermediate phase or the smectic C phase. From a microscopic test of their different textures, this phase exhibited a close similarity to the smectic diethylazoxybenzoate; therefore it was designed as a smectic A′ phase. To confirm this relationship, we prepared a contact specimen of dihexylazoxycinnamate and a mixture of 15.8 mol% diethylazoxybenzoate and 84.2 mol% didodecyloxyazobenzene (cf. Fig. 4 top).

One can suppose that the concentrations in the contact area are on the vertex line of the concentration triangle of that ternary system whose binary border systems are given by Figs 2, 3 and 4 (Fig. 5). We found a complete ternary mixing series between the smectic A′ phase of the binary mixture and the smectic A phase of the azoxycinnamate; the smectic C phases as well as the isotropic-liquid phases were completely miscible. Thus, the miscibility gap between the phases A and A′ in Fig. 4 vanishes on passing over to the ternary system (Fig. 5).

It should be emphasised that the most important result of our observations is that the smectic A phase of diethylazoxybenzoate is *not* completely miscible with the smectic C phase of the didodecyloxyazobenzene.

The miscibility behaviour of the three compounds investigated in their cr.-l. state consequently is completely analogous to the texture results given at the

407

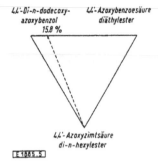

Fig. 5 Concentration triangle. Dashed line is the vertex line.

beginning of this section 3. Those smectic phases of the pure compounds, which must be taken to be related because of their microscopic texture phenomena, exhibit complete miscibility in their binary systems. Immiscibility is found if these texture relationships are not present.

It should be mentioned also that the smectic phases of dioctyloxy-, dinonyloxy- and didecyloxyazobenzene belong to the group of smectic C phases as follows from their miscibility relationships given in Part II.

4.   The system methoxy- and ethoxybenzal-aminoethylcinnamate

Both compounds were investigated, amongst other trimorphic cr.-l. substances, by X-ray scattering by Herrmann[7] (see Part I). Their high temperature smectic modifications as well as the low temperature ones exhibited comparable X-ray patterns. This is also the case for their miscibility as shown by Fig. 6. The smectic modifications, named smectic A and B phases (a justification for these terms is given in Section 5) are completely miscible, as are the nematic and isotropic-liquid phases.

Heterogeneous areas could not be detected for any of the transitions between the liquid phases of the mixtures. The curve for the transition between smectic A and B phases partially lies within the area of the supercooled stable crystalline-solid modification (melting points enclosed in brackets in Fig. 6). We could not obtain a clear picture of the temperature range for the crystalline-solid phase, by contrast with the cr.-l. phases, which is our main interest. Probably there is complete miscibility of the solid ethoxybenzalaminoethylcinnamate and the metastable modification of methoxybenzalaminoethylcinnamate. Also, because of the high viscosity of the smectic B phase, a three-phase coexistence point and thus a miscibility gap in the solid state could have been overlooked.

The transition between the smectic A and B phases took place similarly for the pure compounds and the mixtures, and could also be observed between the

[7] K. Herrmann, Z. Kristallogr., Mineralog., Petrogr. Abt. A 92, 49 (1935).

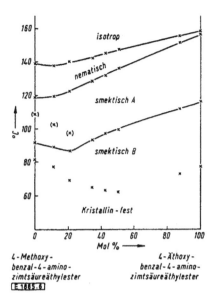

Fig. 6 Phase diagram.

cover slip and object slide for substances previously melted into the isotropic liquid, contrary to Vorländer's statement.[4,5] Decreasing the temperature below the transition, the optically perceivable discontinuities gradually vanished, but returned discontinuously on increasing the temperature.[2]

According to Gabler[8] the high temperature smectic modifications, as well as the low temperature ones of 4-ethoxy- and 4-n-propyloxybenzal-4-aminoethyl-cinnamate are completely miscible. Both smectic modifications of the latter compound thus are to be labelled smectic A and B phases according to our scheme (cf. Introduction).

5. Results from further systems by the contact method

Results obtained by the contact method (see Part II) proved to be very reliable. For the system given in Fig. 4 all 7 three-phase coexistence points and their temperatures were obtained exactly by this method, and the nature of the phase diagram could be seen before detailed investigations were carried out.

In order to gain more information about the miscibility of smectic phases we investigated several binary systems using only the contact method, and without a quantitative determination of the phase diagrams. The miscibility results for cr.-l. phases are given in Table 3 with the two components and their cr.-l. modifications. The polymorphism is shown as a sequence of modifications from top

[8] R. Gabler, Ph.D. Thesis, Leipzig 1939.

to bottom on decreasing temperature. A straight line between the phases indicates *complete* miscibility; the *absence* of the line means *incomplete* miscibility. Isotropic-liquid phases were always miscible. Regarding the data for three-phase coexistence points and schematic phase diagrams we refer to reference.[2]

We designated the high temperature smectic modification of ethoxybenzalamino-ethylcinnamate as a smectic A phase because of its complete miscibility with the smectic diethylazoxybenzoate (system 1 in Table 3). Herrmann[7] obtained identical X-ray results for both A phases; furthermore, the microscope textures are identical.

Neither the high nor the low temperature modification of the ethoxybenzalamino-ethylcinnamate were completely miscible with the smectic C phase of the dialkoxyazoxybenzenes (system 2, 3 and 4, Table 3). As the low temperature modification of the cinnamate ester mentioned above is quite different from the smectic type C, and also because of its texture we identified it as a 'smectic phase B'.

The results of systems 5, 6 and 7 show that the cr.-l. high and low temperature modifications of dihexyl-, didecyl- and didodecylazoxycinnamates are to be assigned

Table 3 Results of the contact method*

Component 1			Component 2
1. 4,4′-azoxydiethyl-benzoate		Nem	4-ethoxybenzal
	SmA —	SmA	4-aminoethyl
		SmB	cinnamate
2. 4,4′-Di-n-dodecyloxy-azoxybenzene		Nem	4-ethoxybenzal
	SmC	SmA	4-aminoethyl
		SmB	cinnamate
3. 4,4′-Di-n-decyloxyazoxy-benzene	Nem —	Nem	4-ethoxybenzal
		SmA	4-aminoethyl
	SmC	SmB	cinnamate
4. 4,4′-Di-n-octyloxyazoxy-benzene	Nem —	Nem	4-ethoxybenzal
		SmA	4-aminoethyl
	SmC	SmB	cinnamate
5. 4,4′-Azoxy-di-n hexylcinnamate	SmA —	SmA	4,4′-Azoxy- di-n-heptyl-cinnamate
	SmC —	SmC	
6. 4,4′-Azoxy-di-n-heptyl-cinnamate	SmA —	SmA	4,4′-Azoxy-di-n-decyl-cinnamate
	SmC —	SmC	
7. 4,4′-Azoxy-di-n-heptyl-cinnamate	SmA —	SmA	4,4′-Azoxy-di-n-dodecyl-cinnamate
	SmC —	SmC	
8. 4,4′-Azoxy-di-n-heptyl-cinnamate	SmA---(SmA)		4,4′-Azoxy-di-ethyl-cinnamate
	SmC		
9. 4,4′-Azoxy-di-ethylbenzoate	SmA---(SmA)		4,4′-Azoxy-di-ethyl-cinnamate
10. 4,4′-Di-n-dodecyloxy-azoxybenzene	SmC	(SmA)	4,4′-Azoxy-di-ethyl-cinnamate

*Editor's note: we have used the modern abbreviations for smectic A (SmA), smectic C (SmC) and smectic B (SmB).

to the smectic A or C types, respectively; the same holds for the assignment of the dihexylester according to Section 3.

Because of the decomposition hazard of the diethylazoxycinnamate at higher temperatures, the results for systems 8, 9 and 10 must be taken as uncertain. Probably, their smectic phases are of type A. There is no indication of demixing in the cr.-l. state in system 9 as supposed by Friedel.[6]

6. Summary and discussion

Generally speaking, the smectic modifications of the compounds investigated proved to be typical concerning their mutual miscibility. Consequently, they should be related to each other in the way shown by Table 4. The full lines indicate, as in Table 3, the existence of complete miscibility between the phases concerned. (As to the dashed line for diethylazoxycinnamate see Section 5.) The notations (H) and (T) refer to high and low temperature modifications.

Our picture of related modifications characterised by complete miscibility within each single group is self-contained. At the moment there is no example of complete miscibility between phases of different groups. This might be expected for the combination of the smectic A and B types as well as for A and C, only by the fact that they appear together in the same substances (alkoxybenzalaminoethyl-cinnamates and dialkylazoxycinnamates).

The relationships given in table 4 can be simply explained in this way, that phases belonging to the same group have a common or very closely related structure. Consequently, they are completely miscible, whilst incomplete miscibility of different phase types is caused by structural differences.

This assignment, based only on miscibility relations, is a broad assertion which deserves further support, possibly from other criteria of isomorphism (see Part I). Phases of the same group possess extensive similarities in their microscopic textures, whilst there are characteristic differences between phases of group A and C as well as between B and C; the differences between phases A and B are smaller [see Section 3 for phases A and C, also[2]].

Furthermore, within the homologous series the complete miscibility runs parallel to the regularities of the transition temperatures of the smectic phases. (A complete investigation of the dependence of the transition temperatures on the chain length in the series of dialkylazoxycinnamate is still to be done.) As far as X-ray data of the mixture components are available [as for the system given in Fig. 6 and system 1 in Table 3] complete miscibility is found if the scattering pictures of the modifications are identical.

With regard to the assumption of a relationship between structure and miscibility, it follows from Herrmann's X-ray results[7] that phases of group A form a smectic structure like that of Friedel (molecular layers with random lateral molecular distances), whilst for phases of group B a smectic structure with a lattice within the molecular layers can be assumed. Phases of type C should possess a common third type of structure, but at present there is no indication of its nature.

Table 4 Smectic phases

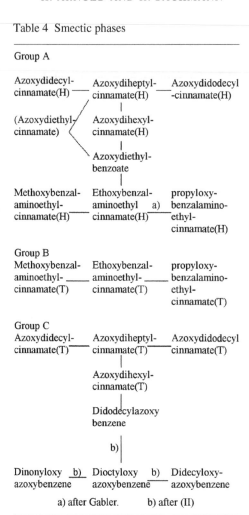

Group A

Azoxydidecyl-   Azoxydiheptyl-   Azoxydidodecyl
cinnamate(H)   cinnamate(H)   -cinnamate(H)

(Azoxydiethyl-   Azoxydihexyl-
cinnamate)   cinnamate(H)

                      Azoxydiethyl-
                      benzoate

Methoxybenzal-   Ethoxybenzal-   propyloxy-
aminoethyl-   aminoethyl-   a)   benzalamino-
cinnamate(H)   cinnamate(H)   ethyl-
                                                  cinnamate(H)

Group B
Methoxybenzal-   Ethoxybenzal-   propyloxy-
aminoethyl- _____   aminoethyl- _____   benzalamino-
cinnamate(T)   cinnamate(T)   ethyl-
                                                 cinnamate(T)

Group C
Azoxydidecyl-   Azoxydiheptyl-   Azoxydidodecyl
cinnamate(T)   cinnamate(T)   cinnamate(T)

                     Azoxydihexyl-
                     cinnamate(T)

                     Didodecylazoxy
                     benzene

                  b) |

Dinonyloxy _b)   Dioctyloxy   b)   Didecyloxy-
azoxybenzene   azoxybenzene   azoxybenzene

a) after Gabler.   b) after (II)

As observed for the smectic phases A′ and A in the system shown in Fig. 4, it can happen that two phases of the same group are not completely miscible. The reason for this behaviour is probably not a difference in their structural type but minor energetic and steric effects. One cannot conclude without further consideration that there is a different structure even if there is not complete miscibility, e.g. of the smectic B and C phases in the systems 2, 3 and 4 of Table 3. These types were never found simultaneously in one and the same substance, as is the case for types A and B as well as for A and C, which can be taken as the most reliable indication of a different structure. For confirmation we must take account of the result of microscopic texture investigations showing characteristic differences.

As a continuation of our work, we have to investigate how far smectic modifications of other chemical compounds, especially of homologous series, can be fitted into

412

this definite pattern, and if further smectic types exist. By including other methods, especially the X-ray technique, assumptions about the relationships between miscibility and phase structure must be proved.

Finally we note that according to our results on mixtures, the transition between two smectic modifications takes place at a three-phase coexistence point if a third phase is involved. This is an argument that these modifications must be distinguished as phases, and that their mutual transformations are of first order.

Horst Sackmann was born in Freiburg im Breisgau, in the Black Forest region of western Germany, in 1921. He would normally have attended his local university in Freiburg, but when Sackmann was eighteen the Second World War began and it was temporarily closed. He instead entered the University of Halle in 1939, where he studied chemistry. This was where his first contact with liquid crystals took place, under Wilhelm Kast, who was professor of physics there. Once the university in Freiburg reopened, he continued his studies in Freiburg until he was drafted the army in 1941. At the end of the war he resumed his studies in Halle, obtaining his Ph.D in 1949 with a thesis on the thermodynamics of mixing.

His habilitation thesis in physical chemistry, which he submitted in 1954, concerned the material properties of spherically shaped molecules. At that stage he decided to extend his studies on the isomorphism of rod-shaped components. In Halle Sackmann found an extensive collection of samples in the Institute of Physical Chemistry, dating right back to the time of Daniel Vorländer.

At that time it was known from Vorländer's and Friedel's results that liquid crystalline compounds could exist as polymorphic modifications. By means of extensive miscibility investigations he was able to throw light on the tangle of various polymorphic liquid crystal modifications, especially for smectics. Heinrich Arnold, his co-author in the paper included in this collection, was his doctoral student in this area.

In 1958 he was appointed to a full professorship in physical chemistry and became Director of the Institute of Physical Chemistry in Halle. His election as Vice-President of the *Deutsche Akademie der Naturforscher Leopoldina*, the oldest academy in the world, in 1973, indicated the esteem in which his peers held him.

After his retirement he continued his research in liquid crystals, as well as actively involving himself in scientific administration. In the period after German reunification in 1989 he played an active role in the political renewal of the University of Halle. He died in 1993.

Heinrich Arnold was born in Merseburg, close to Halle in the eastern part of Germany, in 1933. In 1951 he entered the University of Halle to study chemistry. He obtained his Ph.D degree in 1959 under the supervision of Horst Sackmann. He was Sackmann's first doctoral student in the field of liquid crystal polymorphism, and paper C5 is one of the papers from his Ph.D work.

Afterwards, he focussed his interests on very precise calorimetric measurements of phase transition enthalpies in liquid crystals. One of the by-products of this work was the calorimetric indication of the cholesteric-blue phase transformation, performed in collaboration with Roediger in 1967.

In 1964 he obtained his habilitation degree in physical chemistry. In the same year he was appointed as full professor of chemistry at the Technical University of Ilmenau (Thuringia). Because of political restriction of the East German Government at that time he was not allowed to continue his investigations on liquid crystals and worked on semiconductors and microelectronics as professor of physical chemistry in the Physical Department of Electronic Devices.

He retired in 1998.

C6
Archives for Rational Mechanics and Analysis **28**, 265–83 (1968)

Some Constitutive Equations for Liquid Crystals

F. M. Leslie

Communicated by J. L. Ericksen

1. Introduction

Since their discovery at the end of the last century, the materials called liquid crystals have aroused considerable interest. Briefly, liquid crystals are states of matter which are capable of flow, and in which the molecular arrangement gives rise to a preferred direction. Brown & Shaw [1] discuss many of their properties in an extensive review. A number of attempts have been made to formulate continuum theories to describe properties of these peculiar liquids. The earliest seems to be Oseen's static theory [2], which was later revised by Frank [3]. More recently Ericksen [4] has reformulated these static theories. Anzelius [5] and Oseen [6] have put forward dynamical theories to describe the behavior of liquid crystals. However, Anzelius did not employ concepts now commonly used in continuum mechanics. Also, Oseen's theory is rather general and has not been developed. The static part of this theory has a precedent in the earlier work of Duhem [7]. In recent years, with a view to describing the behavior of these materials, Ericksen [8] has formulated a simple theory of anisotropic liquids. However, as Ericksen pointed out, his theory appeared inadequate for this task, since it did not reduce statically to the Oseen-Frank theory. Also, in Ericksen's anisotropic liquids it is not possible to prescribe boundary conditions for the preferred direction, and these appear to be important in many phenomena.

Motivated by his hydrostatic theory of liquid crystals, Ericksen [9] proposed rather general conservation laws for the dynamical behavior of these liquids. Adopting these laws, Leslie [10] formulated constitutive equations similar to those discussed earlier by Ericksen [8], and essentially recovered Ericksen's theory of anisotropic liquids. Following the scheme proposed by Coleman & Noll [11], Leslie discussed the restrictions placed upon these constitutive equations by the Clausius-Duhem inequality. Here I make broader constitutive assumptions, and obtain a theory, which, apart from minor differences, reduces statically to the theory discussed by Ericksen [4]. Here, however, a generalization of the entropy inequality due to Müller [12] is adopted, *viz*, the entropy flux is not assumed equal to the heat flux divided by temperature. In general, one finds that the entropy flux differs from this quantity in a non-trivial manner. However, in the special case in which the preferred direction is described by a vector of fixed magnitude, the classical result holds.

In the second part of the paper shear flows of these liquids are examined. In the solutions obtained, sufficiently far from the boundary, the preferred direction has the value given by the stable solution of Ericksen's earlier theory [13]. In certain circumstances, the preferred direction has this value throughout the liquid apart from "boundary layers" at the bounding surfaces.

Finally, I compare the theoretical results obtained here with the experimental measurements of PORTER & JOHNSON [14]. In this connection the measurements of SAUPE [15] and MIESOWICZ [16] and the calculations of ERICKSEN [17] are helpful. It is found that the analysis presented here provides a possible explanation of the "adsorption layer" phenomenon mentioned by PORTER & JOHNSON. Also, variations in apparent viscosity are predicted at low rates of shear.

2. Kinematics

The motion of the continuum is referred to a fixed set of Cartesian axes. Let x_i denote the position of a typical particle at current time t, and $\xi_i(\tau)$ its position at a previous time τ, where

$$\xi_i(\tau) = \xi_i(x_1, x_2, x_3, t, \tau), \quad x_i = \xi_i(t), \quad -\infty < \tau \leq t. \tag{2.1}$$

The components of velocity of the particle are $v_i(\tau)$, and

$$v_i(\tau) = \frac{d}{d\tau} \xi_i(\tau),$$

where $d/d\tau$, which denotes differentiation with respect to τ keeping x_i and t constant, is the material time derivative. We employ the notation

$$v_i = v_i(t),$$

and define the tensors A_{ij} and ω_{ij} by

$$2A_{ij} = v_{i,j} + v_{j,i}, \quad 2\omega_{ij} = v_{i,j} - v_{j,i}, \tag{2.2}$$

where $v_{i,j}$ is the partial derivative of v_i with respect to x_j.

Associated with the particle is a director $d_i(\tau)$ which describes the preferred direction, and

$$d_i(\tau) = d_i(x_1, x_2, x_3, t, \tau), \quad -\infty < \tau \leq t. \tag{2.3}$$

The director may vary in magnitude, and it has the dimensions of a length. A director velocity $w_i(\tau)$ is defined by

$$w_i(\tau) = \frac{d}{d\tau} d_i(\tau).$$

It is convenient to employ the notation

$$d_i = d_i(t), \quad w_i = w_i(t),$$

and to introduce the quantities N_i and N_{ij} where

$$N_i = w_i + \omega_{ki} d_k, \quad N_{ij} = w_{i,j} + \omega_{ki} d_{k,j}. \tag{2.4}$$

We consider motions of the continuum which differ from that given by equations (2.1) and (2.3) only by superposed rigid body motions,

$$\xi_i^*(\tau) = c_i^*(\tau) + Q_{ij}(\tau)\{\xi_j(\tau) - c_j(\tau)\},$$
$$d_i^*(\tau) = Q_{ij}(\tau) d_j(\tau), \tag{2.5}$$

where $c_i^*(\tau)$ and $c_i(\tau)$ are vector functions of τ, and $Q_{ij}(\tau)$ is a proper orthogonal tensor function of τ. Above quantities are defined in terms of the original motion. We denote corresponding quantities, derived from the motion given by equations (2.5), by the same symbol with an asterisk attached. One can readily show that

$$A_{ij}^* = Q_{ip}Q_{jq}A_{pq}, \qquad \omega_{ij}^* = Q_{ip}Q_{jq}\omega_{pq} + \Omega_{ij},$$
$$N_i^* = Q_{ip}N_p, \qquad N_{ij}^* = Q_{ip}Q_{jq}N_{pq}, \tag{2.6}$$

where

$$Q_{ij} = Q_{ij}(t), \qquad \Omega_{ij} = Q_{jk}\frac{d}{dt}Q_{ik},$$

and d/dt denotes the material time derivative evaluated at current time t.

3. Conservation Laws and Entropy Inequality

In order to describe the dynamical behavior of liquid crystals, we adopt the conservation laws proposed by ERICKSEN [9]. GREEN, NAGHDI & RIVLIN [18] have discussed these laws in a more general context, and BEATTY [19] considers this topic in a recent paper. An account is also to be found in the work of TRUESDELL & NOLL [20]. For completeness a brief outline of a derivation of the conservation laws is given below.

For a material volume V bounded by a surface A, we assume that

$$\frac{d}{dt}\int_V \rho(\tfrac{1}{2}v_iv_i + \tfrac{1}{2}w_iw_i + U)\,dV$$
$$= \int_V \rho(r + F_iv_i + G_iw_i)\,dV + \int_A (t_iv_i + s_iw_i - h)\,dA, \tag{3.1}$$

and

$$\frac{d}{dt}\int_V \rho w_i\,dV = \int_V (\rho G_i + g_i)\,dV + \int_A s_i\,dA, \tag{3.2}$$

where ρ is density, U internal energy per unit mass, r heat supply per unit mass, per unit time, F_i body force per unit mass, G_i external director body force per unit mass, t_i surface force per unit area, s_i director surface force per unit area, h the flux of heat out of the volume per unit area, per unit time, and g_i an intrinsic director body force per unit volume. Also, for superposed rigid body motions, it is assumed that

$$\rho^* = \rho, \qquad U^* = U, \qquad h^* = h, \qquad r^* = r,$$
$$t_i^* = Q_{ij}t_j, \qquad s_i^* = Q_{ij}s_j, \qquad g_i^* = Q_{ij}g_j,$$
$$F_i^* - \frac{dv_i^*}{dt} = Q_{ij}\left(F_j - \frac{dv_j}{dt}\right), \qquad G_i^* - \frac{dw_i^*}{dt} = Q_{ij}\left(G_j - \frac{dw_j}{dt}\right). \tag{3.3}$$

419

Consider a second motion of the type (2.5) in which

$$c_i^*(\tau) = a_i \tau, \quad c_i(\tau) = 0, \quad Q_{ij}(\tau) = \delta_{ij},$$

where a_i is an arbitrary, constant vector. From equations (2.5) and (3.3), for this second motion, equation (3.1) becomes

$$\frac{d}{dt} \int_V \rho [\tfrac{1}{2}(v_i+a_i)(v_i+a_i) + \tfrac{1}{2} w_i w_i + U] \, dV$$
$$= \int_V \rho [r + F_i(v_i+a_i) + G_i w_i] \, dV + \int_A [t_i(v_i+a_i) + s_i w_i - h] \, dA. \tag{3.4}$$

Hence, from equations (3.1) and (3.4), one finds

$$\frac{d}{dt} \int_V \rho (v_i a_i + \tfrac{1}{2} a_i a_i) \, dV = \int_V \rho F_i a_i \, dV + \int_A t_i a_i \, dA,$$

and, since a_i is arbitrary,

$$\frac{d}{dt} \int_V \rho \, dV = 0, \quad \frac{d}{dt} \int_V \rho v_i \, dV = \int_V \rho F_i \, dV + \int_A t_i \, dA. \tag{3.5}$$

From this one sees that the classical equation of conservation of linear momentum cannot be generalized to include a term corresponding to the intrinsic director body force in equation (3.2). From equations (3.5), with the usual smoothness assumptions,

$$\frac{d\rho}{dt} + \rho v_{i,i} = 0, \quad t_i = \sigma_{ji} v_j, \quad \rho \frac{dv_i}{dt} = \rho F_i + \sigma_{ji,j}, \tag{3.6}$$

where σ_{ji} are the components of the surface force across the x_j-planes, and v_i is the unit normal to the surface. Similarly, from equation (3.2),

$$s_i = \pi_{ji} v_j, \quad \rho \frac{dw_i}{dt} = \rho G_i + g_i + \pi_{ji,j}, \tag{3.7}$$

where π_{ji} are the components of the director surface force across the x_j-planes. Also, from equation (3.1), one may now deduce that

$$h = q_i v_i, \tag{3.8}$$

where q_i is the heat flux across the x_i-planes. With the aid of equations (2.2), (2.4), (3.6), (3.7) and (3.8), after some manipulation, equation (3.1) reduces to

$$\rho \frac{dU}{dt} = \rho r - q_{i,i} + \sigma_{ji} A_{ij} + \pi_{ji} N_{ij} - g_i N_i + \tilde{\sigma}_{ji} \omega_{ij}, \tag{3.9}$$

where

$$\tilde{\sigma}_{ji} = \sigma_{ji} - \pi_{kj} d_{i,k} + g_j d_i. \tag{3.10}$$

Consider now another motion of the type (2.5) in which

$$Q_{ij} = \delta_{ij}, \quad \frac{d}{dt} Q_{ij} = a_{ij}, \tag{3.11}$$

where a_{ij} is an arbitrary, skew symmetric tensor. From equations (2.6) and (3.3), for this motion, equation (3.9) takes the form

$$\rho \frac{dU}{dt} = \rho\, r - q_{i,\,i} + \sigma_{ji} A_{ij} + \pi_{ji} N_{ij} - g_i N_i + \tilde{\sigma}_{ji}(\omega_{ij} + a_{ij}),$$

and thus, using equation (3.9),

$$\tilde{\sigma}_{ji}\, a_{ij} = 0.$$

Hence, since a_{ij} is skew symmetric and arbitrary,

$$\tilde{\sigma}_{ji} = \tilde{\sigma}_{ij}. \tag{3.12}$$

The equation (3.9) reduces to

$$\rho \frac{dU}{dt} = \rho\, r - q_{i,\,i} + \sigma_{ji} A_{ij} + \pi_{ji} N_{ij} - g_i N_i, \tag{3.13}$$

and the equations of motion are therefore $(3.6)_1$, $(3.6)_3$, $(3.7)_2$, (3.12) and (3.13). Apart from differences of notation, these are the equations proposed by ERICKSEN.

In conjunction with the above conservation laws, we consider an entropy production inequality of the form

$$\frac{d}{dt} \int_V \rho\, S\, dV - \int_V \frac{\rho\, r}{T}\, dV + \int_A k\, dA \geqq 0. \tag{3.14}$$

Here S is entropy per unit mass, T temperature, and k the flux of entropy out of the volume per unit area, per unit time. Following MÜLLER [12], we do not set the entropy flux equal to the heat flux divided by temperature. If p_i is the flux of entropy across the x_i-plane, one can deduce from the inequality (3.14) that

$$k = p_i\, v_i.$$

The argument is essentially that given by GREEN & RIVLIN [21]. Hence, since T is positive,

$$\rho T \frac{dS}{dt} - \rho\, r + T\, p_{i,\,i} \geqq 0.$$

If we introduce a Helmholtz free energy F and a vector φ_i, where

$$F = U - TS, \qquad \varphi_i = q_i - T\, p_i, \tag{3.15}$$

with the aid of equation (3.13), the inequality becomes

$$\sigma_{ji} A_{ij} + \pi_{ji} N_{ij} - g_i N_i - p_i T_{,\,i} - \rho \left(\frac{dF}{dt} + S \frac{dT}{dt} \right) - \varphi_{i,\,i} \geqq 0. \tag{3.16}$$

4. Constitutive Equations

Adopting the principle of equipresence [20], we assume that at any particle at time t

$$U,\ S,\ q_i,\ p_i,\ \sigma_{ij},\ \pi_{ij},\ g_i \tag{4.1}$$

are all single-valued functions of

$$\rho, \ T, \ d_i, \ d_{i,j}, \ w_i, \ v_{i,j}, \ T_{,i},$$

evaluated at that particle at time t; equivalently, from equations (2.2) and (2.4), they are functions of

$$\rho, \ T, \ d_i, \ d_{i,j}, \ N_i, \ A_{ij}, \ \omega_{ij}, \ T_{,i}. \tag{4.2}$$

In addition to the conditions (3.3), we assume that S, k, and T are unaffected by superposed, rigid body motions. Hence

$$U^*=U, \quad S^*=S, \quad q_i^*=Q_{ij}q_j, \quad p_i^*=Q_{ij}p_j,$$
$$\sigma_{ij}^*=Q_{ip}Q_{jq}\sigma_{pq}, \quad \pi_{ij}^*=Q_{ip}Q_{jq}\pi_{pq}, \quad g_i^*=Q_{ip}g_p, \tag{4.3}$$

and using equations (2.5) and (2.6), we have

$$\rho^*=\rho, \quad T^*=T, \quad d_i^*=Q_{ip}d_p, \quad d_{i,j}^*=Q_{ip}Q_{jq}d_{p,q},$$
$$N_i^*=Q_{ip}N_p, \quad A_{ij}^*=Q_{ip}Q_{jq}A_{pq}, \quad T_{,i}^*=Q_{ip}T_{,p}, \tag{4.4}$$

but

$$\omega_{ij}^*=Q_{ip}Q_{jq}\omega_{pq}+\Omega_{ij}.$$

Consider now a second motion of the type (2.5) for which equations (3.11) hold. At time t the quantities (4.1) and (4.2) are all unaltered except for ω_{ij}, which may be varied arbitrarily. Consequently ω_{ij} cannot appear explicitly in the constitutive equations, and (4.2) must be replaced by

$$\rho, \ T, \ d_i, \ d_{i,j}, \ N_i, \ A_{ij}, \ T_{,i}. \tag{4.5}$$

On account of equations (4.3) and (4.4), the quantities (4.1) (and also F and φ_i) are hemitropic functions of the variables (4.5).

Our concern here is primarily with liquid crystals of the nematic type. Hence, following FRANK [3], we assume that the constitutive equations are invariant under reflections through planes containing the director. This excludes liquid crystals of the cholesteric type. However, the result of this assumption is that our constitutive functions become isotropic rather than hemitropic, i.e., equations (4.3) and (4.4) hold for both proper and improper orthogonal tensors. Also, we assume that d_i and $-d_i$ are physically indistinguishable, and thus

$$U\to U, \quad S\to S, \quad q_i\to q_i, \quad p_i\to p_i,$$
$$\sigma_{ij}\to\sigma_{ij}, \quad \pi_{ij}\to-\pi_{ij}, \quad g_i\to-g_i, \tag{4.6}$$

if

$$d_i\to-d_i, \quad d_{i,j}\to-d_{i,j}, \quad N_i\to-N_i. \tag{4.7}$$

These assumptions are identical to those made by ERICKSEN [8]. Also, we confine our attention to homogeneous materials.

In order that the number of equations be equal to the number of unknowns, equation (3.12) must hold as an identity. In addition, we consider that the heat supply r, and the body forces F_i and G_i, may be specified arbitrarily. Consequently, at any particle, there is no restriction upon the choice of the quantities (4.5), and their material and spatial derivatives, on account of equations $(3.6)_3$, $(3.7)_2$ and (3.13).

With the above constitutive assumptions, and if we employ equations (2.2), (2.4), $(3.6)_1$ and

$$\frac{d}{dt}(d_{i,j}) = w_{i,j} - v_{k,j}\, d_{i,k}, \qquad \omega_{kj,i} = A_{ki,j} - A_{ji,k},$$

the inequality (3.16) becomes

$$\left(\sigma_{ji} + \rho^2\frac{\partial F}{\partial\rho}\delta_{ij} + \rho\frac{\partial F}{\partial d_{k,j}}d_{k,i}\right)A_{ij} + \left(\pi_{ji} - \rho\frac{\partial F}{\partial d_{i,j}} - \frac{\partial\varphi_j}{\partial N_i}\right)N_{ij}$$

$$-\left(g_i + \rho\frac{\partial F}{\partial d_i}\right)N_i - \left(p_i + \frac{\partial\varphi_i}{\partial T}\right)T_{,i} - \frac{\partial\varphi_i}{\partial d_j}d_{j,i}$$

$$+\rho\left(d_j\frac{\partial F}{\partial d_i} + d_{j,k}\frac{\partial F}{\partial d_{i,k}} + d_{k,j}\frac{\partial F}{\partial d_{k,i}}\right)\omega_{ji} - \frac{\partial\varphi_i}{\partial\rho}\rho_{,i} - \frac{\partial\varphi_i}{\partial d_{k,j}}d_{k,ij} \qquad (4.8)$$

$$-\frac{\partial\varphi_i}{\partial T_{,j}}T_{,ij} - \left(\frac{\partial\varphi_i}{\partial A_{jk}} + d_k\frac{\partial\varphi_j}{\partial N_i} - d_i\frac{\partial\varphi_k}{\partial N_j}\right)A_{jk,i} - \rho\left(\frac{\partial F}{\partial T} + S\right)\frac{dT}{dt}$$

$$-\rho\frac{\partial F}{\partial N_i}\frac{dN_i}{dt} - \rho\frac{\partial F}{\partial A_{ij}}\frac{dA_{ij}}{dt} - \rho\frac{\partial F}{\partial T_{,i}}\frac{d}{dt}(T_{,i}) \geq 0.$$

Since one may choose the material time derivatives of T, N_i, A_{ij}, and $T_{,i}$ arbitrarily and independently of all other quantities appearing in the above inequality, it follows that

$$\frac{\partial F}{\partial T} + S = 0, \qquad \frac{\partial F}{\partial T_{,i}} = 0,$$

$$\frac{\partial F}{\partial N_i} = 0, \qquad \frac{\partial F}{\partial A_{ij}} + \frac{\partial F}{\partial A_{ji}} = 0.$$

Hence

$$F = F(\rho, T, d_i, d_{i,j}), \qquad S = -\frac{\partial F}{\partial T}. \qquad (4.9)$$

Also, since one may choose the vorticity tensor arbitrarily and independently,

$$d_i\frac{\partial F}{\partial d_j} + d_{i,k}\frac{\partial F}{\partial d_{j,k}} + d_{k,i}\frac{\partial F}{\partial d_{k,j}} = d_j\frac{\partial F}{\partial d_i} + d_{j,k}\frac{\partial F}{\partial d_{i,k}} + d_{k,j}\frac{\partial F}{\partial d_{k,i}}. \qquad (4.10)$$

This condition is automatically satisfied since the free energy F is an isotropic function of d_i and $d_{i,j}$ [9].

It is possible to choose the spatial derivatives of ρ, $d_{i,j}$, $T_{,i}$, and A_{ij} arbitrarily and independently (except for the obvious symmetries), and therefore one obtains from the inequality

$$\frac{\partial\varphi_i}{\partial\rho} = 0, \qquad \frac{\partial\varphi_i}{\partial d_{k,j}} + \frac{\partial\varphi_j}{\partial d_{k,i}} = 0, \qquad \frac{\partial\varphi_i}{\partial T_{,j}} + \frac{\partial\varphi_j}{\partial T_{,i}} = 0, \qquad (4.11)$$

$$\frac{\partial\varphi_i}{\partial A_{jk}} + \frac{\partial\varphi_j}{\partial A_{kj}} + d_k\frac{\partial\varphi_j}{\partial N_i} + d_j\frac{\partial\varphi_k}{\partial N_i} - d_i\frac{\partial\varphi_k}{\partial N_j} - d_i\frac{\partial\varphi_j}{\partial N_k} = 0. \qquad (4.12)$$

By use of equation $(4.11)_3$, it is readily shown that φ_i depends linearly upon the temperature gradient. Also, from equation (4.12), one can show that, apart from an arbitrary dependence upon the scalar product $d_i\, N_i$, φ_i is linear in the vector N_i and the rate of strain tensor A_{ij}. Hence φ_i is of the form

$$\varphi_i = \alpha_i + \alpha_{ij}\, T_{,j} + \beta_{ij}\, N_j + \beta_{ijk}\, A_{jk} + \gamma_{ijk}\, T_{,j}\, N_k + \gamma_{ijkp}\, T_{,j}\, A_{kp},$$

where the coefficients may depend upon T, d_i, $d_{i,j}$, and $d_i\, N_i$, and β_{ijk} and γ_{ijkp} are symmetric in their last two indices.

However, employing equation $(4.11)_2$ one can show that i) α_i is at most quadratic in $d_{i,j}$, ii) α_{ij}, β_{ij}, and β_{ijk} are at most linear in $d_{i,j}$, and iii) γ_{ijk} and γ_{ijkp} are independent of the director gradients*. As a result, by use of the work of SMITH [22] and recalling (4.6) and (4.7), it is a straightforward matter to show that

$$\varphi_i = \alpha(d_i\, d_{j,j} - d_j\, d_{i,j}), \tag{4.13}$$

where α is a function of temperature T, the magnitude of the director d, and the product $d_i\, N_i$.

The inequality (4.8) reduces to

$$\left(\sigma_{ji} + \rho^2\, \frac{\partial F}{\partial \rho}\, \delta_{ij} + \rho\, \frac{\partial F}{\partial d_{k,j}}\, d_{k,i}\right) A_{ij} + \left(\pi_{ji} - \rho\, \frac{\partial F}{\partial d_{i,j}} - \frac{\partial \varphi_j}{\partial N_i}\right) N_{ij}$$

$$- \left(g_i + \rho\, \frac{\partial F}{\partial d_i}\right) N_i - \left(p_i + \frac{\partial \varphi_i}{\partial T}\right) T_{,i} - \frac{\partial \varphi_i}{\partial d_j}\, d_{j,i} \geq 0. \tag{4.14}$$

By considering a static, isothermal deformation in which the magnitude of the director does not vary, it follows immediately from the above inequality that the coefficient α, appearing in equation (4.13), must be zero when $d_i\, N_i$ vanishes. Therefore, using the notation

$$D_i = d_i\, d_{j,j} - d_j\, d_{i,j},$$

and assuming sufficiently smooth behavior, we have

$$\varphi_i = [\alpha_0(T, d)\, d_j\, N_j + \alpha_1(T, d, d_j\, N_j)]\, D_i \tag{4.15}$$

where

$$\alpha_1 = o(d_i\, N_i) \qquad \text{as } d_i\, N_i \to 0.$$

Since the tensor N_{ij} may be chosen arbitrarily and independently of the other quantities in the inequality (4.14),

$$\pi_{ji} = \rho\, \frac{\partial F}{\partial d_{i,j}} + \frac{\partial \varphi_j}{\partial N_i}. \tag{4.16}$$

We denote the values of the stress, intrinsic body force and entropy flux in a static, isothermal deformation by σ_{ij}^0, g_i^0 and p_i^0 respectively, and employ the notation

$$\sigma_{ij} = \sigma_{ij}^0 + \sigma_{ij}', \qquad g_i = g_i^0 + g_i', \qquad p_i = p_i^0 + p_i'.$$

---

* Up to this point the results obtained are independent of whether the constitutive functions are isotropic or hemitropic, and of the material property (4.6) and (4.7).

From the inequality (4.14) and the equations (3.12), (4.10), (4.15) and (4.16), it follows that

$$\sigma_{ij}^0 = -\rho^2 \frac{\partial F}{\partial \rho} \delta_{ij} - \rho \frac{\partial F}{\partial d_{k,i}} d_{k,j},$$

$$g_i^0 = -\rho \frac{\partial F}{\partial d_i} - \frac{\partial}{\partial d_k}(\alpha_0 D_j d_i) d_{k,j}, \qquad (4.17)$$

$$p_i^0 = 0.$$

The non-equilibrium parts of the stress, intrinsic director body force and entropy flux must satisfy

$$\sigma_{ji}' A_{ij} - g_i' N_i - \left(p_i + \frac{\partial \varphi_i}{\partial T}\right) T_{,i} - \frac{\partial}{\partial d_j}(\alpha_1 D_i) d_{j,i} \geqq 0. \qquad (4.18)$$

Using the work of SMITH [22], one can obtain general forms for the non-equilibrium parts of the stress, intrinsic body force and entropy flux. However, on account of their complexity, we refrain from doing this. When investigating a particular situation, one can always obtain the relevant forms from SMITH's paper. Also the restrictions arising from the entropy inequality may be found from (4.18).

From equations (4.15), (4.16) and (4.17), for a static, isothermal deformation, the stresses and the intrinsic body force are

$$\sigma_{ij} = -\rho^2 \frac{\partial F}{\partial \rho} \delta_{ij} - \rho \frac{\partial F}{\partial d_{k,i}} d_{k,j},$$

$$\pi_{ij} = \rho \frac{\partial F}{\partial d_{j,i}} + \alpha_0 D_i d_j,$$

$$g_i = -\rho \frac{\partial F}{\partial d_i} - (\alpha_0 D_j d_i)_{,j},$$

and the heat and entropy fluxes are zero. Apart from the terms involving the coefficient α_0, the above expressions are those obtained by ERICKSEN [4]. However, these extra terms do not appear in the equilibrium equations, and are therefore significant only if the director stress is specified as a boundary condition.

5. Incompressible Liquid with Director of Constant Magnitude

When the director is constrained to be of fixed length, it is convenient to absorb its magnitude into other fluid properties, and to consider the vector d_i as a unit vector. Hence, redefining the director body forces and stress, we postulate

$$\frac{d}{dt}\int_V [\rho(\tfrac{1}{2}v_i v_i + U) + \tfrac{1}{2}\rho_1 w_i w_i]\, dV$$
$$= \int_V [\rho(r + F_i v_i) + \rho_1 G_i w_i]\, dV + \int_A (t_i v_i + s_i w_i - h)\, dA,$$

and

$$\frac{d}{dt}\int_V \rho_1 w_i\, dV = \int_V (\rho_1 G_i + g_i)\, dV + \int_A s_i\, dA,$$

where w_i is the material time derivative of a unit vector d_i, and, since the fluid is incompressible, ρ and ρ_1 are constants. The unit vector is dimensionless, and consequently

$$[\rho_1]=ML^{-1}, \quad [G_i]=T^{-2}, \quad [g_i]=ML^{-1}T^{-2}, \quad [s_i]=MT^{-2}.$$

As in section 3, one can derive the equations of motion, and $(3.6)_3$, (3.12) and (3.13) are unaltered, but equations $(3.6)_1$ and $(3.7)_2$ are replaced by

$$v_{i,i}=0, \tag{5.1}$$

and

$$\rho_1 \frac{dw_i}{dt}=\rho_1 G_i+g_i+\pi_{ji,j}. \tag{5.2}$$

The entropy inequality again takes the form (3.16).

The constitutive assumptions are identical to those of the previous section. However, on account of the constraints, it is no longer possible to choose the kinematic quantities completely arbitrarily. Nonetheless, one can proceed as before to show that

$$F=F(T, d_i, d_{i,j}), \quad S=-\frac{\partial F}{\partial T}, \tag{5.3}$$

omitting the explicit dependence upon density. Again F must satisfy equation (4.10). By identical arguments, one can show that

$$\varphi_i=\alpha(d_i d_{j,j}-d_j d_{i,j})$$

where α is a function of temperature T. Consideration of a static, isothermal deformation shows that the inequality requires that α be zero. In this case therefore the classical result holds,

$$\varphi_i=0, \quad p_i=\frac{q_i}{T}. \tag{5.4}$$

Employing equations (5.3) and (5.4), the inequality (3.16) reduces to

$$\left(\sigma_{ji}+\rho \frac{\partial F}{\partial d_{k,j}} d_{k,i}\right) A_{ij}+\left(\pi_{ji}-\rho \frac{\partial F}{\partial d_{i,j}}\right) N_{ij}$$
$$-\left(g_i+\rho \frac{\partial F}{\partial d_i}\right) N_i-\frac{q_i T_{,i}}{T} \geqq 0. \tag{5.5}$$

Since the fluid is incompressible, the stress is indeterminate to an arbitrary pressure, and thus

$$\sigma_{ij}=-p\delta_{ij}+\hat{\sigma}_{ij}, \tag{5.6}$$

where $\hat{\sigma}_{ij}$ is called the extra-stress. The scalar p is called pressure, and is an unknown to be determined from equations $(3.6)_3$. Consider a director stress and an intrinsic body force of the form

$$\pi_{ji}=\beta_j d_i, \quad g_i=\gamma d_i-\beta_j d_{i,j}, \tag{5.7}$$

where γ and β_i are an arbitrary scalar and vector. Since d_i is a unit vector

$$d_i N_i = 0, \qquad d_i N_{ij} + d_{i,j} N_i = 0,$$

and therefore the system (5.7) plays no role in the inequality (5.5), or the equation (3.13). Also equation (3.12) is satisfied identically. The contribution to the equation of motion (5.2) is simply a scalar multiple of the director, and the scalar may be regarded as an unknown to be determined from equations (5.2), which otherwise would be overdeterminate. Hence, we set

$$\pi_{ji} = \beta_j d_i + \hat{\pi}_{ji}, \qquad g_i = \gamma d_i - \beta_j d_{i,j} + \hat{g}_i, \tag{5.8}$$

and call $\hat{\pi}_{ji}$ the extra director stress, $\hat{g}_i$ the extra intrinsic director body force, and γ the director tension. The terms involving β_i contribute nothing new to the equations of motion, but they could be significant if the director stress is specified at the boundary.

Since, apart from the constraint, one may choose N_{ij} arbitrarily and independently in the inequality (5.5),

$$\hat{\pi}_{ji} = \rho \frac{\partial F}{\partial d_{i,j}}. \tag{5.9}$$

We denote the equilibrium values of extra stress, extra intrinsic body force, and heat flux by $\hat{\sigma}_{ij}^0$, $\hat{g}_i^0$ and q_i^0 respectively, and employ the notation

$$\hat{\sigma}_{ij} = \hat{\sigma}_{ij}^0 + \hat{\sigma}_{ij}', \qquad \hat{g}_i = \hat{g}_i^0 + \hat{g}_i', \qquad q_i = q_i^0 + q_i'.$$

From the inequality (5.5) and equations (3.12), (4.10) and (5.9), it follows that

$$\hat{\sigma}_{ij}^0 = -\rho \frac{\partial F}{\partial d_{k,i}} d_{k,j},$$

$$\hat{g}_i^0 = -\rho \frac{\partial F}{\partial d_i}, \tag{5.10}$$

$$q_i^0 = 0.$$

The non-equilibrium parts of the extra stress, extra intrinsic body force and heat flux must satisfy

$$\hat{\sigma}_{ji}' A_{ij} - \hat{g}_i' N_i - \frac{q_i T_{,i}}{T} \geq 0. \tag{5.11}$$

For the reasons given earlier we again refrain from giving the general forms of these quantities.

For an isothermal, static deformation, from equations (5.6), (5.8), (5.9) and (5.10)

$$\sigma_{ij} = -p \delta_{ij} - \rho \frac{\partial F}{\partial d_{k,i}} d_{k,j},$$

$$\pi_{ij} = \beta_i d_j + \rho \frac{\partial F}{\partial d_{j,i}},$$

$$g_i = \gamma d_i - \beta_j d_{i,j} - \rho \frac{\partial F}{\partial d_i},$$

and the heat and entropy fluxes are zero. Apart from the terms involving β_i, these are the expressions obtained by ERICKSEN [4]. However, ERICKSEN seems to have overlooked the possibility of such an additional indeterminacy.

6. Shear Flow

The purpose of this section is to investigate the shear flow of an incompressible liquid with a director of constant magnitude. Motivated to some extent by the work of ANZELIUS [5], we make the following additional constitutive assumptions. In the free energy, we retain only the terms of lowest degree in the director gradients, and therefore choose the form discussed by FRANK [3]. For the non-equilibrium parts of the extra stress and extra intrinsic body force, we assume a linear dependence upon N_i, A_{ij} and $T_{,i}$, and omit terms involving director gradients. These quantities therefore take the forms discussed by ERICKSEN [8] and LESLIE [10].

Our aim is to obtain solutions of the equations (5.1), (3.6)$_3$ and (5.2),

$$v_{i,i}=0, \qquad \rho \frac{dv_i}{dt}=\rho F_i+\sigma_{ji,j},$$

$$\rho_1 \frac{d}{dt} w_i=\rho_1 G_i+g_i+\pi_{ji,j}, \tag{6.1}$$

where, from equations (5.6), (5.8), (5.9) and (5.10),

$$\sigma_{ji}=-p\,\delta_{ij}-\rho\,\frac{\partial F}{\partial d_{k,j}}\,d_{k,i}+\hat{\sigma}'_{ji},$$

$$\pi_{ji}=\beta_j\,d_i+\rho\,\frac{\partial F}{\partial d_{i,j}}, \tag{6.2}$$

$$g_i=\gamma\,d_i-\beta_j\,d_{i,j}-\rho\,\frac{\partial F}{\partial d_i}+\hat{g}'_i,$$

and

$$2\rho F=k_{22}\,d_{i,j}\,d_{i,j}+(k_{11}-k_{22}-k_{24})\,d_{i,i}\,d_{j,j}$$
$$+(k_{33}-k_{22})\,d_i\,d_j\,d_{k,i}\,d_{k,j}+k_{24}\,d_{i,j}\,d_{j,i}, \tag{6.3}$$

$$\hat{\sigma}'_{ji}=\mu_1\,d_k\,d_p\,A_{kp}\,d_i\,d_j+\mu_2\,d_j\,N_i+\mu_3\,d_i\,N_j$$
$$+\mu_4\,A_{ij}+\mu_5\,d_j\,d_k\,A_{ki}+\mu_6\,d_i\,d_k\,A_{kj}, \tag{6.4}$$

$$\hat{g}'_i=\lambda_1\,N_i+\lambda_2\,d_j\,A_{ji}.$$

In the expression (6.3), it is convenient to retain FRANK'S notation for the coefficients. In order to satisfy the equation of motion (3.12) identically, it is necessary that

$$\lambda_1=\mu_2-\mu_3, \qquad \lambda_2=\mu_5-\mu_6. \tag{6.5}$$

The terms involving the free energy satisfy this equation identically on account of equation (4.10). The inequality (5.11) imposes further restrictions upon the coefficients occurring in the expressions (6.4). LESLIE [10] discusses these, and of his results we require

$$\lambda_1 \leq 0. \tag{6.6}$$

Also, for reasons which become apparent below, it is necessary to consider

$$|\lambda_1| \leq |\lambda_2| \neq 0. \tag{6.7}$$

Thermal effects are ignored, and therefore, in the equations (6.3) and (6.4), the dependence of the coefficients upon temperature is not taken into account. Consequently the equations (6.1) become seven equations for the seven unknowns velocity v_i, pressure p, the unit vector d_i, and the director tension γ. The equation (3.13) reduces to an equation for the temperature, once these other quantities have been determined, and therefore we ignore it. Also, we consider situations where the body forces are conservative, and external director body forces are absent. Hence

$$\rho F_i = -\chi_{,i},$$

where χ is a scalar potential.

Our intention is to obtain solutions of the above equations in which the velocity and director components referred to Cartesian axes take the form

$$\begin{aligned} v_x &= u(y), & v_y &= v_z = 0, \\ d_x &= \cos\theta(y), & d_y &= \sin\theta(y), & d_z = 0. \end{aligned} \tag{6.8}$$

The first of equations (6.1) is automatically satisfied, and the remainder reduce to

$$\frac{\partial}{\partial y}\sigma_{yx} - \frac{\partial}{\partial x}(p+\chi)=0, \quad \frac{\partial}{\partial y}(\hat{\sigma}_{yy}-p-\chi)=0, \quad \frac{\partial}{\partial z}(p+\chi)=0, \tag{6.9}$$

$$\frac{\partial}{\partial y}\pi_{yx} + g_x = 0, \quad \frac{\partial}{\partial y}\pi_{yy} + g_y = 0. \tag{6.10}$$

It follows from equations (6.9) that

$$\sigma_{yx} = a y + c, \quad p + \chi = p_0 + a x + \hat{\sigma}_{yy},$$

where a, c and p_0 are arbitrary constants. From the first of these

$$g(\theta)\frac{du}{dy} = a y + c, \tag{6.11}$$

where

$$2 g(\theta) = 2\mu_1 \sin^2\theta \cos^2\theta + (\mu_5 - \mu_2)\sin^2\theta + (\mu_6 + \mu_3)\cos^2\theta + \mu_4. \tag{6.12}$$

After some elimination, the equations (6.10) reduce to

$$2f(\theta)\frac{d^2\theta}{dy^2} + \frac{d}{d\theta}f(\theta)\left(\frac{d\theta}{dy}\right)^2 + \frac{du}{dy}(\lambda_1 + \lambda_2 \cos 2\theta) = 0, \tag{6.13}$$

where

$$f(\theta) = k_{11}\cos^2\theta + k_{33}\sin^2\theta. \tag{6.14}$$

ERICKSEN [23] has given reasons for considering that the function $f(\theta)$ be positive. Also, for a deformation of the type (6.8), the inequality (5.11) reduces to

$$g(\theta)\left(\frac{du}{dy}\right)^2 \geq 0,$$

requiring that $g(\theta)$ is positive. For the solution given below, it is necessary to consider both $f(\theta)$ and $g(\theta)$ to be strictly positive. In addition, we consider the case when the constant a is zero.

Eliminating the velocity gradient from the equations (6.11) and (6.13), one obtains

$$2f(\theta)\frac{d^2\theta}{dy^2}+\frac{d}{d\theta}f(\theta)\left(\frac{d\theta}{dy}\right)^2+c(\lambda_1+\lambda_2\cos 2\theta)/g(\theta)=0. \tag{6.15}$$

The remainder of this section is devoted to two boundary value problems for this equation.

Flow Near a Boundary

We look for a solution of equation (6.15) satisfying the boundary conditions

$$\begin{aligned}u(0)&=0, \quad \theta(0)=\theta_1,\\ \theta(y)&\to\theta_0, \quad y\to\infty,\end{aligned} \tag{6.16}$$

where θ_0 and θ_1 are constants having values between zero and 2π. These conditions correspond to flow near a stationary solid boundary at $y=0$. Since the theory does not distinguish between d_i and $-d_i$, the values θ and $\theta\pm\pi$ are equivalent. It follows from equation (6.15) that

$$\cos 2\theta_0 = -\frac{\lambda_1}{\lambda_2}, \tag{6.17}$$

and, recalling the conditions (6.7), this gives essentially two possible values for θ_0. The trivial cases θ_1 equal to θ_0 or $\theta_0\pm\pi$ are excluded.

One may integrate equation (6.15) to obtain

$$f(\theta)\left(\frac{d\theta}{dy}\right)^2=-c\lambda_2\int_{\theta_0}^{\theta}\frac{(\cos 2\varphi-\cos 2\theta_0)}{g(\varphi)}d\varphi. \tag{6.18}$$

For a solution to exist, it is necessary that the right hand side of this equation be a positive quantity. This condition determines the value of θ_0 which one obtains. With the aid of the condition (6.6), one finds that

i)	$0\leq\theta_0\leq\frac{1}{4}\pi,$	when $c>0,\ \lambda_2>0,$
ii)	$\frac{1}{4}\pi\leq\theta_0\leq\frac{1}{2}\pi,$	when $c<0,\ \lambda_2<0,$
iii)	$\frac{1}{2}\pi\leq\theta_0\leq\frac{3}{4}\pi,$	when $c>0,\ \lambda_2<0,$
iv)	$\frac{3}{4}\pi\leq\theta_0\leq\pi,$	when $c<0,\ \lambda_2>0.$

(6.19)

One obtains the relationship between orientation and distance by integrating equation (6.18), and thus

$$y=\pm\int_{\theta_1}^{\theta}\left[\frac{f(\varphi)}{c\,h(\varphi)}\right]^{\frac{1}{2}}d\varphi, \tag{6.20}$$

where

$$h(\theta)=\lambda_2\int_{\theta}^{\theta_0}\frac{(\cos 2\varphi-\cos 2\theta_0)}{g(\varphi)}\,d\varphi.$$

From equation (6.20), it follows that θ is a monotonic function of y. Also it can be seen that the function $h(\theta)$ tends to zero at θ_0 in such a manner that the integral in equation (6.20) has a singularity there. Consequently the orientation approaches the value θ_0 asymptotically with distance. If desired, one may obtain the velocity by integrating equation (6.11).

A result of some interest is the following. Writing

$$\zeta=(|k_{11}|+|k_{33}|)/|c|,$$

and using equations (6.14) and (6.20), one can show that

$$\lim_{\zeta\to 0} y=0, \qquad \text{provided } \theta\neq\theta_0.$$

This indicates a "boundary layer" type behavior when the length $\zeta^{\frac{1}{2}}$ is sufficiently small.

It is of interest to compare the above results with those of the simpler theories of anisotropic liquids discussed by ERICKSEN [8] and LESLIE [10]. In contrast with these theories, one obtains only one solution of the type considered. However, the asymptotic values given by (6.19) correspond to ERICKSEN'S stable solutions, which is rather a satisfactory state of affairs.

Flow between Parallel Plates

We now look for a solution of equation (6.15) satisfying the boundary conditions

$$u(-h)=0, \quad u(h)=V, \quad \theta(-h)=\theta(h)=0, \tag{6.21}$$

where h and V are constants. This corresponds to flow between two parallel plates at a constant distance $2h$ apart, one of which is at rest, and the other moving with uniform velocity V along a straight line in its own plane. To fix ideas, the orientation at the wall has been set equal to zero, so that the director is parallel to the plates at the boundaries. This is a possible boundary condition for the flow of liquid crystals of the nematic type. Also it is convenient to consider λ_2 positive. One may readily obtain solutions similar to the one given below for negative λ_2, and for other boundary conditions. In view of the conditions (6.21), it is reasonable to assume that

$$\theta(-y)=\theta(y), \qquad \frac{d\theta(0)}{dy}=0, \tag{6.22}$$

and that the constant c is positive.

20*

Upon integration equation (6.15) yields

$$f(\theta)\left(\frac{d\theta}{dy}\right)^2 = c\,\lambda_2 \int_\theta^{\theta_2} \frac{(\cos 2\varphi - \cos 2\theta_0)}{g(\varphi)}\,d\varphi, \qquad 0 \leqq \theta \leqq \theta_2.$$

On account of (6.22), θ_2 is the orientation at $y=0$. It is necessary that θ_2 shall not exceed θ_0, where, recalling the condition (6.6),

$$\cos 2\theta_0 = -\frac{\lambda_1}{\lambda_2}, \qquad 0 < \theta_0 \leqq \tfrac{1}{4}\pi.$$

The trivial case θ_0 equal to zero is excluded. Writing

$$F_{\theta_2}(\theta) = \int_0^\theta \left[\frac{f(\varphi)}{h_{\theta_2}(\varphi)}\right]^{\frac{1}{2}} d\varphi, \tag{6.23}$$

where

$$h_{\theta_2}(\theta) = \lambda_2 \int_\theta^{\theta_2} \frac{(\cos 2\varphi - \cos 2\theta_0)}{g(\varphi)}\,d\varphi,$$

one chooses the constants θ_2 and c so that

$$c^{\frac{1}{2}} h = F_{\theta_2}(\theta_2). \tag{6.24}$$

The required solution is therefore

$$\begin{aligned} c^{\frac{1}{2}}(h-y) &= F_{\theta_2}(\theta), & y &\geqq 0, \\ c^{\frac{1}{2}}(h+y) &= F_{\theta_2}(\theta), & y &< 0. \end{aligned} \tag{6.25}$$

From equations (6.23) and (6.25), it is seen that θ is a monotonic function of y in either half of the channel.

One may obtain the velocity from equation (6.11), and, using conditions (6.21) and (6.22), it follows that

$$u = c \int_{-h}^{y} \frac{ds}{g[\theta(s)]},$$

and

$$V = 2c \int_0^h \frac{ds}{g[\theta(s)]} = 2c^{\frac{1}{2}} \int_0^{c^{\frac{1}{2}} h} \frac{d\xi}{g[F_{\theta_2}^{-1}(\xi)]}.$$

Therefore, with the aid of equation (6.24), one obtains the following relationship between V, h and θ_2,

$$Vh = 2F_{\theta_2}(\theta_2) \int_0^{F_{\theta_2}(\theta_2)} \frac{d\xi}{g[F_{\theta_2}^{-1}(\xi)]}. \tag{6.26}$$

Defining the apparent viscosity η by

$$\eta = \frac{\sigma_{yx}}{\left(\dfrac{V}{2h}\right)} = \frac{2ch}{V},$$

and combining equations (6.24) and (6.26),

$$\eta = F_{\theta_2}(\theta_2) \Bigg/ \int_0^{F_{\theta_2}(\theta_2)} \frac{d\xi}{g[F_{\theta_2}^{-1}(\xi)]}. \tag{6.27}$$

Therefore the apparent viscosity is a function of θ_2, or, by equation (6.26), of the product Vh.

As in the previous section, a "boundary layer" phenomenon can be demonstrated for the solution. The function $F_{\theta_2}(\theta_2)$ becomes infinite as θ_2 approaches the value θ_0. Therefore, setting

$$\zeta' = (|k_{11}| + |k_{33}|)/c\,h^2,$$

from equation (6.24), one may deduce

$$\lim_{\zeta' \to 0} \theta_2 = \theta_0,$$

and, from equations (6.25),

$$\lim_{\zeta' \to 0} (h-y)/h = 0, \quad y > 0, \quad \theta < \theta_2 \leqq \theta_0,$$

$$\lim_{\zeta' \to 0} (h+y)/h = 0, \quad y < 0, \quad \theta < \theta_2 \leqq \theta_0.$$

Consequently, if ζ' is sufficiently small, the orientation will have a value close to θ_0, except for boundary layers at either boundary. Also, from equation (6.27), one can show that

$$\lim_{\zeta' \to 0} \eta = g(\theta_0).$$

Finally, for sufficiently small values of c or h, the orientation is close to the value at the boundaries, since, from equation (6.24),

$$\lim_{\zeta' \to \infty} \theta_2 = 0.$$

In addition, from equation (6.27), one obtains the result

$$\lim_{\zeta' \to \infty} \eta = g(0).$$

7. Discussion

Assuming that the Oseen-Frank theory describes the static behavior of nematic liquid crystals, SAUPE [15] has given estimates of the FRANK constants from measurements of static deformations in the presence of magnetic fields. For p-azoxy-anisole, he estimates the constants k_{11} and k_{33} to be of the order of 10^{-6} dynes. MIESOWICZ [16] has measured the viscosity of this material in the presence of magnetic fields. If our theory is applicable, his measurements, coupled with ERICKSEN's calculations [17], suggest that the viscosity function $g(\theta)$ is of the order of a centipoise. Therefore one would expect boundary layer behavior of the orientation to be insignificant if the product of the shear stress and the square of the gap width is large compared with 10^{-6} dynes, or, alternatively, if the product of the velocity difference and the gap width is large compared with 10^{-4} cm^2/sec. Otherwise, the influence of the boundary upon the orientation will be significant.

In their experiments with p-azoxyanisole, PORTER & JOHNSON [14] found in general that, for shear rates greater than $10^3 \sec^{-1}$, the apparent viscosity was constant at a given temperature. However, in measurements with the thickness of the fluid layer as small as several microns, they found higher values than with larger gap widths, which they attributed to "adsorption layers". This would seem to be in accord with our theoretical predictions. To investigate this further, estimates of the various coefficients in the function $g(\theta)$ would be necessary. These do not appear to be available so that this question is not pursued at present.

Our calculations also predict departures from the limiting apparent viscosity $g(\theta_0)$ at low rates of shear. PETER & PETERS [24] have measured increases in apparent viscosity at low rates of shear, but again the lack of estimates for the constants in our theory prevents a close comparison. Some of this increase one expects is due to the effect discussed here[1], but other reasons have been given such as fractional orientation. Measurements of the apparent viscosity as a function of shear rate for different gap widths would indicate how important the "adsorption layer" phenomenon is at low rates of shear.

Acknowledgment. The work described in this paper was supported by a grant from the U.S. National Science Foundation. Also it is a pleasure to express my thanks to Professor J.L. ERICKSEN for many helpful discussions.

References

1. BROWN, G.H., & W.G. SHAW, The mesomorphic state, liquid crystals. Chem. Rev. **57**, 1049—1157 (1957).
2. OSEEN, C.W., Die anisotropen Flüssigkeiten. Tatsachen und Theorien. Forts. Chemie, Phys. und Phys. Chemie **20**, 25—113 (1929).
3. FRANK, F.C., On the theory of liquid crystals. Discussions Faraday Soc. **25**, 19—28 (1958).
4. ERICKSEN, J.L., Hydrostatic theory of liquid crystals. Arch. Rational Mech. Anal. **9**, 371—378 (1962).
5. ANZELIUS, A., Bewegung der anisotropen Flüssigkeiten. Uppsala Univ. Arsskr., Mat. och Naturvet, 1 (1931).
6. OSEEN, C.W., The theory of liquid crystals. Trans. Faraday Soc. **29**, 883—899 (1933).
7. DUHEM, P., Le potentiel thermodynamique et la pression hydrostatique. Ann. Ecole Norm. (3) **10**, 183—230 (1893).
8. ERICKSEN, J.L., Anisotropic fluids. Arch. Rational Mech. Anal. **4**, 231—237 (1960).
9. ERICKSEN, J.L., Conservation laws for liquid crystals. Trans. Soc. Rheol. **5**, 23—34 (1961).
10. LESLIE, F.M., Some constitutive equations for anisotropic fluids. Quart. J. Mech. Appl. Math. **19**, 357—370 (1966).
11. COLEMAN, B.D., & W. NOLL, The thermodynamics of elastic materials with heat conduction and viscosity. Arch. Rational Mech. Anal. **13**, 167—178 (1963).
12. MÜLLER, I., On the entropy inequality. Arch. Rational Mech. Anal. **26**, 118—141 (1967).
13. ERICKSEN, J.L., Transversely isotropic liquids. Kolloidzeitschrift **173**, 117—122 (1960).
14. PORTER, R.S., & J.F. JOHNSON, Orientation of nematic mesophases. J. Phys. Chem. **66**, 1826—1829 (1962).
15. SAUPE, A., Die Biegungselastizität der nematischen Phase von Azoxyanisol. Z. Naturforschg. **15**a, 815—822 (1960).
16. MIESOWICZ, M., The three coefficients of viscosity of anisotropic liquids. Nature **158**, 27 (1946).
17. ERICKSEN, J.L., Some magnetohydrodynamic effects in liquid crystals. Arch. Rational Mech. Anal. **23**, 266—275 (1966).
18. GREEN, A.E., P.M. NAGHDI, & R.S. RIVLIN, Directors and multipolar displacements in continuum mechanics. Int. J. Engng. Sci. **2**, 611—620 (1965).

---

[1] Compare the remarks of PORTER, BARRALL & JOHNSON [25].

19. BEATTY, M.F., On the foundation principles of general classical mechanics. Arch. Rational Mech. Anal. **24**, 264—273 (1967).
20. TRUESDELL, C., & W. NOLL, The Non-Linear Field Theories of Mechanics. Handbuch der Physik III/3. Berlin-Heidelberg-New York: Springer 1965.
21. GREEN, A.E., & R.S. RIVLIN, The relation between director and multipolar theories in continuum mechanics. Z. ang. Math. Phys. **18**, 208—218 (1967).
22. SMITH, G.F., On isotropic integrity bases. Arch. Rational Mech. Anal. **18**, 282—292 (1965).
23. ERICKSEN, J.L., Inequalities in liquid crystal theory. Phys. of Fluids **9**, 1205—1207 (1966).
24. PETER, S., & H. PETERS, Über die Strukturviskosität des kristallin-flüssigen $p-p'$-Azoxyanisols. Z. Phys. Chemie (N.F.) **3**, 103—125 (1955).
25. PORTER, R.S., E.M. BARRALL, II, & J.F. JOHNSON, Some flow characteristics of mesophase types. J. Chem. Phys. **45**, 1452—1456 (1966).

The University
Newcastle-upon-Tyne

(Received November 7, 1967)

Frank Matthews Leslie was born in 1935 in Dundee, Scotland. He entered Queen's College, Dundee, then part of the University of St Andrews in 1953, and graduated in mathematics in 1957. He was awarded the Ph.D degree in 1961 for a dissertation on non-linear fluid motion at the University of Manchester, where he had already been appointed as  assistant lecturer. Following his Ph.D. he spent a year in America at MIT, before returning to England as a lecturer in mathematics at the University of Newcastle.

The academic year 1966–7 was spent at Johns Hopkins university in Baltimore, Maryland, where he was able to interact with the rational mechanics group of Professor Jerry Ericksen. In the picture, taken in 1977, Leslie (left) can be seen with Ericksen. In 1968 Leslie was appointed to a senior lectureship at the University of Strathclyde in Glasgow; he was promoted to reader in 1971, to a personal chair in 1979 and to an established chair in 1982.

Most of Leslie's best-known work is devoted to the study of liquid crystals. In this book we reprint fundamental papers on the theory of nematic liquid crystals and on the theory of switching in twisted nematic devices. He is also well-known for his continuum theory of smectic C liquid crystals.

Leslie's attachments to Scotland, and in particular to the Scottish National Party, as well as his advocacy of an independent Scotland, were well-known to colleagues and friends. He was awarded many honours, including appointments as visiting professor in the United States, Germany, Italy and Japan, and fellowships of the Royal Societies of Edinburgh and of London.

He was due to retire from the University in September 2000, but unfortunately died on June 15, following complications to what should have been a routine hip operation.

Short Range Order Effects in the Isotropic Phase of Nematics and Cholesterics†

P. G. DE GENNES‡

Department of Physics
Simon Fraser University
Burnaby 2, B. C., Canada

Received September 22, 1970

Abstract—We assume that (1) the local state of order in the isotropic phase is a symmetric traceless tensor $Q_{\alpha\beta}$, proportional to the anisotropic part of a tensor property such as the magnetic susceptibility; (2) the free energy may be expanded in powers of $Q_{\alpha\beta}$ and of its gradients. This allows a unified description covering the anomalous magnetic birefringence, the intensity of light scattering, and the properties of the nematic/isotropic interface. For a cholesteric, although the optical rotation is huge in the ordered phase, we predict that it should *not* be anomalous just above the transition point T_c. We also investigate the dynamics of fluctuations of $Q_{\alpha\beta}$, and discuss the flow birefringence, the frequency width of the Rayleigh scattering, and the attenuation of ultrasonic shear waves, in terms of 3 viscosity coefficients.

1. Introduction

The nematic↔isotropic transition is of first order, but weak.[1] In most thermotropic materials, the isotropic phase still shows some remarkable short range order effects above the transition point T_c; we give here a short list of the relevant experiments:

(a) *magnetic birefringence*[2–4]: the refractive indices $n_{\parallel}$ and $n_{\perp}$ (measured for polarisations respectively parallel and normal to the magnetic field H) differ by an amount

$$n_{\parallel} - n_{\perp} = \alpha(T) H^2 \qquad (1.1)$$

Near T_c, α may be a hundred times larger than in conventional organic liquids. Plots of $1/\alpha$ versus T are roughly linear and extra-

---

† Invited Paper presented at the Third International Liquid Crystal Conference, Berlin, August 24–28, 1970.

‡ Permanent address: Physique des solides, Faculté des Sciences, 91 Orsay, France.

polation suggests that α would diverge at a temperature T^* which is only slightly smaller than T_c ($T_c - T^* \gtrsim 1\,^\circ K$).

(b) *intensity of light scattering I*: although smaller than in the nematic phase[5] I is still significant.[6] It is essentially independent of the scattering angle θ, and the ratio $I_\perp / I_\parallel$ (for the two conventional polarisation set ups) is $\frac{3}{4}$ as it is for a collection of anisotropic scattering objects with random orientations.[6] Since I is independent of θ, the size of these objects, or, more accurately, the "coherence length" $\xi(T)$, is smaller than the optical wavelength. Typically we expect $\xi(T_c)$ to be of order 10 times the molecular length (i.e. $\sim 200\,\text{Å}$): This is still large enough to allow a macroscopic description of fluctuation and correlation effects. One of the aims of the present paper is to give a detailed definition of these coherence lengths.

(c) *flow birefringence*[2]: with a flow velocity v, in the x direction, and under a velocity gradient $\partial v / \partial z$, the isotropic phase becomes birefringent, with two optical axes (1) and (2) at 45° from x and z. The difference in refraction indices $n_1 - n_2$ is proportional to the rate of shear

$$n_1 - n_2 = \tau(T)\frac{\partial v}{\partial z} \qquad (1.2)$$

τ has the dimension of time, and is much larger than in conventional liquids.

The Zvetkov measurements of τ[2] were not very accurate, but, they do show a strong increase of τ when T decreases down to T_c.

(d) *frequency width of the scattered light*: this has been measured with a Fabry–Perot interferometer.[3,6] The observed line is a single Lorentzian, of half width ($\Gamma/2\pi$). Γ is small (in the megacycle range) when $T = T_c$, and increases with T. The presence of only one Lorentzian is remarkable, and will be discussed in detail in section 4.

More generally, the aim of the present paper is to give a unified discussion of short range order effect, in terms of a Landau model for the nematic isotropic transition. The principles have been sketched in an earlier short communication.[7] The first problem is to define an adequate order parameter—here a tensorial object $Q_{\alpha\beta}$, as explained in section 2. Then we assume that the free energy F may be expanded in powers of $Q_{\alpha\beta}$: the symmetry properties of $F(Q)$ then force the transition to be first order. However, just above

438

T_c, the fluctuations may be large. This is discussed in section 3, together with some static applications, including effects (a) and (b) above, plus the properties of the nematic/isotropic interface. Section 4 discusses the dynamics of fluctuations, using the thermodynamics of irreversible processes as a framework, and including the coupling between molecular rotation and flow. Finally we try to extend these considerations to cholesteric materials: the new feature here is the presence of a term proportional to Q grad Q in the free energy. We discuss the physical implications of this unusual term on the static optical properties in section 5.

2. Definition of an Order Parameter

(a) *microscopic approach*: For a system of rod-like molecules, the natural order parameter, in a nematic phase, is

$$S = \langle \tfrac{1}{2}(3 \cos^2 \theta - 1) \rangle \tag{2.1}$$

where θ is the angle between rod axis and nematic axis. For rigid molecules of more general shape, a natural generalisation of (2.1) amounts to use

$$S_{ij}^{\alpha\beta} = \langle \tfrac{1}{2}(3i_\alpha j_\beta - \delta_{ij} \delta_{\alpha\beta}) \rangle \tag{2.2}$$

where i, j, k are three orthonormal vectors linked to the molecule, while α and β are indices referring to the laboratory frame. In a uniaxial nematic of optical axis parallel to z, the only non 0 components of $S_{ij}^{\alpha\beta}$ are the following

$$\left. \begin{array}{l} S_{ij}^{zz} = S_{ij} \\ S_{ij}^{xx} = S_{ij}^{yy} = -\tfrac{1}{2}S_{ij} \end{array} \right\} \tag{2.3}$$

(b) *macroscopic approach*: Many nematic molecules are partly flexible, and different parts of one same molecule would give different S_{ij} tensors. Also, from a thermodynamic point of view, it is preferable to define the amount of order from a macroscopic property, independently of any assumption on the rigidity of the molecules. Consider for instance the anisotropy of the magnetic susceptibility: let us define[8]

$$Q_{\alpha\beta} = \chi_{\alpha\beta} - \tfrac{1}{3}\chi_{\gamma\gamma} \delta_{\alpha\beta} \tag{2.4}$$

$Q_{\alpha\beta}$ is a symmetric, traceless tensor: we call it the *tensor order parameter*. It must be emphasized that any other tensor property (i.e. the dielectric constant $\epsilon_{\alpha\beta}$) could have been used to define $Q_{\alpha\beta}$. In the isotropic phase, where $Q_{\alpha\beta}$ is small, any other anisotropic effect is linear in Q. For instance we may write

$$\delta\epsilon_{\alpha\beta} \equiv \epsilon_{\alpha\beta} - \tfrac{1}{3}\epsilon_{\gamma\gamma}\,\delta_{\alpha\beta} = M_{\alpha\beta\gamma\delta}\,Q_{\gamma\delta} \tag{2.5}$$

The only matrix M relating two symmetric traceless tensors ($\delta\epsilon$ and Q) and compatible with rotational invariance is a multiple of the unit matrix. Thus Eq. (1) reduces to

$$\delta\epsilon_{\alpha\beta} = M\,Q_{\alpha\beta} \tag{2.6}$$

For simple rod-like molecules M is simply the ratio of dielectric/magnetic anisotropies in the ordered phase.

We have chosen here to define $Q_{\alpha\beta}$ through the diamagnetic anisotropy for the following reason: theoretical calculations of $\chi_{\alpha\beta}$ are feasible, since magnetic interactions between different molecules are negligible. For instance, with a rigid molecule, let us assume that we know the suspectibility tensor (per molecule) in the molecular frame χ^{ij}. Then $\chi_{\alpha\beta}$ is simply a superposition of individual responses, and if n is the number of molecules/cm^3, we may write

$$Q_{\alpha\beta} = nS_{ij}^{\alpha\beta}\,\chi^{ij}$$

In a uniaxial nematic, the difference between parallel and perpendicular susceptibilities is

$$\chi_{\parallel} - \chi_{\perp} = Q_{zz} - Q_{xx} = \tfrac{3}{2}S_{ij}\chi^{ij} \tag{2.7}$$

Equation (2.7) gives, for rigid molecules, the link between the microscopic definition of order (via S) and the macroscopic (via Q). If we had chosen to define Q from the dielectric anisotropy, we could not have produced such explicit formulae—the present theory of dielectric constants in liquids being unable to take into account correctly the electric interactions between different molecules.

(c) *biaxial versus uniaxial nematics*: Our definition of the order parameter covers both uniaxial and biaxial nematic systems: the matrix $Q_{\alpha\beta}$, when diagonalized, may have the form

$$Q_{\alpha\beta} = \begin{bmatrix} \dfrac{-Q+P}{2} & & 0 \\ 0 & \dfrac{-Q-P}{2} & 0 \\ 0 & & Q \end{bmatrix} \tag{2.8}$$

Depending on the form of the free energy F as a function of Q and P,

440

the minimum of F may correspond to $P = 0$ (uniaxial nematic) or $P \neq 0$ (biaxial nematic). The possible existence of biaxial phases has been stressed in particular by Freiser.[9]

3. Free Energy and Static Properties

3.1. Landau Expansion of the Free Energy

Let us assume that the free energy F (per unit volume) may be expanded in powers of $Q_{\alpha\beta}$. Then, retaining only terms which have rotational invariance, we may expect the following structure

$$\bar{F} = F_0 + \tfrac{1}{2}A Q_{\alpha\beta}Q_{\beta\alpha} + \tfrac{1}{3}B Q_{\alpha\beta}Q_{\beta\gamma}Q_{\gamma\alpha} + 0(Q^4) - \tfrac{1}{2}Q_{\alpha\beta}H_\alpha H_\beta \quad (3.1)$$

where F_0 is independent of Q. Because Q is symmetric traceless there is only one invariant of order Q^2 (with the coefficient $A/2$) and one invariant of order Q^3 (there would be two invariants of order Q^4). $A(T)$ is expected to be small near T_c. More precisely we can put

$$A(T) = a(T - T^*)^\gamma \quad (3.2)$$

where T^* is a temperature slightly below T_c, and γ an unknown exponent.

($\gamma = 1$ in a mean field theory such as the Maier Saupe theory.)

The last term in (3.1) is the anisotropic part of the diamagnetic energy, in a field H. The presence of a cubic term BQ^3 in (3.1) imposes a first order transition, as explained in Fig. 1. If one assumed that an expansion to order Q^4 is acceptable even in the ordered phase, one could derive formulas for the order parameter just below T_c, the latent heat, etc; however, it is not obvious that the terms of order Q^5, etc., are indeed negligible in the ordered phase. For this reason we shall mainly restrict our attention to the isotropic phase $(T > T_c)$ where Q is indeed small.

3.2. Application to the Magnetic Birefringence

In a non 0 external field, the minimum of F (Eq. 3.1) corresponds to a non 0 Q, i.e. to a finite anisotropy in the optical properties. Minimizing F with respect to $Q_{\alpha\beta}$, and keeping in mind the constraint of 0 trace ($Q_{\alpha\alpha} = 0$) one finds (to order H^2) for the thermal average of Q:

$$\langle Q_{\alpha\beta}\rangle = \frac{1}{2A}(H_\alpha H_\beta - \tfrac{1}{3}H^2\delta_{\alpha\beta}) \quad (3.3)$$

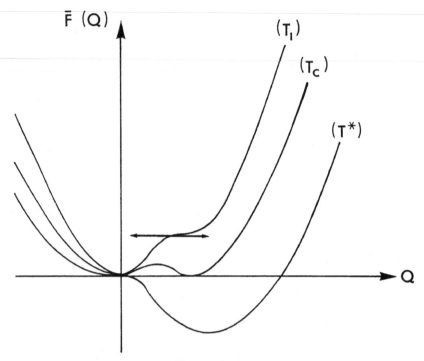

Figure 1. Plot of free energy $\bar{F}$ versus order parameter Q, in zero magnetic field, at various temperatures. T_c is the equilibrium transition point. T^* is the temperature below which the isotropic phase is absolutely unstable. T_1 is the temperature above which the nematic phase is absolutely unstable.

Take for instance H along Z. Then

$$Q_{zz} - Q_{xx} = \frac{H^2}{2A}$$

and from Eq. (2.6), the dielectric anisotropy is

$$\delta\epsilon_{zz} - \delta\epsilon_{xx} = \frac{MH^2}{2A}$$

(where ϵ and M are defined for the frequency of the optical measurement)

Finally the birefringence is obtained by writing $n_\parallel^2 = \epsilon_{zz}, n_\perp^2 = \epsilon_{xx}$

$$n_\parallel - n_\perp = \frac{MH^2}{4A\bar{n}} = \alpha(T)H^2 \tag{3.4}$$

We do not expect any strong variation in M or $\bar{n}$ near T_c, thus

$1/\alpha(T)$ is essentially proportional to $A(T)$. The existing data suggest that $1/\alpha$ is nearly linear in T ($\gamma \cong 1$).

3.3. DEFINITION OF COHERENCE LENGTHS

Let us now add terms in F which can describe situations where the order parameter varies slowly from point to point. In a nematic fluid, the first terms allowed by symmetry are quadratic in the gradients of Q and have the form

$$F_g = \tfrac{1}{2}L_1 \partial_\alpha Q_{\beta\gamma} \partial_\alpha Q_{\beta\gamma} + \tfrac{1}{2}L_2 \partial_\alpha Q_{\alpha\gamma} \partial_\beta Q_{\beta\gamma} \tag{3.5}$$

where $\partial_\alpha \equiv \partial/\partial x_\alpha$. L_1 and L_2 may be called the elastic constants in the isotropic phase. Their number is smaller than in the ordered phase (where there are three constants[10]) because in Eq (3.5) we restrict our attention to terms of order Q^2, i.e. to small Q. To get more information on the meaning of L_1 and L_2 let us consider the case where Q depends only on one co-ordinate, say z. Then

$$F_g = \frac{L_1}{2}[(\partial_z Q_{zz})^2 + (\partial_z Q_{xx})^2 + (\partial_z Q_{yy})^2] + L_1[(\partial_z Q_{xy})^2 + (\partial_z Q_{yz})^2$$

$$+ (\partial_z Q_{zx})^2] + \frac{L_2}{2}[(\partial_z Q_{zz})^2 + (\partial_z Q_{zx})^2 + (\partial_z Q_{zy})^2] \tag{3.6}$$

From Eq. (3.6) we may derive a number of inequalities which must be satisfied by the elastic constants to ensure stability (i.e., F_g must be positive for all distortions of Q).

Let us start with a situation where Q_{xy} is the only non-vanishing component. Then from Eq. (3.6) we get the stability condition

$$L_1 > 0 \tag{3.7}$$

Similarly with Q_{xz} (or Q_{yz}) we obtain

$$L_1 + \tfrac{1}{2}L_2 > 0 \tag{3.8}$$

Considerations involving the diagonal terms Q_{xx}, Q_{yy}, Q_{zz} are more delicate because the sum of these three terms must vanish identically. Let us consider first a case where, at all points, the nematic axis is parallel to $0z$. (Fig. 2a) (at $T = T_c$, this would describe a nematic/isotropic interface with the optical axis normal to the interface). In such a situation we may put

$$Q_{zz} = Q \quad Q_{xx} = Q_{yy} = -\frac{Q}{2} \quad Q_{xy} = Q_{yz} = Q_{zx} = 0$$

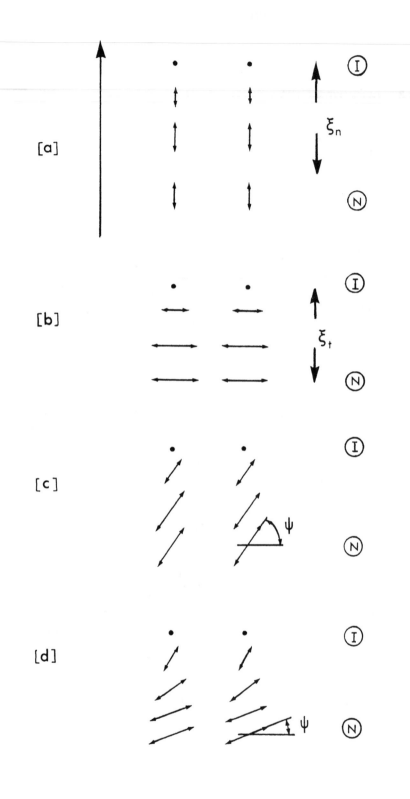

and we get from Eq. (3.6)

$$F_g = \frac{3L_1 - 2L_2}{4} \left(\frac{dQ}{dz} \right)^2 \qquad (3.9)$$

Adding the other terms of order Q^2 from Eq. (3.1) we get for the free energy

$$F = \bar{F} + F_g = \tfrac{3}{4}AQ^2 + \frac{3L_1 + 2L_2}{4} \left(\frac{dQ}{dz} \right)^2$$

$$= \tfrac{3}{4}A \left[Q^2 + \xi_n^2 \left(\frac{dQ}{dz} \right)^2 \right] \qquad (3.10)$$

$$\xi_n^2(T) = \frac{(L_1 + \tfrac{2}{3}L_2)}{A(T)} \qquad (3.11)$$

where we have defined the " normal coherence length " $\xi_n(T)$. ξ_n^2 must be positive = this implies the inequality

$$L_1 + \tfrac{2}{3}L_2 > 0$$

Now let us go the tangential case of Fig. (2b). Here we assume

$$Q_{xx} = Q \quad Q_{yy} = Q_{zz} = -\frac{Q}{2} \quad Q_{xy} = Q_{yz} = Q_{zx} = 0$$

and we obtain for the free energy, to order Q^2

$$F = \tfrac{3}{4}A \left[Q^2 + \xi_t^2(T) \left(\frac{dQ}{dz} \right)^2 \right] \qquad (3.12)$$

where the " tangential coherence length " $\xi_t(T)$ is given by

$$\xi_t^2(T) = (L_1 + \tfrac{1}{6}L_2) \frac{1}{A(T)} \qquad (3.13)$$

We also have the stability requirement :

$$L_1 + \tfrac{1}{6}L_2 > 0$$

but, if the conditions (3.7) and (3.11) are satisfied, the other conditions will also be satisfied. The two lengths ξ_n and ξ_t are expected

---

Figure 2. Possible structures for the nematic/isotropic interface:
(a) normal case; (b) tangential case; (c) conical case with constant angle in the transition layer; (d) conical case with variable angle.

to be comparable in magnitude (for qualitative discussions we will sometimes replace them by a single length $\xi(T)$). Near T_c the coherence lengths are large (but finite) since $A(T_c)$ is small (but non 0). Qualitatively, neglecting all indices, and neglecting the difference between ξ_n and ξ_t) we may say that the QQ correlation function has the form :

$$\langle Q(0) Q(R) \rangle = \frac{\text{const. } k_B T}{L_1 R} e^{-R/\xi} \quad (R \gtrsim \xi) \tag{3.14}$$

At first sight we might hope to measure the coherence lengths ξ_n and ξ_t (or suitable admixtures of these two) by a study of the scattered light intensity I. However, the data of Litster and Stinson[6] show that I is essentially independent of the scattering wave vector q. This means then $q\xi \ll 1$ and in this limit we cannot measure ξ.

The magnitude of the scattered intensity is easily derived for $q\xi \ll 1$. Consider for instance the case where the incident light is polarized along x, while the outgoing light is polarized along y. The corresponding intensity $I_\perp$ is proportional to the average square.

$$I_\perp \cong \langle |\delta\epsilon_{zx}|^2 \rangle = M^2 \langle Q_{zx}^2 \rangle \tag{3.15}$$

To order Q^2 the only terms in the free energy (3.1) involving Q_{zx} are $\frac{1}{2}A(Q_{zx}^2 + Q_{xz}^2) = AQ_{zx}^2$. The gradient terms (3.5) are negligible for $g\xi \ll 1$. Then applying the equipartition theorem gives (per unit volume)

$$A \langle Q_{zx}^2 \rangle = \tfrac{1}{2} k_B T$$

$$I_\perp \cong \langle |\delta\epsilon_{zx}|^2 \rangle = \frac{M^2 k_B T}{2A}. \tag{3.16a}$$

Similarly, when both polarizations are parallel to X, we get

$$I_{\shortparallel} \cong \langle |\delta\epsilon_{zz}|^2 \rangle = \tfrac{2}{3} \frac{M^2 k_B T}{A} \tag{3.16b}$$

Thus the intensities are proportional to $1/A$. This has been verified by Stinson and Litster.[6]

3.4. The Nematic/Isotropic Interface

In the present paragraph we shall assume that the distortion free energy is correctly described by Eq. (3.5), even when the variations of Q take place in a distance comparable to $\xi(T)$. (This is equivalent

446

to a mean field approximation). We can then apply our earlier Eqs. (3.10) and (3.12) to a discussion of the N/I interface, at $T = T_c$. Here Q goes from a finite value (at $z = -\infty$) to 0 (at $z = +\infty$), the variations taking place in a thickness $\xi(T_c)$ which is large compared with the molecular dimensions a.

Consider first the "normal" case where Q has the structure defined immediately before Eq (3.9). We may write the surface tension γ_r in the form

$$\gamma_n = \int_{-\infty}^{\infty} dz \left[\bar{F}(Q) + b_n^2 \left(\frac{dQ}{dz}\right)^2 \right] \tag{3.17}$$

Here $b_n^2 = (3L_1 + 2L_2)/4$ and $\bar{F}(Q)$ is the free energy (3.1), including now all powers of Q. The origin of free energies is such that $\bar{F}(0) = \bar{F}(Q^*) = 0$ where Q^* is the order parameter in the nematic phase, at $T = T_c$. Writing that the form of $Q(z)$ minimizes γ we get the equation

$$2b_n^2 \frac{d^2Q}{dz^2} = \frac{\partial \bar{F}}{0Q} \tag{3.18}$$

This equation has the first integral

$$b_n^2 \left(\frac{dQ}{dz}\right)^2 = \bar{F}(Q) \tag{3.19}$$

where the integration constant must vanish, since both dQ/dz and $\bar{F}(Q)$ are zero far from the transition layer. Inserting Eq. (3.19) into (3.17) we get

$$\gamma_n = 2 \int_{-\infty}^{\infty} b_n^2 \frac{dQ}{dz} dQ = 2b_n \int_0^{Q^*} [\bar{F}(Q)]^{1/2} dQ \tag{3.20}$$

Similarly, for the tangential case, we get

$$\left. \begin{array}{l} \gamma_t = 2b_t \displaystyle\int_0^{Q^*} [\bar{F}(Q)]^{1/2} dQ \\[2mm] b_t^2 = \tfrac{1}{4}(3L_1 + \tfrac{1}{2}L_2) \end{array} \right\} \tag{3.21}$$

Comparing (3.20) and (3.21) we see that, if $L_2 > 0$, b_t is smaller than b_n, and $\gamma_t < \gamma_n =$ the tangential conformation is favoured. On the other hand, if $L_2 < 0$, the normal conformation is favoured.

A word of caution should be inserted at this point: the actual conformation at the interface may be different from the normal and tangential cases considered above. In macroscopic terms, we may say that the preferred angle between the director and the N/I surface need not be 0, or $\pi/2$: it might be some intermediate angle. To compute the interface tension γ for such cases is difficult. We have only carried out a simple variational calculation, assuming that at all points in the transition layer the medium is uniaxial, with an optical axis which is the same everywhere (making a constant angle ψ with the surface—Fig. 2c). Within this approximation we can show that the minimum of γ occurs either at $\psi = 0$ if $L_2 > 0$) or at $\psi = \pi/2$ if $L_2 < 0$), but not at an intermediate ψ value. But a much more elaborate calculation would be required to elucidate this point completely.

4. Dynamics of Fluctuations

4.1. FLUXES AND FORCES

In the nematic phase we know both from theory[5] and from experiment[11] than an orientational fluctuation of wave vector q relaxes in a purely viscous way (no oscillations) with a time constant τ_g given qualitatively by

$$\frac{1}{\tau_g} \simeq \frac{Kg^2}{\eta_{\text{eff}}} \tag{4.1}$$

where K is an average of the Frank elastic constants,[10] and η_{eff} an average viscosity. This result suggests that, above T_c, where the fluid is more disordered, we will also find a strongly dissipative behaviour: such a behaviour may then be analyzed in terms of the thermodynamics of irreversible processes, introducing *fluxes* and *forces*.

(a) *Fluxes*. One group of such fluxes is given by the rate of change in time of $Q_{\alpha\beta}$

$$R_{\alpha\beta} = \frac{\delta Q_{\alpha\beta}}{\delta t} \tag{4.2}$$

The differentiation symbol $\delta/\delta t$ denotes the variation (along one flow line) with respect to the background fluid; in particular, if the fluid is in rotation, the part of the change of Q due to this rotation must be subtracted, as explained in Ref. (7). However, in all what follows, we shall treat v_α and $Q_{\alpha\beta}$ as infinitesimal quantities of first order. Then the difference between $\delta/\delta t$ and the partial derivative $\partial/\partial t$ is of second order and may be neglected:

$$R_{\alpha\beta} \to \frac{\partial Q_{\alpha\beta}}{\delta t} \qquad (4.3)$$

Another group of fluxes which is important here is the hydro-dynamic shear rate tensor

$$e_{\alpha\beta} = \partial_\alpha v_\beta + \partial_\beta v_\alpha \qquad (4.4)$$

We restrict our attention to incompressible flow ($e_{\alpha\alpha} \equiv 0$). As explained in Ref. (5), this appears to be justified because the fluctuations of Q have a frequency spectrum much lower than the sound waves.

(b) *Forces.* The force $\phi_{\alpha\beta}$, conjugate to $Q_{\alpha\beta}$, may be obtained directly from the free energy F (e.g., 3.1, 3.5). We restrict our attention to terms of F which are quadratic on Q, giving linear contributions to ϕ. Furthermore, in most of the applications to be discussed below, the wave vectors q of interest will turn out to be small ($q\xi \ll 1$) and the derivative terms from (3.5) will be negligible; then:

$$\phi_{\alpha\beta} = \frac{-\delta F}{\delta Q_{\alpha\beta}} = -A Q_{\alpha\beta}. \qquad (4.5)$$

The force conjugate to $e_{\alpha\beta}$ is $\frac{1}{2}\sigma_{\alpha\beta}$, where $\sigma_{\alpha\beta}$ is the viscous stress tensor.

Finally the entropy source may be written as a bilinear function of fluxes and forces

$$TS = \phi_{\alpha\beta} R_{\alpha\beta} + \tfrac{1}{2}\sigma_{\alpha\beta} e_{\alpha\beta} \qquad (4.6)$$

Our choice of fluxes and forces purposely omits certain effects (such as temperature fluctuations) which are not anomalously large and not strongly coupled to the order parameter Q.

4.2. Hydrodynamic Equations in the Isotropic Phase

We now write down a phenomenological system of linear equations coupling the fluxes and the forces. Since all these quantities are symmetric traceless tensors of rank 2, the most general form for these equations, which is compatible with rotational invariance and the Onsager relations, is[12]:

$$\tfrac{1}{2}\sigma_{\alpha\beta} = \tfrac{1}{2}\eta\, e_{\alpha\beta} + \mu R_{\alpha\beta} \qquad (4.7)$$

$$\phi_{\alpha\beta} = \mu\, e_{\alpha\beta} + \gamma R_{\alpha\beta} \qquad (4.8)$$

If Q is taken as dimensionless, both R and e have the dimension of frequency. A and σ have the dimension of pressure. The three coefficients η, μ, ν have the dimension of a viscosity.

Imposing that the entropy source (4.6) be positive leads to the inequality

$$2\mu^2 > \nu\eta \tag{4.9}$$

In addition we need the hydrodynamic acceleration equation

$$\rho\frac{dv_\alpha}{dt} \sim \rho\frac{\partial v_\alpha}{\partial t} = \partial_\beta\sigma_{\alpha\beta} - \partial_\beta p \tag{4.10}$$

where ρ is the density and p the scalar pressure, Eqs. (4.5, 8, 9, 10) together with the incompressibility condition ($e_{\alpha\alpha} = 0$) define entirely the problem of small motions.

4.3. Discussion of Experiments

(a) *flow birefringence*: with a flow velocity v along x, and a velocity gradient $e_{xz} = \partial v/\partial z$, we have a steady state ($R_{\alpha\beta} = 0$) corresponding to $\phi_{xz} = \mu\,e_{xz}$ from Eq. (4.8). Inserting Eq. (4.5) for ϕ_{xz} we arrive at

$$Q_{xz} = -\frac{\mu}{A}\frac{\partial v}{\partial z} \tag{4.11}$$

all other components of Q being 0. This implies that two principal axes of the Q tensor (1) and (2) are the bisectors of the x and z axis. The dielectric anisotropy is

$$\left|\epsilon_1 - \epsilon_2\right| = M\left|Q_{11} - Q_{22}\right| = MQ_{xy} \tag{4.12}$$

The difference in refracting indices is

$$\left|n_1 - n_2\right| = \frac{M}{2n}\left|Q_{xy}\right| = \left|\frac{M\mu}{2nA}\right|\frac{\partial v}{\partial z} \tag{4.13}$$

Thus the characteristic time $\tau(T)$ defined in Eq. (1.3) is given by

$$\tau(T) = \frac{M\mu}{2nA(T)} \tag{4.14}$$

In Eq. (4.14) the most important temperature effects come from $A(T)$. The factors M and n are probably temperature insensitive. The friction coefficient μ might vary significantly with T. The data of Zvetkov[3] are not accurate enough to decide on this point.

(b) *inelastic scattering of light—simplified treatment*: For small wave vectors q, it turns out that the characteristic frequencies of the fluctuations of Q are very different from the frequencies associated with v. In such case, the coupling between v and Q is uneffective: in the dynamical Eq. (4.8) for $Q_{\alpha\beta}$ we may neglect the v term. Then we find a simple exponential relaxation for $Q_{\alpha\beta}$, involving one single relaxation rate:

$$\frac{\partial Q_{\alpha\beta}}{\partial t} = -\Gamma Q_{\alpha\beta} \qquad (4.15)$$

$$\Gamma(T) = \frac{A(T)}{\nu} \qquad (4.16)$$

The consequences of Eq. (4.16) have been investigated by Stinson and Litster.[3] Taking $A(T)$ from the magnetic birefringence, they find that the temperature dependence of Γ can be accounted for if one assumes that ν varies in temperature just as the average viscosity η. Cases where Eq. (4.15) is not applicable (because of the coupling between Q and v) will be discussed in paragraph 4.4.

(c) *shear wave attenuation*: Let us assume that the liquid crystal is driven at an angular frequency ω, with shear waves propagating along z, the flow velocity being along x. The shear stress is from Eq. (4.7)

$$\sigma_{xz} = \eta \frac{\partial v}{\partial z} + 2\mu \frac{\partial Q_{xz}}{\partial t} \qquad (4.17)$$

and the equation for the order parameter is:

$$\frac{\partial Q_{xz}}{\partial t} + \Gamma Q_{xz} = -\frac{\mu}{\nu} \frac{\partial v}{\partial x} \qquad (4.18)$$

Replacing $\partial/\partial t$ by $i\omega$ and eliminating Q_{xz} between (4.17) and (4.18) we arrive at a simplified form of the acceleration equation

$$i\omega \rho v = \eta(\omega) \frac{\partial^2 v}{\partial x^2} \qquad (4.19)$$

where the effective viscosity $\eta(\omega)$ is defined by

$$\eta(\omega) = \eta - \frac{2\mu^2}{\nu} \frac{i\omega}{\Gamma + i\omega} \qquad (4.20)$$

Thus $\eta(\omega)$ goes from η (for $\omega \ll \Gamma$) to $\eta - (2\mu^2/\nu)$ (for $\omega \gg \Gamma$) Dispersion anomalies in this frequency range have indeed been observed very recently,[13] but it is not known yet whether the simple form (4.20) accounts for them or not.

The concept of an effective viscosity $\eta(\omega)$ applies only when the penetration depth of the shear waves $(\eta/\rho\omega)^{1/2} = \delta$ is much larger than the coherence length $\xi(T)$. This is indeed correct when $\omega \sim \Gamma$, as can be seen from the following argument:

$$\frac{\xi^2}{\delta^2} = \xi^2 \frac{\rho\Gamma}{\eta} \sim \frac{\rho A L_1}{\eta\gamma A} = \frac{L_1\rho}{\eta\gamma} \tag{4.21}$$

The parameter $L_1\rho/\eta\gamma$ is similar in magnitude and in physical content to the parameter $K\rho/\eta^2$ ($K =$ Frank elastic constant) which is central to the discussion of the fluctuation modes in the ordered phase. As pointed out in Ref. (5) this parameter is of order 10^{-4} or less: thus $\xi/\delta \gtrsim 10^{-2}$ and the concept of an effective viscosity is acceptable.

4.4. COUPLED MODES OF BIREFRINGENCE AND FLOW

We now discuss in more detail the coupled relaxation modes of Q and v, for a Fourier component of given wave vector q. The characteristic frequency associated with v is $\eta q^2/\rho$. The characteristic frequency associated with Q is the Litster width Γ. For small q values, these two frequencies are widely different, and the coupling between v and Q has very little influence on the power spectrum of Q, as measured by light scattering: we made use of this observation in paragraph 3b of this section. Here, we shall focus our attention on the opposite case, where both frequencies are comparable:

$$\Gamma \sim \frac{\eta q^2}{\rho} \tag{4.22}$$

It should be emphasized that (4.22) is compatible with our general assumption $q\xi \ll 1$. In fact, from Eq. (4.16) we see that the condition (4.22) corresponds to

$$q\xi \sim \left(\frac{\eta v}{L_1\rho}\right)^{1/2} \sim 10^{-2}$$

Let us put the z axis of our reference frame along q ($\partial/\partial x \to 0$ $\partial/\partial y \to 0$ $\partial/\partial z \to iq$). Because of the incompressibility condition we have only two non 0 components of v (v_x and v_y). v_x is coupled to Q_{xz} by Eqs. (4.7, 4.8). Similarly v_y is coupled to Q_{yz}. All the other components of Q ($Q_{zz}, Q_{xx}, Q_{xy}, \ldots$) are not coupled to the hydrodynamic flow, and relax with a single relaxation rate Γ. From now on, we shall concentrate on Q_{xz} and v_x. With suitable polarizations, the light scattering experiment (performed at wave vector q and frequency shift ω) measures the quantity:

$$S(q\omega) = \int dt < Q_{xy}(-q, 0) Q_{xy}(qt) e^{-i\omega t} > = -\frac{k_B T}{\pi\omega} IM[\chi(q\omega)]$$

(4.23)

$\chi(q\omega)$ is a response function giving Q_{xy} when the system is submitted to an external perturbation $(H_x H_z)$ modulated at wave vector q and frequency ω. Equation (4.23) is a statement of the fluctuation dissipation theorem for χ. Our use of response functions is similar in spirit to Ref. (5). χ may be derived very directly from Eqs. (4.7, 4.8, 4.10), which give, after inclusion of the external pertubation:

$$iq\frac{\mu}{\nu}v_x + (i\omega + \Gamma)Q_{xz} = \Gamma\chi_0 H_x H_z \qquad (4.24)$$

$$\left(i\omega + \frac{\eta}{\rho}q^2\right)v_x + \frac{2\mu}{\rho}q\omega Q_{xz} = 0 \qquad (4.25)$$

Here $\chi_0 = 1/2A$ is the static susceptibility (Eq. 3.3). Solving Eq. (4.24, 4.25) for $Q_{xz} \equiv \chi(q\omega) H_x H_z$ and taking the imaginary part of χ leads to:

$$S(q\omega) = \frac{k_B T}{2\pi\nu} \frac{\omega^2 + ab\,q^4}{(\omega^2 - \Gamma aq^2)^2 + \omega^2(\Gamma + bq^2)^2} \qquad (4.26)$$

where

$$a = \frac{\eta}{\rho}$$

$$b = \frac{\eta}{\rho} - \frac{2\mu^2}{\nu\rho}$$

Equation (4.26) shows that in general the power spectrum of Q_{xz} is not composed of a single Lorentzian. However, when $aq^2 \ll \Gamma$ we recover a single Lorentzian of width Γ. In the opposite limit $(aq^2 \gg \Gamma)$ we also find a single Lorentzian, but with a modified width:

$$\Gamma' = \frac{a}{b}\Gamma \qquad \left(\frac{\Gamma}{a} < q^2 < \xi^{-2}\right) \qquad (4.27)$$

The more complicated scattering law described by Eq. (4.26) has not been observed by Litster and Stinson [2,6] but it might be worthwhile to search for it with suitable q values and polarization indices.

P.G. DE GENNES

5. Extension to Cholesterics

5.1. STRUCTURE OF THE FREE ENERGY

With an optically active material, it is still possible to define an order parameter $Q_{\alpha\beta}$ by Eq. (2.7). The static magnetic susceptibility tensor is always symmetric, and $Q_{\alpha\beta}$ is symmetric traceless. The expansion of the free energy $\bar{F}$ in powers of Q retains the structure (3.1) and Eq. (3.4) for the magnetic birefringence is always valid. Thus, if T^* is only slightly below T_c, cholesterics will show a large magnetic birefringence anomaly, just like nematics.[14]

New features appear when we investigate the terms involving the gradients of Q in the free energy: symmetry allows for a new term

$$F_c = q_0 L_1 \epsilon_{\alpha\beta\gamma} Q_{\alpha\mu} \partial_\gamma Q_{\beta\mu} \qquad (5.1)$$

In this formula $\epsilon_{\alpha\beta\gamma}$ is the alternant symbol ($\epsilon_{xyz} = 1$, $\epsilon_{xxa} = 0$, etc.) and is antisymmetric with respect to all pairs of indices. The energy density F_c is a pseudoscalar, and its coefficient must vanish in a nematic. But in a cholesteric F_c must be included: the complete free energy contains (a) the term $\bar{F}$ for a uniform Q; (b) the term F_c (linear in grad Q); (c) the usual term F_g quadratic in the gradients (Eq. 3.5). The coefficient in F_c has been written as $q_0 L_1$ where L_1 is still defined by Eq. (3.5), and q_0 has the dimension of an inverse length. Qualitatively we may say that $2\pi/q_0$ is the helical pitch which the isotropic phase would tend to display for T just above T_c. We shall now explore some consequences of the presence of F_c in the free energy.

5.2. INTENSITY OF LIGHT SCATTERING

Let us now discuss the magnitude of the fluctuations of Q for a Fourier component of given wave vector q. As in section 4, we take our z axis parallel to Q. From Eqs. (3.1, 3.5 and 5.1) we get for the free energy associated with this Fourier component, to order Q^2:

$$F(q) = \tfrac{1}{2}(A + L_1 q^2)[\tfrac{3}{2}Q^2 + \tfrac{1}{2}P^2 + 2(|Q_{xy}|^2 + |Q_{yz}| + Q_{zx}{}^2|)]$$
$$+ \tfrac{1}{2}L_2 q^2[Q^2 + |Q_{xz}|^2 + |Q_{zy}|^2]$$
$$+ 2q_0 L_1[P'' Q'_{xy} - P' Q''_{xy} + Q''_{zz}Q'_{yz} - Q'_{zz}Q''_{yz}] \qquad (5.2)$$

In Eq. (5.2) we have used the notation of Eq. (2.8) for the diagonal

454

components of $Q_{\alpha\beta}$. We have also analyzed the Fourier component $Q_{\alpha\beta}(q)$ in its real and imaginary part $Q_{\alpha\beta}(q) = Q'_{\alpha\beta}(q) + iQ''_{\alpha\beta}(q)$. Diagonalizing the quadratic form (5.2) and applying the equipartition theorem we arrive at the following averages:

$$\langle Q_{zz} \rangle \equiv \langle Q^2 \rangle = \frac{2k_\beta T}{3(A + L_1 q^2) + 2L_2 q^2} \tag{5.3}$$

$$\langle |Q_{xy}|^2 \rangle = \tfrac{1}{4} \langle P^2 \rangle = \frac{k_\beta T(1 + \xi_1^2 q^2)}{2A[(1 + \xi_1^2 q^2)^2 - 4q_0^2 q^2 \xi_1^4]} \qquad \xi_1^2 = \frac{L_1}{A} \tag{5.4}$$

$$\langle |Q_{xz}|^2 \rangle = \langle |Q_{yz}|^2 \rangle = \frac{k_\beta T \tilde{A}}{2[\tilde{A}^2 - q_0^2 L_1^2 q^2]} \tag{5.5}$$

$$\tilde{A} = A + (L_1 + \tfrac{1}{2}L_2) q^2$$

The fluctuations of Q_{zz}, as defined by Eq. (5.3), are not different from what they are in the nematic state. But all other averages are modified. Let us focus our attention on $I_{xy}(q) = \langle |Q_{xy}|^2 \rangle$. Examination of Eq. (5.5) shows the following features

–if $q_0 \xi_1(T) < \tfrac{1}{2}$ the maximum of $I_{xy}(q)$ is at $q = 0$

as it is in a nematic

–if $q_0 \xi_1(T) > \tfrac{1}{2}$ the maximum of $I_{xy}(q)$ is at a

finite q (comparable to q_0): we then have a broad scattering peak which is reminiscent of the Bragg peak in the ordered phase.
However $q_0 \xi_1$ cannot become much larger than $\tfrac{1}{2}$: in fact, if $q_0 \xi_1$ would reach the value 1, the fluctuations $\langle |Q_{ky}|^2 \rangle$ would diverge for $q = q_0$. The temperature T^{**}, such that $q_0 \xi_1(T^{**}) = 1$, is thus the temperature below which the isotropic phase is absolutely unstable. A first order transition from isotropic to cholesteric must occur (because of the Q^3 terms in F) at a temperature $T_c > T^{**}$. Thus $q_0 \xi(T_c) < 1$.
In many cases we may in fact have $q_0 \xi(T_c) < \tfrac{1}{2}$: then $I_{xy}(q)$ has a central peak at all temperatures above T_c, and the tendency to build up a spiral is not strongly apparent in the properties of the isotropic phase.
A similar discussion can be carried out for the components Q_{xz} and Q_{yz}, with qualitatively similar conclusions. It can be shown, however, (using the inequalities of section III on L_1 and L_2) that the

455

fluctuations of Q_{xy} are more singular: if we were able to supercool below T_c, the onset of instability would occur (at $T = T^{**}$) through Q_{xy}, while Q_{xz} is still comparatively small.

5.3. Optical Rotation at Long Wavelengths

Cholesterics in their ordered phase show a huge optical rotation: thus, if the cholesteric → isotropic transition is nearly of second order, we might, at first sight, expect a large optical rotation just above T_c. This is not correct, however, as shown by the following argument.

To derive the optical rotation we essentially look at the forward scattering amplitude for a process where a photon is absorbed at point (1), a virtual photon propagates from (1) to (2), and a final photon is emitted from (2). This gives a second order correction to the non local polarisability tensor, proportional to

$$p_{\alpha\beta}(\mathbf{R}) = \langle Q_{\alpha\gamma}(\mathbf{r}_1) T_{\gamma\mu}(\mathbf{R}) Q_{\mu\beta}(\mathbf{r}_2) \rangle$$
$$\mathbf{R} = \mathbf{r}_1 - \mathbf{r}_2 \tag{5.6}$$

Here $T_{\gamma\mu}$ represents the virtual photon propagator, and in the long wavelength limit ($\lambda \gg R$) it essentially describes the field of a static dipole:

$$T_{\gamma\mu} = \frac{1}{R^3}\delta_{\gamma\mu} - \frac{3}{R^5}R_\gamma R_\mu \tag{5.7}$$

The optical rotation is proportional to Ω/λ^2, where Ω is the integral:

$$\Omega = \int d\mathbf{R}\, p_{xy}(\mathbf{R})\, R_z \tag{5.8}$$

Ω may be computed in detail from the Eqs. (5.4 and 5.5). Qualitatively, we may estimate Ω as follows: the chiral part of the $\langle QQ \rangle$ correlation function which occurs in (5.6) is of order

$$\langle Q(0) Q(\mathbf{R}) \rangle_{\text{chiral}} \simeq \frac{k_B T}{L_1 R} e^{-R/\xi} q_0 R \tag{5.9}$$

Inserting this in (5.8) we have

$$\Omega \sim q_0 \int 4\pi R^2\, dR\, \frac{1}{R^2}\frac{k_\beta T}{L_1} e^{-R/\xi} \sim \text{const}\, \frac{k_\beta T}{L_1} q_0 \xi \tag{5.10}$$

We have seen that $q_0 \xi \gtrsim 1$. Thus $\Omega \sim k_\beta T/L_1$ is non singular near T_c. It is our hope that this point will soon be checked experimentally.

6. Concluding Remarks

We know that there are spectacular short range effects in the isotropic phase of nematics, and we know how to correlate them in terms of a small number of phenomenological constants. We do not have, however, any measurement of the coherence lengths ξ_n and ξ_t; light scattering involves wavelengths which are too large; studies on the reflectance of the nematic/isotropic interface might be helpful. Qualitatively, since the enhancement factor found in the magnetic birefringence or the flow birefringence reaches values of order 100, we expect the ratio coherence length/molecular length to be of order $\sqrt{100} \sim 10$.

Another attractive direction for experimental work is the study of *metastable phases*. Is it possible to observe a metastable isotropic phase in the small interval $T^* < T < T_c$, and to follow the fluctuations in such a phase? Also, can one measure the temperature T_1 above which the nematic phase is absolutely unstable? Finally, from the nucleation processes of one phase in the other, can one obtain significant information on ξ_n or ξ_t?

All our analysis has been restricted to properties which are quasi-macroscopic (spatial variations slow on the molecular scale). However, certain microscopic properties can be estimated from it: in particular the Pincus calculation of nuclear relaxation rates[15] may be extended to temperatures above T_c, using equations such as (5.26), suitably generalized to cover situations where $q\xi \sim 1$. A qualitative estimate has been described in Ref. (7), but more detailed calculations are clearly required.

Acknowledgments

The present author is greatly indebted to R. B. Meyer for a number of discussions on pretransitional phenomena in liquid crystals. He would also like to thank D. Litster and A. Arrott for some very stimulating conversations. Part of this work was done during a visit at the General Electric Research and Development Center, and the author would like to thank M. Fiske and C. P. Bean for their hospitality.

REFERENCES

1. A useful list of latent heats and other characteristics of the transition is given in the review by A. Saupe. *Angurandte Chemie* (English Version) **7**, 97 (1968).
2. Zadoc Kahn, J., *Annales de Physique* **6**, 31 (1936), Zvetkov, V., *Acla Physicochemica USSR* **19**, 86 (1944).
3. Stinson, T. W., Litster, J. D. To be published.
4. Allain, Y. Private communication.
5. For an analysis of the scattering in the ordered phase, see Orsay group on liquid crystals, *Journ. Chem. Phys.*, **51**, 816 (1969).
6. Litster, J. D., Stinson, T., *Journ. App. Phys.*, **41**, 996 (1970).
7. de Gennes, P. G., *Physics Letters*, **30A**, p. 454 (1969).
8. Summation over repeated indices is always implied in the equations.
9. Freiser, M. J., *Phys. Rev. Lett.* **24**, 1041 (1970).
10. Frank, F. C., *Faraday Soc. discussions* **25**, 19 (1958). Ericksen, J. L., *Arch. Rat. Mech. Anal.* **10**, 189 (1962).
11. Orsay group on liquid crystals. *Phys. Rev. Letters*, **22**, 1361 (1969).
12. The proof is similar to that leading to Eq. (2.6) and can be found in the book by de Groot: " Thermodynamics or irreversible processes ", North Holland 1954.
13. Private communication from Dr. Candau.
14. Of course the magnitude of the birefringence may be smaller than in nematics, because the anisotropy of the electric polarisability is rather small for cholesterol esters.
15. Pincus, P., *Solid State Comm.* **7**, 415 (1969).

Pierre-Gilles de Gennes was born in Paris in 1932. He studied physics and biology at the École Normale Supérieure in Paris, between 1951 and 1955. Between 1955 and 1959 he worked as a research engineer at the Atomic Energy Centre in Saclay, working mainly on neutron scattering and magnetism. He was awarded the Ph.D degree in 1957. In 1959 he was a postdoctoral visitor to the solid state physics group of C. Kittel in Berkeley (USA), subsequently serving for two years in the French Navy.

In 1961 he was appointed as assistant professor at the Université de Paris-Sud in Orsay, and started a group working in superconductivity. In 1968 he switched to liquid crystals, creating a joint experimental-theoretical group in this area. In 1971 he began work on polymers. Later still he worked in colloid science, the theory of dynamical wetting, adhesion, granular matter and artificial muscles. Apart from his initial work in magnetism and superconductivity, most of his work would now be recognised as belonging to the theory of so-called soft condensed matter, in which he is widely regarded as a pioneer. A distinguishing mark of his work has been a commitment to interdisciplinarity.

In 1971 he became professor of condensed matter physics at the Collège de France. At the same time, between 1976 and 2002, he was director of the École Supérieure de Physique et Chimie Industrielle in Paris.

Over his career he has so far (2002) written 500 research papers, and a number of influential books, including *Superconducting metals and alloys* (Benjamin 1964), *The physics of liquid crystals* (OUP 1976, 2nd edn with J. Prost 1993), *Scaling concepts in polymer physics* (Cornell Univ. Press 1979) and *Simple views on condensed matter physics* (World Scientific 1992).

de Gennes is a member of the French Academy of Sciences, the Dutch Academy of Arts and Sciences, the Royal Society of London, the American Academy of Arts and Sciences, and the National Academy of Sciences, USA. He has received a large number of prizes and awards, including the Holweck Prize from the joint French and British Physical Society; the Ampere Prize, French Academy of Science; the gold medal from the French CNRS; the Matteuci Medal, Italian Academy; the Harvey Prize, Israel; the Wolf Prize, Israel; The Lorentz Medal, Dutch Academy of Arts and Sciences; and polymer awards from both APS and ACS. He received the 1991 Nobel prize in physics.

Section D

THE DEVELOPMENT OF LIQUID CRYSTAL DISPLAY TECHNOLOGY

THE DEVELOPMENT OF LIQUID CRYSTAL DISPLAY TECHNOLOGY

Introduction

The emergence of liquid crystals as new states of matter provided many decades of controversy and intellectual exercise for those involved. While the rhetoric flowed back and forth, it was experiments and demonstrations of liquid crystals that captured the imagination. In particular Lehmann took his travelling show of liquid crystal demonstrations to many centres around Europe, and while he may not always have changed scientific opinions, the demonstrations certainly made an impact. The dramatic effects of liquid crystals are for the most part a consequence of their optical properties, and even Reinitzer (see Part A) had remarked on the unusual colour changes observed on melting or cooling cholesteryl benzoate. Mauguin, as a mineralogist, was primarily interested in the optical properties of liquid crystals, and these early observations were reported in his 1911 article [A9]. Reading this, there is an undisguised excitement for the experiments; the important results are italicised, perhaps to convince both the reader and the author of their authenticity. The optical properties of liquid crystals continued to fascinate experimental scientists, but it required a big step to go from their description and interpretation to applications. Indeed in 1924, Vorländer wrote[1]

> I have been asked if the crystalline-liquid substances may be technically exploited? I do not see any such possibility.

The most important application of liquid crystals is in the area of displays, and for this reason we devote a separate section of this book to them. Astonishingly, for a large majority of the public, the term liquid crystal is only recognised in the context of a display; the discovery of a new state of matter is not a subject of public excitement.

In some respects liquid crystal displays have revolutionised the electronics industry, which since the discovery of the transistor had become firmly based on solid state physics. Solid state technology exploited the electronic effects of semi-conducting materials, and gave rise to the silicon-based industries, as in Silicon Valley. Liquid crystal display devices and other non-display devices use a fluid as their active element, and they are based on molecular effects rather than electronic effects. An important type of device produced by the electronics industry is the user interface, or VDU (visual display unit) as it used to be called. For a long time this was the CRT (cathode ray tube), but electro-luminescent effects in solid state semi-conductors have also been developed into a variety of display devices. The revolution in displays caused by liquid crystals encountered early resistance from the electronics industry, even from some of those companies which were best placed to exploit the new technologies. Liquid crystal devices have required the acquisition of a new knowledge base, new research methods, and above all new production methods, but in spite of these difficulties, the liquid crystal display business has grown from a few million dollars in 1980 to approaching 30 billion dollars in 2002. In the year 2000 approximately 2 billion liquid crystal displays were made, which is nearly one for every third member of the human race.

There are numerous physical effects in liquid crystals that may be exploited in applications, and many of these can form the basis for a display device. Thus there are a number of potential liquid crystal display device technologies. A few of the potential technologies have been tried commercially; some have failed, some have been superseded, and yet others have adapted to become highly successful in the hands of major world electronics companies, most of which are based in the Far East.

However, one particular liquid crystal technology above all others has dominated the recent period of display development. This is the so-called *twisted nematic* or TN display. The manner in which this was discovered, and how the various constituent physics and materials science suddenly came together is a dramatic story. Since the discovery of the twisted nematic display in the early 1970s, there have been competing claims, counter claims and litigation on three continents to establish the rights to this particular invention. But this is not the story we wish to tell here, rather it is the deeper story of how fundamental scientific research over nearly 100 years resulted in the birth of the multi-billion dollar business of liquid crystal displays. The key papers reproduced here have been selected to trace the scientific odyssey that resulted in the emergence of the first nematic liquid crystal displays, and in particular the twisted nematic display. We have added a further two papers which are concerned with display devices based on ferroelectric smectic liquid crystals. While these latter displays have not been commercialised to anything like the extent of the TN device, the scientific ideas behind these new devices are sufficiently ground-breaking to warrant exposure in this volume.

Liquid crystals are anisotropic materials, and their properties are different in different directions. The concept of anisotropic properties is easy to understand for solid crystals, where a simple rotation of the crystal will change its direction, and hence the properties. Indeed the optical properties of liquid crystals convinced Lehmann [A2] that they were crystals, but having such a low surface tension that they could flow and appear fluid-like. We now know that the correct description of liquid crystals is that they are anisotropic liquids, but in liquids it is far from obvious how a direction may be defined, or how it might be changed. This was not a problem for Lehmann, who thought that liquid crystals were crystals, and so could properly have an optic axis which would define a unique direction in the materials.

It was established by Mauguin [A8] and Grandjean [A10] that the optic axis of a liquid crystal film could be aligned on suitable surfaces. Mauguin, clearly an admirer of Lehmann, presented in paper A8:

> ...new results which are a clear confirmation of the conclusions of the distinguished German scientist (Lehmann). Because of their fluidity, these birefringent *liquids* (editor's italics) can give rise to very varied structures. The simplest are optically uniform oriented films, which one obtains on melting the material, with special precautions, between two clean glass surfaces.

It seems that it was Mauguin who persuaded both Lehmann and Friedel that liquid crystals really were liquids. Mauguin [A8] investigated the optical properties of films of azoxyanisole and azoxyphenetole aligned on glass. Films having the optic axis parallel to the glass substrates were prepared by slow crystallisation of the liquid to form large uniformly aligned domains. On reheating, the alignment of the liquid crystal remained. This observation had first been made by Lehmann and interpreted by him as a result of thin membranes of crystal which remained on the glass surfaces. Mauguin confirmed this observation, but also went on to establish unambiguously that the medium between the microscope slide and the cover glass was an optically uniaxial anisotropic liquid. He did this by observing the interference fringes in convergent monochromatic light, and noticed that on raising and lowering the cover glass, the interference fringes moved, but preserved their sharpness. Thus the liquid is preserving its anisotropy, while it is forced to flow. A perpendicular (homeotropic) alignment could also be achieved in thin films using very clean glass surfaces. As a result of these findings, surfaces could be used to control the orientation of the optic axis, and so the properties of liquid crystals could be measured in different directions.

The behaviour of liquid crystals on crystal surfaces was also examined by Mauguin,[2] but the most complete account of such early studies is given by Grandjean [A10]. One reason for these investigations was that the surfaces of crystals had some intrinsic periodicity, and it was thought that any periodicity in liquid crystals or crystalline liquids might be revealed because of commensurate periodicities with the crystal surface. In fact this did not happen, but something surprising did, which was that the optic axis of the liquid crystal was aligned by crystal surfaces. Mauguin – as reported in paper A9 – discovered that an easier and more reliable method of aligning liquid crystal films was to use a magnetic field, and he also demonstrated that an external field could be used to change the orientation of the optic axis of the liquid crystal.

...under the action of a magnetic field, the liquid assumes all the optical properties of a uniaxial crystalline plate, the optic axis of which is now directed along the lines of force.

The modern way of aligning liquid crystal films in displays is to rub the glass. This technique was known to early researchers of liquid crystals, but is not clear exactly when the practice developed. Chatelain describes the method in detail, and we have included two of his papers (B11 and B12) on this topic. He explains the alignment effect as due to charge pairs forming on the glass surface from impurities deposited by the rubbing paper. The electric field generated by these charge pairs causes the alignment of the liquid crystal. It doesn't work with smectic phases because the fields aren't large enough. Chatelain's explanation is no longer accepted, and much work has since been done to provide a more accurate interpretation of alignment by rubbing. Whatever the explanation, a key component of the liquid crystal display had been discovered, namely how to control the alignment of a liquid crystal at a surface. Chatelain was certainly aware of the significance of his work, even if he did not anticipate the display applications:

> ...this study specifies in a complete way the conditions necessary to obtain good samples of nematic liquids oriented by rubbing the support surfaces. ...the straightforward practical consequences identified show that this phenomenon should allow important steps to be made towards a deeper understanding of the nematic phase.

The display technology that we are primarily focussing on is that of the TN display, and the twisted or helical structure of the liquid crystal film is a key aspect of its optical behaviour. In the device, incident plane-polarised light is guided by the twisted structure to emerge as plane-polarised light, but with the plane of polarisation rotated by 90°. Mauguin [A8] was responsible both for discovering the twisted state of nematic liquid crystals, and also for providing an explanation of the optics. Indeed the conditions under which plane-polarised light is guided by a helical director configuration is known as the Mauguin condition.

> There exists for each domain two special linear vibrations which remain linear on propagating through the liquid structure, but which turn during their passage by an angle equal to the angle between the surface films fixed to the glass slides... provided that the product of the birefringence of the lamellae and the helicoidal pitch is large in comparison with the wavelength of light.

In his investigations of the alignment of nematics on glass surfaces, reported in 1911, Mauguin [A8] created the alignment direction for the liquid crystal by allowing large single crystalline domains to form on cooling. These conditioned the glass surfaces, so that on melting the optic axis of the nematic liquid crystal

466

was aligned in the same direction as that of the crystal. Later, Chatelain too discovered that he could construct twisted films, in which the surface alignment was determined by rubbing:

> Thanks to this method, it is easy to create helicoidal films of great extent, which have a predetermined twist. It is sufficient to turn the cover slide in such a way that the direction of rubbing makes the desired angle with the direction of rubbing on the support slide.

Thus a key aspect of the TN liquid crystal display had been demonstrated and understood well in advance of its application. However, to create a display there must be some method of changing its optical appearance to carry an image. Furthermore if it is to be a useful display, then a simple method to erase and rewrite the image is desirable. For most displays, the write/read/erase/write sequence is achieved by application of an electric potential. This is certainly the case with liquid crystal displays, and so it is necessary to be able to change the optical properties of the liquid crystal by application of an electric field. As pointed out above, Mauguin [A9] knew that a magnetic field would align the optic axis of a nematic. Thus he showed how to control and change the optical properties of a nematic film using a magnetic field. It must have been obvious to him that such effects could be achieved also with electric fields, since both the Kerr electro-optic effect and Cotton–Mouton magneto-optic effect in liquids were well-known at the time Mauguin was working. In fact Friedel [B1] reports that he used electric fields to align nematic liquid crystals, but he found that the optic axis aligned perpendicularly to the field direction, in contrast to the observations of Mauguin with magnetic fields. We now know that the orientation of a nematic with respect to an applied field depends on the sign of the electric or magnetic susceptibility anisotropy, and the early liquid crystal materials studied had positive magnetic anisotropies and negative electric anisotropies. The first study of the effect of electric fields on liquid crystals was reported in 1904 by Bredig and Schukowsky,[3] but in 1918 Björnståhl[4] working in Uppsala published the results of a comprehensive investigation of the influence of electric fields on azoxyanisole and anisaldazine. Amongst other effects, he describes electric field-induced turbulence, which above a critical voltage results in strong scattering of light. This observation anticipated by nearly 50 years the discovery of dynamic scattering, which was the basis for the first commercialised liquid crystal displays (see below and papers D2 and D3).

Much of the physics behind today's liquid crystal displays was developed in the 1920s, and the effects of electric and magnetic fields on the optical properties of liquid crystals had been studied in a number of laboratories during the decades at the beginning of the twentieth century. Experiments with a magnetic field required magnets, and these tended to be big and cumbersome. Furthermore if you wanted to study liquid crystals, then not only was a heated stage necessary, but it was also desirable to have something like a microscope to peer into the magnet cavity. It is easy to see that the concept of a field-activated portable liquid crystal display did not fit easily with the laboratory requirements for the experiments. However, it was recognised in the 1930s that the electro-optical properties of fluids, and in particular the Kerr effect, provided a mechanism to translate electrical signals into an optical effect: the basis of a visual display unit. Kerr cell shutters were already used in the burgeoning cinematographic industry. The first patent claiming a possible display application of liquid crystals, granted to the Marconi Wireless Telegraph Company in January 1936, was for an improved Kerr cell light valve. In this patent, reproduced here as paper D1, a cell is described which uses a nematic liquid crystal. The proposed application is as a light modulator, for use in television, facsimile telegraph and other electro-optical translating systems. The claimed novelty of the device is the liquid crystal material, which can operate at much lower voltages than the nitrobenzene-filled devices normally used. The patent is remarkable in that it proposes the use of both thermotropic and lyotropic nematic materials, but the physics base for the patent is very weak. There is no attempt to use aligning glass surfaces to define the on or off states of the device. The on-state is created by applying an electric field to the liquid crystal in a direction perpendicular to the light beam. The field aligns the nematic, and the consequent birefringent film between crossed polarisers at 45° to the field direction will allow light to be transmitted. The authors claim that the transmitted light intensity can be modulated by adjusting the voltage applied to the cell. The patent apparently ignores the contemporaneous literature on the Frederiks transition, which could have been used to produce a much more efficient device than that described in the patent. But it was the first. Because there were no liquid crystals known at this time to be nematic at ambient temperatures, the device incorporated a heating element. In an attempt to avoid this complication, the patent proposes the use of a lyotropic nematic solution, stable at room temperature, formed by an aqueous solution of 9-bromophenanthrene-3 (or 6) sulphonic acid and related compounds. The patent does not record the long-term effect of applying 200 volts to an aqueous solution of an organic salt.

From the late 1930s until 1945, the scientists had, or were forced to have had, things on their mind other than liquid crystals. Although liquid crystals did not have an obvious usage in the conduct of war, displays certainly did, and no doubt much effort was expended in developing the display technology of the age – the cathode ray tube. With its perfection, and subsequent adaptation for television, the CRT seemed to be the final solution to the requirement for easy and fast display of text, graphics and images. Clearly not everyone was satisfied, and in the late 1950s and 1960s programmes were initiated to look for new display technologies. In the United States, the search for new display technologies was based in companies such as Westinghouse, and in particular the Radio Corporation of America Laboratories in Princeton, New Jersey. It was here, in 1963, that an important discovery stimulated the birth of liquid crystal display technology.

Early studies of the effects of electric fields on the optical properties of liquid crystals had been restricted to configurations in which the electrodes, and hence the electric field, were in a plane perpendicular to the direction in which the beam of light was travelling. There was a simple reason for this: the metal electrodes were opaque to light, but this was not the best arrangement[5] to exploit the electro-optical phenomena of liquid crystals. Following the Second World War, electrically conducting transparent glass became available. Indium-tin oxide (ITO) coated glass had been developed as heated windscreen material to prevent icing and misting in aircraft, and it was ideal for liquid crystal experiments. ITO glass allowed electric fields to be applied to the sample, while being able to observe and measure effects in the direction of the field, i.e. through the electrodes.

So it was that Richard Williams in 1963, working at RCA, applied an electric field to a thin film of nematic *p*-azoxyanisole contained between two pieces of glass coated with indium-tin oxide. The surfaces of the glass had not been rubbed, and the film was not observed between crossed polarisers. An article published in the *Journal of Chemical Physics*, and reproduced here as paper D2, records the results:

> As the voltage on the specimen is increased from zero, nothing is observed to happen until, at a fairly sharp threshold field of about 1000 V/cm, a pattern forms in the liquid as illustrated in Fig. 1.

These patterns or domains are now known as Williams domains, but the significance of these results for displays was not revealed in the paper. However the author modestly claims:

> Their existence (domains) and general behaviour suggest several features of the structure of liquid crystals which have not previously been reported.

In the absence of an electric field, the nematic film was optically transparent. However at the threshold voltage for the formation of domains, there was a sharp decrease in the intensity of light transmitted, which further decreased with increase of voltage. This effect was similar to that observed by Björnståhl[4] in 1918, although the sample geometry was different. It was correctly explained in terms of variations in the refractive index of the sample which were caused by the domains and the intrinsic birefringence of the liquid crystal. The inhomogeneous refractive index of the film caused strong scattering of the light, and so the sample was opaque to transmitted light. The origin of the domains was more difficult to explain, and Williams at first thought they were due to ferro-electrically aligned nematic regions with different orientations of the electric polarisation. The relatively low conductivity of the liquid crystal caused Williams to dismiss the possibility of flow or conduction effects. We now know that the formation of Williams domains is an electrohydrodynamic phenomenon, and is observable in nematics which have a positive conductivity anisotropy and a negative dielectric anisotropy. If Williams had used rubbed glass plates, he would have seen that the domains were aligned perpendicular to the rubbing direction, and he might have noticed circulation of dust particles around the axis of the domains, indicating flow. He did however establish that the spatial period of the domains was determined by the thickness of the sample. The mechanism of the effect was satisfactorily explained by Wolfgang Helfrich[6] and Edward Carr,[7] and is sometimes known as the Carr–Helfrich effect, although we shall see that Helfrich has a stronger claim to fame in another area of liquid crystals. The importance of the observations of Richard Williams was that he had found an optical effect with a threshold behaviour, which gave a rapid change in the transmission of light: just the requirements for a display device. It was however another five years before the phenomenon recorded by Williams became the basis for a display.

The intervening years saw the exploitation of another optical effect in liquid crystals, that of thermo-optical scattering.[8] This did not produce a display in the conventional sense, but provided a temperature sensor with a wide range of applications. From the very beginning, starting with the observation of colours from cholesteryl benzoate by Reinitzer, the brightly coloured scattering from cholesteric liquid crystals had been a source of interest and visual excitement. It had been understood from the work of Mauguin, Grandjean and Friedel that the source of the scattering was the helicoidal structures of the chiral nematic phase. In his 1933 paper [B5], Oseen considered the propagation of light through such a structure, and showed that for a wavelength corresponding to the Bragg condition, the medium completely back-scatters circularly polarised light of the same spatial handedness as the structure, whereas a circularly polarised wave of opposite handedness passes through. However it was not until 1951 that H. de Vries[9] formulated a complete theory of the phenomenon. Since the wavelength of the selectively scattered light depends on the pitch of the helical structure, and as this in turn is affected by temperature, a simple indicator could be devised based on the change in wavelength as a function of temperature. The dependence of scattered wavelength on temperature is particularly sensitive just above the transition from the cholesteric phase to a smectic A phase. These surface temperature sensors continue to have a place in the market, with diverse applications in materials testing, novelty thermometers for wine, and monitoring the temperatures of new-born babies.

Returning to the main theme of our story of displays, the electrohydrodynamic phenomenon discovered by Williams was much studied in the 1960s and 1970s. It was not surprising therefore that in 1968 other researchers from RCA, in particular George Heilmeier, were able to report the performance of a liquid crystal display based on the dynamic scattering of light. This resulted from optical turbulence caused by application of electric fields above a critical threshold to thin films of nematic liquid crystals. The first liquid crystal display is described in the paper D3 by George Heilmeier, Louis Zanoni and Lucian Barton. The materials used had changed, and Schiff's bases were reported as yielding the best performance. It had also been realised that electrical conduction played a vital role in the effect, and that it was necessary to have a material of negative dielectric

470

anisotropy. There was still a problem that the materials used were nematic above room temperature,[†] and so if a display was to be constructed it would need internal heating to maintain the liquid crystal in the nematic phase. A partial solution to this problem was provided by Joel Goldmacher and Joe Castellano,[10] also working in RCA, who reported a scattering display utilising Schiff's base mixtures, some of which were nematic down to 22 °C. The fact that Schiff's bases had been identified as good materials for dynamic scattering provided a stimulus for chemists. In 1969 a Schiff's base, N-(p-methoxybenzylidene)-p-n-butylaniline (MBBA) was reported by H. Kelker and B. Scheurle from the Hoechst Company that was nematic over the temperature range 20 °C to 41 °C. Their paper is reproduced as D4. Not only did this enable many more physical experiments to be carried out, without the need for thermostatically controlled heated chambers, it also allowed the realisation of a liquid crystal display that could be produced without the need for an internal heater. For a time, MBBA became the standard material for physical measurements, and also the key component, along with other Schiff's bases of liquid crystal displays. The suppliers of MBBA were Hoechst, a German chemical company based in Frankfurt, and in order to promote their new product, they placed an advertisement which challenged display manufacturers to produce a wall-mounted flat-screen television.

Bald können Sie Ihren Fernseher an den Nagel hängen

Soon you will be able to hang your television on the wall (*ed. lit.* a nail)

The challenge was not well-received, and CRT manufacturers brought a legal action against Hoechst.

However Schiff's bases are chemically unstable, and readily decompose in the presence of water, light, heat, and electric currents. The poor stability of these liquid crystals did not deter some electronics companies who went into production with liquid crystal watches and calculators based on the dynamic scattering effect. In the hands of some companies, this commercial exploitation was premature, and some of the earliest devices had only a short lifetime before the liquid crystal materials were seriously decomposed and the ITO electrodes destroyed by electrochemical processes. This was not an insurmountable problem, and it was possible to prevent deterioration of Schiff's bases in liquid crystal displays by introducing a protective layer of silicon dioxide on the glass. The coating prevented hydroxide ions from the glass substrates reaching the liquid crystal material, and together with the use of glass-to-glass seals allowed the manufacture of long-life displays.[11]

[†] There existed one or two liquid crystals that were nematic at room temperature – e.g. n-nonadi-2,4-ene acid, nematic between 23 °C and 49 °C, discovered by Weygand (C. Weygand, R. Gabler and J. Hoffmann, Z. phys. chem. **B50**, 124–27 (1941)). However these would not have been suitable for display applications.

As it turned out, liquid crystal displays based on dynamic scattering were destined to have a limited history, and did not do a lot to inspire public confidence in this new display technology. There were other more durable effects that could be exploited for displays, and in the late 1960s and early 1970s the search for new liquid crystal devices that could be commercialised intensified, and new players entered the game. Although RCA had created a strong liquid crystal display research group, the company was reluctant to commercialise the devices that came from the laboratories. Liquid crystal displays could provide a challenge to the CRT technology in which RCA had a huge investment, and which was financially very successful. However indirectly RCA was to have a big influence on future developments in liquid crystal displays in two very different ways. The huge royalties being paid to RCA for the shadow mask colour TV tube resulted in the UK Government setting up a Working Party to search for an alternative to the CRT technology. A key figure in the UK in this search for new display technologies was Cyril Hilsum, a senior scientist at the Royal Radar Establishment (RRE), Malvern, Worcestershire. Attention became focussed on liquid crystals, and a new research group on liquid crystal displays, led by Hilsum, was established at RRE – of which more later. But across the Atlantic, the greatest contribution which RCA made to the development of the highly successful TN display was to lose one of its young researchers, Wolfgang Helfrich, to the Swiss company Hoffmann-La Roche based in Basle which was looking for new research projects.

Despite the emergence in the early 1970s of new research groups in Europe directed towards the development of liquid crystal display technologies, there was still a strong research activity in the United States. Other phenomena such as the guest–host effect[12] and the so-called cholesteric–nematic phase change were developed into liquid crystal displays, some of which were commercialised. The guest–host effect used the dichroic properties of dyes dissolved in nematic liquid crystals. The devices were attractive, since they involved a colour change, and they did not require polarisers. The early work of Mauguin and Friedel had established that the cholesteric phase of liquid crystals was one in which the director, or optic axis formed a helical structure. This was capable of rotating the plane of polarisation of incident light, and exhibited selective reflection for wavelengths of light related to the helical pitch. Apart from selective reflection, the cholesteric phase could scatter light strongly, due to the formation of focal conic domains, the similarity of which to smectic phases had been noted by Grandjean and Friedel. The optical appearance of a cholesteric film depended on the treatment of the surfaces of the glass which contained the film. Application of an electric field to the cholesteric phase resulted in a change in optical appearance, and it had been deduced that this was a consequence of the unwinding of the helix by the field; magnetic fields also worked. Since the cholesteric phase without a helix is essentially a nematic phase, the effect of helix unwinding has become known as the cholesteric–nematic phase change effect, despite the fact that inviolable rules of symmetry forbid such a transition. The unwinding of the helix of a cholesteric by an electric field was initially observed by Zocher and Birstein.[13] Many have argued that the unwinding of a cholesteric helix is just the effect exploited in the twisted nematic display. However there are subtle but important differences, and the rôle of surface alignment in twisted nematic devices is not only essential, but changes the physics from the cholesteric–nematic phase change effect. As soon as boundary conditions become important, then the physicist must turn to theory for guidance. Fortunately help was at hand, and in 1970 the Scottish mathematician Frank Leslie produced a firmly based theory of the effects of fields on twisted structures in liquid crystals, which did include the effects of surfaces.

The effects of fields on helical structures had been considered before. In particular it had been demonstrated, both theoretically and experimentally, that applying a field in a direction perpendicular to the helix axis of a cholesteric would result in the distortion of the helix, with an increase in pitch. At some critical field strength the pitch would become infinite, and the director or optic axis was then everywhere parallel to the field direction. Leslie, whose paper is reproduced in D5, considered the case of a twisted nematic configuration contained between two parallel aligning surfaces, the alignment directions of which made an angle of ϕ_o to each other. The angular displacement of the alignment directions created a twisted structure in the cell, with the director varying continuously from one surface to the other. A magnetic field was present in a direction perpendicular to the planes of the confining surfaces and parallel to the axis of twist. Leslie solved the appropriate continuum equations, with the result:

[#]The above analysis points to the existence of a critical field strength H_c, given by

$$\nu l^2 H_c^2 = \alpha_1\left(\frac{\pi}{2}\right)^2 + (\alpha_3 - 2\alpha_2)\phi_o^2$$

provided that the angle ϕ_o is not too large.

This equation shows that the elastic energy stored in a twisted structure, represented by the right-hand side of the equation, can be overcome by application of a magnetic field, above a certain critical field strength given by the left-hand side of the equation. The equation also describes the threshold behaviour of a twisted nematic display, provided that the magnetic field is replaced by an electric field and the dielectric susceptibility anisotropy is used in place of the magnetic anisotropy. Leslie's result was originally presented at the Third International Liquid Crystal Conference held in Berlin during 24–28 August 1970. At this same meeting a paper was presented by John Dreyer, the abstract for which is reprinted below in its entirety.

[#] Here α_1, α_2 and α_3 are the elastic constants for splay, twist and bend, ν is the anisotropy of the magnetic susceptibility and l is the separation of the confining boundaries.

S 1.2

John F. Dreyer
Polacoat, Inc.
9750 Conklin Road
Cincinnati, Ohio, U.S.A., 45242

A LIQUID CRYSTAL DEVICE FOR ROTATING THE PLANE OF POLARIZED LIGHT

A nematic liquid crystal film between two surface oriented plates becomes a device for rotating the plane of polarized light when the plane of the entering polarized light is parallel or perpendicular to the orientation of the first surface. The direction of the plane of polarization of the exiting light is aligned in accordance with the orientation direction of the second surface.

The device described exploits the optical properties of twisted nematic films, which had been discovered by Mauguin more than half a century earlier. The meeting in Berlin would have been attended by many of those searching for a viable liquid crystal display, and they had not been disappointed. Leslie had given the theory for the threshold response of a twisted nematic film, while Dreyer had reminded everyone of its optical properties. Apparently everything was now in place for the physical realisation of the twisted nematic display.

The milestone, which is generally recognised as marking the birth of the twisted nematic display is the Swiss Patent 564 207, which was granted to Martin Schadt and Wolfgang Helfrich of Hoffmann-La Roche Company, Basle, Switzerland with a priority date of 4 December 1970. The device was described in a paper entitled 'Voltage-Dependent Optical Activity of a Twisted Nematic Liquid Crystal', authored by Schadt and Helfrich, and submitted to Applied Physics Letters on 8 December 1970. We reproduce this here as article D6. In the paper, the authors describe the construction and optical characteristics of a twisted nematic cell, and quote Leslie's result for the threshold voltage. It was necessary to use a nematic with a positive dielectric anisotropy, in contrast to the materials required for dynamic scattering devices. Work had been going on behind the scenes to produce more suitable materials, and mesogens with potentially large positive dielectric anisotropy had been prepared[14] by adding cyano groups to the terminal positions of alkoxy Schiff's bases. In fact in September 1968 a patent[15] had been filed from RCA, which reported an electro-optic light modulator using mixtures of such Schiff's bases having positive dielectric anisotropies and nematic ranges from about 30 °C to 90 °C. Since the mixtures readily supercooled, they could be used in devices at room temperature, and Castellano reported on such mixtures at the Berlin Conference in August 1970. The materials used in the first twisted nematic display made by Schadt and Helfrich used alkylcyano-Schiff's bases.

474

Fig. D1 One of the first prototype twisted nematic displays (courtesy of M Schadt). See Plate III.

Although the twisted nematic device is often referred to as the 'Schadt–Helfrich' display, it is clear that a lot of earlier science actually contributed to its realisation. Furthermore a number of research groups had been exploring new ideas for liquid crystal displays, and had come very close to the configuration described in D6. Particular mention should be made of James Fergason who had worked on liquid crystals at Westinghouse, and in 1970 had a position at the Liquid Crystal Institute, Kent State University. In January 1970 Fergason and others from the Liquid Crystal Institute published an article in Electro-Technology[16] on liquid crystals and their applications. In this the authors describe a guest–host display having a twisted nematic configuration, but the device exploits the dichroic properties of the dye and does not have polarisers. Undoubtedly it was not the twisted nematic display of Schadt and Helfrich, but it was similar. In fact Fergason subsequently filed a patent[17] in the United States in April 1971 which describes the twisted nematic display, but Hoffmann-La Roche had the edge and bought the Fergason patent rights. Significant scientific developments are often waiting to happen: the precise time and place being somewhat serendipitous.

It was not obvious at this time that the twisted nematic configuration was necessarily the best choice for a display, and a number of other optical effects were being investigated. One which attracted attention was the electric field-induced deformation of homeotropically aligned films of a nematic (the DAP effect), described[18] in 1971 by Schiekel and Fahrenschon of AEG-Telefunken. Their device, like the dynamic scattering displays, used materials of negative dielectric anisotropy, but the scattering effect was suppressed by using alternating voltages of sufficiently high frequency.

Although prototype liquid crystal displays had been developed in both the United States and Europe, there was some reluctance by display manufacturers in these countries to add liquid crystal devices to their product lists. In Switzerland, the birth-place of the twisted nematic display, manufacturers of watches were not enthusiatic about the appearance of rival digital watches. Despite these difficulties, twisted nematic displays started appearing on the market in the early 1970s, but there were still problems with the liquid crystal materials used. The first room temperature nematics for TN displays were mixtures containing alkylcyano Schiff's bases and cyano-esters. These materials had been developed[19] by Schadt's group at Hoffmann-La Roche, and were used by Japanese manufacturers in the development of liquid crystal watches and other displays. However these displays suffered from the stability problems exhibited by Schiff's bases, which have already been referred to.

The real breakthrough in terms of materials was the discovery and synthesis of a family of cyanobiphenyl derivatives by George Gray. He and his colleagues at the University of Hull, were working under a contract with the UK Ministry of Defence at RRE Malvern, and in conjunction with the new liquid crystal display group led by Cyril Hilsum at RRE Malvern. Gray's research group at Hull had been very active during the 1950s and 1960s on fundamental studies of liquid crystal materials (see Section C), and so were well-placed to respond to the demand for new stable liquid crystal materials having a positive dielectric anisotropy. The materials were the alkyl- and alkoxy-cyanobiphenyls and related 4-n-alkyl-4″-cyano-p-terphenyls, and they provided for the first time stable, colourless room temperature materials of strong positive dielectric anisotropy, which were ideal for use in twisted nematic displays. Paper D7 briefly reports their properties for the first time after patents had been filed, and further details were made public[20] at the American Chemical Society Meeting held in Chicago in 1974. The team at RRE Malvern soon produced suitable mixtures for the displays from the available homologous cyano-biphenyls and – terphenyls which met the exacting demands of the manufacturers in the US and increasingly Japan. Concerning this successful period[21] for the production of new liquid crystal display materials, George Gray recalls:

On a remarkably short time scale, thanks to the efforts of Research Director Ben Sturgeon at the UK Company BDH Ltd, the materials were quickly produced in bulk and sold worldwide. It is interesting to note that around this time (1973) BDH Ltd came under the ownership of E. Merck of Darmstadt, Germany, but the UK Company was left to operate its successful liquid crystal business independently, although there was close collaboration between the research teams in Darmstadt and at Hull University and RRE. By the end of the 1970s, mixtures based on the cyano-biphenyls and –terphenyls were the industry standard for nematic displays. Their great contribution was that at a time when user confidence in twisted nematic liquid crystal displays was faltering because of lifetime problems with materials such as cyano Schiff's bases, they made possible for the first time the manufacture and sale of TN devices that had long lifetimes and were attractive to users, so providing the emerging liquid crystal display industry with the secure

476

base that it needed in order to consolidate before developing in the burgeoning multi-billion dollar display device industry that we have today.

The German company E. Merck itself had a long-standing interest in liquid crystals. This can be traced back to a correspondence between its founder Emanuel Merck and the liquid crystal pioneer Otto Lehmann, in the early years of the twentieth century. In a letter dated 16 June 1905, Lehmann asked Emanuel Merck for permission to announce that the liquid crystalline compounds used by Lehmann would be supplied by Merck. Emanuel Merck promptly replied the following day:

> I must point out that the requirements to be met regarding the purity of your substances are extremely difficult. Consequently, I can only entrust very experienced people with their synthesis. I must reserve a considerable margin concerning the time for delivery and the price. I agree that you may announce the willingness of my company to prepare the compounds as far as possible. Please emphasise that the compounds are not in stock, but will be delivered if possible by special request.

Merck included in their catalogue some of the first liquid crystalline materials (see Fig. D2), such as the nematic material p-azoxyanisole, for many years, but without any significant commercial success. However around 1970 Merck began developing new liquid crystals in their research laboratories at Darmstadt. This work resulted in the discovery by Rudolf Eidenschink and Ludwig Pohl of important new materials, the cyanophenylcyclohexanes[22] and cyanobicyclohexanes,[23] which were the cyclohexane analogues of Gray's materials. The chemical structures of examples of these materials are shown below.

Examples of (a) cyanocyclophenylhexane, (b) cyanobicyclohexane and (c) cyanobiphenyl liquid crystalline compounds

These new liquid crystals from Merck also provided excellent colourless stable nematic materials for displays. Apparently competitors to the biphenyls/terphenyls,

Fig. D2 Advertisement by E. Merck for liquid crystals.

the phenylcyclohexanes and bicyclohexanes were in fact interesting complementary systems, since they had lower birefringences than the Gray/BDH materials, and from the late 1970s these two classes of compounds shared the marketplace. Since the performance of displays is directly linked to the electro-optical properties of the liquid crystal materials, there were strong commercial pressures to produce better and better display mixtures. Hoffmann-La Roche, who held the twisted nematic patents, was fundamentally a chemical company, and as well as manufacturing biphenyls and terphenyls under licence from BDH, they also became strongly involved in the production and the development new liquid crystal materials to improve the performance of displays.

As well as the advancements in materials, the 1970s were a time of rapid development of liquid crystal displays. There was a limit to the complexity of information that could be displayed using simple twisted nematic devices. However an apparently trivial change to the twisted structure, increasing the twist angle from 90° to 270°, opened up the way for much more complex displays such as screens for lap-top computers. In some respects the invention in the early 1980s of the supertwisted nematic display (STN), as these came to be called, was a re-run of the TN story. Just as the first twisted nematic device used the dichroic properties of a dye guest–host material without polarisers, so this patent[24] – filed in June 1982 by Colin Waters, and Peter Raynes from the Royal Signals and Radar Establishment at Malvern – described a dichroic supertwisted nematic device which operated without polarisers and which was capable of a high level of multiplexing.[†] However, this patent also recognised that the device could work without a dye, but with polarisers, although its operation was not described.

> For high birefringence materials and layers of 12 μm or more, a polarisation switch effect (c.f. the $\pi/2$ twisted nematic) is obtainable without a dye by using the cell between two polarisers…

Within a year, a similar patent was filed by Terry Scheffer and Johann Nehring of Brown Boveri Company, Switzerland, but this time the device operated with polarisers, and the optical effect exploited was birefringence rather than dichroism.[25] The display image was still coloured, like the dichroic display, but this was due to the optical properties of the liquid crystal film, which was outside the Mauguin limit for guiding of plane-polarised light.

[†] The term *multiplexing* describes a technique whereby individual display elements are sequentially activated to produce a complex image. The relatively slow response of the eye registers the complete image, despite the fact that its individual elements are being activated at different times. The complexity of an image that may be displayed in this way depends on the sequencing of the electrical pulses that activate the elements of the image, but more importantly on the electro-optical response of the liquid crystal material. Selecting liquid crystals with particular elastic properties enabled the multiplexing level (i.e. the number of image elements that could be activated) to be increased in TN cells. STN cells gave a dramatically increased level of multiplexing.

A key invention[26] from Sharp, Japan, allowed the colour to be eliminated from the STN displays, and they were subsequently developed as lap-top computer screens. Researchers at Sharp had designed a double layer TN cell, in which a second twisted liquid crystal film was included, which compensated the birefringence effects of the first active film, and so eliminated colour from TN displays; it also worked extremely well for STN displays. The first Sharp patent for a double layer compensated TN cell had been filed in Japan on 9 December 1980,[27] a year before the filing date of the US patent.[26] The first page of the priority Japanese patent is reproduced below, together with the figure from the patent, which shows the construction of a compensated TN cell.

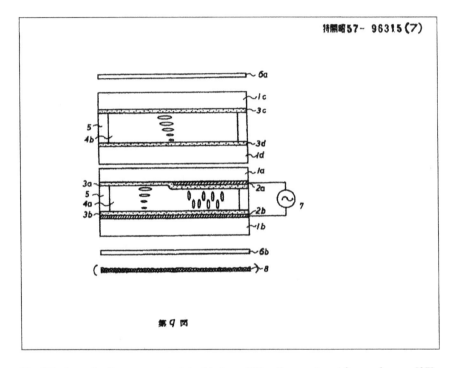

Fig. D3 An optically compensated double layer TN cell, reproduced from reference [27]. The passive layer is represented as 4b.

Liquid crystal computer displays of the twenty-first century use active matrix addressing, whereby each individual pixel of the display is separately addressed by its own transistor. This technology was only developed after 1980, and so is outside the review period of this book.

There have been many refinements to the experimental studies of liquid crystals, and the models and theories used to explain the experiments. In the context of displays, one of the most exciting developments has been the discovery of ferroelectric smectic liquid crystals, and their application in faster displays. Ferroelectric

(19) 日本国特許庁（JP）　　(11) 特許出願公開

(12) 公開特許公報（A）　昭57—96315

(51)Int. Cl.³	識別記号	庁内整理番号	
G 02 F 1/133		7348—2H	(43)公開 昭和57年(1982) 6月15日
	110	7348—2H	発明の数 1
G 09 F 9/00		6865—5C	審査請求 未請求

（全 7 頁）

(54)二層型液晶表示装置

(21)特　　願　昭55—174406

(22)出　　願　昭55(1980)12月9日

(72)発 明 者　船田文明
大阪市阿倍野区長池町22番22号
シャープ株式会社内

(72)発 明 者　松浦昌孝
大阪市阿倍野区長池町22番22号
シャープ株式会社内

(72)発 明 者　和田富夫
大阪市阿倍野区長池町22番22号
シャープ株式会社内

(71)出 願 人　シャープ株式会社
大阪市阿倍野区長池町22番22号

(74)代 理 人　弁理士 福士愛彦

明　　細　　書

1.　発明の名称

二層型液晶表示装置

2.　特許請求の範囲

1.　液晶分子の長軸方向を螺旋状に配向した液晶層を螺旋軸方向に二層積層し、前記液晶層の少なくとも一部に電圧を印加して液晶分子の配向状態を変換する給電手段と前記液晶分子の配向変換を顕視化する偏光手段を配設して成るツイステッドネマティック液晶表示装置に於いて、前記給電手段が実質的に前記液晶層の一方のみに設定され、他方の液晶層を補正板として機能せしめたことを特徴とする二層型液晶表示装置。

2.　各液晶層の最近接液晶分子の長軸方向が相互に略直交状態に設定されている特許請求の範囲第1項記載の二層型液晶表示装置。

3.　各液晶層の層厚dと正常光線及び異常光線に対する屈折率の差△nの積d・△nを略々等しく設定した特許請求の範囲第1項記載の二層型液晶装置。

4.　|d・△n|の値を0.36 μm以上2.0 μm以下に設定した特許請求の範囲第3項記載の二層型液晶表示装置。

3.　発明の詳細な説明

本発明は2枚の基板間に螺旋軸を基板面と垂直な方向にして基板間で液晶分子長軸を実質的に90°捩つたいわゆるツイステッドネマティック電界効果型液晶表示装置に関するものであり、特にその非活性時の表示の色付き現象を軽減化する技術に関するものである。

近年、液晶表示装置の分野に於いても表示情報量の拡大化が要求されるようになり、従来のセグメント型表示からマトリックス型表示へ需要動向が移行しつつある。しかしながら、マトリックス型表示で表示情報量を拡大するためには必然的にいわゆるマルチプレックス駆動の度数（デューティ比）を増加する必要があり、それに伴なつて表示コントラストの低下や視角範囲の狭在化という問題が生じる。この問題を解決する手段の1つとして、液晶層厚dと液晶の複屈折△n（＝n_e

—75—

The first page of Japanese patent 55-174406
Double-layered liquid crystal display
Inventors: F Funada, M Matsuura, T Wada
Assigned to Sharp, Osaka, Japan

materials have many associated interesting properties because of their symmetry. They are pyroelectric, piezoelectric and also exhibit a linear electro-optical response (the Pockels effect). The existence of ferroelectricity in liquid crystals had been the subject of much speculation, and indeed as we have already seen an early theory of nematic liquid crystals by Max Born[28] required ferroelectric polarisation to exist in nematics. It has already been pointed out that ferroelectricity has not so far been observed in nematics, but the situation for smectic liquid crystals is very different.

In 1973 Bob Meyer while on sabbatical at the Laboratoire de Physique des Solides in Orsay, gave a series of lectures at a Summer School in Les Houches on Molecular Fluids. In these lectures Meyer reviewed, along the lines of F.C. Frank, the symmetry arguments applicable to smectic liquid crystals, and realised that molecular polarisation perpendicular to the director should couple to bend polarisation. Since in a chiral smectic C phase there is a spontaneous bend of the tilted director, it was clear that there should also be a spontaneous polarisation in the ground state of a chiral smectic C phase. This was predicted to be in the plane of the smectic layers along the two-fold rotation axis perpendicular to the tilt plane. Since the chiral tilted smectic phases in the bulk have a helicoidal structure in which the molecular tilt direction precesses around the layer normal, the spontaneous polarisation also rotates from layer to layer, and the net polarisation is zero.

In order to test Meyer's idea, the chiral material DOBAMBC (4-decyloxyben-zylidene-4'-amino-2-methylbutylcinnamate) was synthesised by chemists at Orsay, and it proved to have both smectic A and smectic C phases. Samples were prepared in which the smectic layers were parallel to containing glass substrates. Electrodes had been embedded in the sample to give an electric field in the plane of the smectic C layers, and the samples were observed using a polarising microscope as the electric field was increased from zero. In the absence of a field the observed conoscopic image was that of an optically uniaxial material, consistent with the view of a helicoidal structure along the helix axis. As the electric field was increased from zero, the observed image shifted in a direction normal to the field direction indicating that the helix was distorted by the field. At higher fields the helix was completely unwound, giving a biaxial monodomain of the smectic C phase having a uniform tilt direction. Reversal of the field direction resulted in the reversal of the tilt direction, which is consistent with ferroelectric coupling between the spontaneous polarisation and the electric field. The transition between the smectic A phase and the smectic C phase was identified as a Curie point. These observations were described in a paper finally published in 1975, and reproduced here as D8. A consequence of this paper has been a whole new area of experimental and theoretical physics, and there is now a new subject of ferroelectric liquid crystals. In addition to the prediction and observation of ferroelectricity in chiral smectic C phases, Meyer and his co-authors also predicted and observed a tilt induced in the chiral smectic A phase by an electric field applied perpendicular to the layer normal. This effect, known as the electro-clinic effect has also been much studied in the years following Meyer's paper. The

482

conclusion to the paper suggests that the ferroelectric properties of these phases should provide useful probes for the study of phase changes and structure. It turned out that the ferroelectric properties also opened up a number of possibilities for faster liquid crystal displays based on smectic liquid crystals.

As the twisted nematic display became commercialised in the 1970s, its limitations became apparent. In particular the response time was relatively slow (~milliseconds), and this was a consequence of the switching mechanism from the twisted state to the optically non-guiding homeotropic state. Furthermore, TN cells had no memory, and required constant refreshing with an electric field to keep them in the on-state. The discovery of ferroelectricity in chiral smectic C liquid crystals provided an opportunity to make a display which would be both faster and have intrinsic bistability. This was the surface stabilised ferroelectric liquid crystal display reported by Noel Clark and Sven Lagerwall in 1980; their paper is reproduced here as D9.

Clark and Lagerwall discovered that by using very thin samples of liquid crystal (~1 μm) it was possible to produce films of DOBAMBC (the same material used by Meyer) in which the smectic layers were normal to the bounding glass surfaces, and the chirality-induced helix on the smectic C tilt cone was was unwound. By gently shearing the sample in the smectic A phase, a film was produced in which the director was parallel to the plane of the glass substrate. On cooling into the chiral smectic C phase, the liquid crystal adopted in local domains either of two possible surface-stabilised orientations, having ferroelectric polarisations in opposite directions. In this geometry, the symmetry arguments of Meyer show that the spontaneous polarisation is perpendicular to the glass substrates, with the different domains having the polarisation either up or down. With the smectic layers uniformly aligned, the domains were easily visible in a polarising microscope as regions of different contrast, and they were separated by clearly identifiable domain walls. In contrast to the experiments of Meyer, where the field was in the plane of the glass substrates, the substrates used by Clark and Lagerwall had been coated with an electrically conducting layer, so they were able to apply an electric field perpendicular to the substrate surfaces. Application of a voltage to the sample of ferroelectric domains favoured those domains having a polarity of the appropriate sign, and the domain walls moved to increase the favoured domain areas. Reversal of the voltage caused domains of the opposite polarity to be stabilised, and on removal of the voltage the favoured domain structure remained. This feature of bistability together with the very fast switching mechanism promised to revolutionise the liquid crystal display industry.

There were two problems which slowed the introduction of this new technology. Firstly submicron films of liquid crystal were required, raising manufacturing difficulties, and secondly the cell fabrication required alignment control of both the smectic layers and the director orientation. Additionally, smectic layers aligned perpendicular to a substrate surface are very fragile, which made the displays sensitive to mechanical shocks. However all of these problems are minimised in cells of small dimensions, so that the technology has been commercialised in the form of active matrix ferroelectric liquid crystal micro-displays for camera and camcorder displays. The TN display used old physics, and it was really the search for new display ideas coupled with advances in materials that resulted in the successful technology. On the other hand the ferroelectric liquid crystal display used new physics and new materials in a very original conjunction, but it has not yet shared the commercial success of the TN display.

This is a history book, and so we are looking back not forward. However since 1980 there have been massive developments in liquid crystal displays, and there continue to be many new ideas for new display technologies based on liquid crystals. No story of liquid crystal displays would be complete without referring to the rôle played by Japanese companies and scientists in the successful commercialisation and development of these devices.[29] Companies such as Sharp, Seiko, Epson, Hitachi and Toshiba have taken the simple ideas behind the liquid crystal display, and transformed the whole industrial sector of information display. Liquid crystal displays now challenge the traditional vacuum-tube TVs, and the LC displays of today are unrecognisable in comparison with the simple seven segment TN displays that were the first products of the new technology. The story of the twisted nematic display is a special one for liquid crystals: the transformation of an esoteric discovery into a household item. The story could be retold many times in different contexts to remind us of the contributions of fundamental science to our modern life-style.

References

1. D. Vorländer, '*Chemische Kristallographie der Flüssigkeiten*' (Chemical Crystallography of Liquids), Leipzig 1924, p. 89ff.
2. Ch. Mauguin, *Orientation des cristaux liquides par les lames de mica*, Comptes Rendus de l'Académie des Sciences **156**, 1246–7 (1913).
3. G. Bredig and N. Schukowsky, *Prüfung der Natur der flüssigen Kristalle mittels elektrischer Kataphorese*. Berichte der Deutschen Chemischen Gesellschaft, 3419–25 (1904).
4. Y. Björnståhl, *Untersuchungen über anisotrope Flüssigkeiten*, Ann. der Phys. **56**, 161–207 (1918). Björnståhl reports in this paper on a number of electro-optical experiments, and he demonstrates a sudden reduction in transmitted light intensity with increasing applied electric field. This has the appearance of a threshold effect. However, the sample configuration is such that the electric field is perpendicular to the direction of the light beam, and there is no reference to the use of polarisers. The reduction in transmitted light intensity is almost certainly a consequence of electric field-induced turbulence in the sample, but is not quite the same effect as that observed by Williams under better-defined circumstances.
5. The liquid crystal display industry grew up with devices in which the electric field is applied perpendicular to the liquid crystal film, hence the importance of ITO glass. However in 1995 a new method of switching TN displays was invented, In-Plane Switching, which had electrodes in the plane of the liquid crystal film. See e.g.

G. Baur, R. Kiefer, H. Klausmann and F. Windscheid, *In-plane switching: a novel electro-optic effect*, Liquid Crystals Today **5**(3), 13–14 (1995).

6. W. Helfrich, *Conduction-induced alignment of nematic liquid crystals: basic model and stability considerations*, J. Chem. Phys. **51**, 4092–105 (1969).

7. E.F. Carr, *Influence of electric fields an the molecular alignment in the liquid crystal p-(anisalamino)-phenyl acetate*. Mol. Cryst. Liq. Cryst. **7**, 253–68 (1969).

8. See: J.A. Fergason, *Liquid crystals*, Scientific American **211**, August 1964, 76–85, or J.L. Fergason, T.P. Vogl and M. Garbung, *Thermal imaging devices utilizing a cholesteric liquid crystalline phase material*. U.S. Patent 3,114,836; 17 December 1963.

9. H. de Vries, *Rotatory power and other optical properties of certain liquid crystals*, Acta Crystallogr. 4, 219–26 (1951).

10. J.E. Goldmacher and J.A. Castellano, *Electro-optical compositions and devices*. US Patent 3,540,796; 17 November 1970.

11. J. Castellano, private communication, The 100,000th liquid crystal watch containing Schiff's bases made by Fairchild Semiconductor Corporation using this method is still functioning.

12. G.H. Heilmeier and L.A. Zanoni, *Guest-host interactions in nematic liquid crystals. A new electro-optic effect*. Appl. Phys. Lett. **13**, 91–2 (1968).

13. H. Zocher and V. Birstein, *Beitrage zur Kenntnis der Mesophasen* (Zwischenaggregatzustände) *V: Über die Beeinflussung durch das elektrische und magnetische Feld*, Z. phys. Chem. **142A**, 186–94 (1929).

14. J.A. Castellano, J.E. Goldmacher, L.A. Barton and J.S. Kane, *Liquid crystals II. Effects of terminal group substitution on the mesomorphic behaviour of some benzylideneanilines*, J. Org. Chem. **33**, 3501–4 (1969).

15. J.A. Castellano, *Electro-optic light modulator*, US Patent No. 3,597,044; 3 August 1971.

16. J.A. Fergason, T.R. Taylor and T.B. Harsch, *Liquid crystals and their applications*, Electro-Technology, January 1970, pp. 41–50.

17. J.A. Fergason, *Display devices utilizing liquid crystal modulation*, US Patent No. 3,731,986; 8 May 1973 (filed April 22 1971).

18. M.F. Schiekel and K. Fahrenschon, *Deformation of nematic liquid crystals with vertical orientation in electric fields*, Appl. Phys. Lett. **19**, 391–3 (1971).

19. A. Boller, H. Scherrer, M. Schadt and P. Wild, *Low electro-optic threshold in new liquid crystals*, Proc. IEEE **60**, 1002–3 (1972).

20. G.W. Gray, K.J. Harrison, J.A. Nash, J. Constant, D.S. Hulme, J. Kirton and E.P. Raynes, *Stable, low melting nematogens of positive dielectric anisotropy for display devices*, in Liquid Crystals and Ordered Fluids, Vol. 2, eds J.F. Johnson and R.S. Porter (Plenum Press, New York, 1974) pp. 617–43.

21. For an engaging insider's account of the RRE role in the development of liquid crystal devices, see: C. Hilsum, *The anatomy of a discovery – biphenyl liquid crystals*, in *Technology of chemicals and materials for electronics*, Ch.3, ed. E.R. Howells (Ellis Horwood, Chichester 1991).

22. R. Eidenschink, D. Erdmann, J. Krause and L. Pohl, *Substituted phenylcyclohexanes – a new class of liquid crystalline compounds*, Angew. Chem. Int. Ed. Engl. **16**, 100 (1977).

23. R. Eidenschink, L. Pohl, J. Krause and D. Erdmann. SID (Society for Information Display) Digest, 102 (1978).

24. C.M. Waters and E.P. Raynes, *Liquid crystal devices*, UK Patent No. 2123163; 25 January 1984 (filed 27 June 1983).

25. T.J. Scheffer and J. Nehring, *A new, highly multiplexable liquid crystal display*, Appl.Phys.Lett. **45**, 1021 (1984).

26. F. Funada, M. Matsuura and T. Wada, *Interference color compensation double layered twisted nematic display*, US Patent No. 4,443,065; 17 April 1984 (filed 3 December 1981).

27. F. Funada, M. Matsuura and T. Wada, JP Patent No. 55-174406, *Double-layered liquid crystal display*, filed 9 December 1980.

28. M. Born, Sitzungsber. Preuss. Akad Wiss. **30**, 614 (1916). This is the report of the maths and physics session of the Prussian Academy of Sciences dated 25 May 1916.

29. See: H Kawamoto, *The history of liquid crystal displays*, Proc. IEEE **90**, 460–500. This review article provides a comprehensive account of the commercial development of liquid crystal displays from the 1960s to the 1990s.

PATENT SPECIFICATION

Application Date: July 13, 1934. No. 20621/34.
„ „ Aug. 3, 1934. No. 22721/34.

One Complete Specification Left: April 29, 1935.
(Under Section 16 of the Patents and Designs Acts, 1907 to 1932.)

Specification Accepted: Jan. 13, 1936.

441,274

PROVISIONAL SPECIFICATION
No. 20621 A.D. 1934.

Improvements in or relating to Light Valves

We, MARCONI'S WIRELESS TELEGRAPH COMPANY LIMITED, a company organised under the laws of Great Britain, of Marconi Offices, Electra House, Victoria Embankment, London, W.C.2, BARNETT LEVIN, of 104, Leighton Gardens, Kensal Rise, London, N.W.10, and NYMAN LEVIN, of 15, Kyverdale Road, Stamford Hill, London, N.16, both British subjects, do hereby declare the nature of this invention to be as follows:—

This invention relates to light valves and has for its object to provide improved light valves of great sensitivity suitable for use as electro-optical translating devices in television, facsimile telegraph and other systems.

There are many well known systems, notably television and facsimile telegraph systems, wherein it is required to translate received electrical signals into corresponding variations in light. For example, a commonly employed electro-optical translating system in television apparatus is of the polarised light type and comprises a nitrobenzene filled Kerr cell to the electrodes of which received signal (after suitable amplification) are applied and which is interposed in the path of a light beam between Nicol prisms, the arrangement being such that the variation in the ellipticity of the polarisation of the beam in the Kerr cell, as the result of incident electric potentials causes variations in the intensity of the transmitted beam.

One of the most important disadvantages or difficulties met with in known light valves or electro-optical translating devices of the kind depending upon the so-called Kerr effect for their action is that of obtaining sufficient sensitivity while a further difficulty is that large voltages are required to be applied to the cell. For example, in a practical Kerr cell arrangement using nitrobenzene as the bi-refringent material an applied voltage of the order of 4000 was required for a light path only 2 cm. across and with an electrode gap in the cell of only 0.13 cms. Where bi-refringent solids instead of liquids are employed, the voltages required are even higher.

The present invention is based upon the fact that certain substances, when in the nematic phases or phases or states intermediate between the solid state and the liquid state, exhibit a Kerr effect very much greater than that which they exhibit when either in the liquid or in the solid state. For example, whereas the Kerr constant of nitrobenzine is 4×10^{-3}— and nitrobenzine is the material which is at the moment probably the most widely used for light valves of the kind in question—the Kerr constant of ethyl-anisalamino-cinnamate (the formula of which is $CH_3OC_6H_4CH = NC_6H_4CH = CH COOC_2H_5$) is approximately 7 when the material in question is in the nematic phase of its liquid crystalline state.

According to this invention a light valve of the kind utilising the Kerr effect, i.e. of the kind employing bi-refringent material, is characterised in that the bi-refringent material employed is maintained in a phase or state intermediate between the liquid and solid states. The material in question may be almost any substance which can exist in the nematic phase and the particular compound already mentioned is only one of many substances which are utilisable in carrying out the invention. This particular compound is representative of a class of which other examples are methyl-*p*-ethoxybenzalamino-α-methyl-'cinnamate, ethyl - *p* - ethoxybenzalamino-α-methyl-cinnamate, methyl-anisalamino-α-methyl-cinnamate, and ethyl-anisalamino-α-methyl-cinnamate. These various substances differ among themselves as to transparency, temperature of formation of the nematic phase, and other properties and selection among them may, of course, be made to choose that which, having regard to its general properties is most suitable and convenient in any particular case.

[*Price* 1/-]

The invention may be carried into practice in various different ways. Generally speaking the production of bi-refringent material in a phase or state between the liquid and solid states presents no practical difficulties, the practical difficulties being mostly concerned with maintaining the material in the desired state in question. Various expedients may be adopted for maintaining a bi-refringent material in the desired intermediate state between liquid and solid. For example one of the electrodes on which the bi-refringent substance is placed can be arranged to be heated by a wire or winding carrying a heating current, the wire or winding being fixed on an asbestos sheet or other carrier of good heat insulation qualities and being arranged to be controlled by a rheostat inserted in the heating current supply circuit for the purpose of controlling or varying the temperature.

Owing to possible difficulties concerning variations in ambient temperature it is preferable to employ as the bi-refringent substance a material in which the intermediate phase or state in which the Kerr co-efficient is at or near the maximum, occurs over a relatively large temperature range. If the bi-refringent material be so chosen, it may be possible to ignore the possibility of varying ambient temperatures for such ambient temperature changes as may be likely to occur may be between limits which are closer together than the temperature limits between which the material remains in the state in which a high Kerr co-efficient is manifested. Where it is not convenient or practical to choose a material having a relatively large temperature range over which the desired state is maintained, the whole Kerr cell may be enclosed in a lagged or heat insulated chamber and/or any thermostatic control means as known per se may be provided to ensure that a predetermined temperature of the bi-refringent material is maintained irrespective of changes in ambient temperature.

In carrying the invention into practice, the thickness of the layer or layers of bi-refringent material employed should be small for the reason that if thick layers be used there is the danger that different parts of the layer will not be in the same state or phase. In practice it is preferred to use very thin layers of the order of 0.5 mm. in thickness.

In one form of light valve in accordance with this invention two strips of tinfoil are cemented on to a thin glass plate, such as a microscope slide in such a way as to leave a small gap between the two adjacent edges in the centre of the slide. The bi-refringent material to be employed is spread in a thin layer between the two strips of tin foil (which serve as electrodes) and then covered with a similar glass plate. Two sheets of asbestos each with a hole in the centre and each carrying a heating wire or winding are placed one above and one below the glass plates. The assembly as so far described constitutes the complete valve. Light is passed vertically through the hole in the asbestos sheets and the gap between the tin foil electrodes, and of course, any known polarised light optical system, such as is employed with known Kerr cells, is employed in conjunction with this cell. If it be desired to arrange the source of light to give a horizontal light beam and to arrange for the output light to travel horizontally, mirrors or prisms may be used to reflect a horizontal light beam from the source vertically down through the cell and to reflect light which has passed vertically through the cell horizontally away from the cell. Such prisms are preferably constituted by 45° prisms of Iceland Spar silvered on their hypotenuse faces.

It will be realised that with Kerr cells in accordance with this invention, it will be in general impossible to secure a light valve action immediately at any time, and that in general a short time will be necessary for "starting up" in order to allow the bi-refrigerant material to reach the state between liquid and solid at which the maximum Kerr effect is manifested.

It will be appreciated that the present invention, residing as it does in the use of liquid crystals or anisotropic melts provides a substantial improvement as regards sensitivity when compared to known light valves utilising the Kerr effect and it may be stated that with the substance already referred to,—namely ethylanisalaminocinnamate—the maximum voltage to be applied to the cell is only of the order of 50 volts, though the voltage to be applied will depend upon the distance between the electrodes.

Dated the 13th day of July, 1934.

CARPMAELS & RANSFORD.

Agents for the Applicants,

24, Southampton Buildings, London. W.C.2.

PROVISIONAL SPECIFICATION
No. 22721 A.D. 1934.

Improvements in or relating to Light Valves

We, MARCONI'S WIRELESS TELEGRAPH COMPANY LIMITED, a company organised under the laws of Great Britain, of Marconi Offices, Electra House, Victoria Embankment, London, W.C.2, BARNETT LEVIN, of 104, Leighton Gardens, Kensal Rise, London, N.W.10, and NYMAN LEVIN, of 15, Kyverdale Road, Stamford Hill, London, N.16, both British subjects, do hereby declare the nature of this invention to be as follows:—

This invention relates to light valves and has for its object to provide improved light valves of great sensitivity suitable for use as electro-optical translating devices in television, facsimile telegraph and other systems.

There are many well known systems, notably television and facsimile telegraph systems, wherein it is required to translate received electrical signals into corresponding variations in light. For example, a commonly employed electro-optical translating system in television apparatus is of the polarised light type and comprises a nitrobenzene filled Kerr cell to the electrodes of which received signals (after suitable amplification) are applied and which is interposed in the path of a light beam between Nicol prisms, the arrangement being such that the variation in the ellipticity of the polarisation of the beam in the Kerr cell, as the result of incident electric potentials causes variations in the intensity of the transmitted beam.

One of the most important disadvantages or difficulties met with in known light valves or electro-optical translating devices of the kind depending upon the so-called Kerr effect for their action is that of obtaining sufficient sensitivity while a further difficulty is that large voltages are required to be applied to the cell. For example, in a practical Kerr cell arrangement using nitrobenzene as the bi-refringent material an applied voltage of the order of 4000 volts was required for a light path only 2 cm. across and with an electrode gap in the cell of only 0.13 cms. Where bi-refringent solids instead of liquids are employed, the voltages required are even higher.

The specification accompanying our co-pending specification No. 20621/34 describes an invention which is based upon the fact that certain substances, when in the nematic phases or phases or states intermediate between the solid state and the liquid state, exhibit a Kerr effect very much greater than that which they exhibit when either in the liquid or in the solid state, and according to the invention contained in the said co-pending specification a light valve of the kind utilising the Kerr effect, i.e. of the kind employing bi-refringent material, is characterised in that the bi-refringent material employed is maintained in a phase or state intermediate between the liquid and solid states.

The various liquid crystalline substances mentioned in the co-pending specification above referred to, are organic compounds in themselves, that is to say, they are not in combination with other substances. There are two practical difficulties met with in light cells as described in the above mentioned co-pending specification, namely (1) that heating of the cell is necessary and (2) that there is some difficulty in obtaining a high degree of homogeneity.

The object of the present invention is to provide improved light valves of great sensitivity i.e. of sensitivity comparable to that obtained with valves as described in the co-pending specification above referred to, and wherein the above mentioned practical difficulties are avoided.

Certain substances, such as 9-bromo-phenanthrene-3 (or 6)-sulphonic acid, the material known under the registered trade mark "Salvarsan", m^1-nitro-benzoyl -m- aminobenzoyl -2- naphthyl-amino-4.8-disulphonic acid, and the di-sodium salt of m^1-nitrobenzoyl-m-amino-benzoyl-2-naphthylamine-6.8-disulphonic acid, when dissolved in water exist in the mesomorphic state, and that certain concentrations are in the nematic phase of this state.

According to this invention the bi-refringent material of a light valve utilising the Kerr effect is constituted by an aqueous solution of a substance as mentioned in the preceding paragraph in such concentration as to be in the nematic phase.

One form of cell in accordance with this invention is made by cementing together four pieces of glass, for example glass sheets 2 mm. thick, the four pieces being so arranged as to surround a square space having a side of from about 5 to 10 mm. The glass edges constituting one pair of opposite walls of this square

489

space are covered with metal foil, for example tin foil, or are sprayed with metal, for example gold or platinum, the metal coatings forming the electrodes of 5 the cell. The structure thus obtained is cemented on to a glass plate and the bi-refringent material is then poured into the square space which is now closed at the bottom by the plate upon which the 10 structure is cemented. Another glass plate is then placed on top to close in the bi-refringent material.

The light valve thus manufactured is a complete self-contained cell and may be used in any manner known per se in con- 15 junction with a suitable optical system for electro-optical translating purposes in connection with television, facsimile telegraph or other systems.

Dated the 1st day of August, 1934.
CARPMAELS & RANSFORD.
Agents for the Applicants,
24, Southampton Buildings, London,
W.C.2.

COMPLETE SPECIFICATION

Improvements in or relating to Light Valves

20 We, MARCONI'S WIRELESS TELEGRAPH COMPANY LIMITED, a company organised under the laws of Great Britain, of Marconi Offices, Electra House, Victoria Embankment, London, W.C.2, BARNETT 25 LEVIN, of 104, Leighton Gardens, Kensal Rise, London, N.W.10, and NYMAN LEVIN, of 15, Kyverdale Road, Stamford Hill, London, N.16, both British subjects, do hereby declare the nature of this inven- 30 tion and in what manner the same is to be performed, to be particularly described and ascertained in and by the following statement:—

This invention relates to light valves 35 and has for its object to provide improved light valves of great sensitivity suitable for use as electro-optical translating devices in television, facsimile telegraph and other systems.

40 There are many well known systems, notably television and facsimile telegraph systems, wherein it is required to translate received electrical signals into corresponding variations in light. For 45 example, a commonly employed electro-optical translating system in television apparatus is of the polarised light type and comprises a nitrobenzene filled Kerr cell to the electrodes of which 50 received signals (after suitable amplification) are applied and which is interposed in the path of a light beam between Nicol prisms, the arrangement being such that the variation in the ellipticity of 55 the polarisation of the beam in the Kerr cell, as the result of incident electric potentials causes variations in the intensity of the transmitted beam.

One of the most important disadvan- 60 tages or difficulties met with in known light valves or electro-optical translating devices of the kind depending upon the so-called Kerr effect for their action is that of obtaining sufficient sensitivity 65 while a further difficulty is that large voltages are required to be applied to the

cell. For example, in a practical Kerr cell arrangement using nitrobenzene as the bi-refringent material an applied voltage of the order of 4000 was 70 required for a light path 2 cm. across and with an electrode gap in the cell of only 0.13 cms. Where bi-refringent solids instead of liquids are employed, the voltages required are even 75 higher.

The present invention is based upon the fact that certain substances, when in the nematic phase or state which is intermediate between the solid and the liquid 80 states, exhibit a Kerr effect very much greater than that which they exhibit when either in the liquid or in the solid state. For example, whereas the Kerr constant of nitro-benzine is 4×10^{-5} and 85 nitro-benzine is the material which is at the moment probably the most widely used for light valves of the kind in question—the Kerr constant of ethylanisal-amino-cinnamate (the formula of which is 90 $CH_3OC_6H_4CH = NC_6H_4CH = CHCOOC_2H_5$) is approximately 7 when the material in question is in the nematic phase of its liquid crystalline state.

According to this invention a light 95 valve of the kind utilising the Kerr effect, i.e. of the kind employing bi-fringent material which is subjected to varying electric strain to produce varying light effects, is characterised in that the bi- 100 refringent material employed is in the nematic phase or state. The material in question may be almost any substance which can exist in the nematic phase and the particular compound already men- 105 tioned is only one of many substances which are utilisable in carrying out the invention. This particular compound is representative of a class of which other examples are methyl-p-ethoxybenzalamino- 110 a-methyl-cinnamate, ethyl - p - ethoxy-benzalamino-a-methyl-cinnamate, methyl-anisalamino-a-methyl-cinnamate, and

490

ethyl anisalamino-α-methyl-cinnamate. These various substances differ among themselves as to transparency, temperature of formation of the nematic phase,
5 and other properties and selection among them may, of course, be made to choose that which, having regard to its general properties, is most suitable and convenient in any particular case.
10 The invention may be carried into practice in various different ways. Generally speaking the production of bi-refringent material in a phase or state between the liquid and solid states pre-
15 sents no practical difficulties, but in some cases means must be specially provided for maintaining the material in the desired state in question. Various expedients may be adopted for maintain-
20 ing a bi-refringment material in the desired intermediate state between liquid and solid. For example where the material is one which must be maintained at a temperature above normal atmo-
25 spheric temperatures, the necessary heating may be obtained by arranging for one of the electrodes of the cell containing the bi-refringent substance to be heated by a wire or winding carrying
30 a heating current, the wire or winding being fixed on an asbestos sheet or other carrier of good heat insulation qualities and being arranged to be controlled by a rheostat inserted in the heating current
35 supply circuit for the purpose of controlling or varying the temperature.

Owing to possible difficulties concerning variations in ambient temperature it is preferable, in cases where the bi-refrin-
40 gent material selected is one needing to be heated to maintain it in the desired state, to employ as the bi-refringent substance a material in which the intermediate phase or state in which the Kerr
45 co-efficient is at or near the maximum, occurs over a relatively large temperature range. If the bi-refringent material be so chosen, it may be possible to ignore the possibility of varying ambient tem-
50 peratures, for such ambient temperature changes as may be likely to occur may be between limits which are closer together than the temperature limits between which the material remains in the state in which
55 a high Kerr co-efficient is manifested. Where it is not convenient or practical to chose a material having a relatively large temperature range over which the desired state is maintained, the whole
60 Kerr cell may be enclosed in a lagged or heat insulated chamber and/or any thermostatic control means as known per se may be provided to ensure that a predetermined temperature of the bi-refrin-
65 gent material is maintained irrespective

of changes in ambient temperature.

In carrying the invention into practice, the thickness of the layer or layers of bi-refringent material employed should be small for the reason that if thick layers 70 be used there is the danger that different parts of the layer will not be in the same state or phase. In practice it is preferred to use very thin layers of the order of 0.05 mm. in thickness. 75

One form of light valve in accordance with this invention is illustrated in the accompanying figures 1 and 2 which are schematically mutually perpendicular views (not drawn to scale) of the cell. 80 Referring to these figures, two metal foil electrodes 1, 2, each about .002″ thick and 5 mm. wide and separated by a space of about 0.75 mm. are cemented between glass plates 3, 4 each about 1/16th″ 85 thick. The cement is placed above and below the electrodes and round the edges of the plates and the whole cell is baked at a high temperature. The upper glass plate 3 has a small central hole 5 about 90 4 mm. in diameter. A small heating wire 6 is placed between two thin asbestos sheets 7, 8 and held in position under the plate 4 by asbestos strips 9, 10. A small amount of the bi-refringent sub- 95 stance is placed in the hole 5 and heated so that it melts and spreads in a thin layer between the two electrodes, where it remains. When it cools it solidifies in very small crystals, as the cooling will 100 be relatively rapid, and remains as a thin layer. A cell as described in connection with figures 1 and 2 can be readily constructed for a maximum voltage of the order of 200 volts. 105

It will be realised that with a Kerr cell as illustrated in figures 1 and 2 or as so far specifically described, it will be in general impossible to secure a light valve action immediately at any time, 110 and that in general a short time will be necessary for " starting up " in order to allow the bi-refringent material to reach the state between liquid and solid at which the maximum Kerr effect is mani- 115 fested.

The various bi-refringent liquid crystalline substances so far specifically mentioned are organic compounds in themselves, that is to say they are not in com- 120 bination with other substances and they present two practical difficulties when employed for the purposes of this invention, namely (1) that heating of the cell is necessary and (2) that there is some 125 difficulty in obtaining a high degree of homogeneity. The difficulties may be avoided by employing instead of the substances already mentioned substances such as 9-bromophenanthrene-3 (or 6) 130

491

sulphonic acid, the substances known under the Registered Trade Mark salvarsan, m^1-nitrobenzoyl-m-amino-benzoyl-2-naphthylamine-4.8-disulphonic acid and the disodium salt of m^1-nitro-benzoyl- m -aminobenzoyl- 2 -naphthyl-amino-6.8-disulphonic acid which, when dissolved in water exist in the meso-morphic state, and at certain concentra-tions are in the nematic phase of this state, the substance in question being employed in an aqueous solution in such concentration as to be in the nematic phase.

The concentrations at which the sub-stances will exist in aqueous solution in the nematic phase of the mesomorphic state will vary with the nature of the substance. Thus, 9-bromophenanthrene-3 (or 6) -sulphonic acid requires a concen-tration of 13.5 gm. in 100 ccs. of water and the disodium salt of m-nitrobenzoyl-m-aminobenzoyl- 2 -naphthylamine-6.8-disulphonic acid requires a concentration of 5 to 6 gm. in 100 ccs. of water.

Figures 3 and 4 (which are not drawn to scale) of the accompanying drawings, are mutually perpendicular views show-ing schematically a form of cell in accordance with this invention and wherein the bi-refringent material is such as not to require heating. The cell shown in figures 3 and 4 is made by cementing together four pieces 11, 12, 13, 14, of glass, for example glass sheets 2 mm. thick, the four pieces being so arranged as to surround a square space 15 having a side of about 2 mm. The glass edges 11a, 12a constituting one pair of opposite walls of this square space 15 are covered with metal foil, 16, 17, for example tin foil, or are sprayed with metal for example gold or platinum, the metal coatings forming the electrodes of the cell. The structure thus obtained is cemented on to a glass plate 18 and the bi-refringent material is then poured into the square space 15 which is now closed at the bottom by the plate 18. Another glass plate 19, which is removed in figure 3, is then placed on top to close in the bi-refringent material. The plate 19 need be sealed only round the edges.

The light valve thus manufactured is a complete self-contained cell and may be used in any manner known per se in con-junction with a suitable optical system for electro-optical translating purposes in connection with television, facsimile tele-graph or other systems.

It will be appreciated that the present invention, provides a substantial improvement as regards sensitivity when compared to known light valves utilising the Kerr effect. With the substance ethylanisalaminocinnamate the maximum voltage to be applied to the cell is only of the order of 200 volts, though the voltage to be applied will depend upon the dis-tance between the electrodes.

Having now particularly described and ascertained the nature of our said inven-tion and in what manner the same is to be performed, we declare that what we claim is :—

1. A light valve of the kind utilising the Kerr effect and wherein the bi-refrin-gent material employed is in the nematic phase or state.

2. A valve as claimed in claim 1 and wherein the material is one which must be maintained at above normal tempera-ture to be in the nematic phase and wherein heating means are provided for enabling said temperature to be reached and maintained when the valve is in use.

3. A valve as claimed in claim 2 and wherein the heating means includes an electrical heater wire or winding.

4. A valve as claimed in claim 2 and wherein the heating means includes an electrical heater wire or winding arranged to heat one of the electrodes of the valve.

5. A valve as claimed in claim 2 and wherein an enclosed lagged or heat insu-lating chamber is provided to assist in maintaining the requisite temperature.

6. A valve as claimed in claim 2 and wherein thermostatically controlled heater means is provided to maintain the requisite temperature.

7. A valve as claimed in any of the preceding claims and wherein the bi-refringent material is ethylanisalamino-cinnamate, methyl-p-ethoxybenzalamino-α-methyl-cinnamate, ethyl-p-ethoxy-benzalamino-α-methyl-cinnamate, methyl-anisalamino-α-methyl-cinnamate or ethyl-anisalamino-α-methyl-cinnamate or some other compound from the same class.

8. A valve as claimed in claim 1 and wherein the bi-refrigerant material is an aqueous solution of a substance which, when dissolved in water exists in the mesomorphic state, said solution being in such concentration as to be in the nematic phase at ordinary atmospheric tempera-tures.

9. A valve as claimed in claim 1 and wherein the bi-refrigerant material is an aqueous solution of 9-bromophenanthrene-3 (or 6)-sulphonic acid, salvarsan, m^1-nitrobenzoyl - m - aminobenzoyl - 2-naphthylamine-4.8-disulphonic acid or di-sodium salt of m^1-nitrobenzoyl-m-aminobenzoyl - 2 - naphthylamine - 6.8-disulphonic acid in the nematic phase of the mesomorphic state at ordinary atmo-spheric temperatures.

492

10. Light valves substantially as herein described with reference to the accompanying drawings.

Dated the 29th day of April, 1935.

CARPMAELS & RANSFORD,
Agents for the Applicants,
24, Southampton Buildings, London,
W.C.2.

Leamington Spa: Printed for His Majesty's Stationery Office, by the Courier Press.—1936.

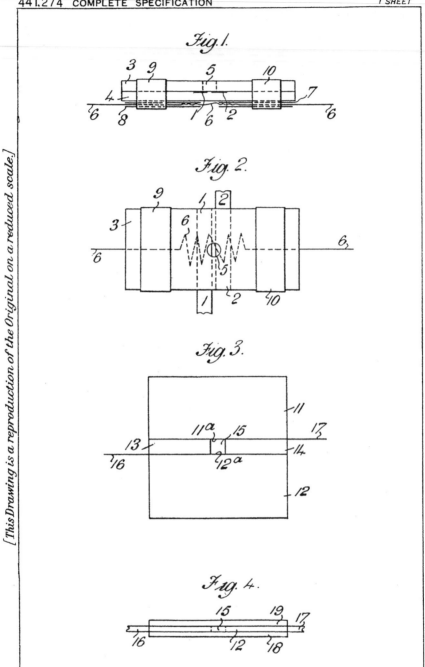

Fig. 1.

Fig. 2.

Fig. 3.

Fig. 4.

Nyman Levin was born in the East End of London in 1906 and received his bachelor and doctoral degrees at Imperial College, London. Between 1930 and 1940 he worked for the research department of the Marconi Company in London. Apart from the liquid crystal project leading to the patent republished here, he was also involved in developing microwave valves. This work led to his secondment, in 1940, to the Admiralty, where he worked for the Royal Naval Scientific Service. He was particularly involved with microwave receivers and radar, and with the development of cross-channel VHF links in preparation for the allied invasion of Europe in 1944.

In 1946 he transferred his loyalties to the Admiralty on a more permanent basis, and was subsequently promoted to Superintendent of the Admiralty Gunnery Establishment (1951). Between 1955 and 1958 he was Chief of Research and Development at Rank Precision Industries (UK). In this post he was responsible for developing of the photocopying technique invented by the Xerox Corporation in the US (at that time known as 'Xerography'), over which Rank had European rights.

In 1958, however, he resigned his post on being appointed as Assistant Director of the Atomic Weapons Research Establishment in Aldermaston (Berkshire., UK), and was promoted to Director the following year. He held this post until shortly before his death, when he was appointed to the UK Atomic Energy Authority.

He died of a heart attack at the age of 58 in 1965.

D2
Journal of Chemical Physics **39**, 384–88 (1963)

Domains in Liquid Crystals

RICHARD WILLIAMS

RCA Laboratories, Princeton, New Jersey

(Received 21 March 1963)

A regular domain pattern has been observed to form in nematic liquid crystals when an electric field is applied. The domains are visible in unpolarized light by either transmission or reflection. Both ac and dc fields produce similar domain patterns and these appear at a rather sharp threshold field of about 1000 V/cm. The general features of the phenomenon have been investigated. It is believed that the domains are due to ordering in the liquid of a kind which has not previously been recognized.

IT has recently been observed[1] that application of a modest electric field to a thin specimen of the liquid crystal, p-azoxyanisole, causes the appearance of a regular pattern in the liquid which is readily visible to the unaided eye. The pattern consists of an array of long parallel regions which are here referred to as "domains." Examples of their appearance are shown in Figs. 1 and 2. Their existence and general behavior suggest several features of the structure of liquid crystals which have not previously been reported. A number of experiments have been done to determine some of the properties of the domains and to try to understand why they form. In what follows, these experiments and an interpretation will be discussed in detail.

EXPERIMENTAL DETAILS

The domains may be observed at any temperature within the range over which p-azoxyanisole forms a nematic liquid crystal. This range is from the melting point, at 117°C, up to 134°C, at which temperature there is a phase transition to the isotropic liquid phase. The solid is melted between two glass plates. Each plate has a transparent tin oxide conductive coating on its inner face and an appropriate lead for making contact to an external circuit (Fig. 3). The spacing between the plates was in the range from 10 to 200 μ for the experiments reported here. It is advantageous to use a strip of conductive coating which is less than the width of the glass plate, as illustrated in Fig. 3. This allows simultaneous observation of adjacent regions of the same specimen with and without an electric field as in Fig. 1. The electric field is perpendicular to the surface of the glass and has an appreciable magnitude only in the square area where the conductive strips on the two plates cross. Either an ac or a dc voltage produces the domains. An ac voltage of 1 kc/sec was used where possible because an ac voltage produces a somewhat more stable pattern and, in addition, any electrochemical deterioration of the liquid is thereby minimized. The electrical resistivity of the liquid used was about 10^9 $\Omega \cdot$cm. This indicates that any ionic impurities which might be present are there only in small amounts. Thus, none of the effects reported here can be due to electro-

chemical deposition of material since the current flow is much too small to produce the effects in the required time. An ac signal of around 10 V is required and this was obtained from an audio oscillator. The liquid-crystal specimen between the glass plates was maintained at the required temperature by supporting it on a transparent heating plate of glass with conductive coating which was connected to a variable transformer. The supporting plate with specimen was mounted on the stage of a low-power microscope and observed by unpolarized transmitted light. Quantitative measurements of transmitted light intensity were made by mounting a photomultiplier on the microscope eyepiece. Some experiments were done with other compounds which form nematic liquid crystals. These were anisaldazine, dibenzal-benzidine and p-methoxycinnamic acid. Similar domains were observed to form for all the compounds studied.

EXPERIMENTAL RESULTS

The most detailed set of experiments was done with p-azoxyanisole. As the voltage on the specimen is increased from zero, nothing is observed to happen until, at a fairly sharp threshold field of about 1000 V/cm, a pattern forms in the liquid as illustrated in Fig. 1. Adjacent areas of the liquid, with and without field, are displayed by looking where one of the strips of conductive coating has its edge within the field of view. Figure 2 shows the domains under somewhat different lighting and illustrates that the detailed appearance of the domains under the microscope is somewhat sensitive to the details of illumination, but that the general features are readily apparent in either case. This photograph was chosen to show certain other properties of the domains which is discussed in a later section.

The domain pattern appears within about 2 msec after the field is applied and disappears within about 20 msec after the field is removed. This was ascertained by displaying, on an oscilloscope, the light intensity transmitted by the specimen as a function of time and switching the voltage source on or off. Considerable scattering of light is caused by the domain pattern and the transmitted light intensity changes during its formation to a stable but different value when the forma-

[1] R. Williams, "Liquid Crystals in an Electric Field," *Science* (to be published).

tion is complete. A similar change occurs when the field is removed and the domain pattern disappears.

The appearance and characteristics of the domains do not change if the transparent tin oxide conductive coating is replaced by a thin semi-transparent layer of evaporated gold or aluminum. Nor do any features of the domain pattern correspond to any observable irregularities or roughness of the solid surface. Thus the domains are a property of the liquid itself, though the solid surface very likely plays an important part since it is known[2] that a wall has an orienting effect in liquid crystals which extends over distances comparable with

FIG. 2. Domains in another specimen with different lighting. Same magnification as in Fig. 1. Note sharply bounded areas with domains perpendicular to those in adjoining areas. Circular spots at right and bottom are air bubbles.

FIG. 1. Domains in p-azoxyanisole liquid crystal. The vertical line about ⅓ of the way in from the right border is the edge of the strip of transparent conductive coating. To the right of this there is no field in the liquid. To the left there is a 1-kc/sec ac field of 2500 V/cm directed perpendicular to the plane of the page. Specimen thickness about 50 μ temperature, 125°C.

the specimen thickness in these experiments. Observation of the domain pattern by monochromatic light shows that the pattern does not change as the wavelength of the light is changed. This demonstrates that the pattern does not result from any kind of interference phenomenon which might give fringes.

When the applied field is increased steadily the domains appear at a threshold field. The change in transmitted light intensity was again used as a criterion for the appearance of the domains. A slowly increasing dc voltage was applied by means of a battery and a motor-driven Helipot potentiometer at such a rate that the threshold field was reached in about 30 sec. Applied voltage was displayed on the horizontal axis of a pen recorder and transmitted light intensity on the vertical axis. A representative tracing is shown in Fig. 4. It is seen that there is a sharp change of light intensity at

the threshold followed by a slower change as the field increases further. The sharp change is due to the formation of domains. The slower change over many volts is due to a stirring action which the field exerts on the liquid. There is first a gentle stirring and finally a vigorous agitation as the field is increased above the threshold value. This stirring action of an electric field on liquid crystals has been noted by earlier workers.[3] Measurements of the threshold field were made for the temperature range 117°–134°C. Within the precision of the measurements, which was about 5%, there was no change of the threshold field with temperature.

There was an upper frequency limit for the applied voltage, above which the domains did not form. This was usually around 20 kc/sec though an occasional specimen showed domain formation up to 100 kc/sec. Below this limiting frequency the appearance of the domains and the threshold field for their formation showed no appreciable dependence on frequency.

A property of the domains which is important to an understanding of their origin is the dependence of the domain size on the specimen thickness. It is evident from photographs of the domains that the characteristic dimension is the width of the long parallel strips and

FIG. 3. Schematic drawing of glass plates with strips of conductive coating. The liquid crystal was melted between the plates as indicated by the crosshatching.

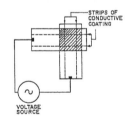

[2] G. W. Gray, *Molecular Structure and Properties of Liquid Crystals* (Academic Press Inc., New York, 1962). This volume summarizes many early references on liquid crystals. The photograph showing "threads" is Plate 3.

[3] V. Zwetkoff; Acta Physicochim. U.R.S.S. 6, 885 (1937).

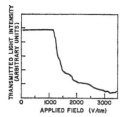

FIG. 4. Recorder tracing showing transmitted light as a function of applied field. The sharp vertical dip at about 1300 V/cm indicates where domains form in the liquid.

not their length. It can also be seen that the regularity of the pattern is sufficient that it is meaningful to characterize a given specimen by a single value of the domain width. It was noticed in initial experiments that the domain width was greater in thick specimens than in thin specimens. To obtain a semiquantitative measure of the dependence of domain width on specimen thickness a wedge-shaped specimen was prepared. The glass plates were mounted at an angle by means of appropriate spacers such that the separation between the plates increased continuously from one end to the other. With liquid crystal filling the space between the plates a field was applied and photographs of the domain pattern were taken. The width of the domains was seen to increase across the picture in the direction of increasing specimen thickness. The picture was divided into equal areas by drawing equally spaced lines perpendicular to the direction of increasing thickness. An average width was estimated for the domains lying in a given area and plotted against the average thickness for this area. Results are shown in Fig. 5. It is seen that there is a strong dependence of domain width on specimen thickness. The procedure used here for obtaining this dependence is, of course, somewhat arbitrary. The important fact is that the domain width increases roughly proportional to the specimen thickness.

INTERPRETATION OF DOMAINS

On the basis of the above data it is possible to set down a plausible qualitative interpretation of the domains. A brief discussion of the structure of nematic liquid crystals will be given first.

The nematic liquid-crystal phase is formed by many organic compounds.[2] The most significant common feature of all these compounds is that they have long, rodlike molecules. In the nematic phase the molecules are packed in such a way that they have complete freedom of rotation only about one axis, which is ordinarily the long axis of the molecule. The long axes of the molecules all lie approximately parallel to an axis which we shall label the z axis. There is an angle θ between the long axis of each molecule and the z axis which varies in a random way from molecule to molecule, but has a small average value. In general, the properties are similar to those of a uniaxial crystal with the z axis

analogous to the optic axis of the crystal. Without special precautions this resemblance may not be apparent since a macroscopic volume of liquid is often broken up into many small volumes in which the z axes are randomly oriented with respect to each other. A measure of the actual degree of order has been given by Maier and Saupe[4] who obtained the quantity, $S=1-3/2\langle\sin^2\theta\rangle$, for p-azoxyanisole. Anisotropies of several experimental quantities including refractive index, infrared and ultraviolet spectra, and diamagnetic susceptibility were compared with a theoretical calculation and gave good agreement. As an example, for 120°C, the value of S is around 0.6. There is an additional orienting influence exerted by the wall of the container which extends over distances which may be comparable to the specimen thickness in the present experiments.[2] This requires some modification of the above picture and is probably important to the present work. It seems likely that the model for the nematic structure is still correct in its general features even for thin specimens and it will be the basis of the ensuing discussion.

To formulate a model for the domains, it is first necessary to establish what it is that makes them visible. They are not due to electrochemical decomposition products. Because of the high resistivity of the liquid, there is not enough current flow in a millisecond to make a visible deposit of any product. From the sharp value of the threshold field it might seem that some kind of reversible phase transition is caused by the electrical field and that the domains are due to segregation of two phases. The phase rule places severe restrictions on the possible models here, but even if a consistent combination of components and phases were present this mechanism could be ruled out. This is because the domains form in milliseconds, and segregation of phases would require material to diffuse distances the order of domain widths, or about 20 μ. Diffusion for these distances in these short times would require diffusion coefficients which would be orders of magnitude larger than those actually observed in liquids.

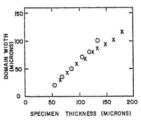

FIG. 5. Domain width as a function of specimen thickness. The circles and crosses indicate data for two different wedge-shaped specimens.

[4] W. Maier and A. Saupe, Z. Naturforsch. 16, 816 (1961); 15, 287 (1960); 14, 882 (1959).

Another complication with this model is that electrically induced phase transitions have rarely (perhaps never) been observed, even in fields much higher than those used here.

Since there are no permanent changes, the observed patterns are likely due to orientation effects. All the observations suggest that what one sees is regions of differing refractive index which are formed in an orderly array when the field is applied. Since the liquid has properties analogous to those of a uniaxial crystal, there are two different refractive indices for light polarized parallel and perpendicular to the z axis. These have been measured,[5] and the values reported for the ordinary and extraordinary ray are 1.561 and 1.849, respectively, for the temperature 117°C. This difference in refractive indices is enough to make adjacent areas of liquid with different orientations clearly visible by transmitted or reflected light. For example, it is nearly twice as great a difference as that between the refractive indices of glass and water.

It is clear that substantial orientation effects at the low fields applied here could arise only by the concerted action of many molecules in response to the field. For an electric field E of 3000 V/cm acting on a dipole moment μ of 1 D at the temperature of our experiments, $\mu E/kT$ is 2×10^{-4}. This is the exponent in the Boltzmann factor which gives the degree of orientation of individual molecular dipoles in the field. The result is a familiar one[6] and corresponds to less than one molecule in a thousand having any substantial orientation produced by the field. For the field to produce an orientation affecting all the liquid it is necessary for a large number of molecules, n, to respond to the field cooperatively. The minimum value which n may have is then given by the criterion $n\mu E/kT>1$. This gives $n>5000$. It seems likely that a much larger value of n is appropriate to the present experiments because the orientation occurs in a short time and viscous forces cannot be neglected. In addition, there is probably not enough molecular order for individual molecular dipoles to add in the most advantageous way as assumed in this argument.

A possible explanation of the effects observed here may be that they are analogous to ferroelectric domains which are somewhat similar in appearance. The thickness effect illustrated in Fig. 5 has an analog in the case of ferroelectric domains.[7] There, too, the domain width increases with increasing specimen thickness. This is attributed to two effects which work in opposite directions: one of these is the energy gained by reversal of the direction of a persistent internal polarization in

[5] P. Chatelain and O. Pellet, Bull. Soc. Franc. Mineral. Crist. **73**, 154 (1950) (cited in Ref. 2).
[6] P. Debye, *Polar Molecules* (Dover Publications Inc., New York; Reprint of the same title which was originally printed by Chemical Catalog Company, 1929.
[7] W. Känzig, *Solid State Physics* (Academic Press, New York, 1957), pp. 5–197; W. J. Merz, Progr. Dielectrics **4**, 101–149 (1962).

FIG. 6. (a) Schematic illustration of the arrangement of rod-like molecules in a nematic liquid crystal. The direction of the z axis is indicated and also the direction of M, the internal polarization postulated in the present discussion. (b) Illustration of the boundary between domains with M oppositely directed in the two domains.

adjacent volume elements of the crystal. This leads to formation of domains in the first place. The presence of internal polarization and the possibility for the polarization to change or reverse its direction are both properties characteristic of ferroelectric crystals. The process leads to an increased energy for the material in the walls where the domains meet, and the two effects combine to make the domain width depend on the specimen thickness.

An adaptation of these ideas will be applied here. It is assumed that the liquid crystal has its z axis parallel or with a substantial component parallel to the glass plates enclosing it. Figure 6(a) indicates the arrangement of long rodlike molecules with the ordering characteristic of the nematic phase. It is further assumed, and this is fundamental to this model, that the combination of the ordering effects of the nematic liquid and the container walls leads to a net dipole moment per unit volume in a direction parallel to the container wall. This is indicated by the arrow labeled M. M is visualized here as the resultant along the z direction of many small contributions from individual molecules which might arise if molecules have approximate but not complete freedom of rotation about the z axis. For any given volume, the value of M must be much smaller than the product of the molecular dipole moment and the number of molecules contained within the volume. If the direction of M in adjacent macroscopic volume elements is antiparallel the free energy will be lower than if they are parallel in analogy with the case of ferroelectricity.[7] This does not lead to any visible structure since the long axes of individual molecules are parallel to the glass for both orientations of M. Figure 6(b) illustrates this situation with oppositely directed values of M in neighboring volume elements. An electric field would act on the polarization M to exert a torque on a volume of the liquid containing a large number of molecules. The field, however, cannot act to rotate individual molecules about an axis perpendicular to the z axis. In the nematic liquid even the effect of thermal energy is not sufficient to produce this kind of rotation and $\mu E/kT \ll 1$. The polarization per unit volume due to added contributions from individual molecules would respond to the field as a torque on the entire volume exactly as a bar magnet responds to a magnetic field. Thus a volume of liquid would rotate in the field when the field strength reached a magnitude great enough to break up the nematic structure at the boundaries of the volume or to overcome any hindrance imposed by the walls. It is believed that this process is responsible for the domains observed. Above the critical field, elements of volume containing many molecules rotate so that the direction of M follows the field as a macroscopic dipole. The region at the boundary where the regions of oppositely directed M meet retains its original orientation since here there is no net torque exerted by the field. In a thin layer where the domains meet it is assumed that the interactions between the oppositely directed values of M is stronger than their interaction with the applied field. Because of the anisotropy of refractive index, the domain structure becomes visible. When the field is removed the original structure is reformed and the domains again become invisible.

This picture fits the data in a general way. It can occur at low applied fields, respond to either ac or dc fields, give the appearance of domains at a threshold field and give a dependence of domain width on specimen thickness. With the data obtained so far it is not possible to be more specific about how an internal polarization might arise or, indeed, how likely such a process is. At this stage the model is proposed chiefly as a concrete guide to suggest further experiments.

In support of the proposed model, there is direct experimental evidence showing orientation of p-azoxyanisole molecules in an ac electric field. X-ray data reported by Kast[8] showed that substantial orientation of molecules occurred at field strengths and frequencies comparable to those used here. Kast's interpretation of these effects is in general agreement with the above discussion.

"THREADS" IN NEMATIC LIQUID CRYSTALS

A useful property of the domains is the fact that they reveal something new about the order present in the nematic structure. The name "nematic" derives from the observation that lines or "threads" are frequently seen on microscopic examination of the liquid. A photograph showing these is given in Ref. 2. They also may be seen in the photograph of Fig. 2 here. It can be seen that there are several areas in the photograph which are separated by sharp boundary lines. Within each area the domains are parallel but in the adjoining area just across the boundary they are still parallel to each other, but all are approximately perpendicular to those in the first area. These sharply bounded areas form mainly when the liquid is cooled down from the isotropic liquid phase. The order revealed by the domains suggests that these bounded areas are analogous to the individual crystallites present in a polycrystalline specimen of an ordinary solid. The boundary lines between the t'crystallites" are the threads ordinarily seen. Thus the threads appear to be crystal boundaries. They are present without the field and remain unchanged as the field is applied and the domains appear. They grow over a period of minutes to larger areas just as a polycrystalline solid changes, under proper conditions, to give larger crystallites. The appearance of these "crystallites" on cooling the liquid down from the isotropic phase is probably a consequence of the space-filling requirements of elongated molecules. On going from the isotropic phase, where individual molecules are randomly oriented, to the nematic phase, where they are all parallel, liquid crystallites nucleating at different points could have their z axes randomly oriented with respect to each other. As they grow, their boundaries would meet. To minimize surface energy, it is likely that adjoining crystallites would orient themselves so that their z axes are either parallel or perpendicular. This would give the densest packing of molecules along the interface. With the z axes parallel the boundary would disappear and they would coalesce to one larger crystallite. With the z axes perpendicular, there would result an aggregate such as that shown in Fig. 2.

It should be mentioned that periodic structures, similar in appearance to those described here, have been reported for solutions of poly-γ-benzyl-L-glutamate.[9] The structure forms without any applied field and forms very slowly. It is attributed to parallel stacking of polymer molecules into layers. Each polymer molecule has an α-helix structure. A small progressive twist of the orientation from layer to layer gives a macroscopic periodicity. It is not clear at this time whether this bears any close relation to the present structure or has only a superficial resemblance.

ACKNOWLEDGMENTS

The author is indebted to R. E. Shrader and to P. Wojtowicz for cooperation and helpful discussions during the present work.

[8] W. Kast, Z. Physik **71**, 39 (1931).

[9] C. Robinson and J. S. Ward, Nature **150**, 1183 (1957); C. Robinson, Trans. Faraday Soc. **52**, 571 (1956).

Richard Williams was born in 1927, and began his career as a physical scientist with an A.B degree in chemistry from Miami University, Oxford, Ohio, followed in 1954 by a Ph.D from Harvard in physical chemistry. In 1958 Williams joined the RCA laboratories, and began work on liquid crystals.

Despite a life-long interest in chemistry, Richard Williams is best known for his fundamental work on electro-hydrodynamics in liquid crystals, and specifically for his discovery of the instability now known as Williams domains. This work introduced the basic display configuration of a liquid crystal thin film sandwiched between glass plates with transparent conductive coatings, and in due course resulted in the invention at RCA of the first liquid crystal display.

Apart from visiting appointments in South America and China, Williams was a member of the technical staff at RCA laboratories until 1991. During this period he worked on various aspects of solid state devices, and is the author or co-author of more than 130 technical articles and patents. Williams is a Fellow of the American Physical Society and a Corresponding Member of the Brazilian Academy of Sciences.

D3
Proc. I.E.E.E. **56**, 1162–71 (1968)

Dynamic Scattering: A New Electrooptic Effect in Certain Classes of Nematic Liquid Crystals

GEORGE H. HEILMEIER, SENIOR MEMBER, IEEE, LOUIS A. ZANONI,
AND LUCIAN A. BARTON

Abstract—A new electrooptic effect in certain classes of nematic liquid crystals is presented. The effect has been termed "dynamic scattering" because scattering centers are produced in the transparent, anisotropic medium due to the disruptive effects of ions in transit. The ions can be produced by field assisted dissociation of neutral molecules and/or Schottky emission processes. The rise times of 1 to 5 ms and decay times of less than 30 ms, together with dc operating voltages in the 10 to 100 V range, make dynamic scattering seem attractive for such applications as alphanumeric indicators, and do not preclude its use in line-at-a-time matrix addressed, real-time displays. Reflective contrast ratios of better than 15 to 1 with efficiencies of 45 percent of the standard white have been demonstrated.

LIQUID CRYSTALS—NEMATIC PHASE

THE FIELD of liquid crystals has been the subject of reviews by Gray [1] and Brown and Shaw [2], hence a comprehensive review will not be attempted here. Nevertheless, some general information necessary to familiarize the reader with this rather exotic field will facilitate an understanding of the new effect and its possible applications.

The term "liquid crystals" is applied to substances whose rheological behavior is similar to that of fluids but whose optical behavior is similar to the crystalline state over a given temperature range. Liquid crystals are by no means rare. Approximately one out of every 200 organic compounds exhibits mesomorphic behavior, although the occurrence in inorganic substances is extremely uncommon. Both aliphatic and aromatic compounds can exist as liquid crystals.

Manuscript received January 16, 1968.
The authors are with RCA Laboratories, Princeton, N. J. 08540

The feature common to molecules exhibiting liquid crystallinity is a planar, rod-like structure. Three classes of mesomorphic behavior exist: smectic, cholesteric, and nematic. As evident in Fig. 1, each is characterized by a specific arrangement of molecules in the fluid. Cholesteric liquid crystals have been used as temperature and vapor indicators [3]. We shall restrict ourselves to a brief discussion of the nematic state which is the particular mesophase exhibiting our electrooptic effect.

The term "nematic" is derived from the Greek word meaning thread. This term describes the thread-like lines which can be seen in this class of liquid crystals under a microscope. The molecules of nematic liquid crystals are arranged with their long axes parallel but not in layer form as is the case with smectic liquid crystals (see Fig. 1). The molecules are free to slide past each other; nevertheless, they remain essentially parallel within a given region. Nematic materials can be aligned by both electric and magnetic fields. This alignment can be essentially complete [4] in contrast to the relatively weak alignment produced by high fields in conventional polar fluids (i.e., one part in 10^4) as predicted by Langevin theory ($\sim \mu E/kT$, $\mu =$ dipole moment, $E =$ electric field). This follows from the cooperative behavior of the molecules in the nematic liquid crystal.

Probably the best known nematic substance and one whose rheological, optical, and electrical properties have been extensively explored is p-azoxyanisole. This material transforms from a crystalline solid into the nematic state at 116°C and becomes an isotropic liquid at 133°C.

Fig. 1. Molecular alignment for various mesophase types.

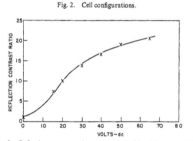

Fig. 2. Cell configurations.

PHENOMENOLOGICAL DESCRIPTION OF THE EFFECT

The structure which we wish to consider is a sandwich cell consisting of a transparent front electrode (i.e., nesa-coated glass) and a specularly reflecting back electrode. This is seen in Fig. 2, cell I. The nematic material is sandwiched between these electrodes using teflon spacers of $\frac{1}{4}$ to 1 mil thickness. Capillary action is sufficient to hold the liquid between the plates for a variety of orientations. In general, some means for maintaining the temperature within the nematic range is required.

In its quiescent state with no field applied, the liquid crystal is essentially transparent. This means that if the specularly reflecting back electrode is faced into a black background, the cell appears black. When a dc field of the order of 5×10^3 V/cm (corresponding to 6 V for a $\frac{1}{2}$ mil thick sample) is applied, the liquid becomes turbulent and scatters light. We have designated this state as the *dynamic scattering mode (DSM)*. In this state the cell appears white. Increasing the field results in increased brightness so that a gray scale is obtainable. Saturation of contrast ratio versus field occurs at about 5×10^4 V/cm. A typical result is shown in Fig. 3. Under dc conditions maximum contrast ratios in excess of 20 to 1 with maximum brightness of the order of 50 percent of $MgCo_3$ (the standard for white) have been obtained under conditions where the specular reflecting back electrode "looks" into a black background.

While we have discussed this effect phenomenologically as a reflective effect, the true operation is due to the scattering of light. It is characteristic of nonconducting scattering centers which are larger than 5 to 10 times the wavelength of the incident light that the scattered radiation is not a function of the incident wavelength [5]. Moreover, the bulk of the radiation is forward scattered rather than back-scattered; thus, a specular reflecting back electrode is necessary to direct this radiation back through the scattering medium to subjectively maximize the effect.

Fig. 3. Reflection contrast ratio versus voltage. Material: C_1-APAPA; cell thickness $\sim\frac{1}{2}$ mil.

Fig. 4. Anisylidene-*p*-aminophenylacetate.

MATERIALS

The materials which have yielded the best performance in the dynamic scattering mode of operation are members of a class of organic compounds known as Schiff bases. These materials, when highly pure, are essentially transparent in the visible and have resistivities of 1 to 5×10^{10} $\Omega \cdot$cm and a dielectric constant of 3.5 at 90°C. We have found the compound APAPA (anisylidene para-amino-phenylacetate), shown in Fig. 4, to be of particular interest in demonstrating the dynamic scattering mode [6]. Its nematic range is from 83°C to 100°C.

DRIVING MECHANISM OF THE DYNAMIC SCATTERING MODE

The turbulence noted in the liquid crystal when subjected to fields above threshold requires that domain-like regions of neutral molecules be set in motion. Scattering is produced

504

by localized variations in the index of refraction. As we have discussed, these regions must be large compared to the wavelength of the incident radiation. Their nature will be explored more fully in another section.

There are several mechanisms by which an electric field can exert a force on a fluid: 1) electrostriction, 2) spatially varying dielectric constant, 3) dielectrophoresis, 4) electrophoresis, 5) electrohydrodynamic effects. Mechanisms 4) and 5) are somewhat related. To review briefly, electrostrictive forces arise from the change in dielectric constant produced by field induced density changes. Dielectrophoresis is defined as the motion of matter caused by polarization effects in a nonuniform field. Electrophoresis is the motion of a charged particle and its associated counterion cloud in an electric field. The counterions themselves are moving, on the average, in the opposite direction under the influence of the applied field. Since they are solvated, they tend to carry with them their associated solvent molecules, so that there is a net flow of solvent in a direction opposite to the motion of any given (solvated) central charge.

What we shall call electrohydrodynamic forces [7], [8] are those mechanical forces which are produced by essentially unipolar electrical conduction. When ions in a partly ionized medium move under the influence of an electric field, friction with the carrier medium transfers momentum to the latter. If only one sign of ion is present, the pressures created can be significant.

We have performed experiments using five basic sample configurations in an effort to determine the mechanism or mechanisms of the DSM. These experimental configurations are shown in Fig. 2. Cell I is similar to the arrangement used in the actual display with $6\mu < d < 25\ \mu$. In cells II and III both electrodes are on the same substrate. The liquid crystal covers both electrodes, and a transparent cover slide with spacers makes up the remainder of the cell. Cell IV is fabricated from glass tubing and filled with nematic materials. The metal screen electrodes are spaced 5 mm apart, and provisions are made for measuring the pressure difference across the cell with electric field applied.

In cell I, for thicknesses less than 25 μ, with one nesa-coated glass electrode and one metal electrode, the DSM is always greater at a given voltage when the metal electrode is negative. In addition, the current-voltage characteristic is not linear, as seen in Fig. 5. A threshold of 5 to 10 kV/cm for the DSM is generally found. Transient measurements, to be described more completely in a later section, were also made using this type of cell. When the cell was subjected to a voltage step, the current, in addition to exhibiting an initial displacement current transient, also possessed a later secondary peak similar to that found for transient space charge limited currents in solids (see Fig. 6) [9]. The time of occurrence of the secondary peak or "cusp" was inversely proportional to the applied voltage, as seen in Fig. 7. This secondary peak was not seen when one of the electrodes was made an ideal blocking contact by covering it with teflon. In addition, no optical effect was seen under these conditions. Unfortunately, measurements as a function of cell thickness were nonreproducible in a consistent manner due

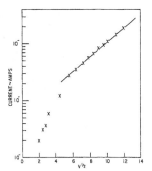

Fig. 5. Current versus voltage sandwich cell. Thickness $\sim \frac{1}{4}$ mil; $T \sim 90°$C.

Fig. 6. Typical current transient.

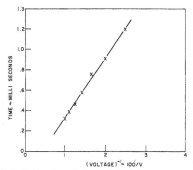

Fig. 7. Position of cusp (time) versus (voltage)$^{-1}$. Material: APAPA, electrode separation $\sim 12\ \mu$.

to the nature of the cell fabrication technique.

When cell III was viewed under the same conditions with crossed polarizers, an optical effect (i.e., change in transmitted light) was seen to initiate at the *negative* electrode and travel to the positive electrode. At threshold field the effect initiated nonuniformly at the negative electrode.

When cell III was viewed under the same conditions as above with the smaller-radius electrode negative, the optical effect initiated at the negative electrode for fields about the

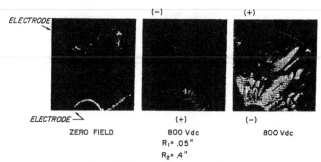

ZERO FIELD   800 Vdc   800 Vdc

$R_1 = .05"$

$R_2 = .4"$

Fig. 8. Cylindrical electrode system—optical effects.

same as those observed for the sandwich cell. This optical effect at threshold also propagates to and terminates at the anode, as seen in Fig. 8. The propagation rate appeared to be nonuniform, slowing noticeably near the anode. In this geometry with the field at the small electrode equal to 5 to 10 kV/cm, that at the anode is much less and substantially *below* threshold.

In cell IV, although the electrode separation is orders of magnitude greater than that of cell I, turbulence in the liquid crystal initiates at roughly the same field strength. The appearance of an optical effect is accompanied by a rise in pressure at the anode. The difference in the height of the column at the anode and at the cathode as a function of field is shown in Fig. 9, along with the current.

Another cell similar to IV without the provisions for measuring pressure was also used. This cell had two solid metal electrodes separated by a metal screen electrode (see Fig. 2, cell V). It was found that the current was 10 to 15 percent higher when the screen electrode was positive and only one of the solid electrodes was used as a cathode, compared to the case with the polarities reversed. Optical effects were also noted in the field free region beyond the screen anode. When both solid electrodes were used as the cathode, the current to the screen anode in the center of the cell was not doubled as might be expected.

It is recognized that the widely different cell and electrode geometries and sizes may themselves help determine which of the several possible effects may be dominant in driving the DSM in a specific experiment. Nevertheless, if we choose to minimize this possibility and assume that there is experimental consistency independent of the geometry, we can summarize our data as follows: 1) The appearance of the optical effect initiates at the negative electrode. 2) It is accompanied by an increase in pressure at the anode and the deviation of the current-voltage characteristic to a more rapid variation with voltage. 3) The magnitude of the optical effect at a given field is dependent on the nature of the negative electrode. 4) The optical effect which initiates at the cathode propagates to the anode, although the field in this region may be far below threshold for the effect.

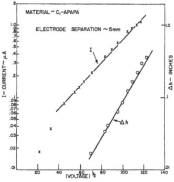

Fig. 9. Current and pressure versus voltage.

5) The experiments with cell V indicate that there is a definite relationship between electrical and optical effects and hydrodynamic effects. 6) Transient measurements indicate that the passage of current is necessary to produce the DSM.

Now let us critically examine these observations. Forces due to electrostriction and/or spatially varying dielectric constant cannot account for the pressure rise at the anode. In addition, these forces would not require current to be passed, which seems necessary for the DSM. Dielectrophoresis requires a nonuniform field. In our planar geometry this could only be produced by space charge. Nevertheless, the experiments with cylindrical geometry electrodes with the high field at the cathode saw the optical effect still propagate *from* cathode to anode, which is contrary to what would be predicted for dielectrophoresis. Other true field effects are also suspect when we note that the optical effect in the aforementioned case propagates through regions where the field is far below normal threshold for the effect. Thus, a model based on dielectrophoresis does not seem consistent with all the experimental data.

The experiments in one way or another seem to eliminate all possibilities except those based on charge transport as the mechanism for the DSM. Electrohydrodynamic effects require charge transport in the liquid crystal to be dominated by one sign of carrier. Positive ions are not felt to be the dominant carriers, since the pressure rise is at the anode, the optical effect initiates at the cathode, and the effect is somewhat sensitive to cathode material. The means of producing transport imbalance include preferential adsorption or accumulation of one type of ion at the walls of the container and/or electrode, mobility differences, and injection of electrons followed by their capture by neutral molecules (forming negative ions).

Fundamental to the experimental results is the requirement that the electrical force present on the space charge in the fluid must be balanced by a pressure gradient given by

$$\nabla p = \rho E \qquad (1)$$

where

p = pressure
ρ = net space charge density
E = electric field strength.

Using Poisson's equation and integrating, the pressure difference between anode and cathode is given by

$$\Delta p = \tfrac{1}{2}\varepsilon_0\varepsilon(E^2_{\text{anode}} - E^2_{\text{cathode}}) = Tg\Delta h \qquad (2)$$

where

T = density of fluid
g = acceleration due to gravity
Δh = difference in height of column between anode and cathode
ε = relative dielectric constant.

If the effects of diffusion are neglected, this equation can be written

$$\Delta p = L \sum_i J_i/\mu_i \qquad (3)$$

where J_i represents the various contributing conduction processes, and L is the electrode spacing. If the dominant conduction mechanism is also the dominant contributor to the pressure generation, a plot of Δp versus V will have the same behavior as J versus V. If this is not the case, the behavior of the two plots will differ. Pressure and current-voltage data are shown in Fig. 9. Note that the slope of the ln (pressure) versus $V^{1/2}$ is greater than ln (I) versus $V^{1/2}$, indicating that we are probably dealing with separate mechanisms in this case. Pickard [10] has examined in detail the variation of pressure with the uniform V/L in conventional liquids in terms of a parameter η which reflects changing space charge conditions. Our data fit the case where η is increasing monotonically to one as the voltage is increased.

As an example of how subtle field differences between anode and cathode in the presence of larger average fields can lead to the observed pressure differences, consider the case shown in Fig. 9.

$$\Delta p = \tfrac{1}{2}\varepsilon_0\varepsilon(E^2_{\text{anode}} - E^2_{\text{cathode}}) = Tg\Delta h \qquad (4)$$

where

T = density $\sim 1.3 \times 10^3$ kg/m^3
g = 9.8 m/s^2
$\varepsilon_0\varepsilon \sim 36 \times 10^{-12}$ F/m
$\Delta h|_{V=10\text{kv}} \sim 2.2 \times 10^{-3}$ m.

Hence,

$$(E^2_{\text{anode}} - E^2_{\text{cathode}}) \sim 1.5 \times 10^{12} \text{ (V/m)}^2. \qquad (5)$$

The square of the average field for the 4 to 5 mm electrode separation in the pressure cell is 4 to 6×10^{12} (V/m)^2.

While the importance of charge in producing dynamic scattering has been established, two major questions remain. 1) What is the origin of the charge? 2) What are the details of the process which produces scattering centers in the liquid? These questions are the subject of the following sections of this paper.

CONDUCTION STUDIES AND THEIR RELEVANCE
TO DYNAMIC SCATTERING

In the previous section it was established that a relationship does exist between dynamic scattering and the production of a pressure difference between anode and cathode when field is applied. Dc current-voltage data for the pressure cell and sandwich cell (see Fig. 2), where the electrode spacings are 15 cm and 6×10^{-4} cm, respectively, are shown in Figs. 9 and 5, respectively. Note that the data give an excellent fit to an $I \propto \exp (V^{1/2})$ law. The slope has a value of 2.48×10^{-3} (MKS) for the sandwich cell and 2.7×10^{-3}(MKS) for the pressure cell. These values do not compare favorably to the theoretical value of

$$e^{3/2}/2kT(\pi\varepsilon\varepsilon_0)^{1/2} \sim 0.88 \times 10^{-3} \text{ (MKS)}$$

for Schottky emission which has the same functional dependence [11]. Measurements could not be made over a significantly wide temperature range to check the temperature dependence of the slope.

Experiments in the sandwich cell geometry with evaporated aluminum, nickel, cobalt, platinum, and chromium cathodes revealed a difference in current of no more than a factor of ten for fields up to 10^5 V/cm and no consistent dependence on cathode work function. The thickness dependence of the current is shown in Fig. 10. Note that the current increases linearly with increasing sample thickness. This is, indeed, contrary to ohmic and Schottky emission behavior where a decrease is expected.

As a final check, a cell with an electrode geometry of a point and a plane was fabricated. No significant variation in current was found at fixed voltage when the polarity of the voltage on the cell was reverse. Schottky emission is dependent on the field at the cathode. In the case of the point cathode, conditions favoring emission certainly existed, yet no variation in current was observed.

We can summarize the experimental conductivity studies by noting that 1) $I \propto \exp kV^{1/2}$ where the experimental value for k is a factor of 2 to 3 times larger than that pre-

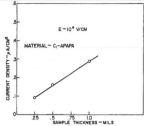

Fig. 10. Current density versus sample thickness
for constant applied field.

dicted by Schottky emission; 2) little or no dependence was found on electrode work function or field configuration; and 3) current *increased* with increasing electrode separation.

The data are consistent with a model based on field assisted molecular dissociation of the liquid or impurities therein. In general, if we assume essentially no space charge, the current is given by

$$J = eE(n\mu_n + p\mu_p) \tag{6}$$

and

$$\nabla \cdot J_n = e\mu_n E \frac{dn}{dx} = cD(E) - \gamma np \tag{7}$$

where

E = average electric field
μ_n, μ_p = negative and positive ion mobilities, respectively
n, p = negative and positive ion densities, respectively
$D(E)$ = field assisted dissociation constant
γ = recombination constant
c = constant.

If, in the spirit of the semiquantitative nature of this analysis, we assume that

E = constant (no space charge)
$n = p$
$\mu_n = \mu_p$
$\gamma = 0$ (direct recombination is small)
$n = 0$ (at cathode),

then (7) can be integrated simply to yield

$$n = \frac{cD(E)}{\mu E e} L$$

where L = electrode spacing. Since we have assumed that $n \sim p$ and $\mu_n \sim \mu_p$,

$$J = 2cD(E)L. \tag{8}$$

It remains to develop a simple expression for the field assisted dissociation constant.

In the absence of field, this has the general form

$$D(0) = D_0 \exp(-U_0/kT) \tag{9}$$

where U_0 = dissociation energy in the absence of field. Using the same general approach of simplified derivations of the Schottky emission equation [11], the dissociation energy is given by the integrated sum of the electrostatic forces

$$U(E) = \int_0^{x_c} \frac{-e\,dx}{4\pi\varepsilon\varepsilon_0 x^2} - \int_0^{x_c} eE\,dx$$

$$= U_0 - \frac{e^2}{4\pi\varepsilon\varepsilon_0 x_c} - eEx_c. \tag{10}$$

Now when $x = x_c$, the net force is zero; hence,

$$x_c = \left(\frac{e}{4\pi\varepsilon\varepsilon_0 E}\right)^{1/2} \tag{11}$$

and the dissociation energy becomes

$$U = U_0 - 2e^{3/2}\left(\frac{E}{4\pi\varepsilon\varepsilon_0}\right)^{1/2}. \tag{12}$$

Substitution of (9) and (12) in (8) yields the current-voltage relationship for the dissociation model

$$J = 2cLD_0 \exp\left\{-U_0 + 2e^{3/2}(E/4\pi\varepsilon\varepsilon_0)^{1/2}\right\}/kT. \tag{13}$$

Note that the slope of a $J \propto \exp V^{1/2}$ plot is twice that of the Schottky emission model, and an *increase* in current with electrode separation is predicted. In addition, no dependence on electrode work function is indicated.

While possessing some similarities with Schottky emission, the field assisted dissociation model offers much better agreement with our experimental data and seems to be the dominant conduction process. This is not to say that it is the only active conduction mechanism in our experiments. The initiation of an optical effect at the negative electrode followed by its propagation to the anode could be due to a field nonuniformity or an emission process. Thus, while field assisted dissociation may dominate the conduction process, it is still possible that an emission process which is enhanced by high fields at the cathode may be a factor in the dynamic scattering. This emission process would seem to have much in common with Schottky emission, and indeed I-V data in agreement with this mechanism were found in *one* sample. This finding leads us to suspect that the dissociating species is an impurity molecule.

A MODEL FOR DYNAMIC SCATTERING BASED ON SHEAR INDUCED ALIGNMENT BY IONS IN TRANSIT

Previous experiments have emphasized the importance of ions in transit in producing dynamic scattering in certain classes of nematic liquid crystals. In further work it was found that little or no dynamic scattering is present in nematic systems whose molecular dipole moment lies along the molecular axis, although resistivities may be quite comparable to those for which the effect is observed in the APAPA family. Such a material is *p-n* ethoxybenzylidene-*p'*-aminobenzonitrile (PEBAB), shown in Fig. 11. We have also found that dynamic scattering can be produced without an electric field by moving one electrode along the other in the sandwich cell geometry, thus subjecting the liquid

Fig. 11. *p-n*-ethoxybenzylidene-*p'*-aminobenzonitrile.

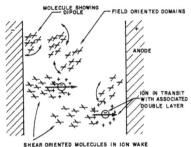

Fig. 12. Molecular model of the dynamic scattering mode.

crystal to a mechanical shear. The shear in this case is roughly $1000\ \text{s}^{-1}$.

In an effort to explain these varied experimental observations, we wish to propose a model based on the alignment of swarms due to the shear induced in the liquid crystal by the ions in transit. This alignment, which is dynamic in nature, tends to align the *molecular axis* along the direction of ion transit (see Fig. 12). The electric field, on the other hand, tends to align the *dipole moment* along the direction of the applied field. In molecules where the dipole moment does not lie along the molecular axis, each of these two forces tends to produce a different alignment, and since the swarms are highly birefringent, the conditions for light scattering exist between the differently oriented regions. In the case where the dipole moment and molecular axis lie in the *same* direction, the field induced alignment and the shear induced alignment are in the same direction, and scattering centers are not dramatically evident.

Consider a rod-shaped domain in a shear field $g = dv_x/dy$. The major axis of the domain lies at an angle θ with respect to the x-axis. If the swarm axis is not aligned in the direction of flow, the ends of the swarm will experience a transverse component of velocity:

$$v_t = 1/2Lg \sin^2 \theta \qquad (14)$$

where $L =$ swarm length. Neglecting inertial effects, the angular velocity of the swarm axis is simply

$$\dot{\theta} = g \sin^2 \theta. \qquad (15)$$

The probability Φ of finding a swarm in any given orientation will be inversely proportional to the rotational speed; hence,

$$\Phi \dot{\theta} = \text{constant}. \qquad (16)$$

The rotational thermal motion of the swarms tends to counteract the hydrodynamic orientation effects. This thermal component of the motion gives rise to an orientational

diffusion of the swarms in the direction corresponding to a decrease of $\Phi(\theta)$. When this component of motion is added to (16), we obtain

$$\Phi(\theta)\dot{\theta} - D_r \frac{d\Phi}{d\theta} = \text{constant}. \qquad (17)$$

Since $\dot{\theta}$ must be proportional to the velocity gradient or shear g, Φ becomes a function of the ratio of the velocity gradient and the rotational diffusion constant D_r. Hence,

$$\Phi = f(\alpha)$$

$$\alpha = g/D_r.$$

Equation (17) can be solved by series techniques. Two special cases are of interest.

Case 1

Let

$$\alpha < 1$$

$$\Phi = 1 + \sum_1^\infty \alpha^m g_m(\theta).$$

Substituting in (17) and equating coefficients yields

$$\Phi = 1 + \frac{\alpha}{4} \sin 2\theta + \cdots. \qquad (18)$$

The alignment tendency is weak but tends to have a maximum at $\theta = \pi/4$.

Case 2

Let

$$\alpha \gg 1$$

$$\Phi = \sum_1^\infty \alpha^{-n} g_n(\theta).$$

Again substituting in (17) and equating coefficients, we obtain

$$\Phi = \frac{\text{constant}}{\alpha \sin^2 \theta} + \cdots. \qquad (19)$$

In this case the swarm axes tend to align along the flow axis, and the distribution function is sharpened and peaked, compared to Case 1, about $\theta = 0$.

Previous experiments on dynamic scattering have revealed the importance of ions in transit in producing the effect. We now consider some of the consequences of this motion. An ion in solution tends to move with its own characteristic velocity v_0 which is independent of other ions in solution. However, the ion atmosphere, being of opposite sign, will tend to move in the opposite direction. This introduces a retarding effect on this central ion. Each element in the ion atmosphere is acted upon by a force per volume which is the product of the space charge density and the field. The space charge is given by

$$\rho = \frac{e\lambda^2 \exp(\lambda a - \lambda r)}{4\pi(1 + \lambda a)r} \qquad (20)$$

where

$$a = \text{ion radius}$$

$$(\lambda)^{-1} = \text{Debye length}.$$

Hence,

$$\text{force/volume} = \rho E$$

and

$$\text{force} = \int 4\pi r^2 \rho E \, dr.$$

This force produces a drag on the central ion corresponding to a velocity v_d. If we assume that the ions are essentially spherical and that the Stokes law holds,

$$v_d = \frac{F}{6\pi\eta r}$$

$$dv_d = \frac{dF}{6\pi\eta r} = -\frac{E\lambda^2 e \exp(\lambda a - \lambda r)}{6\pi\eta(1 + \lambda a)} dr \qquad (21)$$

where η = coefficient of viscosity. Integrating from the closest distance of approach, a, to infinity yields

$$v_d = -\frac{Ee\lambda}{6\pi\eta(1 + \lambda a)}. \qquad (22)$$

For dilute solutions $\lambda a \ll 1$ (low conductivity)

$$v_d = -\frac{Ee\lambda}{6\pi\eta}.$$

The net velocity of the central ion is

$$v = v_0 - |v_d| = \frac{eE}{6\pi\eta}\left(\frac{1}{a} - \frac{1}{\delta}\right) = \frac{eE}{6\pi\eta}\left\{\frac{\delta - a}{a\delta}\right\} \qquad (23)$$

where $\delta = \lambda^{-1}$ = Debye length. The motion of the central ion and the counterion cloud tends to shear the fluid surrounding the ion. Solely for the sake of a rough quantitative estimate, we assume that the shear is given by

$$g \sim \frac{v_0 - |v_d|}{\delta}. \qquad (24)$$

In our materials the Debye length is of the same order as the average distance between ions.

In nematic liquid crystals we have found that the ion mobility is $\sim 10^{-4}$ cm^2/V·s. For the fields of roughly 10^4 V/cm which are necessary for dynamic scattering, this corresponds to $\Delta v \sim 1$ cm/s. Debye lengths for the materials which are of interest are usually $\sim 0.5\,\mu$. Hence,

$$g \sim 2 \times 10^4 \text{ s}^{-1}.$$

As we have previously shown, the orientation of swarms in a shear flow is governed by the parameter α.

$$\alpha = g/D_r = \text{shear/rotary diffusion coefficient}.$$

The rotary diffusion can be estimated if we assume that the axial ratio of the swarm is approximately unity (spherical). According to Morawetz [12],

$$D_r = kT/8\pi\eta R^3 \qquad (25)$$

where

$$R = \text{swarm radius} \sim 0.1\,\mu$$

$$\eta = \text{coefficient of viscosity} \sim 3 \times 10^{-3} \text{ (MKS)}.$$

For our case

$$D_r \sim 50.$$

Thus,

$$\alpha = g/D_r \sim 400.$$

Our previous calculations indicated that appreciable alignment along the direction of flow was obtained for $\alpha \gg 1$; hence, it appears that the swarms are capable of being aligned by the shear induced by ions in transit. When we pass to the isotropic state the molecules are no longer associated, and the size of the orienting species is that of the individual molecule ($R \sim 20$ Å). Hence,

$$D_r \sim 6.6 \times 10^6 \text{ s}^{-1}$$

and

$$\alpha \sim 3.3 \times 10^{-3}.$$

This value of α is too small to produce alignment and hence no dynamic scattering is observed in the isotropic state.

Some of the assumptions made in the course of this crude analysis are open to question, i.e., 1) swarm orients as a rigid body, 2) the simplified Stokes model of viscous friction, 3) the spherical shape used in estimating D_r, and 4) the use of the Debye length in calculating the shear. However, these calculations serve only to suggest the plausibility of a hydrodynamic orientation model for dynamic scattering and do not purport to establish it unequivocally. It is to be noted that a circulatory force on the liquid crystal which could generate an apparent turbulence can be predicted from (1) provided

$$\vec{\nabla} \times \vec{F} = \vec{\nabla} \times \rho\vec{E} = \vec{\nabla}\rho \times \vec{E} \neq 0.$$

Thus, a space charge gradient in a direction which is not parallel to the applied field is required. That such a gradient could exist in a fluid exhibiting optical and electrical anisotropy does not seem too remote.

TRANSIENT BEHAVIOR

Using the sandwich cell configuration, the response of the liquid crystal to a voltage step function was observed. The optical response was recorded using a 931A photomultiplier and oscillosope. Rise times of the order of 1 to 5 ms are possible with dc voltages in the 50 to 100 V range corresponding to fields in the vicinity of 10^4 V/cm. Both the rise and decay times are a function of temperature and sample thickness. While our data are limited to a relatively narrow range, behavior consistent with the following relationships has been found for APAPA:

$$\tau_{\text{rise}} \propto I^{-1}$$

and also

$$\tau_{\text{decay}} \propto \rho^{1/2} L^2$$

where

τ_{rise} = rise time
τ_{decay} = decay time
I = current
L = sample thickness
ρ = resistivity.

In view of the crucial role which ion transport plays in the DSM, it is not surprising that the rise time for the effect should be related to the number of ions and their velocity or, as we have seen, the current. Note that the decay time (defined as the time required to go from 90 percent to 10 percent of the initial level) seems to be proportional to the square of the sample thickness and the square root of the resistivity. The shape of the decay curve could not be accurately measured over more than an order of magnitude; nevertheless, it did *not* fit an exp k/t or ln (kt) type of behavior. This would seem to eliminate models based on classical diffusion and mechanisms in which the rate is determined by past events. A much better fit to the decay curve was obtained using a $(1-kt^{1/2})$ law suggested by Williams [13].

Using the hydrodynamic model presented in the previous section, the relaxation of dynamic scattering would seem to be closely related to a molecular reorientation process. Williams [13] has suggested that reorientation may initiate at the surface and propagate at a diffusion controlled rate. There is ample evidence for the importance of surface effects in the orientation and electrical behavior of nematic liquid crystals [14]. Williams assumes that the scattering intensity is proportional to the sample thickness. When the field is removed the scattered light takes the following form:

$$I_s(t) = \beta[L - 2x(t)] \qquad (26)$$

where

L = sample thickness
x = distance from the sample wall.

This model is shown schematically in Fig. 13. In the simplest diffusion model (diffusion rate independent of the particular wall)

$$x(t) = D^{1/2}t^{1/2}. \qquad (27)$$

Thus,

$$I_s(t) = I_0\left(1 - \frac{2D^{1/2}t^{1/2}}{L}\right) \qquad (28)$$

which can be functionally fitted to our decay transients. Using our previous definition of decay time $(0.9I_0 - 0.1I_0)$

$$\tau = 0.8L^2/4D. \qquad (29)$$

This model agrees with the experimentally determined thickness dependence.

It is also possible to infer the resistivity dependence from the Einstein relation for the diffusion coefficient by a some-

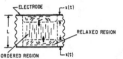

Fig. 13. Simplified model of the relaxation process.

what circuitous route. The diffusion coefficient is assumed to be proportional to a mobility which is, in turn, inversely proportional to the effective radius and viscosity of the diffusing species. This behavior of the mobility[1] follows from the Stokes model for viscous friction. If it is further assumed that the effective radius is a function of the Debye length ($r \propto l_{\text{Debye}} \propto \rho^{1/2}$), an inverse square root dependence of the diffusion coefficient on resistivity follows directly.

Equation (29) can be directly checked with experiment. The value for the decay time of a 1/2 mil thick cell at 90°C is 30 ms. Using (29), this value implies a diffusion coefficient of 10^{-5} cm²/s. On the basis of an Einstein model, the value of mobility should be

$$\mu = De/kT \approx 4.8 \times 10^{-4} \text{ cm}^2/\text{V} \cdot \text{s}.$$

As previously noted, a cusp was observed in the current transient (Fig. 6) which possessed much of the character of the familiar space charge limited current transient in solids [9]. If the transient is treated in this manner, one can calculate the mobility from the position of the cusp (Fig. 7). For our materials this value is approximately 0.3×10^{-4} cm²/V·s, which is a typical value for ions in solution. Thus, the agreement between theory and experiment seems reasonably good in view of our crude model.

There are, however, certain inconsistencies in our attempt to relate the diffusion coefficient to the resistivity through the Debye length. According to the Stokes model for the ion mobility, one calculates (using the macroscopic viscosity) an effective radius in the 10 to 100 Å range for the mobilities which we have inferred from our data, while the Debye length is of the order of 0.1 to 0.5 μ. One can speculate on whether it is valid to use the experimental macroscopic viscosity in computing the mobility. Perhaps the motion is determined by a much smaller microscopic viscosity. There are dramatic reports in the literature in which the experimental mobility remained essentially constant while the macroscopic viscosity changed by orders of magnitude [15]. It is evident that more work must be done in this area if we are to understand the detailed effects of the microscopic viscosity on the mobility and, hence, the dependence of decay time on resistivity.

EVALUATION OF A STATIC LIQUID CRYSTAL DISPLAY
BASED ON DYNAMIC SCATTERING

It is of interest to subjectively evaluate images produced by dynamic scattering. Using the sandwich cell geometry, the image was defined on the transparent electrode by a

---

[1] Mobility is inversely proportional to the product of effective radius and effective viscosity.

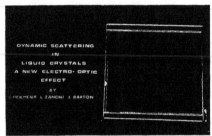

(a) Left: lettered card. Right: liquid crystal panel (3½ by 4 inches). $V=0$.

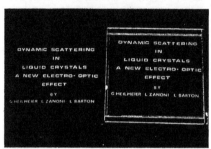

(b) $V=45$ V dc.

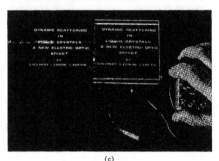

(c)

Fig. 14. Static image produced by dynamic scattering—
qualitative evaluation.

transparent photoresist process. The photoresist is a much
better dielectric than the liquid crystal; hence, field appeared across the liquid only in the regions from which the
photoresist had been removed. Fig. 14 shows the quality
of the image formed by the liquid crystal display compared
to that of a printed page. The liquid crystal panel was
roughly 3½ by 4 inches, and the excitation was 45 V dc.

The maximum contrast ratio on this panel was better than
20 to 1, and the areas of maximum brightness were 40 percent of the standard white ($MgCO_3$).

DYNAMIC SCATTERING AND TRANSMITTED LIGHT

While this paper has dealt almost exclusively with the
reflective mode of operation, the effects can also be observed in transmitted light. In this case, both electrodes in
the sandwich cell geometry are transparent. With no field
applied, light is transmitted by the cell. In its excited state
the material scatters the incident light, thus reducing the
amount of transmitted light. On-to-off ratios in excess of
20 to 1 have been obtained in this scheme.

CONCLUSIONS

A new reflective effect in certain classes of nematic liquid
crystals has been discovered. The effect has been related to
the disruptive effects of ions in transit through the aligned
nematic medium which results in the formation of localized
scattering centers. The ions can be produced by field assisted
dissociation of neutral molecules and/or Schottky emission
processes. The rise times of 1 to 5 ms and decay times of
less than 30 ms, together with dc operating voltages in the
10 to 100 V range, make *dynamic scattering* seem attractive
for such applications as alphanumeric indicators and do not
preclude its use in line-at-a-time matrix addressed, real-
time displays. Reflective contrast ratios of better than 15 to 1
with efficiencies of 45 percent of the standard white have
been demonstrated.

ACKNOWLEDGMENT

The authors wish to thank J. Goldmacher and J. Castellano, who conducted the materials research connected with
this project; J. Van Raalte, B. J. Lechner, D. Kleitman, and
W. Moles, who provided support and insight into many
problems relating to practical display applications; and
Dr. R. Williams, who contributed many helpful discussions
and insights into the field of liquid crystals.

REFERENCES

[1] G. W. Gray, *Molecular Structure and Properties of Liquid Crystals*.
London: Academic Press, 1962.
[2] G. H. Brown and W. G. Shaw, *Chem. Rev.*, vol. 57, p. 1049, 1957.
[3] J. L. Fergason, *Sci. Am.*, vol. 211, p. 77, 1964.
[4] R. Williams and G. H. Heilmeier, *J. Chem. Phys.*, vol. 44, p. 638,
1966.
[5] H. van de Hulst, *Light Scattering by Small Particles*. New York:
Wiley, 1957.
[6] G. Hansen, unpublished dissertation, Halle, Germany, 1907.
[7] A. P. Chattock, *Phil. Mag.*, vol. 48, p. 401, 1899.
[8] O. Stuetzer, *J. Appl. Phys.*, vol. 30, p. 984, 1959.
[9] A. Many et al., *J. Phys. Chem. Solids*, vol. 22) p. 285, 1961; also P.
Mark and W. Helfrich, *Z. Phys.*, vol. 166, p. 370, 1962.
[10] W. F. Pickard, *J. Appl. Phys.*, vol. 34, p. 246, 1963.
[11] See, for example, A. Van der Ziel, *Solid State Physical Electronics*.
Englewood Cliffs, N. J.: Prentice Hall, 1957.
[12] H. Morawetz, *Macromolecules in Solution*, New York: Interscience,
1965.
[13] R. Williams, private communication.
[14] See, for example, O. Pellet and P. Chatelain, *Bull. Soc. Franc.
Mineral.*, vol. 73, p. 154, 1950; R. Williams, *J. Chem. Phys.*, vol. 39,
p. 384, 1965; and R. Williams and G. H. Heilmeier, *J. Chem. Phys.*,
vol. 44, p. 638, 1966.
[15] R. Taft and L. E. Malm, *J. Phys. Chem.*, vol. 43, p. 499, 1939.

George H. Heilmeier was born in Philadelphia in 1936. He took his BS degree at the University of Pennsylvania, and then his Master's and Ph.D at Princeton. After completing his graduate studies. Heilmeier joined RCA Laboratories where he worked on various electronic and electro-optic devices, becoming Head of Solid State Device Research in 1966.

His work on electro-optic effects in liquid crystals resulted in the first liquid crystal displays. For this achievement RCA awarded in 1968 the prestigious David Sarnoff award to the team of George Heilmeier, Joseph Castellano, Joel Goldmacher, Louis Zanoni and Lucian Barton. From 1970 to 1977 Heilmeier was with the US Department of Defense, and became the Director of Defense Advanced Research Projects Agency. During this period he was twice awarded the Department of Defense Distinguished Civilian Service Medal. From 1977, Heilmeier's career was in the senior management of Texas Instruments and Bellcore (later Telcordia Technologies), of which he became Chairman and Chief Executive.

Over his distinguished career in telecommunications, George Heilmeier has received many honours and awards, and has served on numerous advisory committees for government and academic institutions. In 1996 he received the John Scott Award for Scientific Achievements from the City of Philadelphia for his pioneering work in the development of liquid crystal displays. Previous winners of the Scott Award include Albert Einstein, Guglielmo Marconi, Marie Curie, the Wright brothers and Thomas Edison. Heilmeier is a fellow of the American Academy of Arts and Sciences and a Fellow of the Institution of Electronic and Electrical Engineers; he holds the IEEE medal of honor, the highest award of the Institute, for his work on liquid crystal displays.

He retired from Bellcore in 1997. However, he continues to contribute to the telecommunications industry as Chairman Emeritus of Telcordia Technologies.

Louis Zanoni was born in 1933 in Trenton, New Jersey, USA. At the age of 18, he joined the US Navy, where he worked as a radio operator in communications at a US Naval air station in Italy. After four years in the Navy, Zanoni attended the RCA Institute in New York City and studied radio and television broadcasting.

After graduating from the RCA Institute, Zanoni worked for RCA Laboratories in Princeton, New Jersey. It was during this period that he worked on the development of liquid crystal devices and co-wrote the paper included in this volume. In 1970 Zanoni left RCA to co-found the Optel Corporation. Here he participated in the design of the first LCD watch. In 1976, in collaboration with his wife Mary, he founded Zantech Inc, an electronic watch consulting company. In 1993 he founded WZBN TV-25. This a local TV station and is an outgrowth of the electronic watch business.

During his period at RCA and subsequently, Zanoni published widely in the technical literature, and has numerous patents in the field of electro-optic devices and circuits. While at Zantech, Louis and Mary Zanoni published two books on Quartz and digital watch repair. Currently (2002), although officially retired, Zanoni still participates in the growth and development of the WBZN TV-25.

D4
Angewandte Chemie, International Edition **8**, 884–85 (1969)

A LIQUID-CRYSTALLINE (NEMATIC) PHASE WITH A PARTICULARLY LOW SOLIDIFICATION POINT

by

H. Kelker and B. Scheurle[*]

We have prepared a chemically homogeneous, stable, nematogenic substance that is still liquid at room temperature: *N*-(*p*-methoxybenzylidene)-*p*-*n*-butylaniline (*2k*). This melts at 20 °C to a mobile, turbid, pale yellowish liquid with all the properties of a nematic phase; its clarification point is approximately 41 °C. Comparison with homologous and isomeric compounds[1] enables its transition temperatures and other physical properties to be discussed in relation to constitution. The optical and electrooptical properties that characterize some higher-melting liquid-crystalline substances are also observed with (*2k*).

Our investigations started from the finding that in the azomethine series (*1*) there is a very obvious melting-point minimum at the butyric ester (R^1=*n*-propyl)[2]. The melting point of the butyrate is 50 °C and the clarification point 112 °C.

$$H_3C-O-\langle\bigcirc\rangle-CH=N-\langle\bigcirc\rangle-O-CO-R^1$$

(1), R^1 = n-Alkyl

According to *Weygand* alkyl chains are all to be regarded as "wing groups"; examples are *n*-alkyldiphenyl-pyridazines and *p*-butylbenzoic acid.

[*] Dr. H. Kelker and Dr. B. Scheurle Farbwerke Hoechst AG 623 Frankfurt/Main 80 (Germany).

[1] D. *Vorländer*: Chemische Kristallographie der Flüssigkeiten. Akademische Verlagsgesellschaft, Leipzig 1924; C. *Weygand*: Chemische Morphologie der Flüssigkeiten und Kristalle. Handu. Jahrbuch der chem. Physik. Vol. 2, Section IIIC; G. W. *Gray*: Molecular Structure and the Properties of Liquid Crystals. Academic Press, New York 1962; W. *Kast* in *Landolt-Börnstein*, 6th Edit., Vol. II/ 2a, p. 288.

[2] H. *Kelker* and B. *Scheurle*, J. Physique, in press.

515

Comparison of compounds containing an azoxy, azo, or azomethine group shows that the homologous azomethines have the lowest clarification point. There are, however, no general rules for melting-point series, least of all for the initial members of homologous series. We synthesized azomethines of series *(2)* by the known method, condensation of an aromatic aldehyde with an alkylaniline. [*(2a)*[3] and *(2b)*[4] are known.] From the series of solidification and clarification points it could not be deduced with certainty that the butyl derivatives *(2k)* and *(2l)* would have the desired properties[**].

$$R^1 -\!\!\bigcirc\!\!- CH\!=\!N -\!\!\bigcirc\!\!- R^2 \quad (2)$$

	R^1	R^2	M.p. (°C)	Transition point (°C), behavior
(a)	CH_3O	H	60–63 [3]	isotropic
(b)	CH_3O	CH_3	92–93 [4]	38 (monotr. nemat.)
(c)	CH_3O	C_2H_5	57	28 (monotr. nemat.)
(d)	CH_3O	n-C_3H_7	42	57 (enantiotr. nemat.)
(e)	C_2H_5O	CH_3	94	80 (monotr. nemat.)
(f)	C_2H_5O	C_2H_5	67	70 (enantiotr. nemat.)
(g)	C_2H_5O	n-C_3H_7	76	97 (enantiotr. nemat.)
(h)	CH_3	CH_3	92–93	isotropic
(i)	C_2H_5	CH_3	49	isotropic
(j)	n-$C_9H_{19}O$	n-C_3H_7	51	74/84 (smectic I/II)
(k)	CH_3O	n-C_4H_9	20	41 (enantiotr. nemat.)
(l)	C_2H_5O	n-C_4H_9	36	80 (enantiotr. nemat.)
(m)	CH_3	C_4H_9O	80	71 (monotr. nemat.)

Compounds *(2k)* and *(2l)* were prepared by condensing p-methoxy- and p-ethoxy-benzaldehyde with p-n-butylaniline. After the usual working up procedure, *(2l)* was recrystallized several times from ethanol; *(2k)* was obtained in about 80% yield by distillation in a high vacuum. *Gabler*[5], who briefly mentions this type of compound, prepared only members with longer alkoxy chains R^1 (C_8 and C_9) and $R^2 < C_4H_9$, and these also showed as few peculiarities as did the Schiff base *(2m)* which is isomeric with *(2k)*.

Since *(2k)* is readily accessible, physical properties can be studied for this compound which would be hard to study without special apparatus (thermostat) for higher-melting compounds. For instance, we were able to demonstrate the

[3] O. Anselmino, Ber. dtsch. chem. Ges. *40*, 3473 (1907).

[4] O.J. Steinhart, Liebigs Ann. Chem. *241*, 338 (1887).

[**]*Note added in proof*: The next higher homolog, N-(p-methoxybenzylidene)-p-n-pentylaniline, melts at 38 °C to a nematic phase whilst the clarification point is 58 °C.

[5] R. Gabler, Dissertation, Universität Leipzig 1939.

"dynamic scattering" effect[6] for *(2k)*, the phenomenon that forward scattering of incident light is increased in an electric field; this effect is utilized industrially for display of numbers, *etc.* By use of *(2k)* it is also possible to demonstrate strain figures on plastics or glass at room temperature. Here we are clearly dealing with "epitaxy" effects such as have been utilized in the manufacture of polarization films[7].

(2k) can also be used as solvent for *Saupe*'s technique of NMR spectroscopy[8] (direct spin-spin coupling in the nematic phase).

Finally it was of interest to prepare cholesteric mixed phases containing *(2k)* and to study their possible use in thermotopography[9,10]. It is known that nematic phases become cholesteric on addition of optically active cholesteric substances. Use of *(2k)* had the result that the typical cholesteric reflection color appeared but that this was not shifted towards the violet on increase in temperature as is characteristic of many mixtures and for mixtures of various proportions. In mixtures of *(2k)* with cholesteryl benzoate, methyl carbonate, nonanoate, or oleate, a cholesteric phase is formed on cooling below the clarification point; this shows a reflection color which depends on the composition but which remains almost unchanged over a wide temperature range. The color shifts slightly towards the violet before it vanishes at 0 °C. These experiments were carried out on cover slips with the reverse side blackened and placed on a Kofler hot stage.

The mixtures behave like binary mixtures when $\beta > \beta_0$[9]. [β_0 is the composition of a binary cholesteric mixture at which the relation between reflection color and temperature changes sign.] Above β_0 the wavelength of the reflected light shifts to red with increase in temperature, and to blue with decrease in temperature. Whereas the mixtures of *(2k)* and cholesteryl oleate are stable for weeks, the three other esters rapidly crystallize out.

Received: August 15, 1969; revised: October 1, 1969 [Z 91 IE]

German version: Angew. Chem. *81*, 903 (1969)

[6] G. H. *Heilmeier*, L. A. *Zanoni*, and L. A. *Barton*, Appl. Physics Letters *13*, 46 (1968); R. *Williams*, J. chem. Physics *39*, 384 (1963).

[7] J. F. *Dreyer*, US-Pat. 2524286 (1950); US-Pat. 2544659 (1951), both for Polacoat Comp.

[8] A. *Saupe*, Angew. Chem. *80*, 99 (1968); Angew. Chem. internat. Edit. *7*, 97 (1968).

[9] J. *Adams*, W. *Haas*, and J. *Wysocki*, Physic. Rev. Letters *22*, 92 (1969).

[10] J.L. *Fergason*, Sci. American *211*, 77 (1964).

Hans Kelker was born in 1922 in a small village in Saxony in Germany. After army service he studied chemistry at the Technical University of Hanover from 1948 to 1953. In 1955 he obtained his Ph.D thesis in physical chemistry under the supervision of Professor Hans Braune. His subsequent employment was at the Hoechst company in Frankfurt. In 1960 he became director of the Analytical Institute at Hoechst, retaining this post until his retirement in 1988.

In addition to his work at Hoechst, in 1970 he was awarded the habilitation degree in analytical chemistry from the University of Frankfurt. In 1975 he was appointed as Honorary Professor at that university.

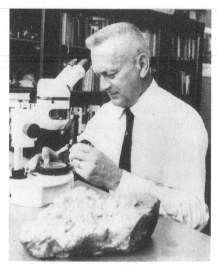

(Photo courtesy of Lutz Rohrschneider, Münster).

He was particularly known for his introduction of physical methods into instrumental analysis. He was the first to use liquid crystals as stationary phases in gas chromatography. His joint paper with Bruno Scheurle included in this volume describing the synthesis of MBBA, the first stable room-temperature nematic phase, was carried out as part of this programme.

Kelker was also known for his work for the liquid crystal community. In 1980, together with Rolf Hatz, he edited the *Handbook of Liquid Crystals*, which long remained *the* international standard work. Beginning in 1971, he was much involved with the annual Liquid Crystal Workshops in Freiburg.

His other special interest was the history of liquid crystals, and his work has significantly influenced that of the present authors. He made detailed studies of the early period of liquid crystal research in the immediate aftermath of the work of Reinitzer and Lehmann. This led to a number of articles and two books with P.M. Knoll, one a biography of Otto Lehmann, and the other a discussion of early Franco-German links in the liquid crystal field. He also established an extensive collection of exhibits on liquid crystal history which is now a part of the Bunsen archive in Giessen.

He died in 1992 in Freiburg.

D5

Molecular Crystals and Liquid Crystals **12**, 57–72 (1970)

Distortion of Twisted Orientation Patterns in Liquid Crystals by Magnetic Fields†

F. M. LESLIE

Mathematics Department
University of Strathclyde
Glasgow, Scotland

Received October 15, 1970

Abstract—At rest between parallel plates, cholesteric liquid crystals commonly exhibit a characteristic twisted orientation pattern, the axis of twist being perpendicular to the plates. Also, this orientation pattern appears possible in nematic liquid crystals, since it is consistent with continuum theory[1] for both types of liquid crystal. This paper discusses the influence upon such a twisted orientation pattern of a magnetic field perpendicular to the plates. If one employs a free energy of the form discussed by Frank,[2] the continuum theory equations have solutions relevant to this situation. For nematic liquid crystals, provided that the twist is not too large, our analysis suggests that no distortion of the orientation pattern occurs until the magnetic field strength exceeds a critical value which varies with the amount of twist. For cholesteric liquid crystals, there seems to be two possibilities depending upon the relative magnitudes of two Frank constants. Either distortion always occurs above a critical field strength, or does so only when the distance between the plates is sufficiently small.

1. Introduction

A number of interesting experiments in liquid crystal theory employ solid boundaries and external magnetic fields as competing influences upon the orientation of the large elongated molecules which occur in these liquids. For a nematic liquid crystal at rest in a small gap between parallel plates, suitable prior treatment of the solid surfaces leads to a uniform orientation pattern either parallel or perpendicular to the plates (see for example Chatelain[3]). Application of a magnetic field to such thin films of nematic liquid crystal leads to three important experiments. Two employ a parallel orientation pattern

† Presented at the Third International Liquid Crystal Conference in Berlin, August 24–28, 1970.

519

with the field either normal or parallel to the plates, but always perpendicular to the molecular axes, and the third uses a perpendicular orientation pattern with the field parallel to the plates. In each case there is no appreciable distortion of the orientation pattern until the magnetic field strength exceeds a critical value, which depends upon the liquid crystal, the distance between the plates, and the arrangement employed. Continuum theory predicts such behaviour, and in this way Saupe[4] obtains estimates for three parameters in the Frank energy.[2] More recently, Dafermos[5] and Ericksen[6] present detailed theoretical investigations of these experiments. Also, Leslie[7] shows that the theory predicts similar behaviour in other situations.

A sample of cholesteric liquid crystal at rest between parallel plates commonly has a twisted or helical orientation pattern. In this arrangement, the orientation of the molecular axes is everywhere parallel to the solid surfaces, being constant in any plane parallel to the plates, but varying uniformly with distance normal to the surfaces. As one traverses the gap between the plates, the ends of the molecules therefore trace out a helix, whose pitch is a characteristic of a given cholesteric liquid crystal. Employing the theory of Frank[2] and ignoring the influence of boundaries, de Gennes[8] and Meyer[9] discuss the application of magnetic fields to this type of liquid crystal. When the field is perpendicular to the helical axes, de Gennes finds that the pitch increases with field strength until at a critical value the molecules align parallel to the field throughout the sample. The experiments of Durand et al.[10] and Meyer[11] confirm this behaviour. De Gennes[8] and Meyer[9] also discuss the case when the field is parallel to the helical axes, but in the absence of boundary constraints consider the possibility that the sample reorientate so that the field is again perpendicular to the helical axes. However, Meyer also discusses the possibility of distortion of the helical structure.

In this paper we consider a liquid crystal at rest between parallel plates exhibiting the twisted orientation pattern described above, and employing continuum theory discuss the application of a magnetic field perpendicular to the plates. For nematic liquid crystals, if one assumes that the solid surfaces control the orientation of the molecules in contact, it seems possible to obtain such a twisted orientation pattern from the uniform parallel orientation pattern mentioned earlier simply by rotating one plate through an angle about its normal. For this type of liquid crystal, provided that the twist is not too large, our analysis suggests that no distortion occurs until one exceeds a critical field strength, which depends upon the amount of twist and the distance between the plates. With data from the experiments described above, one can predict this critical value, and this appears to offer a useful test for consistency between theory and experiment.

For cholesteric liquid crystals, we assume that the molecules align parallel to the boundaries such that the surfaces exert no normal couple stress upon the liquid crystal. Leslie[12,13] makes this assumption for this type of liquid crystal in other problems, and his predictions based upon it appear compatible with observations. As for the nematic case, our analysis suggests that no distortion of the orientation pattern can occur until the magnetic field strength exceeds a critical value. However, depending upon the relative magnitudes of two material parameters in the Frank energy, distortion may occur only when the distance between the plates is sufficiently small. Assuming that theory and experiment are consistent, it seems possible to obtain information regarding material parameters from these experiments.

2. Continuum Theory

This section presents a brief summary of the equations given by Ericksen[14] and Leslie[12,15] to describe static, isothermal behaviour of liquid crystals. It is convenient to choose a set of right-handed Cartesian axes, and to employ Cartesian tensor notation.

As is commonly done, one describes the orientation of the molecular axis by a unit vector d_i, and assumes that d_i and $-d_i$ are physically indistinguishable. The relevant equations are a balance of forces

$$t_{ij,j} + F_i = 0, \qquad (2.1)$$

t_{ij} representing the stress tensor, and F_i the body force per unit volume, and a further balance law

$$s_{ij,j} + g_i + G_i = 0, \qquad (2.2)$$

where s_{ij} is an orientation stress tensor, and g_i and G_i are intrinsic and extrinsic orientation body forces per unit volume respectively. Associated with the orientation stress tensor is a couple stress tensor l_{ij} given by

$$l_{ij} = e_{ipq} d_p s_{qj}. \tag{2.3}$$

The constitutive equations for the stress tensors and the intrinsic orientation body force are

$$t_{ij} = -p\delta_{ij} - \frac{\partial W}{\partial d_{k,j}} d_{k,i} + \alpha\, e_{jkp} (d_p d_i)_{,k}, \tag{2.4}$$

$$s_{ij} = d_i \beta_j + \frac{\partial W}{\partial d_{i,j}} + \alpha\, e_{ijk} d_k, \tag{2.5}$$

$$g_i = \gamma\, d_i - (d_i \beta_j)_{,j} - \frac{\partial W}{\partial d_i} - \alpha\, e_{ijk} d_{k,j}. \tag{2.6}$$

The undetermined pressure p in Eq. (2.4) arises on account of the assumed incompressibility, and similarly the scalar γ and the vector β_i in Eqs. (2.5) and (2.6) stem from the constraint that the vector d_i has fixed magnitude. Also, the function W is the Helmholtz free energy per unit volume, and depends only upon the vector d_i and its gradients. Here, we adopt the form due to Frank,[2]

$$2W = \alpha_1(d_{i,i})^2 + \alpha_2(\tau + d_i\, e_{ijk}\, d_{k,j})^2$$
$$+ \alpha_3\, d_i\, d_j\, d_{k,i}\, d_{k,j} + (\alpha_2 + \alpha_4)\, [d_{i,j} d_{j,i} - (d_{i,i})^2]. \tag{2.7}$$

In the present context, the coefficients in the above expression and α are constants. For nematic liquid crystals, the coefficients α and τ are both zero.

In this paper, we consider body forces arising from an applied magnetic field H_i and gravity. Consequently, accepting the estimates of Ericksen[14]

$$F_i = -\chi_{,i} + \{(\nu_1 - \nu_2)\, H_j d_j d_k + \nu_2 H_k\}\, H_{k,i}, \tag{2.8}$$

$$G_i = (\nu_1 - \nu_2)\, H_j d_j H_i, \tag{2.9}$$

where χ is the gravitational potential, and ν_1 and ν_2 are the magnetic susceptibilities parallel and perpendicular to the molecular axis respectively. As in earlier work, we assume that ν_1 and ν_2 are positive constants, and that the former is the larger.

With the aid of Eqs. (2.5), (2.6) and (2.9), Eq. (2.2) becomes

$$\left(\frac{\partial W}{\partial d_{i,j}}\right)_{,j} - \frac{\partial W}{\partial d_i} + \gamma d_i + (\nu_1 - \nu_2)\, H_j\, d_j\, H_i = 0. \tag{2.10}$$

Also, when the external body forces take the form (2.8) and (2.9), it is straightforward to show that Eq. (2.1) integrates with the aid of Eq. (2.10) to yield

$$p = p_0 - \chi - W + \tfrac{1}{2}[(\nu_1 - \nu_2)\,(H_j\,d_j)^2 + \nu_2 H_j H_j], \tag{2.11}$$

where p_0 is an arbitrary constant. Consequently, below we seek solutions of Eq. (2.10).

3. The Solution

Referred to a set of right-handed Cartesian axes, consider a solution of Eqs. (2.7) and (2.10) of the form

$$d_x = \cos\theta(z)\,\cos\phi(z), \quad d_y = \cos\theta(z)\,\sin\phi(z), \quad d_z = \sin\theta(z), \tag{3.1}$$

with

$$H_x = 0,\ H_y = 0,\ H_z = H, \tag{3.2}$$

where H is a constant. After some manipulation to eliminate the scalar γ the Eq. (2.10) reduces to

$$f(\theta)\frac{d^2\theta}{dz^2} + \tfrac{1}{2}\frac{d}{d\theta}f(\theta)\left(\frac{d\theta}{dz}\right)^2 - \tfrac{1}{2}\frac{d}{d\theta}g(\theta)\left(\frac{d\phi}{dz}\right)^2$$

$$- 2\alpha_2\tau\sin\theta\cos\theta\,\frac{d\phi}{dz} + \nu H^2\sin\theta\cos\theta = 0, \tag{3.3}$$

and

$$g(\theta)\frac{d^2\phi}{dz^2} + \frac{d}{d\theta}g(\theta)\frac{d\theta}{dz}\frac{d\phi}{dz} + 2\alpha_2\tau\sin\theta\cos\theta\,\frac{d\theta}{dz} = 0, \tag{3.4}$$

in which

$$f(\theta) = \alpha_1\cos^2\theta + \alpha_3\sin^2\theta, \quad g(\theta) = (\alpha_2\cos^2\theta + \alpha_3\sin^2\theta)\cos^2\theta, \tag{3.5}$$

and

$$\nu = \nu_1 - \nu_2. \tag{3.6}$$

Eq. (3.4) promptly integrates to yield

$$g(\theta)\frac{d\phi}{dz} - \alpha_2\tau\cos^2\theta = k, \tag{3.7}$$

where k is an arbitrary constant. Below, we examine solutions in which both θ and ϕ vary. For these, multiply Eq. (3.3) by the derivative of θ and Eq. (3.4) by the derivative of ϕ, add and integrate to obtain

$$f(\theta)\left(\frac{d\theta}{dz}\right)^2 + g(\theta)\left(\frac{d\phi}{dz}\right)^2 + \nu H^2 \sin^2\theta = c, \qquad (3.8)$$

where c is an arbitrary constant. From Eqs. (3.7) and (3.8), it is straightforward to determine derivatives of θ and ϕ as functions of θ, and the solution follows readily by further integration. However, since the boundary conditions considered differ for nematic and cholesteric liquid crystals, it is necessary to discuss these cases separately.

4. Nematic Liquid Crystals

For this type of liquid crystal, material symmetries require that in Eqs. (2.4)–(2.7)

$$\alpha = 0, \quad \tau = 0. \qquad (4.1)$$

Also, the argument of Ericksen[16] leads one to impose the conditions

$$\alpha_1 > 0, \quad \alpha_2 > 0, \quad \alpha_3 > 0. \qquad (4.2)$$

We consider a layer of liquid crystal lying between parallel plates which coincide with the planes $z = 0$ and $z = 2l$. As remarked earlier, there is evidence that a solid surface imposes a particular orientation upon the molecules of a nematic liquid crystal immediately adjacent to it. Further, this is controllable to some degree by the prior treatment given to the surface. Consequently, it appears physically relevant to discuss the boundary conditions

$$\theta(0) = 0, \ \theta(2l) = 0, \ \phi(0) = -\phi_0, \ \phi(2l) = \phi_0, \qquad (4.3)$$

where ϕ_0 is an arbitrary positive constant. In the absence of a reasonable alternative, we assume that the magnetic field does not alter the orientation at the solid interface. However, Saupe[4] suggests that this may be open to question on occasion.

When the magnetic field is absent, one solution of Eqs. (3.3) and (3.4) which meets the boundary conditions (4.3) is

$$\theta = 0, \quad \phi = \phi_0(z - l)/l. \qquad (4.4)$$

Provided that ϕ_0 is sufficiently large, it appears to be possible to construct other acceptable solutions in which θ varies. For the present, we disregard these and assume that the solution (4.4) represents the initial orientation pattern. This presumably places some restriction on the range of the parameter ϕ_0.

When the field is present, the twisted orientation pattern (4.4) remains a solution of Eqs. (3.3) and (3.4), but it is natural to examine another possibility. Consider solutions of the type

$$\theta(z) = \theta(2l - z), \quad 0 \leqslant z \leqslant l, \qquad (4.5)$$

with

$$\theta(l) = \theta_m, \quad \frac{d}{dz}\theta(l) = 0, \qquad (4.6)$$

θ_m being a constant to be determined. On account of the symmetry of the problem, one may choose θ_m positive without loss of generality. In this event, Eqs. (3.7) and (3.8) at once yield

$$g(\theta)\frac{d\phi}{dz} = k, \qquad (4.7)$$

and

$$f(\theta)\left(\frac{d\theta}{dz}\right)^2 = \nu H^2\left(\sin^2\theta_m - \sin^2\theta\right) + k^2\left(\frac{1}{g(\theta_m)} - \frac{1}{g(\theta)}\right). \qquad (4.8)$$

From Eq. (4.7), it follows that

$$\phi(z) = -\phi(2l - z), \quad 0 \leqslant z \leqslant l, \qquad (4.9)$$

and in particular that

$$\phi(l) = 0. \qquad (4.10)$$

The relevant solution is therefore given by

$$z = \int_0^\theta \left[\frac{f(\psi)}{\nu H^2(\sin^2\theta_m - \sin^2\psi) + k^2(1/g(\theta_m) - 1/g(\psi))}\right]^{1/2} d\psi, \quad 0 \leqslant z \leqslant l, \qquad (4.11)$$

and

$$\phi = -\phi_0 + \int_0^\theta \left[\frac{f(\psi)}{\nu H^2(\sin^2\theta_m - \sin^2\psi) + k^2(1/g(\theta_m) - 1/g(\psi))}\right]^{1/2}\frac{k\,d\psi}{g(\psi)}, \quad 0 \leqslant z \leqslant l, \qquad (4.12)$$

provided that the constants θ_m and k satisfy

$$l = \int_0^{\theta_m} \left[\frac{f(\theta)}{\nu H^2(\sin^2\theta_m - \sin^2\theta) + k^2(1/g(\theta_m) - 1/g(\theta))}\right]^{1/2} d\theta, \qquad (4.13)$$

and

$$\phi_0 = \int_0^{\theta_m} \left[\frac{f(\theta)}{\nu H^2(\sin^2\theta_m - \sin^2\theta) + k^2(1/g(\theta_m) - 1/g(\theta))} \right]^{1/2} \frac{k\,d\theta}{g(\theta)}. \quad (4.14)$$

The last two equations therefore determine the parameters θ_m and k for a given field strength H. Alternatively, one may regard them as equations which give the field strength H and the parameter k as functions of θ_m.

If one makes a change of variable

$$\sin\lambda = \frac{\sin\theta}{\sin\theta_m}, \quad (4.15)$$

Eqs. (4.13) and (4.14) become

$$l = \int_0^{\pi/2} \left[\frac{f(\theta)}{\nu H^2 - k^2 F(\theta, \theta_m)/g(\theta)\,g(\theta_m)} \right]^{1/2} \frac{d\lambda}{\cos\theta}, \quad (4.16)$$

$$\phi_0 = \int_0^{\pi/2} \left[\frac{f(\theta)}{\nu H^2 - k^2 F(\theta, \theta_m)/g(\theta)\,g(\theta_m)} \right]^{1/2} \frac{k\,d\lambda}{\cos\theta\,g(\theta)}, \quad (4.17)$$

where

$$F(\theta, \theta_m) = \frac{[g(\theta_m) - g(\theta)]}{(\sin^2\theta_m - \sin^2\theta)}$$

$$= \alpha_3 - 2\alpha_2 - (\alpha_3 - \alpha_2)(\sin^2\theta + \sin^2\theta_m). \quad (4.18)$$

From these, it follows that

$$\lim_{\theta_m \to 0} k = \alpha_2 \phi_0/l, \quad \lim_{\theta_m \to 0} \nu l^2 H^2 = \alpha_1(\pi/2)^2 + (\alpha_3 - 2\alpha_2)\phi_0^2, \quad (4.19)$$

the latter assuming that

either $\qquad \alpha_3 \geqslant 2\alpha_2,\qquad$ or $\qquad \phi_0^2 \leqslant \alpha_1 \pi^2/4(2\alpha_2 - \alpha_3). \quad (4.20)$

With the notation

$$\beta = \sin^2\theta_m, \quad (4.21)$$

differentiation of expressions (4.16) and (4.17) leads to

$$2l\left(\frac{dk}{d\beta}\right)_{\beta=0} = (\alpha_3 - 2\alpha_2)\phi_0, \quad (4.22)$$

$$2\nu l^2 \left(\frac{dH^2}{d\beta}\right)_{\beta=0} = \alpha_3(\pi/2)^2 - (\alpha_3^2 - \alpha_3\alpha_2 + \alpha_2^2)\phi_0^2/\alpha_2. \quad (4.23)$$

Consequently, if the condition (4.20) holds, and also

$$\phi_0^2 < \alpha_3\alpha_2 \pi^2/4(\alpha_3^2 - \alpha_3\alpha_2 + \alpha_2^2), \quad (4.24)$$

one has the result that

$$\frac{dH}{(d\beta)}_{\beta=0} > 0. \tag{4.25}$$

The conditions (4.20) and (4.24) are therefore necessary to ensure that the field strength increases monotonically with θ_m in the interval zero to $\pi/2$. However, it is beyond the scope of the present investigation to determine sufficient conditions for this.

Given the existence of more than one solution, following Dafermos[5] we select on stability grounds that which minimizes the energy function

$$\epsilon = \int_V [W - \tfrac{1}{2}\{(\nu_1 - \nu_2)(H_i d_i)^2 + \nu_2 H_i H_i\}] dV, \tag{4.26}$$

where V is the volume of liquid crystal. If one denotes by ϵ_0 the value of the above integral when the solution (4.4) occurs, and introduces an energy difference

$$\Delta = \epsilon - \epsilon_0, \tag{4.27}$$

Eqs. (2.7), (3.1), (3.2), (3.5), (3.6) and (4.1) yield

$$\Delta = \tfrac{1}{2} A \int_0^{2l} \left[f(\theta) \left(\frac{d\theta}{dz}\right)^2 + g(\theta) \left(\frac{d\phi}{dz}\right)^2 - \nu H^2 \sin^2 \theta - \alpha_2 \phi_0^2/l^2 \right] dz, \tag{4.28}$$

where A is the area of the plates. Hence, with the aid of Eqs. (4.5), (4.7) and (4.8), one obtains

$$\Delta = A \int_0^l [\nu H^2 (\sin^2 \theta_m - 2\sin^2 \theta) + k^2/g(\theta_m) - \alpha_2 \phi_0^2/l^2] dz, \tag{4.29}$$

and changes of variable employing Eqs. (4.8) and (4.15) lead to

$$\Delta = A \int_0^{\pi/2} \frac{[f(\theta)]^{1/2} [\nu H^2 \sin^2 \theta_m \cos 2\lambda + k^2/g(\theta_m) - \alpha_2 \phi_0^2/l^2]}{\cos \theta [\nu H^2 - k^2 F(\theta, \theta_m)/g(\theta) g(\theta_m)]^{1/2}} d\lambda. \tag{4.30}$$

By differentiation of this expression and the relations (4.16) and (4.17), one can show that

$$\left(\frac{d\Delta}{d\beta}\right)_{\beta=0} = 0, \quad 4l \left(\frac{d^2\Delta}{d\beta^2}\right)_{\beta=0} = -A[\alpha_3 (\pi/2)^2 - (\alpha_3^2 - \alpha_3 \alpha_2 + \alpha_2^2) \phi_0^2/\alpha_2]. \tag{4.31}$$

c

Therefore, provided that the conditions (4.20) and (4.24) hold,

$$\epsilon > \epsilon_0 \tag{4.32}$$

for values of θ_m in some neighbourhood of zero.

The above analysis points to the existence of a critical magnetic field strength H_c, given by

$$\nu l^2 H_c^2 = \alpha_1 (\pi/2)^2 + (\alpha_3 - 2\alpha_2) \phi_0^2, \tag{4.33}$$

provided that the angle ϕ_0 is not too large. For field strengths below this value, the twisted orientation pattern occurs, and above this value there is distortion of this orientation pattern. As Saupe[4] demonstrates, it is possible to estimate the three Frank constants appearing in the expression (4.33) by means of the experiments mentioned earlier, allowing one to predict the critical field strength for the present situation. This seems to offer a worthwhile opportunity to compare theory and experiment. Further, it could provide a means of assessing the range of applicability of the Frank free energy. As Ericksen[16] discusses, this energy essentially represents an expansion about a uniform orientation pattern, and therefore should be appropriate to describe small departures from such an orientation pattern, this being the case in the experiments employed by Saupe.[4] Here, however, the initial orientation pattern is non-uniform, the degree of non-uniformity being given by the ratio ϕ_0/l. Consequently, the extent of agreement between the relationship (4.33) and experiment as the distance between the plates decreases should give some indication of the range of validity of this form of energy.

5. Cholesteric Liquid Crystals

For this type of liquid crystal, the static form of the theory of Leslie[12] differs from that due to Ericksen,[14] the terms with coefficient α in Eqs. (2.4)–(2.6) appearing only in the former. These arise in a natural way in Leslie's derivation, but the need to include them in the theory remains to be established. Since their absence leads to some simplification, we initially omit them. In this event, application of Ericksen's[16] reasoning to the commonly observed twisted orientation pattern leads to

$$\alpha_1 > 0, \quad \alpha_2 > 0, \quad \alpha_3 > 0, \quad \alpha_2 + \alpha_4 = 0, \quad \tau = \frac{2\pi}{\zeta}, \tag{5.1}$$

where ζ denotes the pitch of the helix.

Here, the liquid crystal again lies between parallel plates coincident with the planes $z = 0$ and $z = 2l$. On this occasion, however, the boundary conditions for the molecular orientation are less clear, and would seem to depend upon the nature of the material forming the plates. One reasonable proposition follows from the common occurrence of the characteristic twisted orientation pattern, namely that at a solid boundary the molecular axes align parallel to the surface such that the normal component of couple stress is zero. Leslie[12,13] employs this condition, and his conclusions based upon it appear consistent with observations. From Eqs. (2.3), (2.5), (2.7), (3.1) and (3.5), the relevant component of couple stress is

$$l_{zz} = g(\theta)\frac{d\phi}{dz} - \alpha_2 \tau \cos^2 \theta, \tag{5.2}$$

and therefore the boundary conditions considered below are

$$\theta(0) = 0, \quad \theta(2l) = 0, \quad \frac{d}{dz}\phi(0) = \tau, \quad \frac{d}{dz}\phi(2l) = \tau. \tag{5.3}$$

One solution of Eqs. (3.3) and (3.4), compatible with the boundary conditions (5.3), is the characteristic twisted orientation pattern,

$$\theta = 0, \quad \phi = \phi_1 + \tau z, \tag{5.4}$$

where ϕ_1 is an arbitrary constant. When the magnetic field is absent, one can show that this is the only solution satisfying the conditions (5.3). However, when the field is present, another is possible.

As in the previous section, consider solutions in which

$$\theta(z) = \theta(2l - z), \quad 0 \leqslant z \leqslant l, \tag{5.5}$$

with

$$\theta(l) = \theta_m, \quad \frac{d}{dz}\theta(l) = 0, \tag{5.6}$$

θ_m again being a positive constant. In view of the boundary conditions (5.3), Eq. (3.7) takes the form

$$\frac{d\phi}{dz} = \frac{\alpha_2 \tau}{h(\theta)}, \quad h(\theta) = \alpha_2 \cos^2 \theta + \alpha_3 \sin^2 \theta. \tag{5.7}$$

With this result and the conditions (5.6), Eq. (3.8) becomes

$$f(\theta)\left(\frac{d\theta}{dz}\right)^2 = \nu H^2(\sin^2\theta_m - \sin^2\theta) + \alpha_2^2\tau^2\left(\frac{\cos^2\theta_m}{h(\theta_m)} - \frac{\cos^2\theta}{h(\theta)}\right). \qquad (5.8)$$

The appropriate solution is therefore

$$z = \int_0^\theta \left[\frac{f(\psi)}{\nu H^2(\sin^2\theta_m - \sin^2\psi) + \alpha_2^2\tau^2(\cos^2\theta_m/h(\theta_m) - \cos^2\psi/h(\psi))}\right]^{1/2} d\psi,$$

$$0 \leqslant z \leqslant l, \qquad (5.9)$$

provided that the constant θ_m satisfies

$$l = \int_0^{\theta_m} \left[\frac{f(\theta)}{\nu H^2(\sin^2\theta_m - \sin^2\theta) + \alpha_2^2\tau^2(\cos^2\theta_m/h(\theta_m) - \cos^2\theta/h(\theta))}\right]^{1/2} d\theta,$$

$$(5.10)$$

this last equation relating θ_m and the field strength H.

The substitution (4.15) reduces Eq. (5.10) to

$$l = \int_0^{\pi/2}\left[\frac{f(\theta)}{\nu H^2 - \alpha_3\alpha_2^2\tau^2/h(\theta)h(\theta_m)}\right]^{1/2}\frac{d\lambda}{\cos\theta}, \qquad (5.11)$$

from which one readily obtains

$$\lim_{\theta_m\to 0}\nu H^2 = \alpha_1\left(\frac{\pi}{2l}\right)^2 + \alpha_3\tau^2. \qquad (5.12)$$

Also, with the notation (4.21), differentiation of the expression (5.11) leads to

$$2\nu\left(\frac{dH^2}{d\beta}\right)_{\beta=0} = \alpha_3\left[(\pi/2l)^2 + 3(\alpha_2 - \alpha_3)\tau^2/\alpha_2\right], \qquad (5.13)$$

and therefore

$$\left(\frac{dH}{d\beta}\right)_{\beta=0} > 0, \qquad (5.14)$$

provided that,

either $\qquad \alpha_2 \geqslant \alpha_3, \qquad$ or $\qquad l^2 < \dfrac{\alpha_2\pi^2}{12(\alpha_3 - \alpha_2)\tau^2}. \qquad (5.15)$

To discriminate between the solutions, we examine their energies as defined by Eq. (4.26). Denoting that of the solution (5.4) by ϵ_0, and that of the second by ϵ, let

$$\Delta = \epsilon - \epsilon_0, \qquad (5.16)$$

and Eqs. (2.7), (3.1), (3.2), (3.5) and (3.6) combine to yield

$$\Delta = \tfrac{1}{2} A \int_0^{2l} \left[f(\theta) \left(\frac{d\theta}{dz} \right)^2 + g(\theta) \left(\frac{d\phi}{dz} \right)^2 - 2\alpha_2 \tau \cos^2 \theta \frac{d\phi}{dz} \right.$$

$$\left. + \alpha_2 \tau^2 - \nu H^2 \sin^2 \theta \right] dz, \qquad (5.17)$$

A being the area of the plates. With the aid of Eqs. (5.5), (5.7) and (5.8), one obtains

$$\Delta = A \int_0^l \left[\nu H^2 (\sin^2 \theta_m - 2 \sin^2 \theta) \right.$$

$$\left. + \alpha_2 \tau^2 \left\{ \alpha_2 \frac{\cos^2 \theta_m}{h(\theta_m)} - 2\alpha_2 \frac{\cos^2 \theta}{h(\theta)} + 1 \right\} \right] dz, \qquad (5.18)$$

and proceeding as before it follows that

$$\Delta =$$

$$A \int_0^{\pi/2} \frac{\left[f(\theta) \right]^{1/2} \left[\nu H^2 \sin^2 \theta_m \cos 2\lambda + \alpha_2 \tau^2 \left\{ \alpha_2 \dfrac{\cos^2 \theta_m}{h(\theta_m)} - 2\alpha_2 \dfrac{\cos^2 \theta}{h(\theta)} + 1 \right\} \right]}{\cos \theta \left[\nu H^2 - \alpha_3 \alpha_2^2 \tau^2 / h(\theta) \, h(\theta_m) \right]^{1/2}} \, d\lambda.$$

$$(5.19)$$

Differentiation of this last expression leads to

$$\left(\frac{d\Delta}{d\beta} \right)_{\beta=0} = 0, \quad 4 \left(\frac{d^2 \Delta}{d\beta^2} \right)_{\beta=0} = -A l \alpha_3 \left[\left(\frac{\pi}{2l} \right)^2 + 3(\alpha_2 - \alpha_3) \frac{\tau^2}{\alpha_2} \right], \qquad (5.20)$$

and thus, if the condition (5.15) holds,

$$\epsilon < \epsilon_0 \qquad (5.21)$$

for values of θ_m in a neighbourhood of the origin.

For this type of liquid crystal, therefore, the above analysis points to the occurrence of a critical field strength H_c, given by

$$\nu H_c^2 = \alpha_1 \left(\frac{\pi}{2l} \right)^2 + \alpha_3 \tau^2, \qquad (5.22)$$

provided that the condition (5.15) is satisfied. For field strengths below this value, the magnetic field does not influence the twisted orientation pattern, but above this value distortion occurs. With the aid of the relations (5.1), the inequality in condition (5.15) is equivalent to

$$l < \left[\frac{\alpha_2}{3(\alpha_3 - \alpha_2)} \right]^{1/2} \frac{\zeta}{4}. \qquad (5.23)$$

Consequently, if the Frank constant α_3 is larger than α_2, this requirement places a severe restriction on the distance between the plates. However, if α_2 is greater than α_3 measurements when l is large compared with the characteristic pitch ζ should provide an estimate of α_3.

When one retains the terms with coefficient α in Eqs. (2.4)–(2.6), the argument preceding the conditions (5.1) requires modification. In this case, Leslie[13] provides motivation for

$$\alpha_1 > 0, \quad \alpha_2 > 0, \quad \alpha_3 > 0, \quad \tau - \frac{\alpha}{\alpha_2} = 2\pi/\zeta, \tag{5.24}$$

ζ again being the pitch of the characteristic helical arrangement.

With the retention of these terms, the component of couple stress normal to the plates is

$$l_{zz} = g(\theta)\frac{d\phi}{dz} + (\alpha - \alpha_2\tau)\cos^2\theta, \tag{5.25}$$

and therefore it is necessary to replace the boundary conditions by

$$\theta(0) = 0, \quad \theta(2l) = 0, \quad \frac{d}{dz}\phi(0) = \tau - \frac{\alpha}{\alpha_2}, \quad \frac{d}{dz}\phi(2l) = \tau - \frac{\alpha}{\alpha_2}. \tag{5.26}$$

One solution of Eqs. (3.3) and (3.4) subject to the above boundary conditions is

$$\theta = 0, \quad \phi = \phi_1 + \left(\tau - \frac{\alpha}{\alpha_2}\right)z, \tag{5.27}$$

where ϕ_1 is an arbitrary constant. In addition, if one considers a solution of these equations satisfying conditions (5.5), (5.6) and (5.26), Eqs. (3.7) and (3.8) become

$$g(\theta)\frac{d\phi}{dz} = \alpha_2\tau\cos^2\theta - \alpha, \tag{5.28}$$

and

$$f(\theta)\left(\frac{d\theta}{dz}\right)^2 = \nu H^2(\sin^2\theta_m - \sin^2\theta) + \frac{(\alpha_2\tau\cos^2\theta_m - \alpha)^2}{g(\theta_m)}$$
$$- \frac{(\alpha_2\tau\cos^2\theta - \alpha)^2}{g(\theta)}. \tag{5.29}$$

The solution is therefore

$$
z = \int_0^\theta \left[\nu H^2(\sin^2\theta_m - \sin^2\psi) + \frac{(\alpha_2\tau\cos^2\theta_m - \alpha)^2}{g(\theta_m)} - \frac{(\alpha_2\tau\cos^2\psi - \alpha)^2}{g(\psi)} \right]^{1/2} d\psi,
$$

$$
0 \leqslant z \leqslant l, \qquad (5.30)
$$

provided that θ_m satisfies

$$
l = \int_0^{\theta_m} \left[\nu H^2(\sin^2\theta_- - \sin^2\theta) + \frac{(\alpha_2\tau\cos^2\theta_m - \alpha)^2}{g(\theta_m)} - \frac{(\alpha_2\tau\cos^2\theta - \alpha)^2}{g(\theta)} \right]^{1/2} d\theta.
$$

$$
(5.31)
$$

Here, one notes the possibility of a solution of this type when the field is absent.

With the substitution (4.15), the expression (5.31) becomes

$$
l = \int_0^{\pi/2} \left[\frac{f(\theta)}{\nu H^2 - G(\theta,\theta_m)/g(\theta)\,g(\theta_m)} \right]^{1/2} \frac{d\lambda}{\cos\theta}, \qquad (5.32)
$$

where

$$
G(\theta,\theta_m) = [\alpha_3\alpha_2^2\tau^2 - 2\alpha_2\tau\alpha(\alpha_3 - \alpha_2)]\cos^2\theta\,\cos^2\theta_m
$$
$$
+ \alpha^2[(\alpha_3 - 2\alpha_2) + (\alpha_2 - \alpha_3)(\sin^2\theta + \sin^2\theta_m)]. \qquad (5.33)
$$

From Eq. (5.32), one readily finds that

$$
\lim_{\theta_m \to 0} \nu H^2 = \alpha_1 \left(\frac{\pi}{2l}\right)^2 + \alpha_3 \left(\tau - \frac{\alpha}{\alpha_2}\right)^2 + 2\alpha\left(\tau - \frac{\alpha}{\alpha_2}\right), \qquad (5.34)
$$

and that

$$
2\alpha_2\nu\left(\frac{dH^2}{d\beta}\right)_{\beta=0} = \alpha_2\alpha_3 \left(\frac{\pi}{2l}\right)^2 + 3\alpha_3(\alpha_2 - \alpha_3)\left(\tau - \frac{\alpha}{\alpha_2}\right)^2
$$
$$
+ 6\alpha(\alpha_2 - \alpha_3)\left(\tau - \frac{\alpha}{\alpha_2}\right) - 3\alpha^2. \qquad (5.35)
$$

Consequently, the retention of the terms with coefficient α does not necessarily alter the nature of some of our earlier results. However, when they are present, the stability argument employed above appears to require modification, and it is beyond the scope of this paper to consider this question further. It is perhaps worth noting that the existence of a critical field strength given by Eq. (5.34) offers some prospect of obtaining information on the coefficient α.

Finally, as Ericksen[16] points out, the form of energy employed in this paper for cholesteric liquid crystals is open to criticism in that it is an expansion about a uniform orientation pattern rather than about the characteristic twisted orientation pattern. In a recent paper, Jenkins[17] proposes an energy for cholesteric liquid crystals which remedies this. However, predictions based on this energy have yet to be determined.

REFERENCES

1. Ericksen, J. L., *Appl. Mech. Rev.* **20**, 1029–32 (1967).
2. Frank, F. C., *Disc. Faraday Soc.* **25**, 19–28 (1958).
3. Chatelain, P., *Bull. Soc. Franc. Mineral. Crist.* **77**, 323–52 (1954).
4. Saupe, A., *Z. Naturforsch.* **15a**, 815–22 (1960).
5. Dafermos, C. M., *S.I.A.M. J. Appl. Math.* **16**, 1305–18 (1968).
6. Ericksen, J. L., *Z. Angew. Math. Phys.* **20**, 383–8 (1969).
7. Leslie, F. M., *J. Phys. D.* **3**, 889–97 (1970).
8. De Gennes, P. G., *Solid State Commun.* **6**, 163–5 (1968).
9. Meyer, R. B., *Appl. Phys. Letters* **12**, 281–2 (1968).
10. Durand, G., Leger, L., Rondelez, F., and Veyssie, M., *Phys. Rev. Letters* **22**, 227–8 (1969).
11. Meyer, R. B., *Appl. Phys. Letters* **14**, 208–9 (1969).
12. Leslie, F. M., *Proc. Roy. Soc.* **A307**, 359–72 (1968).
13. Leslie, F. M., *Mol. Cryst. and Liquid Cryst.* **7**, 407–20 (1969).
14. Ericksen, J. L., *Arch. Rat. Mech. Anal.* **9**, 371–8 (1962).
15. Leslie, F. M., *Arch. Rat. Mech. Anal.* **28**, 265–83 (1968).
16. Ericksen, J. L., *Phys. of Fluids* **9**, 1205–7 (1966).
17. Jenkins, J. T., *J. Fluid Mech.* (to appear).

Frank Matthews Leslie was born in 1935 in Dundee, Scotland. He entered Queen's College, Dundee, then part of the University of St Andrews in 1953, and graduated in mathematics in 1957. He was awarded the Ph.D degree in 1961 for a dissertation on non-linear fluid motion at the University of Manchester, where he had already been  appointed as assistant lecturer. Following his Ph.D. he spent a year in America at MIT, before returning to England as a lecturer in mathematics at the University of Newcastle.

The academic year 1966–7 was spent at Johns Hopkins university in Baltimore, Maryland, where he was able to interact with the rational mechanics group of Professor Jerry Ericksen. In the picture, taken in 1977, Leslie (left) can be seen with Ericksen. In 1968 Leslie was appointed to a senior lectureship at the University of Strathclyde in Glasgow; he was promoted to reader in 1971, to a personal chair in 1979 and to an established chair in 1982.

Most of Leslie's best-known work is devoted to the study of liquid crystals. In this book we reprint fundamental papers on the theory of nematic liquid crystals and on the theory of switching in twisted nematic devices. He is also well-known for his continuum theory of smectic C liquid crystals.

Leslie's attachments to Scotland, and in particular to the Scottish National Party, as well as his advocacy of an independent Scotland, were well-known to colleagues and friends. He was awarded many honours, included appointments as visiting professor in the United States, Germany, Italy and Japan, and fellowships of the Royal Societies of Edinburgh and of London.

He was due to retire from the University in September 2000, but unfortunately died on June 15, following complications to what should have been a routine hip operation.

D6
Applied Physics Letters **18**, 127–28 (1971)

VOLTAGE-DEPENDENT OPTICAL ACTIVITY OF A TWISTED NEMATIC LIQUID CRYSTAL

M. Schadt and W. Helfrich

Physics Department, F. Hoffmann-La Roche & Co. AG., Basel, Switzerland

(Received 8 December 1970)

A new electro-optical effect in twisted nematic liquid crystals is described which allows variation of the rotation of linearly polarized light continuously from 0° to 90°. It requires lower voltages than other electro-optic effects.

It is well known that the orientation pattern of nematic liquid crystals can be influenced by weak magnetic and electric fields. The response is the basis of some interesting electro-optical effects known as domain formation,[1] dynamic scattering,[2] and guest-host interaction.[3] To observe these, a thin layer of a suitable nematic liquid crystal is placed between two glass plates provided with a conductive coating. In the present note we report on a fourth electro-optical effect of this general type. Like the third, it involves realignment by dielectric torques. In contrast to the others, the new effect requires the orientation pattern to be twisted in the fieldless state.

The orientation pattern which we used corresponds to the planar texture of cholesteric liquid crystals. It is obtained if the alignment is parallel to the walls but differs in direction on the surfaces. With a twist angle of 90° and a cell thickness of typically 10^{-3} cm the resulting pitch is 4×10^{-3} cm. This being much larger than the wavelength of light, one may expect the polarization plane of linearly polarized light travelling normal to the glass plates to rotate with the unique axis.[4] Accordingly, a 90° twist should lead to a 90° rotation.

An applied voltage can influence the orientation pattern, provided the material has a positive dielectric anisotropy ($\epsilon_{\parallel} > \epsilon_{\perp}$). At high enough fields the alignment in the bulk will be practically parallel to the field. For light travelling normal to the cell there remains only a very weak optical anisotropy which arises from the residual influence of the fixed wall alignment. At the walls there will be transition regions with strong splay and bend but without twist. Consequently, the rotation of the light will be weak or absent. The

two extreme cases, i.e., orientation patterns with twist only or no twist at all, are sketched in Fig. 1. Mixed patterns and intermediate rotation angles are expected to occur at lower electric fields.

We chose as liquid crystal material $n(4'$-ethoxybenzylidene) 4-amino-benzonitride (PEBAB) which is nematic between 106 and 128 °C and has a very large positive dielectric anisotropy[5] with

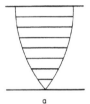

a

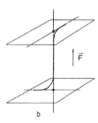

b

FIG. 1. (a) 90° twist in a nematic liquid crystal; (b) splay and bend instead of twist in a strong electric field $\vec{F}$.

$\epsilon_\parallel \sim 21$ and $\epsilon_\perp \sim 7$. We also used a room-temperature mixture of

$$R- \hexagon -CH = N- \hexagon -CN$$

with $R_1 = n\text{-}C_4H_9\text{-}O\text{-}$, $R_2 = n\text{-}C_9H_{13}\text{-}O\text{-}$, and $R_3 = n\text{-}C_7H_{15}\text{-}COO\text{-}$ in molar proportions of 1:1:1, which is nematic between 20 and 94 °C. The desired wall orientation was obtained by the well-known method of rubbing the electrodes with a cotton swab. (The molecules of the liquid crystal align parallel to the direction of rubbing.) A twist of 90° was achieved by melting the material between "crossed" electrodes. The cell was placed between two polarizers which were parallel or perpendicular to the direction of rubbing and illuminated at normal incidence. When no voltage was applied to the electrodes the liquid crystal rotated the light by 90°. Applying dc or ac voltages we found that the rotatory power of the cell can practically be reduced to zero. The effect showed no hysteresis and the cell returned to its initial state when the voltage was turned off.

The angle of rotation as a function of voltage for a mixed liquid crystal is shown in Fig. 2(a) for a $10\text{-}\mu$-thick sample. Also plotted for the same driving frequency of 1 kHz is the transmission with parallel polarizers [Fig. 2(b)]. The brightening effect has a threshold of only 3 V (rms) and is roughly 90% complete at 6 V. The behavior in PEBAB is frequency-independent from about 0.1 Hz to 80 kHz. At higher frequencies turbulences begin to show, destroying the optical homogeneity of the cell. The threshold for PEBAB is even lower (1 V at 1 kHz) and the brightening effect is almost complete at 3 V. Under dc operation, higher voltages were required for the brightening effect to occur, the threshold being at 2.5 V for PEBAB and 6 V for the mixed liquid crystal. Also, the appearance of the cell was not as homogeneous as with ac. This may have been due to electrohydrodynamic effects. The voltage needed for a given brightness fell slightly with increasing temperature, but the change was less than 10% over the entire nematic range.

A formula for the threshold field of deformation of slightly twisted nematic liquid crystals was given by Leslie[6] for the case of magnetic fields. Rewriting his formula for the electric threshold voltage V_c and for dielectric constants instead of susceptibilities one gets

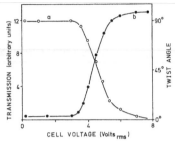

FIG. 2. (a) Rotation angle of linearly polarized light versus voltage for a nematic mixed liquid crystal at room temperature (at 1 kHz); (b) transmission versus voltage with parallel polarizers.

$$(4\pi)^{-1}(\epsilon_\parallel - \epsilon_\perp) V_c^2 = k_{11}(\pi/2)^2 + (k_{33} - 2k_{22})\varphi_0^2$$

in cgs units. Here k_{11}, k_{22}, and k_{33} are the elastic moduli for splay, twist, and bend, respectively, φ_0 is the twist angle which was $\pi/2$ in our experiments. It is seen that the threshold voltage can be made very small (and in fact, zero) if $k_{33} - 2k_{22} < 0$. The threshold reduction is not surprising since realignment releases the elastic energy associated with the initial twist. This may explain why we observed a threshold voltage smaller than the 5–8 V common in the other electro-optical affects. The new effect which has a thickness-independent threshold voltage like the other three promises interesting device applications.

We would like to thank H. Scherrer for having synthesized the nematic liquid crystals used in our experiments.

[1] R. Williams, J. Chem. Phys. 39, 384 (1963).
[2] G. H. Heilmeier, L. A. Zanoni, and L. A. Barton, Proc. IEEE 56, 1162 (1968).
[3] G. H. Heilmeier and L. A. Zanoni, Appl. Phys. Letters 13, 91 (1968).
[4] C. Mauguin, Bull. Soc. Franç. Minéral. 34, 71 (1911); H. de Vries, Acta Cryst. 4, 219 (1951).
[5] G. H. Heilmeier, L. A. Zanoni, and J. E. Goldmacher, in Liquid Crystals and Ordered Fluids, edited by J. F. Johnson and R. S. Porter (Plenum, New York, 1970).
[6] F. M. Leslie, Molecular Cryst. Liquid Cryst. (to be published).

Martin Schadt was born in Liestal, Switzerland, in 1938. He entered the University of Basel in 1960 to study physics. He received his master's degree in 1964 and his doctorate in solid state physics in 1967. From 1967 to 1969 he was a postdoctoral fellow at the Canadian National Research Council (NRCC) in Ottawa. There he continued his research on organic single crystal semiconductors and invented and patented the first solid state, blue light emitting organic diode (OLED).

In 1969 Schadt returned to Switzerland working first for the Laboratoire Suisse de Recherche Horlogère in Neuchâtel. Between 1970 and 1994 he was Research Scientist and Project Leader in the Central Research Department of Hoffmann-La Roche Inc., Basel, Switzerland. The paper republished in this volume is a product of Schadt's collaboration with Wolfgang Helfrich at Hoffman-La Roche at the beginning of this period. Schadt then started to investigate correlations between liquid crystalline molecular structures, material properties and electro-optical effects with the aim of finding display-specific design criteria for new liquid crystals to render the twisted nematic effect and subsequent field-effects applicable and to advance liquid crystal displays. His interdisciplinary approach amalgamated physics and chemistry and established Roche as a major liquid crystal manufacturer. Under his leadership (1987–1994) the Roche Liquid Crystal Research Department pioneered research on photo-alignment of liquid crystal displays and liquid crystalline polymer thin-films on single substrates.

In the early 1990s Hoffmann-La Roche focused on its pharmaceutical business, and in 1994 its nematic chemical patent portfolio was sold to E. Merck of Darmstadt. This portfolio consisted of over 65 US patents and equivalents in ten other countries, including among others, the alkenyl liquid crystals. At the same time, the interdisciplinary research and development company ROLIC Ltd., headed by Schadt, was formed. ROLIC's business basis has been a broad and rapidly expanding intellectual property portfolio of basic device and material patents. These include novel LC-technologies which ROLIC's scientists had started to invent in the late 1980s, and which are being licensed worldwide. Photo-alignment has since been advanced by ROLIC and its licensees to an enabling technology in various fields. Examples include: optical alignment and patterning of liquid crystal displays; functional optical thin-films for increased brightness and angle of view of displays; and novel optical security elements for banknotes. Martin Schadt was Chief Executive Officer (CEO) at ROLIC until his retirement in October 2002. He is now Chief Technology Officer (CTO) of the company and remains a member of the board of directors.

Following the Schadt-Helfrich invention of the twisted nematic effect (TN-LCD), Martin Schadt has been granted more than 100 US patents. A majority of these patents have become the basis for new products. He has published over 170 research papers in the field of applied physics, mainly oriented towards correlations between molecular structures, material properties and electro-optical effects, but with some also in the biophysics area. He has been the recipient of a number of prestigious prizes in information display and applied physics. Martin Schadt has maintained a strong involvement in the liquid crystal community, including conference organisation and journal board membership.

Wolfgang Helfrich was born in 1932. From 1951 to 1958 he studied physics at the universities of Munich, Tübingen and Göttingen. In 1961 he obtained his Ph.D degree at the Technical University of Munich, where he was employed as Research Assistant until 1964. From 1964 to 1966 he was Postdoctoral Fellow and Assistant Research Officer at the National Research Council, Ottawa, Canada. In 1966 he returned to the Technical University of Munich where he was awarded the habilitation degree in physics in 1967. Until this time his research activities concerned the structure and physical properties of organic crystals.

From 1967 he turned to another scientific topic and begun to work in the field of liquid crystals as a member of the Technical Staff at the RCA Laboratories, Princeton, NJ. In 1970 he moved to Hoffmann-La Roche, Basle, Switzerland, where initially he remained in the liquid crystal field. It was at Hoffmann-La Roche that he and Martin Schadt developed the nematic twisted nematic cell described in he paper in this collection. Towards the end of his period at Hoffmann-La Roche his interests moved towards the study of lipid membranes.

In 1973 Helfrich was appointed as professor of physics at the Free University of Berlin. In Berlin his principal activities concerned the structure of membranes and lyotropic liquid crystalline systems. His work on the fluctuation-induced interaction between membranes is particularly well known. He has been awarded several prizes for his scientific work. The development the twisted nematic cell was recognised by the award of the Robert-Wichard-Pohl-Prize of the German Physical Society to Schadt and Helfrich in 1996.

D7
Electronics Letters **9**, 130–31 (1973)

NEW FAMILY OF NEMATIC LIQUID CRYSTALS FOR DISPLAYS

Indexing terms: Liquid crystals, Electro-optical effects, Display equipment

New organic compounds of positive dielectric anisotropy giving nematic liquid crystals at room temperature, either as the pure compounds or in admixture, are described. These materials are of improved stability, and function well in electro-optical devices based on the twisted-nematic or cholesteric-nematic phase-change effects.

Room-temperature liquid crystals currently command interest because of their potential use in display devices.[1] The materials available until now show a number of undesirable properties, and there has been a great demand for improved compounds. In this letter, we suggest that the major deficiencies are inherent in the structures of the compounds currently used. Our reasoning leads us to propose a completely new family of low-melting-point materials, which give liquid crystals of the nematic type suitable for displays, and we report encouraging preliminary results.

The structures of the materials in current use may be generalised and represented simply, as shown below:

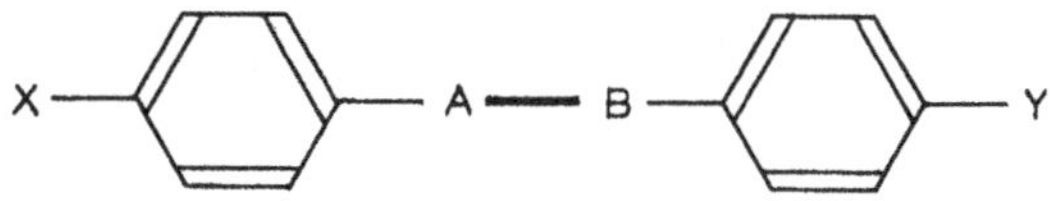

The significant feature is the central unit A—B, which links the two phenyl rings. The molecules of a liquid crystal must be elongated, and, except in certain esters where A—B is CO.O, the link in the A—B unit is generally a double or triple bond, to confer rigidity on the molecule. Typical examples of compounds with a double-bond linkage (CH=N) are MBBA and EBBA, the Schiff's bases used in many commercial devices. We believe that the chemical and photochemical instability observed in materials of the type represented above are directly

543

connected with their structures, since chemical attack and isomerisation may occur at the A—B linkage. For example, Schiff's bases are readily hydrolysed, even by traces of moisture, yielding potentially toxic amines; and the α chlorostilbenes, an alternative family of compounds recently proposed as preferable to compounds such a MBBA, decompose in ultraviolet radiation. The most stable compounds with this structure are those which belong to the family of azoxy compounds [where A—B is N=N(O)], but these have the disadvantage of being yellow.

We reasoned that elimination of the central unit A—B would give compounds with more stability and less colour. The resulting biphenyl system, is a well-known one for liquid crystals, but previous examples have all shown high melting points, and the compounds have therefore been ignored for display applications. We thought that the use of simpler terminal groups X and Y than had been used earlier might give materials which would form liquid crystals at room temperature. The preparative techniques, which were conventional, will be described in full elsewhere. Here we give some of our results, to show that our line of approach has proved promising.

Table 1 gives details of four compounds, all with CN as the substituent Y in the second structure.

All four compounds are colourless and chemically stable. They were purified by column chromatography and distillation, and checked for purity by elemental analysis, mass spectrometry and thin-layer chromatography. Controlled alignments in thin films of the nematic melts were obtained without difficulty.

Compound A is particularly valuable for the preparation of test cells, because its nematic melt supercools for long periods at room temperature (4 months so far), either in closed sample tubes or in test cells. The low-frequency relative permittivities parallel to, and perpendicular to, the long molecular axes are, respectively, $\varepsilon_{11} \simeq 17$ and $\varepsilon_{\perp} \simeq 6$. The positive dielectric anisotropy ($\varepsilon_a \simeq +11$) indicates the suitability of the compound for use in twisted nematic[2] and phase-change[3] electro-optical devices. As detailed elsewhere,[4] the use of this compound in such devices leads

Table 1

Material	X substituent	Melting point	Nematic clearing point
		deg C	deg C
A	$n-C_5H_{11}$	22.5	35
B	$n-C_6H_{13}$	13.5	28
C	$n-C_7H_{15}$	28.5	42
D	$n-C_6H_{13}O$	58.0	76.5

to favourably low thresholds of 1.1 V r.m.s. for the twisted nematic effect, and of 4.5×10^5 V/m r.m.s. for the phase-change effect; in the latter case, a mixture containing 10% by weight of cholesteryl chloride in compound A was used. For the first time, therefore, experimental studies of the twisted nematic electro-optical effect may be made at room temperature with a pure, stable nematogen.

The stability of these compounds has been estimated by several methods. For example, the nematic clearing temperature for compound A has remained constant at 35 °C for samples stored for 3 months in screw-capped bottles, and also for samples left open to the atmosphere for several weeks. This contrasts strongly with MBBA, for which the nematic clearing temperature falls by as much as 3 degC from the value of 47.5 °C when purified material is exposed for a few minutes to the laboratory atmosphere. Thin-layer chromatography and chemical analysis also show that there is complete stability and lack of contamination over periods of several weeks. MBBA, on the other hand, again shows clear evidence of instability and contamination after exposure to the atmosphere for a short time. Indeed, MBBA reacts with the ostensibly inert supports, such as silica gel used in thin-layer chromatography, owing to cleavage of the A—B linkage. The electrical resistivity of compound A has been measured both to direct current and to 100 Hz alternating current. Such measurements require very careful interpretation if any quantitative conclusions are to be made, but qualitative trends in the resistivity with time show that compound A is indeed more stable to atmospheric contamination than MBBA. Finally, exposure of compound A to an ultraviolet source for 4 h produced little effect on the compound. This contrasts strongly with the α-chlorostilbenes, which show large decreases in the resistivity and nematic clearing temperature, and eventually give a red colouration when irradiated by the same source for a few seconds.

By employing mixtures of compounds A with other similar compounds, lower-melting-point materials may, of course, be obtained. For example, a mixture of 83.5 mol% of compound A and 16.5 mol% of compound D melts at 15 °C and remains nematic until 42 °C. This stable eutectic mixture may be cooled to, and maintained at, −55 °C without crystallisation occurring. The performance of the mixture in display devices is similar to that of compound A.

Similarly, a mixture of 56 mol% of compound A and 44 mol% of compound C melts at 0.5 °C and is nematic until 37 °C; the nematic melt is induced to crystallise only with great difficulty.

We have described above some of the properties of four biphenyl compounds. Many other compounds of a structurally related kind and binary and ternary mixtures have also been prepared, and shown to have favourable properties. Among the compounds are some of negative dielectric anisotropy. It is therefore clear that the phases given by such compounds* are potentially valuable additions to the range of liquid crystals available for display devices.

* A patent application relating to these materials has been made.

Acknowledgment: The authors wish to thank J. Kirton, E.P. Raynes and their colleagues at the Royal Radar Establishment, Great Malvern, Worcs., England, who obtained the data on the electrical and electro-optical properties of the compounds.

21st February 1973[†]

G.W. GRAY
K.J. HARRISON
J.A. NASH
Department of Chemistry
University of Hull
Hull, Yorks. HU6 7RX, England

References

1. GRAY, G.W.: 'Synthetic chemistry related to liquid crystals'. 4th International liquid crystal conference, Kent State University, Ohio, USA, 1972 (to be published in *Mol. Cryst. & Liquid Cryst.*).
2. SCHADT, M., and HELFRICH, W.: 'Voltage-dependent optical activity of a twisted nematic liquid crystal', *Appl. Phys. Lett.*, 1971, **18**, pp. 127–128.
3. WYSOCKI, J.J., ADAMS, J., and HAAS, W.: 'Electric-field-induced phase change in cholesteric liquid crystals', *Phys. Rev. Lett.*, 1968, **20**, pp. 1024–1025.
4. ASHFORD, A., CONSTANT, J., KIRTON, J., and RAYNES, E.P.: 'Electro-optic performance of a new room-temperature nematic liquid crystal', *Electron. Lett.*, 1973, **9**, pp. 118–120.

[†] It should be noted that the letter was received on the same date as the companion letter by Ashford *et al.*, *Electron. Lett.*, 1973, **9**, pp. 118–120.

George William Gray was born in Edinburgh, Scotland, in 1926. He entered the University of Glasgow in 1943, graduating with a B.Sc. in Chemistry in 1946. In the same year he was appointed to an assistant lectureship in chemistry at the University of Hull in the North of England, obtaining an external Ph.D from the University of London in 1953. At Hull he was successively promoted to Lecturer, Senior Lecturer, Reader in Organic Chemistry, then to a Personal Chair in Organic Chemistry in 1978 and finally to the G.F. Grant chair in Chemistry in 1984. In 1990 he left Hull for a post as Research Coordinator at Merck UK in Poole on the English South Coast, before retiring in 1993. The internationally recognised Hull liquid crystal group he created continues today under the leadership of John Goodby.

Gray has published more than 350 papers and patents, two books and several edited books over his career. His liquid crystal work began in the late 1940s when the subject was out of fashion and apparently of purely academic interest. By synthesising very large numbers of new liquid crystal materials, he gradually established relationships between molecular structure and liquid crystal behaviour. These enabled him in later years to produce the cyano-biphenyl/-terphenyl materials enabling the manufacture of the first high quality and long-lifetime twisted nematic displays. These very important materials provided the liquid crystal display industry with the secure base from which it expanded so dramatically. They also provided physical scientists with stable materials for reliable physical measurements and observations.

Gray was the first to observe the cubic D phase in the 1950s. His work on smectics played a major role in establishing that single pure materials could exhibit several different smectic phases, and his materials provided the earliest examples of smectic F and I phases. With Goodby he produced the current classification system for smectics. This distinguishes the true smectics (A, Hexatic B, C,F,I) from the crystal smectics (B,E,G,H,J,K). Throughout the 1980s he maintained a strong interest in materials for display applications, including contributing a range of appropriate ferroelectric smectic C* materials for experimentation.

In addition to his research work, Gray has played an active role in public scientific life. He was British Editor of the journal *Molecular Crystals Liquid Crystals* over the period 1979–91, Editor of the journal *Liquid Crystals* in the period 1992–2002, and has also sat on many national committees concerned with Chemistry and Liquid Crystal Science.

The role played by Gray in the development of the modern liquid crystal display industry has been publicly recognised by his election as Fellow of the Royal Society of London (1983), of the Royal Society of Edinburgh (1989), and the award of CBE (1991). He has received a large number of honorary degrees and prizes, including most notably the Kyoto International Gold Medal (1995). He has received the Karl Ferdinand Braun Gold Medal of the Society for Information Display (1996), and the Freedericksz Medal of the Russian Liquid Crystal Society. In 1995 he was elected as a Foreign Member of the Japanese Academy of Engineering, and in 2001 as an Honorary Member of the Royal Irish Academy of Dublin. At the time of writing (2002) he remains scientifically active as a research consultant.

D8
Journal de Physique **26**, L69–71 (1975)

FERROELECTRIC LIQUID CRYSTALS

R. B. MEYER (*), L. LIÉBERT, L. STRZELECKI and P. KELLER

Université Paris-Sud, Physique des Solides (**), 91405 Orsay, France

(*Reçu le 29 novembre 1974, accepté le 9 janvier 1975*)

Résumé. — Un argument général de symétrie est présenté et des expériences sur le p-décyloxybenzy-lidène p′-aminocinnamate de méthyl-2 butyle sont décrites démontrant que les smectiques C et H chiraux sont ferroélectriques. Quelques propriétés de cette nouvelle famille de ferroélectriques sont discutées.

Abstract. — A general symmetry argument is presented, and experiments on newly synthesized p-decyloxybenzylidene p′-amino 2-methyl butyl cinnamate are described, demonstrating that chiral smectic C and H liquid crystals are ferroelectric. Some of the properties of this new class of ferro-electrics are discussed.

In spite of speculation on the possibility of ferro-electric liquid crystalline phases [1], there has never been a compelling fundamental reason for, or expe-rimental demonstration of the existence of ferro-electricity in these systems. In this letter we show by symmetry that smectic C and H liquid crystals composed of chiral molecules must have a sponta-neous polarization. The synthesis of a new material exhibiting these phases is reported, and experiments are described which determine the existence and approximate magnitude of the spontaneous polari-zation. Some of the unusual properties of these fluid ferroelectrics are discussed.

In a smectic C liquid crystal, rod-like molecules are arranged in layers, with the long molecular axes parallel to one another and tilted at an angle θ from the layer normal. Each layer is a two-dimensional liquid. In the smectic H phase (also called tilted B), the molecular layers are crystalline; there remains some question about the degree of correlation of the lattices on different layers. These properties are well established by x-ray [2] and optical studies [3].

Both these phases have monoclinic symmetry, the point group for which contains only a two-fold rotation axis parallel to the layers and normal to the long molecular axis, a reflection plane normal to the two-fold axis, and a center of inversion. However, if the phase is composed of chiral molecules (not

superposable on their mirror image) then the mirror plane and the center of inversion are eliminated. The remaining single two-fold axis allows the exis-tence of a permanent dipole moment parallel to this axis.

The typical liquid crystal molecule has a permanent dipole moment and is of low enough symmetry so that it has only two degenerate minimum energy positions in a monoclinic environment, connected by the two-fold rotation. If the molecule is non-chiral, the mean orientation of its permanent dipole must be normal to the two-fold axis. However, the skewed form of a chiral molecule forces the permanent dipole to have a component parallel to the two-fold axis. If all the molecules are identical this produces a net polarization of at most a few Debye per molecule. A racemic mixture, however, will have no net pola-rization.

Two effects tend to reduce the magnitude of the spontaneous polarization. First, the coupling of the molecule to the monoclinic environment may be weak, so that the molecule is almost freely rotating about its long axis. Second, if the chiral part of the molecule is only weakly coupled to the polar part, then internal molecular rotations may reduce the polarization.

With these considerations in mind, a new chiral material, p-decyloxybenzylidene p′-amino 2-methyl butyl cinnamate (DOBAMBC), was synthesized. By analogy with similar molecules it was expected to have a smectic C phase, and because the chiral 2-methylbutyl group is next to the polar ester group, it was hoped that the problem of internal rotation would be minimized. The thermal phase diagram

(*) Alfred P. Sloan Foundation Research Fellow. Present address, Division of Engineering and Applied Physics, Harvard University, Cambridge, Massachusetts, U.S.A. 02138.

(**) Laboratoire associé au CNRS.

was determined by polarized light microscopy and x-ray studies :

$$\text{Crystal} \xrightarrow{76°} \text{Smectic C} \xrightarrow{95°} \text{Smectic A} \xrightarrow{117°} \text{Isotropic}$$
$$\searrow{}^{63°}$$
$$\text{Smectic H}$$

The smectic C → smectic A transition appears, by differential scanning calorimetry and optical measurements, to be second order, with θ going continuously to zero, while the other transitions are first order.

In both the C and H phases, a helicoidal structure is observed, in which the molecular tilt direction precesses around the normal to the layers, with a pitch of several microns. There are two causes for this helix. First, as in a cholesteric liquid crystal, there is a chiral component to the intermolecular interactions which induces a spontaneous twist in the ground state structure [4]. Second, the same symmetry argument that predicts a spontaneous polarization also requires in the ground state a spontaneous bending curvature [5]. In fact, this helicoidal structure is a state of uniform torsion and bending of the molecular alignment. (This was expected for the smectic C phase [6], but it raises fundamental questions about the H phase to which we will return later) ; because the helix results from two different interactions, it should be possible to find conditions where, in a pure material, the helix disappears (pitch → ∞), while the spontaneous polarization does not. Another way to get the same situation could be to mix different materials, compensating for the pitch but not for the local dipole.

To detect the spontaneous polarization, the electro-optical properties of DOBAMBC were examined. First, samples were prepared between glass plates treated with hexadecyltrimethyl ammonium bromide, which induces the smectic planes to lie parallel to the glass, in the smectic A phase. Upon cooling into the C phase, this ordering is preserved. The electric field was applied parallel to the layers between a pair of 200 μm diameter copper wires imbedded in the sample 1.5 mm apart ; the wires also served as spacers for the glass.

For these samples, the conoscopic image was observed, using crossed linear polarizers. With no field, due to the presence of the helix a uniaxial interference figure was seen, centered on the layer normal, and consisting of a series of concentric rings, just as in the A phase, and an extinction cross, the center part of which is more or less distinct, depending on the molecular tilt angle. With a small applied field, this figure shifts, without apparent distortion, in a direction normal to the field. This effective rotation of the macroscopic optical axis is linear in the field, changing direction when the field is reversed. At higher fields, the helicoidal structure is completely unwound, with the molecular tilt direction uniformly oriented normal to the field, producing the typical biaxial interference figure of a monodomain smectic C sample. Again, reversing the field reverses the tilt direction.

This behavior is consistent with ferroelectric coupling. At high field the polarization is uniformly aligned, producing the observed monodomain structure. The low field behavior results from distortion of the helix, in which regions of favorable polarization grow at the expense of the intervening regions of opposite polarization. This shifts the mean optical axis toward the high field direction. The helix serves as an ideal ferroelectric domain structure, although it arises from local, not long range, interactions.

For quantitative measurements, the ferroelectric coupling must be separated from the ordinary dielectric coupling due to the anisotropy of the dielectric constant [7]. Since the ferroelectric coupling requires the tilt direction to rotate with the field, this response is damped by the rotational viscosity of the fluid, and in practice disappears above a few hundred Hertz. The dielectric coupling, which is quadratic in field, produces a static response which persists at high frequency.

By preparing samples between glass slides coated with transparent tin oxide electrodes, a focal conic texture is obtained in which the applied field is again parallel to the smectic layers, in some regions. In these regions, the helix produces a series of parallel stripes, allowing direct measurement of the pitch and the critical field E_c for unwinding the helix. The high frequency value of E_c was typically at least a factor of 20 higher than the d.c. value, meaning that the dielectric coupling can be ignored at low frequency. The ferroelectric coupling results in a critical field $E_c = \pi^4 K/(4\ lP)$, in cgs units, with l the full pitch of the helix, P the polarization, and K a torsional elastic constant for the helix [8]. Using measurements at 86 °C, $E_c = 2\ 400$ V/cm, and $l = 1.5$ μm. Estimating $K \simeq 10^{-6}$ dyne, then $P \simeq 125$ statcoul./cm^2, or about 0.25 Debye/molecule, which is not unreasonable ; estimating the anisotropy of the dielectric constant $E_a/4\ \pi \sim 1$, the electric polarization induced by E_c (~ 8 cgs) is much smaller than our estimated value for P, showing the consistency of our interpretation. E_c varies from 600 V/cm at 94 °C to 6 500 V/cm at 63.5 °C. The pitch varies from 3 μm at 94 °C to 1 μm at 73 °C, below which it was too small to be measured in this experiment. Unfortunately, K is not known, so that P cannot be determined.

The properties described above are the most important for classifying a material as a ferroelectric, namely the presence of a spontaneous polarization which is easily oriented by an applied field. The analogy with crystalline ferroelectrics extend further than this. The smectic A ↔ smectic C transition of DOBAMBC is actually a Curie point.

This transition is driven by intermolecular forces producing the tilt, and not by the ferroelectric coupling [1, 9]. This is confirmed by comparing the transition temperature T_c of the chiral and racemic versions of DOBAMBC, which differ by less than 1 °C. The situation is then fundamentally different from what happens in usual ferroelectric transitions in solids, where dipole-dipole interactions are directly responsible for the phase change. By symmetry, then, the polarization P is proportional to θ, going to zero continuously at T_c. The linear coupling of P and θ also produces a piezoelectric effect in the smectic A phase. Shear of the smectic layers over one another produces tilt [10] which in turn produces a polarization transverse to the shear. Conversely, an electric field in the plane of the layers produces a polarization and therefore a tilt normal to the field. The latter effect is easily observed in DOBAMBC, although quantitative measurements have not yet been made. As the Curie point is approached from above, the piezoelectric coefficients diverge, which also leads to a divergent dielectric constant.

The molecular tilt mode plays the role of a soft optical phonon at the transition. In this case it is an overdamped mode, and its viscous relaxation frequency tends toward zero at the transition. This should be observable in electro-optic or dielectric constant measurements.

Ferroelectric liquid crystals should have a number of unusual properties. An ordinary smectic C material exhibits curvature elasticity for spatial gradients of the tilt direction, much like a nematic liquid crystal [4, 11]. The presence of the ferroelectric polarization fundamentally alters this situation, since any divergence in P produces space charge and long range coulomb interactions. A bending mode of the tilt direction, with wave vector q parallel to the smectic layers, is a divergence mode of P. The electrostatic energy of this mode is independent of q, in contrast to the elastic energy which varies as q^{-2}. The electrostatic interaction therefore makes this a non-hydrodynamic mode with finite relaxation frequency at zero wave vector. The role of ionic impurities will further complicate the problem.

Another unusual effect will be flow induced polarization. As pointed out by L. Leger, shear of the layers over one another will distort the helix, producing a preferred alignment of the molecules and therefore a polarization transverse to the flow. A similar distortion and polarization can be induced by a magnetic field obliquely oriented to the layer normal. We are grateful to I. W. Smith for a very useful discussion of such an experiment.

The smectic H phase presents fundamental structural problems. The observed helicoidal structure is incompatible with a three dimensional lattice, suggesting that the two dimensional lattices in the smectic layers rotate with the tilt direction in the helix [12]. This requires weak coupling between the layers, and perhaps an equilibrium network of screw dislocations. The electro-optical effects seen in the C phase are observable in the H phase as well, but with a much longer relaxation time, varying from about a tenth of a second near the smectic H-smectic C phase change to several seconds at lower temperatures. The mechanism of the apparently uniform rotation of the tilt direction in response to an applied field may involve the motion of dislocations within the lattice of each layer.

In conclusion, we have demonstrated the existence of a new class of ferroelectric materials. Their unusual combination of fluid and ferroelectric properties leads to some novel phenomena. The ferroelectric and piezoelectric properties of these materials should provide useful probes for the study of their phase changes and their basic structure.

The authors are grateful to J. Doucet for performing x-ray measurements. R. B. Meyer wishes to thank G. Durand for a critical reading of the manuscript, R. Ribotta and Y. Galerne for help with experimental problems, I. W. Smith for both experimental assistance and numerous valuable discussions of this project.

References

[1] McMillan, W. L., *Phys. Rev.* A 8 (1973) 1921.
[2] Doucet, J., Lambert, M., Levelut, A. M., *J. Physique Colloq.* 32 (1971) C 5A-247.
De Vries, A., Proc. of Fifth Intl. Conf. on Liquid Cryst., 1974, *J. Physique Colloq.* 36 (1975) C 1, to be published.
[3] Taylor, T. R., Fergason, J. L., Arora, S. L., *Phys. Rev. Lett.* 24 (1970) 359.
[4] Frank, F. C., *Disc. Faraday Soc.* 25 (1958) 19.
[5] Meyer, R. B., *Lectures in Theoretical Physics* (Les Houches) 1973, to be published.
[6] Helfrich, W., Oh, C. S., *Mol. Cryst. & Liquid Cryst.* 14 (1971) 289 ;
Urbach, W., Billard, J., *C. R. Hebd. Séan. Acad. Sci.* 272 (1972) 1287.
[7] Cheung, L., thesis, Harvard Univ., 1973 ;
Gruler, H., Sheffer, T. J., Meier, G., *Z. Naturforsch.* 27a (1972) 966.
[8] The calculation is formally identical to that of De Gennes, P. G., *Solid State Commun.* 6 (1968) 163.
[9] De Gennes, P. G., *C. R. Hebd. Séan. Acad. Sci.* 274B (1972) 758.
[10] Brochard, F., thesis, Université Paris-Sud, Physique des Solides (1974).
[11] De Gennes, P. G., *The Physics of Liquid Crystals* (Oxford Univ. Press, London) 1974, chapter 7.
[12] This point was clarified in a discussion with De Gennes, P. G.

Robert B. Meyer was born in 1943. He obtained a bachelor degree in physics (1965) and a doctorate in applied physics (1970) from Harvard University. He worked at Harvard (1970–8), including a spell as a visitor at Chalmers University of Technology in Gothenburg, Sweden (1977), and as Joliot-Curie professor at ESPCI, Paris (1978). He was then appointed as associate professor at Brandeis University, Waltham, Massachusetts (1978–85). In 1985 he was promoted to full professor at the same institution.

Meyer's research has concerned various aspects of the physics and chemistry of liquid crystals, and includes both theory and experiment. He has been involved in fundamental studies of liquid crystal ordering in a variety of systems, electric and magnetic field effects, defect structures, phase changes, and the relationship between molecular structure and novel macroscopic properties such as flexoelectricity and ferroelectricity. In recent years his research has concentrated on polymer based liquid crystals, colloidal liquid crystals formed from virus particles, self-assembling liquid crystal systems, dynamical systems based on nonlinear phenomenology of liquid crystals, and smart materials and structures. He has published over a hundred papers and also holds a number of patents. He is widely thought of as one of the leading workers in the liquid crystal field at the present time.

D9
Applied Physics Letters **36**, 899–901 (1980)

Submicrosecond bistable electro-optic switching in liquid crystals

Noel A. Clark [a] and Sven T. Lagerwall

Department of Physics, Chalmers Technical University, Goteborg, Sweden

(Received 3 March 1980; accepted for publication 13 March 1980)

Ferroelectric smectic C (FSC) liquid crystals are used in a simple new geometry that allows the spontaneous formation of either of two surface-stabilized smectic C monodomains of opposite ferroelectric polarization. These domains are separated by well-defined walls which may be manipulated with an applied electric field. The resulting electro-optic effects exhibit a unique combination of properties: microsecond to submicrosecond dynamics, threshold behavior, symmetric bistability, and a large electro-optic response.

PACS numbers: 61.30. − v, 78.20.Jq

In the smectic C (SC) liquid-crystalline phase molecules form a layered structure with the average orientation of the molecular long axes, denoted by the unit vector director $\hat{n}$ tilted at an angle Ω_0 to the layer normal ($\hat{z}$ axis). The director $\hat{n}$ exhibits a continuous degeneracy in its azimuthal orientation, lying on a cone coaxial with $\hat{z}$ (Fig. 1). The azimuthal orientation is thus readily manipulated by external forces or fields. Guided by an elegant physical argument, Meyer *et al.*[1] showed that, for suitably constructed chirally asymmetric molecules, the SC structue will be ferroelectric, with a macroscopic electric dipole density **P** locally normal to $\hat{n}$ and lying in the (x, y) plane of the layers. In principle, FSC's yield, as a result of **P·E** torques, a strong linear coupling of $\hat{n}$ to applied electric field **E**.[1,2] However, this linear coupling is eliminated on a macroscopic scale by an additional consequence of the molecular chirality, namely that $\hat{n}$ and **P** spiral about the $\hat{z}$ axis to form an effectively antiferroe-

lectric helical structure. We demonstrate here that it is possible to use surface interactions to suppress the antiferroelectric helix in a geometry that simultaneously provides (i) the stabilization of either of two domains of opposite ferroelectric polarization separated by domain walls (a situation not permitted in the bulk because of the rotational degeneracy of $\hat{n}$); (ii) convenient **E** field selection of molecular orientation via domain-wall manipulation; (iii) a large optical response (equivalent to the rotation of a uniaxial dielectric of refractive index anisotropy $\Delta n \sim 0.2$ through an angle $2\Omega_0 \sim 4545$

Figure 1 illustrates the essential features of our geometry. The SC is confined between flat plates which are treated so that the director at a surface is constrained to lie in the plane of the surface ($\gamma_1 = 0$), but with no strong tendency for a particular orientation in the surface plane (γ_2 free). For samples having the layers normal to the plates, this boundary condition requires a disclinated texture in order to be

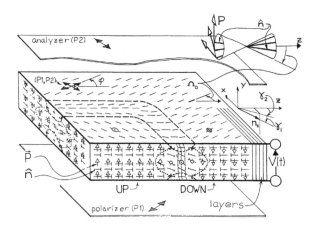

FIG. 1. Schematic of the sample geometry showing UP and DOWN domains and a domain wall (—| indicates a molecule whose right end projects outward). In the regions bounded by the dashed lines (-----) and the plates the SC tilt angle Ω is less than Ω_0. This represents the core of the disclination and, in fact, may be very small. There may also be small layer compression effects. The polarizer (P1) and analyzer (P2) are crossed, with the polarization direction at an angle φ to the $\hat{z}$ axis. For $\varphi = \Omega_0$, light traversing the polarizer-sample-analyzer sandwich will be extinguished in the DOWN state and transmitted in the UP state. A positive applied voltage moves the domain wall so that the DOWN region grows. Domain walls having the opposite helix sense are also possible and are observed.

[a] Permanent address: Dept. of Physics, Univ. of Colorado, Boulder, CO 80309.

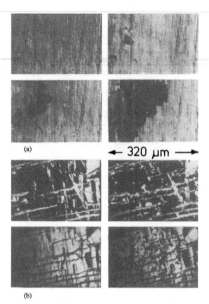

(a)

← 320 μm →

(b)

FIG. 2. Transmission micrographs (resolution ~1 μm): (a) Domain-wall migration in a 1.5-thick SC DOBAMBC sample at $T = 84$ °C. The sample has been oriented by shear such that the layers are horizontal and normal to the page. The vertical texture is caused by remaining weak smectic layer undulations. The analyzer is horizontal and the polarizer is vertical making a small angle with the layer normal ($\varphi \approx 1°$). Voltage sequence (left to right and down): 0, 0.5, 1.0, 1.5 V. The transmitted light contrast between UP (bright) and DOWN (dark) domains is evident. This contrast reverses as the sample is rotated through $\varphi = 0$ and at $\varphi = 0$ only the domain walls are visible, dark, and sharp on a light background. Note the closed domain-wall loops, indicating the presence of domain walls of both signs, and the coalescence of two loops to make one. (b) Sample appearance with $\varphi = \Omega_0$ under conditions of bistable operation. Vertical texture is as in 2(a), and horizontal lines are thin homeotropic regions resulting from the orienting shear. These serve to nucleate and terminate domain-wall motion. Voltage sequence: 1, (top left) -5 V; 2, (top right) 0 V; 3, (bottom left) $+5$ V; 4, (bottom right) 0 V.

of a field favoring the UP orientation produces torques in the wall which induce wall motion that expands the UP region and vice versa. Possible structures for domain walls parallel and normal to the SC layers are shown in Fig. 1. Note that the domains in our case are stabilized by surface mechanical forces and not by bulk forces, distinguishing it from that of crystalline ferroelectrics.

The optical effects associated with domain-wall motion arise from the different director orientations in the UP and DOWN states. Although weakly biaxial,[4] the SC may be taken for present purposes to be uniaxial, with the optic (high index) axis along $\hat{n}$. The simplest geometry has the sample between crossed polarizers with $\hat{n}$ parallel to the polarization direction in the DOWN state ($\varphi = \Omega_0$, Fig. 1) leading to extinction of light passing through the polarizers and sample (DOWN = OFF). In the UP state the polarization will make an angle $2\Omega_0$ with the optic axis and a fraction of the incident optical power F will be transmitted, with

$$F = F_0[\sin(4\Omega_0) \sin(\pi \Delta n \, d / \lambda)]^2.$$

Here Δn is the refractive index anisotropy, λ the vacuum optical wavelength, and F_0 the parallel polarizer transmis-

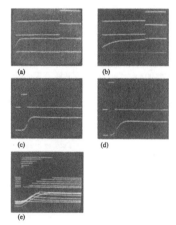

(a)   (b)

(c)   (d)

(e)

Fig. 3. Response of the optical transmission of a SC HOBACPC sample at $T = 68$ °C to pairs of opposite-polarity rectangular voltage pulses, spaced by 70 ms. VERT, photodiode output or applied voltage; HORIZ, time. (a) The top trace represents the optical response to above-threshold pulses showing switching to the OFF state at the beginning of the trace and back to the ON state 70 ms later. The ON pulses are above threshold for all traces. Top to bottom: response to OFF pulses of fixed width ($\tau = 1.4$ μs) and variable amplitude $A = 44, 37, 29$, and 26 V (10 ms/div). (b) Photodiode response to OFF pulses of fixed amplitude ($A = 44$ V) and variable width: $\tau = 2.2, 1.1, 0.9$, and 0.6 μs (top to bottom), again the ON pulses are above threshold (10 ms/div). (c) Dynamic optical response showing applied pulse (top) and photodiode output (bottom): LEFT (2 μs/div): $A = 20$ V, $\tau = 2.0$ μs; RIGHT (1 μs/div): $A = 40$ V, $\tau = 1.0$ μs. (d) Dynamic optical response showing a series of b actinic pulses (top) of varying width for which $A = 1.5$ V and $20 < \tau < 240$ μs, and photodiode response (bottom); 50 μs/div. Traces have been displaced vertically to avoid overlap. For $A = 1.5$ V full latching occurs for $\tau > 200$ μs.

compatible with the bulk $\hat{n}$, P helix. Since the energy required to unwind the helix decreases as the sample thickness d, this boundary condition will suppress the $\hat{n}$, P helix[3] for sufficiently thin samples ($d \lesssim P$, the helix pitch). In the absence of the helix there are two stable, equal energy configurations of the SC, illustrated in Fig. 1 for the idealized case of complete freedom of angle γ_2. For FSC's these two types of $\hat{n}$, P monodomains will possess opposite P normal to the plates, and will be denoted as the "UP" and "DOWN" states. The UP and DOWN states are structurally identical, differing only by a π rotation about the $\hat{z}$ axis (symmetric bistability). Adjacent UP and DOWN regions in a sample will be separated by well-defined domain boundaries (Fig. 1) which are π inversion walls in the $\hat{n}$, P field. The application sion. $F = F_0$ can be achieved for $\Omega_0 \gtrsim 23°$ [this condition is met for DOBAMBC[5] for $T \lesssim 85$ °C (Refs. 2, 4)], and $d > \lambda / 2\Delta n$, implying $d > 2.5\lambda \sim 1.2\,\mu m$ ($\Delta n \sim 0.2$ for DOBAMBC[4]).

We now turn to our experimental results. The bounding plates were glass, coated with semitransparent conductive ($100\,\Omega$ /cm^2) SnO$_2$ layers. The SnO$_2$ surfaces were cleaned of contaminants and dust with spectrographic acetone and placed together without spacers (overlap area $= 6 \times 6$ mm). The sample material was introduced between them by capillary suction from the isotropic phase, resulting in samples which were slightly wedged, typically varying from 0.5 to 3 μm in thickness. The compounds used in this study were optically active DOBAMBC[5] and HOBACPC.[5] Observations of the SA textures obtained upon cooling from the isotropic and obtained as a result of a gentle shearing of the bounding plates suggest the operative boundary conditions for these compounds to be those described above ($\gamma_1 = 0$, γ_2 free). The overall behavior of the two compounds was qualitatively similar except as noted below.

In the SC phase the helix was absent and the electro-optic response was characterized by the motion of resolution-limited domain walls separating regions having apparent optic axes (as determined by the angle φ of the crossed polarizer-analyzer required for extinction) oriented at angles $\varphi \sim \pm \Omega_0$ from the layer normal. The domain walls thus separate regions of nearly opposite polarization. Contrast ratios were limited by the degree of layer orientation achieved, with 20:1 typical over millimeter square areas for $|\varphi| = \Omega_0$.

Figure 2(a) shows typical domain-wall appearance in the SC phase. With a slowly varying voltage the detailed motion of the domain walls was followed and indicated their strong interaction with defects in the layer structure, surface imperfections and scratches, and subresolution defects. this observation prompted a search for bistable operation, which was indeed found, as indicated by Fig. 2(b), showing a sample in either the UP or DOWN state for zero applied field. Bistability was further studied by applying pairs of opposite polarity rectangular voltage pulses of selectable amplitude A, width τ, and time separation to the sample. The sample transmission was monitored by passing a 2-mW He-Ne laser through the microscope to a photodiode. Figure 3 shows typical results obtained on a 1.5-μm-thick[6] HOBACPC sample at $T = 68$ °C for pairs of pulses separated by 70 ms. Figures 3(a) and 3(b) show respectively the optical response to pulses of fixed width ($\tau = 1.4\,\mu s$ and varying amplitude, and to pulses of fixed amplitude ($A = 44$ V) and varying width. For $A\tau$ sufficiently large, the optical response is bistable, with the (+ , −) pulse latching the monitored area ($200 \times 200\,\mu m$) into the (ON, OFF) state. The bistable latch-

ing exhibits a relatively sharp threshold, going from zero to saturated memory response for a less than 25% change in $A\tau$. The dynamic behavior of the optical response to a pulse [c.f. Fig. 3(c)] is characterized by a rise time τ, which depends on pulse amplitude, increasing from a minimum of 1 μs for $A = 20$ V to 4 ms at $A = 0.2$ V. In general, for latching to occur, A and τ must be such that the saturated optical response is achieved during the applied pulse. An exception to this occurs for short, high-voltage pulses ($A > 20$ V) for which, although the optical response does not decrease below 0.9 μs, it does exhibit "inertia," continuing to saturation after termination of the pulse. This minimum response time is comparable to the RC time constant of the sample sandwich. Full switching could be achieved with pulse widths τ_l over the range $A = 55$ V, $\tau_l = 500$ ns ($A\tau_l = 25$ V μs) to $A = 0.2$ V, $\tau_l = 4$ ms ($A\tau_l = 800$ V μs). This general trend and fast response is expected from the simplest theoretical estimate: $A\tau_l \sim \eta d /P \sim 10^{-4}$ V s, where is an orientational viscosity, although the predicted $\tau_l \sim A^{-1}$ dependence is obeyed only at the higher voltages ($A > 10$ V). The full response, once attained, was stable over periods of at least several hours. The dynamic response to fast-rise-time pulses was homogeneous (i.e., independent of the size L of the sample area monitored for $L > 5\,\mu m$), reflecting the nucleation and motion of many domain walls. Results in DOBAMBC were similar, with rise times and requisite pulse widths two to three times longer, presumably a result of the smaller value of P in this material.[7]

To conclude, the effects described here are of potential use where electro-optic effects having fast response and/or built-in memory are required (for example, matrix-addressed video display). This work was supported by the Swedish Natural Science Research Council and the Swedish Board of Technical Development.

[1]R. B. Meyer, L. Liebert, L. Strzelecki, and P. Keller, J. Phys. Lett. **36**, L69–71 (1975).

[2]P. Martinot-Lagarde, J. Phys. (Paris) **37**-C3, 129–132 (1976).

[3]The suppression of the helix for the stronger boundary condition $\gamma_1 = 0$, $\gamma_2 = \Omega_0$ to form a unique stable SC monodomain has been demonstrated [M. Brunet and C. Williams, Ann. Phys. 3, 237–248 (1978)].

[4]S. Garoff, Ph.D. thesis, Harvard University, 1977 (unpublished).

[5]DOBAMBC is decyloxybenzylidene p'-amino 2 methyl butyl cinnamate. See Ref. 1; HOBACPC is hexyloxybenzylidene p'-amino 2 chloropropyl cinnamate. See P. Keller, S. Juge, L. Liebert, and L. Strzelecki, C. R. A. S. 282C, 639-641 (1976). Both compounds have isotropic-SA-SC-SF phases.

[6]Sample thickness was determined using the Newton color sequence.

[7]P. Martinot-Lagarde, J. Phys. (Paris) Lett. **38**, L17-19 (1977).

Noel Clark was born in 1940 in Cleveland, Ohio, where he received his B.S. and M.S. physics degrees in physics at John Carroll University. He received his Ph.D from MIT in 1970 and spent the next seven years in the Harvard liquid crystal group with Peter Pershan and Bob Meyer, first as a research fellow and then as junior faculty. He moved to Boulder in 1977 where he is currently Professor of Physics at the University of Colorado.

He has worked in many areas of soft condensed matter/complex fluid physics, mostly liquid crystals and colloids. Research highlights include initiating, while at Harvard, the modern study of ultrathin freely suspended liquid crystal films and a career-long interest in their properties; the discovery, with Sven Lagerwall, of the surface stabilized ferroelectric liquid crystal structure; and the observation, with the Boulder team, of macroscopic chiral domains in a fluid liquid crystal of achiral bent-core molecules.

In 1984 he co-founded Displaytech, Inc., currently the world's leading maker of ferroelectric liquid crystal devices and materials. His work is the most cited among experimentalists who have worked in liquid crystals.

Sven T. Lagerwall was born in 1934 in Göteborg (Gothenburg) in southern Sweden. He studied at Chalmers University of Technology in his hometown and graduated in Nuclear Engineering in 1958. He then continued his studies at the Hahn-Meitner Institute for Nuclear Research in Berlin-Wannsee and obtained his Ph.D from the Technical University in Berlin in 1964. In 1966 he became lecturer in physics at Chalmers and for a number of years was particularly active in teaching at both undergraduate and graduate levels.

In 1972 he decided to switch his research field to liquid crystals and spent more than a year in the Orsay Liquid Crystal Group in Paris. In 1974 he started his own activities at Chalmers, first in the hydrodynamics of nematics, then dealing with defects (dislocations and disclinations) in smectics, in particular in smectic C, together with R.B. Meyer. In 1978 he started a collaboration with N.A. Clark on chiral smectics which led to the discovery of the surface-stabilized ferroelectric structure in 1980. His group in Göteborg and Clark's group in Boulder, Colorado, have been closely collaborating since then.

Lagerwall became full Professor of Physics at Chalmers in 1986. He has published some 200 papers, about 30 patents and several books. He is a member of the Royal Swedish Academy of Sciences as well as the Royal Swedish Academy of Engineering Sciences.

Section E

LYOTROPIC AND POLYMERIC LIQUID CRYSTALS

LYOTROPIC AND POLYMERIC LIQUID CRYSTALS

Introduction

So far in this narrative, the story of liquid crystals has been followed chronologically, and we have focussed on those materials that form liquid crystals as a result of a change of temperature. However, as we have already seen in Section B (p. 156), A.S.C. Lawrence had specifically noted in 1933 that the action of thermal energy on a solid crystal resembles that of a solvent, by virtue of the fact that the lattice forces of the molecules in a crystal are 'loosened'. In the early history of liquid crystal research it had been observed that certain compounds, on the action of water, form states lying conceptually between solid crystals and true (i.e. isotropic) solutions. Such compounds, described as amphiphilic, exhibit a polar, hydrophilic head and a hydrophobic tail, and can incorporate considerable amounts of water or other solvents into their crystal structure, forming intermediate or mesostates. In the 1933 London Faraday Discussion (see Section B), Lawrence described these different types of intermediate states as lyotropic mesophases. However, considerably before this Otto Lehmann had been aware of the early experiments on such systems, in particular ammonium oleate, and as we shall see, he was keen to apply his ideas to them. Such water-containing materials were also interesting because they embraced the components of living systems, and we have seen in Section A (p. 18) that Lehmann was attracted by the idea of '*flüssige Kristallen*' as the link between inanimate and animate systems.

In fact, liquid crystals are widespread in living systems, and were observed even before the concept of organised fluids was introduced by Lehmann. Some liquid crystals from biology have structures that resemble lyotropic mesophases, and others derive from biomacromolecules. In recent years liquid crystalline phases have been identified in a wide variety of polymers, and polymer liquid crystals have emerged as important new materials. These aspects of liquid crystal history are much more complex than the topic of thermotropic liquid crystals, which occupies four fifths of this book. Our examination of lyotropic, biological and polymeric liquid crystals in this final chapter embraces chemistry, colloid science, biology, biophysics and materials science. We need therefore to follow a number of apparently different threads, but ultimately these will be seen to be united by the concept of liquid crystals. The study of lyotropic, biological and polymeric liquid crystals is of profound importance for understanding physical processes in biology, and in addition such systems are providing new materials with remarkable properties. Research will increasingly focus on these areas in the future.

We have seen that the detection of *thermotropic* liquid crystals can be exactly fixed to the year 1888, when Friedrich Reinitzer observed 'two melting points' on heating derivatives of cholesterol [see paper A1]. It was the pioneering idea of Otto Lehmann to interpret Reinitzer's observation in terms of the existence of materials combining the fluidity of liquids with the anisotropy of solid crystals, which he named 'flowing crystals' [see paper A2]. However, results from the field of life science obtained in the middle of the nineteenth century revealed observations in biological systems which unquestionably indicate liquid crystalline properties. In 1854 the famous German pathologist Rudolf Virchow described his microscopic observations of the behaviour of nerve fibres in water.[1] He named the resulting structures *myelins*, an example of which is given in Fig. E1 (a wooden printing block from Lehmann[4], 1904).

Only a few years later, Ch. Freiherr von Mettenheimer (1824–98), using polarising microscopy, found that the myelin forms[2] were optically birefringent. Mettenheimer, who became a medical doctor in Frankfurt/Main in 1847, subsequently worked as an ophthamologist and also as the physician to the philosopher Arthur Schopenhauer. Of course at that time nobody was aware of the liquid crystalline nature of the myelins. In fact the complicated structure of myelins was not understood until 1977, when studies by Alfred Saupe, whom we have already met, were published.[3]

Fig. E1 Myelin observed as early as 1854 by R. Virchow[1] (Otto Lehmann, Flüssige Kristalle, Leipzig 1904, p. 254, Fig. 468).

Extensive studies on myelin were carried out by Georg Quincke, Professor of Physics at Berlin (see Section A). He worked on interfaces and growth processes of highly dispersed systems, and in 1894 published a detailed paper[5] entitled: 'On the spontaneous formation of hollow bubbles, foam and myelin forms by alkali oleates and related phenomena, especially of protoplasma'. He produced birefringent myelin-like structures from oleic acid, ammonia and water and interpreted his observations in terms of solid soap crystals included in tubes of thin oleic acid films. These he likened to the diffusion tubes formed by deposited membranes, as in the plant-like structures that grow when water glass (silicic acid) is put into water. In this paper Quincke also explained Lehmann's observations on Reinitzer's cholesteryl derivatives in the same way, i.e. that the birefringent structures observed were solid crystals coated by films of liquids. Lehmann repeated and extended Quincke's experiments on myelin, and showed that the fibres have anisotropic cores, and must be non-equilibrium structures, since they were of limited stability. One year after Quincke's paper, Lehmann published his results[6] entitled 'On contact movements and myelin forms'. In his famous monograph on liquid crystals, published[7] in 1904, Lehmann clearly established the connection between myelin fibres and '*flüssige Kristalle*':

> Following Quincke, the myelin-like structures were prepared by surrounding rigid soap crystals with a thin film of oleic acid. Water diffusion through this film causes the included volume to increase, and the oleic acid bubble grows, forming e. g. cylindrical tubes. I repeated Quincke's experiments, and it turned out that the myelin forms are nothing but flowing crystals. Because of differences in the surface tension at different areas, they had been distorted into strange forms similar in appearance to organic shapes.

The relation to living systems may be seen in Fig. E2, an original photograph of Otto Lehmann, showing worm-like liquid crystalline structures.

As an example of myelin formation, we reproduce in Fig. E3 an illustration from Lehmann's book showing the growth of myelin forms into a water phase from oleic acid and ammonium oleate.

In this particular system three components are necessary for myelin-like structures to form, although sometimes, for example in the lecithin-water system, only two components are sufficient. From his studies Lehmann had identified these myelin-like systems as a special type of liquid crystal nowadays known as 'lyotropic'. However, as we indicated in Section B (p. 147), Daniel Vorländer recorded strong reservations about Lehmann's interpretation.

At the beginning of the twentieth century, studies of liquid crystals provided a new perspective to the ongoing debate between scientists and philosophers over the origin of life and spontaneous generation (*Urzeugung*). In 1906 Otto

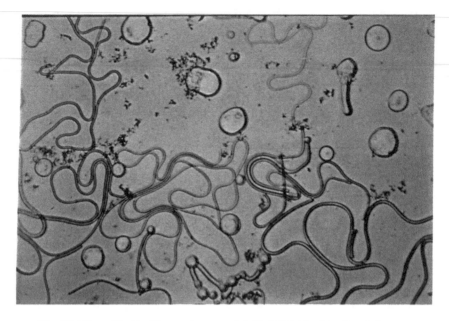

Fig. E2 Worm-like liquid crystal texture, labelled 'Trichites' by Otto Lehmann.

Lehmann published a two part paper in Biologisches Zentralblatt[8] on what he called *apparently living crystals*, where, in addition to ammonium oleate, he mentioned:

> ... lecithin (isolated from eggs and obtained from E. Merck, Chemische Fabrik, Darmstadt) is a component of protoplasma. When combined with water, lecithin forms a myelin structure, which develops in some respects like the growth of a living organism.

Lehmann[8] regarded myelin forms as 'artificial cells with liquid-crystalline membranes', and regarding Quincke's investigations,[5] he attributed a liquid-crystalline structure to all semi-permeable membranes. Nowadays, the important role of myelin in living systems is rather well understood, although it took more than 100 years after their first observation to reach this understanding. Nerve fibres are coated by myelin sheaths, and these in turn permit nerve impulses to travel at relatively high speeds of about 100 meters per second. Thus the fast response of the human brain relies crucially on myelin. Lehmann suggested[8] that the force of muscle contraction could be explained by the contraction forces which appear during enantiotropic transformation between polymorphic modifications.

In 1906 Otto Lehmann gave a lecture at the 78th *Tagung der Gesellschaft Deutscher Naturforscher und Ärzte* (Meeting of the Society of German Natural

564

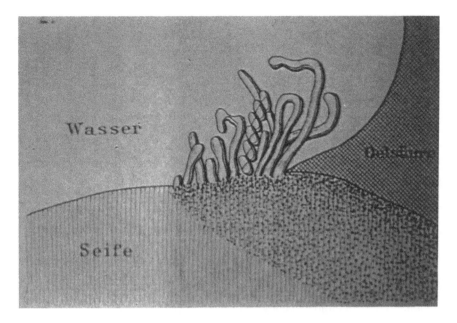

Fig. E3 Myelins growing into the water phase from oleic acid and soap (Otto Lehmann, Flüssige Kristalle, Leipzig 1904, p. 255, Fig. 471).

Scientists and Physicians) in Stuttgart entitled 'Liquid and apparently living crystals'. The manuscript was later published as a monograph, but now with the new and more provoking title '*Flüssige Kristalle und Theorien des Lebens*' (Liquid Crystals and Theories of Life). In this work, he discussed the monistic doctrine propagated by the famous natural philosopher Ernst Haeckel (1834–1919). We mentioned this briefly in Section A as a possible impediment to the acceptance of Lehmann's ideas, but it must be emphasised that Lehmann did not take a clear-cut position on monism.[10]

Haeckel had been Professor of Zoology at the University of Jena since 1862, and had been a childhood friend of the physicist Georg Quincke. In Haeckel's theory, atoms in living systems are endowed with 'atomic souls', which interact just so long as the entity under consideration remains alive. The death of any given living system involves a breakdown of the interaction between these atomic souls. Lehmann rejected the existence of atomic souls in inorganic compounds, e.g. solid crystals, and took a negative view of spontaneous generation because 'relations between atomic souls cannot occur spontaneously'. Ernst Haeckel, however, believed Otto Lehmann's liquid crystals to be a linking element between inorganic and organic materials, and emphasised this point of view in the course of his intensive correspondence with Lehmann from 1917 to 1919 (see also Haeckel's book *Crystal souls* mentioned in Section A). Haeckel considered the vigorous and permanent spontaneous motions of myelin forms, which he named 'life phenomena of rheocrystals', as strong evidence for his ideas. Haeckel concluded:

> By objective critical comparison of spherical myelin forms with simple probiontics (Chroococcus), the traditional and artificial separation between inorganic and organic nature is finally removed.

On the other hand, Lehmann[9] was at pains to point out that his

> apparently living crystals are obviously not to be understood as real living beings.

but he was hopeful that his investigations on liquid crystals would help the understanding of the interaction between forces in living organisms.

Although many had dallied with the notion of biological liquid crystals, it was not until the appearance in 1936 of the paper by Bawden, Pirie, Bernal and Fankuchen, 'Liquid Crystalline Substances from Virus-infected Plants', that liquid crystals in biology were given a firm basis. This highly significant paper on living systems is included in our collection as the paper E2. In a certain sense a virus may be regarded as a link between living and inorganic systems, because it is well known that virus materials can be obtained in crystalline forms. We recall Haeckel's vision of 'living crystals' mentioned above, a controversial discussion revived by the properties of viruses. The authors of paper E2 were a distinguished team from Cambridge, which included both virologists and crystallographers. Such interdisciplinary teams were rare at the time, but perhaps came about in this instance because Pirie and Bernal had formed a friendship through the Cambridge Communist Party. They isolated a protein from the sap of tobacco plants infected by the so-called tobacco mosaic virus (TMV). This protein does not occur in healthy tobacco saps and consequently must come from the virus. The Cambridge team demonstrated that neutral aqueous solutions of this protein exhibit liquid crystalline properties. They found that there are three strains of the TMV which show the same physical properties in aqueous solutions, namely birefringence, anisotropy of flow, orientation by electric fields, all properties typical of liquid crystals. From X-ray investigations on dry gels obtained from the aqueous solutions, the authors derived a rod-like molecular shape for the virus proteins, which then organise in solution to give a liquid crystal. Unfortunately the authors were not able to identify the mesophase type. In TMV solutions there is no polydispersity, i.e. all virus particles are of the same length, and TMV solutions became a standard system to test theoretical models for the organisation of long rods in solution (see the paper E4 by Onsager).

Another group of materials, namely soaps and related compounds, had also been identified as liquid crystals by Lehmann, who wrote in his book '*Flüssige Kristalle*' that:

ordinary potash soap used in each household is nothing but an aggregate of flowing crystals.

Despite this, we have seen that it took nearly 30 years for Lehmann's view to become accepted. In the intervening period, there had been many studies of colloidal systems, such as sols and gels formed by amphiphilic substances, including soaps, in water. The focus of this work was to try to understand the apparently complex structures formed by soaps and similar compounds in water, and the relationship of these to liquid crystals was largely ignored. Of course soap had already been around for a long time. As long ago as 2500 BC, the Sumerians used soaps for washing textiles and for medical and cosmetic applications. The soaps were prepared by thermal reaction of oils and potash, and in the fourth century AD the term 'saponarius' was used to label people responsible for the preparation of soaps. By the nineteenth century, industrial soap production had developed, and soap boilers were already observing qualitative differences between isotropic solutions of soap in water and those at higher concentrations, which we now know to be liquid-crystalline. Furthermore they had detected two (lyotropic) mesophases having different viscosities and named them 'middle' and 'neat' soaps, but it was not until the middle of the twentieth century that the structures of these liquid crystal phases were finally established.

Before continuing the story of lyotropic liquid crystals, we must be reminded that there is a significant difference between thermotropic and lyotropic liquid crystals with respect to their local structure. We recall the strong arguments in the 1920s and 1930s (and recorded in Section B) over the existence of swarms in thermotropic mesophases. While such local aggregates do not contribute to thermotropic liquid crystals, for which the building blocks are individual molecules, they are important for the formation of lyotropic liquid crystals. Thus lyotropic mesophases form in a stepwise fashion, starting with microscopic organisation, via a mesoscopic structure, to a macroscopic assembly. The first process is the concentration-dependent self-organisation of amphiphilic molecules in solution to aggregates. This takes place above a certain molecular concentration, named the 'critical micelle concentration (cmc)'. The micelles form an isotropic solution but on increasing concentration a second process occurs, and eventually the aggregates (micelles) organise to form orientationally ordered systems, namely lyotropic liquid crystalline phases.

The earliest paper we reproduce (in part) in this section as E1 concerns studies of these organising systems, and it was published in 1912 by Richard Zsigmondy and his research student Wilhelm Bachmann. The authors considered the gels formed primarily by soaps, but also mentioned those formed by silicic acid, and gelatine. The new and pioneering technique for optical investigations of colloidal systems used in this work was ultramicroscopy. At that time, standard optical microscopy had too low a resolution to detect the very small colloid particles. However, Zsigmondy's striking idea was to use the Tyndall effect[11] for indirect visualisation of the particles by scattered light. His invention was recognised by the award of the Nobel prize for Chemistry in 1925.

The phases studied by Zsigmondy and co-workers were described by the authors in E1 as gels. At the time of this publication, the term 'gel' did not have a precise meaning in the contemporary literature, but the distinguishing feature of 'gel' phases was an increased viscosity. The results reported in E1 were for mixtures having soap concentrations of less than 15 per cent. We now know (for example see Fig. 1 in E5 for potassium palmitate,) that the samples studied by Zsigmondy were middle phases, having a hexagonal structure and characterised by an enhanced viscosity.

The main importance of paper E1 is its critical discussion of the different theories about the structure of the gel phases at that time. The earliest hypothesis was the micellar theory of Carl von Nägeli, Professor in Munich, reported in his book[12] entitled 'The Theory of fermentation – a contribution to molecular physiology'. He assumed that molecules aggregated to form assemblies for which he introduced the term 'micelles'. The micelles are separated by water but can organise to create a network, and so result in the formation of a gel. By observation in polarised light, von Nägeli detected birefringence for these micellar solutions, and as a consequence he proposed a crystal-like nature for the micelles:[13]

By means of polarised light the micelle turned out to be small birefringent crystals.

Another theory of gel formation was put forward[5] by Georg Quincke in 1894 (cf. also Ref. 8 in E1). He proposed that soap gels are foams consisting of aqueous liquid and separating surfaces of liquid oleic acid or solid soap lamellae linked to each other. Then in 1898, Bütschli derived his honeycomb theory, based on microscopic studies. These suggested that gels consist of a honeycomb structure of solid material, e.g. gelatine, filled with water (cf. Ref. 10 in E1). The ultramicroscopic investigations presented in paper E1 clearly support Nägeli's micellar theory but not the foam and honeycomb theories. An open question, however, remained at that time: were the gels crystalline or amorphous?

A large part of paper E1 concerns morphological studies of soap solutions and soap gels carried out using the ultramicroscope. These are results from the Ph.D thesis of W. Bachmann, who is a co-author, and whose thesis title constitutes a subtitle for the paper; we have omitted the very detailed experimental section of the paper. At the end of the paper, the authors mention that some of their data agree with the soap crystallisation temperatures reported in 1899 by F. Krafft.[14] In a concentration/temperature diagram of lyotropic systems the Krafft point (or Krafft temperature T_K) is the intersection of the cmc (critical micelle concentration) curve with the solubility curve. The term Krafft point is nowadays standard, and was first introduced for soap gels in the paper E1.

568

For the most part, scientists working in the field of colloids during the first decades of the twentieth century did not recognise (or ignored) the liquid crystalline character of their systems. An exception is Hans Zocher, who is also firmly established in liquid crystal history as a pioneer of the continuum theory of torsional elasticity in liquid crystals, eponymously referred to as the Frank–Oseen–Zocher theory. But Zocher's first encounter with anisotropic fluids seems to have been through an experimental study of vanadium oxide and iron oxide sols. This work,[15] entitled 'On formation of spontaneous structures in sols' was published in 1925 in volume 147 of the *Zeitschrift fur Anorganische Chemie*. In this Zocher pointed out for the first time the strong relation between the anisotropic structures he observed in these sols and those in thermotropic liquid crystals. Part of the first page of this paper is reproduced below, and our translation follows.

H. Zocher. Über freiwillige Strukturbildung in Solen.   91

Über freiwillige Strukturbildung in Solen.

(Eine neue Art anisotrop flüssiger Medien.)

Von H. ZOCHER.

Mit 3 Figuren im Text und 4 Tafeln.

Eines der interessantesten Probleme der Kolloidlehre ist das von der Reichweite der Kräfte, die von Phasengrenzen ausgehen (1). Diese Kräfte sind zweierlei Natur. Erstens handelt es sich um molekulare Wirkungen, die man meistens als chemischer Natur, als Wirkung von Valenzkräften anzusehen geneigt ist. Äußerungen dieser Kräfte sind die Oberflächenspannung und die Adsorption neutraler Molekeln, z. T. wohl auch der Ionen. Aber auch zwischen den Kolloidteilchen nimmt man derartige Kräfte an, und zwar als das treibende Moment bei der Koagulation (2). Die zweite Art von Kräften sind die gröber elektrischen, die die elektrokinetischen Potentiale, z. T. wohl auch die Adsorption der Ionen und die Stabilität der

---

One of the most interesting problems in colloid science is the range of forces extending from phase boundaries[1]. These forces are of two types. First, there are molecular effects, which one normally tends to think of as chemical in nature, resulting from the effects of valence forces. Examples of these forces are the surface tension, the adsorption of neutral molecules, as well as sometimes probably the adsorption of ions. These forces, however, are also believed to occur between colloid particles, as the particular factor which drives coagulation[2]. The second type of force is stronger and electrical in nature. This causes the electrokinetic potentials, and sometimes also the adsorption of ions and the stability of colloidal solutions. Now, molecular theoretical considerations indicate that the range of molecular forces, and hence also the thickness of interfaces and adsorption layers, cannot be much larger than molecular dimensions. However, by contrast, in order to understand a number of observations on extremely thin layers, on semipermeability, on the electrokinetic potentials, on coagulation etc., one must suppose the layers to be considerably thicker. The observations on spontaneous structure in sols reported in this paper demonstrate unambiguously the existence of long-range forces. These forces can be either attractive or repulsive.

Stable structures, not destroyed by molecular motions, can in general exist in sols. This may at first seem strange. However, the discovery of liquid crystals by LEHMANN has already demonstrated that stable structures are possible, even in pure single-phase media. Moreover, these structures can be very complex. It is evident that the BROWNIAN motion of the structure-forming particles in sols cannot be very strong. Moreover, in sufficiently concentrated sols with higher viscosity this is absolutely the case. Thus, it is known that the BROWNIAN motion in gelatine solutions is rather small. In vanadium pentoxide sols[3] this author has observed that both the streaming-induced parallel orientation of oblate colloid particles, and the optical anisotropy, vanish increasingly slowly for older and more concentrated sols. In aged sols (about 1%) these properties persist; there is no longer any disorientation. In the same system Szegvari[4] demonstrated that there remains a very weak BROWNIAN motion consisting of oscillations around equilibrium positions. The existence of stable structures is probably the result of this behaviour.

[1] H. Freundlich Kapillarchemie, Leipzig 1922, p.419 f.
[2] R. Zsigmondy, Kolloidchemie, Leipzig 1918, p.69.
[3] H. Zocher, Z. phys. Chem. **98** (1922), 293.
[4] A. Szegvari, Z. phys. Chem. **112** (1924), 295.

In the following section of the paper, Zocher gives experimental details of the optical anisotropy observed in the polarising microscope of vanadium pentoxide and hydrated iron oxide sols. As an example we reproduce in Fig. E4 a microphoto of vanadyl pentoxide sol between crossed Nicols.

Zocher described the Brownian motion observed in vanadium pentoxide sols as follows:

> Rotating an anisotropic area into the extinction position, one observes a vigorous flickering. This is caused by a back and forth rotation of the anisotropic rods out of the extinction position. This is the same behaviour as observed by Friedel (paper B1 ed.) for several mesomorphic substances (liquid crystals).

And as a consequence of his optical observations on the hydrated iron oxide sols, Zocher suggested a periodic layer arrangement of the disc-like colloid particles in layers:

> They exhibit a brilliant nearly metallic strong green iridescent lustre. ...Obviously, the layer period is just an integral multiple of the half wavelength of green light.

At the end of his paper, Zocher emphasises a strong relation between the behaviour of his sols and (thermotropic) liquid crystals. He notes:

> Finally, I draw attention to a remarkable analogy. The similarity of the permanently birefringent sols to liquid crystals — FRIEDEL's mesomorphic substances — is so great that at first sight they could easily be confused. It is particularly unexpected that each of the observed sol

Fig. E4 Microphoto of vanadyl pentoxide sol between crossed Nicols (H. Zocher, Z. anorg. Chem. **147**, 91 (1925).

structures (*i.e. vanadium pentoxide and hydrated iron oxide, ed.*) corresponds to one of the principal types of mesomorphic state classified by Friedel (paper B1; *ed.*). The nematic type is formed by molecules with a parallel orientation of their axes. In the smectic type the molecules not only have their axes oriented parallel to each other, but in addition they exhibit regular spacings in this axis direction. There is a complete analogy between the first type and the vanadium pentoxide structures, and between the second type and the hydrated iron oxide structures.

Zocher was clearly a versatile and talented researcher. Having established the link between certain colloidal systems and thermotropic liquid crystals, he moved on to more detailed investigations of thermotropic materials. Zocher's contributions in that area have already been discussed in Section B of this book, and one of his important papers [B7] was presented at the Faraday Discussion on Liquid Crystals held in London in 1933. This meeting has figured strongly in the history of thermotropic liquid crystals, but there were also papers on lyotropic systems related to soaps and living systems. We have already noted the paper[16] presented by Lawrence, entitled 'Lyotropic Mesomorphism'. In his eight page paper Lawrence discussed soap systems and emphasised:

> ...the confusion concerning the forms in which soap-water systems exist....

A few years earlier, J.W. McBain,[17] working at the University of Bristol, had established phase diagrams for various soap/water systems, and had observed that there were several hydrated phases between the isotropic solution and solid crystalline soap. Lawrence in his paper at the Faraday Discussion pointed out the problems of applying the phase rule to these systems, and, referring to the work of McBain, presented a list of possible phases.

Curd soap = lamellar crystals
Neat soap = smectic liquid crystal
Middle soap = also liquid crystalline
Isotropic liquid

In summary, Lawrence stated that the interest of his results:

> ...is not in the light that they throw upon liquid crystals generally, but rather in that the formation of liquid crystals from solution points to structure in the liquid from which they appear.

A further paper[18] on lyotropic systems by Friedrich Rinne from Freiburg im Breisgau was included in the Faraday Discussion, but it was not presented as Rinne died one month before the Meeting. The title of the paper was 'Investigations and Considerations Concerning Paracrystallinity', and was concerned with the lyotropic system bromo-phenanthrene sulphonic acid in water. Remember (paper D1), this lyotropic liquid crystal was subsequently proposed as a suitable material for a liquid crystal display. However our interest in Rinne's paper was his discussion of the nomenclature of liquid crystals. It was a pity that neither he, nor Friedel, who was terminally ill, were at the meeting to give voice to the arguments. There is no doubt that the discussion would have been lively, since Friedel was well known to have been intolerant of contradictory views, especially where *his* liquid crystals were concerned.

Rinne had strong objections to the terms 'liquid crystal' (due to Otto Lehmann), 'mesomorphism' (proposed by G. Friedel) and 'mesophase' (coined by Hans Zocher). Instead of these Rinne introduced the term 'Paracrystal' for systems having one- or two-dimensionally periodic structures instead of the three-dimensional space lattice in crystals. Because of the close relation of crystals and paracrystals he wanted to bring out this natural correspondence in their description, while noting the peculiarities in structural dimensions by the prefix 'para'. Thus in the lyotropic system he reported (bromo-phenanthrene sulphonic acid in water), he observed two liquid-crystalline phases which he named 'α-paracrystalline (nematic)' and 'β-paracrystalline (smectic)'. Rinne's paper is also of interest in the context of biological liquid crystals, and a section of the paper is devoted to 'Organic Paracrystals' (Bioparacrystals). Using microscopy and X-ray measurements he investigated 'Lipoid Drops' (following up Mettenheimer's observations on myelin forms in 1858 – see above), 'Sperms', 'Muscles' and 'Nerves' and presented some interesting photographs for these systems.

At this point in our story of lyotropic liquid crystals, i.e. in the mid-1930s, various lyotropic mesophases had been identified, and the connection with micellar solutions recognised, if not fully understood. A breakthrough was to come, as in many other areas of science, through the application of X-ray diffraction, and our next paper E3 from Joachim Stauff reports his X-ray experiments on soap solutions. However before describing Stauff's results, we should recall what had been proposed concerning the nature of micellar solutions. As early as 1879, von Nägeli had detected birefringence in micellar solutions. We have already encountered Carl von Nägeli in our discussion of paper E1, and in his book '*Theorie der Gärung*', he writes:[19]

> The micelles in solution are much less mobile than solute molecules. As a result they can associate with each other very easily. I name these aggregates 'Micellverbände' [micelle associates], and using this, several characteristic properties of micellar solutions can be explained.... Micelles are organised either in parallel layers or in spherical shells around a common centre or in cylinder casings around a common axis.

For gels such as those found in silicic acid solutions, he proposed a network structure formed by micelle chains with wide meshes in which water is included.

Although von Nägeli was far away from the concept of lyotropic liquid crystals he did anticipate that micelles can be organised in different structures, these we nowadays call lamellar or hexagonal phases.

Thus it was clear that lyotropic mesophases developed from the organisation of micelles. In the early 1930s it was generally accepted that amphiphilic molecules aggregate in solution above the so-called critical micelle concentration (cmc) to form micelles. However, the structure of micelles was still an open question at that time, and a matter of controversial discussion. As early as 1920 McBain[17] had derived a theory of soap solutions from measurements of conductivity and vapour pressure. This proposed the existence of two different types of micelle: one type carrying charges, and another uncharged which was essentially a colloidal particle. On the other hand, Hartley[20] assumed the existence of only one spherical type of micelle having paraffin chains in the core and adopting structures as in liquid paraffin. But others, such as Heß et al.,[21] asserted that the paraffin chains inside the micelle should be ordered as in crystalline soap. All this was imaginative speculation, and the matter could only be resolved through the application of the relatively new technique of X-ray diffraction to these organised fluids.

Paper E3 of our collection, by Joachim Stauff, reports the first systematic X-ray investigation on soap solutions. There is a detailed discussion of the different contemporary theories of micelle structure. Stauff concluded from his measurements, that there are two different types of micelle, the proportion of which depends on concentration. He named them '*Groß- und Kleinmizellen*' (Large and Small Micelles). His results indicated that Large Micelles were only present at high soap concentrations, and were formed by aggregation from Small Micelles, and not from single soap ions. He deduced that Large Micelles had a liquid-crystalline structure, analogous to that of smectic phases (see Fig. 3 in E3), with liquid-like paraffin chains. Stauff also observed Small Micelles in the concentration range that had been studied by Hartley,[20] and their structure seemed to be spherical, in agreement with Hartley's proposal. Soap solutions consisting of Large Micelles exhibited flow birefringence and probably had a layered structure; in modern nomenclature[22] they would be designated as 'lamellar phase L_{α}'. Stauff's work is typical of much research on colloidal systems, then and now, in that the close relation to liquid-crystalline phase behaviour has been either missed or ignored.

With the close of the 1930s decade, micellar (e.g. soaps and the like) and rod-like (TMV or inorganic sols) lyotropic liquid crystals had been identified, but there was a complete lack of theories to describe their structure. By contrast a mean-field theory of thermotropic liquid crystals had already been formulated (if not read!) by Grandjean (C1), and a nearly complete continuum theory was available from Oseen (B5) and Zocher. There had been attempts to describe the stability of lyophobic colloidal suspensions by Langmuir[23] and Hamaker[24], Derjaguin and Landau[25] and also later by Verwey and Overbeck[26] in 1948. The theory developed by Derjaguin and Landau and independently by Verwey and Overbeck has become known as DLVO theory. It considers the interaction between van der Waals attractive forces and electrical double-layer repulsion of charged particles, and has since been applied to micellar solutions.[27] However, theories for the ordering of non-spherical particles in solutions were not available. It was not until 1949 that Onsager published a theory, reproduced here as E4, describing the interaction of rigid colloidal particles, which enables the prediction of phase transitions from isotropic to anisotropic phases in lyotropic systems. However it seems likely that the work must have been completed at least seven years earlier, for one item in Onsager's bibliography is a 1942 reference to a paper in Physical Review. Onsager was notoriously wayward, perhaps even secretive, in his publication habits, and he may only have published his work in 1949 under pressure from colleagues who wished to use it.

Under the by-line *Proceedings of the American Physical Society*, the 1942 reference turns out to be the Minutes of a meeting of the New England section of the American Physical Society, which had taken place at Trinity College, Hartford, Connecticut, on 24 October 1942. There were a number of relatively anodyne invited talks at this meeting, headed by an address by H.G. Houghton of MIT entitled 'The training of meteorologists for the armed services'. As the war was raging at its fiercest at that point, no doubt that talk was received with interest by the 'at least forty five' members whom the minutes record as being in attendance. The minutes further record that there were six contributed papers, numbers 1 and 6 of which were submitted by Onsager, coming from Yale University in nearby New Haven.# With hindsight it is possible to realise that three out of the other four papers were indirectly concerned with atomic bomb work, although this would not have been evident at the time. Luckily the abstracts of the contributed papers are given in the brief report of the meeting. Onsager's first paper is entitled 'Anisotropic solutions of colloids'. As far as one can gather from the brief abstract, all the main results in the 1949 paper are included, including the ratio of concentrations at equilibrium and predicted osmotic pressure in the anisotropic phase.

In paper E4, Lars Onsager makes the first significant contribution to the theory of lyotropic liquid crystals. This work is a theoretical *tour de force*, significant both for its author's mathematical power, and for his ability to use concepts in theoretical physics. The article has several basic inputs. One is the classic Debye–Hückel theory for the distribution of ions near charged particles. This leads to a formula for the repulsive force between parallel plates, and Onsager quotes classic formulae for this. He then extends this, inexactly but perfectly adequately, to obtain a formula

---

Although it is only indirectly relevant to progress in liquid crystal physics, readers may be interested in Onsager's other paper, entitled *Crystal statistics*. Although its title conceals it, this paper nevertheless is one of the most important in the history of theoretical physics, for it reports on the exact evaluation of the partition function of the two-dimensional Ising model, and predicts a logarithmic singularity in the specific heat close to the "Curie point" (the quotation marks are Onsager's). This solution begins the study of critical phenomena, for it carries within it the seeds of the failure of the Landau–Ehrenfest classification of phase transitions. In the 1970s and 1980s a large part of the theoretical effort in the liquid crystal field would be concerned with the properties of phase transitions between liquid crystalline phases.

for the repulsion between cylindrical particles with given mutual inclination and distance of closest approach.

Having established this, he goes on to discuss the statistical mechanics of interacting particles. This is a very difficult problem. In the low-density limit, the pressure or free energy of such a system can be expanded in terms of a density expansion. The leading term is the perfect gas — the non-interacting particle — term. The subsequent terms, in increasing powers of the density or concentration, correct this. The expansion is known as a *virial expansion*, first used by Kamerlingh-Onnes,[28] but made systematic by Ursell and Mayer[29] in terms of so-called cluster integrals. These cluster integrals, which in the general case involve (Boltzmann-factor) weighted integrals over the possible positions of N nearest neighbour particles of a given particle, take into account the potential energy between particles in a systematic way. It was this expansion that Onsager used; after enormous mathematical effort, most of which was relegated to an appendix, he obtained a virial expansion for a system of stiff rod-like particles of given sizes and with given orientational probability distribution functions.

At this stage Onsager stealthily inserts the most important part of his calculation. The cluster integrals for the exact inter-particle potentials, including all the Debye–Hückel corrected electrostatic repulsions, are so difficult that any normal theoretical physicist would give up. However, by supposing that the second virial term is the most important, and by comparing the form of this term with the form of the electrostatic repulsion, Onsager was able to assert that the main effect of this repulsion is very short-range. To begin with this offends theoretical sensibilities, because we know that true electrostatic forces are long-range, and in particular sufficiently long-range that the cluster-expansion approach does not work. But the screened case is different, and localises the interaction. The double layers must not overlap, and the size of the double layers measures the degree of screening. In Onsager's own words there is 'padding' around the original rod-like molecules; essentially the ionic double layer causes the effective diameter of the particles to increase. Now all the cluster terms only depend on volume exclusion, and become possible (though not easy!) to calculate. Thus the theory now applies the effect of excluded volume alone to the organisation of long hard particles (length >> diameter) in solution; there are no attractive forces. Using the virial expansion Onsager was able to compute a free energy as a function of rod concentration for the anisotropic phase.

The problem remained to pick out the anisotropic solution, if there was one, and to determine at what rod concentration the transition from the isotropic to the anisotropic solution would occur. Onsager solves the variational problem of minimising the free energy with respect to the orientational distribution function by using a very unpleasant but just tractable trial function for this function. Finally he is able to compute a free energy as a function of rod concentration for the anisotropic phase, and it turned out that rigid rod solutions form an anisotropic (nematic) phase if the ratio of the total co-volume to the solution volume exceeds a critical value. At this point the problem is solved, although Onsager

576

never uses the term order parameter, apparently being unfamiliar with Tsvetkov's work (paper C2), which had been published after he derived the result. Thus he does not comment that at the transition the order parameter $\overline{P}_2$ jumps from a value of zero to 0.84 (i.e. almost complete order). As rigid rods represent an athermal system there is no temperature dependence of the results. Furthermore the theory does not apply to anisotropic phases of polydisperse solutions because this would require different angular distribution functions for each particle.

As formulated, Onsager's theory was restricted to dilute solutions and rigid particles. Many developments of the theory have appeared since his original formulation, and these have been excellently reviewed by Vroege and Lekkerkerker.[30] A significant early reformulation of Onsager's theory was given by Isihara[31] in 1951, and later a key paper on the application to flexible chains was published by Khokhlov and Semenov.[32] At the time that Onsager published his seminal work, suitable experimental data to test the theory was not available, but eventually nematic TMV solutions provided a test-bed for his model.[33,34]

Another approach to the theory of the organisation of rigid particles in solution was developed by Paul Flory in 1956 using a lattice model. Variants of the lattice model had been used for many years to describe a variety of phenomena in condensed matter physics.[35–37] In general, the advantage of lattice model is that it simplifies the effects of interparticle interaction sufficiently so that it is possible to construct a physically sensible fluid theory even beyond the low density or (in the case of a solution) the low dilution limit. For hard anisotropic fluids Onsager's model only applies in this limit, because only the second virial term is included. Flory was a powerful exponent of the lattice method, and applied it in the early 1940s to a variety of polymer fluid problems. In paper E5, entitled 'Phase equilibria in solutions of rod-like particles', he extended lattice theory ideas to deal with athermal rod-like molecules. The athermality is important, in that particular molecular conformations are either *allowed* or *forbidden*, and this simplifies the counting of states that is at the heart of the lattice model. This state-counting was what made the lattice method tractable when it was not yet possible to treat analogous continuous systems.

In his theory, Flory imagined the rod-like particles to be divided into segments which were located at the points of a lattice. The theory predicts a temperature-independent phase transition from a disordered to an ordered phase at concentrations larger than those found by Onsager. Flory was well aware that his theory is the lattice equivalent of the Onsager problem, and indeed compares his results with the Onsager result.*

Flory's model can also be applied to highly ordered phases of concentrated solutions. It is interesting to note that the paper[38] which precedes E5 in the Proceedings of the Royal Society is also by Flory. In that paper, Flory discusses a

---

* A comparison of the Onsager and Flory theories appears in the seminal text by P.-G. de Gennes and J. Prost, *The Physics of Liquid Crystals* (2nd Edn, Oxford, 1993) pp. 59–66.

lattice model of the of semi-flexible chain molecules, in which local orientational correlations play an important role. With the benefit of hindsight, such a model might also describe a liquid crystalline phase, but in the paper Flory concentrated more on how crystallisation might occur in an irregular polymeric fluid.

In the years that followed, new liquid crystalline rigid rod systems became available with which to test the theories. As early as 1956, it had been found by Conmar Robinson[39] that solutions of poly-γ-benzyl-L-glutamate (PBLG) in organic solvents organise to a helical structure, and form (lyotropic) cholesteric (chiral nematic) phases with exceptional optical properties. Ten years later the same author published[40] a careful study of the concentration dependence of the mesophase/isotropic phase equilibrium as a function of the axial ratio of the PBLG molecules. The results are compared with Flory's predictions, and showed the correct trends, but

...a considerable difference from Flory's calculated values.

Now let us briefly summarise the situation around the late 1950s as it related to lyotropic liquid crystals. Sophisticated theories of the organisation of rigid rods in solution had been devised, but had not been tested. Certainly liquid crystal behaviour in suitable systems such as TMV solutions and inorganic sols had been seen, but not sufficiently characterised to provide suitable data with which to test the theories. The theoretical approaches developed by Onsager [E4], Isihara[31], and Flory [E5] were for rigid particles, and, at least in their original form, could not be applied to describe the behaviour of flexible micelles in solution or lyotropic mesophases. A variety of mesophases had been identified in these soap-like systems, but their structures had not been firmly established, and there were no quantitative theories.

The next major step forward in the story is the unequivocal determination of the structures of the lyotropic mesophases. We recall that the so-called *middle phase* and the *neat phase* had been observed, depending on concentration, in aqueous soap solutions. In 1948 Doscher and Vold[41] had proposed a model of the neat phase in which layers of soap molecules are bound to each other by water molecules. They also suggested that the middle phase had a lamellar structure. In the same year, Marsden and McBain[42] carried out X-ray investigations of aqueous solutions of dodecyl sulfonic acid and triethanolamine laurate which exhibit the same phase sequence as simple soap solutions. For the neat phase they derived a 'lamellar non-expanded phase', and the middle phase was suggested to exhibit a hexagonal lattice of long fibres separated by water. Finally in 1960 much more detailed and convincing X-ray results on the structure of soap-like mesophases were published by Luzzati, Mustacchi, Skoulios and Husson from the University of Strasbourg; their paper is reproduced here as paper E6.

The importance of the paper E6 is that it established for the first time models for the 'classical' middle and neat phases of soap/water systems. Luzzati *et al.* derived from their X-ray data on the middle phase, a structure consisting of an hexagonal array of cylinders of indefinite length, separated by water (see Fig. 3a of paper E6). This phase is now termed[22] the *hexagonal phase H_1*. For the neat phase they derived a structure consisting of a stack of parallel and equidistant flat layers: this is now named the *lamellar phase* L_α (see Fig. 3b in E6). For both phases the authors deduced that the hydrocarbon chains had a chaotic distribution of conformations, similar to a liquid hydrocarbon. This they demonstrated by comparing the X-ray scattering of liquid tetradecane with that from sodium myristate /water (see their Fig. 2). In paper E6 the work of Stauff (E3) is dismissed as being only of historical interest, and he is criticised for missing the different structures of the middle and neat phases. However Stauff had not discussed the phase behaviour of his soap solutions, and so perhaps can be forgiven for not identifying the hexagonal middle phase. Undoubtedly his preliminary studies gave a clear indication of the bilayer structures in soap solutions, which through self-organisation result in the wide variety of lyotropic mesophases.

As well as establishing the structures of the middle and neat phases, the Strasbourg group discovered four new lyotropic mesophases. Careful X-ray measurements on their soap solutions as functions of concentration and temperature, revealed four intermediate liquid crystalline phases between the regions of the neat and middle phases. This important observation contradicted previous suggestions from other authors (e.g. Vold[43]), who had described the intermediate regions as a mixture of neat and middle phases.

The new phases were described as follows: (i) A *deformed middle phase* in which the hexagonal array becomes orthorhombic with elliptically deformed cylinders. (ii) A *rectangular phase* the model of which is given in Fig. 3c of their paper. (iii) A *Complex hexagonal phase* consisting of hollow cylinders with water inside and hydrocarbon chains on the outside.(see Fig. 3d of E6). (iv) Finally a *cubic phase* was discovered, the structure of which was derived by the authors as a face-centred cubic lattice. In all of the new phases, the paraffin chains were found to be liquid-like. Luzzati and his co-workers had identified and determined the structures for a range of different lyotropic mesophases in soap/water systems, which eventually would have their analogues in thermotropic liquid crystals. There was however one notable exception – no lyotropic nematic mesophase, and it was not until 1967 that this was detected.[44]

The serious problem in carrying out the X-ray investigations reported in E5 was the difficulty of obtaining samples with only a single phase in the intermediate region instead of two coexisting phases. However using a skilful procedure, Luzzati *et al.* were able to identify the diffraction patterns of each phase separately. The paper also reports typical optical textures observed in the polarising microscope for the neat (*lamellar*) and middle (*hexagonal*) phases which permits an identification of these phases. However, the textures of the intermediate phases (except the cubic phase which of course was optically isotropic) are similar to that of the neat phase and were not different enough to permit identification (see Fig. 4–6 in paper E5).

In concluding this section on more complex types of liquid crystal, we turn our attention to polymers. Although the idea of polymers or macromolecular structures had been established in the 1920s and 1930s, in particular by the Nobel Prize Winner Hermann Staudinger, the detection of thermotropic liquid crystalline phases in polymer melts was not straight forward. There was a vague hint from Vorländer[45] in 1936 that liquid crystalline properties may be exhibited by polymers, since he reported high-molecular weight diazo-amino compounds and linear polyazobenzenes having liquid crystalline-like properties accompanied by decomposition. He wrote:

> (In the case of para-disubstituted benzene derivatives)...one can study the change of molecular properties if one lengthens the molecules in one dimension. Will we perhaps arrive at a new state...(which might be called)...supracrystalline-liquid?

Polymers are formed either naturally or artificially by the combination of smaller molecular units, usually in a regular sequence. If the molecular units, or monomers as they are sometimes called, are joined end to end, then the resulting polymeric molecule is quite properly described as a chain molecule. We have already discussed Flory's contributions (represented by paper E5) to the theory of lyotropic phase formation, but the paper[38] which preceded E5 in the issue of the Proceedings of the Royal Society was entitled 'Statistical thermodynamics of semi-flexible chain molecules'. Since this paper provided the mathematical basis for Flory's theory of 'Phase equilibria in solutions of rod-like particles', it might be argued that Flory had anticipated the formation of main chain polymer liquid crystals. He predicts a first order transition between the disordered (isotropic liquid) state, to an ordered state. In the former disordered state the polymer segments deviate from a linear configuration, while in the ordered state the polymer segments within a single chain are parallel to each other (i.e. rod-like), and the chains themselves are also parallel. This sounds like a liquid crystal phase, but Flory did not refer specifically to such a mesophase, and was more interested in the subsequent crystallisation process.

Basically, two different types of liquid crystalline structures are possible in polymers. Monomer units that promote liquid crystal behaviour can be part of the polymer chains, forming so-called *main chain liquid crystal polymers*, or they can be laterally attached to the backbone chain of the polymer, resulting in *side chain liquid crystal polymers*. First attempts to synthesise liquid crystalline main chain polymers started at the beginning of the 1970s. In 1972 Yeh,[46] from diffraction and microscopic studies on polymers like polystyrene, had obtained strong evidence that there is:

> ...a liquid-crystalline type of chain-packing order.

It was observed that fibres spun from anisotropic polymer melts or concentrated solutions possess outstanding mechanical properties, in particular high tensile strengths.[47] A well-known example is a polyamide fibre with the trade name

Kevlar, for which fibres are obtained by spinning from a nematic solution of the polymer in concentrated sulphuric acid. In 1975 Roviello and Sirigu published a paper on 'Mesophase structures in polymers'. This contains the first convincing proof for liquid crystalline phases in melts of polymeric esters, and must be taken as the pioneering publication on liquid crystalline main chain polymers. The paper is reproduced here as E7. The authors measured a considerable heat of transition for the phase transformation into the isotropic melt, and by polarising microscopy they detected a Schlieren texture (Fig. 2 in E7), due either to nematic or smectic C phases. Shortly after, Jackson and Kuhfuss[48] published a paper in which they reported the synthesis and mechanical properties of co-polymerised esters for injection moulding. They did not give any details of the liquid crystalline behaviour of their materials, but only at the very end of their publication do they claim that:

> ...these polyesters are the *first* thermotropic liquid crystal polymers to be recognised.

It seems that they were not aware of the paper by Roviello and Sirigu, although the actual timing of the discovery of liquid crystal polymers is obscured by key patents[47,49] which had been filed in the early 1970s.

Early attempts to prepare liquid crystalline side chain polymers focussed on the polymerisation of anisotropically ordered molecules (e.g. Tanaka *et al*.[50]). Thus the starting materials were nematic phases of compounds in which the mesogenic unit is directly connected to the polymerisable group. In most cases only amorphous polymers were obtained, although an exception due to Perplies *et al*.[51] was a liquid crystal polymer obtained by polymerisation of acryl-Schiff's bases in the nematic state. Those authors observed a polymethacrylate ester having a smectic A phase, but the nematic phase was lost during polymerisation. X-ray investigations showed that the polymer main chains are packed in between the smectic layers.

The reason why most attempts to obtain liquid crystalline side chain polymers failed is that in all cases the molecular group that promoted liquid crystallinity was directly attached laterally to the polymer main chain. Thus, a statistical distribution of chain conformations prevented the orientational ordering of these liquid crystal groups. The problem was solved by a simple idea first introduced by Finkelmann *et al*. in 1977 at an American Chemical Society symposium in Chicago.[52] The full paper was published in *Makromolekulare Chemie* in 1978, and is reproduced here as E8. The idea was to decouple the mesogenic side groups from the polymer backbone by introducing flexible spacers. The authors used alkyl chains as spacers with two to twelve methyl groups, and in this way the polymer main chain could remain in a liquid state having a statistical distribution of chain conformations, while the mesogenic side groups could become ordered. This novel concept suddenly opened up a number of possibilities for the synthesis of many liquid crystalline polymers having nematic and smectic phases depending on the spacer length. It was also possible to synthesise cholesteric (chiral nematic) copolymers[52] exhibiting selective reflection bands in the visible, like the well-known cholesteric phases of monomers (see also references[53,54]). The spacer concept even allowed the mesogenic groups to adopt a cubic lattice structure, and so led to the synthesis[55,56] of liquid crystalline side-chain polymers with blue phases.

By analogy with lyotropic monomers, lyotropic polymers are also amphiphilic; they possess hydrophilic head groups and hydrophobic tails. There are lyotropic main chain polymers formed by head to tail cross-linking of suitable amphiphilic monomers as well as lyotropic side chain polymers formed either by tail to tail or head to head cross-linking. Additionally, block copolymers exhibit lyotropic liquid crystalline phases. As noted above, liquid crystallinity in lyotropic polymer solutions had already been observed in 1956 by Robinson[39] for PBLG in organic solvents, while aqueous solutions of the natural biopolymer TMV had also been shown to be liquid crystalline. There are a few papers dealing with lyotropic polymers in the early 1960s[57,58] describing hexagonal and lamellar phases. A more detailed investigation on block copolymers in several organic solvents was carried out by Douy *et al.*[59] in 1969 who observed lyotropic lamellar, hexagonal and cubic phases independent of the nature of the solvent. However, systematic studies of lyotropic liquid crystalline polymers began in the 1980s with Finkelmann's and Rehage's investigations[60] on the relationship between molecular architecture of amphiphilic polymers and their lyotropic phases. Similar investigations were carried out by many other groups, and extend up to the present.[61]

In this section on the history of liquid crystals we have referred to those important papers which have strongly contributed to our understanding of lyotropic and polymer liquid crystals. But we have also highlighted ideas which have survived and stimulated investigations and developments up to the present time. In the context of polymer liquid crystals, we draw attention to liquid crystalline elastomers, which carry great promise for many applications. Elastomers are cross-linked polymer networks (e.g. rubber) with remarkable mechanical properties. By introducing liquid crystallinity into elastomers, it is possible to create materials having the anisotropic properties of liquid crystals, which can be changed by small mechanical forces. Clearly these changes enter through the elastic properties of liquid crystals, and as we have already seen in Sections B and C an understanding of their elastic behaviour was at the core of theories for liquid crystals. In a certain sense de Gennes in his paper[62] published in 1975 predicted the possible existence of nematic elastomers, and the first liquid crystalline elastomer was described in 1981 by Finkelmann *et al.*[63] Several liquid crystalline phases such as nematic, cholesteric, smectic A and smectic C* as were identified in these elastomers, and it was found that they could be oriented by external mechanical forces. Furthermore mechanical strains can cause a significant shift of the phase transition temperatures, such as that between the nematic and isotropic phases.[64] Since the discovery of elastomers having liquid crystalline order, there has been a considerable interest in these materials, and this is currently an important area of research.

Like many stories of fantasy, our narrative finishes with the prospect of future exciting discoveries and new adventures for the characters. Except that ours has been a story of reality, the characters have been real people, and their adventurous discoveries have contributed to improving the lives of everyone. The progress of science as recorded in the published literature is largely impenetrable to all but the dedicated, focussed and, almost by definition, boring expert, but we hope that the story we have recounted here of liquid crystals, or *crystals that flow*, shows that science is a supremely human endeavour.

References

1. R. Virchow, *Ueber das ausgebreitete Vorkommen einer dem Nervenmark analogen Substanz in den thierischen Geweben*, Archiv für pathologische Anatomie und Physiologie und für klinische Medizin, **6**, 562–572 (1854); Rudolf Ludwig Carl Virchow (1821–1902) was a doctor, pathologist, anthropologist and later became a politician, when he founded the Prussian Progressive Party in opposition to Bismarck. Virchow's fame in anthropology results from his misidentification of Neanderthal skeletons as the bodies of Napoleonic soldiers.

2. C. Mettenheimer, *Zwei Mittheilungen über das Myelin* (Scientific notes), Correspondenz-Blatt des Vereins für gemeinschaftliche Arbeiten zur Förderung der wissenschaftlichen Heilkunde **31**, 467–471 (1858); **24**, 331 (1857).

3. A. Saupe, *Textures, deformations and structural order of liquid crystals*, J. Colloid and Interface Sci. **58**, 549–58 (1977).

4. O. Lehmann, *Molecülphysik I*, 1888, p. 522.

5. G. Quincke, *Über die freiwillige Bildung von hohlen Blasen, Schaum und Myelinformen durch ölsaure Alkalien und verwandte Erscheinungen, besonders des Protoplasmas*, Ann. Phys. Chem. **53**, 593–631 (1894).

6. O. Lehmann, *Über Contactbewegung und Myelinformer*, Ann. Phys. Chem. **56**, 771–88 (1895).

7. O. Lehmann, *Flüssige Kristalle* (Wilhelm Engelmann, Leipzig, 1904), p. 254.

8. O. Lehmann, *Scheinbar lebende Kristalle, Pseudopodien, Cilien und Muskeln*, Biolog. Zentralblatt **28**, 481–9, 513–26 (1906). The idea of liquid crystal phase transitions being responsible for muscle activity proved not to be true. Artificial muscles can probably be constructed using liquid crystalline elastomers, (see e.g. H. Wermter and H. Finkelmann, e-Polymers, No.013, 1–13 (2001).

9. O. Lehmann, *Flüssige und scheinbar lebende Kristalle* , Physikal. Zeitschr. **7**, 789–93 (1906).

10. Monism (from Greek *monos* = unit) is a philosophical doctrine which denies any separation between mind and matter.

11. The Tyndall effect is the scattering of light by particles which are too small to be visible to the naked eye. Depending on the nature of the particle, the wavelength of light preferentially scattered may depend on the particle size.

12. C. von Nägeli, *Theorie der Gärung – Ein Beitrag zur Molekularphysiologie* (R. Oldenbourg, München, 1879).

13. It is now clear that, given the optical microscopes available to him, von Nägeli could not possibly have observed individual micelles. Rather, what he did see was the birefringence exhibited by the micellar solutions. Suggestions of a liquid-crystalline

structure for micelles were made in 1939 by Stauff (see paper E3, Fig. 3) and by Hartley, see e.g. G.S. Hartley, Kolloid-Zeitsch. **22**, 88 (1939).

14. F. Krafft *Ueber die Kristallisationsbedingungen colloidaler Salzlösungen*, Ber. Dt. Chem. Ges. **32**, 1596–608 (1899).

15. H. Zocher, *Über freiwillige Strukturbildung in Solen*, Z. anorgan. Chem. **147**, 91 (1925).

16. A.S.C. Lawrence, *Lyotropic mesomorphism*, Trans. Far. Soc. **29**, 1008–15 (1933). Although only 30 at the time of the Faraday Discussion, Stuart Lawrence had already written a popular book entitled 'Soap Films', which was to earn him the life-long nickname of 'Soapy' Lawrence. The early part of his career had been as research assistant to Sir James Dewar and Sir William Bragg at the Royal Institution, London, and we have already speculated on the latter's role as '*eminence grise*' in promoting the subject of liquid crystals in the English-speaking world through the Faraday Discussion in 1933. At the time of this Meeting, Lawrence had moved to Cambridge as assistant to Professor (later Sir) Eric Rideal in the newly formed Department of Colloid Chemistry. During the war years Lawrence was employed as scientist in the Royal Navy, where among other tasks he dealt with the problem of de-icing warships. Leaving the Royal Navy in 1946 as Commander, Lawrence returned to a distinguished academic life in the Chemistry Department of the University of Sheffield, where he remained until his death in 1971.

17. J.W. McBain and C.S. Salmon, *Colloidal electrolytes. Soap solutions and their constitution*, J. Amer. Chem. Soc. **42**, 426–60 (1920).

18. F. Rinne, *Investigations and considerations concerning paracrystallinity*, Trans. Far. Soc. **29**, 1016–31 (1933). Despite the crusading efforts of Rinne's former student, R. Hosemann, who vainly struggled for its adoption right up to the end of the twentieth century, the *paracrystallinity* nomenclature introduced in Rinne's paper was never generally accepted.

19. C. von Nägeli, op. cit. pp. 100, 102.

20. G.S. Hartley, *Aqueous Solutions of Paraffin Chain Salts* (Paris, 1937).

21. K. Heß, W. Philippoff and H. Kiessig, *Viskositätsbestimmungen, Dichtemessungen und Röntgenuntersuchungen an Seifenlösugen*, Kolloid-Zeitschr. **88**, 40–51 (1939).

22. See: K. Hiltrop, in: H. Stegemeyer (ed.), Topics in Physical Chemistry, Vol. 3, Liquid Crystals, Darmstadt, New York 1994, p. 143.

23. I. Langmuir, *The role of attractive and repulsive forces in the formation of tactoids, thixotropic gels, protein crystals and co-acervates*, J. Chem. Phys. **6**, 873–96 (1938).

24. See e.g. H.C. Hamaker, Physica **4**, 1058 (1937); E.J.W. Verwey and H.C. Hamaker, *The role of the forces between the particles in electrodeposition and other phenomena*, Trans. Faraday Soc. **36**, 180–6 (1940).

25. B. Derjaguin and L. Landau, Acta Physicochem. USSR **14**, 633 (1941), B.V. Derjaguin, *On the repulsive forces between charged colloid particles and on the theory of slow coagulation and the stability of lyophobic sols*, Trans. Faraday Soc. **36**, 203–15 (1940).

26. E.J.W. Verwey and J.T.G. Overbeek, *Theory of stability of lyophobic colloids*, Elsevier, Amsterdam 1948.

27. J.N. Israelachvili, *Intermolecular and surface forces* (Academic Press, London, 1985).

28. H. Kamerlingh Onnes, Proc. Sec. Sci. Kon. Ned. Akad. Weten. Amsterdam **4**, 125 (1901).

29. J.E. Mayer and M. Goeppert-Mayer, *Statistical Mechanics* (Wiley, New York, 1940).

30. G.J. Vroege and H.N.W. Lekkerkerker, *Phase transitions in lyotropic, colloidal and polymer liquid crystals*, Rep. Prog. Phys. **55**, 1241–309 (1992).

31. A. Isihara, *Theory of anisotropic colloidal solutions*, J. Chem. Phys. **19**, 1142–7 (1951).

32. A.R. Khokhlov and A.N. Semenov, *Liquid crystalline ordering in the solution of partially flexible macromolecules*, Physica **A112**, 605–14 (1982).

33. S.D. Lee and R.B. Meyer, *Computations of the phase equilibrium, elastic constants and viscosities of a hard-rod nematic liquid crystal*, J. Chem. Phys. **84**, 3443–8 (1986).

34. S. Fraden, G. Maret, D.L.D. Caspar and R.B. Meyer, *Isotropic-nematic phase transition and angular correlations in isotropic suspensions of Tobacco Mosaic Virus*, Phys. Rev. Lett. **63**, 2068–71 (1989).

35. W.L. Bragg and E.J. Williams, *The effect of thermal agitation on atomic arrangement in alloys*, Proc. Roy. Soc. **A145**, 699–730 (1934).

36. H.A. Bethe, *Statistical theory of superlattices*, Proc. Roy. Soc. **A150**, 552–75 (1935).

37. J.E. Lennard-Jones and A.F. Devonshire, *Critical and cooperative phenomena*, Proc. Roy. Soc. **A169** 317–38 (1939).

38. P.J. Flory, *The statistical mechanics of semi-flexible chain molecules*, Proc. Royal Soc. London A **234**, 60–72 (1956).

39. C. Robinson, *Liquid-Crystalline Structures in Solutions of a Polypeptide*, Trans. Far. Soc. **52**, 571–92 (1956).

40. C. Robinson, *The cholesteric phase in polypeptide solutions and biological structures*, Mol. Cryst. **1**, 467–94 (1966).

41. T. Doscher and R.D. Vold, *Colloidal Structures in Binary Soap Systems*, J. Phys. Coll. Chem. **52**, 97 (1948).

42. S.S. Marsden Jr. and J.W. McBain, *Diffraction in aqueous systems of dodecylsulphonic acid*, J. Amer. Chem. Soc. **70**, 1973–4 (1948).

43. R.D. Vold and M.J. Vold, *Thermodynamic behaviour of liquid crystalline solutions of sodium palmitate and sodium laurate in water at 90°*, J. Amer. Chem. Soc. **61**, 37–44 (1939).

44. K.D. Lawson and T.J. Flautt, *Magnetically oriented lyotropic liquid crystalline phases*, J. Amer. Chem. Soc. **89**, 5489–91 (1967).

45. D. Vorländer, '*Über Suprakrystalline organischen Verbindungen*', Naturwiss. **24**, 113 (1935).

46. G.S.Y. Yeh, *Morphology of amorphous polymers and effects of thermal and mechanical treatments on morphology*, Pure and Appl. Chem. **31**, 65–89 (1972).

47. S.L. Kwolek, US Patent 3.671.542 (1972).

48. W.J. Jackson, Jr. and H.F. Kuhfuss, '*Liquid Crystal Polymers. I. Preparation and Properties of p-Hydroxybenzoic Acid Copolyesters*', J. Polymer Sci., Polymer Chem. Ed. **14**, 2043 (1976).

49. H.F. Kuhfuss and W.J. Jackson, US Patent 3.778.410 (1973), US Patent 3.804.805 (1974).

50. Y. Tanaka, S. Kabaya, Y. Shimura, A. Okada, Y. Kurihara and Y. Sakakibara, *Polymerizaton of mesomorphic monomers. I. Bulk polymerization of cholesteryl methacrylate*, J. Polymer Science, Polymer Letters **10**, 261–4 (1972).

51. E. Perplies, H. Ringsdorf and J.H. Wendorff, *Polyreaktionen in orientierten Systemen, 5. Polyacryl- und – methacryl-Schiffsche Basen mit flüssig-kristallinen Eigenschaften*, Ber. Bunsenges. Phys. Chem. **78**, 921–3 (1974).

52. H. Finkelmann, H. Ringsdorf, W. Siol and J.H. Wendorff, *Enantiotropic (Liquid Crystalline) Polymer Synthesis and Models*, ACS Symposium Ser. **74**, 22–32 (1978).

53. V.P. Shibaev, N.A. Platé and Y.S. Freidzon, *Thermotropic Cholesterol-Containing Liquid Crystalline Polymers*, ACS Symposium Ser. **74**, 33–55 (1978).

54. A. Blumstein, Y. Osada, S.B. Clough, E.C. Hsu and R.B. Blumstein, *Liquid Crystalline Order in Polymers and Copolymers with Cholesteric Side Groups*, ACS Symposium Ser. **74**, 56–70 (1978).

55. H. Stegemeyer, H. Onusseit and H. Finkelmann, *Formation of a blue phase in a liquid-crystalline side chain polysiloxane*, Makromol. Chem. Rapid Commun. **10**, 571 (1989).

56. J.M. Gilli, M. Kamayé and P. Sixou, *Quenched Blue Phase, Below the Glass Transition of a Side Chain Polysiloxane: Electron Microscope Studies*, Mol. Cryst. Liq. Cryst. **199**, 79 (1991).

57. F. Husson, H. Mustacchi and V. Luzzati, *La structure des colloides d'association. II Description des phases liquid-crystallines de plusieurs systemes amphiphile-eau: amphiphiles anioniques, cationiques, non-ioniques*, Acta Cryst. **13**, 668–77 (1960).

58. A. Skoulios and G. Finaz, *Structure of associated colloids. VII Amphipathic character and mesomorphic phases of block co-polymers of styrene and ethylene oxide*. J. Chimie Physique **59**, 473–80 (1962).

59. A. Douy, R. Mayer, J. Rossi and B. Gallot, *Structure of liquid crystalline phases from amorphous block co-polymers*, Mol. Cryst. Liq. Cryst. **7**, 103–26 (1969).

60. H. Finkelmann, B. Lühmann and G. Rehage, *Phase behaviour of lyotropic liquid crystalline side chain polymers in aqueous solutions*, Colloid Polymer Sci. **260**, 56 (1982).

61. K. Kratzat, in: H. Stegemeyer (ed.), *Lyotrope Flüssigkristalle*, Darmstadt 1999, Chapter 3, p. 59.

62. P.-G. de Gennes, *Réflexions sur un type de polymères nématiques*, C.R. Acad. Sci. Paris, B **281**, 101 (1975).

63. H. Finkelmann, H.-J. Kock and G. Rehage, *Investigations on Liquid Crystalline Polysiloxanes 3. Liquid Crystalline Elastomers – A New Type of Liquid Crystalline Material*, Macromol. Chem., Rapid Commun. **2**, 317 (1981).

64. J. Schätzle, W. Kaufhold and H. Finkelmann, *Nematic elastomers: The influence of external mechanical stress on the liquid-crystalline phase behaviour*, Makromol. Chem. **190**, 3269 (1989).

E1
Kolloid-Zeitschrift **11**, 145–57 (1912)

Zeitschrift für Chemie und Industrie der Kolloide

("Kolloid - Zeitschrift")

Wissenschaftliche und technische Rundschau
:: für das Gesamtgebiet der Kolloide ::

Herausgegeben von

Priv.-Doz. **Dr. Wolfgang Ostwald** in Leipzig, Brandvorwerkstraße 77

| Erscheint monatlich 1 mal | Verlag von THEODOR STEINKOPFF Dresden und Leipzig | Preis für den Band M. 16.— |

Ueber Gallerten.

Von R. Zsigmondy und W. Bachmann. (Eingegangen am 2. August 1912)

Ultramikroskopische Studien an Seifenlösungen und -gallerten.

Von W. Bachmann.

Die Erstarrung von Lösungen sogenannter lyophiler Kolloide (z. B. Gelatine, Agar-Agar, Kieselsäure) führt bei hinreichender Konzentration der kolloid gelösten Substanz bekanntlich zu einer Gallerte. Es ist für eine solche Gallerte charakteristisch, daß schon geringe Mengen der gallertbildenden Stoffe ihre Existenz ermöglichen. So genügt im Falle der Gelatine, des Agar-Agar und der Kieselsäure oft (unter günstigen Bedingungen) schon ein Gehalt von etwas weniger als 1 Proz. der genannten Stoffe, um aus deren Lösungen bei der Gelatinierung eine zusammenhängende Gallerte entstehen zu lassen.

Bei verschiedenartigstem chemischen Bau und verschiedenen Entstehungsbedingungen[1] weichen die Gallerten auch in ihrem physikalischen Verhalten oft nicht unerheblich voneinander ab. So ist namentlich der Verlauf der Flüssigkeitsabgabe und -aufnahme bei den quellbaren Gallerten [Gelatine, Agar-Agar, Stärke[2]] unterschiedlich von demjenigen der weniger oder nicht quellbaren Gele (Kieselsäure) u. a. m. Eingehende Untersuchungen in dieser Richtung sind von J. M. van Bemmelen[3] am Kieselsäure-, Eisenoxydhydrogel u. a., von A. Ra-

kowski[4] an der Stärke vorgenommen worden. — Zur Erklärung des merkwürdigen Verlaufs seiner Geldampfspannungsisothermen adoptierte J. M. van Bemmelen gewisse Vorstellungen über die Struktur (den geometrischen Bau oder die morphologische Beschaffenheit) des Kieselsäuregels, die im wesentlichen der Mizellartheorie K. v. Nägeli's entlehnt waren. Hier ergab sich also beispielsweise die Möglichkeit, Struktureigenheiten und -änderungen der Gele für den Verlauf der Ent-, Wieder- und Wiederentwässerung verantwortlich zu machen. Die Erkenntnis der morphologischen Beschaffenheit der Gallerten ist aber nicht nur zur Klärung physikochemischer Fragen zu wünschen, sie ist auch von Bedeutung für die Beurteilung des Wesens der Gele überhaupt, der Art und Weise ihrer Bildung, ihrer verschiedenen Konsistenz ("schleimig-fadenziehend" bis "halbfest-elastisch" mit allen Uebergängen) und nicht zum mindesten für biologische Probleme, wie den Bau des Plasmas usw. Besonders dieser letztere Umstand gab gerade den Vertretern biologischer Fachrichtung Anlaß zu theoretischen Spekulationen über den Bau der Gallerten. Bei der Schwierigkeit, dieses Problem direkt experimentell, etwa auf optischem Wege, in Angriff zu nehmen, begnügte man sich zunächst mit der reinen Theorie. Und hier erwies sich die Anschauung K. v. Nägeli's[5] als von Be-

---

[1] Gelatinierung bei Temperaturänderung (Glutin, Agar-Agar), "spontane" Koagulation (Kieselsäure), Temperaturkoagulation (Eiweiß), Koagulation durch Zusatz chemischer Reagenzien (Kieselsäure) u. a. m.

[2] A. Rakowski, Koll.-Zeitschr. 9, 225—230 (1912).

[3] Zeitschr. f. anorg. Chem. 13, 233 ff. (1897); 18, 98 ff. (1898); 20, 185 ff. (1899); s. a. J. M. v. Bemmelen, Die Absorption (Dresden 1910), 196, 324, 370.

[4] L. c.

[5] K. v. Nägeli, Theorie der Gärung (München 1879), 102.

E1
Kolloid-Zeitschrift **11**, 145–57 (1912)

ON GELS

by

R. Zsigmondy and W. Bachmann

(received 2 August 1912)

Ultramicroscopic studies on soap solutions and soap gels

by

W. Bachmann

It is known that solidification of solutions of so-called lyophilic colloids (e.g. gelatine, agar-agar, silicic acid) results in a gel if the concentration of the colloidal solute is sufficient. Such a gel can exist even with rather small amounts of the gelling substance. In the cases of gelatine, agar-agar, and silicic acid sometimes less than 1 per cent is sufficient to form a continuous gel.

Because of the very different chemical structures and different formation conditions,[1] the physical properties of the gels vary considerably from one another. In particular, the process of release and uptake of liquid is different for gels capable of swelling [gelatine, agar-agar, starch[2]] from the gels less capable or incapable of swelling (silicic acid). Detailed investigations of this were carried out by J.M. van Bemmelen[3] for gels of silicic acid, iron hydrogel and others, and by A. Rakowski[4] for starch.

To explain the curious behaviour of the vapour pressure isotherms of his gels, J.M. van Bemmelen made certain assumptions about the structure (geometry or morphologic constitution) of the silicic acid gel, which were largely taken from K. v. Nägeli's micellar theory. Using this, it was possible to relate the properties and changes of the gel structure to the process of removal and reabsorption of water. The knowledge of the morphological constitution of the gels is not only desirable

---

[1] Gelification with temperature variation (gelatine, agar-agar), 'spontaneous' coagulation (silicic acid), temperature induced coagulation (protein), coagulation by addition of chemical reagents (silicic acid), etc.

[2] A. Rakowski, Koll.-Zeitschr. **9**, 225–30 (1912).

[3] Zeitschr. f. anorg. Chem. **13**, 233 ff. (1897); **18**, 98 ff. (1898); **20**, 185 ff. (1899); cf. J.M. v. Bemmelen, Die Absorption (Dresden 1910), 196, 234, 370.

[4] Op.cit.

in order to clarify physico-chemical questions. It is also important for the assessment of the nature of gels in general, the way they form, their different consistencies ('slimy and thread-like' to 'semisolid and elastic' with all states in between), and last but not least for biological problems such as plasma structure. The latter problems especially gave biologists reason to speculate theoretically about the gel structure. Because of the difficulty in solving this problem directly by experiment, e.g. by optical methods, one has to be content for the moment with pure theory. In this regard K. v. Nägeli's[5] concept became important. He considered the gels to consist of micelles (molecular complexes) possessing crystalline properties.[6] The micelles are separated from each other by a water layer. In the gels they adhere to each other, preserving their individuality, and form networks, in which water is included and 'will be fixed by molecular attraction, not completely, but in a more or less mobile state'.[7]

In an ingenious way v. Nägeli anticipated a lot of things, which have been proved by recent investigations using improved facilities.

Subsequently, v. Nägeli's micellar theory has often been attacked. There were attempts to obtain insight into the gel structure by experiments. However, all these hypotheses were based either on the generalisation of some experimentally derived facts or by analogy.

G. Quincke[8] established his foam theory of the gels, which transferred macroscopic or microscopic experiences about deposited membranes[9] to the field of the fine submicroscopic and amicroscopic* discontinuities found in colloidal solutions and gels. O. Bütschli's 'honeycomb theory' of gels[10] demonstrates how a genuinely critical researcher, who is totally convinced of the truth of his scientific conviction, is able to promote his theory to a universal level of importance by an enormous amount of factual material.

Based on extensive microscopic investigations O. Büschli tried to obtain evidence for his idea of 'honeycombed' and 'foamy' gel structures (gelatine, agar-agar, silicic acid, respectively). Recent investigations, however, have not verified Bütschli's structure theory of gels.

In contradiction to O. Bütschli, the dry (glass-like) gel of silicic acid does not exhibit a honeycomb structure but must be seen as a conglomerate of small SiO_2 particles[11] containing amicroscopic cavities which are connected to each other. The honeycomb structure observed microscopically by O. Bütschli is a figment and is temporarily caused by a heterogeneous accumulation of liquid in this system of cavities.

[5] K. v. Nägeli, Theorie der Gärung (München 1879), 102.

[6] Formerly, M.L. Frankenheim believed that both gels and glasses were crystalline; cf. M.L. Frankenheim, Die Lehre von der Kohäsion (Breslau 1835).

[7] K. v. Nägeli, op. cit.

[8] Cf. Ann. d. Phys., 4. Folge, 9, 793 ff., 969 ff. (1902).

[9] Op.cit. 793 ff.

Editors note: we interpret amicroscopic as meaning below the resolution of a microscope.

[10] O. Bütschli, Untersuchungen über Strukturen (Leipzig 1898).

[11] See R. Zsigmondy; Zeitschr. F. anorg. Chem. 71, 361 (1911).

The structure of more concentrated gelatine gels (more than 6 per cent) is also amicroscopic. Their truly heterogeneous constitution can be proved first of all by the separation of the gels into two components[12] by squeezing, and furthermore optically by the polarisation of the bluish low intensity Faraday–Tyndall cone[13] (slit-ultramicroscope).

Diluted gels (1 to 6 per cent) of gelatine (and agar-agar) and of silicic acid show considerable discontinuities in the ultramicroscope,[14] which can be differently interpreted, at least from an optical point of view. They appear as densely packed whitish particles of alternating shape and medium brightness, and as diffraction fringes and distinct microparticles. From an optical point of view these diffraction fringes can be consistent with optically denser and optically thinner material.[15] For the moment it cannot be decided if the observed heterogenities can be regarded as small gelatine grains, surrounded by a layer of water (grain structure) or as water (the phase having a low gelatine content,) surrounded by walls rich in gelatine (honeycomb structure). Because of facts which will be given below, the grain structure of gelatine, agar-agar and silicic acid gels has been taken as much more plausible.

In a letter to one of us (*Ed.: Zsigmondy*) Wolfgang Ostwald, referring to J. Babinet's theorem, emphasises that this grain structure is unprovable, because the small black borders around the grains mentioned above can be taken also as a gelatine rich phase surrounding, like honeycombs, the phase with low gelatine content (water, dispersant). As pointed out several times, from an optical point of view there is no objection to this interpretation. In this case honeycombs would be present in 1 to 6 per cent gels, which would be much smaller than predicted by O. Bütschli's theory.[16]

---

[12] See O. Bütschli, Ueber den Bau quellbarer Körper (Göttingen 1896), 22–7.

[13] See W. Bachmann, Zeitschr. F. anorg. Chem. **73**, 140/141, 151/152 (1911).

[14] Zeitschr. F. anorg. Chem. **73**, 138–52 (1911).

[15] Cf. R. Zsigmondy, Uber die Struktur des Gels der Kieselsäure. Theorie der Entwässerung, Zeitschr. F. anorg. Chem. **71**, 359 (1911), where it was pointed out that the scattering fringes (in dry silicic acid gel) may be caused by denser as well as from less denser silicic acid agglomerations, finally also by larger cavities in the amicroscopic SiO_2 agglomerate.

[16] O. Bütschli has shown that the gel structures of gelatine caused by coagulating agents become finer on decreasing water content of the gelatine, and coarser on decreasing gelatine concentration: Untersuchungen über Strukturen (1898), 172. [Concerning the relation between water content and size of the ultramicroscopic structure the same resulted for unchanged aqueous gels (above 1 per cent); see W. Bachmann, Zeitschr. f. anorg. Chem. **73**,140 (1911).] Following O. Bütschli the honey-combs in (10 per cent) alcoholic gels should exhibit the same dimensions as in unchanged aqueous gels of the same concentration. The honeycomb diameter is about 0.7–0.8 μm (in 10 per cent gel; op. cit. 154, 172). In our experiments we found 1 μm and more (sometimes 3–4 μm). These dimensions remain uncertain because the borders of the often rather irregular "honeycombs" could not be detected perfectly which is also (the case for) O. Bütschli's investigations. The honeycomb diameter in a 1–2 per cent gel (which was not investigated by O. Bütschli) is expected to be considerably larger even than 0.8 μm. Actually, the heterogeneities observed in such unchanged gels are obviously much finer than 0.8 μm. Sometimes their dimensions cannot be detected with sufficient certainty. From this, Bütschli's theory cannot be verified; otherwise the honeycomb diameter must be assumed to be considerably finer than postulated by O. Bütschli.

The assumption of a honeycomb structure, however, does not hold when considering gelatine flakes, which are formed in 0.5 to 1 per cent gelatine solutions during 'de-emulsification', the structure of which is identical to that of compact gels. Figures 1 and 2 in the text[17] show gel flakes on a dark background. The dark, nearly optically transparent, space corresponds to the squeezable liquid;

Fig. 1 'Flocculation' in an aging aqueous gelatine solution (0.5 per cent). Slit-ultramicroscope. Objective D 'Star'. Compensated ocular 18. Magnification about 2000 times.

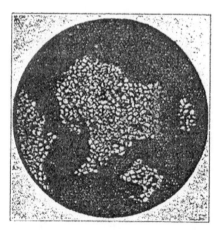

Fig. 2 Single gelatine flocs in a 0.5 per cent aqueous gelatine sample. Cardioid ultra-microscope. Apochromate 3 mm. Compensated ocular 18. Magnification 2166 times.

[17] Fig. 1 in the text was drawn by Mr. cand. jur. Doering, Fig. 2 by W. Bachmann by means of the Abbe drawing instrument.

several floating brighter particles are necessarily the gelatine rich phase. This is due to the edges of the flakes which jut out into the liquid and are almost totally surrounded by it. Thus, the denser constitution of the bright particles is proved.

> A short section of the paper containing more details of the experiments has been omitted, since it does not add to the conclusions of the paper.

This structure can be observed for up to about 6 per cent gels. The differences appearing with increasing concentration consist only of increasingly dense packing and less brightness of the particles, in so far as they are still distinguishable. Larger irregular water-filled cavities (optically transparent) which can often be observed in 1 to 2 per cent continuous gels become rarer on increasing concentration. Or in other words: the space becomes fuller and fuller of gel particles, leaving no room for such cavities.

From this, a close relation becomes obvious between the structure of concentrated (up to 6 per cent) and dilute gels (0.5 to 1 per cent) for which the grain structure must be postulated (see above).

Ten per cent gels cannot be differentiated in the ultramicroscope.[18] Taking Büschli's honeycomb structure as a basis, one should observe very distinct heterogeneities in the ultramicroscope. Gelatine walls with a separation of 0.7 μm should be obvious under any circumstances in the ultramicroscope. The objection that the difference in the refractive indices between the gelatine rich phase and water (gelatine poor phase) is too small to allow a visualisation is invalid, because a 10 per cent gel shows a distinct diffraction fringe against water. Grinding such a gel to a fine mash and suspending it in water, one can even observe flaky assemblies of fine submicron particles, which contrast distinctly with each other and with the optically homogeneous dark liquid.

> A further short section of the paper containing more details of the experiments has been omitted, since it does not add to the conclusions of the paper.

In the slit ultramicroscope undifferentiated gels show a pale-bluish strongly polarised light cone. This provides evidence for the very fine heterogeneous structure. Following Büschli's hypothesis, however, the gels should contain much coarser structures.

The formation conditions of the (diluted) gels [apparent coagulation of submicron (or amicron) particles to 'visible gel elements'[21]] as well as the linear polarisation

[18] W. Bachmann, Zeitschr. f. anorg. Chem. **73**, 140/141 and 151 (1911).
[21] Zeitschr. f. anorg. Chem. **73**, 140/141 and 151 (1911).

of light which these 'gel grains' scatter indicate a structure of the latter of still finer scale [amicrons].[22]

All these results provide evidence for Nägeli's concept of the gel structure. Bütschli's hypothesis, however, would require considerably smaller dimensions of the foam with respect to the honeycomb structure. However, the suitability of such a hypothesis to describe gels would demand further proofs in the future.

As the present paper has shown for various gel structures, the possibility has been pointed out that honeycomb structures may really exist. Only for gels of gelatine, agar-agar, and silicic acid and – as will be shown below – for soaps, Büschli's honeycomb theory has not been proved to be valid.[23]

Consequently, we must prefer v. Nägeli's idea. However, we emphasise that a final verification of Nägeli's theory in which a crystalline nature is attributed to the ultramicron particles (micelles, micellar assembly) has yet to come.

Recently, in particular the Russian researcher P.P. v. Weimarn[24] tried to verify the crystallinity of gels through extensive investigations. He investigated especially the structure of deposited membranes, obtained by pouring together highly concentrated solutions of the reaction components. He provided us with the likely proof that initially undifferentiated membranes decompose[25] after a more or less longer time into small particles having a crystalline structure.[26] Based on this, he argued that the undifferentiated membranes also consist of densely packed crystallites[27] from which those decomposition particles should be formed by molecular crystallisation or crystallisation of aggregates (P.P. v. Weimarn).

However objections to these arguments consider that the formation of small crystallites can also be caused by transformation of amorphous substances into crystalline ones, as well as by growth of amicroscopic crystallites.[28] P.P. v. Weimarn compares these gel-like deposited membranes, with respect

---

[22] Op.cit. 141.

[23] The authors concept verified many times for gels, as cited above, cannot be generalised in the way that O. Bütschli's honeycomb theory has been disproved; cf. R. Zsigmondy, Zeitschr. f. anorg. Chem. **71**, 356–77 (1911), especially summary 376/377; and W. Bachmann, Zeitschr. f. anorg. Chem. **73**, (1911), generalised summary 169–71; and R. Zsigmondy, Kolloidchemie (Leipzig 1912), 151, 154, 251 and so on.

[24] P.P. v. Wiemarn, cf. this author's papers in Koll.-Zeitschr. **2** ff.; further Grundzüge der Dispersiodchemie (Dresden 1911) and so on.

[25] This process becomes obvious in its first stage by the appearance of a clouding of the films.

[26] On this proof method cf. Grundzüge der Dispersoidchemie, supplement I, 115; further: Zur Lehre von den Zuständen der Materie, II. part, Koll.-Zeitschr. **3**, 281 ff (1908) and so on.

[27] Cf. Grundzüge der Dispersoidchemie 56 ff.

[28] Cf. Wo. Ostwald, Grundriß der Kolloidchemie, 1. Edition (1909), 51; 2. Edition, 1st part (1911), 71; C. Doelte, Koll.-Zeitschr. **7**, 29–34 (1910).

to their structure, with typical gels of lyophilic colloids which should also be crystalline.[29]

> Another short section of the paper containing more details of the experiments has been omitted, since it does not add to the conclusions of the paper.

Consequently, there is still no unimpeachable proof of crystallinity for gels; on the other hand it must be emphasised that the statement that gels consist of amorphous particles is also unproved. This remains an open question and at present one can only give probable arguments for both points of view.

However, there are investigations of another class of gels, which without any doubt are formed by crystallisation processes: the soap gels. These typical gels are formed on solidification of aqueous solutions of alkaline salts of higher saturated and unsaturated fatty acids. Their ultramicroscopic structure and their gel-like behaviour contradicts the generalised assumption that foam or honeycomb structures are in all cases characteristic of gels.

> The following 7 pages contain a very detailed description of experiments with and preparation of various alkaline salts of fatty acids together with analytical data. This part of the present paper is from Bachmann's Ph.D. thesis and attributed to the subtitle (see above): 'Ultramicroscopic studies on soap solutions and gels by W. Bachmann'.

As can be seen, these data, and also others reported here, vary considerably even in the case of the same samples which were repeatedly investigated. In particular, the temperature at which various turbidities appear seems to depend on uncontrolled fortuitousness. Even the glass walls may have an effect on the precipitation of the first 'crystalline turbidity', since we observed that the precipitation was retarded – sometimes very considerably – in Jena glass in contrast to common glasses. However, this effect deserves further investigation and should be mentioned only with reservations.

The temperatures at which a 'strong' turbidity appears roughly agree with F. Krafft's crystallisation temperature of soaps.[63] This "strong" turbidity corresponds ultramicroscopically to the appearance of a second solid phase (cf. Table) whereas the so-called 'crystalline' turbidity represents the first solid phase (cf. Table).

[29] Koll.-Zeitschr. **2**, 370 (1908).
[31-62] omitted.
[63] cf. F. Krafft, B.Dt.Chem.Ges. **32**, 1596–608 (1899).

Table

Per cent aqueous solutions		Appearance of crystalline turbidity	Onset of viscosity increase	Sudden appearance of strong turbidity	Strong viscosity increase	Thermal arrest	Remarks
Sodium Oleate	5	–	Marked at onset (+30 °C)	before solidification	during solidification	–1.85°	supercools to –3 °C
	10	–				+2.15 to 5.2°	0 °C; ice forms
	15	–				+5.15 to 6.2°	supercools to +4.5 °C
Potassium Oleate	10	–	Marked at onset (+30 °C)	–	–	–	–
	20						
Sodium Palmitate	10	88 – 87 °C	~54 °C	~55 – 50 °C	~50 °C	49.7°	
	20	75 °C	–	–	–	53.6°	supercools to 52.4 °C
Potassium Palmitate	10	96 – 95 °C	23 °C	~23 – 22 °C*	22 – 20 °C	19.9°	–
Sodium Stearate	5	above 80 °C	–	62 – 55 °C	45 – 43 °C	44.5 – 43°	–
	15	–	–	–	–	50.1 – 49.3°	supercools to 0.5 – 1 °C
Potassium Stearate	5	~80 °C	~40 °C	60 – 53 °C*	36 °C	~33°	–
	15	–	–	–	–	40.5°	–

*The ('strong') turbidity appearing at these temperatures sometimes was less strong and less significant than for sodium salts. In the case of potassium palmitate it was not always observable with certainty or often only at a much lower temperature, i.e. shortly before the thermal arrest.

Finally, we present some thermal data of soap solutions in the following table. All data represent mean or limiting values of several observations. 'Thermal effects' (or thermal arrests) are based on numerous solidification curves (cooling curves) which will be reported elsewhere.

The 'strong' increase of viscosity most probably coincides with the formation of threads etc. The thermal effects are below the onset temperatures for crystallisation. The main part of the heat of crystallisation is only liberated and is significant if the rate of crystallisation becomes a maximum.

In a few cases we succeeded in filtering off very small amounts of the first 'crystalline' turbidity (of the potassium salts at about 60) and determined (after drying) their melting points, as follows: Crystalline turbidity of potassium palmitate: m.p. 95–103, Crystalline turbidity of potassium stearate: m.p. 110–115. From this, one must conclude that the 'crystalline' turbidity is due to a substance containing an excess of free fatty acid in an adsorbed state[64].

---

[64] The neat salts were unmelted still at 230 °C but mostly decomposed. Cf.: F.G. Donnan and A.S. White, The System: palmitic acid – sodium palmitate; Trans. Chem. Soc. **99**, 1668–79 (1911).

Richard Adolf Zsigmondy was born in Vienna in 1865, the son of a prominent dentist. He studied chemistry at the Technische Hochschule in Vienna and subsequently organic chemistry at the University of Munich, where he obtained his Ph.D degree in 1889. He obtained his habilitation degree in 1893 from the Technische Hochschule in Graz, and became lecturer there.

From 1897 to 1900 he worked in the Schott und Genossen company in Jena on problems of colloidal gold. These were pioneering investigations in the field of colloids, e.g. his detection of protective colloids. He discovered how to prepare gold hydrosols. Together with the optician H. F. W. Siedentopf of the firm Zeiss-Jena, he developed the ultramicroscope (1902–3). This instrument enabled the detection of otherwise unobservable small colloidal particles, which became visible as light spots because of their light scattering. It was now possible to measure, albeit indirectly, particles with dimensions as low as $1.5\,\mu$. Zsigmondy exploited his invention himself in his colloidal studies, and the method became widely used in the field.

In 1907 he was appointed as full Professor and Director of the Institute of Chemistry at the University of Göttingen. He kept this position until his retirement in February 1929.

Amongst his colloidal studies in Göttingen were fundamental researches on the gel structure in soap solutions. These later on became well-known as lyotropic systems. He published a very influential textbook on colloid science, as well as a number of other books. For his invention of the ultramicroscope and his pioneering colloidal work Zsigmondy was awarded the Nobel prize in 1925.

He died only a few months after his retirement in September 1929.

Wilhelm Bachmann was born in 1892 in Kassel and studied chemistry at the universities of Munich and Göttingen. He carried out research in Göttingen under Professor Zsigmondy's supervision, and obtained his PhD in 1911. The paper included in this volume formed part of that Ph.D work. He later obtained the habilitation degree in Göttingen in physics in 1916. In the same year together with Zsigmondy he invented their well-known membrane filter.

After 1917 he worked as a director in chemical industry for the de Haen Company and at the same time served as professor at the Technical University of Hanover, giving lectures in colloid chemistry. Over a ten-year period before his death he also served as Managing Secretary of the Deutsche Bunsen-Gesellschaft. He died during a hunting accident in 1933 at the early age of 41.

E2

Nature **138**, 1051–52 (1936)

Letters to the Editor

The Editor does not hold himself responsible for opinions expressed by his correspondents. He cannot undertake to return, or to correspond with the writers of, rejected manuscripts intended for this or any other part of NATURE. *No notice is taken of anonymous communications.*

NOTES ON POINTS IN SOME OF THIS WEEK'S LETTERS APPEAR ON P. 1060.

CORRESPONDENTS ARE INVITED TO ATTACH SIMILAR SUMMARIES TO THEIR COMMUNICATIONS.

Liquid Crystalline Substances from Virus-infected Plants

STANLEY[1] has described the preparation of a crystalline protein possessing the properties of tobacco mosaic virus from the sap of infected tobacco and tomato plants. The crystals were small needles made

FIG. 1. Wake of a goldfish swimming in a dilute solution of protein from infected sap (observed between crossed nicols). *Photo by Ramsey and Muspratt.*

by precipitation with acid ammonium sulphate. We have confirmed these results, but have found that by further purification the protein in neutral aqueous solution can be obtained in liquid crystalline states.

The sap of tobacco and tomato plants infected with strains of tobacco mosaic virus, after clarification by centrifuging, contains five to ten times as much protein as sap from similarly treated uninfected plants. This extra protein can be precipitated from dilute salt solutions at around pH 3·4 and from neutral solutions with from 10 to 12 per cent ammonium sulphate. The protein in uninfected sap is not precipitated under these conditions. Using these properties, 1–2 gm. of protein can be isolated from a litre of sap ; the yield varying with the age of the infected plant and the duration of infection. No enzyme preparation has yet been found which can attack the protein at an appreciable rate, and some impurities can be conveniently removed by incubation with trypsin.

We have worked with three strains of tobacco mosaic virus, those causing common tobacco mosaic, aucuba mosaic and enation mosaic. No gross chemical or physical differences have been found between the three proteins isolated, but each reproduces its characteristic disease when inoculated to susceptible plants. Differences in detail, which will be described in a later publication, have been noticed. Solutions of the protein are antigenic, and the antisera produced in rabbits give specific precipitates with 1 c.c. of solution containing 10^{-7} gm. The three proteins are serologically related. Plants inoculated with 1 c.c. containing 10^{-9} gm. usually become infected, and occasional infections have followed inoculations of 10^{-11} gm.

These proteins, when precipitated with acid and dried, have the usual analytical figures : C 51 per cent, H 7·1 per cent, N 16·7 per cent. The sulphur contents vary from 0·2 to 0·7 per cent and there is

0·5 per cent phosphorus and 2·5 per cent carbohydrate. The last two constituents can be isolated as nucleic acid of the ribose type from protein denatured by heating. Neutral aqueous solutions of the proteins are almost colourless and faintly opalescent. The protein sediments in a centrifugal field of 23,000 times gravity at about 5 mm. per hour and is deposited as a very viscous layer containing up to 30 per cent of dry matter. Highly purified protein solutions, if stronger than 2 per cent, separate into two layers on standing. The lower layer, which may be water clear, is liquid crystalline. The birefringence depends on the concentration, but may rise to 0·002. The upper layer shows, on gentle agitation, the phenomenon of anisotropy of flow. This was noticed by Takahashi[2] in clarified sap. We have, however, been unable to confirm his later claim that the effect could be observed in healthy tobacco sap. The orientation will persist for several seconds in purified solutions, but if the solutions are impure or dilute the persistence is much shorter. This effect is illustrated in Fig. 1, where the wake of a goldfish swimming in a dilute solution is clearly shown ; the cross of isocline will be noticed in the eddies.

Solutions can readily be orientated by electric currents, but not by magnetic fields of 6,000 gauss. The lower layer orientates itself by flow like the top layer and also tends to orientate parallel to glass or air surfaces. On flowing through small capillary tubes it can be thrown into a state of reversed spiral vibration[3]. When small portions of lower layer are in equilibrium with the top liquid, they appear as

FIG. 2. Highly purified protein from infected sap, when stronger than 2 per cent, separates into two layers on standing. If there is very little of the lower layer, it arranges itself in spindles. Photographed between crossed nicols, × 30.

spindle-shaped bodies (see Fig. 2) which are characterized by an approximately constant meridional curvature. Large spindles are consequently nearly spherical and small ones practically linear. On standing there appear holes, filled with top liquid, of

precisely the same spindle shape. All these phenomena indicate the presence of long fibres.

If any of these liquids be allowed to dry, a film
forms on the surface consisting of three sharply
distinguishable layers. The first formed is an extremely
soft gel well orientated and with much higher birefringence, 0·007, than the liquid. The outer part of
this wet gel shrinks by 50 per cent and forms a layer
of higher refractive index, but lower birefringence,
0·003. Finally, on placing in air, extensive cracks
appear in this dry gel and it becomes translucent
and rather feebly doubly refractive. The dry gel is
usually well-orientated but in certain conditions
shrinks longitudinally to give a herring bone pattern,
which indicates that little longitudinal shrinkage of
the fibres themselves takes place.

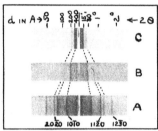

FIG. 3. X-ray patterns of dry gel (*A*), wet gel (*B*)
and 13 per cent solution (*C*).

The X-ray patterns of these different forms show
remarkable similarities and differences. For the large
angle scattering there appears to be little difference,
except in general intensity, between patterns given
by all the different forms from the top liquid,
orientated by flow in a Lindemann glass capillary,
to the dry gel and the 'crystals', from ammonium
sulphate solutions. This pattern is therefore entirely
due to the protein molecules themselves and may be
called the intramolecular pattern. It appears to have
about the same order of complexity as that produced
by feather keratin, with a repeat unit in the fibre
direction of $3 \times 22 \cdot 2 \pm 0 \cdot 2$ A.

Near the centre of the photographs, spots were
observed which varied from one preparation to
another. A camera of 40 cm. length was constructed
employing an X-ray beam of copper-$K\alpha$ radiation
with an angular width of 3′ and capable of showing
spacings up to 1,200 A. All the intermolecular
pattern so far observed lies in the equatorial plane,
that is, refers to the relative sideway positions of
the rod-like molecules. The dry gel gives the most
definite pattern (Fig. 3) showing five lines (see table)
corresponding to the first five possible reflections of
a hexagonal close-packing with intermolecular
distance of $152 \cdot 0 \pm 0 \cdot 5$ A. The wet gel gives three
distinct lines which seem to correspond to hexagonal
close-packing with intermolecular distance of 210 A.,
but this varies slightly with the composition. The
bottom layer liquid gives a pattern of three lines ; two
sharp lines the position of which depends on concentration, indicating a mean molecular distance ranging
from 300 A. to 470 A., and a diffuse line at a spacing
of 100 A. which is independent of the concentration
and is thus plainly of intramolecular origin. The
pattern of the liquid suggests that there exists a quasi-
regular Debye-Hückel arrangement of parallel rod-like

charged molecules in the solution. The sharp difference
between the wet and dry gel can best be explained
at the present stage by postulating somewhat
triangular molecules that fit together in two ways,
containing different amounts of water. The reflection
at 100 A. supports this hypothesis. So far it has
proved impossible to get any indication of intermolecular reflections in the other directions of the
molecule, though any regularity of less than 1,200 A.
would have been detected. This suggests strongly
that the arrangement is regular only in cross-section
and that no molecular sheets are formed.

OBSERVED INTERMOLECULAR SPACINGS IN ANGSTROMS.

$h k i l$	$10\bar{1}0$	$11\bar{2}0$	$20\bar{2}0$	$12\bar{3}0$	$30\bar{3}0$
Dry gel : obs.	131·8	75·75	65·90	49·75	43·5
calc.	131·8	76·00	65·90	49·75	43·9
Wet gel	188	106	93		
Liquid : 23 per cent	300	175			
13 ,, ,,	397	225			

From the results already gained, it is legitimate
to make certain conclusions as to the nature of the
protein molecules. First, the molecules seemed to be
identical in cross-section. Secondly, each molecule
has quasi-regular structure and thus may be considered to be built up of sub-units of approximately
the same character. The physical properties of the
substance can best be explained by postulating rod-
shaped molecules. The minimum cross-section area
of these is 20,100 sq. A. for the dry gel. The molecular
length is more uncertain. The extreme character
of the orientation phenomena and the X-ray data
point to a minimum length of not less than ten times
the width, or greater than 1,000 A. This gives a
minimum molecular weight in reasonable agreement
with Svedberg's[4] estimate of 17×10^6, though there
is nothing to show that the lengths of the molecules
are uniform.

These results have a certain intrinsic interest, but
this would naturally be greatly enhanced could it be
shown that these rods are in fact virus particles.
This conclusion seems to us both reasonable and
probable, but we feel that it is still not proved, nor
is there any evidence that the particles we have
observed exist as such in infected sap.

Note added Dec. 3.

Wyckoff and Corey have published[5] an X-ray study
of the ammonium sulphate crystals of tobacco
mosaic and aucuba protein. Their measurements
of the intramolecular spacings obtained with un-
orientated material agree with ours, notably the lines
they record at 11·0, 7·44, 5·44 and 3·7 A. correspond
to our measurements of the planes (0006, 9, 12, 18)
respectively.

F. C. BAWDEN.

Rothamsted Experimental Station,
  Harpenden.

N. W. PIRIE.

  Biochemical Laboratory,
  Cambridge.

J. D. BERNAL.

Crystallographic Laboratory,   I. FANKUCHEN.
  Cambridge.
  Nov. 17.

[1] *Phytopathology*, **26**, 305 (1936).
[2] *Science*, **77**, 26 (1933).
[3] van Iterson, *Proc. Roy. Akad. Wetensch.*, **37**, 367 (1934).
[4] *J. Amer. Chem. Soc.*, **58**, 1863 (1936).
[5] *J. Biol. Chem.*, p. 51, Nov. 1936.

Frederick Charles (Fred) Bawden was born in 1908 in Devon, England. In 1926 he entered Emmanuel College at Cambridge University on a scholarship, studying natural science with an emphasis on botany. He subsequently obtained the Diploma in Agricultural Science, graduating in 1930. He then joined the new Potato Virus Research Station in Cambridge, remaining until 1936, when he was appointed as Virus Physiologist at Rothamsted Experimental Station in Hertfordshire, working in the Department of Plant Pathology. He was promoted to head of the Plant Pathology Department in 1940, Deputy Director of Rothamsted in 1950, and Director in 1958.

Bawden was the author of some 200 scientific papers, many of which were in common with N.W. Pirie (see below). His first paper with Pirie was published in 1936, and the last after his own death. His work ranged over all aspects of the science of viruses. Apart from the X-ray structural work included in this volume, and general work on the physiology of viruses, he is known for his work on virus multiplication, immunological work, work on virus classification, and work on intracellular structures.

Bawden was very active in scientific politics and international affairs. The Royal Society elected him to a fellowship in 1949 and he was knighted in 1967. He sat on many committees, received honorary degrees from the Universities of Hull, Bath, Reading and Brunel. A number of Scientific Academies overseas elected him to honorary membership. He died in 1972.

Norman Wingate Pirie (usually known as Bill) was born in Sussex, England in 1907. His family was Scottish, and he spent most of his early childhood in Stirlingshire in Scotland, where his family house was located. His father was a well-known painter, and later became, as Sir George Pirie, president of the Royal Scottish Academy.

Pirie entered Emmanuel College, Cambridge in 1925 to study natural science, graduating in 1929 with a specialism in biochemistry. He worked in the Cambridge Biochemistry Department until 1940, then switching to Rothamsted (like Bawden, in whose group he worked, as a Virus Physiologist). Between 1947 and his formal retirement in 1972 he headed the Rothamsted biochemistry department. After his retirement his scientific work continued more or less unabated.

Pirie's early work in Cambridge concerned the purification of sulphur compounds and their metabolism in dogs. Apart from the work on plant viruses celebrated in this volume, he is known for work on leaf protein and world nutrition, early work on the biochemistry of reproduction and contraception, the origin of life and the stability of β-carotene. He was initially sceptical of the pre-eminent role of DNA in inheritance, arguing that the structure of complex lipids would eventually turn out to be important in the carrying of information in living beings.

Bill Pirie was elected a Fellow of the Royal Society in 1949, was awarded its Copley medal in 1971, and was also elected as Fellow of the New York Academy of Sciences in 1963. He was known for his multifarious left-wing and trade union activities, and his outspoken atheism, some of which may explain why, notwithstanding his scientific eminence, he was not further decorated with more orthodox honours.

He died at the age of 89 in 1997.

John Desmond Bernal was born in Nenagh, County Tipperary, Ireland (then part of the United Kingdom) in 1901. His iconoclastic career seems to have begun at birth, for he was the son of an Irish Catholic gentleman farmer (seemingly of Sephardi Jewish origin) and an American protestant heiress. After an English private school education, some of which was spent in a strictly Roman Catholic environment, he entered Cambridge University in 1919, obtaining a degree in physics in 1923. Having published a paper on theoretical aspects of crystal physics, he was invited to by W.H. Bragg (the elder) to perform graduate work on X-ray crystallography at the Royal

Institution in London. In 1927 he was appointed as lecturer in structural crystallography in the Physics Department in Cambridge. After ten years in Cambridge, he moved to Birkbeck College, London, as Professor of Physics and subsequently (from 1963) crystallography.

Bernal was one of the major figures in X-ray crystallography in the twentieth century. In addition, he exhibited polymathic scientific interests and degree of learning, earning him the nickname 'Sage'. Apart from the work on the structure of viruses included in this volume, his main scientific achievements included: the determination of the shape of the cholesterol molecule (discussed already by Reinitzer in 1888), a prediction that at sufficiently high pressure all matter should ultimately become metallic as a result of electron orbital overlap, a determination of the structure of water, and suggestions concerning the chemical processes involved in the origin of life. He also played a major role in the International Union of Crystallography, and was president 1963–6.

Bernal was also noted for his other interests. He wrote in all ten books, including the immensely influential *The social function of science* (1939). In related vein were *Science and industry in the nineteenth century* (1954), *Science in history* (1954) and *The extension of man — a history of physics before 1900* (1972). These works emphasise his interest in science as a human endeavour. He was converted from Catholicism to what many of his colleagues regarded as an uncritical Marxism while still an undergraduate, and this commitment stayed with him the rest of his life, governing much of his activity. His left wing activity put him in touch with Bill Pirie, enabling the fruitful collaboration leading to the publication of the paper on TMV structure in this volume. He was instrumental in the founding of the World Federation of Scientific Workers (1946), but later defended the Lamarckian theories of the Soviet geneticist T.D. Lysenko and wrote an uncritical hagiographic obituary of Stalin. He was awarded the Order of Lenin by the Soviet authorities in recognition of his left-wing activity.

In 1942 Bernal was invited to investigate the effects of bombing on civilian populations. This led to a wide-ranging involvement in war work which was widely appreciated in the highest circles. Bernal was particularly involved in the planning of the Anglo-American invasion of Normandy in 1944, and was physically present on D-day.

Bernal was elected to the Royal Society in 1937, and was awarded the Royal Medal of the Royal Society (its highest medal) in 1945. His most enduring scientific heritage will be, not his own work, but that of the pupils and younger colleagues whom he inspired to work chiefly on the structure of biological macromolecules. One of these was the American Isidore Fankuchen, his collaborator on paper E2. Several others, including Dorothy Crowfoot Hodgkin, Max Perutz, John Kendrew and Aaron Klug, were later to win Nobel prizes. No discussion of Bernal's magnetic personality can fail also to mention the almost irresistable attraction he had for women, leading to numerous long-lasting affairs.

Bernal suffered a stroke in 1963 which affected his ability to work. He retired in 1967, and died in 1971.

Isidore Fankuchen (almost universally known to his friends as 'Fan') was born in Brooklyn, New York in 1904. He came from a poor family, and initially did not go to college, but rather set up a radio repair and installation store. He was thus able to follow courses at Cooper Union, from which he graduated with a B.S. degree in 1926. After Cooper Union, Fankuchen attended graduate school at Cornell University, obtaining the Ph.D degree under the supervision of C.C. Murdock in 1933. He then spent 1934–6 in Manchester in the  laboratory of the Nobel-prize-winning crystallographer W.L. Bragg, before migrating to Cambridge, where he worked on X-ray crystallography under the supervision of J.D. Bernal. In 1938 when Bernal was appointed to the chair in crystallography at Birkbeck College, Fankuchen followed him to London.

Shortly after the outbreak of war in 1939, Fankuchen returned to the U.S. to Harvard Medical School as a National Research Fellow. In 1941 he obtained an academic post at the University of Minnesota Medical School. The following year he returned to Brooklyn as adjunct professor in crystal chemistry in the

Brooklyn Polytechnic Institute. He was promoted to associate professor in 1945 and Professor of Applied Physics in 1947.

Fankuchen was major apostle of the then newly-born science of X-ray crystallography. His contribution to the TMV problem was to design an ingenious monochromator enabling a narrow angle diffraction experiment to be carried through successfully, with sufficiently great intensity for sensitive observations to be made. He played a major role in the international organisation of crystallography, was an editor of *Acta Crystallographica*, and organised the fabrication of X-ray equipment in the US.

He died of cancer at the age of 59 in 1964.

E3
Kolloid-Zeitschrift **89**, 224–33 (1939)

Aus dem Kaiser-Wilhelm-Institut für physikalische Chemie und Elektrochemie, Berlin-Dahlem.

Die Mizellenarten wässeriger Seifenlösungen.

Von *Joachim Stauff (Berlin).*   (Eingegangen am 13. September 1939)

1. Einleitung.

Die besten Modellsubstanzen zur Untersuchung des Verhaltens von Mizellkolloiden in wässerigen Zerteilungen sind seit jeher die Seifen und ähnlich gebaute Stoffe gewesen. Einer ihrer wichtigsten Vorzüge ist ihre Fähigkeit, in Wasser in Ionen zu zerfallen. Dadurch ist es möglich gewesen, die weitläufigen experimentellen und theoretischen Erfahrungen der wässerigen Elektrolytlösungen auf sie anzuwenden. Es ist daher durchaus verständlich, daß die meisten Theorien der Konstitution von Seifenlösungen aus Ergebnissen elektrochemischer Untersuchungen gewonnen worden sind.

So ist die klassische Theorie der Seifenlösungen von Mc Bain[1] im wesentlichen durch die Deutung der Konzentrationsabhängigkeit ihrer Leitfähigkeit auf dem Boden der klassischen Arrhenius'schen Elektrolyttheorie entstanden. Durch Kombination dieser Messungen mit Messung der Dampfdrucke kam Mc Bain zu der Ansicht, daß zwei verschiedene Mizellenarten in den Seifenlösungen vorhanden sein müssen, die er mit „Ionenmizelle" und „Neutralkolloid" bezeichnete. Erstere sollte durch Aggregation von Ionen, letztere durch Aggregation von undissoziierten Molekülen entstanden sein. Durch eine große Zahl experimenteller Untersuchungen haben Mc Bain und Mitarbeiter versucht, diese Anschauungen zu stützen, vor allem die Existenz der beiden verschiedenen Mizellenarten direkt nachzuweisen[2].

Die Theorie von Mc Bain wie auch der Nachweis der beiden Mizellenarten wurde jedoch von Hartley[3], zum Teil wegen der experimentellen Voraussetzungen, zum Teil wegen der Deutung der Ergebnisse stark angezweifelt. Er begründete daher eine neue Theorie der Seifenlösungen, die mit der Annahme nur einer einzigen Mizellenart auskommt. Während die Bildung der beiden Mizellenarten nach Mc Bain etwa in dem Gebiet einsetzt, in welchem die Leitfähigkeitskurve ein Minimum aufweist, sollen nach Hartley bereits bei viel geringeren Konzentrationen Mizellen auftreten. Die stärkste Stütze für diese Anschauungen besteht vor allem in den vielseitigen Erscheinungen, die bei der sogenannten „kritischen Konzentration" von Lottermoser und Püschel[4] erstmalig beobachtet worden sind. Bei dieser Konzentration treten ganz plötzliche Veränderungen verschiedener Eigenschaften der Lösungen auf, die nur durch eine Aggregation der langkettigen Ionen nebst gleichzeitiger Assoziation der Gegenionen erklärt werden können. Ein weiterer Vorteil der Hartley'schen Theorie ist die Rolle, die den elektrostatischen Kräften zwischen den Ionen zugewiesen wird; diese Wechselwirkung ist in der Mc Bain'schen Theorie zum größten Teil unberücksichtigt geblieben. Daher kann Hartley die Ergebnisse von Mc Bain auch mit einer einzigen Mizellenart deuten.

Die aus den Paraffinkettenionen der Seifen sich zusammensetzende Mizelle soll nach Hartley[5] kugelförmig sein; die zusammenhaftenden Paraffinketten bilden das als flüssig anzusehende Innere[6], während die ionisierten Säure- oder Basenreste an der Mizelloberfläche mit dem Wasser in Berührung stehen und dem Gebilde eine gewisse Stabilität verleihen. Ein Teil der Ladungen ist durch assoziierte Gegenionen praktisch neutralisiert.

Da diese Mizelle einen Gleichgewichtszustand darstellen sollte, folgert Hartley, daß die Größe der Mizelle einen konstanten, von der Konzentration unabhängigen Wert haben müßte. Durch Diffusionsmengen konnten Hartley und Runnicles[7] zeigen, daß im Gebiet zwischen der kritischen Konzentration und etwa dem Leitfähigkeitsminimum der Radius der Mizellen fast unverändert 25—27 Å, das sind etwa 4—5 Å mehr als die Länge des untersuchten Paraffinkettenions, beträgt.

Die Hartley'sche Theorie stellt somit im ganzen gesehen eine Ablösung der Mc Bain'schen Theorie dar, da sie imstande ist, alle elektro-

[1] J. W. Mc Bain und C. S. Salmon, J. Amer. chem. Soc. **42** (1920) 46; J. W. Mc Bain, M. E. Laing und A. F. Titley, J. chem. Soc. London **115**, 1292 (1919).
[2] Mc Bain und Jenkins, J. chem. Soc. London **121**, 2325 (1922); Mc Bain und Tsun Hsien Lin, J. Amer. chem. Soc. **53**, 59 (1931).
[3] G. S. Hartley, Aqueous Solutions of Paraffin-Chain Salts (Paris 1937).

[4] A. Lottermoser und F. Püschel, Kolloid-Z. **63**, 175 (1933).
[5] G. S. Hartley, loc. cit.
[6] Dies kann nach Hartley aus der Lösungsfähigkeit der Seifenlösungen für in Wasser unlösliche organische Substanzen geschlossen werden.
[7] Hartley und Runnicles, Proc. Roy., London Ser. A. **168**, 420 (1938).

E3
Kolloid-Zeitschrift **89**, 224–33 (1939)

From the Kaiser-Wilhelm-Institute of Physical Chemistry and Electrochemistry,
Berlin-Dahlem

ON MICELLAR TYPES OF AQUEOUS SOAP SOLUTIONS

by

Joachim Stauff (Berlin)

(Received 13 September 1939)

1. Introduction

Soaps and similar structured compounds have been known for a long time to be the best model substances for the study of the behaviour of micellar colloids in aqueous deflocculation. Their ability to split into ions in water is one of their most important advantages. This has made it possible to apply the extensive experimental and theoretical knowledge of aqueous electrolyte solutions to micellar colloids. It is thus understandable that most of the theories of the constitution of soap solutions have been obtained from results of electrochemical investigations.

By the same token, the classical theory of soap solutions by McBain[1] was developed mainly from the interpretation of the concentration dependence of the conductivity using the classical Arrhenius theory of electrolytes. Combining these measurements with those of the vapour pressure, McBain reached the view that two different micelle types must be present in soap solutions. These he named 'ion micelle' and 'neutral colloid'. The former must be produced by ion aggregation, the latter by aggregation of undissolved molecules. McBain and coworkers tried to provide evidence for these ideas from a large number of experiments, especially to prove directly the existence of both types of different micelles.[2]

Hartley,[3] however, strongly distrusted McBain's theory, as well as the proof of both types of micelles, partly because of the experimental conditions, and partly

---

[1] J.W. McBain and C.S. Salmon, J. Amer. Chem. Soc. **42**, 46 (1920); J.W. McBain, M.E. Laing and A.F. Titley, J. chem. Soc. London **115**, 1292 (1919).

[2] McBain and Jenkins, J. chem. Soc. London **121**, 2325 (1922); McBain and Tsun Hsien Lin, J. Amer. Chem. Soc. **53**, 59 (1931).

[3] G.S. Hartley, Aqueous Solutions of Paraffin-Chain Salts (Paris 1937).

because of the interpretation of the results. He therefore put forward a new theory of soap solutions which assumed only a single type of micelle. According to McBain the formation of both micelle types begins roughly in the region where the conductivity curve exhibits a minimum. But according to Hartley, micelles should occur at much lower concentrations. The strongest support for this idea has been given above all by numerous observations at the so-called 'critical concentration', first observed by Lottermoser and Püschel.[4] At this concentration various properties of the solutions suddenly change. This can only be explained by the aggregation of long-chain ions, together with a simultaneous association of counter ions. A further advantage of Hartley's theory is the role of electrostatic interactions between the ions. This interaction was not adequately considered in McBain's theory. Consequently, Hartley is able to interpret McBain's results assuming only a single micelle type.

According to Hartley[5] a micelle which is composed of ionic paraffin chains of the soap should be spherical. The interacting paraffin chains then form the centre,[6] which may be regarded as liquid, whilst the ionised acid or base groups are in contact with water at the micelle surface giving a certain stability to the whole structure. To a partial extent, the charges are almost neutralised by associated counter ions.

As such a micelle should be in an equilibrium state, Hartley concludes that the micelle dimension must have a constant, concentration-independent value. Using diffusion measurements, Hartley and Runnicles[7] were able to show that the micelle radius is nearly constant at 25 to 27 Å in the region between the critical concentration and around the conductivity minimum. This value exceeds the length of the ionic paraffin chain under discussion by about 4 to 5 Å.

Thus, because it is capable of making all the electrolytic phenomena self-consistent, Hartley's theory generally represents an improvement of McBain's theory.

These ideas, however, to some extent contradict both the behaviour of soap solutions at higher concentrations[8] and results obtained by non-electrochemical methods.

Thiessen and coworkers[9] established in several investigations that larger anisometric particles already occur at intermediate concentrations. Thiessen[10] concluded that particles of different dimensions are present in the solutions, with the dimensions and number of particles depending on concentration and temperature.

---

[4] A. Lottermoser and F. Püschel, Kolloid-Z. **63**, 175 (1933).

[5] G.S. Hartley, op. cit.

[6] According to Hartley this can be concluded from the solvent power of soap solutions for organic compounds which are insoluble in water.

[7] Hartley and Runnicles, Proc. Roy. Soc. London, Ser. A. **168**, 420 (1938).

[8] Hartley believes that a totally different behaviour of soap solutions can be observed only at very high concentrations.

[9] P.A. Thiessen and E. Triebel, Z. phys. Chem. Abt. A, **156**, 309 (1931); P.A. Thiessen and R. Spychalski, Z. phys. Chem. Abt. A, **156**, 435 (1931); P.A. Thiessen, Z. phys. Chem. Abt. A, **156**, 457 (1931).

[10] P.A. Thiessen in R. Zsigmondy, Kolloid-chem. **2**, 167 (Leipzig 1927).

From the viscosity of concentrated solutions, Lawrence[11] also concluded that larger, anisometric particles occur. He therefore assumed that a kind of secondary aggregate was formed from Hartley's spherical micelles, which exhibited an extended shape.

Using X-ray investigations of concentrated, flowing soap solutions, Heß and Gundermann[12] also established a flow anisotropy. They explained this phenomenon by the existence of crystalline, lamellar particles.

The contradiction with the Hartley theory is significant insofar as it is based only on the presence of spherical micelles of constant dimensions. This state alone is assumed to be a (concentration-independent!) minimum in the potential energy. However, the experimental results obtained for concentrated solutions are striking evidence that at least in these cases other states must play a considerable rôle.

The question now arises whether this discrepancy may be resolved, and whether there is a simple relation between the conflicting properties. Only two possibilities appear to answer the question:

1. At any time there is a unique particle type of uniform structure. Consequently, the anisometric particle observed at higher concentrations must result from a continuous change of dimension and size, starting with the sphere at lower concentrations.
2. Additionally, in addition to the Hartley particles, there is a second type of particle, with another dimension, size and structure. These particles must be in equilibrium with the Hartley particles, and at higher concentrations the equilibrium must shift very much in their favour.

In a preliminary communication[13] we interpreted the results of X-ray investigations and the measurements of the rate of nuclei formation in sodium palmitate solutions at 75°. We assumed that in these solutions a second, larger, particle type must be present in addition to the Hartley micelles. We proposed naming this micellar type 'Large Micelles', with the Hartley particles as 'Small Micelles'. The Large Micelles, however, should not be identical with McBain's 'neutral colloid', because they are neither uncharged nor crystalline, as we shall demonstrate in this paper.

Meanwhile, Heß, Philippoff and Kiessig[14] have also published X-ray investigations, together with studies of viscosity, density and refractive indices on sodium decanoate and sodium dodecyl sulfate. From these experiments they determined three solution states for these substances. Since the first solution state is seen as

[11] A.S.C. Lawrence, Trans. Faraday Soc. **31**, 189 (1935).

[12] K. Heß and Gundermann, Ber. Dtsch. Chem. Ges. **70**, 1800 (1937).

[13] J. Stauff, Naturwiss. **27**, 213 (1939).

[14] K. Heß, W. Philippoff, H. Kiessig, Kolloid-Z. **88**, 40 (1939).

a real molecular solution, their results confirm our theory of Large and Small Micelles.

In this paper we will present detailed results of X-ray investigations on two different soap solutions, sodium tetradecyl sulfate and sodium laurate.

2. Experimental

The X-ray photographs of the soap solutions were carried out using a heatable X-ray chamber with a copper block as sample holder, which was equipped with a vertical bore for the Mark tube and could be illuminated perpendicularly by the X-rays. The block was adjusted on a self-made Laue camera.[15]

Because it was necessary to measure the X-ray intensities, a NaCl single crystal was adjusted between sample and film with the 100 plane perpendicular to the primary beam. Using this, a Laue pattern was obtained together with the sample pattern. This procedure offers the advantage that the pattern could be measured photometrically at the same time as the NaCl reflections.

The solutions were prepared by weighing the substance in the required amount of water in a sealed glass tube, which was raised to higher temperatures. The high concentration solutions were annealed for many days until they became totally homogeneous.

The photometrical measurements were performed with a Zeiss scanning photometer.

Diffraction analyses were performed for several solutions of sodium tetradecyl sulphate and sodium laurate at room temperature and at 75 °C.

At room temperature the solutions are opaque and cloudy, containing the substance in the form of a dispersed precipitate. At 75 °C, however, the solutions are totally optically clear and contain the substance in deflocculated colloidal form, as is well known.

A few experimental details related to the X-ray pictures have been omitted here

The results obtained are given in Table 1.

3. Structure of the diffracting substance

The first results for soap solutions were published by Krishnamurti,[16] who had already pointed out the remarkable similarity between the diagrams and those of paraffins in the liquid state. Heß and Gundermann[17] investigated solutions of

[15] unpublished.
[16] K.Krishnamurti, Indian J. Physics 3, 209, 307 (1928).
[17] Heß and Gundermann, op. cit.

Table 1

Conc.	20°			75°	
	$d_1/\text{Å}$	$d_2/\text{Å}$	$d_3/\text{Å}$	$d_1'/\text{Å}$	$d_2'/\text{Å}$
			Na-tetradecylsulfate		
0.1 N	4.6	4.0	–	–	–
0.25 N	4.6	3.99	38.2	4.6	–
0.4 N	4.55	4.00	39.0	4.6	–
0.6 N	4.57	4.03	38.3	4.62	~60–65
40%	4.53	3.95	38.5	4.58	55.0
60%	4.55	3.98	38.0	4.57	48.2
80%	4.57	3.98	38.1	4.59	42.4
100%	4.55	3.98	38.2	–	–
Mean	4.56	3.98	38.3	4.59	
			Na-laurate		
0.2 N	–	4.05	–	–	–
0.4 N	4.73	4.05	29.5	4.57	–
0.6 N	4.72	4.03	30.0	4.58	~40–42
1.0 N	4.70	4.05	29.5	4.55	39
35%	4.74	4.08	29.8	4.62	35.6
70%	4.73	4.04	29.7	4.58	32.0
100%	4.73	4.04	29.8	–	–
Mean	4.73	4.04	29.8	4.58	

sodium oleate and observed the same diffraction rings. They discussed the X-ray interference in terms of crystalline colloidal particles in the solution. In the paper by Heß, Philippoff and Kiessig[18] the same origin is assumed. Their opinion is confirmed by the sharpness of the interference rings.

In view of the fact that Hartley[19] considered the micelles of soap solutions as liquid and structureless, it seemed to be worthwhile to investigate this question in some more detail.

A decision could best be taken by comparing the diagrams of soap solutions with that of solid or molten soap. However, the soaps partially decompose during melting. They are therefore not entirely appropriate for investigations. So, for comparison, a substance with a closely similar structure was used. It turned out that the pattern for palmitic acid at 75 °C was quite similar to those of the soap solutions. This could best be shown by the photometer curves.

In Fig. 1 the photometer curves of the soap deflocculations at 20 °C and 75 °C are compared with those of liquid palmitic acid.

[18] Heß, Philippoff and Kiessig, op. cit.
[19] Hartley, op. cit.

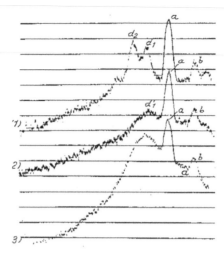

Fig. 1 Photometer curves of 60 per cent sodium tetradecyl sulphate solutions at 20°
(1) and 75° (2) and of liquid palmitic acid at 75° (3). d_1, d_2, and d_1' denote the
interference maxima of the substances under study, a and b result from the test
substance NaCl. Distance substance – film: 23.3 mm.

At 20 °C the curves correspond totally with the solid crystallised soap. This is
in accord with the results of Thiessen and Spychalski[20] who found by X-ray
investigation of gel fibers of sodium palmitate and sodium stearate that the
structural elements of solid/water mixtures consist of a crystalline anhydrous
substance.

At 75 °C [curve (2)] the similarity of the patterns for the soap solutions and the
liquid fatty acids is quite obvious. In both cases the position and half-width of the
intensity maximum are nearly equal.

It is known from the investigations of Stewart and Morrow[21] concerning the
structure of liquids with extended molecules (paraffins, long-chain alcohols and
fatty acids) that the molecules are oriented parallel in certain regions of the
liquid. This orientation causes a definite X-ray interference which corresponds to
the average lateral molecular separation. It is remarkable that this distance of
about 4.6 Å is almost independent of the molecular length. Moreover, the presence
or absence of an end group is also not relevant.

A further diffraction feature associated with the molecular length occurs in the
case of alcohols and fatty acids, but not for paraffins. Its half-width is considerably
larger in liquid fatty acids than in soap solutions at 75 °C. Morrow's[22] intensity

[20] Thiessen and Spychalski, op. cit.
[21] G.W. Stewart and R.M. Morrow, Physical Rev. **30**, 232 (1927).
[22] R.M. Morrow, Physical. Rev. **31**, 12 (1928).

curve for liquid undecanoic acid exhibits a half-width of about 1.5° whilst the values for a 35% sodium laurate solution are 0.6° at 75 °C and 0.3° at 20 °C which have been derived from the photometer curves of Fig. 2. Obviously, the half-width of the solution at 75 °C is in between those of the liquid fatty acid and the crystalline substance.

In this state the molecules seem to possess an order over larger regions with respect to the long direction than in liquid fatty acids. Hence, the regions of order must be anisometric.

In summary, we reach the following view. The X-ray interference of the solutions at 75 °C were caused by colloidal micelles because the solutions are optically transparent. Within the micelles the paraffin chains of the molecules are oriented parallel, similar to liquid fatty acids. The average lateral distance between these chains is the same as for all structurally similar substances. With respect to their long direction, the packing of the molecules in micelles is more highly ordered than in liquid fatty acids. The molecules cannot be shifted arbitrarily with respect to each other but must be arranged as shown in Fig. 3. This situation would correspond to the smectic state of crystalline-liquid phases. Hence, the micelle structure should be considered as crystalline-liquid.

The solid anhydrous soap is still totally crystalline at 75 °C. Its structure must be preserved when deflocculated into micelles, because the number of unit cells in the particle is still large with respect to the particle surface, as will be shown presently. From a purely thermodynamic point of view it is inconceivable

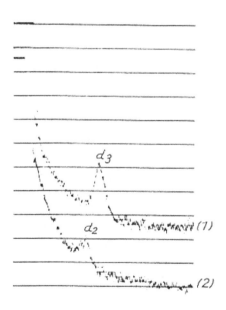

Fig. 2 Photometer curves of 35 per cent sodium laurate solutions at 20° (1) and 75° (2). Distance substance – film: 137.5 mm.

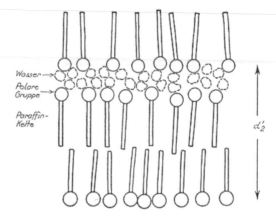

Wasser →
Polare Gruppe →
Paraffin- Kette

d_2'

Fig. 3 Sketch of the molecular arrangement in the micelles. d_2' corresponds to the distances given in Table 1.

that a particle consisting of solid anhydrous soap should not be crystalline at these temperatures.

The existence of at least a crystalline-liquid state can be supposed only because the Large Micelle consists of hydrated soap material. This also follows from the magnitude of the longitudinal distance, which should actually correspond to twice the molecular length.[23]

It can be seen from Table I that the distances d_2' at 75° increase with decreasing concentrations. If one accepts the value of 16.6 Å for the sodium laurate molecule 12×1.27 Å $= 15.2$ Å $+ 1.4$ Å for the NaO group[24], the double chain length amounts to 33.2 Å. At lower concentrations the values of sodium laurate considerably exceed this distance, if one neglects the value of 70 per cent.[25] No values of the length for the sodium tetradecyl sulfate molecule are known; however, the distances of 50 to 60 Å should significantly exceed twice the molecular length. The lattice extension can only be caused by the water penetration between the polar layers of the lattice, as has been schematically demonstrated in Fig. 3. Heß and Gundermann[26] also explain the magnitude of these distances by the presence of water.

The increase in distance on decreasing concentration of the solution can be simply explained by the increased water content of the polar layers on increasing the total water content. Furthermore, no diffraction from the crystalline substance is detectable in highly concentrated solutions at 75 °C. Thus, a water content of 20 per cent in this substance must already be sufficient to produce a totally homogeneous hydrated phase. This confirms that in the presence of 20 per cent water

[23] In crystal lattices the soaps form double molecules which adjoin each other with their polar groups. Cf. P.A. Thiessen and J. Stauff, Z. physik. Chem. **176**, 397 (1936).

[24] P.A. Thiessen and J. Stauff, op. cit.

[25] The long chains apparently are inclined with respect to the basic plane also in the hydrated phases.

[26] Heß and Gundermann, op. cit.

at 75 °C the anhydrous crystalline body cannot be thermodynamically stable. The micellar substance obviously also consists of hydrated soap molecules.

This is in agreement with the results of McBain and coworkers,[27] who established phase diagrams for various soap-water systems. Accordingly, there are several hydrated phases between the 'isotropic solution' and the solid, crystalline soap. This follows especially from the vapour pressure measurements of soap-water systems performed by Vold and Ferguson[28] and Vold and Vold.[29] The different thermodynamic behaviour of these phases must be due in some way to the different amounts of water inserted into the soap lattice. Vold and Vold[29] gave three possible explanations for this:

1. The water molecules are situated between the polar groups of the soap molecules.
2. They fill any lattice defects of the crystalline substance.
3. There are fibrous systems of soap colloids containing water bound at their surfaces.

In view of the present results, the second possibility can be definitely excluded, for the soap structure is totally changed with respect to the solid soap. In a system with defects, however, only a broadening of the interference lines without any positional change would occur. The line width would also be different at different concentrations. This is also not the case.

Vold's third possibility, however, is only a special case of either of the others, in which the soap with one of either possible structures is still fibrously defloccu-lated. Hence, strictly speaking, only the first possibility can be considered, in which the water is locally bound to the polar groups. Only this is in agreement with the present results.

This hydrate theory is compatible with the properties of solid soaps. Soaps form so-called ionic layer lattices. Because of these layers they are extremely stable up to high temperatures, in contrast to the molecular lattices of pure fatty acids. If this ionic layer lattice is broken up by inserting water molecules, the lattice forces causing the high stability are strongly reduced. The lattice now behaves in a similar way to a corresponding molecular lattice, e.g. that of a fatty acid. Thus, the molecular thermal motion at 75 °C is already sufficient to produce a state of reduced ordering. In the absence of water, this only occurs at much higher temperatures.

[27] J.W. McBain in Jerome Alexander's Colloid Chemistry (New York 1926).

[28] R.D. Vold and R.H. Ferguson, J. Amer. Chem. Soc. **60**, 2066 (1938).

[29] R.D. Vold and M.J. Vold, J. Amer. Chem. Soc. **61**, 37 (1939).

4.   Identification of large micelles

To decide if the X-ray interferences of colloid particles are caused by the enlargement of the Hartley micelles, or are caused by another larger type of particle, we proceeded as follows.

The intensities of the data at various concentrations were compared with the intensity standard of NaCl. The resulting relative intensity as a function of concentration is shown in Figs. 4 and 5.

Generally, the intensity of the diffracted X-rays is always proportional to the amount of the radiated substance. In the case of deflocculation of a nearly insoluble substance in water, the intensity is expected to be proportional to the amount of the dispersed substance.

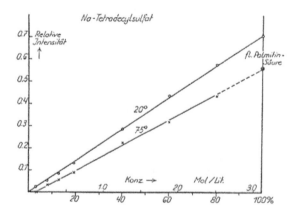

Fig. 4   Concentration dependence of the intensity of the reflections d_1 and d_1' of sodium tetradecyl sulfate. The 20° curve corresponds to d_1, the 75° curve to d_1'.

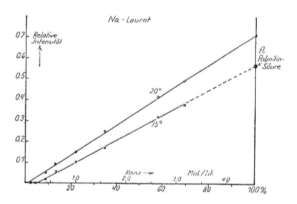

Fig. 5 Dependence of the intensity of the reflections d_1 and d_1' on the concentration of sodium laurate.

The molecular solubility of tetradecyl sulfate at 20 °C is about 0.003 mol per litre, i.e. 0.1 per cent; of sodium laurate 0.025 mol per liter, i.e. 0.6 per cent. These quantities are already outside the measurement accuracy and can be neglected. The molecular solubility of sodium tetradecyl sulfate at 75 °C is about 0.0031 and that of sodium laurate about 0.029 moles per litre as we have determined elsewhere.[30,31] Because of this small difference the molecular solubility even at 75 °C can be neglected.

As shown by Figs. 4 and 5, there is a direct proportionality between intensity and concentration only in the data obtained at room temperature. For tetradecyl sulfate the straight line runs through the origin, and at 100 per cent corresponds to the intensity of the solid substance. For sodium laurate, however, the influence of the considerably higher solubility is obvious. The data at 75 °C show another result. In fact, a straight line is also obtained, but this line does not run through the origin and cuts the abscissa. Below this intercept, X-ray interferences could not be detected even using very long exposure times. Furthermore, it is remarkable that at 100 per cent this line passes through the intensity for liquid palmitic acid. This is evidence that the scattering powers of tetradecyl sulfate, laurate, and palmitic acid for X-rays at 75 °C do not differ significantly from each other, and also that the state of order with respect to the lateral distances of all three substances is nearly equal.

Thus, the disappearance of X-ray interference patterns at a definite concentration is not caused by insufficient resolution of the X-ray diffraction method. The interference patterns due to tetradecyl sulfate and laurate disappear at quite different concentrations in spite of their equal scattering power. According to Heß, Philippoff and Kiessig[32] the scattering patterns of sodium caproate and dodecyl sulfate disappear at about 22 to 25 per cent. It cannot be supposed that the scattering powers of these substances differ significantly from those of ours. Thus the interference patterns disappear at a precise concentration, specific to each substance. The reason for this can only be that the scattering substance concerned itself disappears.

The question asked above was as follows. Are the particles producing the X-ray interferences identical with the Hartley micelles, do they develop from these, or is there another type of particle? The considerations above enable this question to be answered. In the first case, the intensity-concentration line at 75 °C should cut the abscissa approximately at the critical concentration of the substance concerned. It should thus exhibit the same intersection point as the intensity line at 20 °C, because of the small difference between molecular solubility and the critical concentration (see above). As this is not the case, one must assume that there is a different type of particle present from the Hartley micelle. It has been already

[30] J. Stauff, Z. physik. Chem. A.

[31] This fact rebuts the objection of Heß, Philippoff and Kiessig (op. cit.) against the comparison of soap systems at 20 °C and 75 °C, as in this case different solubilities should be present. The small difference of solubility, however, does not play any rôle.

[32] Heß, Philippoff, and Kiessig, op. cit.

proposed[33] to name this particle type the 'Large Micelle' and the Hartley particle the 'Small Micelle'.

The different dimensions of the two particles are as follows. Using diffraction measurements, Hartley and Runnicles[34] determined the micellar radius above the critical concentration up to the conductivity minimum. They obtained values of about 25 to 27 Å and observed that the radius practically does not change with concentration. In the X-ray data, periodicities corresponding to about 40 to 60 Å were observed, depending on the substance and concentration. Those periodicities, however, can only occur if there are several lattice planes of this distance in a particle. That means that the particle diameter must at least amount to a multiple of this distance. It then follows that these Large Micelles must exhibit a considerably larger diameter than Hartley's Small Micelles. For this reason alone they cannot be identical with the Hartley particles.

The exact determination of the concentrations at which the Large Micelles are first formed can be better achieved by extrapolation of the intensity-concentration curves than by visual observation, whether or not there is an interference. By counting and averaging the individual peaks of the photometer curve it is possible to measure the intensity maximum with an accuracy of 10 to 20 per cent even for weak intensities. Hence the values at low concentrations are relatively reliable.

5. Conclusion

The assumption of the presence of Small and Large Micelles in soap solutions has not only been derived from X-ray investigations, but also from the concentration dependence of the curve of rate of nuclei formation, as has been already pointed out in a preliminary communication.[35] Meanwhile, this theory has been confirmed by the investigations of Heß, Philippoff, and Kiessig,[36] who obtained qualitatively the same results in X-ray studies, which subsequently were supported by viscosity and density measurements. However, there is a large difference in their interpretation. They attribute a crystalline structure to the Large Micelles and regard the concentration range of Small Micelles as an intermediate state.

Philippoff and Kiessig[37] observed X-ray interference also in benzene-containing soap solutions. From the increase of the distances d_2' they concluded that benzene was incorporated into the micelles between the adjacent CH_3 groups of the paraffin chains. Also in this case the appearance of X-ray interferences was taken as evidence generally for the crystalline structure of the micelles.

However, this assumption is by no means justified. The broadening of the interference rings of the soap solutions with respect to the solid soap cannot be

[33] J. Stauff, op. cit.

[34] G.S. Hartley and Runnicles, op. cit.

[35] J. Stauff, op. cit.

[36] Heß, Philippoff, and Kiessig, op. cit.

[37] Philippoff and Kiessig, Naturwiss. **27**, 593 (1939).

exclusively related to a colloidal deflocculation of the crystalline soap substance. This is the result of the following considerations:

1. If the micelle consists of a crystalline, colloid-dispersed substance, a coincidence of the rings could occur because of a large change in the half-width of the X-ray interferences. The resulting new ring should be positioned in between both rings of the coarse-crystalline substance. This is not the case, as follows from comparing d_1 and d_2 with d_1' (cf. Table I).
2. The half-width of the X-ray interference always reaches its largest value when the substance is fluid. The values of less strongly dispersed crystalline forms must consequently always be smaller. The diagrams of the soap solutions at 75 °C, however, exhibit practically the same half-width as the liquid palmitic acid.

According to the results given above, the structure of the Large Micelles can be considered with certainty as crystalline-liquid.

However, the range of the Small Micelles was investigated by Hartley[38] and other authors in such an accurate manner that there can be no doubt about the existence of the defined state of Small Micelles and its special behaviour.

Recently, McBain[39] also has accepted the existence of a critical concentration for micelle formation. He has checked his results of osmotic pressure of the soap solutions. In so doing, he also confirmed his theory that generally *two* micelle types must exist in the solutions. This is in agreement with the viewpoint developed in this paper. However, he assumed that the type of micelle which he calls a 'neutral colloid' should not possess any kind of charge. This cannot be correct for the following reasons:

Small Micelles carry charges on their surface. According to Hartley[40] the separation of these charges is given by the molecular distance, of about 5 Å on average. These charges are formed by dissociation of the polar groups of the soap molecules and can be neutralised by associated counter ions to a degree, which depends on concentration.[41] According to our results the distance between two molecules in Large Micelles is about 4.5 Å, and hence practically equal to the distance in Small Micelles.

There is no reason to assume that a soap molecule at the surface of a Large Micelle behaves in principle in any other way than at the surface of a Small Micelle. The possibility of ion formation is available in both cases, although there may exist differences in the degree of association of counter ions, and in the way the surrounding ion atmosphere forms. The Large Micelle certainly cannot be regarded as uncharged, and it therefore has nothing to do with McBain's

[38] G.S. Hartley, op. cit.

[39] J.W. McBain, J. physik. Chem. **43**, 458 (1939).

[40] G.S. Hartley, loc. cit.

[41] The association is a pure electrostatic one as was shown by high frequency conductivity measurements by G. Schmidt and E. Larsen, Z. Elektrochem. **44**, 651 (1938).

neutral colloid. The difference between the two micelle types only applies to their dimension and structure, but not to their electrical properties.

The structure of the Large Micelles leads to an explanation of the osmotic behaviour of the solutions found in the results of McBain and Salmon.[42] Because of the layers (cf. Fig. 3) larger amounts of counter ions must be incorporated inside the micelle, i.e. most of the inside of the micelle consists of neutral substances.

From his measurements McBain draws the conclusion that in the solutions one part of the substance must be osmotically and electrically inactive. However, one need not require that this part be formed from a special type of uncharged colloidal particles, and one cannot justify their separate existence. During the formation of Large Micelles one part of the substance also becomes osmotically and electrically inactive because undissociated molecules accumulate inside the micelle.

Large Micelles cannot be formed simultaneously with Small Micelles simply from dissolved ions, but must evolve from Small Micelles by aggregation, and be in equilibrium with the latter. This assumption is supported by the way in which the Large Micelles are formed, which does not occur at a precisely defined concentration but over a rather broad concentration range. Otherwise there should exist a critical concentration, similar to the formation of Small Micelles, at which the solution properties change discontinuously. However McBain's ideas, by which both micelle types should exist simultaneously in comparable amounts over a broad concentration range, cannot be correct. The intensity-concentration curves clearly show that the proportion of Small Micelles rapidly decreases on increasing the concentration, so that at high concentrations almost exclusively Large Micelles must be present.

The Large Micelles are thus identical with those particles which cause the flow birefringence. As a consequence their shape must be anisometric. Thus, the contradiction between Hartley's theory and the results of Thiessen and coworkers, as well as with Lawrence and also with Heß and Gundermann (c.f. above) has been clarified.

Finally, it is worthwhile to note that the formation of Large Micelles always begins in the range in which one will also find the minimum of the conductivity curve. This fact suggests the interpretation of the subsequent increase of the conductivity curve as due to the formation of Large Micelles.

As already mentioned, a certain number of ions disappear by neutralisation in the micelle interior. In this way, firstly the concentration of the counter ions is reduced, and secondly the number of micelles, i.e. also the micelle concentration, decreases. From the experience of modern electrolytic theories, such a process should always cause an enhancement of the activity coefficients. However, the degree may be modified in individual cases. At all events this enhancement causes an increase of the equivalent conductivity. The concentration decrease resulting from Large Micelle formation does not enter this quantity, because the equivalent conductivity is always related to the total concentration.

[42] J.W. McBain and C.S. Salmon, op. cit.

Summary

Summarising, it follows from the preceding remarks:

1 From X-ray studies in different soap solutions, it has been established that larger colloidal particles appear in sodium tetradecyl sulfate at 0.1 M and sodium laurate at 0.2 to 0.25 M. These particles have been named Large Micelles, in contrast to the smaller colloidal aggregates below the concentrations given above.

2 The structure of the Large Micelles has been shown to be crystalline-liquid, and these micelles consist of a soap/water system. According to the X-ray results the water is situated *between* the polar groups of the soap molecules.

3 These results enable the contradictions between the results and theory of Hartley to be reconciled with those of McBain, Thiessen and others. With respect to the state of charge, there is only one micelle type consistent with Hartley's view. However, regarding the dimension, size, and space filling, two different types of micelles exist, consistent with Zsigmondy's opinion.

4 The formation of Large Micelles can furthermore explain both the minimum in the conductivity curves of the soap solution and its osmotic behaviour.

* * *

I am grateful to Prof. Dr P.A. Thiessen for supporting this work and for stimulating discussions. I also thank the Deutsche Forschungsgemeinschaft for supplying apparatus, and the Henckel & Cie. Company for donating materials.

Joachim Stauff was born in Berlin in 1911. He studied chemistry at the Universities of Berlin, Göttingen, and Munich, obtaining his Ph.D thesis in Berlin in 1936. Between 1936 and 1940 he worked as a senior scientist at the Kaiser-Wilhelm-Institute of Physical Chemistry in Berlin-Dahlem. During this period he made intensive and pioneering studies of the structure of micelles in soap solutions. His paper reprinted in this volume is a product of this period.

In 1940 he moved to the Institute of Physical Chemistry at the University of Frankfurt/Main, and from there obtained the habilitation degree. From 1942 to 1945 he was a member of the Institute of Physical Chemistry at the Technical University of Darmstadt.

After the end of the war till 1953 he was employed at the Hoechst Company at Frankfurt. In 1953 he was appointed Professor at the University of Frankfurt, and in 1966 he was promoted to Professor of Colloid Chemistry and Director of the Institute of Physical Biochemistry.

He retired in 1976. Nevertheless he continued work as a member of a research group at the Technical University of Darmstadt till 1982. Apart from his work in micelle colloids and soap solutions, his scientific interests included photochemistry, luminescence, and protein chemistry.

E4

Annals of the New York Academy of Sciences **51**, 627–59 (1949)

THE EFFECTS OF SHAPE ON THE INTERACTION
OF COLLOIDAL PARTICLES

By Lars Onsager

Sterling Chemistry Laboratory, Yale University, New Haven, Connecticut

Introduction. The shapes of colloidal particles are often reasonably compact, so that no diameter greatly exceeds the cube root of the volume of the particle. On the other hand, we know many colloids whose particles are greatly extended into sheets (bentonite), rods (tobacco virus), or flexible chains (myosin, various linear polymers).

In some instances, at least, solutions of such highly anisometric particles are known to exhibit remarkably great deviations from Raoult's law, even to the extent that an anisotropic phase may separate from a solution in which the particles themselves occupy but one or two per cent of the total volume (tobacco virus, bentonite). We shall show in what follows how such results may arise from electrostatic repulsion between highly anisometric particles.

Most colloids in aqueous solution owe their stability more or less to electric charges, so that each particle will repel others before they come into actual contact, and effectively claim for itself a greater volume than what it actually occupies. Thus, we can understand that colloids in general are apt to exhibit considerable deviations from Raoult's law and that crystalline phases retaining a fair proportion of solvent may separate from concentrated solutions. However, if we tentatively increase the known size of the particles by the known range of the electric forces and multiply the resulting volume by four in order to compute the effective van der Waal's co-volume, we have not nearly enough to explain why a solution of 2 per cent tobacco virus in 0.005 normal $NaCl$ forms two phases.

General Kinetic Theory and Conventions. Some care is needed when we apply the general principles of statistical thermodynamics to solutions of colloidal particles. On one hand, any force acting on a particle of whatever size is important as soon as the work of the force is comparable to kT. On the other hand, the presence of one colloidal particle will usually affect the free energy of dilution of the electrolyte present by a large multiple of kT. This difficulty must be circumvented by all theories and experiments pertaining to the distribution of colloidal particles. One suitable piece of experimental apparatus is an osmometer whose membrane is impermeable to the colloidal particles, but permeable to all small molecules and ions of the electrolytic solvent. The osmotic pressure measured across such a membrane will be exactly proportional to the number of particles if the solution behaves like an ideal gas. The analogy can be extended to real gases and real solutions, whereby the gas pressure still corresponds to osmotic pressure.

The imperfection of an ideal gas can be computed when we know the forces between the molecules for every configuration. For that purpose, we have to evaluate the integral

$$B(T) = \int e^{-u/kT} \, d\tau / N!$$

(1)

where u stands for the potential of the forces and $d\tau$ denotes a volume element in configuration-space. The free energy of the gas in terms of this integral is

$$F(N, V, T) = N\mu_0(T) - kT \log B(N, V, T)$$

(2)

where the additional function $\mu_0(T)$ depends only on the temperature and does not enter into the computation of the pressure,

$$P = -(\partial F/\partial V)_{N,T} = kT(\partial \log B/\partial V).$$

(3)

The osmotic properties of a colloidal solution can be computed by a similar procedure. What we need to know initially is the *potential*

$$w((q_1), (q_2), \ldots (q_N))$$

of the *average forces* which act between the particles in a configuration described by the sets of coordinates $(q_1), (q_2), \ldots (q_{N_p})$ of particles $1, 2, \ldots N_p$. In general, it is necessary to specify the orientations of the particles as well as the positions of their centers, and the work against the corresponding torques must be included in w.

With

$$B_\nu(N_p, V, T) = \int e^{-w/kT} \, d\tau / N_p!$$

(4)

we have then

$$F \text{ (solution)} - F \text{ (solvent)}$$

$$= N_p \, \mu_p^\circ \, (T, \text{solvent}) - kT \log B_p \left(N_p, V, T \right).$$

(5)

Here, the difference between "solution" and "solvent" means that the former contains colloidal particles, and we compare solutions of different colloid concentrations $c = (N_p/V)$ always in "dialytic" equilibrium across an osmometer membrane with a "solvent" of constant composition. The proportions of ions and molecules present between the particles in the colloidal solution may differ from those in the "solvent" as we have defined it. This complication can hardly be avoided if we want simple relations and *precise interpretation of practicable experiments.*

With these conventions the analog of EQUATION 3,

$$P = kT(\partial \log B_p/\partial V),$$

(6)

is valid for the osmotic pressure and

$$\mu_P = \mu_P^0 - kT(\partial \log B_p/\partial N_p) \tag{7}$$

for the chemical potential of the colloidal particles.

Moreover, the conditions for coexistence of two phases are simply

$$P = P' \tag{8a}$$

$$\mu_p = \mu_p' \tag{8b}$$

The assumed dialytic equilibrium takes care of all small molecules and ions.

Electric Forces. According to theories developed by Helmholtz, Lamb, and Smoluchowski, the speed of migration of a colloid in an electric field is quantitatively related to the potential difference between the first mobile layer of liquid in contact with the particle and the bulk of the solution. It is customary to specify the electric charges of particles indirectly in terms of this so-called ζ-potential. The theory is still somewhat incomplete as regards cases where the thickness of the electric double layer is of the same order of magnitude as the dimensions of the particle; a factor variable between the limits of unity and 3/2 then enters into the interpretation.

With slight approximations, the general kinetic theory for the distribution of ions near charged particles leads to the well-known Poisson-Boltzmann differential equation for the electric potential

$$\nabla^2 \psi \equiv \frac{\partial^2 \psi}{\partial x^2} + \frac{\partial^2 \psi}{\partial y^2} + \frac{\partial^2 \psi}{\partial z^2} = -\frac{4\pi}{D} \sum_j n_j e_j e^{-e_j \psi/kT} \tag{9}$$

where e_1, e_2, $\cdots$ denote the charges of ions present in concentrations n_1, n_2, $\cdots$ (in the solution or, rather, in a "solvent" maintained in dialytic equilibrium), and D denotes the dielectric constant. Whenever the condition $|e_j \psi| \ll kT$ is satisfied for all kinds of ions present, EQUATION 9 may be replaced by

$$\nabla^2 \psi = \kappa^2 \psi$$

$$\kappa^2 = (4\pi/DkT) \sum_j n_j e_j^2. \tag{10}$$

The normal gradient of ψ at the surface of the particle is related to the charge density on the particle. We have to expect an implicit boundary condition determined by the adsorption and surface ionization in equilibrium with ions present at the surface in local concentrations $n_1 \exp(-e_1 \zeta/kT)$, $n_2 \exp(-e_2 \zeta/kT)$, etc. Since the kinetics of the surface ionization is rarely known, the relation

$$\psi = \zeta = \text{constant}; \quad \text{(at surface)} \tag{11}$$

has often been assumed, regardless of modifying factors, although systematic variations of ζ with electrolyte concentration, etc., should be expected and have been demonstrated in some cases. We shall not pursue these questions, because the expected variations of ζ will have but little effect on the forces between the particles.

In one-dimensional cases, $\psi = \psi(x)$, EQUATION 9 is generally soluble by quadratures. However, even the simplest case of a binary electrolyte between parallel plates, both maintained at the potential ζ, leads to elliptic integrals (Langmuir, 1938), and the resulting exact formula for the force is fairly involved (Verwey and Overbeek, 1948). Their approximation

$$K(x) = 16\, n\, kT(\tanh(e_1\zeta/4kT))^2\, e^{-\kappa x} \tag{12}$$

for the force per unit area between two parallel plates separated by a distance x is valid for not too small distances and will suffice as a basis for discussion. We note that the force decreases exponentially and that the screening constant κ, given by EQUATION 10, depends only on the ionic strength of the solvent. Moreover, for any fixed distance d between the plates, the force approaches a finite limit with increasing particle potential ζ. These two features are general.

We may use the result of EQUATION 12 to estimate the force between two infinite cylinders of the same diameter d crossing at an angle γ in such a manner that the mantles are separated by a distance x_0 between the points of closest approach.

We choose Cartesian coordinates in a plane parallel to the axes of both cylinders and identify points on the cylinder mantles by the coordinates (y, z) of their projections upon that plane. Then, the distance between points on the two cylinder mantles with the same (y, z) coordinates will be:

$$x(y, z) = x_0 + d - (\tfrac{1}{4}\, d^2 - y^2)^{1/2} - [\tfrac{1}{4}\, d^2 - (y\cos\gamma - z\sin\gamma)^2]^{1/2}.$$

If we allow the approximation

$$x(y, z) \sim x_0 + (y^2/d) + ((y\cos\gamma - z\sin\gamma)^2/d),$$

and compute the local force density $K(x)$ according to EQUATION 12, an elementary integration yields for the total force

$$\text{Average Force} = (\pi d/\kappa \sin\gamma)K(x_0) \tag{13}$$

and we obtain for the potential w of the average force

$$w/kT = (d/q \sin\gamma)[\tanh(e_1\zeta/4kT)]^2 e^{-\kappa x_0} \tag{14}$$

where we use the abbreviation

$$q = e_1^2/2\, DkT = z_1^2 \times 3.56 \times 10^{-8}\ \text{cm.} \tag{15}$$

and $e_1 = -e_2$ denote the charges, $z_1 = -z_2$ the valences of the ions in the solvent; the numerical value refers to water at 25°C.

As an example, we may consider two perpendicular cylindrical particles of $d = 150$ A. U., $\zeta = 0.15$ volt, in a 0.005 mol NaCl solution, whereby $1/\kappa = 43$ A. U. We find

$$w/kT = 34.5\, e^{-\kappa x_0} \tag{16}$$

which equals $e^{-c} = 0.561$ at a distance $x_0 = \delta = 4.12/\kappa = 184$ A. U.

Here, we have neglected the divergence of the electric force-lines, which must be quite appreciable because δ is by no means small, compared to d. We apply a correction of the right order of magnitude if we multiply w by the factor $d/(d + x_0)$; according to the corrected formula we then find $w = 0.561\, kT$ at a distance of about 151 A.U.

It will be evident that over a considerable range of particle diameters and orientations and over a wide range of concentrations of electrolyte, the effective range of the electrostatic repulsion will be a modest multiple of the screening distance $1/\kappa$. While exact computations are not available, there can be little doubt about the orders of magnitude involved.

One further observation is in order: unless the electric double layers of three particles overlap *in the same region*, the repulsive forces are additive. When $\kappa d \gg 1$, the exceptional configurations are just about impossible; but even under much less stringent conditions very few of them can occur. On the strength of these estimates, we shall treat the electrostatic repulsion as an additive *short range effect*. For very low concentrations of ions such that κd is small, our procedure may be unreliable. On the other hand, we shall make no allowance for differential van der Waal's attraction. This omission would tend to become particularly serious for high concentrations of ions and low ζ-potentials, under conditions approaching those which lead to flocculation of the particles.

Imperfect Gas Theory. We proceed to evaluate the configuration integral of Equation 4 according to the general method developed by Mayer and Mayer. Assuming additive forces:

$$w = w_N((q_1), \cdots (q_N)) = \sum_{i<j} w_{ij}\, ; \tag{17}$$

$$w_{ij} = w_2((q_i), (q_j))$$

we put

$$\Phi_{ij} = \Phi_{ij}((q_i), (q_j)) = e^{-w_{ij}/kT} - 1. \tag{18}$$

In order to avoid confusion with a distribution-function f, we write Φ_{ij} for the functions which Mayer and Mayer denote by f_{ij}. Their notation sometimes implies the hypothesis that w_{ij}, and with it Φ_{ij}, depends only on the distance between two particles. Their specialization is not essential and their method is valid, with obvious pertinent modifications, for the more general case with which we have to deal.

Upon suitable rearrangement of the sum

$$e^{-w/kT} = 1 + \sum_{i>j} \Phi_{ij} + \sum \Phi_{ij} \Phi_{i'j'} + \cdots \qquad (19)$$

which now constitutes the integrand of EQUATION 4, Mayer and Mayer obtain an expansion for the integral in terms of the irreducible cluster integrals

$$\beta_1 = \frac{1}{V} \int \Phi_{12} \, d\tau_1 \, d\tau_2 \qquad (20)$$

$$\beta_2 = \frac{1}{2V} \int \Phi_{12} \, \Phi_{23} \, \Phi_{31} \, d\tau_1 \, d\tau_2 \, d\tau_3$$

and these furnish the first two correction terms to the ideal gas laws in the expansion

$$\log B_p = N_p \{ 1 + \log (V/N_p) + \tfrac{1}{2}\beta_1(N_p/V) + \tfrac{1}{3}\beta_2(N_p/V)^2 + \cdots \}. \qquad (21)$$

Similarly, for a solution which contains $N_1, \cdots, N_s, \cdots$ particles, of different types $1, \cdots, s, \cdots$, respectively, we have

$$\log B_p = \sum_s N_s (1 + \log (V/N_s)) + \frac{1}{2V} \sum_{s,s'} \beta_1(s, s') N_s N_{s'}$$

$$+ \frac{1}{3V^2} \sum_{s,s',s''} \beta_2(s, s', s'') N_s N_{s'} N_{s''} + \cdots . \qquad (22)$$

The arguments of the cluster integrals indicate that the functions Φ_{12}, $\Phi_{23}, \cdots$ involve the interaction potentials w appropriate to pairs of particles from the sets of types (s, s'), (s, s', s''), etc.

In EQUATION 21, the generalized volume elements $d\tau_i$ are ordinary volume elements whenever the forces are central, so that w_{ij} and with it Φ_{ij} depend only on the distance between the two particles involved; but we shall be very much interested in the mutual orientations of the particles. In dealing with *isotropic* solutions, we have two alternative procedures at our disposal. The first method is to include an averaging over orientations (Ω) in the definitions of volume elements, thus

$$d\tau_i = dV_i \, d\Omega_i \bigg/ \int d\Omega_i. \qquad (23)$$

With particles of axial symmetry, it is, of course, enough to specify the directions of the symmetry axes, so that, for a cylindrical particle, we may let $d\Omega$ be an element of solid angle including the direction a_i of the cylinder axis:

$$d\tau_i = dV_i \, d\Omega_i / 4\pi. \qquad (23a)$$

The second method is more general, in that it applies to anisotropic phases without periodic structure, in other words, liquid crystals of the *nematic* type. For the purpose of computing B_p, we then introduce the artifice that we treat particles of different orientation as particles of different kinds. The distribution of particles among different orientations is determined by the condition that B_p must be a maximum. Incidentally, the convention that the terms log (V/N) are now formed separately for each "kind" of particles makes due allowance for the entropy of "mixing" (Gibbs Paradox). On the other hand, we must remember that the generalized volume in space and orientation available to a particle of orientation restricted to an element of solid angle $d\Omega$ is only $Vd\Omega$, rather than $4\pi V$ for a particle of unrestricted orientation. Thus, when we divide the total of all directions in space among elements of solid angle $\Delta\Omega_1, \cdots, \Delta\Omega_\nu, \cdots, \Delta\Omega_s$ surrounding the directions $a_1, \cdots, a_\nu, \cdots, a_s$, respectively, these will have populations of particles which we shall denote by

$$\Delta N_\nu = N_p f(a_\nu)\Delta\Omega_\nu \; ; \nu = 1, 2, \cdots, s, \tag{24}$$

whereby, of course,

$$\sum_1^s \Delta N_\nu = N_p \sum_\nu f(a_\nu)\Delta\Omega_\nu = N_p. \tag{24a}$$

With this notation, EQUATION 21 is generalized as follows:

$$\log B_p = \sum \Delta N_\nu (1 + \log (V\Delta\Omega_\nu/4\pi\Delta N_\nu))$$

$$+ \frac{1}{2V} \sum_{\nu,\nu'} \beta_1(a_\nu, a_{\nu'})\Delta N_\nu N_{\nu'} \tag{25}$$

$$+ \frac{1}{3V^2} \sum_{\nu,\nu',\nu''} \beta_2(a_\nu, a_{\nu'}, a_{\nu''})\Delta N_\nu \Delta N_{\nu'} \Delta N_{\nu''} + \cdots.$$

Here, the cluster integrals of EQUATION 20 are computed for fixed orientations a_1, a_2, viz. a_1, a_2, a_3, of the particles involved. Replacing the sums by integrals in terms of the distribution-function $f(a)$, the integral of which is now normalized:

$$\int f(a) \, d\Omega(a) = 1, \tag{26}$$

we may write EQUATION 23 in the form

$$\log B_p = N_p \bigg\{ 1 + \log (V/N_p) - \int f(a) \log (4\pi f(a)) \, d\Omega(a)$$

$$+ (N_p/2V) \int\int \beta_1(a, a')f(a)f(a') \, d\Omega \, d\Omega' \tag{27}$$

$$+ (N_p^2/3V^2) \int\int\int \beta_2(a, a', a'')f(a)f(a')f(a'') \, d\Omega \, d\Omega' \, d\Omega'' + \cdots \bigg\}.$$

The Cluster Integrals. When the forces are repulsive at all distances, we have $w \geq 0$ everywhere, whence the functions Φ_{ij}, defined by EQUATION 19, satisfy the inequalities

$$-1 \leqq \Phi_{ij} \leqq 0 \tag{28}$$

everywhere. In this case, the first two cluster integrals β_1 and β_2, defined by EQUATION 21, are necessarily negative, because the integrands are formed from one and three negative factors, respectively. The former has a particularly simple geometrical meaning in the ideal case of "hard" particles, which repel each other at contact but do not interact otherwise. In this case, we have

$$w_{ij} = +\infty; \quad \Phi_{ij} = -1; \quad \text{(particles intersecting)}$$
$$w_{ij} = 0; \quad \Phi_{ij} = 0; \text{ otherwise}$$

and $(-\beta_1)$ then equals the volume which is denied to particle j by the condition that it must not intersect particle i. For a pair of spheres of radius r, the excluded volume is obviously a sphere of radius $2r$. This leads to the familiar result first derived by Boltzmann, that the van der Waal's "covolume" (per particle) equals four times the volume of one spherical particle

$$b = -\tfrac{1}{2}\beta_1 = 4(4\pi r^3/3) = 4v_p$$

The analogous problem for two cylinders of lengths of l_1, l_2 and diameters d_1, d_2 is solved in the Appendix; we reproduce here the result (from A 11)

$$-\beta_1(\gamma) = (\pi/4)d_1 d_2(d_1 + d_2) \sin \gamma \tag{30}$$
$$+(\pi/4)(l_1 d_1^2 + l_2 d_2^2) + (\pi/4)(l_1 d_2^2 + l_2 d_1^2)|\cos \gamma|$$
$$+(l_1 + l_2)d_1 d_2 E(\sin \gamma) + l_1 l_2(d_1 + d_2) \sin \gamma,$$

where $E(\sin \gamma)$ denotes the complete elliptic integral of the second kind

$$E(\sin \gamma) = \int_0^{\pi/2} (1 - \sin^2 \gamma \sin^2 \phi)^{1/2} d\phi. \tag{30a}$$

For special orientations or dimensions, the formula simplifies more or less. The following cases are instructive:

$\gamma = 0$: $\qquad -\beta_1 = (\pi/4)(l_1 + l_2)(d_1 + d_2)^2$ (31a)

$\gamma = 0$: $\quad l_1 = l_2 = l; \quad d_1 = d_2 = d: \; 8(\pi/4)l d^2$ (b)

$\gamma = \pi/2$: $l_1 l_2(d_1 + d_2) + (l_1 + l_2 + d_1 + d_2)d_1 d_2 + (\pi/4)(l_1 d_1^2 + l_2 d_2^2)$

$l_1 = l_2; \quad d_1 = d_2$: (c)

$(2 l^2 d + (\pi/2)d^3) \sin \gamma + \{(\pi/2)(1 + |\cos \gamma|) + 2 E(\sin \gamma)\}l d^2$, (d)

$l_1 = l_2 = 0$: $\qquad (\pi/4)d_1 d_2(d_1 + d_2) \sin \gamma$ (e)

$l_1 = d_2 = 0$: $\qquad (\pi/4)l_2 d_1^2 |\cos \gamma|$ (f)

$l_2 = d_2 = 0$: $\qquad (\pi/4)l_1 d_1^2$ (g)

$l_1 \gg d_1 + d_2 \ll l_2$: $\qquad l_1 l_2(d_1 + d_2) \sin \gamma.$ (h)

Case (b) yields 8 times the volume of one particle, as for spheres. This is generally true for centrosymmetrical convex particles in parallel orientation. Most of the others explain themselves. We call attention to the idealized cases (e) and (f), where the particles have a mutual covolume although neither has any volume, and to the case (h), which shows that the ratio (covolume/volume) for long needles is (length/diameter) rather than (4/1).

The theory of isotropic solutions involves a simple average of EQUATION 30 over all directions in space:

$$-\bar{\beta}_1(l_1, d_1; l_2, d_2)$$

$$= 2b = -\int \beta_1(\gamma_{12}) \, d\Omega_2/4\pi$$

$$= -\frac{1}{2}\int \beta_1(\gamma) \sin \gamma \, d\gamma$$

$$= (\pi/4)^2 d_1 d_2 (d_1 + d_2) + (\pi/4)(l_1 d_1^2 + l_2 d_2^2)$$
$$+ (\pi/8)(l_1 d_2^2 + l_2 d_1^2) + (\pi^2/8)(l_1 + l_2)d_1 d_2$$
$$+ (\pi/4)l_1 l_2(d_1 + d_2). \tag{32}$$

For details of the integration, we again refer to the Appendix (A 14). For particles of equal diameters $d_1 = d_2 = d$, EQUATION 32 simplifies

$$-\bar{\beta}_1(l_1, l_2) = 2b_{12} = \tfrac{1}{2}\pi d\{l_1 l_2 + \tfrac{1}{4}(\pi + 3)(l_1 + l_2)\,d + \tfrac{1}{4}\pi\,d^2\} \tag{33}$$
$$= 1.5708 \, d(l_1 l_2 + 1.5354(l_1 + l_2)\,d + 0.7854\,d^2), \cdot$$

and, when the lengths as well as the diameters are equal, it simplifies still a little further

$$- \bar{\beta}_1 = \tfrac{1}{2}\pi d(l^2 + \tfrac{1}{2}(\pi + 3)l\,d + \tfrac{1}{4}\pi\,d^2). \tag{34}$$

It is interesting to examine the ratio of covolume to volume as a function of the ratio (l/d) according to EQUATION 34. The ratio

$$-\bar{\beta}_1/2(\pi/4)\,d^2l = b/(\pi/4)\,d^2l = b/v_p$$

becomes in various limiting cases

$$b/v_p \sim l/d; \quad (l \gg d)$$

$$b/v_p = \text{minimum} = \pi^{1/2} + \tfrac{1}{2}(\pi + 3) = 4.843; \quad (l = (\pi/4)^{1/2}\,d)$$

$$b/v_p \sim (\pi/4)\,d/l; \quad (l \ll d).$$

When the dimensions are about equal the ratio is not much more than 4, but for highly anisometric particles, whether needles or pancakes, (b/v_p) is about equal to the ratio of the long to the short dimension.

While the evaluation of the first cluster integral β_1 defined by EQUATION 20 proved perfectly feasible, the integral β_2 depends on three directions, and to compute it exactly would be an extremely tedious task at best. For that reason we shall be content to estimate the order of magnitude of β_2.

The value for spheres of equal diameters was computed by Boltzmann; in that case, one finds

$$-\beta_2 = (15/64)\beta_1^2.$$

This result gives us the right order of magnitude of the ratio (β_2/β_1) for isometric particles in general. Where anisometric particles are concerned, we must distinguish between slender rods and thin plates. For the latter case, a little experimentation with various orientations will show that in most cases where two plates intersect each other, the volume within which a third plate of comparable diameter will intersect the other two simultaneously will be a sizable fraction of the volume within which it will intersect a given one of the others. Accordingly, barring special orientations, we have the result

$$\beta_2/\beta_1 = O(-\beta_1); \text{ (spheres, cubes, plates).} \tag{35}$$

For the slender rods, we obtain the same result only if the three rods are nearly coplanar, whereby the admissible deviation in angle is of the order (d/l). Otherwise, it is easily seen that if we look at a pair of interesting rods along the direction of a third, the projection of their intersection upon the plane normal to the axis of the third rod will be (at most):

$$d_1 d_2/\sin \phi_3 ,$$

where ϕ_3 is the angle between the projections of two rods, alias the angle between the planes containing the pairs of directions $(\mathbf{a}_1, \mathbf{a}_3)$ and $(\mathbf{a}_2, \mathbf{a}_3)$, respectively. Or, considering the spherical triangle whose corners have the directions $\mathbf{a}_1, \mathbf{a}_2, \mathbf{a}_3$, ϕ_3 is the angle at the third corner. If we denote the three sides of this triangle (angles between the directions pairwise) by γ_{12}, γ_{23} and γ_{31}, we arrive at the following estimate for the second cluster integral:

$$-\beta_2 = (d_1 + d_2)(d_2 + d_3)(d_3 + d_1)\{O(l^2 d) + l_1 l_2 l_3(\sin \gamma_{12}/\sin \phi_3)\};$$

$$(\phi_3 > d/l). \tag{36}$$

By the theorem of sine proportions, valid for spherical triangles, the quotient of the two sines is a symmetrical function of the three directions. The angle ϕ_3 vanishes (or equals π) whenever the three directions $(\mathbf{a}_1, \mathbf{a}_2, \mathbf{a}_3)$ are coplanar, whereby $\mathbf{a}_1$ and $\mathbf{a}_2$ are normally not parallel. For such directions, the estimate (36) becomes infinite; as we have mentioned above, the estimate (35) is then valid instead.

In computing the average of β_2 over all combinations of directions $(\mathbf{a}_1, \mathbf{a}_2, \mathbf{a}_3)$ we find that the combinations of directions which are coplanar within an angle $\pm\phi_m$ do form a fraction of the order ϕ_m of the total. Thus, when we integrate EQUATION 36 over all other orientations the average of the sine ratio will be of the order

$$-\log \phi_m + \text{const.}$$

When we put $\phi_m = d/l$ and substitute the estimate (35) for coplanar orientation, the added term does not change the order of magnitude, and we still obtain

$$-\bar{\beta}_2 = O(d^3l^3(\log (l/d) + \text{const})). \tag{37}$$

No concerted effort has been made to render this estimate more definite. As it is, the result

$$\bar{\beta}_2/(\bar{\beta}_1)^2 = O((d/l) \log (l/d)) \tag{38}$$

will offer some justification for the procedure which we shall, perforce, adopt in the following, where order corrections which depend on β_2 and higher cluster integrals will be neglected altogether. If we talk about "concentrated" solutions whenever $\beta_1 N/V$ is of the order unity or greater, then we may hope that our results will describe fairly concentrated *isotropic* solutions of rod-shaped particles reasonably well. The results for anisotropic solutions will be somewhat doubtful in all cases, and more so the more concentrated the solutions. Where plate-like particles are concerned, our approximations will introduce more serious errors, and we can hardly hope for more than that our result will describe concentrated solutions of such particles qualitatively rather than quantitatively.

We shall inquire, next, about the effects due to the finite range of the electrostatic repulsion between the particles.

We have mentioned before that the force (per unit area) between two parallel plates varies exponentially with the distance, and that the law of force for a different geometry is only modified by the effects due to divergence of the electric force-lines. The data needed for an exact prediction of the forces are not available, and even if we had them it would be a difficult and laborious task to compute the forces. But fortunately the resulting uncertainty will not, as a rule, count for much in the computation of the cluster integrals. Only the cases where very few cations or very few anions are present (or very few of either sign) might well require careful separate analysis. The most important modification of our previous results (EQUATIONS 30 and 31 a–h) will occur for long rods (EQUATION 31 h), in which case the effect of the electrostatic repulsion will be equivalent to an increase of the effective *diameter*. A similar increase of the effective *length* will cause a relatively insignificant increase of the covolume, unless the concentration of electrolyte is so low that κl is of the order unity, in which case, the problem of the repulsive forces must be reconsidered as a whole.

For the law of force between two cylindrical particles of the same diameter $d_1 = d_2 = d$, whose mantles are separated by a distance x, we now assume

$$w/kT = A(\gamma)e^{-\kappa x} \tag{39}$$

(of EQUATION 14 and pertinent discussion). More precisely, we assume that w has the value given by EQUATION 39 whenever the two cylinders cross, in the sense that the projections of their axes upon the plane parallel to both intersect. Such configurations yield the leading term

$$(d_1 + d_2)l_1 l_2 \sin \gamma = 2d l_1 l_2 \sin \gamma$$

of the excluded volume for long cylinders (the central parallelopiped of the solid figure illustrated in FIGURE 7, and compare EQUATION 30). The first cluster integral β_1 is defined by EQUATIONS 18 and 20. Assuming $w = w(x)$ as given by EQUATION 39 for all "crossed" configurations, the part of β_1 due to such configurations is simply

$$\beta_1 \text{ (crossing)} = 2l_1 l_2 \left\{ -d + \int_0^\infty (e^{-w(x)/kT} - 1) \, dx \right\} \sin \gamma. \tag{40}$$

The consequent correction to the effective diameter, assuming EQUATION 39, is accordingly

$$\int_0^\infty (1 - e^{-Ae^{-\kappa x}}) \, dx = \int_0^A (1 - e^{-u}) \, du/\kappa u$$

$$= \kappa^{-1} \left(\log A + C + \int_A^\infty e^{-u} \, du/u \right). \tag{41}$$

Here C denotes Euler's constant

$$\int_0^1 (1 - e^{-u}) \frac{du}{u} - \int_1^\infty e^{-u} \frac{du}{u} = C = -\Gamma'(1) = 0.5772 \cdots. \tag{41a}$$

For reasonably large values of A the exponential integral in EQUATION 41 may be neglected, and we get simply

$$-\beta_1 (\text{crossing}) = 2l_1 l_2 \, d_{\text{eff}}(\gamma) \sin \gamma$$

$$= 2l_1 l_2 \{d + \kappa^{-1}(C + \log A(\gamma))\} \sin \gamma. \tag{42}$$

Thus, the effective diameter equals the actual diameter increased by the distance δ at which the condition

$$w(\delta) = kT e^{-C} = 0.561 \, kT \tag{43}$$

is satisfied.

In deriving this important rule, we have made certain physical assumptions and mathematical approximations; but, as long as κd is reasonably small (less than unity), the errors thus incurred ought to be very modest and EQUATION 42 should describe a good estimate. Moreover, the result is rather insensitive to modifications of the geometry, so that the required modifications of the first four terms of EQUATION 30 may be estimated in a similar manner. We shall be content to observe that the increments of the effective lengths and diameters are of the same order of magnitude for all terms. This observation together with EQUATIONS A 14 and A 15 should be helpful in case an estimate of the end-corrections for long rods (which we shall neglect in the following) should be desired.

In EQUATION 42 we have indicated that the force constant A is expected to vary with the angle of intersection γ. A precise specification of that variation is contained in our EQUATION 13, and, in spite of the several approximations involved in the derivation, the relation

$$A(\gamma) = A(\pi/2)/\sin \gamma \qquad (44)$$

ought to be very nearly true, with the one exception that, for small angles $\gamma < d/l_{12}$ the factor $1/\sin \gamma$ must be replaced by a smaller number of the order l_{12}/d, where l_{12} denotes the length of the overlap between the two cylinders.

If we disregard the exception just mentioned and neglect terms which represent end-wise approach of the particles, (corresponding to the first four terms in EQUATION 30), we may write EQUATION 42 in the form

$$-\beta_1(\gamma) = 2l_1l_2\{[d + \kappa^{-1}C + \kappa^{-1}\log A(\pi/2)]\sin \gamma$$
$$- \kappa^{-1}(\sin \gamma)\log(\sin \gamma)\}. \qquad (45)$$

The average of the excluded volume over all orientations equals

$$-\bar{\beta}_1 = 2b_{12} = -\int \beta_1(\gamma)\, d\Omega/4\pi = -\frac{1}{\pi}\int_0^\pi \beta_1(\gamma)\sin\gamma\, d\gamma$$
$$= (\pi/2)l_1l_2(d + \bar{\delta}),\, . \qquad (46)$$

with

$$\bar{\delta} = \kappa^{-1}[C + \log A(\pi/2) + \log 2 - \tfrac{1}{2}] = \kappa^{-1}[0.7704 + \log A(\pi/2)]. \quad (47)$$

The integral

$$\int_0^\pi \log(\sin \gamma)\sin^2 \gamma\, d\gamma,$$

which enters into the computation, might seem difficult, but it is easily computed from the Fourier series

$$-\log|2\sin \gamma| = \cos 2\gamma + \tfrac{1}{2}\cos 4\gamma + \tfrac{1}{3}\cos 6\gamma + \cdots.$$

Concerning the absolute value of the force-constant A, we refer back to EQUATION 14 with attendant discussion and references.

Our results (33) and (46) for straight rods should apply without change to bent rods and *flexible chains* as long as they are not so tightly coiled that multiple contacts between pairs of different chains will be common. The statistics of such multiple contacts has not been investigated. In addition, it stands to reason that, when the particles are so slender as to be very flexible, the effective range of the electrostatic repulsion will constitute the main part of their effective diameters.

We shall deal summarily with the case of thin, plate-shaped particles. According to EQUATION 30, the mutual excluded volume for a pair of such particles is practically independent of their thickness, barring only nearly parallel orientations of the particles. Then, if only

$$\kappa d \gg 1, \tag{48}$$

which condition excludes very low concentrations of electrolyte at the most, an increase of the effective diameter by a distance δ, determined according to EQUATION 43, will make very little difference. For the case of very low electrolyte concentration, the question of the forces would seem to require a more careful analysis than we have available at present.

Thermodynamic Properties of Isotropic Solutions. We shall be generally content with the first order corrections to the laws of ideal solutions. Accordingly, we abbreviate the expansion (21) as follows

$$\begin{aligned}
\log B_p &= N_p(1 + \log (V/N_p) + \tfrac{1}{2}\beta_1(N_p/V)) \\
&= N_p(1 + \log (V/N_p) - b(N_p/V)),
\end{aligned} \tag{48}$$

whereby, for a monodisperse solution of rod-shaped particles of length l and diameter d according to EQUATION 46

$$b = (\pi/4)l^2(d + \delta), \tag{49}$$

including a correction δ for the "padding" due to an ionic double layer. For a solution of plate-shaped particles EQUATION 48 is also valid over a more limited range of concentrations with the different value

$$b = (\pi/4)^2 d^3 \tag{50}$$

for the covolume, as given by EQUATION 34 when specialized to the case $l = 0$.

The variation of the free energy with the particle concentration can be obtained by substituting the result (48) in EQUATION 5. We are particularly interested in the derived quantities. We obtain from EQUATIONS 6 and 7, respectively,

$$P = kT\{(N_p/V) + b(N_p/V)^2\} \tag{51}$$

for the osmotic pressure and

$$\mu_p = \mu_p^0 + kT\{\log (N_p/V) + 2b(N_p/V)\} \tag{52}$$

for the chemical potential. These correspond to well-known formulas in the theory of gases; the salient point of the present theory is that the covolumes may be much greater than the actual volumes of the particles. We also get reasonably simple results for a polydisperse solution which contains rod-shaped particles of various lengths $l_1, \cdots, l_s, \cdots$, but of identical diameters $d_1 = \cdots = d_s = \cdots = d$, and otherwise sufficiently similar, so that the effective diameter for any pair of particles is always $d + \bar{\delta}$. For this purpose, we substitute the covolumes given by EQUATION 46 in the more general formula (22), which yields

$$
\begin{aligned}
\log B_p &= \sum_s N_s(1 + \log (V/N_s)) - (\pi(d + \bar{\delta})/4V) \sum_{s,s'} N_s N_{s'} l_s l_{s'} \\
&= \sum_s N_s(1 + \log (V/N_s)) - (\pi/4)(d + \bar{\delta})(L^2/V),
\end{aligned}
\tag{53}
$$

for the configuration integral, with the abbreviation

$$
L = \sum_s N_s l_s \tag{54}
$$

for the *sum of the lengths of all particles present*. The osmotic pressure is accordingly

$$
P = kT \left\{ \sum_s (N_s/V) + (\pi/4)(d + \bar{\delta})(L/V)^2 \right\}, \tag{55}
$$

and we get

$$
\mu_s = \mu_s^0 + kT \log (N_s/V) + 2kT(\pi/4)(d + \bar{\delta})(L/V)l_s \tag{56}
$$

for the chemical potential of the particles of length l_s.

The corresponding formulas for a polydisperse solution of circular plate-shaped particles are almost equally simple. We specialize EQUATION 32 to the case $l_1 = l_2 = 0$:

$$
-\bar{\beta}_1(0, d_1; 0, d_2) = 2b = (\pi/4)^2 d_1 d_2(d_1 + d_2);
$$

then, with the abbreviations

$$
D = \sum_s N_s d_s
$$

$$
A = (\pi/4) \sum_s N_s d_s^2 \tag{57}
$$

for the sums of all the diameters viz. areas of all particles present, we obtain

$$
\log B_p = \sum_s N_s(1 + \log (V/N_s)) - (\pi/4)(DA/V) \tag{58}
$$

$$
P = kT \left\{ \sum_s (N_s/V) + (\pi/4)(DA/V^2) \right\} \tag{59}
$$

$$
\mu_s = \mu_s^0 + kT \log (N_s/V) + kT[Ad_s + D(\pi/4) d_s^2](\pi/4V). \tag{60}
$$

These results for plates might well have qualitative rather than quantitative significance. While certain colloids (bentonite) are known to consist of sheet-like particles, it is not known whether the outlines of the sheets are regular curves or polygons that might be reasonably approximated by circular disks. Nevertheless, it seems worth pointing out, that for a given total area of the particles, both terms in the formula (59) for the osmotic pressure do increase as the degree of dispersion increases. (When all particles are cut into quarters the sum of the diameters is doubled and the total number of particles is increased by a factor of four.)

Returning to our result (55) for the osmotic pressure of *rod-shaped* particles, we note that the absolute value of the second term, which represents the deviation from the value appropriate to ideal solutions, is quite *independent of the subdivision into individual lengths.* (On the other hand, a lengthwise splitting of the particles, if possible, will increase both terms in EQUATION 55.) On this basis, we should be prepared to find that the osmotic pressures of *flexible* chain-like particles in concentrated solutions may be practically independent of the subdivision of the chains. For rigid rod-shaped particles, we do not anticipate this phenomenon (in the isotropic phase, anyway), because, as we shall show next, such solutions will form an anisotropic phase as soon as the ratio (total covolume/volume) exceeds a certain critical value.

Anisotropic Solutions. We shall investigate the possibility that a solution of rod-shaped particles may form a nematic *liquid crystal* in which the distribution of orientations of the particles is anisotropic, while the distribution of the particles in space is homogeneous, and does not exhibit the periodic variation of density which characterizes solid crystals (periodicity in three dimensions) and *smectic* liquid crystals (periodicity in one dimension). We shall show that the concentration of particles need not be so very large (in terms of actual volume occupied) before the isotropic solution becomes unstable, relative to an anisotropic phase of the nematic type. Whether the latter will be stable, relative to other types of anisotropic phases, is a question which involves much more difficult computations, and we shall not try to settle it.

We introduce a distribution-function $f(\mathbf{a})$ for the directions $\mathbf{a}$ of the axes of the cylindrical particles, normalized according to EQUATION 26. When we neglect the terms which depend on β_2 and higher cluster integrals in the expansion given by EQUATION 27, we arrive at the following formula for the configuration-integral

$$\log B_p = N_p\{1 + \log (V/N_p)\} - \int f(\mathbf{a}) \log 4\pi f(\mathbf{a})\, d\Omega(\mathbf{a})$$

$$+ (N_p/2V) \int \beta_1 (\cos^{-1} (\mathbf{a}\cdot\mathbf{a}'))f(\mathbf{a})f(\mathbf{a}')\, d\Omega\, d\Omega'. \tag{61}$$

The function $f(\mathbf{a})$ is implicitly determined by the condition

$$B_p = \text{maximum}, \tag{62}$$

(subject to the restriction (26)).

We shall introduce convenient abbreviations for the two functionals which enter into EQUATION 61:

$$\sigma(f) = \int f(\mathbf{a}) \log 4\pi f(\mathbf{a}) \, d\Omega(\mathbf{a}), \tag{63}$$

$$-2b\rho(f) = \bar{\beta}_1 \rho(f) = \int \beta_1 \left(\cos^{-1}(\mathbf{a} \cdot \mathbf{a}')\right) f(\mathbf{a}) f(\mathbf{a}') \, d\Omega \, d\Omega'; \tag{64}$$

where, in conformity with EQUATION 46, we understand:

$$-2b = \bar{\beta}_1 = \int_0^{\pi/2} \beta_1(\gamma) \sin \gamma \, d\gamma. \tag{64a}$$

In addition, we shall denote the concentration of particles by

$$c = (N_p/V). \tag{65}$$

In this shorthand, the condition (62) becomes

$$\sigma(f) + bc\rho(f) = \text{minimum}, \tag{66}$$

wherein f is subject to the restriction

$$\int f \, d\Omega = 1. \tag{26}$$

The value of the minimum required by the condition (66) determines the free energy of the system according to EQUATIONS 5 and 61:

$$F(\text{solution}) - F(\text{solvent}) = N_p \mu_p^0 - kT \log B_p$$
$$= N_p \mu_p^0 + N_p kT \{\log c - 1 + \sigma(f) + bc\rho(f)\}. \tag{67}$$

We may apply Lagrange's method to the problem (66), thus

$$\delta\sigma(f) + bc\delta\rho(f) - \lambda\delta \int f \, d\Omega = 0. \tag{68}$$

The usual manipulations lead to the non-linear integral equation

$$\log(4\pi f(\mathbf{a})) = \lambda - 1 + c \int \beta_1(\mathbf{a}, \mathbf{a}') f(\mathbf{a}') \, d\Omega'. \tag{69}$$

EQUATION 69 is satisfied by every function which renders the functional of the problem (66) stationary; the true solution of (66) is included among these. The constant function

$$f = f_0 = 1/4\pi, \tag{70a}$$

which describes the isotropic distribution, is always a solution of EQUATION 69, with

$$\sigma(f_0) = 0; \quad \rho(f_0) = 1; \quad \lambda = 1 + 2bc \tag{70b}$$

On the other hand, in order to show that for sufficiently large values of c the solution (70) will not be the true solution of the problem (66), we only have to find some function f_1 such that $\sigma(f_1)$ is finite and $\rho(f_1) < 1$; then, when we take c large enough, the inequality

$$\sigma(f_1) + bc\rho(f_1) < \sigma(f_0) + bc\rho(f_0) = bc$$

can certainly be satisfied.

According to our previous considerations, the function $-\beta_1(\gamma)$ is an increasing function of $(\sin \gamma)$, so that a trial function f_1 with the required properties can be constructed very simply as follows: We choose an angle γ_1 such that

$$-\beta_1(\gamma) < -\bar{\beta}_1; \qquad \gamma < 2\gamma_1,$$

a preferred direction $\mathbf{a}_0$ and the following trial function

$$f_1(\mathbf{a}) = 0; \qquad |(\mathbf{a}_0 \cdot \mathbf{a})| < \cos \gamma_1$$

$$f_1(a) = 1/4\pi(1 - \cos \gamma_1); \qquad \cos \gamma_1 < |(\mathbf{a}_0 \cdot \mathbf{a})| < 1.$$

Some of the unwanted solutions of EQUATION 69—possibly all of them—may by interpreted as solutions of a modified variation problem:

$$\left.\begin{array}{c} \displaystyle\int f \, d\Omega = 1 \\[2mm] \sigma(f) = \sigma_1 \geqq 0 \end{array}\right\} \text{ prescribed}$$

$$\rho(f) = \text{minimum} = \rho_m(\sigma_1). \tag{71}$$

This leads again to EQUATION 68 with the difference that c is interpreted as a Lagrange multiplier on par with λ.

The second restriction in the problem (71) is in effect no different than the inequality

$$\sigma(f) \gtreqqless \sigma_1, \tag{71a}$$

because the function $\beta(\gamma)$ in EQUATION 64 is continuous (less would suffice). In consequence, if we know one function $f(\mathbf{a})$, which realizes a certain value of ρ, we can always find another which realizes very nearly the same value of ρ, but gives us a *greater* value of σ. All we have to do is introduce a very rapid local fluctuation of $f(\mathbf{a})$. This reasoning leads to the inequality

$$(\rho_m(\sigma_2) - \rho_m(\sigma_1))/(\sigma_2 - \sigma_1) \leqq 0; \tag{71b}$$

in words: the minimum of ρ is a never-increasing function of σ.

One way to solve the problem (66), at least in principle, is to solve the more general problem (71) first for all values of σ. For greater flexibility, we may describe the resulting relation between ρ_m and σ in parameter form

$$\sigma = \sigma(\alpha); \qquad \rho_m = \rho(\alpha); \tag{72a}$$

then the solution of EQUATION 66 must satisfy the condition

$$\sigma'(\alpha) + bc\rho'(\alpha) = 0, \tag{72b}$$

and, in order that the state thus described be stable

$$\sigma''(\alpha) + bc\rho''(\alpha) > 0. \tag{72c}$$

According to these results, if the function

$$(-d\rho_m/d\sigma) = -\rho_m'(\sigma) \tag{73}$$

is a steadily decreasing function of σ, then the transition from the isotropic to the anisotropic phase will be continuous and take place at the concentration given by the condition

$$1 + bc(d\rho_m/d\sigma)_{\sigma=0} = 0. \tag{74}$$

On the other hand, if the function (73) increases for small values of σ, reaches a maximum (as it must because $\rho > 0$), and decreases thereafter, then the isotropic solution will become unstable towards finite disturbances at some concentration lower than that required by EQUATION 74. In this case, the anisotropic phase will always possess a finite degree of anisotropy. Moreover, there will be a pair of concentrations for which the two phases can coexist; a solution of concentration intermediate between these will separate into two phases.

It is possible to show by rather general qualitative reasoning that the second alternative—a discontinuous transition—must be realized when an anisotropic solution is formed. We may as well assume that the anisotropic phase has cylindrical symmetry around some preferred direction $\mathbf{a}_0$; this restriction is unimportant, because it allows the distribution-function to contain spherical harmonics of all (even) orders. The odd orders are excluded if we assume that the solution is not polar in the crystallographic sense (seignette-electric):

$$f(\mathbf{a}) = f(-\mathbf{a}).$$

Under these assumptions, $f(\mathbf{a})$ may be developed in a series of even Legendre polynomials

$$4\pi f(\mathbf{a}) = 1 + 5A_2 P_2(\mathbf{a} \cdot \mathbf{a}_0) + 9A_4 P_4(\mathbf{a} \cdot \mathbf{a}_0) + \cdots \tag{75}$$

Moreover, since the homogeneous quadratic functional defined by EQUATION 64 is invariant against all rotations of the frame of reference, the Legendre polynomials are its eigenfunctions and an expansion of the type

$$\rho(f) = 1 - B_2 A_2^2 - B_4 A_4^2 - \cdots \tag{76}$$

is valid. Only even powers of A_2, A_4, $\cdots$, (in fact, only the second powers), occur in EQUATION 76. However, when we substitute the expansion (75) in EQUATION 73, we get

$$\sigma(f) = (5/2)A_2^2 + (9/2)A_4^2 + \cdots$$
$$- (25/21)A_2^3 - (45/7)A_2^2 A_4 - \cdots$$
$$+ (125/28)A_2^4 + \cdots \tag{77}$$

Linear terms do not occur in this expansion either, but *every cubic term is present*. The critical condition defined by EQUATION 74 is fulfilled by the smallest value (c_0) of c, which causes any one of the coefficients in the expansion

$$((5/2) - B_2 bc)A_2^2 + ((9/2) - B_4 bc)A_4^2 + \cdots$$

to vanish. It does not matter which one vanishes first; for the sake of argument let us assume that it is the coefficient of A_2^2. Then, if we take

$$f(\mathbf{a}) = 1 + A_2 P_2(\mathbf{a} \cdot \mathbf{a}_0)$$

we have the expansion (convergent for $|A_2| < 1$):

$$\sigma(f) + bc_0\rho(f) = bc_0 - (25/21)A_2^3 + (125/28)A_2^4 + \cdots$$

For finite, not too large, positive values of A_2, the sum of this expansion certainly takes values smaller than bc_0. This means that the isotropic solution becomes unstable towards finite disturbances at some concentration lower than that required by EQUATION 74 for a continuous transition, so that a *discontinuous transition* must take place at some lower concentration.

We shall try next, to get some idea about the distribution of orientations of the particles in the anisotropic solutions, and to estimate the thermodynamic functions for these solutions. The variation problem (66) is best attacked directly: the technique is to construct plausible trial functions with as many variable parameters as one can handle conveniently; the parameters are then adjusted so as to approach the required minimum as closely as possible.

As regards the general nature of the distribution, we note that $-\beta_1(\mathbf{a}, \mathbf{a}')$ has a minimum when the directions $(\mathbf{a}, \mathbf{a}')$ are parallel and a maximum when they are perpendicular. The smallest possible value of $\rho(f)$ is, therefore, attained when all particles have exactly the same orientation; according to our somewhat approximate formula (45), the function $\rho(f)$ then vanishes. For this singular distribution, however, $\sigma(f)$ becomes infinite. As a compromise, we have to expect a distribution which is more or less concentrated around a preferred direction $\mathbf{a}_0$, with this direction as an axis of symmetry. EQUATION 69 indicates, in a general way, how the density will decrease with the angle

$$\theta = \cos^{-1}(\mathbf{a} \cdot \mathbf{a}_0) \quad ; \tag{78}$$

for large angles an exponential function of $(\sin \theta)$ is thus indicated. A trial function of the type

$$\text{constant} \times (\cosh(\alpha \sin \theta))^{-n} \tag{79}$$

would therefore seem promising, and tentative computations for $n = 3$ gave encouraging results, but this lead was abandoned on account of the effort involved. The simpler function

$$f(\mathbf{a}) = (\alpha/4\pi \sinh \alpha) \cosh (\alpha(\mathbf{a} \cdot \mathbf{a}_0)) = (\alpha/4\pi \sinh \alpha) \cosh (\alpha \cos \Theta) \quad (80)$$

decreases rather too rapidly for large angles, and it contains but one parameter. It was, nevertheless, adopted as the best tractable function. Even so, according to EQUATION 67, the function to be minimized for the problem (66) is the *free energy* itself, so that our results for this important thermodynamic function ought not to be very much in error.

Further, to simplify the computations, we shall allow the approximate description

$$-\beta_1(\gamma) \sim 2l_1 l_2(d + \delta) \sin \gamma = (8/\pi)b \sin \gamma \quad (81)$$

of the more complicated function given by EQUATION 45. When EQUATIONS 80 and 81 are substituted in EQUATION 64, the resulting integral can be evaluated in terms of elementary functions, together with a Bessel function of order 2, and imaginary argument. For the definition (B 18) and properties of this function, and for the details of the integration, we refer to the Appendix, Section 2. We quote here the result for a monodisperse solution

$$\rho(\alpha) = 2(\sinh \alpha)^{-2} I_2(2\alpha) \quad (82)$$

(cf. EQUATION B 17). The evaluation of the integral in EQUATION 63 for the function (80) is elementary and yields

$$\sigma(\alpha) = \log(\alpha \coth \alpha) - 1 + (\sinh \alpha)^{-1} \tan^{-1}(\sinh \alpha). \quad (83)$$

The power series

$$4\rho(\alpha) = 4 - (\alpha^4/90) + (2\alpha^6/945) - (71\alpha^8/226800) + \cdots ;$$
$$(|\alpha| < \pi) \quad (84)$$
$$\sigma(\alpha) = (\alpha^4/90) - (2\alpha^6/810) + (108\alpha^8/226800) - \cdots ; \ (|\alpha| < \pi/2),$$

illustrate our consideration of continuous vs. discontinuous transition (when an expansion of the type (75) is constructed for the function (80), the coefficient of P_2 is of the order (α^2)). However, the values of α which correspond to stable anisotropic solutions are far outside the limits of convergence of the series (84). Fortunately, the required values of α are so large that the asymptotic representations

$$\sigma(\alpha) \sim \log \alpha - 1 , \quad (85)$$
$$\rho(\alpha) \sim 4(\pi\alpha)^{-1/2} \{1 - 30(32\alpha)^{-1} + 210(32\alpha)^{-2} + 1260(32\alpha)^{-3} + \cdots\}, \quad (86)$$

(cf. B 20), are eminently suitable for computation. The condition (72b) for internal equilibrium becomes, after rearrangement,

$$(\pi\alpha)^{1/2} = 2bc\{1 - 90(32\alpha)^{-1} + 1050(32\alpha)^{-2} + 8820(32\alpha)^{-3} + \cdots \}. \quad (87)$$

The asymptotic behavior of α for high concentrations is evident from the inverted series

$$\alpha \sim (4/\pi) \ (bc)^2 - (45/8) + O((bc)^{-2}). \tag{88}$$

As a measure of the spread in angle, we may compute the mean square of $\sin(\Theta/2)$, if we understand by Θ, (this time), the angle between the direction a of a particle and the nearer of the two directions $(+a_0, - a_0)$. With this convention, we have

$$\overline{(2 \sin (\Theta/2))^2} = (2/\alpha) \coth \alpha \sim 1/\alpha. \tag{89}$$

The standard deviation of the angle Θ is, therefore, about proportional to $\alpha^{-1/2}$ or, if we disregard the higher terms in EQUATION 87, inversely proportional to the concentration:

$$\overline{(2 \sin (\Theta/2))^2} \sim (\pi/4) \ (bc)^{-2} \tag{89a}$$

Combination of EQUATIONS 85, 86, and 87 yields for high concentrations

$$bc\rho \sim 2 + 75\pi(8bc)^{-2} + \cdots \tag{90}$$

$$\sigma \sim \log(4/\pi) - 1 + 2 \log(bc) - 90\pi(8bc)^{-2} + \cdots . \tag{91}$$

The free energy is given by EQUATION 67; in view of EQUATION 72b we derive therefrom the following simple *general* formula for the *osmotic pressure*

$$P = -(\partial F/\partial V)_{N_,} = kTc(1 + bc\rho). \tag{92}$$

In particular, if we take EQUATION 90 seriously

$$P = kT \ c(3 + 75\pi(8bc)^{-2} + \cdots). \tag{93}$$

On this basis, the osmotic pressure should be just a little greater than three times the ideal pressure.

The simple results (89a) and (93), however, depend on a rather severe over-simplification of the physical picture. The approximation (81) for EQUATION 45 tends to overestimate the deviation of the angles for high concentrations. The several approximations and simplifications which enter into EQUATION 93 would tend to make this an underestimate, possibly a very bad one; it is barely conceivable that our neglect of the attractive van der Waal's (dispersion) forces might in some cases bring about a measure of compensation.

For the lowest concentrations at which the anisotropic solutions can exist, the approximations which lead to EQUATIONS 85 and 86 may still be quite tolerable. In those cases it is best to solve EQUATION 87 numerically, because the expansion is only semi-convergent and the inversion (88) aggravates its tendency to diverge. In constructing tables or graphs there is, of course, no need for the inversion; it is just as well to compute σ, ρ and c all as functions of the parameter α.

646

We still have to compute the conditions for equilibrium of the two liquid phases. The osmotic pressure of the anisotropic solution is given by EQUATION 92. The chemical potential is computed according to the definition (7) by differentiation of EQUATION 67, whence in view of EQUATIONS 65 and 72b:

$$\mu_p = (\partial F/\partial N_p)_V = \mu_p^0 + kT\{\log c + \sigma + 2bc\rho\}. \qquad (94)$$

The equilibrium conditions (8) require that the functions given by (92) and (94) be equated with the corresponding functions (51) and (52) for the isotropic phase. The two concentrations c_a and c_i are thus determined by the two equations

$$c_a + bc_a^2\rho = c_i + bc_i^2$$
$$\log c_a + \sigma + 2bc_a\rho = \log c_i + 2bc_i,$$

which we may write in terms of the total covolumes as follows:

$$bc_a(1 + bc_a\rho) = bc_i(1 + bc_i)$$
$$\log(bc_a) + \sigma + 2bc_a\rho = \log(bc_i) + 2bc_i. \qquad (95)$$

Here, σ and ρ are functions of (bc_a) described implicitly by EQUATIONS 85, 86, and 87. The following results were obtained by numerical solution of the system of equations:

$$
\begin{array}{llr}
\alpha = 18.584 & bc_i = 3.3399 & \\
\rho = 0.49740 & bc_a = 4.4858 & \\
\sigma = 1.9223 & \rho bc_a = 2.2313 & (96) \\
\multicolumn{3}{c}{c_a/c_i = 1.343.}
\end{array}
$$

The standard deviation of $\sin(\Theta/2)$ given by EQUATION 89 corresponds to an angle $\Theta = 13.3°$.

It is a matter of interest to see how the expansions (88), (90), and (91) work out in the worst possible case. The values obtained from these formulae, as abbreviated, for the case $bc_a = 4.486$,

$$\alpha = 19.9; \quad \rho bc_a = 2.184; \quad \sigma = 2.021,$$

may be compared with the accurately computed values (96).

It would take us too far to develop a theory for the anisotropic phase of a polydisperse solution. The difficulty is that long rods will be more perfectly oriented than short rods, so that one has to compute a whole set of mutually dependent distribution-functions, one for each size of particles. Moreover, each composition of the anisotropic phase presents a separate problem of this type. Nevertheless, the mathematical analysis in the Appendix has been kept as general as was feasible, in order to facilitate computations for polydisperse systems.

647

It is possible to foresee that when two phases are formed, the longest particles will collect preferentially in the anisotropic phase, and that the total concentrations in each phase will vary with the ratio of the volumes.

Possibly, the best experimental tests of the present theory will consist in measurements of light scattering. As is well known, the light scattering per particle is inversely proportional to (dP/dc). However, when the longest dimensions of the particles are comparable to that of the measuring light, this simple relation applies only to *scattering at small angles*. We have, in effect, shown that the presence of one particle reduces the density of scattering matter up to a distance which equals the length of a presumptive neighbor, so that the *phase relations* of the light waves scattered from *different particles* must be considered in the interpretation of the large angle scattering.

On the basis of our results (92) and (51), the *anisotropic solution* ought to *scatter more light at small angles* than the isotropic phase in equilibrium with it. Predictions for large angles must await a mathematical analysis of the optical problems; the distribution of scattering matter around any one particle obviously depends on the degree of orientation of the particles.

Appendix
The Mutual Excluded Volume of Two Cylinders

We first compute the excluded volume $-\beta_1(0, d_1; 0, d_2)$ for two circular plates of vanishing thickness and diameters $d_1 = 2r_1; d_2 = 2r_2$. Besides, this is one of the interesting limiting cases.

Let the first plate be fixed with its center at the origin and let its normal form an angle γ with the y axis, in the (y, z) plane. We allow the second plate to move, but we keep its orientation constant, so that its normal is always parallel to the y axis. When we require that the two plates must not intersect, what is the volume inaccessible to the center of the second plate?

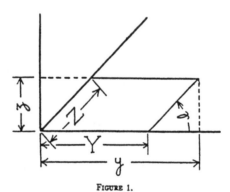

FIGURE 1.

Some of the analysis becomes a little simpler if we replace the coordinates (y, z) by the pair (Y, Z) referred to skew axes parallel to the plates 2, 1 respectively:

$$y = Y + Z \cos \gamma \qquad (A 1)$$
$$z = Z \sin \gamma$$

Then the intersection of the excluded region with the plane

$$z = Z \sin \gamma = \text{const.}$$

is bounded by the curve formed by the center of circle 2 as this circle rolls on a chord AB of the circle 1 (see FIGURE 2).

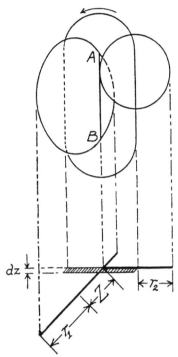

FIGURE 2.

The length of the chord equals

$$2(r_1^2 - Z^2)^{1/2}.$$

The area of the intersection is, evidently,

$$A(Z) = 4(r_1^2 - Z^2)^{1/2} r_2 + \pi r_2^2$$

(rectangle + two half circles), and the excluded volume is, accordingly,

$$-\beta_1(0, d_1; 0, d_2) = \int A(Z) \, dz = \int A(Z) \, dZ \sin \gamma$$

$$= 2\pi r_1 r_2 (r_1 + r_2) \sin \gamma \qquad (A\ 2)$$

$$= (\pi/4) \, d_1 \, d_2 (d_1 + d_2) \sin \gamma .$$

The excluded region, illustrated in FIGURE 3, is bounded by four planes and by the fourth degree surface described by the equations

$$(r_1^2 - Z^2)^{1/2} + (r_2^2 - Y^2)^{1/2} = \pm x. \qquad (A\ 3)$$

(Here and in the following we always take positive roots). This surface joins the planes

$$Z = \pm r_1$$

$$Y = \pm r_2 \qquad (A\ 4)$$

along the circles

$$x^2 + Y^2 = r_2^2$$
$$x^2 + Z^2 = r_1^2,$$

respectively, in such a manner that the normal directions are continuous except at the points

$$x = 0, \quad Y = r_1, \quad Z = r_2$$

where two such circles touch one another.

A curve bounding the intersection of the excluded region with a plane $x = $ const. is described by EQUATION A 3 for $x = $ const, or it consists of segments of curves described thus alternating with segments of the straight lines described by EQUATION A 4, according to the value of x. The cross-section for $x = 0$ is simply the parallelogram given by EQUATION 4. The four possible cases are illustrated in FIGURE 3.

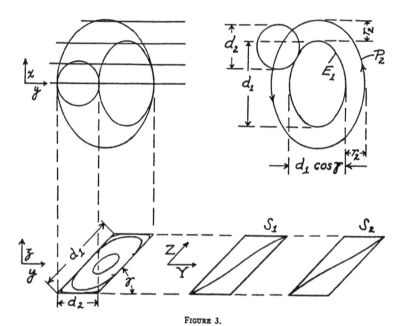

FIGURE 3.

On each curve of this type there is a pair of points for which

$$y = Y + Z \cos \gamma = \text{extremum } (S_2)$$

and another pair for which

$$z' = Y \cos \gamma + Z = \text{extremum } (S_1).$$

The loci of such points are space-curves S_1, S_2 on the surface of the excluded volume; their projections on the (y, z) plane are described by the equations

$$(Z/Y)(r_1^2 - Z^2)^{-1/2}(r_2^2 - Y^2)^{1/2} = \begin{cases} \cos \gamma; & (S_2) \\ 1/\cos \gamma; & (S_1) \end{cases} \qquad (A\ 5)$$

The significance of S_2 will be clear from the observation that when the center of plate 2 is on S_2, not only does this plate make rim-to-rim contact with plate 1, but (in addition), a cylinder raised perpendicularly on the rim of plate 2 also just touches the rim of plate 1. Similarly, when the center of plate 2 is on S_1, then the rim of plate 2 just touches a cylinder raised perpendicularly on the rim of plate 1.

The curve S_2 separates the two parts of the surface of the excluded volume which are seen from opposite directions along the normal of plate (2). Its projection P_2 on the (x, y) plane (parallel to plate 2), which delimits the projection of the excluded volume, may be described as the locus of points whose distance from the nearest point of the projection of plate (1) is precisely r_2. The projection of plate (1) on the plane of plate (2) is an ellipse E_1 of semi-axes r_1, $r_1 \cos \gamma$; if a circle of radius r_2 rolls on this ellipse its center traces P_2.

The area $A(P_2)$ can be evaluated by a general method applicable to rolling-figures of continuous tangent. We let R denote the curve described by the center of a circle of radius r as the circle rolls on closed curve C. The algebraic sum of the curvatures of the circle and the curve C must be positive (convex) everywhere on C. The area between C and R is easily found as follows (FIGURE 4):

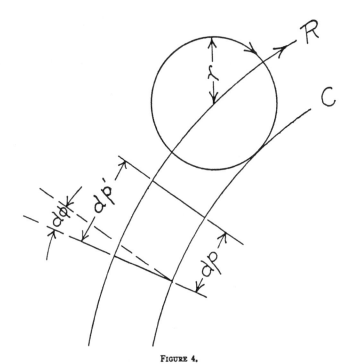

FIGURE 4.

Consider the area contained between two normals of directions $\varphi, \varphi + d\varphi$, the segments $dp(C)$ of the perimeter of C, and the segment

$$dp'(R) = dp(C) + r d\varphi$$

of the perimeter of R. The area of the trapezium thus formed equals

$$\tfrac{1}{2} r (dp + dp') = r dp + \tfrac{1}{2} r^2 d\varphi.$$

Integrating around C we find simply

$$A(R) - A(C) = \int r \, dp(C) + \tfrac{1}{2} r^2 \int d\varphi = r p(C) + \pi r^2 \qquad (A\ 6)$$

where $p(C)$ denotes the circumference of C, and $A(C)$ its area. For the ellipse E_1 described above we have

$$A(E_1) = \pi r_1^2 \mid \cos \gamma \mid \tag{A 7}$$

$$p(E_1) = 4r_1 E(\sin \gamma) = 4r_1 \int_0^{\pi/2} (1 - \sin^2 \gamma \sin^2 \varphi)^{1/2}\, d\varphi$$

where the customary notation $E(\sin \gamma)$ is used for a complete elliptic integral of the second kind. We substitute these results in (A 6) along with $r = r_2$ and obtain for the projection P_2:

$$A(P_2) = \pi r_1^2 \mid \cos \gamma \mid + 4r_1 r_2 E(\sin \gamma) + \pi r_2^2 \tag{A 8}$$

Similarly, of course,

$$A(P_1) = \pi r_1^2 + 4r_1 r_2 E(\sin \gamma) + \pi r_2^2 \mid \cos \gamma \mid . \tag{A 9}$$

Actually, the profile areas (A 8) and (A 9) are the only properties of S_1 and S_2 which will enter into our final result for the excluded volume of two cylinders, but some additional analysis has been included as an aid to visualization.

So far, we have considered the thin plates of diameters d_1, d_2. Now, let us replace the second by a cylinder of length l_2, diameter d_2. Our solution for the case $l_2 = 0$ has an

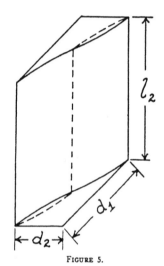

FIGURE 5.

important connection with the more general problem: As we consider the two halves of the surface illustrated in FIGURE 3 separated by the curve S_1, the front half is the locus for the center of the rear end of cylinder (2) at closest approach to plate 1, and the rear half of the surface is the locus for the front end-center of the cylinder for the opposite type of end-wise contact. The corresponding loci for the center of cylinder (2) are displaced by distances $\pm \frac{1}{2} l_2$ in the direction of the cylinder axis. If the two end surfaces formed by these loci are joined rim to rim by a cylindrical mantle parallel to the axis of cylinder (2), then that mantle is the locus for the center of (2) when this cylinder makes lateral contact with the rim of plate (1).

The excluded volume for this case is the same as for $l_1 = l_2 = 0$, plus that of a cylinder of length l_2 and orthogonal section P_2:

$$-\beta_1(0, d_1, l_2, d_2) = -\beta_1(0, d_1, 0, d_2) + l_2 A(P_2). \tag{A 10}$$

The end-faces of the inserted cylindrical piece are parallel, so that the computation of the volume is not affected by their complicated shape.

The final generalization to the case $l_1 > 0$ proceeds in a similar manner: The body illustrated by FIGURE 5 is cut as indicated by the dotted line. The cut, most simply taken perpendicularly to the plane of the projection, follows the separated halves of the curve S_1 and the median plane of the cylinder inserted in the previous step. A second cylindrical

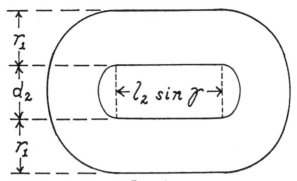

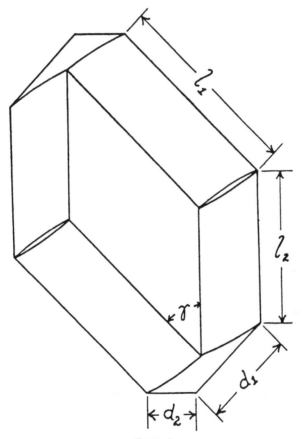

FIGURE 7

piece of length l_2 parallel to the normal of the plate (1) and cross-section, as indicated by FIGURE 6, is inserted. This time the added volume is

$$l_2 A(P_1) + 2\, l_1 l_2 (r_1 + r_2) \sin \gamma,$$

653

and the volume of the resulting domain, illustrated in FIGURE 7, equals

$$-\beta_1(l_1, d_1; l_2, d_2; \gamma) = (\pi/4)d_1 d_2(d_1 + d_2) \sin \gamma$$

$$+l_2\{(\pi/4)d_2^2 + d_1 d_2 E(\sin \gamma) + (\pi/4)d_1^2 | \cos \gamma | \} \qquad (A\ 11)$$

$$+ l_1\{(\pi/4)d_1^2 + d_1 d_2 E(\sin \gamma) + (\pi/4)d_2^2 | \cos \gamma | \} + l_1 l_2(d_1 + d_2)\sin \gamma.$$

The integrations involved in the computation of the average

$$\int \beta_1(\gamma) \, d\Omega/4\pi = \tfrac{1}{2} \int_0^\pi \beta_1(\gamma) \sin \gamma \, d\gamma$$

should be obvious for all terms except those which involve the elliptic integral $E(\sin \gamma)$. These too, reduce to a very simple form by a suitable substitution. By the definition of the complete elliptic integral we have

$$4E(\sin \gamma) = 4 \int_0^{\pi/2} (1 - \sin^2 \gamma \sin^2 \psi)^{1/2} \, d\psi = \int_0^{2\pi} (1 - \sin^2 \gamma \sin^2 \psi)^{1/2} \, d\psi$$

Hence

$$4 \int_0^\pi E(\sin \gamma) \sin \gamma \, d\gamma = \int_0^\pi \sin \gamma \, d\gamma \int_0^{2\pi} d\psi (1 - \sin^2 \gamma \sin^2 \psi)^{1/2} \qquad (A\ 12)$$

By the substitution

$$\sin \gamma \sin \varphi = \cos \vartheta$$

$$\cos \gamma = \sin \vartheta \sin \varphi,$$

(which may be interpreted as a change of polar coordinates in space), the integral (A 12) becomes

$$4 \int_0^\pi E(\sin \gamma) \sin \gamma \, d\gamma = \int_0^\pi \sin \vartheta \, d\vartheta \int_0^{2\pi} d\varphi \sin \vartheta = \pi^2 \qquad (A\ 13)$$

With the aid of this result and obvious elementary integrations we find from EQUATION A 11:

$$- \int \beta_1(l_1, d_1; l_2, d_2; \gamma) \, d\Omega/4\pi = (\pi/4)^2 d_1 d_2(d_1 + d_2)$$

$$+ (\pi/4) (l_1 d_1^2 + l_2 d_2^2) + (\pi/8) (l_1 d_2^2 + l_2 d_1^2) \qquad (A\ 14)$$

$$+ (\pi^2/8)(l_1 + l_2) d_1 d_2 + (\pi/4) l_1 l_2 (d_1 + d_2)$$

For a pair of cylinders of lengths l_1, l_2, capped by hemispheres of diameters d_1, d_2, the computation of the mutual excluded volume is quite analogous to the preceding. Several details are much simpler. For the case $l_1 = l_2 = 0$, the result (A 2) is replaced by the volume of a sphere of diameter $d_1 + d_2 = 2d$. The profiles (P_1, P_2) are replaced by circles of radius d (compare A 8, A 9). The assembled final result for capped cylinders is

$$-\beta_1(\gamma) = (4\pi/3)d^3 + \pi d^2(l_1 + l_2) + 2dl_1 l_2 \sin \gamma. \qquad (A\ 15)$$

Here, the averaging over directions involves only the simple integral

$$\int \sin \gamma \, d\Omega/4\pi = \pi/4. \qquad (A\ 16)$$

The Mean Covolume for Anisotropic Solutions

We shall show how the multiple integral

$$\bar{\beta}_{1\rho}(f_1, f_2) = \int f_1(a_1) f_2(a_2) \beta_1 (\cos^{-1} (a_1 \cdot a_2)) \, d\Omega(a_1) \, d\Omega(a_2), \qquad (B\ 1)$$

where a_1, a_2 denote variable unit vectors which specify the orientations of particles, can be reduced to a single integral when the distribution-functions are of the special type

$$f_s(a) = (\alpha_s/4\pi \sinh \alpha_s) \cosh \alpha_s(a \cdot a_0), \tag{B 2}$$

symmetrical about the discretion a_0 (axis of the liquid crystal). Concerning β_1 we assume only

$$\beta_1(\gamma) = \beta_1(\pi - \gamma) = F(\sin\gamma) \tag{B 3}$$

at this stage, although in the end we shall introduce the approximation

$$\beta_1(\gamma) = \beta_1(\pi/2)\sin \gamma, \tag{B 4}$$

derived in the text.

Consider the integral

$$J = \int \cosh (\alpha_1(a_1 \cdot a_0) + \alpha_2(a_2 \cdot a_0)) F(\sin\gamma) \, d\Omega_1 \, d\Omega_2 ;$$

$$\cos \gamma = (a_1 \cdot a_2). \tag{B 5}$$

The value of this integral is not affected by the substitution

$$a_2 \to - a_2$$

which changes the first factor of the integrand into

$$\cosh (\alpha_1(a_1 \cdot a_0) - \alpha_2(a_2 \cdot a_0)).$$

The arithmetic mean of the two integrals involves the factor

$$\cosh (\alpha_1(a_1 \cdot a_0)) \cosh (\alpha_2(a_2 \cdot a_0))$$

instead, and by comparison with EQUATIONS. B 1, B 2, we readily verify the identity

$$J = (4\pi)^2 (\sinh \alpha_1 \sinh \alpha_2/\alpha_1\alpha_2)\bar{\beta}_1\rho(f_1, f_2). \tag{B 6}$$

We proceed to evaluate J.

For the direction a_1, we next introduce polar coordinates (Θ_1, ϕ_1) referred to a_0, but the direction a_2 we shall specify in terms of polar coordinates (γ, ϕ) referred to the direction a_1, such that $\phi = 0$ for $a_2 = a_0$. Then

$$(a_0 \cdot a_1) = \cos \Theta_1$$

$$(a_0 \cdot a_2) = \cos \Theta_2 = \cos \Theta_1 \cos \gamma + \sin \Theta_1 \sin \gamma \cos \phi \tag{B 7}$$

$$d\Omega_1 = \sin \Theta_1 \, d \Theta_1 \, d \phi_1$$

$$d\Omega_2 = \sin \gamma \, d\gamma \, d\phi.$$

Since the integrand does not depend on ϕ_1, we integrate at once over this variable and get

$$J = 2\pi \int \cosh (\alpha_1 \cosh \Theta_1 + \alpha_2 \cos \Theta_2)F(\sin \gamma) \sin \Theta_1 \, d\Theta_1 \sin \gamma \, d\gamma \, d\phi. \tag{B 8}$$

The limits of the variables are

$$0 < \Theta_1 < \pi; \quad 0 < \gamma < \pi; \quad 0 < \phi < 2\pi.$$

Now we replace the two variables Θ_1 and ϕ by the following substitution:

$$\cos \Theta_1 = \sin \chi \cos (\psi + \eta(\gamma)) \tag{B 9}$$

$$\sin \Theta_1 \cos \phi = \sin \chi \sin(\psi + \eta (\gamma))$$

$$\tan \eta(\gamma) = \alpha_2 \sin \gamma/(\alpha_1 + \alpha_2 \cos \gamma)$$

$$\partial(\Theta_1, \phi)/\partial(\chi, \psi) = \sin \chi/\sin \Theta_1$$

655

whereby the integral takes the form

$$J = 2\pi \int \cosh \left([\alpha_1^2 + \alpha_2^2 + 2\alpha_1 \alpha_2 \cos \gamma]^{1/2} \sin \chi \cos \psi\right) F(\sin \gamma) \sin \gamma \, d\gamma \sin \chi \, d\chi \, d\psi, \tag{B 10}$$

the integral to be taken between the limits

$$0 < \gamma < \pi; \quad 0 < \chi < \pi; \quad 0 < \psi < 2\pi. \tag{B 10a}$$

After a final substitution

$$\sin \chi \cos \psi = \cos \mu$$

$$\cos \chi = \sin \mu \cos \xi \tag{B 11}$$

$$\partial(\chi, \psi)/\partial(\mu, \xi) = \sin \mu / \sin \chi,$$

we can integrate over μ and ξ under the integral sign as follows

$$\int_{\mu=0}^{\pi} \sin \mu \, d\mu \int_{\xi=0}^{2\pi} \cosh \left([\alpha_1^2 + \alpha_2^2 + 2\alpha_1\alpha_2 \cos \gamma]^{1/2} \cos \mu\right) d\xi$$

$$= 4\pi [\alpha_1^2 + \alpha_2^2 + 2\alpha_1\alpha_2 \cosh \gamma]^{-1/2} \sinh \left([\alpha_1^2 + \alpha_2 + 2\alpha_1\alpha_2 \cos \gamma]^{1/2}\right)$$

$$= (- 4\pi/\alpha_1\alpha_2 \sin \gamma) \frac{\partial}{\partial \gamma} \cosh \left([\alpha_1^2 + \alpha_2^2 + 2 \alpha_1\alpha_2 \cos \gamma]^{1/2}\right).$$

The result

$$J = (- 8\pi^2/\alpha_1\alpha_2) \int_{\gamma=0}^{\pi} \frac{\partial}{\sin \gamma \, \partial \gamma} \left\{\cosh \left([\alpha_1^2 + \alpha_2^2 + 2\alpha_1\alpha_2 \cos \gamma]^{1/2}\right)\right\} F(\sin \gamma) \sin \gamma \, d\gamma \tag{B 12}$$

may be integrated by parts and we finally obtain

$$J = (8\pi^2/\alpha_1\alpha_2) \{2 \sinh \alpha_1 \sinh \alpha_2 F(0)$$

$$+ \int_{\gamma=0}^{\pi} \cosh \left([\alpha_1^2 + \alpha_2^2 + 2\alpha_1 \alpha_2 \cos \gamma]^{1/2}\right) dF(\sin \gamma)\} \tag{B 13}$$

or in view of (B 6)

$$2 \sinh \alpha_1 \sinh \alpha_2 \{\beta_1 \rho(f_1, f_2) - \beta_1(0)\}$$

$$= \int_{\gamma=0}^{\pi} \cosh \left([\alpha_1^2 + \alpha_2^2 + 2\alpha_1 \alpha_2 \cos \gamma]^{1/2}\right) d \beta_1(\gamma). \tag{B 14}$$

For identical particles, $\alpha_1 = \alpha_2 = \alpha$,
we have

$$(\alpha^2 + \alpha^2 + 2\alpha^2 \cos \gamma)^{1/2} = 2\alpha \cos \tfrac{1}{2}\gamma. \tag{B 15}$$

Moreover, if we adopt the approximation

$$F(\sin \gamma) = \beta_1(\gamma) = \beta_1(\pi/2)\sin \gamma = -(8/\pi)b \sin \gamma \tag{B 16}$$

(see text), the integral (B 14) can be expressed in terms of a Bessel function as follows

$$2(\sinh \alpha)^2 \beta_1 \rho(f) = \int_{\gamma=0}^{\pi} \cosh (2\alpha \cos \tfrac{1}{2} \gamma)\beta_1(\pi/2) \cos \gamma \, d\gamma$$

$$= \pi\beta_1(\pi/2)I_2(2\alpha) = - 8b \, I_2(2\alpha) \tag{B 17}$$

with the standard notation

$$I_2(2\alpha) = - J_2(2i\alpha) = \sum_{n=0}^{\infty} (\alpha^{n+2}/n!(n + 2)!) \tag{B 18}$$

for the Bessel function of order 2. The integral (B 17) is but a variant of the standard integral definition of the Bessel function

$$\pi I_2(2\alpha) = \int_{x=0}^{x=\pi} \cosh (2\alpha \cos x) \cos (2x) \, dx,$$

the last step being justified by the observation that the integrand remains unchanged when the argument x is replaced by $\pi - x$.

In all cases encountered in the present work, the argument of the Bessel function will be either zero (isotropic solution) or else so large that a few terms of the asymptotic expansion

$$2 I_2(2\alpha) \sim (\pi\alpha)^{-1/2} e^{2\alpha} \left\{ 1 - \frac{3 \cdot 5}{1! \, 16\alpha} + \frac{1 \cdot 3 \cdot 5 \cdot 7}{2! \, (16\alpha)^2} \right.$$
$$\left. - \frac{(-1) \cdot 1 \cdot 3 \cdot 5 \cdot 7 \cdot 9}{3! \, (16\alpha)^3} + \cdots \right\} \tag{B 19}$$

will suffice for computation. The corresponding formula for the mean effective excluded volume is

$$-\tilde\beta_1 \rho(f) \sim 8b(\pi\alpha)^{-1/2} \left\{ 1 - \frac{30}{32\alpha} + \frac{210}{(32\alpha)^2} + \frac{1260}{(32\alpha)^3} + \cdots \right\}. \tag{B 20}$$

In the general case $\alpha_1 \neq \alpha_2$, even though we adopt the approximation (B 16), the integral of (B 14) can no longer be expressed in terms of simple known functions. An asymptotic expansion analogous to (B 20) has been obtained by the usual procedure: A new variable t is introduced by the substitution

$$\alpha_1^2 + \alpha_2^2 + 2\alpha_1\alpha_2 \cos \gamma = (\alpha_1 + \alpha_2 - t)^2; \tag{B 21}$$

the hyperbolic function is approximated by an exponential and the factor

$$dF(\sin \gamma)/dt = \cos \gamma \, (d\gamma/dt)$$

by an abbreviated power series in t; finally the range of integration over t is extended to the interval $(0, \infty)$. The following generalization of (B 20) results

$$-\tilde\beta_1 \rho(f_1, f_2) \sim 8b_{12}(\alpha_1 + \alpha_2)^{1/2}(2\pi\alpha_1\alpha_2)^{-1/2} \left\{ 1 - \frac{3}{8}\left(\frac{1}{\alpha_1} + \frac{1}{\alpha_2} + \frac{1}{\alpha_1 + \alpha_2} \right) \right.$$
$$\left. + \frac{15}{128}\left[\frac{8}{\alpha_1 \alpha_2} - \left(\frac{1}{\alpha_1} + \frac{1}{\alpha_2} + \frac{1}{\alpha_1 + \alpha_2} \right)^2 \right] + \cdots \right\}. \tag{B 22}$$

By this general technique, it is also possible to deal with the more accurate description of the covolume function given by EQUATION 45 in the text. Some terms involving the factor $\log t$ then occur after the substitution (B 21); but the integrals which correspond to these terms are easily evaluated:

$$\int_0^\infty e^{-t} t^n (\log t) \, dt = \frac{\partial}{\partial n} \int_0^\infty e^{-t} t^n \, dt = \Gamma'(n + 1).$$

For the present work, however, it did not seem worth while to complete this computation.

Bibliography

DERJAGUIN, B. 1940. On the repulsive forces between charged colloid particles and on the theory of slow coagulation and stability of lyophobic sols. Trans. Faraday Soc. **36**: 203.

LANGMUIR, I. 1938. The role of attractive and repulsive forces in the formation of tactoids, thixotropic gels, protein crystals and coacervates. J. Chem. Phys. **6**: 873.

ONSAGER, L. 1942. Anisotropic solutions of colloids. Phys. Rev. **62**: 558.

VERWEY, E. J. W., & J. Th. G. OVERBEEK. 1948. *Theory of the Stability of Lyophobic Colloids.* Elsevier Publishing Co. Amsterdam & New York.

Lars Onsager was born in Oslo, Norway, in 1903. He entered the Norges tekniske høgskole in 1920 as a student of chemical engineering, graduating in 1925. He spent the years 1926–8 in Zurich as a student of Peter Debye and Erich Hückel, working on the theory of electrolytes. In 1928 he spent one term at the Johns Hopkins University in Baltimore and was then appointed to a junior faculty post at Brown University, Rhode Island. In 1933 he received a fellowship at Yale University, being awarded a Ph.D from the same university in 1935 for a thesis on deviations from Ohm's law in weak electrolytes. He was subsequently promoted, becoming Professor of Theoretical Chemistry in 1945.

Onsager is perhaps best known for his contributions to the foundations of irreversible thermodynamics, leading to the so-called Onsager coefficients and relations. However, in his own words, over the years, the subjects of his interest came to include colloids, dielectrics, order-disorder transitions, metals and superfluids, hydrodynamics and fractionation theory.

He is generally regarded as one of the leading figures in twentieth century theoretical chemistry. He was widely honoured for his work, and was awarded the Nobel Prize in Chemistry in 1968. He retired in 1972 and died in 1976.

E5

Proceedings of the Royal Society of London A **234**, 73–89 (1956)

Phase equilibria in solutions of rod-like particles

By P. J. Flory†

Department of Chemistry, The University, Manchester

(Communicated by G. Gee, F.R.S.—Received 13 July 1955)

A partition function for a system of rigid rod-like particles with partial orientation about an axis is derived through the use of a modified lattice model. In the limit of perfect orientation the partition function reduces to the ideal mixing law; for complete disorientation it corresponds to the polymer mixing law for rigid chains. A general expression is given for the free energy of mixing as a function of the mole numbers, the axis ratio of the solute particles, and a disorientation parameter. This function passes through a minimum followed by a maximum with increase in the disorientation parameter, provided the latter exceeds a critical value which is 2e for the pure solute and which increases with dilution. Assigning this parameter the value which minimizes the free energy, the chemical potentials display discontinuities at the concentration at which the minimum first appears. Separation into an isotropic phase and a somewhat more concentrated anisotropic phase arises because of the discontinuity, in confirmation of the theories of Onsager and Isihara, which treat only the second virial coefficient. Phase separation thus arises as a consequence of particle asymmetry, unassisted by an energy term.

Whereas for a large-particle asymmetry both phases in equilibrium are predicted to be fairly dilute when mixing is athermal, a comparatively small positive energy of interaction causes the concentration in the anisotropic phase to increase sharply, while the concentration in the isotropic phase becomes vanishingly small. The theory offers a statistical mechanical basis for interpreting precipitation of rod-like colloidal particles with the formation of fibrillar structures such as are prominent in the fibrous proteins.

The asymmetry of tobacco mosaic virus particles (with or without inclusion of their electric double layers) is insufficient alone to explain the well-known phase separation which occurs from their dilute solutions at very low ionic strengths. Higher-order inter-action between electric double layers appears to be a major factor in bringing about dilute phase separation for these and other asymmetric colloidal particles bearing large charges, as was pointed out previously by Oster.

† Fellow of the John Simon Guggenheim Memorial Foundation, 1954. Permanent address: Department of Chemistry, Cornell University, Ithaca, N.Y.

INTRODUCTION

The equilibrium properties of dilute solutions of highly asymmetric particles have been treated by Onsager (1949) and by Isihara (1951). The elaborate methods which they employed depend essentially on the derivation of the excluded volume for a pair of such particles (rods, ellipsoids, or disks) as a function of their relative orientation. Second virial coefficients were obtained by averaging the excluded volume over all orientations, or, for an ordered arrangement of the particles, over their equilibrium distribution of orientations, this distribution being determined by the condition of minimum free energy.

Both Onsager and Isihara concluded that solutions of sufficiently asymmetric particles, e.g. long rods, should exhibit phase separation, even in the absence of repulsive forces between the particles. One of the coexisting phases should contain a low concentration of unoriented particles, and the other a somewhat greater concentration of partially oriented particles. These conclusions were reached without considering higher terms in the virial expansion, evaluation of which would have been prohibitively difficult by an extension of their methods. The results obtained are therefore restricted to low concentrations. Treatment of phase separation without the aid of higher terms leaves much to be desired.

A greatly simplified treatment of solutions of rod-like solute molecules, or particles, is given in the present paper. The treatment depends on estimation of the number of configurations, or volume available in configuration space, for solute molecule $j+1$ when j molecules have previously been assigned to the solution. This quantity is evaluated by an adaptation of the lattice model which is applicable to any specified distribution of orientations of the particles. While the method is an approximate one, it offers the distinct advantage of freedom from restrictions on the concentration range to which it is applied.

THE FREE ENERGY OF MIXING

An expression for the partition function for an isotropic solution consisting of n_1 approximately spherical solvent molecules and n_2 rigid rod-like solute molecules of the same diameter and x times as long, the latter displaying no preferred orientations, may be obtained from (1) of the preceding paper (Flory 1956) by equating the interaction (χ_1) and internal flexibility (f) parameters to zero. The result is

$$Q_M = q_1^{n_1} q_2^{n_2} [(n_1 + xn_2)! / \{n_1! \, n_2! (n_1 + xn_2)^{(x-1)n_2}\}] (\tfrac{1}{2}z)^{n_2}, \qquad (1)$$

where q_1 and q_2 are the internal partition functions for molecules 1 and 2, and z is the lattice co-ordination number. If, however, the rod-like solute molecules were constrained to occur with their axes exactly parallel to one another, the number of arrangements would be that for a binary system of n_1 and n_2 particles, respectively, situated on a linear lattice, with the result that

$$Q_M = q_1^{n_1} q_2^{n_2} (n_1 + n_2)! / (n_1! \, n_2!), \qquad (2)$$

which corresponds of course to the ideal mixing law. This situation can be expected to occur only in the limit of very long rods at high concentration. On the other hand, the state of complete disorientation represented by (1) may prevail only for short rods or at low concentrations. An expression is required for the partition function for states of intermediate order between the extremes represented by (1) and (2).

Let j solute molecules be assigned locations in the volume to be occupied by the solution. The specified distribution of their orientations will be assumed to be symmetrical about an axis. We shall estimate the number ν_{j+1} of situations available to an additional molecule $j+1$ oriented at an angle ψ_{j+1} to this axis. A limitation of the conventional lattice model for the treatment of this problem is its inability to accommodate the rod-like molecule in a continuously varying range of orientations. We propose to circumvent this difficulty by replacing each solute molecule by several submolecules joined together in the manner shown in figure 1b. Thus, a molecule i inclined at an angle ψ_i to the orientation axis (figure 1a) will be divided into $y_i = x \sin \psi_i$ submolecules, each containing x/y_i segments and requiring there-

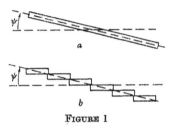

FIGURE 1

fore x/y_i vacant lattice cells. The submolecules are arranged parallel to the orientation axis and joined laterally near their ends (figure 1b). The artificialities of this modification are scarcely more serious than those of the lattice model itself, and the latter is known to give a generally reliable semi-quantitative description of polymer solutions.

Finally, we assume that the distribution of submolecules in any given row of lattice cells parallel to the orientation axis is random, being uninfluenced by circumstances in neighbouring rows. This corresponds to the approximation used in previous treatments (Flory 1942, 1944) of polymer solutions, variously termed 'crude' and 'zero order', but over which so-called higher approximations have thus far failed to show any advantage.

Having acknowledged the various approximations, we proceed in conventional fashion as follows. The number of sites available to the first segment of molecule $j+1$, consisting of $y_{j+1} = x \sin \psi_{j+1}$ submolecules, will be $n_0 - xj$, where n_0 is the total number of lattice sites. The probability that each succeeding site is vacant in the row will be given, assuming, as stated above, a random distribution of submolecules and vacant sites, by the *mole fraction* of vacant sites in this distribution,† i.e. by the ratio of vacant sites to the sum of vacant sites and submolecules. Hence,

† That the mole fraction is required here follows immediately from consideration of the vacancies and the submolecules as two sets of particles arranged in random linear sequence. The probability that any particle of the arrangement is succeeded in a given direction by a vacancy obviously equals the mole fraction of vacancies.

the probability that the $(x/y_{j+1})-1$ sites required for the remaining segments of the first submolecule are vacant will be

$$\left[(n_0-xj)/(n_0-xj+\sum_{i=1}^{j}y_i)\right]^{(x/y_{j+1})-1}.$$

The probability of vacancy in the lattice site for the first segment of the second submolecule will be $(n_0-xj)/n_0$. A factor like that given above will be required for remaining segments of the second submolecule, etc. Combining these factors, we obtain

$$\nu_{j+1}=(n_0-xj)\left[\frac{n_0-xj}{n_0-xj+\sum_{i=1}^{j}y_i}\right]^{(x-y_{j+1})}\left[\frac{n_0-xj}{n_0}\right]^{y_{j+1}-1}.$$

Introduction of y for the average $j^{-1}\sum_{i=1}^{j}y_i$, and combination of factors yields

$$\nu_{j+1}=(n_0-xj)^x/(n_0-xj+yj)^{x-y_{j+1}}n_0^{y_{j+1}-1}. \tag{3}$$

Let the distribution of orientations of the solute molecules be represented by the numbers n_k whose directions occur within the elements of solid angle $\delta\omega_k$. The size of each such element may be chosen to be the same and equal to the range of solid angle assumed for the reference state of so-called perfect order, whose partition function is q_2 per molecule. The *a priori* probabilities for the various orientation states thus defined are identical, and hence may be omitted from the partition function. The latter for the athermal solution will be given by

$$Q_M=q_1^{n_1}q_2^{n_2}\prod_{j=1}^{n_2}\nu_j/\prod_k n_k!. \tag{4}$$

If it is agreed to preserve the specified angular distribution of molecular axes throughout the process of addition of the solute molecules signified by the first product, the average y introduced in (3) remains fixed. Moreover, the geometric mean of ν_j over the distribution may be used in (4), hence the y_{j+1} in (3) should be replaced by their arithmetic mean y. Substitution of (3) in (4) and conversion of products to factorials then yields

$$Q_M=q_1^{n_1}q_2^{n_2}[(n_0-xn_2+yn_2)!/\{(n_0-xn_2)!\,n_0^{(y-1)n_2}\prod_k n_k!\}]$$

or

$$Q_M=q_1^{n_1}q_2^{n_2}[(n_1+yn_2)!/\{n_1!\,n_2!(n_1+xn_2)^{(y-1)n_2}\}]n_2!/\prod_k n_k!. \tag{5}†$$

---

† Equation (5) may be derived alternatively by first dissociating each molecule into the numbers of submolecules required by its orientation. These are distributed at random over the rows of lattice cells running parallel to the orientation axis. The number of configurations for the dissociated system of submolecules is then multiplied by the probability that the submolecules are properly situated in adjacent rows for their combination into the analogues of rods (figure 1b) having the specified orientation.

A result which is at least approximately identical with (5) is obtained by another method which, like the one presented in the text above, involves the derivation of ν_{j+1}. Previously added molecules are resolved into submolecules on rows of lattice cells taken parallel to the axis of molecule $j+1$, and the expected number of situations available to molecule $j+1$ is calculated in the foregoing manner.

Correspondences between (5) on the one hand and (1) and (2) on the other are readily demonstrated. In the case of perfect orientation $y = 1$, the last factor in (5) is unity, and (5) reduces identically to (2).

The correspondence of (5) to (1) for the isotropic solution is less obvious. According to the conventional lattice treatment, each solute molecule may choose between $\frac{1}{2}z$ orientations. The factor $(\frac{1}{2}z)^{n_2}$ in (1) represents the number of configurations arising from these orientations, and it corresponds therefore to the factor $n_2!/\prod_k n_k!$ in (5). For the unoriented system the latter will be much greater than $(\frac{1}{2}z)^{n_2}$ however, owing to the assumption in the present treatment of a distribution of directions which is essentially continuous (if y_k is allowed to assume non-integral values). Reduction of the other factors occurring in (5) to those in (1) requires the substitution $y = x$. The average over all directions for the unoriented system yields $y = (\frac{1}{4}\pi)x$. The discrepancy is not important, however, for we shall be concerned principally with situations in which $y < \frac{1}{2}x$. Moreover, in treating equilibrium properties of the partially ordered phase it will not be necessary to attach precise physical significance to y. We shall merely consider y, therefore, as an index of disorientation which varies from one for perfect order to x for complete disorder. A correspondence with (1) for the unoriented solution is thus established.

In order to render (5) in a more tractable form, it is desirable to relate the last factor to the parameter y. Obviously this factor will increase with y. If we assume a density distribution over solid angle which is uniform out to some angle ψ' and zero beyond, then $n_2!/\prod_k n_k!$ will be proportional to ω^{n_2}, where ω is the solid angle within $\psi < \psi'$. Furthermore, ω is proportional to $y^2 \equiv \overline{(\sin\psi)^2}$ to a good approximation for $\overline{\sin\psi} < \frac{1}{2}$; even at $\overline{\sin\psi} = \frac{1}{4}\pi$ (random distribution) the error in the latter proportionality is less than 30%. For the distribution specified above, therefore, $n_2!/\prod_k n_k!$ will be approximately proportional to y^{2n_2}. The same approximation should be valid for any actual distribution likely to be encountered, at least over the range of orientation $y/x = \overline{\sin\psi} < \frac{1}{2}$ of major interest. Retention of the equivalence of (5) to (1) in the limit of perfect orientation designated by $y = 1$ requires that the constant of proportionality be unity; it is otherwise of no importance to the results which follow. Hence the relation

$$n_2!/\prod_k n_k! \cong y^{2n_2} \qquad (6)$$

will be adopted.

We shall wish to examine the effects of small interactions between the solute particles; hence it is desirable to incorporate the heat of mixing ΔH_M into the expression for the free energy. If the forces are not sufficiently large to vitiate seriously the assumed randomness for a specified degree of orientation, the factors in (5) for the number of configurations may be retained. Moreover, the heat of mixing may then be approximated satisfactorily by an expression of the van Laar form, i.e. ΔH_M may be taken to be proportional to the product of the number of molecules of one type and the volume fraction (v_1) of the other. We thus introduce

$$\Delta H_M = RT\chi_1 n_2 v_1, \qquad (7)$$

663

where $RT\chi_1$ is the energy change per segment on transferring a solute molecule from the pure solute to the infinitely dilute solution.† The parameter χ_1 may be more generally interpreted as a free energy of interaction if desired (Guggenheim 1948; Flory & Krigbaum 1950; Flory 1953).

The expression (7) should hold irrespective of the force law, provided merely that the forces are sufficiently small to fulfil the condition set forth above. Introduction of the effects of interactions in this simple way renders the subsequent results inapplicable to highly charged colloid particles in media of low ionic strength. Even in this case, however, dilute solutions could be treated approximately by assigning dimensions to the particle which include the electric double layer, after the manner set forth by Onsager (1949). The restriction is less serious if the forces between the solute particles are attractive (positive heat of mixing), for in this case the equations to be derived lead straightforwardly to intense precipitation even for comparatively low values of χ_1.

Thus, from (5), (6) and (7) we obtain for the free energy of mixing

$$\Delta G_M/kT = n_1 \ln v_1 + n_2 \ln v_2 - (n_1 + yn_2) \ln [1 - v_2(1 - y/x)] \\ - n_2[\ln(xy^2) - y + 1] + \chi_1 x n_2 v_1, \quad (8)$$

where v_1 and v_2 are volume fractions. When $y = 1$, this equation reduces to that for a regular solution (ideal entropy); when $y = x$ it yields a result which is identical with the free energy of mixing for a polymer solution, except for the disorientation entropy term, which is then represented by $-n_2 \ln x^2$ (see preceding paper).

EQUILIBRIUM DEGREE OF DISORDER

Equation (8) offers an expression for the free energy as a function of the composition and of the disorientation index y. Since no external restraints apply to the latter, it will be assumed to adopt the value which minimizes ΔG_M. Differentiating (8) with respect to y, equating to zero, and solving the resulting expression for v_2, we obtain

$$v_2 = [x/(x-y)][1 - \exp(-2/y)]. \quad (9)$$

This equation possesses two solutions y for a given v_2 and x provided these quantities are sufficiently large. The lower solution y corresponds to a minimum in ΔG_M and the other one to a maximum. Equating dv_2/dy to zero, we obtain

$$x = y^* + \tfrac{1}{2}y^{*2}[\exp(2/y^*) - 1], \quad (10)$$

where y^* is the value of y at the minimum concentration v_2^* for which (9) yields solutions. By series expansion

$$y^* = \tfrac{1}{2}(x-1) - (1/3y^* + 1/6y^{*2} + ...). \quad (10')$$

For large values of x the series of terms in $1/y^*$ may be neglected, and for smaller x the solution is readily obtained by iteration. The value y^* ($\simeq \tfrac{1}{2}x$) sets an upper

† Strictly speaking, we should replace v_2 by a surface fraction
$$(xz - 2x + 2)\, n_2/[zn_1 + (xz - 2x + 2)\, n_2]$$
(see Guggenheim 1944). This refinement, however, is not warranted.

limit for the disorientation in a stable anisotropic phase. As we show later on, phase separation causes the maximum attainable y to be considerably less than y^*.

Substitution of (10) in (9) yields

$$v_2^* = 1 - (1 - 2/y^*) \exp(-2/y^*). \qquad (11)$$

For $v_2 < v_2^*$ the mixing free energy ΔG_M decreases monotonically with y, hence a necessary condition for existence of a stable anisotropic phase is $v_2 > v_2^*$. Series expansion of the exponential in (11) followed by substitution from (10') for y^* with neglect of the series in $1/y^*$ leads to

$$v_2^* \simeq 1 - [1 - 4/(x-1)]^2 + \dots .$$

Numerical calculations show the semi-empirical approximation

$$v_2^* \simeq (8/x)(1 - 2/x) \qquad (12)$$

to be accurate within 2 % for $x > 10$. Thus, the minimum concentration required for stable anisotropy increases inversely as the axis ratio x for large x.

Of course, the minimum in the free energy occurring at the value of y which is the lower solution of (9) for given x and v_2 represents the most stable arrangement only if at this minimum the free energy of mixing is lower than at any other value of y in the physically significant range, and in particular only if it is lower than ΔG_M for $y = x$, which we take to be the upper limit on y for complete disorder. For v_2 values only slightly greater than v_2^*, this condition will not be fulfilled. Numerical calculations show the range of metastability to be small. Moreover, intervention of phase separation (cf. seq.) eliminates the region of metastable $y < x$ for all x almost to its limiting value, which we consider next. It should be borne in mind, nevertheless, that absolute stability of anisotropy within the phase requires a slightly greater concentration than the v_2^* set forth above.

In the absence of diluent ($v_2 = 1$) according to (11), $y^* = 2$. Substituting these values in (9), we find $x = 2e = 5.44$ as the minimum value of x required for a stable anisotropic state for the pure solute. A somewhat greater value of x is required, however, before the free energy of the anisotropic state becomes less than that of the totally disordered state. Assigning the latter a free energy given by (8) with $y = x$, we find by numerical calculation that the minimum x for absolute stability is about 6.7. A smaller maximum value of y (e.g. $\frac{1}{4}\pi$) would yield a lower critical x. Errors arising from the lattice treatment itself may be expected to cause the result to be too large. Hence, we conclude that a length-to-diameter ratio of about 2e should be sufficient to cause spontaneous ordering of the phase. It will be observed that the value of y is in any case considerably less than x, i.e. even at the minimum molecular length for stable anisotropy the degree of ordering is pronounced.

According to the inequality (7) of the preceding paper, the minimum value of x for stability of an isotropic phase *relative to the completely ordered phase* is specified for stiff rods ($f = 0$) by

$$(x-1) - \ln(\tfrac{1}{2}zx) = 0. \qquad (13)$$

Solutions for $z = 6$ and 12 are $x = 3.29$ and 4.24, respectively. The treatment of that paper was developed primarily for very long chains, and fails when applied to shorter chains because of the restriction of the orientations in the isotropic state to $\frac{1}{2}z$ directions (with allowance for the symmetry number, 2) on the one hand, and of the disregard of anisotropic states of partial disorder on the other. Revision of the lattice co-ordination number to achieve consistency with the present treatment would increase the limiting asymmetry for $v_2 = 1$ to $x = 7$ to 8. Correction for the second deficiency mentioned above would lower the result by an unspecifiable amount.

CHEMICAL POTENTIALS

By substituting (9) in (8) the following more convenient expression is obtained for the free energy of formation of the phase with equilibrium disorder (i.e. incomplete disorder):

$$\Delta G_M/kT = n_1 \ln v_1 + n_2 \ln v_2 + 2n_1/y - n_2[\ln (xy^2) - y - 1] + \chi_1 x n_2 v_1. \tag{14}$$

It is to be understood that here and in subsequent equations y is to be assigned its equilibrium value given by the lower solution of (9).

Taking the derivative with respect to n_1, we obtain

$$(\mu_1 - \mu_1^0)/RT = \ln (1 - v_2) + (1 - 1/x) v_2 + 2/y + (n_2 - 2n_1/y^2 - 2n_2/y)(\mathrm{d}y/\mathrm{d}n_1)_{\mathrm{eq.}} + \chi_1 v_2^2,$$

where the n's represent moles instead of numbers of molecules. Evaluation of $(\mathrm{d}y/\mathrm{d}n_2)_{\mathrm{eq.}}$ from (9) yields

$$(\mu_1 - \mu_1^0) RT = \ln (1 - v_2) + [(y - 1)/x] v_2 + 2/y + \chi_1 v_2^2. \tag{15}$$

The corresponding chemical potential for the isotropic phase, obtained from (8) with $y = x$, is the well-known result (Flory 1942, 1953)

$$[(\mu_1 - \mu_1^0)/RT]_{y=x} = \ln (1 - v_2) + (1 - 1/x) v_2 + \chi_1 v_2^2. \tag{16}$$

The similarly derived chemical potentials for the solute, relative to the completely ordered polymer as reference state, are

$$(\mu_2 - \mu_2^0)/RT = \ln (v_2/x) + (y - 1) v_2 + 2 - \ln y^2 + \chi_1 x (1 - v_2)^2 \tag{17}$$

and $\qquad [(\mu_2 - \mu_2^0)/RT]_{y=x} = \ln (v_2/x) + (x - 1) v_2 - \ln x^2 + \chi_1 x (1 - v_2)^2 \qquad (18)$

for ordered and for isotropic solutions, respectively. The latter equation, like (16), is equivalent to that for an ordinary polymer solution except for the additive term $-\ln x^2$ which replaces the free energy of disorientation†.

Chemical potentials calculated from (15) to (18) in conjunction with (9) are shown in figure 2 for athermal solutions $(\chi_1 = 0)$ of rods characterized by $x = 100$. The solvent and solute chemical potentials for the ordered, or anisotropic, solution pass through a maximum and a minimum, respectively, before reaching the dilution below which (9) no longer yields a mathematical solution. For all lower concentrations the free energy decreases monotonically with y, hence only the isotropic state $y = x$ can exist. The location of the maximum and minimum lies within the concentration range, indicated by dashed curves, of metastable anisotropy (see above)

---

† Equation (18) differs from (16) of the preceding paper with $f = 0$ only in terms for the contribution from random orientation, the previous treatment yielding $-\ln (\frac{1}{2}z)$ in place of $-\ln x^2$.

where the free energy for $y = x$ is actually lower than that at the value of y corresponding to the lower solution of (9). Moreover, as will be shown presently, the metastable region is bridged by phase separation, i.e. the region is one of metastability also with respect to conjugate phases having compositions outside this region. The same applies for all values of x at which anisotropy is stable and separation into two phases is predicted.

The curves for isotropic solutions, which occur on the left-hand side of figure 2, are continued only to the limit, $v_2 = 0.0935$, of absolute stability of complete disorder. These curves are identical with those given by ordinary polymer solution theory, except for a vertical displacement in the ordinate scale for $(\mu_2 - \mu_2^0)/RT$

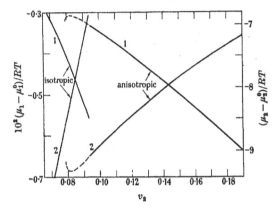

FIGURE 2. Chemical potentials for the solvent ('1', left-hand ordinate scale) and solute ('2', right-hand ordinate scale) for $\chi_1 = 0$ and $x = 100$. Regions of metastable order are shown by broken lines.

required by the difference in additive terms (see above). The appearance of the discontinuity at which the chemical potentials change abruptly owing to transformation to a state of equilibrium anisotropy is of fundamental importance. Disregarding the metastable region, the chemical potentials then change monotonically with further increase in concentration. The rates of change, however, are much less than for the extrapolations of the curves for the isotropic state, i.e. much less than for ordinary polymer solutions.

The basis for these differences between the chemical potentials for the isotropic and the anisotropic states is manifest in (15) to (18). To begin with, there are positive contributions to the chemical potentials for the anisotropic state from terms originating in the sacrifice of disorientation entropy, i.e. from the term $2/y$ in (15) and from the replacement of $-\ln x^2$ in (18) by $2 - \ln y^2$ in (17). Opposing these increases are terms representing the diminished competition in the ordered state for sequences of sites suitable for occupancy by a solute particle. These terms, $[(y-1)/x]v_2$ and $(y-1)v_2$, respectively, are smaller than the corresponding terms for the isotropic state. The changes in the chemical potentials with concentration are moderated by the combined effects of the smaller coefficients of these terms in

v_2 and by the decrease in y with concentration, i.e. by the interplay of the normal effect of an increase in concentration in decreasing the expectancy of vacancies and by the concurrent, and opposite, effect of increase in order with concentration. In short, the entropy of mixing function changes toward that for an ideal solution as the concentration increases, and y consequently decreases.

Chemical potentials for solutions which are not athermal will be discussed in a later section.

PHASE EQUILIBRIUM; ATHERMAL SOLUTIONS

It is obvious from the nature of the discontinuity shown in figure 2 that separation into two phases, one isotropic and the other ordered, is required. From the phase equilibrium conditions $\mu_1' = \mu_1$ and $\mu_2' = \mu_2$, where accents are adopted to denote

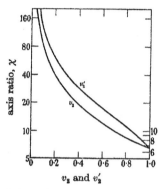

FIGURE 3. Concentrations of phases in equilibrium in relation to x for athermal mixtures.

TABLE 1. SUMMARY OF RESULTS CALCULATED FOR ATHERMAL
EQUILIBRIUM BETWEEN ISOTROPIC AND ORDERED PHASES

x	v_2	v_2'	v_2'/v_2	xv_2	y	y/x
6·702	1·000	1·000	1·000	6·70	1·115	0·1665
8	0·856	0·936	1·093	6·85	1·308	0·1635
10	0·706	0·848	1·201	7·06	1·61	0·161
20	0·3790	0·5405	1·426	7·58	3·34	0·167
30	0·2599	0·3883	1·494	7·80	5·15	0·172
50	0·1597	0·2458	1·539	7·98	8·85	0·177
100	0·0806	0·1248	1·551	8·06	18·65	0·186
200	0·0404	0·0630	1·557	8·08	38·22	0·191

the more concentrated (in this case, ordered) phase in accordance with previous custom, we obtain with the aid of preceding equations

$$\ln(1-v_2) + (1-1/x)\,v_2 + \chi_1 v_2^2 = \ln(1-v_2') + [(y-1)/x]\,v_2' + 2/y + \chi_1 v_2'^2 \quad (19)$$

and

$$\ln(v_2/x) + (x-1)\,v_2 - \ln x^2 + \chi_1 x(1-v_2)^2$$
$$= \ln(v_2'/x) + (y-1)\,v_2' + 2 - \ln y^2 + \chi_1 x(1-v_2')^2. \quad (20)$$

Numerical solution of these equations, along with (9), for $\chi_1 = 0$ and various values of x yields the results summarized in table 1 and shown graphically in figure 3.

668

By substituting $v_2 = v'_2 = 1$ in (9) and (20) it is possible to show that the two curves meet on the right-hand ordinate axis at $x = 6\cdot702$, which may be taken to represent the minimum value of x for co-existence of two phases. Values of other parameters at this limit are given in the first row of table 1. This minimum value is very nearly coincident with the minimum x calculated (see above) for *absolute* stability of anisotropy in the pure solute, though greater than the value $x = 2e$ required for the appearance of a minimum in the free energy with y.

With increase in x the concentrations of both phases diminish. The concentration of the ordered phase is never much greater than that of the isotropic conjugate phase. The ratio of their concentrations (table 1) appears to approach $1\cdot56$ as a limit with increase in x. This is somewhat greater than the value $1\cdot343$ deduced by Onsager (1949) for rods of unspecified length-to-diameter ratio. At large x we find $v_2 \cong 8/x$ and $v'_2 \cong 12\cdot5/x$. Onsager obtained the lower values $v_2 = 3\cdot34/x$ and $v'_2 = 4\cdot48/x$, where x is taken to be the ratio of length to diameter for the rod. Isihara (1951) found the similar result $v_2 = 3\cdot4/x$. The comparison of these results will be considered further in the Discussion.

The ratio y/x for the ordered phase is almost constant (table 1) except for a gradual increase at large x. It is always small, indicating a fairly high degree of order in the anisotropic phase. Attempts to derive analytically these asymptotic relations implied by the numerical calculations have not succeeded.

Absolute values of the chemical potentials for the unoriented phase depend, of course, on the value assigned to y in the limit of complete disorder. Use of $y = (\frac{1}{4}\pi)x$ instead of $y = x$ brings about a small increase in $\mu_1 - \mu_1^0$ and a corresponding decrease in $\mu_2 - \mu_2^0$. Changes in the equilibrium compositions are insignificant, however. The general features indicated for phase separation in a binary athermal system containing rod-like particles appear to be genuine and not an artifact of the approximations employed.

NON-ATHERMAL SOLUTIONS

The energy of mixing has been introduced in coarse approximation only. Its effect on the phase equilibrium is found to be trivial when negative, and so very pronounced when positive as to dwarf any aberrations consequent upon the approximation.

Attention has already been directed to the small rate of decrease of the chemical potential of the solvent for the anisotropic solution. Even a small positive parameter χ_1 has a marked effect on the course of μ_1 with v_2, as is illustrated in figure 4 for $x = 100$. The outstanding feature is the emergence of a maximum at higher concentration. It appears first at $\chi_1 = 0\cdot055$ (not shown in figure 4) and becomes large beyond $\chi_1 = 0\cdot070$.

The presence of the maximum, together with the preceding minimum, raises the possibility for co-existence of two *anisotropic* phases. Indeed, application of the conditions for phase equilibrium, $\mu_1 = \mu'_1$ and $\mu_2 = \mu'_2$, to (15) and (17) for two concentrations v_2 and v'_2 yields mathematical solutions for $0\cdot055 < \chi_1 < ca. 0\cdot08$. At $\chi_1 = 0\cdot060$, for example, the conjugate anisotropic phases in equilibrium have volume fractions $0\cdot26$ and $0\cdot70$. Solution of (19) and (20) yields the further conjugate pair at volume fractions 0·0778 and 0·144. The primary free-energy function from which the various chemical potentials are derived by straightforward operations thus leads *to the existence of two heterogeneous regions*: one involves an isotropic phase ($v_2 = 0·0778$) in equilibrium with a dilute anisotropic phase ($v_2' = 0·144$), and the other a pair of anisotropic phases.

As χ_1 is further increased the intermediate homogeneous region ($0·144 < v_2 < 0·26$) diminishes. The two heterogeneous regions merge at $\chi_1 = 0·070$, hence for $\chi_1 \geqslant 0·070$ equilibrium prevails between the (dilute) isotropic phase and the concentrated, highly anisotropic phase. The two heterogeneous regions then overlap, and therefore merely represent metastable equilibria†. At the critical point (at constant pressure)

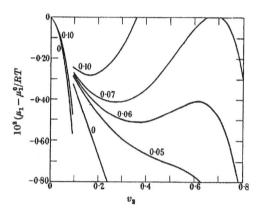

FIGURE 4. Solvent chemical potentials for $x = 100$ and positive energies of mixing indicated by values of χ_1 given with each curve. Curves for isotropic solutions are to the left of the discontinuity; those for ordered solutions are to the right.

$\chi_1 = 0·070$, the stable co-existence of three phases ($v_2 = 0·077, 0·150, 0·804$, respectively) is thus predicted. The phase diagram is shown in figure 5.

At $\chi_1 \cong 0·09$, the composition of the concentrated phase reaches $v_2 = 0·873$, where according to (9) y falls to unity. Use of the phase equilibrium equations (19) and (20) together with the auxiliary equation (9) in the same manner for larger values of the parameter χ_1, and hence for higher concentrations in the anisotropic phase, would involve values of $y < 1$. In order to preserve consistency with the identification of $y = 1$ as the limit for perfect orientation, we have therefore considered y to be constant and equal to unity for all values of $\chi_1 > 0·09$ in calculating the uppermost portions of the curves in figure 5. Substitution of $y = 1$ into (8) reduces it to the expression for a regular solution, and the chemical potentials are given by (17) and (18) of the preceding paper. The equations for phase equilibrium accordingly become

$$\ln(1-v_2) + (1-1/x)\,v_2 + \chi_1 v_2^2 = \ln\left[(1-v_2')/(1-v_2'+v_2'/x)\right] + \chi_1 v_2'^2 \qquad (21)$$

† Beyond $\chi_1 = 0·08$ the equations fail to yield mathematical solutions corresponding to the two metastable equilibria, owing to the presence of the discontinuity. These solutions for $\chi_1 > 0·07$ are, of course, unimportant physically.

670

and

$$\ln(v_2/x) + (x-1)\,v_2 - \ln x^2 + \chi_1 x(1-v_2)^2$$
$$= \ln[(v_2'/x)/(1-v_2'+v_2'/x)] + \chi_1 x(1-v_2')^2, \quad (22\dagger)$$

in which 'polymer' chemical potentials for the dilute phase are equated to those of regular solution theory for the concentrated phase.

Equations (21) and (22) have been used in place of (19) and (20) for $\chi_1 > 0.09$. Neglect of this revision would not, however, have altered seriously the form of the curves. With increase in χ_1 beyond 0.07, v_2 falls rapidly and v_2', already large, increases gradually. At $\chi_1 = 0.25$, $v_2 = 3.0 \times 10^{-6}$ and $v_2' = 0.963$.

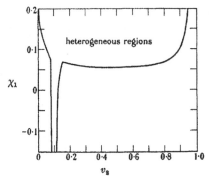

FIGURE 5. Compositions of phases in equilibrium for $x = 100$ and values of the interaction parameter plotted on the ordinate. The curve for isotropic solutions is on the left, and that for the ordered phases on the right, with the range of heterogeneity occupying the region between.

The weak decrease of $(\mu_1 - \mu_1^0)/RT$ with v_2 for the anisotropic phase is strengthened by $\chi_1 < 0$. Although the discontinuity remains, the difference between chemical potentials on either side of it is diminished. The concentrations of the phases consequently shift toward each other as χ_1 decreases. The change is gradual, however, and a heterogeneous zone exists even for very large negative interactions.

No great importance is attached to the shallow minimum occurring in the phase diagram (figure 5) and representing co-existence of two anisotropic phases. It is conceivable that it would disappear with refinement of the theory, although no basis for such a prediction is apparent.

Of far greater importance is the sudden shift in the concentration v_2' from ca. 0.15 to > 0.9 brought about by a comparatively small change in χ_1. This phenomenon may be seen to follow as a direct consequence of the small changes in the partial entropies with concentration in the anisotropic phase. As explained earlier, this feature of the partial entropies finds explanation in the effect of orientation and its change with concentration. The basis for a rapid, if not sudden, change in the

† Equations (21) and (22) are identical with (19) and (20), respectively, of the preceding paper, provided in the latter f is equated to zero and $\ln(\tfrac{1}{2}z)$ is replaced by $\ln x^2$.

character of the phase diagram with increase in χ_1 is thus comprehensible in terms of the physical factors involved. It seems justified to conclude therefore that the reality of the predicted phenomenon transcends possible inaccuracies of the theory itself.

The values of the interaction parameter considered are small compared to those commonly encountered in polymer + solvent systems. Thus, the limit for incidence of phase separation for a solution of a random coiling polymer is $\chi_1 = 0.50$ at infinite chain length. Reckoned on the basis of energy per solute molecule (instead of per segment), the interaction energies appear much larger. At $\chi_1 = 0.10$, for example, the transfer of a solute molecule from pure solute to an infinitely dilute solution requires $x\chi_1 kT$, or $10kT$. A considerable entropy is required to maintain a solution under such circumstances. The polymer mixing law offers a large entropy associated with distribution of solvent and solute molecules, in effect, over the entire volume (Flory 1953). The mixing entropy for rigid rods at equilibrium disorder approaches more nearly the much lower ideal mixing entropy. Hence, the rigid rod having $x = 100$ 'precipitates' at $\chi_1 = 0.10$, whereas more than five times this value would be required even for coacervation (i.e. phase separation about a low critical concentration) for a solution of a random chain polymer of the same size.

Specific interaction forces may be expected to produce further ordering of the solute particles in the 'precipitated' phase in view of its high concentration and the essentially complete parallel order which will be required for its existence. If the particles are of uniform structure, transformation to crystalline regularity, for example, would require only a small further sacrifice of entropy. It is intended in this paper merely to point out that supplementation of the consequence of high asymmetry by a comparatively low interaction energy is *sufficient* to bring about generation of a concentrated phase and near exhaustion of the solute from the dilute phase in equilibrium. The precipitation may, of course, be intensified by forces which are operable only for more highly ordered arrangements, and hence are not represented in the van Laar term for the energy.

Irrespective of the nature of the forces involved, if the energy of interaction is positive and sufficiently large, precipitation may be considered to yield a phase which is essentially pure and crystalline, in so far as thermodynamic equilibrium is concerned. Under such circumstances phase equilibrium equation (21) may be abandoned, and v_2' set equal to unity in (22), which readily reduces to an expression for equilibrium between a dilute and a pure phase.

Discussion

In applying the lattice method to polymer solutions the assumption is made that centres of neighbours to a given segment of a macromolecule may occur only within certain specified volume elements. This assumption fails unless the solvent molecules and polymer segments are interchangeable in their spatial requirements, a condition rarely fulfilled even approximately. In actual solutions, centres of segments which are near neighbours to a given segment will occur over a considerable range of volume in what would otherwise be the first co-ordination sphere.

The solute molecule then effectively excludes others from a volume somewhat greater than its actual volume used in the lattice treatment.

In the limit of complete breakdown of lattice-like regularity, the solvent may be treated as a continuum, and virial coefficients for the expansion of π/RT in powers of the concentration may be deduced (in principle at least) by the well-known method of evaluating integrals of the cluster expansion. The second coefficient obtained in this way (i.e. by averaging the excluded volume for a pair of particles over all orientations) for unoriented rod-like particles by Zimm (1946), Onsager (1949) and Isihara (1950) is given in the present notation by $A_2 = x^2 v_s/M^2$, where $x v_s$ is the molar volume of a solute molecule. This is exactly twice the value obtained by series expansion of (16) with $\chi_1 = 0$. The deficiency of the lattice model pointed out above leads therefore to an error in the second virial coefficient which at most requires correction by a factor of two.

A better representation of dilute solutions might be achieved from the lattice treatment by ascribing a somewhat enlarged cross-sectional domain to the solute particles. Overlapping of such domains at higher concentrations would require use of an effective domain which decreases in size with the concentration. We prefer to avoid the resulting complications at the possible expense of improved accuracy.

As was pointed out earlier in this paper, the present theory predicts athermal phase separation at concentrations about twice those given by Onsager (1949) and Isihara (1951) from consideration of second virial coefficients. The deficiency of the lattice treatment discussed above probably causes the equilibrium concentrations calculated for a given asymmetry (or the asymmetry calculated for a given concentration) to be somewhat too large. On the other hand, the concentration in the anisotropic phase may be shown to be too great to justify the disregard of higher virial coefficients necessitated by insuperable mathematical difficulties in the methods of Onsager and Isihara. Thus, the covolume is $x^2 v_s$ per mole (Zimm, Onsager, Isihara) and the sum of the covolumes of the solute particles in unit volume is $x v_2$. In the isotropic phase co-existing in equilibrium with its conjugate anisotropic phase this product exceeds 8 for large x according to the present theory, and about 3·4 according to Onsager (presumably for large x). The neglect of higher virial coefficients may lead to errors more serious under these conditions than those arising from the limitations of the lattice treatment.[†]

The separation of tobacco mosaic virus (t.m.v.) solutions into two phases at concentrations of ca. 2 % (Bawden & Pirie 1937; Bernal & Fankuchen 1941; Oster 1950) has been attributed to asymmetry of the rod-like t.m.v. particles (Onsager 1949; Isihara 1951). Their measured ratios x are about 15 to 20 (Oster 1950). According to the present treatment, athermal phase separation for particles with $x = 20$ should commence at $v_2 = 0.38$. The results of Onsager and Isihara would predict $v_2 = 0.17$, assuming, however, that an axis ratio of 20 is sufficient for application of their results for very long rods. At the low ionic strengths, ca. 10^{-3} to 10^{-4}M, at which the phase separation is observed, the thickness of the electric double layer

† It is worth noting in this connexion that the chemical potential deduced for the solvent in the anisotropic phase according to the present treatment is not amenable to the usual virial expansion. This is evident from the form of the curve in figure 2.

surrounding the particle exceeds its diameter (*ca.* 150Å). Treating the double layers as integral with the particles (Onsager 1949), a fivefold increase in effective diameter is required (Oster 1950); the axis ratio will be decreased by the same factor, and the effective concentration will be increased by its square. We then have a solution of particles with less than the critical asymmetry (2e) for stable anisotropy even in the absence of diluent. On the other hand, at the point of phase separation the t.m.v. particles and their double layers occupy about 50% of the volume. It is doubtful that the approximate method for treating the double layers about asymmetric particles is valid under these conditions. In any event, the importance of electrical forces in bringing about separation of a dilute anisotropic phase from an even more dilute solution of highly asymmetric particles (Langmuir 1938) is clearly indicated, for the effect with t.m.v. vanishes according to Oster (1950) at salt concentrations exceeding 5×10^{-3}M. The phase separation which occurs in t.m.v. is not adequately explained by consideration of particle asymmetry alone.

Similar considerations may apply to the separation of anisotropic phases, or tactoids, from dilute solutions of inorganic colloidal particles, such as V_2O_5, which are both asymmetric and highly charged (Freundlich 1937; Langmuir 1938; Watson, Heller & Wojtowicz 1949).

Results of the present paper should apply to the aggregation of highly elongated colloidal particles. In particular, they afford a method for treating fibrillar aggregation of helical protein molecules such as those of collagen and myosin. It is to be noted that the tedious problem of calculating the forces operating between *a pair* of particles as a function of distance and orientation is here avoided. Indeed, in so far as equilibrium states are concerned such calculations become irrelevant; we require instead a quantity which is more readily accessible, namely, the free energy of a system of many particles as a function of their concentration and average orientation.

The author takes pleasure in acknowledging his gratitude to Professor G. Gee, F.R.S., and his colleagues for their hospitality during the academic year 1954–5, which it was the author's privilege to spend in residence at Manchester University. He is particularly indebted to Professor Gee and Dr J. S. Rowlinson, of the Chemistry Department, and to Dr S. Levine, of the Department of Applied Mathematics, for helpful discussions and criticisms pertaining to this paper.

REFERENCES

Bawden, F. C. & Pirie, N. W. 1937 *Proc. Roy. Soc.* B, **123**, 274.
Bernal, J. D. & Fankuchen, I. 1941 *J. Gen. Physiol.* **25**, 111.
Flory, P. J. 1942 *J. Chem. Phys.* **10**, 51.
Flory, P. J. 1944 *J. Chem. Phys.* **12**, 425.
Flory, P. J. 1953 *Principles of polymer chemistry*, Ithaca, N.Y.: Cornell University Press.
Flory, P. J. 1956 *Proc. Roy. Soc.* A, **234**, 60.
Flory, P. J. & Krigbaum, W. R. 1950 *J. Chem. Phys.* **18**, 1086.
Freundlich, H. 1937 *J. Phys. Chem.* **41**, 1151.
Guggenheim, E. A. 1944 *Proc. Roy. Soc.* A, **183**, 213.

Guggenheim, E. A. 1948 *Trans. Faraday Soc.* **44**, 1007.
Isihara, A. 1950 *J. Chem. Phys.* **18**, 1446.
Isihara, A. 1951 *J. Chem. Phys.* **19**, 1142.
Langmuir, I. 1938 *J. Chem. Phys.* **6**, 873.
Onsager, L. 1949 *Ann. N.Y. Acad. Sci.* **51**, 627.
Oster, G. 1950 *J. Gen. Physiol.* **33**, 445.
Watson, J. H. L., Heller, W. & Wojtowicz, W. 1949 *Science*, **109**, 274.
Zimm, B. H. 1946 *J. Chem. Phys.* **14**, 164.

Paul Flory was born in Sterling, Illinois, U.S.A. in 1910. He graduated from Manchester College, Indiana in 1931, and then studied for a doctorate in physical chemistry at Ohio State University, which was awarded in 1934. He then joined the Central Research Department of the DuPont Company, where he first became interested in polymerisation while working in the group headed by W.H. Carothers, inventor of nylon. Between 1937 and 1939 he worked at the Basic Science Research Laboratory of the University of Cincinnati. During the Second World War, he worked on synthetic rubber in industry, at Standard Oil (now Exxon) (1940–3), and later at the Research Laboratory of the Goodyear Tire and Rubber Company (1943–8).

In 1948 he was invited by P.J.W. Debye (who had moved before the war from Berlin to Cornell) to join the Department of Chemistry at Cornell University. In 1957 he moved to the Mellon Institute in Pittsburgh and then in 1961 to Stanford University.

Flory is one of the grand figures of polymer science in the twentieth century. To him we owe much of the important understanding of the statistical mechanics of polymeric and macromolecular systems. In particular we might mention: the effect of volume exclusion on polymeric configuration leads to the well-known Flory exponent for molecular size as a function of molecular weight; the theta point at which volume exclusion effects are neutralised; and the Flory-Huggins theory for the free energy of macromolecules in solution.

Flory's books, *Principles of Polymer Chemistry* (1953), and *Statistical Mechanics of Chain Molecules* (1969) were particularly influential. He received numerous honours, most notably the Nobel prize in chemistry in 1974. He died in 1985.

E6

Acta Crystallographica **13**, 660–67 (1960)

La Structure des Colloïdes d'Association. I. Les Phases Liquide–Cristallines des Systèmes Amphiphile-Eau

Par V. Luzzati, H. Mustacchi, A. Skoulios et F. Husson

Centre de Recherches sur les Macromolécules, 6, rue Boussingault, Strasbourg, France

(Reçu le 10 juillet 1959)

The structures of the liquid–crystalline phases of amphiphile–water systems have been investigated by X-rays diffraction and polarizing microscope observations. Some new phases have been discovered. The structures of all the phases are described, and several geometric parameters are determined. It is shown that an essential feature common to all phases is the disordered liquid-like configuration of the paraffin chains.

Introduction

Le terme amphiphile désigne une famille de substances dans lesquelles existent, à l'intérieur d'une même molécule, deux régions douées de solubilités très différentes, et suffisamment éloignées entr'elles pour se comporter de manière indépendante, pour autant que le permettent les liens qui les réunissent. En général, ces deux régions sont formées respectivement par un groupement hydrophile, et une ou plusieurs longues chaînes paraffiniques. On connaît un grand nombre de telles substances; on peut mentionner les savons, sels alcalins d'acides gras, les différents produits dits 'détergents', 'tensio-actifs', 'émulsionnants' ainsi que certaines molécules complexes qu'on rencontre dans les organismes vivants (glycérolipides, phospholipides, etc. ...).

En présence d'eau, un amphiphile a un comportement singulier, car la grande solubilité de son extrémité hydrophile doit se concilier avec l'insolubilité du reste de la molécule. Pour analyser les propriétés d'un tel système, il convient de se référer à son diagramme de phases. A titre d'exemple, nous avons porté dans la Fig. 1 le diagramme relatif au système palmitate de potassium-eau, emprunté à un mémoire de McBain & Lee (1943). On y reconnaît différents domaines:

Domaine de la solution isotrope

C'est la solution embrassant toutes les proportions de savon et d'eau, pour des températures suffisamment élevées. Les propriétés physiques de la solution isotrope changent brusquement à une concentration particulière, et ces changements indiquent qu'il se forme des aggrégats de molécules ou d'ions, appelés micelles. Cette concentration, qui a reçu le nom de concentration micellaire critique, dépend de la nature de l'amphiphile, mais sa valeur est toujours très faible (inférieure à 0,5% en amphiphile). Toutes les propriétés qui sont liées au nombre de particules en solution sont affectées par ce phénomène d'association (pression osmotique, conductivité, tension superficielle, volume spécifique, viscosité, indice de réfraction). De même, la dissolution, dans les solutions aqueuses d'amphiphile, de substances organiques insolubles dans l'eau, commence à cette concentration, pour ensuite croître régulièrement.

Ce domaine micellaire est optiquement isotrope au repos.

Domaine du coagel et du gel

C'est apparemment un domaine à deux phases, qui se trouve situé dans la partie inférieure du diagramme, et qui représente peut-être une succession d'équilibres entre le savon cristallin et la solution très étendue de savon. Lorsqu'on refroidit une solution de savon, avec précaution, à l'abri de tout germe, on obtient parfois un gel clair, transparent, conservant toutes les propriétés de la solution, sauf évidemment la fluidité. Ce gel est un état métastable et se transforme plus ou moins rapidement en coagel.

Domaine des phases liquide cristallines

Il se place entre les deux courbes T_c et T_i du diagramme. La courbe T_c représente les températures de disparition des cristallites du coagel par chauffage; la partie horizontale de cette courbe représente ce que l'on appelle la 'température de Krafft'; c'est la température à laquelle un savon passe rapidement en solution dans l'eau, par élévation de température; pour un savon donné, cette température est très voisine du point de fusion de l'acide gras correspondant. La courbe T_i est la courbe à laquelle apparaît par refroidissement, à partir de la solution isotrope, une phase anisotrope.

On distingue en général deux phases bien définies.

(a) *Phase médiane*. — Ainsi dénommée parce qu'elle se rencontre aux concentrations moyennes. Elle se présente comme très visqueuse, presque plastique, conservant cet état jusqu'à la température à laquelle elle se transforme en solution isotrope.

(b) *Phase lisse*. — Elle est plus concentrée en savon que la précédente, mais elle se présente toutefois sous un aspect beaucoup plus fluide, pratiquement liquide.

E6

Acta Crystallographica **13**, 660–67 (1960)

THE STRUCTURE OF ASSOCIATION COLLOIDS

I. The liquid-crystalline phases of amphiphile-water systems

by

V. Luzzati, H. Mustacchi, A. Skoulios and F. Husson

Centre for Research on Macromolecules, 6 rue Boussingault, Strasbourg, France

(received 10 July 1959)

The structures of the liquid-crystalline phases of amphiphile-water systems have been investigated by X-ray diffraction and polarizing microscope observations. Some new phases have been discovered. The structures of all the phases are described, and several geometric parameters are determined. It is shown that an essential feature common to all phases is the disordered liquid-like configuration of the paraffin chains.

Introduction

The term amphiphile denotes a family of compounds in which the same molecule contains two regions with very different solubilities. These regions are sufficiently separated from each other that, apart from the constraints imposed by the bonds which connect them, they behave in an independent way. In general, the two regions are formed respectively by a hydrophilic group, and one or more long paraffin chains. A large number of such substances are known. Examples are: soaps, alkali metal salts of fatty acids, various substances known as 'detergents', 'surfactants', or 'emulsifiers', as well as certain complex molecules one comes across in living organisms (glycolipids, phospholipids, etc.).

In the presence of water, the high solubility of its terminal hydrophilic group must be reconciled with the insolubility of the rest of the molecule. As a consequence amphiphiles display remarkable behaviour. In order to analyse the properties of such a system, it is appropriate to refer to its phase diagram. By way of an example, we have given in Fig. 1 the phase diagram for the potassium palmitate/water system, taken from a paper by McBain & Lee (1943). In it one recognises different regions:

677

Isotropic solution region

This solution goes across the whole soap-water concentration range at sufficiently high temperatures. The physical properties of the isotropic solution change suddenly at a particular concentration. The nature of these changes indicate that the solution forms molecular or ionic aggregates, which are known as micelles. This concentration has been given the name critical micelle concentration. Its value depends on the nature of the amphiphile, but its value is always very small (less than 0.5% amphiphile). All the properties which depend on the number of particles in solution are affected by the association phenomenon (e.g. osmotic pressure, conductivity, surface tension, specific volume, viscosity, refractive index). In these amphiphilic aqueous solutions organic substances which are normally insoluble in water begin to dissolve at exactly this concentration, and their solubility increases steadily.

The micellar region is optically isotropic at rest.

Region of co-gel and gel

This is apparently a two-phase region, and is situated in the lower part of the diagram. It represents a succession of equilibrium states between the crystalline soap and the strong soap solution. When a soap solution is carefully cooled, avoiding the presence of seeds, one sometimes obtains a clear transparent gel. This has all the properties of the solution, apart obviously from fluidity. This gel is a metastable state which transforms more or less rapidly into a co-gel.

Liquid crystalline region

This exists between the two curves T_c and T_i in the phase diagram. The curve T_c represents the temperatures at which the crystallites of the co-gel disappear on heating. The horizontal part of the curve represents what is known as the 'Krafft temperature'. This is the temperature at which a soap passes rapidly into aqueous solution when temperature is increased. For a given soap, it is very close to the melting point of the corresponding fatty acid. The curve T_i is where an anisotropic phase appears on cooling from the isotropic solution.

In general one can distinguish two well-defined phases.

(a) *Middle phase* – So-called because it appears at intermediate concentrations. It is a very viscous, almost plastic phase, and maintains that state up to the temperature at which it transforms to an isotropic solution.
(b) *Neat phase* – This occurs at higher soap concentrations than the previous case, but it always appears to be a very fluid, almost liquid phase.

The region between the middle phase and neat phase appears to be formed from a mixture of the two bordering phases. In reality this region is much more complex and we shall return to this point further below.

The phase diagrams of all amphiphiles are similar to that for potassium palmitate. The relative importance of the different regions can however vary considerably from one case to another, as can the transition temperatures between phases. In some systems described in the second paper in this series, we have found all the phases which appear in Fig. 1, but in other systems we have found only a single liquid crystalline phase. However it is likely, as in these cases, that a complete study of the phase diagram as a function of concentration and temperature would result in a picture similar to that given in Fig. 1, but with the temperature scale shifted.

The aim of our work has been to establish the structure of different liquid crystalline phases of binary amphiphile-water systems. In two recent notes we have already described the liquid crystalline structures of the saturated water-soap systems of sodium and potassium (Luzzati *et al.*, 1957, 58).

We propose to complete these results here. We describe in this paper the properties common to liquid crystalline phases of amphiphile-water binary systems, and in particular their structure. In a later note we shall describe the individual properties of several systems with different types of amphiphile.

Historical background

In what follows we shall above all be interested in the structural aspects of the liquid crystalline phases. It is therefore appropriate to review the different crystallographic studies as well as the observations from the polarising microscope relevant to this topic. However, to begin with we must consider the structure of the micellar phase, as many researchers believe that this structure is fundamental to all binary amphiphile-water structures.

Although the existence of micelles has been universally accepted, the shape, dimensions and nature of the micelle have provoked much controversy. The two principal points of view on the structure of micelles in aqueous phases have been put forward are those of Hess (1939) and McBain (1950) on the one hand, and of Hartley (1936) on the other.

Hess and McBain propose that the organisation of the amphiphilic molecules in a micelle is analogous to that which one comes across in crystalline soaps. The micelle, known as 'lamellar', is formed from an assembly of layers of amphiphilic molecules, separated from one another by water. The paraffin chains are parallel to each other, and the polar groups are in contact with water.

By contrast, Hartley (1936) proposed that the organisation of the paraffin chains in the micelle is close to that which exists in liquid paraffin. Under these conditions the micelle adopts a spherical shape.

In the study of concentrated amphiphilic phases (middle and neat phases), most workers have followed the point of view of Hess and McBain on the molecular organisation within the micelles. Implicit in the structures that they propose is the idea that the amphiphilic chains are 'crystalline'.

Stauff (1939) was the first to use X-rays to study soap solutions as a function of concentration and at several temperatures. He examined the variation of Bragg spacing associated with a single diffraction line, measured at four or five different concentrations (e.g. for sodium laurate, 13, 22, 35 and 70%). His conclusion was that the middle and neat phases are formed from elementary lamellar micelles of constant thickness. These were separated by water layers whose thickness is concentration-dependent. We remark that these conclusions are based on the partial interpretation of incomplete and imprecise experimental data and are incorrect. At the same time we should not forget that this was the first systematic study of the subject. Stauff missed the fact that the structures of the middle and neat phases are different and that several other phases exist between them. This work is thus only of historical interest.

Dervichian and Lachampt (1945) proposed a model similar to that of Stauff.

Doscher and Vold (1948) made a detailed study of the sodium stearate-water binary system and also arrived at similar conclusions. In the view of these authors the neat phase is formed from a assembly of soap layers, bound to each other by water molecules. The rigid soap molecules are tilted at a definite angle with respect to the planes of the layers. The structure of the neat phase was derived from that of anhydrous soaps, with the water dissolving in the soap in order to separate the crystalline structure. The middle phase, according to Doscher & Vold, is merely a micellar phase, of the lamellar type, in which the organisation is more extended. We shall see later that our experimental results do not agree with this model.

We also mention an important study of Marsden & McBain (1948), who used X-rays to examine the dodecylsulphonic acid-water and triethanolamine laurate-water systems. These systems are analogous to the soap-water systems, but have the advantage that they exhibit liquid crystalline phases at ordinary temperatures. Marsden and McBain propose the same structure for the middle phases of the two systems. Their view of the phase structure was that they were formed by long fibres, arranged in a hexagonal lattice, with water or an ionic atmosphere between the fibres. The neat phases of the two systems were thought to be slightly different. The dodecylsulphonic acid-water system was to be formed by an arrangement of layers with a concentration-independent inter-layer distance. McBain called this phase 'lamellar non-expanding'. The triethanolamine laurate-water system was also thought to be formed by an arrangement of layers; the distance between these layers varied as the inverse of the concentration. This would then be a 'lamellar expanding' phase (see McBain & Marsden 1948).

These authors themselves found some of their structural models quite difficult to explain. In fact if one agrees with McBain that the paraffin chains are only slightly flexible, it is difficult to conceive of a fibrous structure in which the paraffinic chains present their polar ends to the water. Furthermore, in the 'lamellar non-expanding' structural type, the distance between the layers must not change with concentration. As a result, either the water must be elsewhere, in another non-identified phase, or it must penetrate into the interior of the paraffinic molecular bundle.

All these difficulties spring from the fact that they consider that the molecular organisation within the different colloidal particles is similar to that occurring in a crystal. On the other hand, if we take Hartley's point of view that the paraffin chain structure is quasi-liquid, then many of the experimental results can be explained and interpreted much more easily.

Observation using the polarising microscope plays an equally important role in the development and confirmation of the structural models we have described. In fact, it was in mixtures of soap and water that Lehmann (1889) had made observations of liquid crystal textures. These same substances were also studied by Friedel (1922) in his study of the optical properties of mesomorphic stases.

Several researchers, most notably McBain *et al.* (1926) and Lawrence (1938), have found in the neat phase textures which reminded them of those described by Friedel as characteristic of the smectic stase. This has supported the hypothesis of the 'lamellar' structure. On the other hand, some have tried to interpret the middle phase textures as those of a smectic because of the analogous structures which are supposed to exist in neat and middle phases. However, until the recent work of Roosevear (1954), there had not been a detailed comparative study of the two phases. Roosevear showed systematic differences between the two phases, and presented results for characteristic middle phase textures. These results led Roosevear to suppose that the structures are different. The optical properties of the middle phase are not in fact compatible with a smectic structure.

Experimental section

X-ray diffraction

The study of amphiphile-water systems by diffraction of X-rays poses a number of experimental problems.

Most of the diffracted X-rays correspond to fairly large spacings (60 to 20 Å), and have a low intensity. It is therefore necessary to use a scattering chamber which allows the experimenter to go down to small scattering angles ($2\theta \cong 1°$) without sacrificing too much ray intensity. Furthermore, it is frequently found that several narrow diffraction lines are concentrated in the central region of the diffraction pattern. In order to separate these, it is necessary that the spread of the beam due to collimation should be negligible. In addition we have had to study samples containing water at temperatures which occasionally reached 100 °C, and which were in vacuum. The sample ports must therefore be completely tight.

We have resolved these problems in a satisfactory way using a Guinier-type focussing chamber, operating *in vacuo*, equipped with a bent quartz monochromator. The focus of the X-ray tube is sufficiently sharp for it to be easy to isolate the copper $K\alpha_1$ reflection by mechanically adjusting the monochromator. We also used a Philips diffractometer with a Geiger counter. For this we have adapted a bent quartz monochromator; the samples were examined in transmission.

The samples are placed in the metallic chamber between two sheets of mica, which is transparent to X-rays. The temperature is raised using electric heaters, and maintained at constant temperature to ±2 °C using an automatic system.

We do not describe the details of these experimental set-ups. This has been done elsewhere (Mustacchi 1958).

In what follows, all the concentrations are measured as amphiphile weight per gram of mixture.

Polarising microscope

We made use of a Leitz polarising microscope equipped with a hot-stage. The samples have been examined between microscope slides and cover slips at normal temperature. For heating purposes we have chambers similar to those used in the X-ray experiments. In that case the windows were formed from sheets of phlogopite mica, which are uniaxial.

General considerations common to all phases

Appearance of diffraction patterns

We consider the phase diagram of a binary soap-water system, for example that of the potassium palmitate-water system, shown in Fig. 1. Using X-rays at a single fixed temperature above the curve T_c, we increase soap concentration and examine different soap-water mixtures.

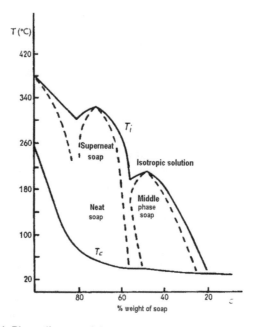

Fig. 1 Phase diagram of the water-potassium palmitate system.

682

Up to a concentration of the order of 30% one remains in the isotropic solution. In this region the diffraction patterns are exclusively formed by diffuse bands. If the concentration of soap increases, one passes the curve T_i and enters the anisotropic phase region. The diffraction patterns differ between the middle phase, the neat phase, or even the region between the middle and neat phases. However, no matter what zone is under consideration, the patterns are constructed from a series of fine distinct lines at small angles. It must be emphasised that the lines are as narrow as the profile of the incident beam. In general between 2 and 7 lines can be distinguished in a region of the diffraction pattern corresponding to diffraction angles smaller than 10°. In addition to these lines, one observes a band of weak intensity which is always situated around $s = 1/4.5 \text{ Å}^{-1}$ ($s = 2\sin\theta/\lambda$, θ: the Bragg angle and λ: wavelength), and a band characteristic of water situated around $s = 1/3.2 \text{ Å}^{-1}$.

For a defined region of the phase diagram, that is to say within the existence boundaries of one phase, the diffraction patterns show the same characteristics and evolve continuously with the concentration.

Chaotic structure of the paraffin chains

The temperature dependence of the diffraction patterns has enabled us to propose a structure for the paraffin chains.

For all water-amphiphile mixtures in the liquid crystalline region, temperature change is accompanied by a shift in the small angle diffraction lines. However, so long as the system stays within a single phase, the ratio of the spacings of the different lines remains constant. More precisely, one notices that *the Bragg spacings of the different lines decrease when the temperature increases.*

This phenomenon is completely general. We have shown it to occur in all amphiphiles studied and in all phases in the liquid crystalline region. For a particular amphiphile and given concentration, the coefficient of linear dilation (d is the Bragg spacing of any line)

$$\alpha = \Delta d/(d\Delta T) \qquad (1)$$

is constant over the range of temperature studied. In all cases its value lies between -8×10^{-4} and -2.5×10^{-3} (see Mustacchi, 1958).

The sign of the coefficient is already surprising. Furthermore, its absolute value is abnormally large for a solid state material, and corresponds rather to a gas.

A phenomenon in many respects analogous to that which we shall describe can be found in some polymers. For example (see Treloar, 1949) when a rubber probe is suspended and stretched by applying a weight, its length is observed to decrease with increasing temperature. Both in sign and absolute value, the coefficient of linear dilatation $dl/l\Delta T$ is similar to that found for soaps,. This phenomenon is explicable in terms of the chaotic arrangement of the chain macromolecules. At each point in time the length of the probe is determined by the equilibrium between the exterior force which tends to orient the macromolecule and the thermal disorder

683

which tends to disorient them. Raising the temperature shifts the equilibrium towards disorder. This is translated into a macroscopic contraction of the probe. The contraction seems not to be accompanied by a volume contraction. On the contrary, the coefficient of volume expansion of such a probe is positive.

One can account for this phenomenon by using a statistical model of a chain, which is partially extended as a result of applying equal but opposite forces at each end. In fact the calculation shows (see for example Treloar, 1949) that for an increase of the temperature of ΔT degrees, the distance between the ends of the chain changes from l to $l[1 - \Delta T/T]$. This yields

$$\alpha = \Delta l/(l\Delta T) = -1/T \tag{2}$$

For $T \cong 350$, $\alpha \cong -2.9 \times 10^{-3}$.

The analogy between this phenomenon and that described for the amphiphiles, just as the agreement between the theoretical value of α (2) and that observed experimentally, is striking. This suggested to us a structural model, motivated by the example of chain statistics.

We suppose, as is usual, that in the binary mixtures the paraffinic part of the amphiphile and the water occupy two distinct regions, each of dimensions large in comparison with atomic distances. The surface of separation is then lined by the hydrophilic groups of the amphiphilic molecules. We suppose as well that the hydrocarbon chains adopt a chaotic configuration similar to that in rubber. However, since one of the ends of each chain is constantly coupled to the interface, it is evident that the elongation direction of the chains is on average perpendicular to the surface of separation. One can therefore localise a force in the interface which tends to extend each chain. In fact, the hydrophilic groups are sufficiently far apart from each other that a force of attraction is exerted between them.* Now each hydrophilic group is accompanied by a constant volume of hydrocarbon. The mutual approach of the polar groups cannot therefore itself produce a reduction in the ratio between the exterior surface and the volume of the paraffinic regions. The nature of the paraffinic regions is characteristic of each phase. The surface to volume ratio can therefore only get smaller if the smallest dimension in these regions gets bigger (see Luzzati et al., 1958). This will tend to extend the hydrocarbon chains. In the middle phase this smallest dimension is the diameter of the cylinder, whereas in the neat phase it is the thickness of the double layer.

The thermal movement which tends to disorient the chains opposes this force. The thermodynamic treatment of this model is then analogous to that of a rubber probe, at least to a first approximation, and leads to the coefficient (2) above.

The disordered structure of the paraffin chains in the different liquid crystal phases is furthermore experimentally confirmed by the following observation. This

---

* If the polar groups move apart, the paraffinic regions come into contact with the water. This opposes the low affinity between the two regions and gives rise to the attraction force between the polar groups. By contrast, if the distance between the polar groups were very small, electrostatic repulsion forces would result.

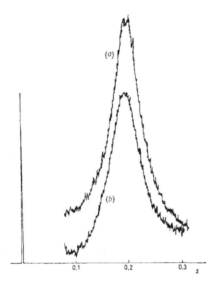

Fig. 2 Microdensitometer trace of the diffuse band characteristic of the liquid state of paraffinic chains. (*a*) Tetradecane at 100 °C. (*b*) Sodium myristate: $c=0.68$, $t=100$ °C.

is that the diffuse band observed in the part of the diffraction pattern corresponding to around 4.5 Å is identical to that given by a liquid hydrocarbon with a chain length comparable to that of the amphiphile. This is shown in Fig. 2.

In the structural models which we propose below, we suppose that the structure of the paraffin chains is chaotic. We shall often describe it as liquid. By this we mean that the organisation of the molecules is closer to the disordered configuration one finds in liquid paraffin than it is to the regular arrangement characteristic of crystalline structures. Moreover, a rapid phase change occurs when the temperature is changed. This indicates that at equilibrium there is certainly a liquid-like structure with substantial mobility, as opposed to a glass constrained in one non-equilibrium configuration.

Structure and texture of the liquid crystalline phases

In each amphiphile-water system, it is possible to identify several types of X-ray diffraction pattern. It is easy to attribute each of these to a phase. Furthermore one finds the same types of patterns in different systems.

In all the systems studied we have only identified a small number of diffraction patterns. Each pattern is characteristic of one of the types of structure we shall describe here. In a later note we will analyse the phase diagrams of several amphiphiles, referring to the structures described below.

We also describe here the textures observed using a polarising microscope, which are common to all amphiphiles.

Middle phase

The diffraction patterns characteristic of this phase have a series of sharp lines for which the Bragg spacings are in the ratios:

$$1:1/\sqrt{3}:1/\sqrt{4}:1/\sqrt{7}$$

Outside these lines one can only distinguish two diffuse bands characteristic of the paraffin chains and water.

The diffraction patterns correspond to an hexagonal array of cylinders of indefinite length, the axis of each cylinder being an axis of sixfold symmetry.

Taking account of these geometrical constraints and the liquid-like structure of the paraffin chains, we have proposed a structural model (Fig. 3) for this phase. This consists of a collection of cylinders of amphiphile of indefinite length, arranged at the nodes of a two-dimensional hexagonal lattice and separated from each other by water. Each cylinder is uniformly filled with the paraffin chains, and the hydrophilic ends are found at the surface of the cylinders, in contact with water.

The dimension d (Fig. 3) of the hexagonal lattice is determined by measuring the spacing of the diffraction lines. The concentration c and the partial specific volume of the amphiphile v_a are known. One can then calculate the diameter d_a of the cylinders (Fig. 3) and the surface S available on average to each hydrophilic group at the water-amphiphile interface.

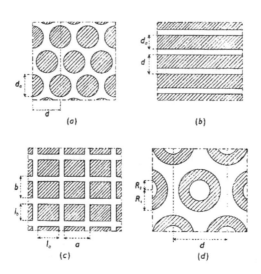

Fig. 3 Schematic representation of the structures. (*a*) Middle phase. (*b*) Neat phase. (*c*) Rectangular phase. (*d*) Complex hexagonal phase. The hatched region represents the zones occupied by amphiphile.

$$(3)$$

$$d_a = d\left[\frac{2\sqrt{3}}{\pi}\frac{1}{1 + v_e/v_a \cdot (1-c)/c}\right]^{\frac{1}{2}}$$

$$S = \frac{4M_a v_a}{d_a N}10^{24} \qquad (4)$$

In these equations, M_a is the molecular mass of the amphiphile, N is Avogadro's number, v_a and v_e the specific volume of the amphiphile and water.

Neat phase

A series of narrow lines of which the Bragg spacings are in the ratios:

$$1:\frac{1}{2}:\frac{1}{3}:\frac{1}{4}$$

is characteristic of the diffraction pattern of this phase. Outside the lines one can distinguish the two diffuse bands of the 'liquid' paraffins and the water.

This diffraction pattern corresponds to a stack of parallel and equidistant flat lamellae of indefinite size.

In the middle phase structure we separate the regions occupied by the water and by the amphiphile. We propose the structural model represented schematically in Fig. 3.

By an analogous calculation to that for the middle phase one can determine the amphiphile and water layer thicknesses, and the surface available on average for each hydrophilic group (Fig. 3):

$$d_a = \frac{d}{[1 + v_e/v_a \cdot (1-c)/c]} \qquad (5)$$

$$d_e = d - d_a \qquad (6)$$

$$S = \frac{2M_a v_a}{d_a N}10^{24} \qquad (7)$$

Intermediate phases

Some authors (Vold, 1939), have considered the region of the diagram intermediate between the neat and middle phases to be a demixing zone. In this region the system would be a two-phase mixture of the neat and middle phases. In general, this hypothesis is contradicted by the diffraction patterns. In fact, these are complex and cannot be interpreted as the superposition of the patterns from the neat and middle phases.

They contain several new lines which cannot be identified without supposing the existence of intermediate phases. However, as in the case of the neat and middle phases, the patterns contain in their central area a collection of sharp lines and a diffuse band in the region of $s = 1/4.5 \text{ Å}^{-1}$. This band and the variation of the positions of the lines as a function of temperature indicate once more that the structure of the paraffin chains is liquid.

Each phase in the intermediate region only exists in a narrow band of concentrations, and the equilibria are established slowly. As a result the experimental study of the intermediate region is often delicate. It was thus unusual to be able to prepare samples containing a single phase. More often we obtained two coexisting phases (it was not unusual even to find three phases in a single sample, most likely resulting from inhomogeneities). However, we have examined a large number of samples with very similar compositions, and compared the intensity of the different lines. In this way we have been able to identify the diffraction patterns of each phase.

With the amphiphiles studied, and under our experimental conditions, we have been able to distinguish four phases in the intermediate region. The position of the phase boundaries has not been determined with great precision, principally because of inhomogeneities. These four phases are the only ones which we have seen under our experimental conditions. This in no way excludes the existence of other phases, especially at higher temperatures.

Deformed middle phase

This phase exists in certain systems, between the middle phase and the complex hexagonal phase. It is characterised by two lines positioned on each side of the first line of the middle phase. It seems to exist in a very narrow region, because we have only seen it accompanied by adjacent phases.

It is possible that this phase structure develops out of that of the middle phase by a deforming the lattice, making the hexagonal lattice become orthorhombic. This is probably accompanied by a deformation of the cylinders, the sections of which become elliptical. However the experimental data are insufficient for this structure to be unambiguously confirmed.

Rectangular phase

The diffraction pattern characteristic of this phase is formed by two lines and their higher orders, in addition to the diffuse bands of water and liquid paraffin. We have seen up to six lines, for which the Bragg spacings are:

$$a, b, 2a, 2b, 3a, 3b$$

The ratio between a and b depends on the nature of the amphiphile (*c.f.* Table 1) which indicates that the lattice depends on at least two independent parameters.

Moreover, in each system the relative intensity of the two series of lines is independent of the concentration. This demonstrates that there is indeed a single phase and not a mixture of two lamellar phases.

Table 1 Rectangular phase

Ratio of the spacings of the first two lines			
Amphiphile	t (°C)	c	a/b
Potassium oleate	20	0.64	1.28
Sodium oleate	65	0.56	1.19
Potassium laurate	20	0.61	1.43
Potassium palmitate	70	0.60	1.26

Such a diffraction pattern is obtained when the diffracting material is organised into two series of equidistant parallel planes, whatever the angle between the two sets of planes. The structural model shown in Fig. 3 satisfies these conditions; we have arbitrarily made the planes perpendicular to each other. In this model the amphiphiles fill the parallelepipeds, which have a rectangular section and are of indefinite length.

One can calculate from a,b and the concentration c, the volume occupied in the unit cell by the amphiphile, and the mean surface area S available to a hydrophilic group at the interface. We have proposed that the rectangular section of the parallelepipeds is homothetic to the unit cell. In this case we can calculate the lengths l_a and l_b of the sides (Fig. 3)

$$S = 2\frac{l_a + l_b}{l_a \cdot l_b} \cdot \frac{M_a v_a}{N} \cdot 10^{24} \tag{8}$$

$$l_a = a\left[\frac{1}{1 + v_e/v_a \cdot (1 - c)/c}\right]^{\frac{1}{2}} \tag{9}$$

$$l_b = l_a \times b/a \tag{10}$$

Complex hexagonal phase

We have been able to identify up to 6 diffraction lines characteristic of this phase. The ratios of the Bragg spacings are:

$$1:1/\sqrt{3}:1/\sqrt{4}:1/\sqrt{7}:1/\sqrt{9}:1/\sqrt{12}$$

As in the case of the middle phase, these patterns are those of a two-dimensional hexagonal lattice. However the lattice parameter is much larger than that of the middle phase, (about double, c.f. Table 2), which indicates that the structure is more complex than previously described.

Knowing the concentration and the parameter d, one can calculate the volume fraction of the unit cell which is occupied by amphiphile. We have considered

Table 2 Comparison between the lattice parameters for the middle and complex hexagonal phases

Amphiphile	$t(°C)$	Middle phase		Complex hex. phase	
		c	$d/Å$	c	$d/Å$
Potassium palmitate	100	0.51	50	0.57	105
Potassium stearate	100	0.44	60	0.62	114
Potassium myristate	100	0.49	46	0.56	93
Sodium palmitate	100	0.48	52	0.53	108
Sodium stearate	100	0.42	63	0.52	124
Potassium oleate	20	0.56	52	0.68	109
Sodium oleate	65	0.52	54	0.60	107
Sodium lauryl sulphate	70	0.60	43	0.68	92

several structural models compatible with this data. Our choice has been guided by the agreement between the calculated and observed intensities.*

The most satisfying model is shown in Fig. 3. It is formed by a collection of hollow cylinders of amphiphile, with water filling the internal cavity and external space.

Cubic phase

We have observed up to four lines characteristic of this phase: the ratios of the Bragg spacings are:

$$1: \sqrt{3}/\sqrt{4}:\sqrt{3}/\sqrt{8}:\sqrt{3}/\sqrt{11}$$

The positions of these lines are those of a face-centred cubic lattice (111, 200, 220, 311).

The cubic symmetry is confirmed by polarising microscope observations. In fact, alone amongst all the phases of the liquid crystalline region, the cubic phase is optically isotropic (see below).

Since a face-centred cubic lattice corresponds to one of the close-packing arrangements of identical spheres, it seems reasonable to propose that in this phase the paraffin chains are organised in spheres, surrounded by water. Knowing the composition of the system and the lattice parameter one can calculate the radius R of the spheres:

$$R = a\left[\frac{3}{16\pi[1 + v_e/v_a \cdot (1-c)/c]}\right]^{\frac{1}{3}} \tag{11}$$

* We will give the crystallographic verification in the following note of this series.

Polarising microscope observations

Some X-ray scattering and polarising microscope observations performed on one and the same sample have allowed us to establish a connection between structure and texture. Thanks to the experience acquired in this way, it became possible to recognise certain phases in the microscope. The study of each new system then began by systematic observations with the polarising microscope. This was followed by X-ray diffraction experiments carried out only on particularly interesting samples.

We summarise the principal observations relating to each phase.

Neat phase – The textures observed are all typical of smectic phases (Fig. 4) (Friedel, 1922). In all cases, one observes focal conic textures. One frequently

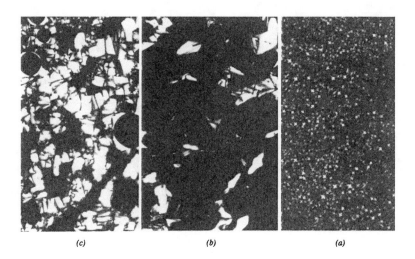

(c)   (b)   (a)

Fig. 4 Textures in the smooth phase (Arkopal 9, $c = 0.65$, $t = 22°$(×80) (a) Units. (b) Bâtonnets. (c) Mosaic.

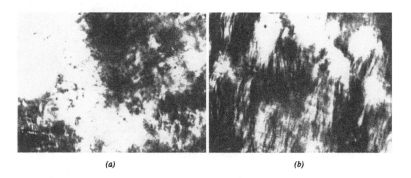

(a)   (b)

Fig. 5 Middle phase textures (a) Potassium oleate, $c = 0.31$, $t = 22°$(×80). (b) Sodium oleate $c = 0.35$, $t = 22°$(×80).

691

comes across positive or negative units, mosaic textures, and oily streaks (Roosevear, 1954). In certain systems, islands of neat phase appear in the midst of the isotropic solution in the form of 'bâtonnets'. Finally the neat phases form large uniaxial regions when they are oriented by shear between the slide and the cover slip.

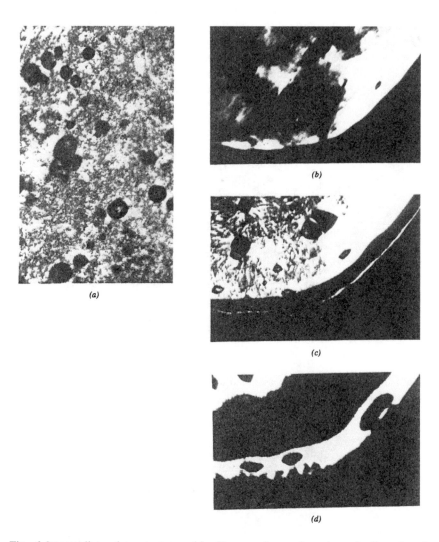

(a)

(b)

(c)

(d)

Fig. 6 Intermediate phase textures. (a) Hexagonal complex phase (sodium lauryl sulphate ($c = 0.65$, $t = 70°$) (×80). (b)-(c)-(d) Series of pictures through the edge of a sample of a cubic region resulting from the progressive evaporation of a middle phase (cetyl trimethylammonium bromide). (b) Homogeneous middle phase. (c) An isotropic region – the cubic phase – appears compressed between two birefringent regions, which are respectively the middle and neat phases. (d) As in (c), but the isotropic region becomes dominant.

Middle phase – The middle phase is strongly birefringent, but generally has a diffuse appearance, without a specific texture. One observes cloudy formations, often combined with fine streaks (Fig. 5). It is in this form that the middle phase precipitates in the midst of the isotropic solution. Nevertheless, in samples which otherwise show by X-ray diffraction a strongly microcrystalline texture, we have sometimes observed 'angular' and 'fan-shaped' geometric textures (Roosevear, 1954).

Intermediate phases – The textures of complex hexagonal, rectangular, and deformed middle phases are very different from those of the neat phase, and generally are close to the non-geometric textures of the middle phase (Fig. 6). We have not found any criterion which enables them to be differentiated from the middle phase, or even to distinguish one (intermediate) phase from another.

The cubic phase, as we have already indicated is the only phase which is optically isotropic. It appears as black regions, between crossed nicols, which sometimes extend throughout the sample. Fig. 6 shows its appearance as a drop of the middle phase of cetyl trimethylammonium bromide is progressively evaporated at 70°.

References

Dervichian, D. & Lachampt, F. (1945). *Bull. Soc. Chim., Paris*, **12**, 189.

Doscher, T. & Vold, R. (1948). *J. Phys. Coll. Chem.* **52**, 97.

Friedel, G. (1922). *Ann.de Phys., Paris*, **18**, 300.

Hartley, G.S. (1936). *Aqueous Solutions of Paraffin Chain Salts*. Paris: Hermann.

Hess, K., Philipoff, W. & Kiessig, H. (1939). *Kolloid-zschr.* **88**, 40.

Lawrence, A.S.C. (1938). *J. Roy. Micr. Soc.* **58**, 40.

Lehmann, O. (1889). *Z. Phys. Chem.* **4**, 462.

Luzzati, V., Mustacchi, H. & Skoulios, A. (1957). *Nature, Lond.* **180**, 600.

Luzzati, V., Mustacchi, H. & Skoulios, A. (1958). *Disc. Faraday Soc.* **25**, 43.

McBain, J.W. & Elford, W.J. (1926). *J. Chem. Soc.* **1**, 421.

McBain, J.W. & Lee, W.W. (1943). *Oil and Soap.* **20**, 17.

McBain, J.W. & Marsden, S.S. Jr. (1948). *Acta Cryst.* **1**, 270.

McBain, J.W. (1950). *Colloid Science*. Boston: Heath.

Marsden, S.S. Jr & McBain, J.W. (1948). *J. Amer. Chem. Soc.* **70**, 1973.

Mustacchi, H. (1958). Thèse. Université de Strasbourg.

Roosevear, F. B. (1954). *J. Amer. Oil. Chem. Soc.* **12**, 628.

Stauff, J. (1939). *Kolloidzschr.* **89**, 224.

Treloar, L.R.G. (1949). *The physics of rubber elasticity*. Oxford: University Press.

Vold, R.D. & Vold, M.J. (1939). *J. Amer. Chem. Soc.* **61**, 37.

Vittorio Luzzati was born in 1923 in Genova (Italy). In 1938 his family, which was of Jewish extraction, moved to Argentina. In 1947 he graduated from the University of Buenos Aires as a mechanical engineer, and then in the same year he settled in France, where he has lived since 1947. He has spent the whole of his career within the French *Centre National de la Recherche Scientifique*.

From 1947 to 1952 he worked at the *Laboratoire Central des Services Chimiques de l'Etat*, in Paris. His main interests were the structure analysis of simple organic and inorganic compounds, and some statistical aspects of structure analysis. In 1953 Dr D. Harker invited him to join the Protein Structure Project that he led at the Polytechnic Institute of Brooklyn (New York, NY, USA), where he first met biological mac- romolecules. On his return to France in 1954, he joined the *Centre de Recherches sur les Macromolécules*, in Strasbourg. At that time he was interested in biological problems, such as the structure of DNA, RNA, proteins and nucleoproteins in solution, and lipid polymorphism. In 1963 he moved to the *Centre de Génétique Moléculaire* at Gif-sur-Yvette. His main interests there were lipid-polymorphism, the X-ray scattering approach to the structure of biological systems (RNA, serum lipoproteins, proteins-detergent complexes) in solution and protein crystallography. He retired in 1991.

In 1972 he was awarded the Prix Robin of the French Physical Society and in 1983 the Prix Léopold Meyer of the French Academy of Science. He remains active (December 2001) in the capacity of Emeritus Investigator.

Antoine Skoulios was born in 1934. He joined the CNRS (National Scientific Research Centre) in 1955 and was supervised by Vittorio Luzzati at the Centre for Macromolecular Science in Strasbourg. He gained his Ph.D in 1959 with a thesis on liquid crystalline structure obtained with pure sodium soaps and with binary mixtures of sodium soaps and hydrocarbons. He was the leader of the X-ray group in this Centre from 1963 to 1987, when he moved into the new Institute for Physics and Chemistry of Materials, also in Strasbourg. His main contributions are based on the concept of the amphiphatic character that he used extensively to describe the mesomorphic structures of thermotropic and lyotropic liquid crystals: soaps, polysoaps, calamitic and discotic liquid crystals. This concept was generalised to the case of block copolymers to describe their different lamellar, columnar and cubic phases.

Apart from the ground-breaking work included in this volume, Skoulios has made wide-ranging contributions to the study of lyotropic mesophases not only in amphiphiles, but also in block copolymers, and has been able to determine the nature of the crystalline state in these materials.

He retired in 1999.

E7
Journal of Polymer Science B: Polymer Letters Edition **13**, 455–63 (1975)

MESOPHASIC STRUCTURES IN POLYMERS. A PRELIMINARY ACCOUNT ON THE MESOPHASES OF SOME POLY-ALKANOATES OF p,p'-DI-HYDROXY-α,α'-DI-METHYL BENZALAZINE

The concepts of liquid crystallinity have been used in relation to polymers essentially in three different ways: (a) – To semiordered structures of mesophasic character have been ascribed some morphological aspects of amorphous polymers (1). (b) – A lyotropic liquid crystal phenomenology has been observed in solutions of polymers like polypeptides, typically poly-γ-benzyl-L-glutamate (2), or some polyesters of alkoxybenzoic acids (3) or some block copolymers like polystirene-polyoxyethylene (4). (3) – Some experimental work has been reported on polymerization in mesophasic media (5-10). Particular attention was paid to possible influences on molecular structure (tacticity) and molecular weight, the mesophasic medium being constituted either by the same monomers or by a substance not directly involved in the polymerization. Much of the experimental work focused on polymers containing the cholesteryl or cholestanyl group in the side chain.

The resulting picture of polymer formation is sometimes contradictory and by no means conclusive.

We have undertaken to systematically study the possibility of obtaining polymeric substances that in the "solid" state markedly show the morphological characters of mesophases.

Molecular aggregations of mesophasic type have been noted on solvent cast films of some synthetic polypeptides, particularly polyglutamates (11-12). A striking example of further induced superstructure is shown by films of poly-γ-benzyl-L-glutamate case in a magnetic field (13,14).

Many of the interesting properties of liquid crystals stem from their being liquid, nonetheless an extended mesophasic morphology should be a source of interesting optical and mechanical properties in materials even in the absence of fluidity.

An obvious approach to the problem of obtaining such polymers is to polymerize monomers that possesses the qualities for self-induced thermotropic mesophases. Alternatively the mesophase-originating atomic grouping could be built with the polymerization itself.

A second point arises concerning the position where the relevant atomic grouping has to be inserted relatively to the polymer chain. Most of the known experimental work deals with polymers having the relevant atomic grouping inserted in the side chain.

As to the kind of mesophase that one could expect to be developed from a particular chemical composition and molecular structure, the same criteria that hold for low molecular weight thermotropic liquid crystals could be tentatively followed. In particular, for a polymer whose chain is formed by an ...ABAB... sequence of atomic groupings, which as to steric encumbrance and nonbonded interactions with neighboring chains are almost interchangeable, a nematic (if any) mesophase can presumably be expected. When the polymer chains are formed by different atomic grouping which are more strictly homophilic a smectic mesophase should be more probable.

On the basis of these hypotheses we have attempted to study and to synthetize polymers of the following formula:

The present paper deals with the preparation and partial characterization of the unfractionated polymers.

Experimental

The p,p'-di-hydroxy-α,α'-di-methyl benzalazine:

has been prepared from p-hydroxyacetophenone and hydrazine sulfate as reported in (15).

Melting point and NMR spectrum confirm the nature and purity of the prepared compound.

The acyl chlorides: $ClOC(CH_2)_nCOCl$ n = 6, 8, 10 have been prepared by standard procedures from the corresponding dicarboxylic acids (Fluka) and were vacuum distilled.

Polymerization. Polyesters of the given formula have been prepared in the following way:

A chloroformic solution of acyl chloride has been added, at room temperature, to a water solution of p,p'-di-hydroxy-α,α'-di-methyl benzalazine, sodium hydroxide, and benzyltriethylammonium chloride. After stirring six min in a

TABLE I

Phase Transition Temperatures and Enthalpy Changes (1)

	(2)	T(I)	ΔH(I)	T(II)	ΔH(II)
	h[a]	483	2.10		
	c	453	1.57		
P12	h[b]	$\left\{\begin{array}{l} 476 \\ 483 \end{array}\right.$	0.71 1.32	514	1.90
	c[c]	454	1.55	489	1.80
	h[d]	479	2.10	512	1.82
P10	(3)	476	1.90	529	2.33
P 8	(3)	511	2.53	568	2.68

[1]Temperatures are in K and are reproducible within 1%; enthalpies are given in Kcal mol^{-1} and are reproducible from sample to sample within 10%.

[2]h = heating run, c = cooling run, Letters in parentheses refer to Figure 1.

[3]From first heating run measurements.

[4]Substantial decomposition starts at $\simeq$ 580 K for P12 and P10 and at $\simeq$ 590 K for P8.

blender, n-heptane was added. The precipitate was filtered and washed with diethylether, ethylalcohol-diethylether, and water to finally eliminate Cl$^-$.

The quantitative elemental analysis gives the following results: (The polymers from here on in will be known as P12, P10, and P8—according to the number of carbon atoms in the aliphatic chain.)

P12— C% (72.70 calc.– 72.48 found); N% (6.06 calc.– 6.02 found);
      H% (7.36 calc.– 7.49 found)

P10— C% (71.89 calc.– 71.62 found); N% (6.45 calc.– 6.48 found);
      H% (6.91 calc.– 6.83 found)

P 8 C% (70.94 calc.– 70.89 found); N% (6.89 calc.– 6.77 found);
      H% (6.40 calc.– 6.49 found).

Solubility and viscosity. The polymers obtained are sparingly soluble in solvents like methyl- and ethylacetate, N,N dimethylformamide and pyridine. A good solubility has been observed in chloroform or in chloroform-phenol or 1,2 dichloroethane-phenol mixtures. Viscosity measurements in phenol containing mixtures show that the polymers are slowly degraded.

Reproducible viscosity measurements have been made with an Ubbelhode viscometer using chloroform as solvent.

TABLE II

X-ray Powder Diffraction Data for P12[a]

Original	Therm. Treated[b]
26.1 (w)	c
7.08 (mw)	5.37 (w)
4.80 (s)	4.80 (s)
4.13 (vs)	4.33 (vs)
	3.87 (s)
3.27 (m)	3.27 (m)
2.88 (w)	2.88 (w)
2.62 (w)	2.43 (w)
2.34 (w)	

[a] Spacings are given in Å, intensities are in relative arbitrary scale.
[b] The sample was previously brought to the liquid anisotropic phase: $483 \leqq T \leqq 500$ K.
[c] Not measurable by photographic means.
[d] The long spacing is 24.4 Å for P10 and 22.7 Å for P8.

At a temperature of $26.10 \pm 0.02°C$ the following limiting viscosity numbers $[\eta]$ have been extrapolated:

$[\eta]$ P12 = 1.01 dl g^{-1}; $[\eta]$ P10 = 0.52 dl g^{-1}
$[\eta]$ P 8 = 0.79 dl g^{-1}.

No satisfactory reproducible molecular weight data could be obtained as yet from osmometric measurements on the unfractionated polymers.

Thermal analysis. A DSC analysis was performed on several samples of the polymers using a Perkin-Elmer DSC-1 apparatus. The relevant results are shown in Table I.

The samples were examined under dry nitrogen flow. The phase-transition temperatures refer to the endotherm maximum point. For the evaluation of the transition enthalpies an indium sample has been used as reference.

Thermal polarizing microscopy. For the microscopic analysis, a Leitz polarizing microscope equipped with heating stage and photographic camera was used. Samples were examined at different heating and cooling rates and for several cycles.

All the examined samples show a first transition to an anisotropic liquid

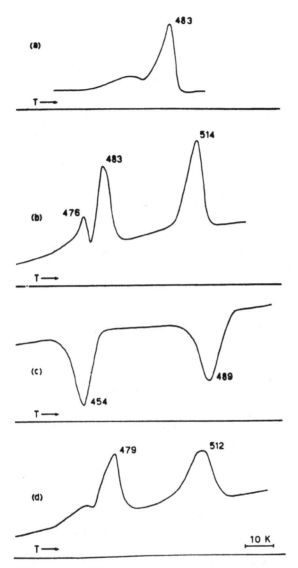

Fig. 1. DSC curves for P12: (a) first run, melting endotherm; (b) same sample, second run, melting endotherm and liquid crystal-isotropic liquid transition endotherm; (c) same sample, cooling run; (d) same sample, third heating run.

phase and a successive transition to an isotropic liquid. The phenomena observable on the microscope are very well reversible—provided that the decomposition temperature of the sample is not reached (580 K for P12 and P10, 590 K for P8).

X-ray analysis. X-ray powder spectra of the polymers with no previous

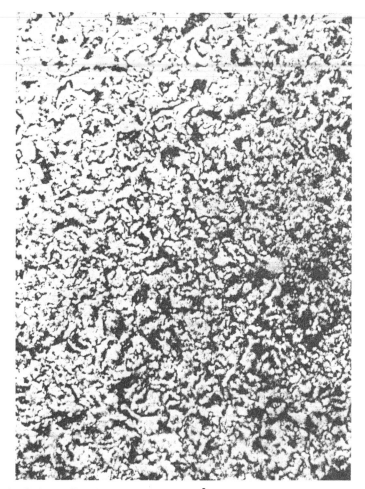

Fig. 2. Polymer P10 liquid crystal at 210°C on cooling. Crossed polarizers.

thermal manipulation were obtained photographically on the cylindrical camera of a Nonius Weissenberg apparatus. The CuK α radiation was used. The
low angle portion of the spectrum was taken on a Philips powder goniometer
and graphically recorded. The FeK α radiation was used. In both cases the
samples were slightly cold-pressed.

The spectra (the wide angle portion) of the thermally treated samples were
recorded by photographic means. The thermal treatment of the samples was
made on the DSC apparatus.

Some X-ray data for P12 are given in Table II.

Discussion

The X-ray analysis shows the polymers to be at least partially crystalline.

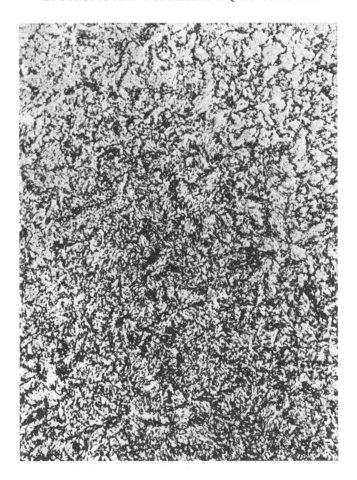

Fig. 3. Polymer P10 solid at room temperature. Same sample as in Fig. 2. Crossed polarizers.

The diffraction patterns depend upon the thermal history. The diffraction patterns obtained from the original samples and those obtained from samples previously melt to the anisotropic liquid phase are different. This is shown in Table II for P12; P10 and P8 show a similar behavior.

Rapid cooling from the isotropic liquid phase restores the original structure.

A parallel behavior is detectable from the DSC analysis (Table I and Figure 1). The melting endotherm of the original sample is broadly shouldered and it is resolved in two distinct endotherms in the successive runs provided that the transition to the isotropic liquid is not obtained. In the latter case, the melting endotherm of the following run is restored to its original form.

Polymers melt to give an anisotropic liquid with the morphological characters, as detectable from the microscopic observations, of a liquid crystalline nematic or a smectic C phase in the Schlieren texture (Figure 2).

703

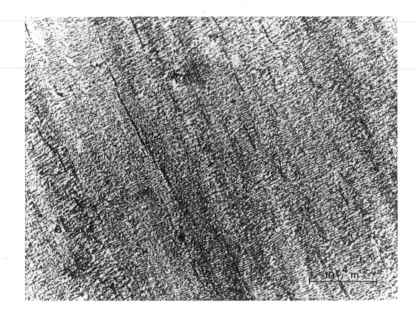

Fig. 4. Polymer P8 solid at room temperature. A shear stress was applied during rapid cooling from the anisotropic liquid phase. Crossed polarizers.

The high enthalpic change involved in the liquid crystal-isotropic liquid transition as compared to the melting enthalpy should indicate in favor of a smectic phase.

By slow cooling ($\sim 1°C/min$) of the anisotropic liquid trace of the threading is preserved down to room temperature in the solid phase (Figure 3).

It seems as if the linear singularities of the liquid crystal phase may act as nucleation lines for crystallization in such a way that the microcrystals ordering in the solid phase keeps some memory of the original semi-ordered liquid matrix.

A much more regular ordering of the microcrystals can be induced by shearing the liquid crystal phase on cooling (Figure 4).

Further work is in progress to characterize selected fractions of the original polymers.

We are grateful to Drs. S. di Martino and L. Pecoraro for helping in the preparative work and also to Drs. R. Palumbo and G. Maglio, and Prof. P. Corradini for their interest.

References

(1) G. S. Y. Yeh, Pure and Appl. Chem., 31, 65, (1972).
(2) C. Robinson, Mol. Crystals, 1, 467, (1966).

(3) V. N. Tsvetkow, E. I. Riumtsev, I. N. Shtennikova, E. V. Korneeva, B. A. Krentsel, and Y. B. Amerik, Europ. Polym. J., 9, 481 (1973).

(4) A. Douy, R. Mayer, J. Rossi, and B. Gallot, Mol. Crystals Liquid Crystals, 7, 103, (1969).

(5) Liquid Crystals, 3, part II, page 1041 f. Report on the 3rd Int. Liq. Cryst. Conf. (1970). G. H. Brown, M. M. Labes editors, Gordon and Breach Sc. Pub., (1972).

(6) W. J. Toth and A. V. Tobolsky, Polym. Lett., 8, 289, (1970).

(7) A. C. de Visser, K. de Groot, J. Feyen and A. Bantjes, J. Pol. Sc. A-1, 9, 1893, (1971).

(8) A. C. de Visser, K. de Groot, J. Feyen and A. Bantjes, Polym. Lett., 10, 851 (1972).

(9) H. Kamogawa, Polym. Lett., 10, 7, (1972).

(10) H. Saeki, K. Iimura, and M. Takeda, Polym. J., 3, 414, (1972).

(11) G. L. Wilkes, Mol. Crystals Liquid Crystals, 18, 165 (1972).

(12) G. L. Wilkes and B. The Vu, Polym. Sci. Technol., 1, 39, (1973).

(13) E. T. Samulski and A. V. Tobolsky, Macromolecules, 1, 555, (1968).

(14) G. L. Wilkes, Polym. Lett., 10, 935 (1972).

(15) G. Lock and K. Stach, Beric., 77/79, 293 (1944/46).

Antonio Roviello
Augusto Sirigu

Istituto Chimico della Università
Via Mezzocannone 4
Napoli, Italy

Received February 26, 1975
Revised March 20, 1975

Antonio Roviello was born in 1948 at Ariano Irpino, South Italy. He studied chemistry at the University of Naples and graduated there in Industrial Chemistry in 1972. He became Assistant Professor in chemistry at the University of Naples in 1976, was promoted to Associate Professor in 1982, and is at present Professor of General and Inorganic Chemistry in the Department of Chemistry at the same university.

His scientific activity started with the study of relationships between structure and properties of macromolecular systems capable of generating thermotropic liquid crystalline phases. In 1975, together with his senior colleague A. Sirigu, he described for the first time the thermotropic behaviour of some linear polyesters showing liquid crystalline properties within an accessible range of thermodynamic stability. This is the work included in this collection. His present activity is focussed on the study of crosslinkable liquid crystalline polymers with non-linear optical properties and long-time stability.

Augusto Sirigu was born in Italy in 1939. After graduating in chemistry at the University of Cagliari, he became, in 1966, an Assistant Professor at the University of Naples. In 1980, he became a Full Professor in Chemistry at the University of Naples, where he has remained. His research activity has been concerned with the synthesis and structural characterization of materials, the latter mainly using single crystal X-ray diffraction methods. His interest in low molar mass and polymeric liquid crystals began in the 1970s, and the paper with his student A. Roviello reproduced here describes the first results on the identification of liquid crystal phases in main chain polymers. More recent interests have focussed on the synthesis and characterization of polymeric materials exhibiting nonlinear optical properties.

E8

Makromolekulare Chemie **179**, 273–76 (1978)

Makromol. Chem. **179**, 273–276 (1978)

Institut für Organische Chemie der Universität Mainz,
J. J.-Becher-Weg 18–20, 6500 Mainz, W.-Germany
*Deutsches Kunststoff-Institut, Schloßgartenstr. 6 R,
6100 Darmstadt, W.-Germany

Short Communication

Model Considerations and Examples of Enantiotropic Liquid Crystalline Polymers

Polyreactions in Ordered Systems, 14[*)]

Heino Finkelmann, Helmut Ringsdorf, and Joachim H. Wendorff*

(Date of receipt: September 19, 1977)

In the last few years studies concerning liquid crystal polymers have become of increasing interest because of their theoretical and technological aspects[1]. As starting compounds, mesogenic polymerizable monomers were used which have the typical structure[2] of conventional liquid crystals. For polymers formed by condensation reactions, e.g. building the low molecular weight mesogenic monomer into the main chain, nematic as well as smectic polymer phases have been described[3].

Within the scope of vinyl polymers with mesogenic side chains only some enantiotropic liquid crystalline polymers have been obtained[4], although numerous experiments have been made. In most cases it was only possible to freeze in the liquid crystalline structure, which was irreversibly lost above the glass transition temperature[5].

In conventional liquid crystalline phases the motions of a molecule are restricted only by the anisotropic interactions with its neighbours. This leads to the formation of the orientational long range order. If the mesogenic groups are directly fixed like side chains to the main chain, the ability to move and orient is changed drastically. In the liquid state of the polymer, the tendency of statistical chain conformation hinders an orientation of the side chains. If the anisotropic interactions of side chains are strong enough to form the mesophase, a liquid crystalline structure can nevertheless be formed, but only in accordance with the limited motions of the main chain. Previous results show that smectic structures are favored where the side groups are ordered in planes. An example is the polymer of 4-ethoxy-N-(4-methacryloyloxybenzylidene)aniline[6]. X-ray investigations indicate that the main chain is ordered in planes within the smectic layers.

In addition to the mesogenic structure of the side chain, the connection of the polymer main chain to the side chain is also crucial for the existence of the different liquid crystalline polymer phases. For a systematic synthesis of enantiotropic liquid crystalline vinyl polymers we started from the following model considerations:

In the liquid state the motions of the polymer main chain have to be decoupled from those of the anisotropically oriented mesogenic side chains. The decoupling should be possible, if flexible spacer groups are inserted between the main chain and the rigid mesogenic side chains.

[*)] Part 13: cf. D. Naegele, H. Ringsdorf, J. Polym. Sci., in press.

MAIN CHAIN

SPACER GROUP

MESOGENIC GROUP

On the basis of this model it is expected that

1) in spite of the statistical conformation of the main chain an anisotropic orientation of the mesogenic side chain is possible,

2) the main chain has little or no restriction on the orientation of the mesogenic side chains.

For the synthesis of suitable monomers phenyl esters of benzoic acid were chosen as the mesogenic groups. They exhibit mesophases in spite of the large variation of p- and p'-substituents[7]. Between the polymerizable group and the mesogenic group, spacer groups of varying length n were inserted. The monomers have the general structure 1[8].

SPACER   MESOGENIC GROUP

$R = H, CH_3; R'$: cf. Tab. 1

They were synthesized by standard methods via the following scheme

$$HO-(CH_2)_n-Hal + HO-\!\!\bigcirc\!\!-COOH \xrightarrow{NaOH} HO-(CH_2)_n-O-\!\!\bigcirc\!\!-COOH$$

The properties of some analytically and spectroscopically characterized monomers are summarized in Tab. 1. The monomers were radically polymerized in solution and were purified through reprecipitation. In accordance with the model considerations nearly all synthesized polymers exhibit enantiotropic liquid crystalline properties. Structures and phase transitions of some prepared polymers are summarized also in Tab. 1. The liquid crystalline phases were identified by X-ray analysis, DSC and with the polarization microscope.

The precipitated polymers or the polymers cast from solution as films are amorphous. On heating above the glass transition temperature an anisotropic liquid crystalline polymer melt is formed. Sharp reversible phase transitions "liquid crystalline-liquid crystalline" and "liquid crystalline-isotropic" are observed. The transition enthalpies for some polymers are

Tab. 1. Structures and properties of spacer carrying liquid crystalline monomers and polymers from 1 (with $R = CH_3$)

No.	n	R'	Monomer phase transition in °C	Polymer phase transition[a] in °C	$\Delta H/(J/g)$
1	2	OCH_3	k 69 i	g 101[b] n 121 i	2,3
2	2	OC_3H_7	k 67 i	g 120[e] sm 129 i	9,2
3	2	CN	k 84 i	amorphous	—
4	2	OC_6H_{13}	k 59 i	g 100[c] sm 140 i	11,3
5	2	$C_6H_4-OCH_3$	k 108 n 211 i	[d] n 177 i	—
6	3	C_6H_5	k 105 i	[d] sm 170 n 187 i	—
7	3	$C_6H_4-OC_2H_5$	k 123 n 202 i	g 120[b] sm 300 i	—
8	6	OCH_3	k 47 i	g 95[b] n 105 i	2,1
9	6	OC_6H_{13}	k 47 n 53 i	g 60[f] sm 115 i	15,5
10	6	C_6H_5	k 64 sm 68 n 92 i	g 130[d] sm 164 n 184 i	—

[a] Using the same nomenclature as in the characterization of liquid crystalline phase transitions, we propose an analogous description for polymers:
g = glass transition temperature T_G
n, sm = transitions to liquid crystalline nematic or smectic polymer phase, resp.
i = transition to isotropic melt of the polymer.
[b] g determined by DSC measurement.
[c] g determined as softening points with the polarization microscope.
[d] g not yet determined.

shown in Tab. 1. As the transition from the liquid crystalline phase to the glass phase takes place upon cooling, no change in texture is observed under the polarization microscope.

In accordance with the behaviour of conventional liquid crystals, nematic or smectic phases were formed by the polymers depending on the chain length of the spacer group and the substituent R'. Thus, nematic phases were observed with short substituents, while smectic phases were observed with long substituents (Tab. 1). For the present kinetic investigations it is of special importance that no phase separation takes place between the monomer and polymer during bulk polymerization, except for the polymerization of a nematic monomer to a smectic polymer.

By introduction of flexible spacer groups between the polymer main chain and the mesogenic side chain, it was possible to synthesize liquid crystalline polymers. They exhibit enantiotropic nematic or smectic phases depending on the substituent. The model consideration of decoupling of the polymer main chain and the mesogenic side chain is thereby confirmed. Further investigations will be made to confirm the general validity.

[1] B. A. Krentsel, Polymerization of Organized Systems, edited by H.-G. Elias, Midland Macromolecular Monographs, Vol. 3, p. 117 (1977); V. P. Shibaev, Vysokomol. Soedin., Ser. A, **9**, 923 (1977)

[2] H. Kelker, R. Hatz, Ber. Bunsenges. Phys. Chem. **78**, 819 (1974)

[3] A. Roviello, A. Sirigu, J. Polym. Sci., Polym. Lett. Ed. **13**, 455 (1975); W. J. Jackson, H. F. Kuhfuss, J. Polym. Sci., Polym. Chem. Ed. **14**, 2043 (1976)

[4] F. Cser, K. Nyitrai, Magy. Chem. Foly. **82**, 207 (1976); V. P. Shibaev, J. S. Freidzon, N. A. Plate, Dokl. Akad. Nauk, SSSR **227**, 1412 (1976); E. Perplies, H. Ringsdorf, J. H. Wendorff, J. Polym. Sci., Polym. Lett. Ed. **13**, 243 (1976)

[5] H. J. Lorkowski, F. Reuther, Plaste Kautsch. **2**, 81 (1976); A. Blumstein, R. B. Blumstein, S. B. Clough, E. C. Hsu, Macromolecules **8**, 73 (1975); Y. B. Amerik, I. I. Konstantinow, B. A. Krentsel, E. W. Malachajow, Vysokomol. Soedin., Ser. A **9**, 2591 (1967)

[6] E. Perplies, H. Ringsdorf, J. H. Wendorff, Ber. Bunsenges. Phys. Chem. **78**, 921 (1974)

[7] R. Steinsträsser, Z. Naturforsch., Teil B: **27**, 774 (1972)

[8] Patent Pending, No. P 2722 5894, BRD (1977)

Joachim Wendorff was born in Naugard (Pommern, formerly Germany) in 1941. He studied physics from 1961 to 1966 at the Philips University in Marburg, where he obtained his Ph.D degree in 1968. He then spent a postdoctoral period in the USA at the University of Massachusetts, Amherst. Between 1972 and 1976 he was a research assistant at the University of Mainz, from where he received his habilitation degree in 1982.

In 1976 he was appointed Head of the Physics Department at the Deutsche Kunststoff-Institut in Darmstadt. In 1991 he returned to the Philips University, Marburg as Professor of Physical Chemistry. In addition to his university position he was Director of the Centre of Material Science in Marburg from 1992 to 1998.

His research interests now concentrate on opto-electronic properties of polymer liquid crystals, glass forming processes, and polymers for medical applications.

Heino Finkelmann was born in Gronau in the lower Saxonian area of Germany in 1945. He initially studied as a chemical engineer at the Scientific Technical Academy at Isny, qualifying in 1966, and receiving the Ing. Grad in 1969. During this period he spent some time with the Unilever Research Lab. and with British Petroleum in Hamburg.

Then he studied chemistry at the Technical University Berlin and began to work on liquid crystals under Horst Stegemeyer, first at the Technical University in Berlin (1969–74), and then at the University of Paderborn (1974–5), where he received his Ph.D in 1975.

He then switched to macromolecular chemistry, working under Professor Ringsdorf at Mainz (1976–8). Here the work included in this volume was carried out. His next move was to the Technical University in Clausthal (1978–84). In 1984 he received his habilitation degree and in the same year was appointed as Full Professor and Director of the Institut für Makromolekulare Chemie at the University in Freiburg, a post he still retains (2004).

Finkelmann has received wide professional recognition. He sits or has sat on the editorial board of several journals, was awarded an honorary doctorate by the University of Toulouse (2004), has served twice on the Board of Directors of the International Liquid Crystal Society, and has received a number of prizes including the Gay-Lussac-Humboldt prize (2000) and the Europhysics prize (2003) together with Mark Warner. He served as Chairman of the German Liquid Crystal Society until 2003.

711

Helmut Ringsdorf was born in 1929 in Giessen. He studied chemistry at he Universities of Frankfurt, Darmstadt and Freiburg where he worked as the last Research Assistant of the Nobel prize winner Hermann Staudinger. In 1958 he obtained his Ph.D from the University of Freiburg.

After a post-doctoral period with H.F. Mark in Brooklyn, he joined the University of Marburg in 1962, obtaining his habilitation there in chemistry in 1967. He worked as Professor of Polymer Science in Marburg until his appointment as Professor of Organic Chemistry at the University of Mainz in 1971. Since his retirement in 1994 he has been active as adjunct and visiting professor in Changchun (China), London and Los Angeles.

His widespread activities in polymer science have been recognised by several awards, including the Staudinger Award of the German Chemical Society and the Humboldt Award (Paris) and by three honorary doctorates (Université de Paris-Sud, Dublin, and ETH Zürich).

INDEX OF BIOGRAPHIES

INDEX OF BIOGRAPHIES

Vorländer, Daniel, 1867–1941   88

Wendorff, Joachim, 1941–   710
Williams, Richard, 1927–   502

Zanoni, Louis, 1933–   514
Zocher, Hans Ernst Werner, 1893–1969   302
Zsigmondy, Richard Adolf, 1865–1929   598

714

AUTHOR INDEX

SUBJECT INDEX

723

733